Metal Casting
Theory and Practice

Metal Casting
Theory and Practice

Jagadeesha T.
Department of Mechanical Engineering
National Institute of Technology
NIT Calicut, Kerala

I.K. International Pvt. Ltd.
NEW DELHI

Published by
I.K. International Pvt. Ltd.
4435-36/7, Ansari Road, Daryaganj
New Delhi-110 002 (India)
E-mail: info@ikinternational.com
Website: www.ikbooks.com

ISBN 978-93-86768-67-4

Reprint 2023

Published by Krishan Makhijani for I.K. International Pvt. Ltd., 4435-36/7, Ansari Road, Daryaganj, New Delhi–110 002.

This book is most affectionately dedicated to

Babuji Maharaj
Founder President, Shri Ramchandra Mission
Shahjahanpur, Uttar Pradesh, India.

Preface

Casting is one of the most commonly used manufacturing processes in industries. It is almost impossible to make a product like automobile, aircraft, etc., without cast components. Many different metals, ranging from aluminium, cast iron, bronze, steel, zinc and brass, are used for casting in the foundries. This book aims to serve as a bridge between the study of the processes and their applications in production industries.

Compared with the classical method of writing a book, here two similar fundamental processes, namely theory and practice, are blended and explained in a different way so that the reader gets holistic approach to casting. First, basic processes are discussed, followed by the special processes and design of systems at the end. Profusely illustrated, it is a comprehensive source of technical metal casting information that foundry engineers and managers, process engineers, and anyone, who has interest in or needs to know about foundry operations and products. It features practical, understandable explanations, even for complex processes, includes clear descriptions of how each of current and recently developed casting processes work.

Author's wide experience in industries made him to include industry data, foundry practices, real-time industrial applications and problems in this book. It is suitable for students, foundry supervisors, engineers and managers. The author also focuses on many new developments in binder technology, different sand control tests and core making. The book also discusses Moulds, Mould Making and the Principles of Gating and Risering. Testing Methods and Metal Pouring is also covered. Solidification of Castings, Dressing of Castings, Important and Special Casting Techniques, Design of Castings, Casting Defects, Production of Iron Castings, Copper, Steel, Aluminium, Zinc Foundry Practices are some of the other topics in the book.

Health and Safety Measures for a Foundry Environment, Preventive Maintenance in Foundries, Application of Computers in the Foundry Industry, Heat Treatment of Castings, Testing, Inspection, and Quality Control in Foundries, Salvage of Defective Castings, Material Handling/Robotics in Foundry and Sand Reclamation are also included.

A sound basis of fundamental theory is presented with up-to-date technical information, with the help of simple line sketches. An attempt has been made to treat the subject matter so that it is attractive and modern. This textbook is suitable for both undergraduate and postgraduate students. A one-term college course of three hours lecture per week and four hours of home assignments is sufficient to cover the entire content. The book conforms to the latest syllabus and exam pattern.

Most of the books available today on metal casting grossly neglect the mathematical treatment of casting. Here, author included complete mathematics of riser design, gating design, solidification and cooling, etc. Several GATE and IES exam questions are solved in this textbook, making it suitable for the readers who take up competitive exams.

Dr Jagadeesha T.
jagdishsg@nitc.ac.in

Acknowledgements

During the course of development of this textbook, I benefitted from the comments and suggestions from several PSG Tech Coimbatore of my friends, clients and associates. I am grateful to all of them.

Acknowledgements are due to several of my teachers, ex-colleagues and team leaders, both in India and abroad. These include: Ashok Kumar, Head (Operations) and Yogesh Kale, Head (Design), Haritha Grammer (TVS-Suzuki Group), Hosur; Louis Kim, Head (Thin Films), Chartered Semiconductors Manufacturing Ltd., Singapore; Guruvaiah, Head (Electric Controls), Shiv Kumar (Fluid Power Controls) and Trivedi (Machine and Tool Design), TELCO, Pune; Ricky Tan, Director (Automation), Md. Johari Head (FMS), IBM Singapore Pvt. Ltd.; Dr. N. Sreenath, Program Director, NTU, Singapore; Kenny Kwan, Head (R&D) ASM, Singapore; Joseph Ong, Head, (CVD) and Stanley Teo, Head (PVD), ST Microelectronics, Singapore; James Lee (Director) APP Systems and Services, Singapore; Young Yee, Head (Thin Films Process) Applied Materials, Korea; Young Yap, Director (Process), Applied Materials, Singapore; Ganesan, Manager, TATA Motors, Pune; Dr. Subbaraya C.K., Vice Chancellor of AIT, Karnataka; Dr. Manoj Gupta, National University of Singapore; Prof. Seung Bok Choi, Director (Smart Materials), Inha University, Korea; and Dr. Radhakrishan, Director (Advanced Studies), PSG Tech Coimbatore.

I would like to express my deep sense of gratitude to our beloved director Sivaji Chakravarthy, a dynamic administrator, who provided an ideal learning atmosphere at NIT Calicut. I am indebted to Dr. Rajendra Kumar P.K., Professor and Head of Mechanical Engineering Department and Prof. S Ashok, Dean (R&D) for their encouragement and support. My gratitude is due to materials and manufacturing group members. My special thanks to Dr. M. A. Joseph, Dr. S. Jayraj, Dr. Srinu, Dr. K. Sekar and Prof. Ghulam Jilani for their support and motivation.

I would like to thank Vasanthi, my wonderful wife and my son, Ramkumar Jagadeesha, for their love, patience and endless sacrifices. My special thanks are due to my sister Prabhavathi T., without whose comforting presence and encouragement, this book would have remained merely a good intention.

I welcome feedback and suggestions for the improvement of this book from readers. While every attempt has been made to eliminate errors, inadvertent lapses are likely. I would be grateful if you would point them out to me. My email id is jagdishsg@nitc.ac.in

Jagadeesha T.

Acknowledgements

Contents

CHAPTER 1

Introduction to Casting Processes

1.1 INTRODUCTION

Useful engineering parts are made from metals and alloys by metal forming. Many ways of forming metals include casting, forging, welding, electroforming, powder metallurgy or a combination of these methods. The choice of a process, for a given component, depends on the material of which it is to be made; on its size, shape, and the required precision; and on how many are to be made—in short, on the design requirements.

A process has certain attributes; they describe the things the process can do. The common process attributes are: size, shape, complexity, precision, surface roughness, etc. A design for a component or assembly calls for a certain profile of these attributes, not for a process per se; any process which meets the profile will do. The problem, then, is that of matching the attribute profiles of available processes to that specified by the design. Process selection involves finding the best match between the process attributes and design requirements. This depends on how material, shape and processing interact with each other (Figure 1.1).

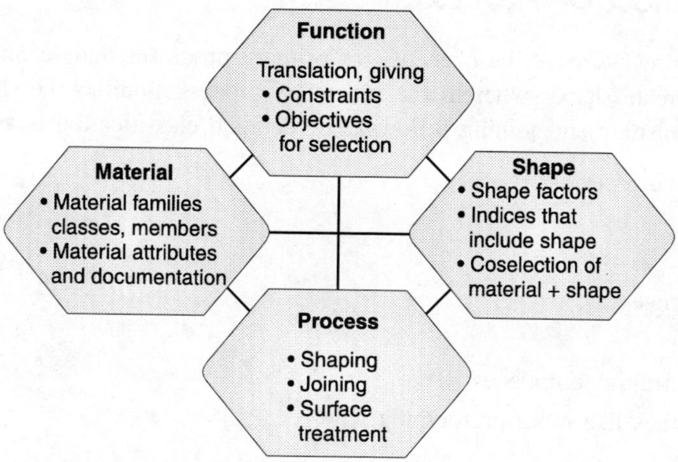

Figure 1.1 Processing selection depends on materials and shape. The "process attributes" are used as criteria for selection.

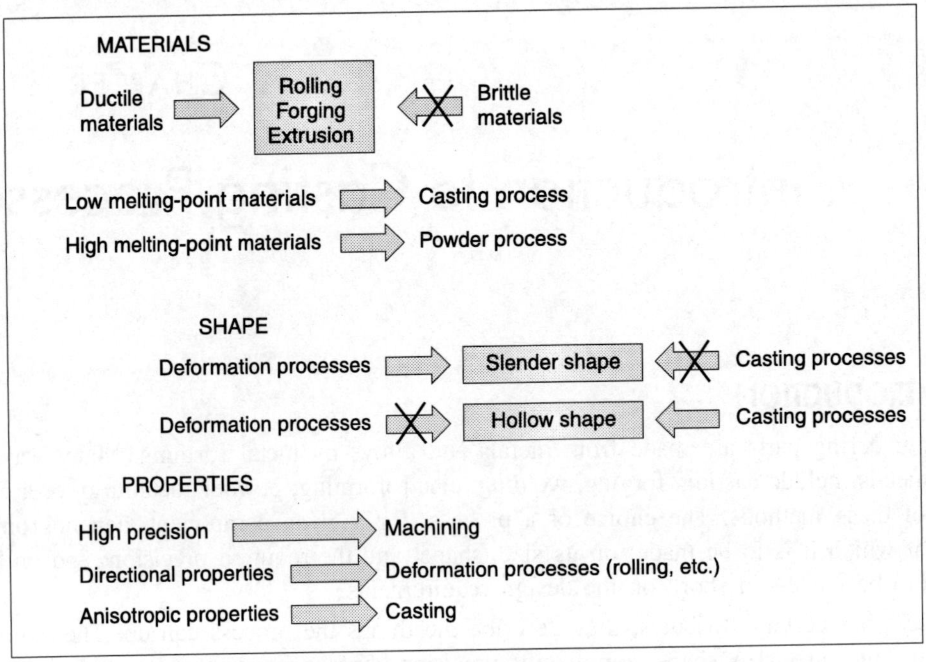

Figure 1.2 Process selection based on process attributes

Selection of the processes depends on type of material, melting point of the material to be processed, shape of the component and precision required. Figure 1.2 gives process selection based on process attributes.

1.2 CLASSIFICATION OF PROCESSES

The manufacturing processes can be classified as primary processes and secondary processes. Primary processes create shapes whereas the secondary process modifies the shape. After secondary processes, finishing and joining follows. The detailed classification is as follows:

Primary shaping processes

- Casting
- Moulding
- Deformation process
- Powder process
- Methods for forming composites
- Special techniques like rapid prototyping

Secondary shaping processes

- Machining
- Heat treatment

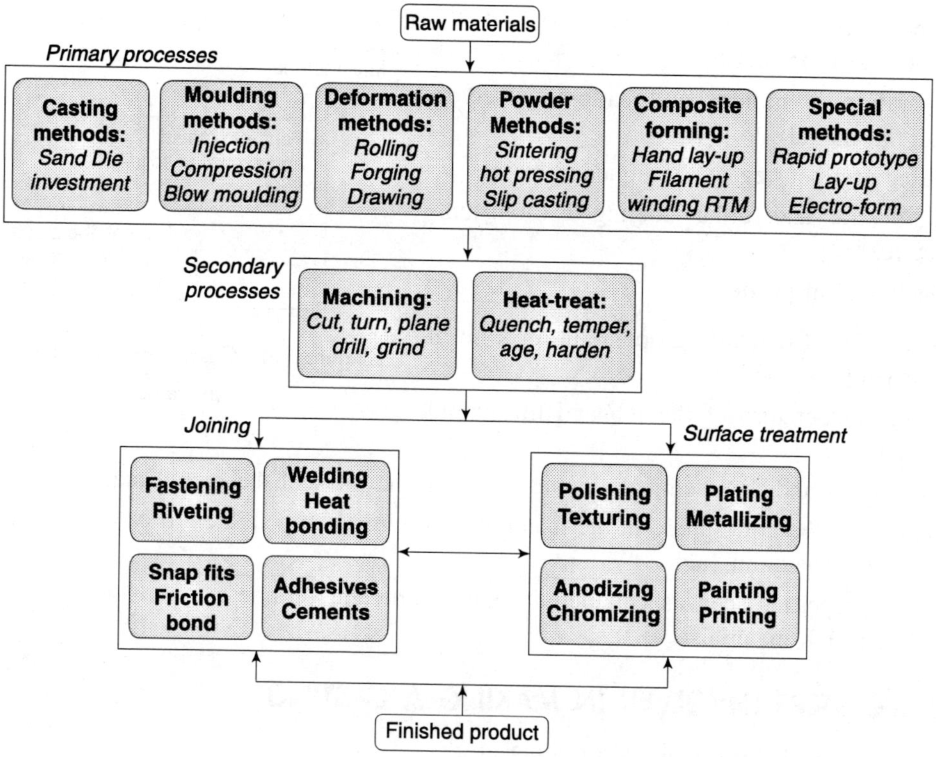

Figure 1.3 Detailed classification of the process

Machining involves chipping from cast or worked material to get the product shape. Machining is a high energy process and it is not an economical process of getting final shape. Another method of getting the finished product is by powder metallurgy process which involves die pressing and sintering of powdered parts. Production of powders of metals and alloys is a costly process.

Casting is a process of introducing molten metal into a mould cavity and allowing it to solidify. Casting is a very versatile process capable of being used in mass production in large components, with intricate designs. Casting is one of the many forming processes and it is one of the cheapest methods of giving desired shape. In this process, the material is first liquefied by properly heating it in a suitable furnace. Then, the liquid is poured into a mould cavity, trimmed, and cleaned to shape. Foundry is a basic industry and its product castings enter every field in which metals serve man.

Definition: Casting is one of oldest manufacturing processes in which a liquid material is usually poured into a mould, which contains a hollow cavity of the desired shape, and then allowed to solidify. The solidified part is also known as a casting, which is ejected or broken out of the mould to complete the process. Figure 1.4 shows the schematics of casting as metal forming processes.

It is clear from the definition of the process that a successful casting operation needs knowledge in the following areas:

(i) Preparation of moulds and patterns (used to make the mould).
(ii) Melting and pouring of the liquefied metal.
(iii) Solidification and further cooling to room temperature.
(iv) Defects and inspection.

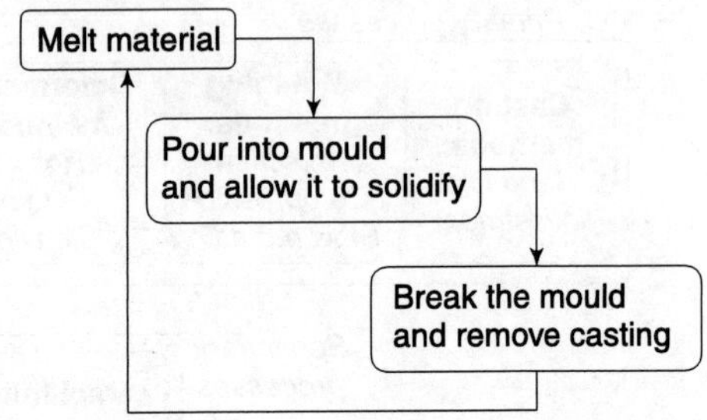

Figure 1.4 Casting as metal forming processes

The suitability of the casting operation for a given material depends on the:

(i) melting temperature of the job and the mould materials,
(ii) suitability of and the chemical reaction between the job and the mould materials,
(iii) solubility of the atmosphere in the material at different temperatures to be encountered in the casting operation, and
(iv) thermal properties such as conductivity and coefficient of linear expansion of both the mould and job materials.

1.3 BASIC STEPS INVOLVED IN MAKING A CASTING

The basic steps involved in making a casting are:

1. Pattern making
2. Core making
3. Moulding
4. Melting of metal and its pouring
5. Fettling and finishing
6. Inspection and quality control

1. Pattern making

The pattern is a physical model of the end product to be produced by metal casting process. Patterns are required to make mould cavities. It is an approximate replica to make mould cavities. It is designed and prepared as per the drawing of the casting received from the planning section and according to the moulding process to be employed. If the casting is to be hollow, additional patterns, referred to as coreboxes, are required.

A mould is made by packing moulding sand around the pattern. The mould is usually made in two parts so that the pattern can be withdrawn. In horizontal moulding, the top half is called the cope, and the bottom half is called the drag. In vertical moulding, the leading half of the mould is called the swing, and the back half is called the ram. When the pattern is withdrawn from the moulding material (sand or other), the imprint of the pattern provides the cavity when the mould parts are brought together. The mould cavity, together with any internal cores as required, is ultimately filled with molten metal through sprue and ingates to form the casting.

2. Core making

If casting is to have internal surfaces, a core must be included in the mould. Cores are made using coreboxes. Cores are usually made of sand, which are placed into a mould cavity to form the interior surfaces of the casting and sometimes the external surfaces as well of the casting. Thus, the void between the mould and core eventually becomes the casting.

3. Mould making

The cavity in the sand mould is formed by packing sand around a pattern, then separating the mould into two halves and removing the pattern. Adequate gating system and feeding system are required in the mould to direct the liquid metal into the mould cavity and to feed the solidification shrinkage. Mould assembly is shown in Figure 1.5.

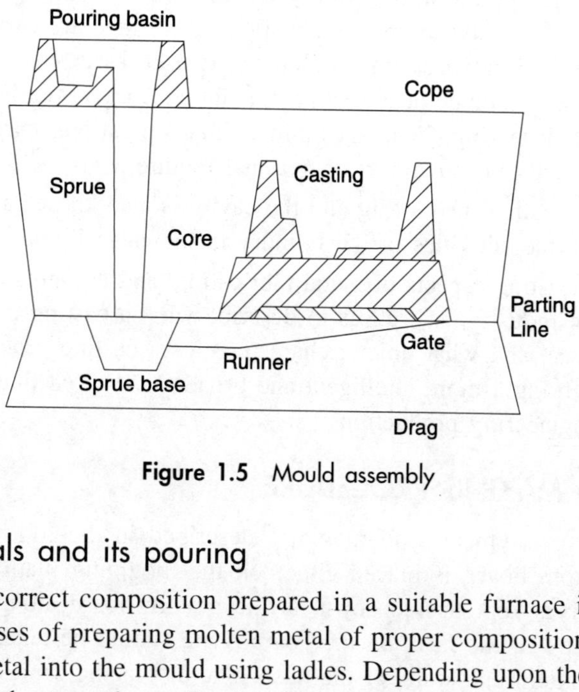

Figure 1.5 Mould assembly

4. Melting of metals and its pouring

The molten metal of correct composition prepared in a suitable furnace is referred to as melting. It involves processes of preparing molten metal of proper composition and temperature and pouring the molten metal into the mould using ladles. Depending upon the requirement, molten metal may or may not be treated or refined before pouring. The mould is then allowed to cool so that the liquid metal solidifies. The castings are finally extracted by breaking the mould. This operation is known as shake out.

5. Fettling and finishing

Fettling refers to all operations necessary for the removal of sand and oxide scale adhered to the casting surface, core and excess metal (fins and other projections, gating system, feeder) from the casting.

After the solidification of casting in the mould, the moulding boxes are opened and casting is knocked out from the sand aggregate. The gates and risers that constitute the gating/feeding system are removed. The adhering sand, scale, parting fins, and other foreign material are removed before the casting is ready for shipment or other processing. The solidified casting needs cleaning as the burnt sand aggregate sticks on the casting surface. Also, vent wires, scales, etc., are removed from the casting.

Finishing of castings is the final stage in cleaning of castings. Typical finishing operations include polishing and other operations (shot blasting, etc.) to improve surface finish and appearance of castings and different surface treatments (painting, electroplating, galvanising, heat treatment, etc.) given to impart special properties to castings.

6. Inspection and quality control

Inspection is necessary to check for defects in the casting as well as to ensure that the casting has the dimensions specified on the drawing and/or specifications. Inspection for internal defects may be involved, depending on the quality specified for the casting. In case any defects or shortcomings are observed, salvage and rectification of castings are carried out to save the casting. The inspected and accepted casting sometimes is used as is, but often it is subject to further processing which may include heat treatment, painting, rust preventive oils, other surface treatment (e.g., hot-dip galvanizing), and machining. Final operations may include electrodeposited plated metals for either cosmetic or operational requirements.

The solidified object is called the casting and the cavity is also known as mould. Figure 1.6 shows basic steps and all the activities of the foundry is shown in Figure 1.7

This book treats only casting aspects of metal fabrication, and attempts to delineate the process as a basic branch of industry, as well as to describe what a casting is, how it is made by different specific processes, and what unique characteristics a casting process has. It is hoped that this knowledge will bring a more intelligent and broader use of casting, improved designs, and more dependable engineering production.

1.4 SAND MOULD MAKING PROCEDURE

The procedure of making a typical sand mould is described in the following steps (refer to Figure 1.8). First a bottom board is placed either on the moulding platform or on the floor. The drag moulding flask is kept upside down on the bottom board along with the drag part of the pattern at the centre of the flask on the board. Care should be taken to provide enough clearance between the pattern and walls of the flask. The dry facing sand is sprinkled over the board and pattern to provide a nonsticky layer. Freshly prepared moulding sand of requisite quality is now poured into the drag and on the pattern to a thickness of 30 to 50 mm. Rest of the drag flask is filled with the backup sand and uniformly rammed to compact the sand. The ramming of sand should be done properly so as not too compact it too hard, which makes the escape of the gases difficult, nor too oloose so that mould would not have enough strength. After the ramming is over, the excess sand in the flask is removed by scaring using a flat bar to the level of the flask edges.

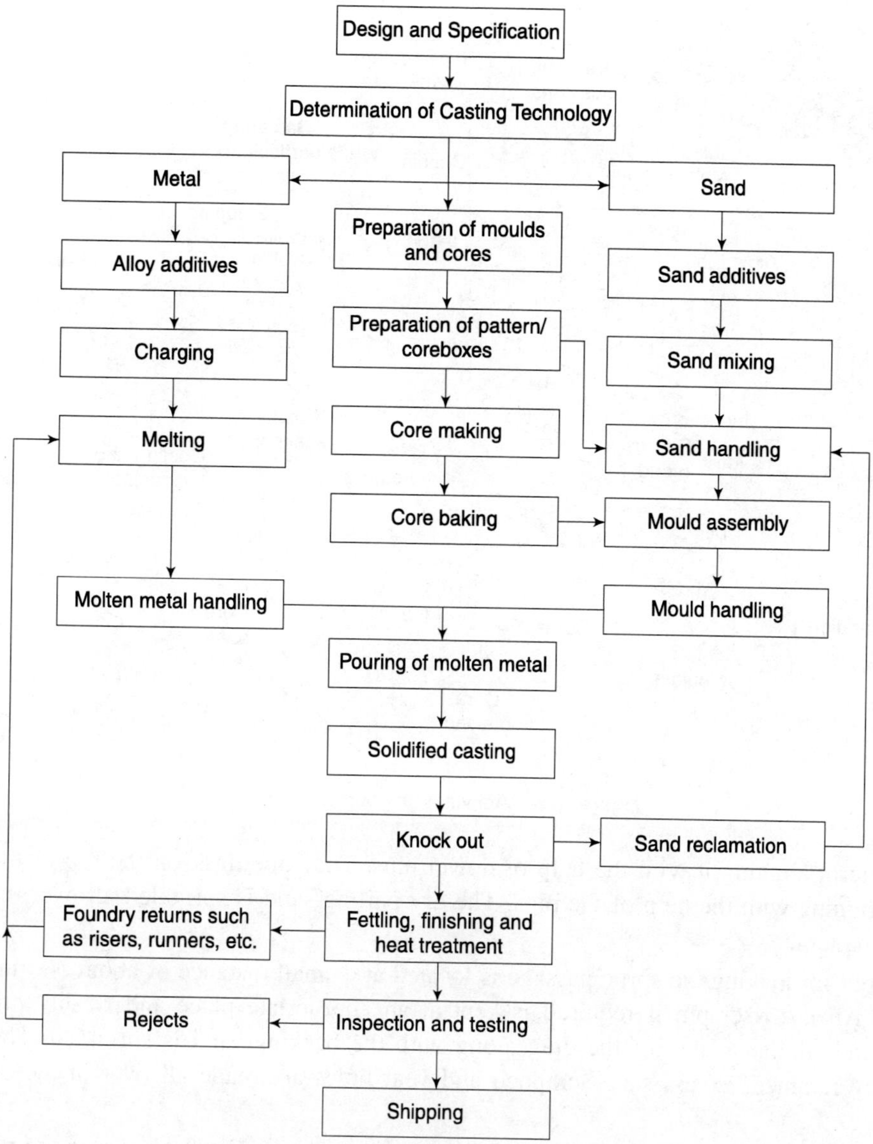

Figure 1.6 Basic steps in casting processes

Now, with a vent wire of 1 to 2 mm with a pointed end, vent holes are made in the drag to the full depth of the flask as well as to the pattern to facilitate removal of gases during casting solidification. This completes the preparation of the drag.

The finished drag is now rolled over to the bottom board exposing the pattern. Using a stick, the edges of sand around the patterns are repaired and cope half of the pattern is placed over

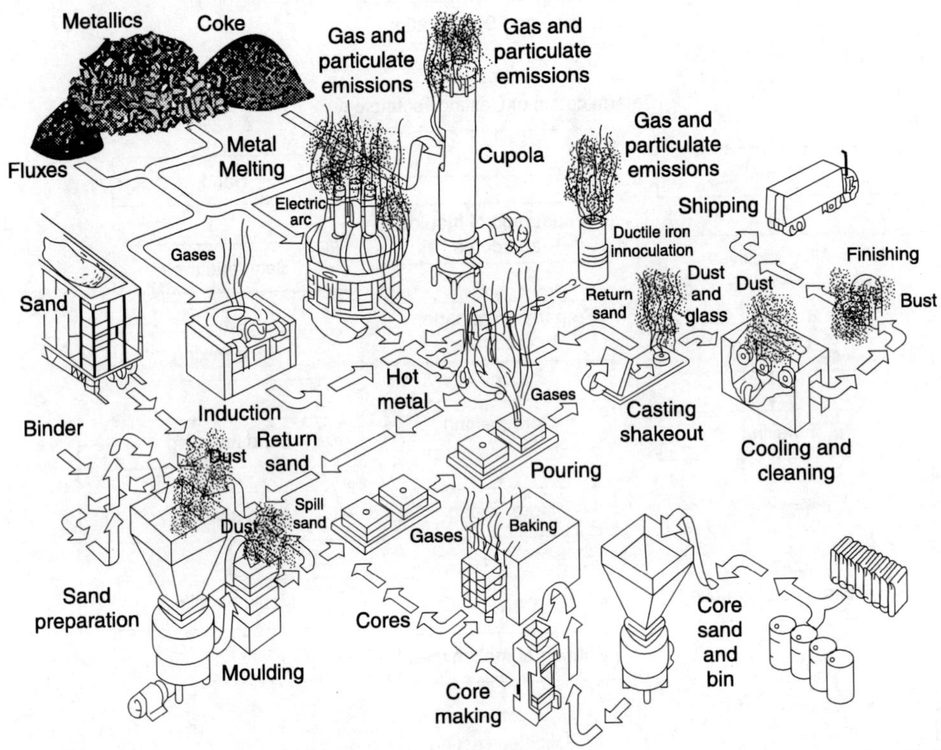

Figure 1.7 Activities in foundery

the drag pattern, aligning it with the help of dowel pins. The cope flask on the top of the drag is located aligning with the help of the pins. The dry parting sand is sprinkled all over the drag and on the pattern.

A sprue pin for making the sprue passage is located at a small distance of about 50 mm from the pattern. Also, a riser pin if required is kept at an appropriate place and freshly prepared moulding sand similar to that of the drag along with the backing sand is sprinkled. The sand is thoroughly rammed, excess sand scrapped and vent holes are made all over in the cope as the drag.

The sprue pin and the riser pin are carefully withdrawn from the flask. Later the pouring basin is cut near the top of the sprue. The cope is separated from the drag and any loose sand on the cope and drag interface of the drag is blown off with the help of bellows. Now the cope and the drag pattern halves are withdrawn by using the draw spikes and rapping the pattern all around to slightly enlarge the mould cavity so that walls are not spilled by the withdrawing pattern. The runners and gates are cut in the mould carefully without spoiling the mould. Any excess or loose sand found in the runners and mould cavity is blown away using the bellows. Now the facing sand in the form of a paste is applied all over the mould cavity and the runners which would give the finished casting a good surface finish. A dry sand core is placed using

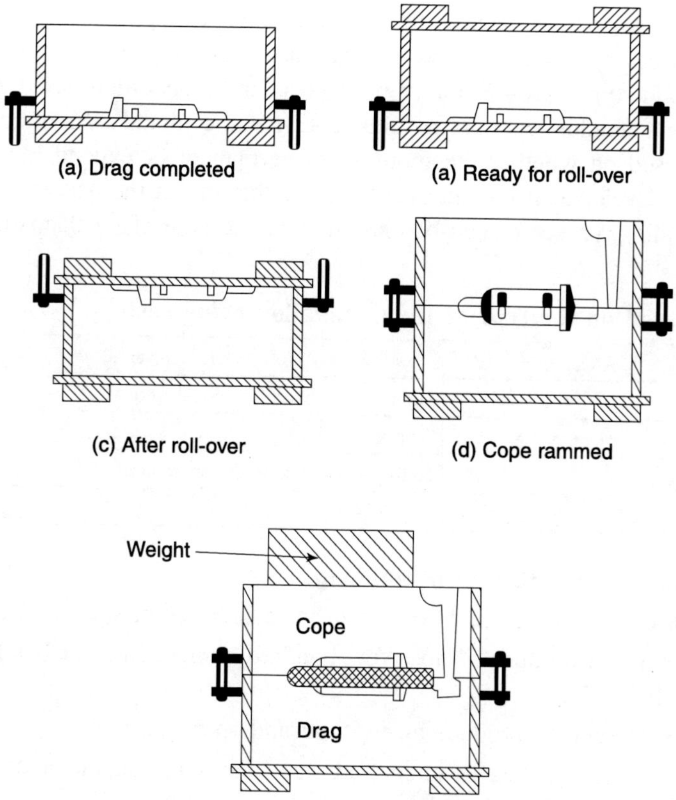

Figure 1.8 Sand mould making process

a corebox. After suitable baking, it is placed in the mould cavity as shown in Figure 1.3. The cope is replaced on the drag taking care of alignment of the two by means of pins. The mould now, as shown in Figure 1.2, is ready for pouring.

1.5 CLASSIFICATION OF FOUNDRIES

Castings are produced in specialized factories called foundries. Over 70% of all metal castings are produced via a sand casting process. Sand casting, also known as sand moulded casting, is a metal casting process characterized by using sand as the mould material.

The term moulding method refers to the process of making the mould and the materials used. The term casting method conveys a broader meaning, often including the moulding methods, the methods of introducing the metal into the mould cavity, or all the processes used in making the casting. Moulding methods and materials encompass a wide range of options. Some moulding (casting) processes are centuries old, while others did not exist until two generations ago. The selection of the correct one to use in a specific casting application depends on the alloy, the dimensional tolerances desired, the quality of the casting, and the price the customer is willing to pay. Most metals and alloys can be cast using any of the methods. There are, however, limitations in some cases. Foundry processes can be classified based on the moulding medium, such as sand moulds, ceramic moulds, and metallic moulds. Moulding methods may also be classified based on whether the moulds are permanent or expendable. Similarly, sub-classification can be developed from patterns, (i.e., whether or not the patterns are expendable). For permanent moulding, processes can be classified by the type of mechanism used to fill the mould.

(i) Classification based on the type of the metal/alloy being cast

Ferrous foundries	*Nonferrous foundries*
Steel foundries	Light metal foundries for Al and Mg
Grey Iron	Copper, brass, and bronze foundries
Malleable iron foundries	Lead, tin and, zinc based foundries
Ductile iron foundries	

(ii) Classification based on nature of production

(a) Jobbing foundries produce castings of small batch production sizes like two pieces.

(b) Mechanized foundries produce 50 to a few hundred components per batch. Good product quality is possible.

(c) Mass production foundries produce a few thousands per run.

(d) Captive foundries are attached to integrated steel plants for maintenance and spares of a particular steel plant.

(iii) Classification based on casting process

(a) Sand casting foundry: sand is used as moulding material. Each mould produces only one casting. It caters for small- and large-sized castings.

(b) Die casting foundry: metallic moulds are used. Production rate and quality is high.

(c) Investment casting foundry: wax and ceramic coating is used as mould material. Very complex shapes can be produced in large numbers using this method.

(iv) Classification based on moulding and casting processes

A. Sand system

Clay/water-bonded sand moulding

- Greensand moulding
- Floor and pit moulding
- Loam moulding

Dry sand moulding

- Skin dried moulding

Resin bonded sand moulding

- No bake process
- Cold box process
- Hot box process
- Warm box process

Sodium silicate processes

Precision casting

Unbounded sand moulding

- Lost foam process
- Vacuum moulding

B. Non-Sand system

Ceramic moulding

Plaster moulding

Graphite moulding

Metal or permanent mould process

- Gravity die casting
- Low pressure die casting
- High pressure die casting

Centrifugal casting

- Vertical centrifugal casting
- Horizontal centrifugal casting

Hybrid processes

- Squeeze casting
- Semisolid metal casting (rheocasting)

Each of the above processes will be discussed in great details in subsequent chapters. Greensand moulding, using expendable mould/permanent pattern is the most commonly used in industry. Investment casting or lost foam casting processes are expendable mould/expendable pattern processes. The investment casting process is one of the oldest casting processes (it has been used for more than 6000 years), while the lost foam process has been used commercially for less than 20 years. Permanent mould processes involve the use of metallic (usually ferrous) or solid graphite moulds. On volume basis, die casting and permanent mould casting are the most important. The classification is given in Figure 1.9.

1.5.1 Sand Moulds

Silica sand, either bonded or unbonded, is the most popular moulding media. Sand has the advantage of being widely available, inexpensive, able to withstand the elevated temperatures involved in casting most alloys, and is easily recycled. In addition, it does not react with most

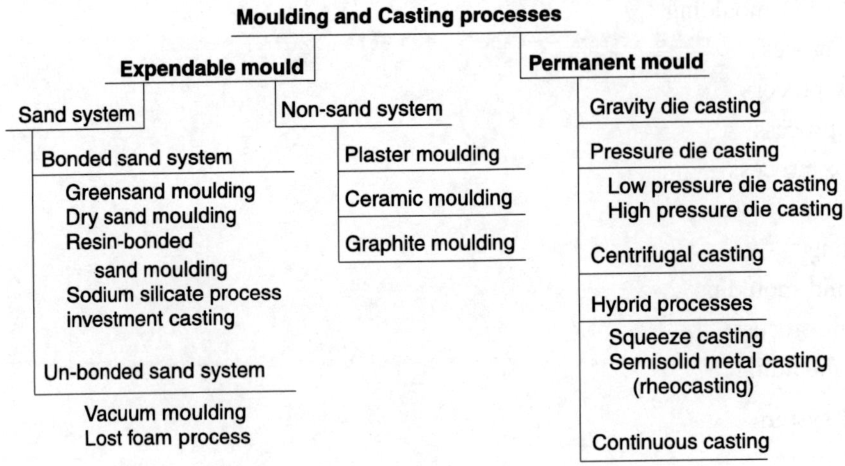

Figure 1.9 Classification of moulding and casting processes

metals and alloys, and it can be bonded with clay, resins, or silicate compounds. Because sand moulds are permeable, the air displaced by the metal when it enters the mould can escape through the mould.

Sand-based moulding processes have certain features in common:

- The use of pattern and/or coreboxes
- An aggregate mixture, comprising a granular refractory material and binder
- A means of forming the aggregate mixture around the pattern
- Hardening of the aggregate (or developing bond) while in contact with the pattern
- Withdrawal of pattern from the hardened aggregate mould
- Assembly of mould and core pieces to make a complete mould to pour into the liquid metal.

Bonded Sand Moulds

In bonded sand moulds, the sand particles are held together with chemical bonds and electrostatic bonds. The bonds that are developed must be easily broken when the casting is solid so that the casting can be removed from the mould.

Common bonded sand moulding processes are:

- Greensand moulding
- Dry sand moulding
- Resin-bonded sand moulding
- Sodium silicate-bonded sand moulding
- Cement-bonded sand moulding
- Investment (or precision) casting processes

(i) Greensand moulds

Greensand moulds are the most widely used of all sand moulds. They are made of sand, bonded by a mixture of clay and water. Other materials may be added to the sand to control its strength and prevent metal penetration. The sand is called "greensand" because of the presence of moisture; it is not baked or fired. Its actual colour is usually black. The advantages and disadvantages are given below.

Advantages	*Disadvantages*
Flexible, can be reused	Creates defects
Rapid and repetitive	Bad dimensional accuracy
Can be mechanised	Some casting designs (example large casting, long sections, thin sections, intricate designs) require other methods.
Cheaper	

Two important varieties in the greensand moulding processes are floor and pit moulding and loam moulding. **Floor and pit moulding** is used to produce castings too large for a flask and the mould is made in the sand bed. The pattern is set in the pit in the position in which the casting is to be poured, and sand is rammed or tucked under and around the sides of the pattern. The cope is finished off with cores or with sand rammed in a cope flask. In the **loam mould process**, a sub-structure is made of brick, wood and other materials to produce the approximate contour of the heavy casting. Slurry made of loam (50% sand and 50% clay mixture) and water is daubed over the base structure and then worked into a rough approximate shape. The mould is brought to the pattern dimensions by using skeleton pattern, sweeps, or templates. The mould is then dried thoroughly by torch or hot air.

(ii) Dry sand moulds

Dry sand moulding is sometimes used to make large castings, although resin-bonded moulds have largely taken over this process. In dry sand moulding, the water is removed from the mould prior to pouring. The bond is created both by using bentonites, and also through the use of other organic binders, such as dextrin, molasses, glutrin, or pitch. The moulds are baked in ovens prior to pouring to drive out the moisture. Dry sand moulds are generally stronger than greensand, and they usually have better dimensional stability than greensand in large castings. As an alternative to drying the entire mould, only the surface of the casting cavity may be dried. These "**skin-dried moulds**" are made from the same sand mixtures as dry sand moulds, but only the surface that will contact the molten metal is dried to a depth of 6 to 12 mm.

(iii) Resin-bonded sand moulds

1-2% (weight) chemical binder (polymer resin) and dry sand are mixed together and formed around a pattern or placed in a corebox. The resins may be cured by heat, or it may be cured by the setting of the resin at room temperature. To accelerate the curing of the resin, solid/liquid/gaseous catalysts may be added.

Advantages

1. Good surface finish.
2. Greater longevity and high strength of the mould.

Disadvantages

1. Limited bench life due to moisture absorption.

Over the last thirty years, a number of resins have been developed as binders for sand, primarily for making cores, although they are also used as moulds. These binder systems generally fall into one of the following categories: no-bake, cold box, protein, hot/warm box, and shell (croning). In these processes, sand and binder are mixed together and formed around a pattern or placed in a corebox. The resin may be cured by heat, or by setting of the resin. To accelerate the curing of the resin, catalysts may be added, either in solid, liquid, or gaseous form.

No-bake processes

The term "no-bake" generally refers to all of the binder systems in which a gas catalyzes the resin while the sand is in contact with the corebox or pattern. The binders cure at room temperature by the addition of an acid catalyst, and they have reaction products, such as nitrogen gas and water. Originally it referred only to the phenolic-urethane-amine process, which is still the most popular core resin. In addition to the phenolic urethane process, other no-bake binder systems include furan acid catalyzed, phenolic acid catalyzed, ester-cured, and alkyd-urethane binders.

Cold box processes

The term "cold box" generally refers to processes in which the sand, binder, and catalyst are mixed together before the sand mixture is introduced into the corebox or pattern; in this case curing begins immediately, and working time is limited. Cold box processes have become popular in high production foundries. The original process is the phenolic urethane process. It is similar to the phenolic urethane no-bake process described above, except that the catalyst is introduced as a vapour after the core is blown, producing a rapid cure, which enhances production rates.

Hot box process

In this process, the binder-sand mixture is wet. A liquid thermosetting binder and a latent acid catalyst are mixed with dry sand and blown into a heated corebox. Upon heating, the catalyst releases acid, which induces rapid cure; therefore the core can be removed in 10 to 30 seconds. After the cores are removed from the pattern, the cure is complete, as a result of the exothermic reaction and the heat absorbed by the core. Large cores require post-curing in an oven. Hot box resins are classified simply as furans or phenolics. Furans contain furfuryl alcohol, and phenolics are based on phenol, while furan-modified phenolics have both. All conventional hot box binders contain urea and formaldehyde. Hot box pattern temperatures are 230 to 290°C, and the patterns are designed for use between 220 and 245°C.

Warm box processes

Warm box resins are minimum-water (%) furfuryl-alcohol-type resins formulated for a nitrogen content of less than 2.5%. Because the resin/sand mix exhibits a high degree of rigid thermoset properties when fully cured, little or no post-strip distortion or sagging occurs. High hot and cold tensile properties are characteristic of warm box sands and generally permit a binder level between 0.8 and 1.8%, or about 20% less than the conventional hot box resin content. Warm box catalysts are copper salts based primarily on aromatic sulfonic acids in an aqueous methanol solution. The catalysts are unreactive with the resin at room temperature but form strong acids when heated. Pattern temperatures range from 150 to 230°C; the process was designed for use at 190°C.

Shell moulding

Shell moulding or "croning process" uses very fine sand coated with a thermosetting resin. The sand is placed in contact with a heated pattern to produce a hardened layer of sand, known as a shell, of only about 10 mm thick. The very fine sand produces an excellent surface finish on the casting. Use of sand of such fineness would not be possible in a conventional mould because mould permeability would be too low; however, in a thin shell mould, it is not a problem. Major steps involved in this process are shown in Figure 1.10.

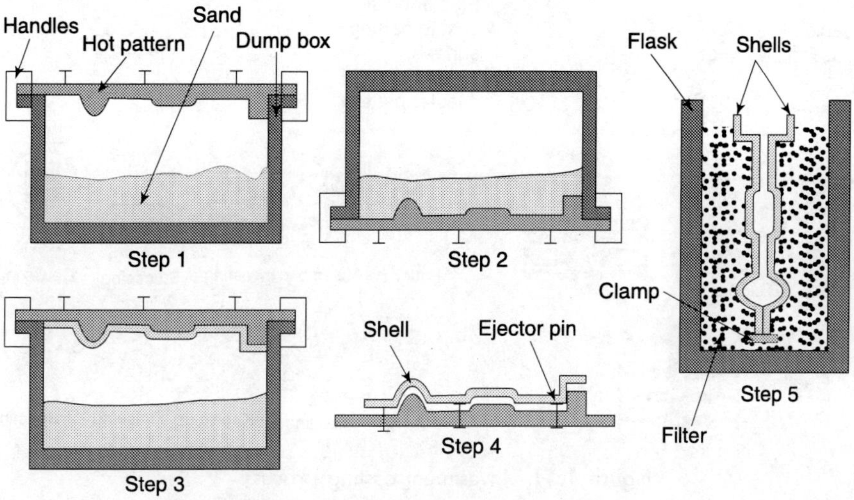

Figure 1.10 Shell moulding process

(iv) Sodium silicate-bonded sand (CO_2 process)

Sand can be bonded using sodium silicate ("waterglass"). Sodium silicate binders are odourless and do not produce volatile organic compounds when metal is poured into moulds. As in resin-bonded sands, the sand is mixed (care should be taken not to overmix the sand, which will produce a crumbly mixture), placed around the pattern, and vibrated to compact it. The sand is then gassed with carbon dioxide gas (hence the name "CO_2 process"), which hardens the mould almost immediately.

(v) Cement-bonded sand moulds

Cement-bonded moulding sand is a mixture of sand, 8 to 12% high early-strength hydraulic cement, and 4 to 6% water. This sand develops great hardness and strength by the setting action of Portland cement.

(vi) Investment (or precision) moulding

In this process, a pattern of an expendable material such as wax or plastic (polystyrene) is used. The investment mould is prepared by surrounding an expendable pattern with refractory slurry that sets at room temperature. The pattern is melted out or burnt off leaving the mould cavity. Major steps are shown in Figure 1.11.

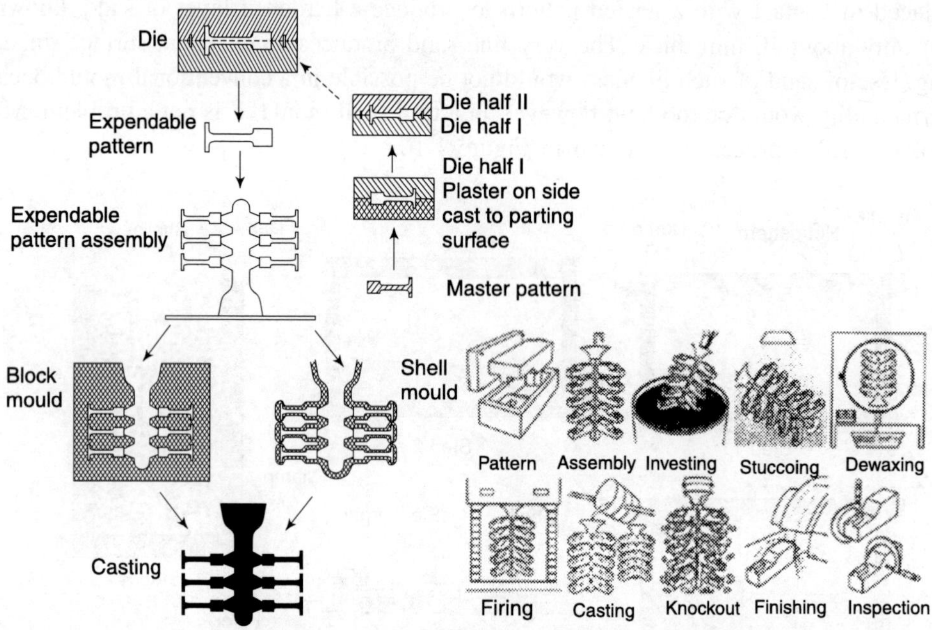

Figure 1.11 Investment casting process

The major steps involved are:

1. Tooling and pattern making: A tool is built to customer provided specification. Cold wax is then injected into the tool to create a wax pattern/prototype.
2. Pattern assembly: The wax patterns are assembled onto the sprue.
3. Dipping and coating: Successive layers of ceramic and stucco are applied to sprue assembly to form a hard shell.
4. Dewaxing and firing: The moulds are flash fired to remove the wax and sprue materials and then heated to 1800°C and placed on a sand bed, ready for pouring.

5. Casting: Molten metal up to 3000°C is poured into the hollow mould and then cooled.
6. Knock out: The ceramic shell is broken off and the individual castings are cut away.
7. Finishing: Excess metal is removed, surfaces are finished and castings are heat treated.

Parts are tested and shipped to the customer.

Unbonded sand moulds

It is not always necessary to use a binder with sand. In unbonded sand moulds, the pattern form is held either by the pattern itself during pouring, or by the use of a vacuum. Unbonded sand moulds have the advantage because they need no binders, mulling equipment, or mould additives.

Common unbonded sand moulding processes are:

- Lost foam moulding
- Vacuum moulding

(i) Lost foam casting

In the lost foam process, patterns are made of expanded polystyrene or other foamable plastics or blend of plastics. The patterns are attached to a gating system, also made of foamed plastic, placed in a flask, and sand is vibrated around them. The molten metal is poured directly onto the plastic, which melts or burns. This method eliminates the use of sand binders, additives, and cores, and it is capable of making some shapes that could be made no other way. Figure 1.12 shows the important steps to make lost foam casting.

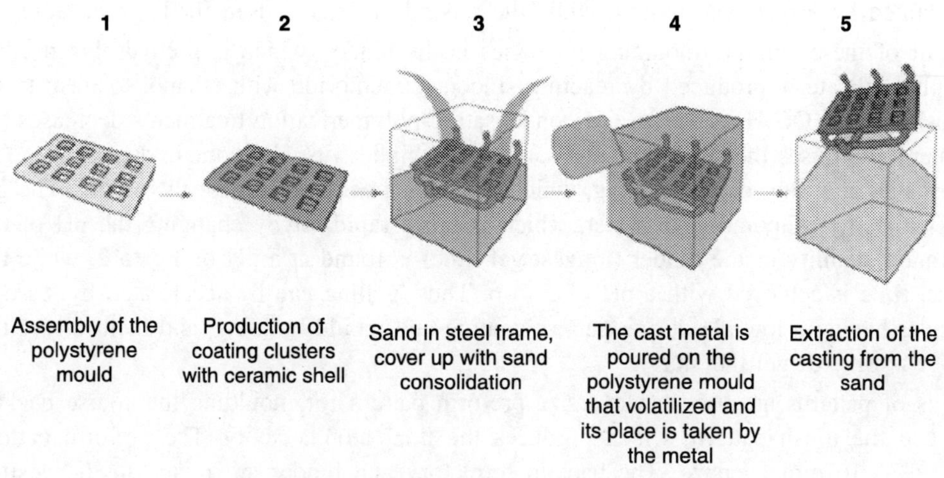

Figure 1.12 Steps involved in lost foam castings

(ii) Vacuum moulding (V-process)

In this binderless process, the sand is positioned between two sheets of thin plastic and held in place during pouring by application of a vacuum. This moulding method is especially useful in making large flat castings, such as sounding boards for pianos. Cope and drag halves of the mould are assembled while maintaining the vacuum, and the casting is poured. After the casting cools, the vacuum is released, and the loose sand falls away from the casting.

1.5.2 Non-sand Moulds

A. Ceramic Moulding

The moulding aggregate used to produce ceramic mould consists of the following: fine grained refractory materials (silica, zirconia, calcined high-alumina mullite, etc.) and chemical binders (ethyl silicate, water, alcohol and hydrochloric acid). The slurry thus produced is poured around the pattern and allowed to form a gel in 4–7 minutes and the pattern is then removed. The mould is then fired to ignite alcohol and bonding develops. They are used primarily to produce forging and extrusion tools and dies; food service equipment; glass moulds; valves for the food processing, chemical, and pharmaceutical industries; and marine hardware. Dimensional control (0.03 mm/cm) and surface finish are similar to that of investment castings, but the size and weight of castings made in ceramic moulds is generally greater than those made in investment castings. A number of ceramic moulding methods are developed based on proprietary processes. Two such important processes are the Shaw process and the unicast process.

The **Shaw process** is actually two processes. One process consists of a mould with a monolithic structure produced from slurry having a fine aggregate. In the second, the mould is a composite. The fine aggregate slurry is used only to form the mould/metal interface; it is backed with less expensive coarser slurry. The first process is used for small parts and has generally been supplanted by investment casting, while the second process is used for large castings.

The heart of these ceramic moulding processes is the binder, which is prehydrolyzed ethyl silicate. Ethyl silicate is produced by reacting silicon tetrachloride with ethanol to form tetraethylorthosilicate, $Si(OC_2H_5)_4$. The use of condensation-polymerization treatments decreases the water content and raises the silica content; commercial binders usually contain 40% silica. The binder forms a gel during mould making, which provides green strength for the mould. The gel is formed either by evaporation of water, which is fairly rapid, or by changing the pH of the gel. Maximum stability of the binder (longest gel time) is found at a pH of 1.5 to 2, while the shortest gel time is achieved with a pH of 5 to 6. Thus, gelling can be accelerated by starting with a binder having a low pH during pattern coating and mould building, and then raising the pH when it is time to gel the slurry.

Two sets of patterns are used, an oversize preform pattern for moulding the coarse backup material, and the finish pattern, which produces the final mould cavity. The preform pattern is about 2.5 to 10 mm oversize. The backup refractory and binder are mixed to the desired consistency, rammed or vibrated over the preform pattern, and gelled by gassing with carbon dioxide. The backup mould aggregate is usually a coarse-grained low-cost refractory, such as an aluminous fireclay, and the backup binder is usually sodium silicate.

The ceramic facing is usually made using fine zircon, alumina, or mullite flour, mixed with hydrolyzed ethyl silicate and often a gelling agent. The slurry is poured over the pattern through pouring gates in the backup mould. The slurry may be degassed in a vacuum chamber before pouring to remove any air bubbles. It gels within a few minutes after pouring, forming a rubber-like consistency, which permits patterns having no draft to be stripped successfully. After the mould is stripped, it is ready for stabilization. In the Shaw process, stabilization is accomplished by burning off the alcohol with a torch. During burning, microcraze cracks form in the mould, which enhances its permeability and facilitates shakeout. The mould is then baked to remove all remaining volatiles. Moulds may be preheated for pouring or poured at room temperature.

The **unicast process** differs from the Shaw process principally in the method of mould stabilization. In the unicast process, the ethyl silicate mould is stabilized by immersing in a liquid before burn-off. The liquid may be boiling water, or other organic liquids, such as kerosene or acetone. Using a liquid stabilization process reduces any warping of the mould that might occur as a result of the thermal gradients established during torch firing. Another difference between the unicast and Shaw processes relates to the sequence followed in preparing composite moulds. In the Shaw process, preparation of the coarse mould backup precedes pouring of the fine ceramic facing; in the Unicast process this sequence is reversed. Other ceramic moulds may be made by bonding the refractory aggregate with calcium or ammonium phosphate. These moulds are made by a dry pressing method in which the moulding mixture containing 4 to 9% moisture is pressed in dies at pressures of 14 to 140 MPa. After pressing, moulds are stripped and fired at temperatures between 900°C and 1300°C.

B. Plaster Moulding

Plaster moulds are used for non-ferrous castings having smoother as-cast surfaces (50 to 125 root mean square), better dimensional accuracy (about 0.008-0.01 inch per inch), and thinner sections then can be produced using sand casting techniques. The plaster casting process is used to produce near-net shape castings such as impellers, tire moulds, match plate patterns for foundries, and electronic equipment housings.

In the conventional plaster mould process, the principal mould ingredient is calcium sulfate. Other materials may be used to enhance mould properties, such as green strength and permeability. These materials include cement, talc, fiberglass, sand, clay, and other materials. Calcium sulfate exists as two different hydrates and in an anhydrous form. Its dihydrate is known as gypsum ($CaSO_42H_2O$). Above 128°C, gypsum loses three-fourths of its water to become plaster of paris, which loses the rest of its water at 163°C to become calcium sulfate,

$$CaSO_42H_2O + \text{Heat} = CaSO_4.\ 1/2\ H_2O + 3/2\ H_2O$$

To produce a plaster mould, 100 parts of gypsum or special metal casting plaster is added to 160 parts of water to form a slurry, which is immediately poured over a pattern or into a core-box, where it sets, forming a solid mould or core of gypsum with free water distributed in the mould. The pattern is normally coated with a release agent before the slurry is poured. Mechanical vibration may be used to eliminate pinholes in the mould while the slurry is being poured over the pattern. The mould is then dried at temperatures (around 200-425°C) high enough to produce calcium sulfate and to burn out any residual pattern materials (such as wax) that may have been used. The drying process should also remove all moisture without mould cracking.

Mould sections thus produced are very fragile and require care in assembling. For this reason, plaster mixes for metal casting ordinarily contain 20 to 30% talc to prevent mould cracking. Asbestos, silica flour is also used, besides chalk, to control the contraction characteristics and setting time. Patterns and coreboxes are usually made of aluminium. However, flexible rubber or urethane patterns may be used to produce moulds having negative draft or very complex sections.

Plaster moulds have low heat capacity, which means metal flows well in them to fill thin sections, but solidification rates are low. For mushy freezing alloys, this means that, because cooling rates are low, properties will generally be lower than those in castings made in sand moulds. Plaster moulds also have low permeability, so vacuum or pressure assists are often used to help fill thin sections. Silver, gold, aluminium, magnesium, copper, and alloys of these metals, particularly brass and bronze may be poured in plaster, but magnesium alloys generally are not because of the possibility of a violent reaction between the molten magnesium and any free water in the mould. Ferrous metals are not poured into plaster moulds because iron reacts with the sulfate to produce very poor surfaces. Plaster mould casting is not favoured for large castings because of its weakness and for high melting point materials because of its destruction at high temperature.

A variation on plaster moulding is the **Antioch process**. In this process, after the moulds have set, they are stripped from the pattern and steam autoclaved for 8 to 10 hours at a pressure of two atmospheres (205 kPa). During this time, partial dehydration occurs and recrystallisation produces a granular structure in the mould having improved permeability. The moulds are then

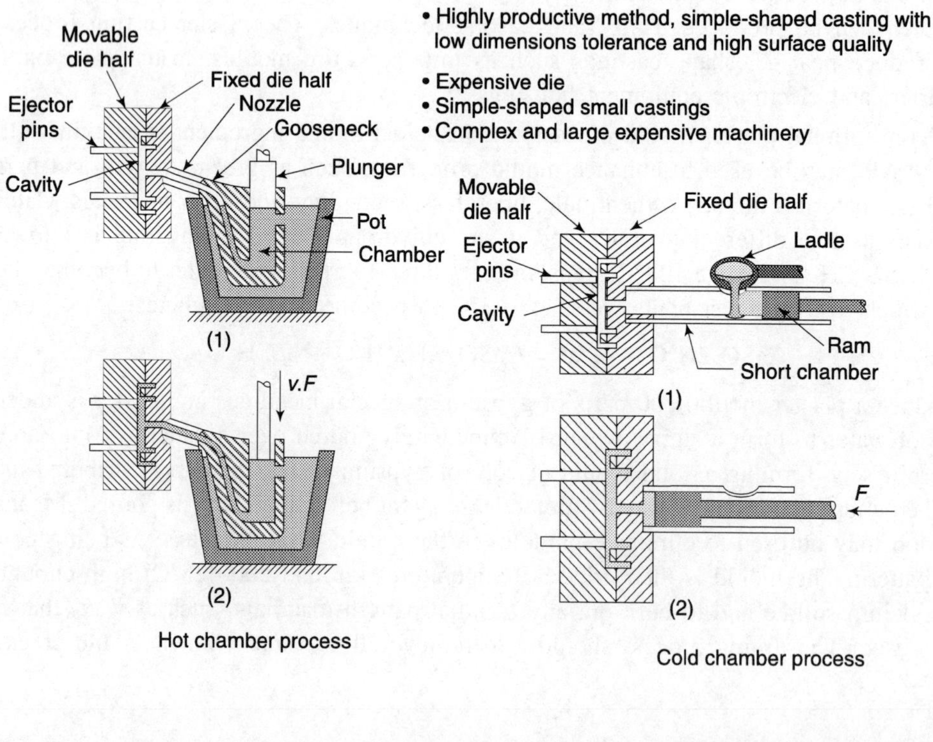

dipped in water and allowed to stand for 10 to 14 hours to complete the granulation reaction. Moulds must then be oven dried to remove the combined and excess water.

C. Graphite Moulding

These types of moulds are inert in nature and are used mainly for casting reactive metals like titanium. The moulds produced are either expendable or permanent in nature. About 70% graphite grains (AFS 85) and 30% binders (pitch, cement, starch and water) are mixed to form the expendable mould. The mould is prepared by squeezing the aggregate around the pattern at about 60-120 psi. It is then dried and fired at a reducing temperature at 1800-2000°F to form a solid mass. In order to produce a permanent mould, a solid block of graphite is machined to produce the cavity.

Usually, ethyl silicate is used as mould coat and the liquid metal is poured into the graphite mould under vacuum to prevent contamination of the metal. The casting thus produced has excellent surface finish and no machining is necessary.

D. Metal Moulds (Figure 1.13)

Moulds which can be used many times are made of metal, usually grey cast iron or steel, though sometimes of bronze. The mould cavity (or die cavity) in this type of permanent mould is often cast to its rough contour and then machined to its finished dimensions. Gating system as well as mould cavity is machined. The machined mould makes it possible to obtain very good surface finish and dimensional accuracy in the castings. Aluminium, magnesium, zinc,

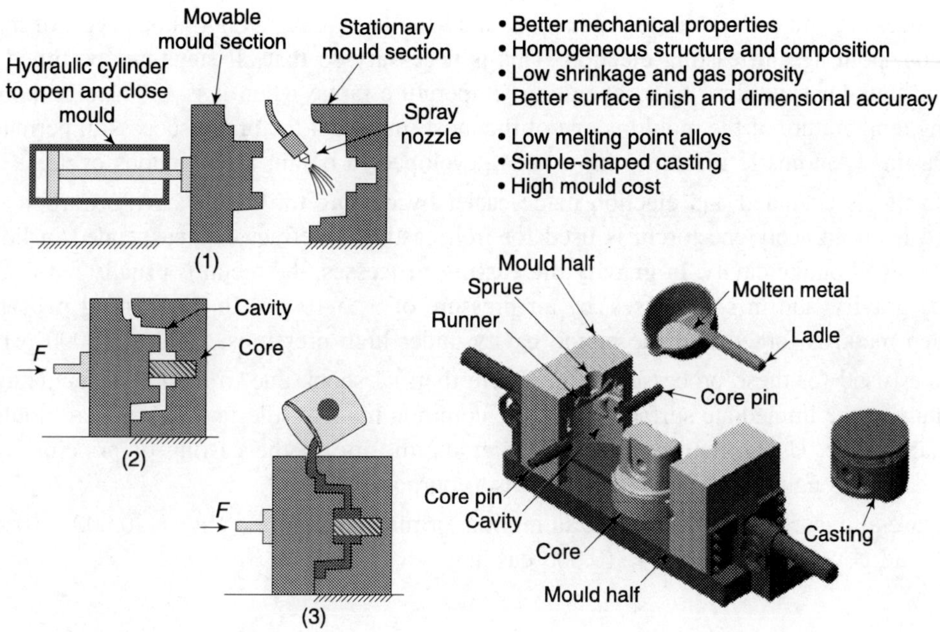

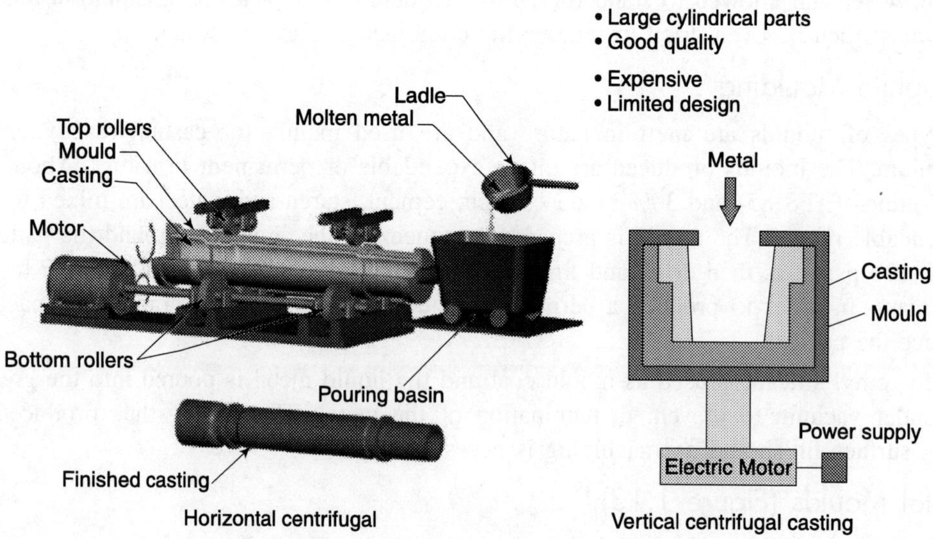

Figure 1.13 Casting process using metal moulds

lead, copper-based alloys, and cast irons are the principal alloys to cast. The extremely high temperatures of casting and consequent mould erosion usually make it unsuitable for most steel castings.

The process is limited to volume production, and usually requires a continuous cycle of mould preparation, pouring and casting ejection. This is necessary so that all steps can be timed and the mould thus kept within a fixed operating temperature range when it is about to be poured. Operating temperature of the mould is one of the most important factors in successful permanent mould casting. Automatic machines have been developed to obtain a continuous cycle.

Mould life is extended and ejection made easier by coating the mould cavity. Carbon soot, deposited from an acetylene torch, is used for iron castings. Refractories suspended in liquids may be sprayed on the cavity. In **gravity die casting processes**, the metal is usually fed into the mould by gravity, and in some cases, by air pressure of 3 to 10 psi. In **die casting processes**, the molten metal is forced into the mould cavity under high pressures, 1000 to 10000 psi.

The dies used for these processes suffer severe thermal shock due to the rapid fluctuation in temperature at the immediate surface. Thus, the common mode of die failure in metal moulding is thermal fatigue. Using conventional die materials, die life in the casting of principal types of alloy can be roughly indicated for comparison purposes.

Zinc base alloys: > 100,000 castings, aluminium or magnesium base alloys: 20,000 – 100,000 castings, and copper base alloys: < 10,000 castings

Specification	*Sand casting*	*Investment casting*	*Permanent mould casting*
Process	Sand moulds are produced around a permament pattern that is withdrawn to leave a cavity. Molten metal is poured into the moluld and solidified. Mould (and core) is broken to retrieve the casting.	A ceramic shell (investment) is slip cast metal cast into the investment that is broken up to remove the casting.	Molten metal poured into a metailic mould where it solidifies.
Shape	Mainly soild components but complex internal shapes produced using friable cores. Very large and small castings possible but thin sections difficult.	Best for relatively small, complex 3D components. Re-entrant angles possible.	Mostly used for small, simple shapes with only simple coring.
Materials	All metals excluding refractory and reactive alloys (e.g., titanium)	Suitable for most metals. Reactive metals can be cast under vacuum.	Mainly used for light alloys. Steel and cast irons are also possible.
Cycle time	Usually long as limited by rate of heat transfer out of the casting. Using multiple moulds increases production rate.	Limited by rate of heat transfer out of the casting. Production rates low because of process complexity. Increased by using multiple moulds and patterns.	Limited by rate of heat transfer across the interface. Production rate can be increased by using multiple moulds.
Quality	Surface texture poor. Porosity endemic. Non-metailic inclusions difficult to control.	Surface texture good. Higher mould temperatures decrease porosity but produce coarse microstructures.	Surface texture is good. Porosity unavoidable but can be minimized by slower mould filling to reduce turbulence.
Flexibility	Patterns are cheap and easy to make.	Moderately high because of the ease of production patterns.	Negligble setting up time for manual operation. Mould making is relatively difficult.
Material utilization	Up to 50% of casting in runners and feeders. Both mould and scrap metal can be directly recycled.	Near-net shape process with little material contained in feeding systems. Wax recycled, investment lost.	Rarely better than 60% utilization. Scrap in the runners and feeders can be directly recycled.
Operating cost	Very low as pattern costs are low and making it relatively easy.	Equipment costs can be high especially where reactive alloys are concerned. Labour costs are high due to the many stages in the process.	Equipment costs can be limited to mould and melt preparation apparatus.

Specification	*Pressure die casting*	*Centrifugal casting*	*Full mould casting*
Process	Molten metal is forced into a water-coated metal mould (die) through a system of sprues and runners. The metal solidifies rapidly and the casting is removed with its sprue and runners.	Molten metal is introduced into a sand or copper-lined, cylindrical steel mould that is rotated about its long axis, distributing the metal over its inner surface.	A refractory coating is applied to a volatile or combusible pattern that is used in a sand mould. The pattern is destroyed by the molten metal.
Shape	Used for complex shapes and thin sections. Cores must be simple and retractable.	Technique used to produce relatively long, hollow objects (e.g., pipes) without the need for cores.	Very complex 3D shapes possible.

Materials	High fluidity requirement means low melting temperature eutectics usually used (e.g., Al-Si). Hot chamber method restricted to very low melting temperature alloys (e.g. Mg and Zn).	Metals excluding refractory and reactive metals.	Non-refractory metals with casting temperatures high enough to vaporise the pattern.
Cycle time	Solidification time is typically < 1s so cycle is controlled by time taken to fill the mould and remove casting.	Determined by the rate of introduction of metal into the mould and the rate of solidification of the metal. The latter is lowered for sand-lined moulds.	Long, due to process complexity. Multiple moulds increase production rate.
Quality	Good surface texture but tolerable mould filling produces high degree of internal porosity.	Porosity and non-metallic inclusions migrate towards the inner surface because of their low density, giving a high quality outer surface.	Normal sand casting defects. Surface texture similar to that of pattern.
Flexibility	High tool changing time and set up time.	Setting up times are relatively short.	Ideal for the manufacture of one-offs.
Material utilization	Near-net shape process but some scrap in sprues, runners and flash which can be directly recycled.	Absence of runners and risers leads to near 100% use of material.	Pattern material entirely wasted. Metal usage poor due to runners, etc.
Operating cost	High, since machine and moulds are expensive.	Equipment is relatively simple and slightly costly. Increased complexity of water-cooled copper-lined moulds are more costly.	All equipment involved are rudimentary and process is very cheap to operate.

Summary of Casting Processes, their advantages and limitations

Process	*Advantages*	*Limitations*
Sand	Almost any metal is cast; no limit to size, shape or weight; low tooling cost.	Some finishing required; somewhat coarse finish; wide tolerances.
Shell mould	Good dimensional accuracy and surface finish; high production rate.	Part size limited; expensive patterns and equipment required.
Expandable pattern	Most metals are cast with no limit to size; complex shapes.	Patterns have low strength and can be costly for low quantities.
Plaster mould	Intricate shapes; good dimensional accuracy and finish; low porosity.	Limited to nonferrous metals; limited size and volume of production; mould making time relatively long.
Ceramic mould	Intricate shapes; close tolerance parts; good surface finish.	Limited size.
Investment	Intricate shapes; excellent surface finish and accuracy; almost any metal cast.	Part size limited; expensive patterns, moulds, and labour.
Permanent mould	Good surface finish and dimensional accuracy; low porosity; high production rate.	High mould cost; limited shape and intricacy; not suitable for high melting point metals.
Die	Excellent dimensional accuracy and surface finish; high production rate.	Die cost is high; size limited; usually limited to nonferrous metals; long lead time.
Centrifugal	Large cylindrical parts with good quality; high production rate.	Equipment is expensive; part shape limited.

General Characteristics of Casting Processes

Process	*Typical Materials Cast*	*Weight (kg)*		*Typical Surface Finish (μm, Ra)*
		Minimum	Maximum	
Sand	All	0.05	No limit	5–25
Shell	All	0.05	100+	1–3
Expandable pattern	All	0.05	No limit	5–20
Plaster	Nonferrous (Al, Mg, Zn, Cu)	0.05	50+	1–2
Investment	All (High melting pt.)	0.005	100+	1–3
Permanent mould	All	0.5	300	2–3
Die	Nonferrous (Al, Mg, Zn, Cu)	<0.05	50	1–2
Centrifugal	All	–	5000+	2–10

Process	*Porosity**	*Shape Complexity**	*Dimensional Accuracy**	*Section Thickness (mm)*	
				Minimum	Maximum
Sand	4	1–2	3	3	No limit
Shell	4	2–3	2	2	—
Expandable pattern	4	1	2	2	No limit
Plaster	3	1–2	2	1	—
Investment	3	1	1	1	75
Permanent mould	2–3	3–4	1	2	50
Die	1–2	3–4	1	0.5	12
Centrifugal	1–2	3–4	3	2	100

*Relative rating: 1 best, 5 worst.

Note: These ratings are only general; significant variations can occur, depending on the methods used.

1.6 HISTORY OF CASTING

Metal casting is the oldest and one of the most widely used manufacturing techniques and antedates 4000 B.C. With the advent of copper age in the history of mankind, casting made from metals have become integral parts in the life of human beings and essential in the progress of civilization. Even though gold was first used by prehistoric man around 8000 B.C. it was mainly used for ornamental purposes. Metal casting was developed based on fire-using technologies (or pyrotechnologies). Initially, pyrotechnologies were used to improve workability of stone, make plaster by burning lime, and produce ceramics by firing the clay. Metal working, however, started 10,000 years ago when the earliest metal objects are found to be wrought rather than cast. Evidence the of such metal objects is found in the form of decorative pendants and beads which were formed by hammering the native copper. Table 1.1 summarizes the advancements in metal working over these 10,000 years. Metal casting, on the other hand, dates back to 5000 and 3000 B.C. which refers to Chalcolithic period during which metals were melted for castings together with the experimentation of smelting copper. Initial moulds were made from smooth textured stones, resulting in fine cast products which could be witnessed in the museums and archeological exhibitions. Both single and multifaceted (carved on both sides of a rectangular piece of stone) moulds were developed from stones to produce castings that are not necessarily flat. However, multifaceted moulds were more popular for being portable and economical in terms of utilizing a suitable piece of stone. Followed by Chalcolithic period, the Bronze Age started in Near East around 3000 B.C. during which alloying elements were added to copper. The first bronze alloy reported was a mix of copper (Cu) with 4–12% per cent of arsenic (As), thus forming a silvery appearance of the cast surface as a result of inverse segregation of the arsenic-rich low-melting phase to the surface. Next, 5–10% of tin (Sn) was added to copper which was advantageous to lower the melting point, improve strength, deoxidize the melt, and produce a fine and easily polished cast surface capable of reproducing the features of the mould with exceptional fidelity, often desirable for art castings. The Bronze Age lasted for 1500 years during which mankind had its first exposure to elemental ores such as tin, copper and silver. Also, lost wax castings of small parts of bronze and silver were reported between 3000 and 2500 B.C. in the Near East region. The earliest known casting in existence is a copper frog, probably cast in 3200 B.C. in Mesopotamia region. The intricate design and three-dimensional characteristics suggest that it was created using sand casting process instead of using a permanent stone mould. The beginning of Bronze Age in Far East was about 2000 B.C., a millennium after its emergence in the Near East region. Casting was forming process in China with little evidence of other metal working operations before 500 B.C. The complexity of antique cast bronze ritual vessels suggested that they must have been produced using lost wax casting method. The use of cast iron in Chinese statuary began in 600 B.C. High phosphorus and sulfur contents in cast iron resulted in melting temperature comparable to bronze together with unusual fluid behaviour in the molten state. In Europe, cast iron was introduced as a casting alloy between 1200 and 1450 A.D. Between 15th and 18th centuries, some other important developments in the casting industry were included but not limited to introduction of sand as mould material in 16th centuries, production of cast iron with coke as fuel at Coalbrookdale (1709 A.D.), casting of the Great Bell of the Kremlin (1735 A.D.), and the development of cast steel by Benjamin Huntsman (1740). The cast iron produced at Coalbrookdale was first used as an important structural material in building the famous iron bridge and in other architectural applications during that period.

Although the development until 18th century was significant, cast materials and casting technologies progressed substantially from 19th century onwards. A new method of centrifugal casting was developed in 1809 A.D. in England. In 1818 A.D. first U.S. cast steel was produced by the crucible process at Valley Forge foundry. The extraction of aluminium from aluminium chloride is also reported in the same century. The first moulding machine was developed by S. Jarvis Adams Co. in Pittsburgh and was available in markets by 1837 A.D. The urge to further develop the casting process led to the development of a die casting machine to supply rapidly cast lead type for newspaper in 1849 A.D. Examination of castings was started by the development of metallography in 1863 A.D. which enabled foundrymen to polish, etch and physically analyze the castings through optical microscopy. Next, sand blasting was developed for large castings in 1870 A.D. By 1876 A.D., aluminium was started to be used as cast material and its first architectural application was reported in 1884 A.D. when a cast aluminium pyramid was mounted on the tip of Washington Monument. Some other developments during the 19th century are oven for core drying (1887 A.D.), non-art application of lost wax casting method to produce dental inlays (1897 A.D.), mould with bonded sand for salt-water piping system (1898 A.D.), and electric arc furnace for commercial production of castings (1899 A.D.). In early 1900s, first patent for low pressure permanent mould casting was issued in England. American foundry Association (AFS) produced rail wheels by centrifugal casting process for the first time. The advancement in the field continued with a die casting machine patented by H.H. Doehler patents in 1905. Heat treatment and artificial ageing were proposed in 1907 to improve the properties of cast aluminium alloys. In 1910, jolt squeezing machining became possible through the development of match plates. Another important development in 1915 was the experimentation on bentonite clay due to its exceptional high green and dry strength. In addition to that, furnaces for nonferrous melting were also developed during 1910s. The quality of casting was first examined through X-ray radiography in 1925 after which all military aircraft castings had to pass X-ray inspection for acceptance by 1940. The development of mathematical relationships between casting geometry and solidification time was established by Chvorinov in 1940. Also, statistical process control was started to be used for casting quality control and assurance in 1940s in the United States. The research on binders resulted in first heat-reactive, chemically-cured binder in Germany in 1944 for rapid production of mortar and artillery shells during World War II. By 1948, ductile iron was not just limited to laboratory castings and started to be used as a cast material in industrial applications.

In order to increase the mould hardness (density), high pressure moulding was experimented in 1950s. Hot box system to prepare and cure the cores simultaneously was introduced in 1953 thereby eliminating the need of dielectric drying ovens. In mid-1950s, squeeze casting process originated in Russia. In addition to that a full mould process was developed in 1958, known as lost foam casting, where the pattern and gating systems were carved from expanded polystyrene (EPS) and placed into a green sand mould. During 1960s, Furan was developed as a binder to be used in core production. Also shell flake resin was introduced in 1963 and it eliminated the need for different solvents. In 1968, a new method called "cold box process" was developed for mass production of cores in foundries. The late 20th century, i.e., 1970–1999 brought more advancements to metal casting such as development of vacuum forming or the V-process method in 1971 to produce moulds using unbonded sand by using vacuum. In 1974, Fiat developed an in-mould process for ductile iron treatment. During 1980s, it was started to investigate the casting processes computationally an example of which is the development and commercialization of a solidification software. In late 1980s, casting solidification software gained acceptance in foundries resulting in optimization of quality and cost of casting process in virtual reality. An important development during this period was 3D visualization techniques followed by the rapid prototyping and CAD/CAM technologies in 1988 which significantly reduced the time of tool development.

Plasma ladle refining (melting and refining in one vessel) was introduced for the first time in 1993 at Maynard Steel Casting Company in Milwaukee, Wisconsin. In order to cast large components through lost foam castings, a low-expansion synthetic mullite sand was patented by Brunswick Corp. in 1994. Microstructure simulations in mid-1990s enabled designers to analyze effects of metallurgy and predict and control mechanical properties of cast components. Cast metal matrix composites (MMCs) were used in automobile applications such as brake rotors for the first time in 1996. In the same year, General Motors Corp. developed a non-toxic and recyclable, water-soluble biopolymer-based core sand binder. Casting simulation developed further by the end of 1990s by stress and distortion simulation. As a result, generation and distribution of residual stresses in the cast component could be well understood which allowed to control casting distortion, reduce residual stresses, minimize defects such as hot tears and cracking, minimize mould distortion and improve mould life. In 2001, a physics-based software was developed by NASA and Department of Energy/OIT capable of predicting the filling of EPS patterns and sand cores when process variables are changed. In 2006, a new approach was developed which emphasized on developing an accept/reject criteria for castings by integrating casting simulation with mechanical performance simulations. Integrated simulations are currently being researched with an aim to improve integrity and quality, which eventually result in reliable operation of cast parts in service.

In India copper and bronze ages started quite early and metallurgy of casting was quite advanced in the earliest known history of India, that is in Indus valley civilization. People at that time knew how to work with close moulds, lost wax process and used cast tools, weapons, ornaments and utensils made of bronze, tin, gold, silver, copper and lead. They were smelted in the brick lined furnaces and poured into crucibles. Most probably these metals in crucibles were remelted in hearths for casting purposes.

History of iron making in India is still obscure and no definite data is available. While the wrought iron worked by hammering was well known before 1000 B.C., the evidence of the use of the iron casting is only available from 1000 B.C. Thus, iron casting came into regular use only after Alexander's invasion of India around 3000 B.C. Well-known iron pillar near Qutab Minar in Delhi is made from wrought iron which was given shape by forging. In India, iron casting may be said to have started around 200 B.C.

A brief summary of history of metal casting technology is given in Table 1.1.

Table 1.1

Date	*Development*	*Location*
9000 BC	Earliest metal object of wrought native copper	Near East
6500 BC	Earliest life size statues of plaster	Jordan
5000-3000 BC	Chalcolithic period: melting of copper, experimentation with smelting. A frog was made from copper metal in Mesopotamia	Near East
3000-1500 BC	Bronze age: arsenical copper and tin bronze alloys	Near East
3000-2500 BC	Lost wax casting of small objects	Near East
2000 BC	Bronze age	Far East

600 BC	Cast iron. Cast iron plowshares poured in China in 233 BC	China
500 AD	Cast crucible steel was produced	India
1200-1450 AD	Introduction of cast iron.	Europe
1455	Cast iron pipe to transport water was used in Dillenburg castle.	Germany
1480	The Vannoccio Biringuccio "father of the foundry industry," in Italy is the first man to document the foundry process.	Italy
1709	The first foundry flask for sand and loam moulding was created by Englishman, Abraham Darby. Cast iron was produced with coke as fuel at Coalbrookdale.	Europe
1735	Great bell of the Kremlin cast	Russia
1740	Cast steel developed by Benjamin Huntsman	England
1779	Cast iron used as architectural material, Iron bridge Gorge	England
1807	AG Eckhardt developed the centrifugal casting process	England
1818	Production of cast steel by crucible process	USA
1837	Development of first moulding machine	USA
1849	Development of die casting machine	-
1863	Metallography of casting surfaces	England
1870	Development of sand blasting of large castings	USA
1876	Aluminium is used as cast material	USA
1887	Development of oven for core drying	-
1896	American Foundrymen's Association (now American Foundrymen's Society) was formed	USA
1898	Mould development with bonded sand	England
1899	Electric arc furnace was used for commercial production of castings	France
1897	B F Philbrook of Iowa who rediscovered the investment casting process thought the roots of investment casitng can be traced when bronze dancing girl was found at Mohenjo-daro around 3000 BC	USA
1900	Low pressure permanent mould casting	England
1907	Heat treatment and artificial ageing to improve cast aluminium alloys	Germany
1910	Matchplates and jolt squeezing machines	-
1925	X-ray radiography for casting quality control	USA
1940	Chvorinov's rule	-
1944	First heat reactive chemically cured binder	Germany
1947	The shell process was invented by J Croning during World War II	Germany
1948	Ductile iron casting in industrial applications	USA
1950	High pressure moulding and squeeze casting moulding	Russia
1958	H F Shroyer was granted a patent for full mould process	-
1960	Development of furan as core binder	-

1968	The cold box process was introduced by L. Toriello and J. Robins for high production coremaking.	-
1969	The Japanese developed vacuum-process moulding. This method uses unbounded sand and the sand was bound by vacuum.	Japan
1971	Rheocasting was developed by MIT	USA
1974	In-mould process for ductile iron treatment by Fiat	USA
1980	Development and commercialization of solidification software	-
1988	Rapid prototyping and CAD/CAM technologies	-
1993	Plasma ladle refining (melting and refining in one vessel)	USA
Mid 1990s	Microstructure simulations	-
1996	Cast metal matrix composites in automobile application	England
End 1990s	Stress and distortion simulation of castings	-
2001	Software development by NASA and department of Energy/OIT	USA
2006	Casting simulations coupled to mechanical performance simulations	

	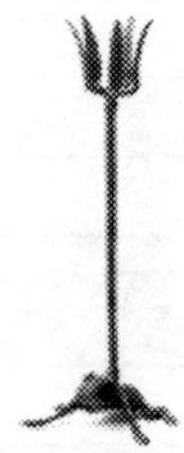		16th Century AD 
Mould prepared by smooth textured stone with an axe	Frog (3000 B.C.)	Cat and Buddha (5th century, China)	Seated Buddha (Thailand) 16th century

Metal casting history (India)

- **3000 B.C.** Earliest castings include the 11 cm high bronze dancing girl found at Mohenjo-daro.
- **2000 B.C.** Iron pillars, arrows, hooks, nails, bowls and daggers were found in Delhi, Roopar, Nashik and other places.
- **500 B.C.** Large scale state-owned mints and processes of metal extraction and alloying have been mentioned in Kautilya's *Arthashastra*
- **500 A.D.** Cast crucible steel was first produced in India.

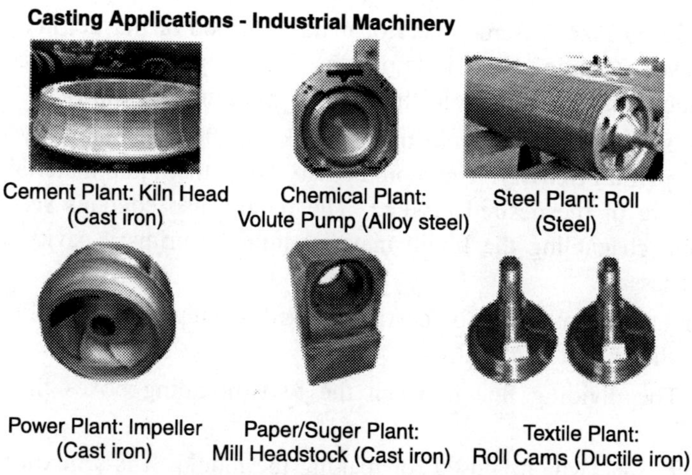

Steel Foundry railway workshop (W. Bengal, 1897)

Cast steel mill house, 198 tonnes

1.7 CASTING TERMS

Before going into the details of the process, defining a number of casting words would be appropriate. (Figure 1.14)

1. **Flask or moulding box** A frame made of metal or wood or plastic, in which the mould is formed. Lower moulding flask is known as drag, upper moulding flask as cope, and intermediate moulding flask, used in three-piece moulding, is known as cheek.
2. **Mould** An assembly of two or more metal blocks, or bonded refractory particles (sand) consisting of a **primary cavity:** The mould cavity holds the liquid material and essentially acts as a negative of the desired product. The mould also contains **secondary cavities** for pouring and channeling the liquid material into the primary cavity and to act as a reservoir, if necessary.
3. **Pattern** A replica of the part to be cast; it is used to prepare the mould cavity. Patterns are made of either wood or metal.
4. **Parting line** The dividing line between the two moulding boxes that makes up the mould.
5. **Moulding sand** Sand which is used for making the mould. It is a mixture of silica sand, clay, and moisture in appropriate proportions. The moulding sand must possess various properties such as permeability, flowability, cohesive strength, etc.
6. **Facing sand** In order to give a better surface finish to the casting, a small amount of fine carbonaceous material, known as facing sand is usually sprinkled on the parting surfaces of the moulding boxes.
7. **Core** The part of mould, made of sand, used to create openings and various shaped cavities in the castings.
8. **Pouring basin** A funnel-shaped cavity at the top of the mould into which the molten metal is poured.
9. **Sprue** The passage through which the molten metal flows from the pouring basin, and reaches the mould cavity. It controls the flow of metal into the mould.
10. **Runner** The channel through which the molten metal is carried from the sprue to the gate.
11. **Gate** A passageway through which the molten metal enters the mould cavity.
12. **Chaplets** Chaplets are used to support the cores inside the mould cavity. They are used to prevent the core against buckling or metallostatic pressure.
13. **Riser** The shapes of the risers are like a sprue, which are placed at that part of the casting which solidified in the last. The riser takes care of the shrinkage of the solidifying metal.
14. **Vent** Small opening in the mould to facilitate escape of air and gases.
15. **Drag** The bottom half of a horizontally parted mould.
16. **Cope** The top half of a horizontally parted mould.
17. **Draft** Slight taper given to a pattern to allow drawing from the sand.

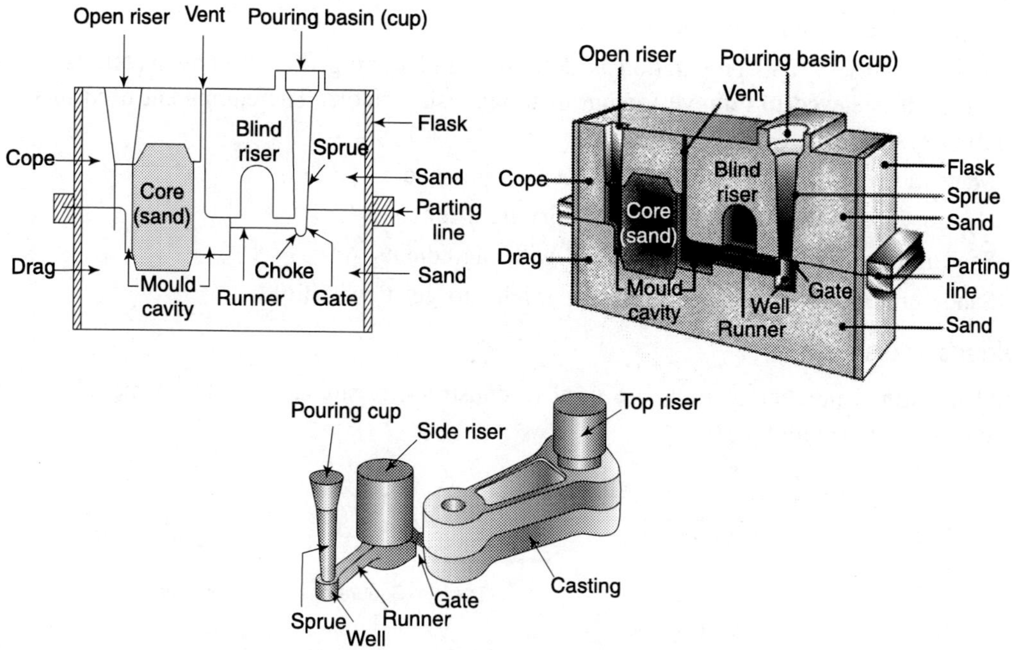

Figure 1.14 Mould section showing some casting terms

1.8 TOOLS AND EQUIPMENT FOR SAND MOULDING

The tools and equipment used for sand moulding process are described below:

1. Shovel: The shovel hand moulding tool is used for mixing and tempering moulding sand for moving the sand from the pile to the flask (Figure 1.15(a)).
2. Hand riddle: Foundry riddles are used by moulder to shift the foundry sand to ensure an even deposit of fine moulding sand over the pattern. Riddle frames are generally round, and their size is denoted by the diameter of the frame and the coarseness of the wire screen (Figure 1.15(b)).
3. Muller: This is a motor operated machine having two stone wheels revolving at slow speed, inside the cylindrical vessel. This machine can perform mixing and coating of the moulding sand with clay and moisture.

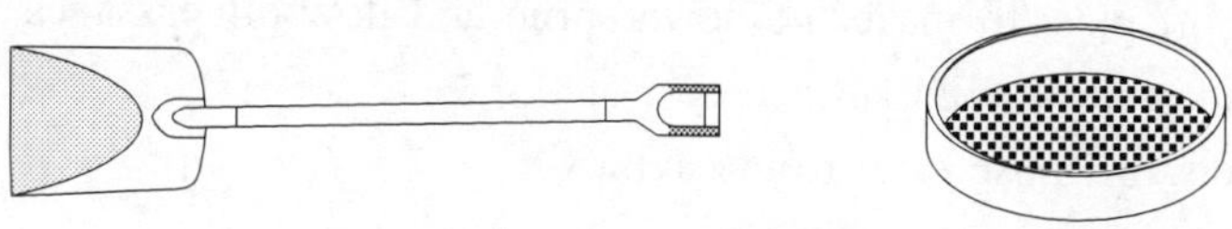

(a) Shovel (b) Riddle

Figure 1.15

Tools for manual sand preparation

In small foundries, sand preparation is done by hand mixing of moulding materials using shovel. Then it is sieved to remove foreign materials using riddle. The equipment used in sand preparation system are:

1. Magnetic separator to remove iron pieces from sand.
2. Gyratory riddle to sieve the pieces to dry hard lumps.
3. Sand muller to mix and coat the sand with ingredients.
4. Sand aerator to separate coarse sand particles to get flowability.

Moulder's tools

In hand moulding, number of tools are used to construct, repair and finish the mould. These are detailed below (Figure 1.16).

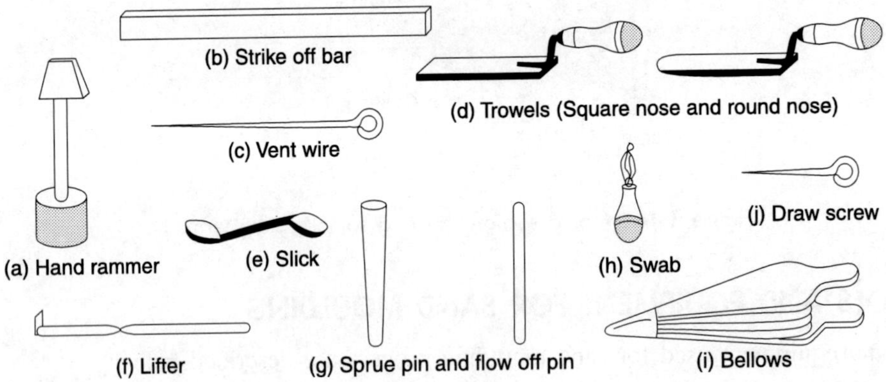

Figure 1.16

(a) Hand rammer: to compact sand by ramming
(b) Strike off bar: a metal bar to remove excess sand over box
(c) Vent wire: for making openings in mould called vent
(d) Trowels: to finish corners
(e) Slick: to repair and finish surface
(f) Lifters or cleaners: to finish vertical surfaces
(g) Sprue and flow off pins: to make the down sprue and flow off passages
(h) Swab: a brush to water mould surface
(i) Bellows: for blow off loose sand from cavities
(j) Draw screw: to remove wooden pattern
(k) Moulding board: to support flask and patterns while moulding

Mould flasks

Mould flasks are boxes to hold, support and handle sand moulds without distortion during moulding and casting. A mould flask is usually of two-piece type, namely, cope at the top and drag at the bottom. The seam between them is the parting line (plane) of the mould. When more than two box pieces are used, the additional middle boxes are called cheeks.

Mould flasks are made of wood, cast iron, steel or aluminium. Wooden flasks are used for small-scale production of light casting. Short-life is expected for wooden flasks. Cast iron flasks are mostly made and used in grey iron foundries. Steel flasks, fabricated with carbon steel sections are strong and durable. Metal flasks are commonly used for mass production of medium and heavy castings. Aluminium flasks are light in weight, and hence can be easily carried in hand moulding. They have less stiffness and strength compared to cast iron and steel.

Moulding flasks can be classified according to their construction as:

1. Rigid or tight flasks
2. Removable or snap flasks

Rigid or tight flasks are mould boxes used to enclose the sand mould till solidification and cooling of the casting. Figure 1.17 shows rigid or tight flasks for small-sized castings made of wood and metal. If the size of box is large, crossbars are added to the box for better reinforcement and holding of moulding sand. Figure 1.18 gives the examples of reinforced large-sized flasks of wood and metal. In large-sized metal flasks, holes are provided on the side walls for the removal of gases during casting.

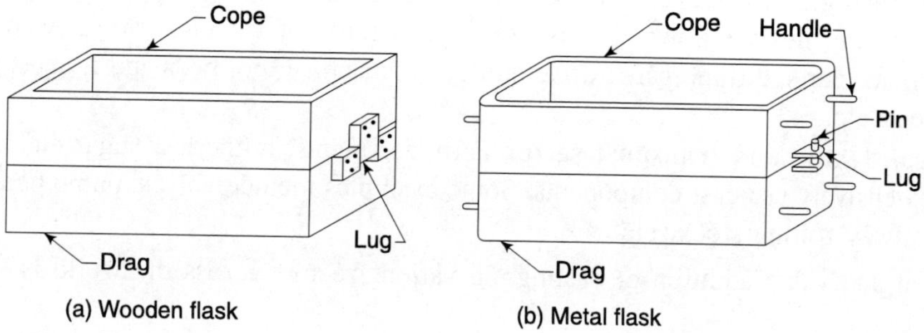

Figure 1.17 Rigid or tight flask (a) wood (b) metal

1.9 APPLICATIONS OF CASTING

Casting industry, popularly known as foundries, plays a key role in all the major sectors of our economy. Today, its applications include automotive components, aerospace components, heavy machinery components, locomotives, etc. Some of the important industrial sectors using casting are:

1. **Mining and mineral processing sector** Large number of cast components can be found in excavators, earth moving equipment, conveyors, crushers, grinders, track pads, shovel teeth, bowl, mantel liners, hammers, jaw plates, etc.

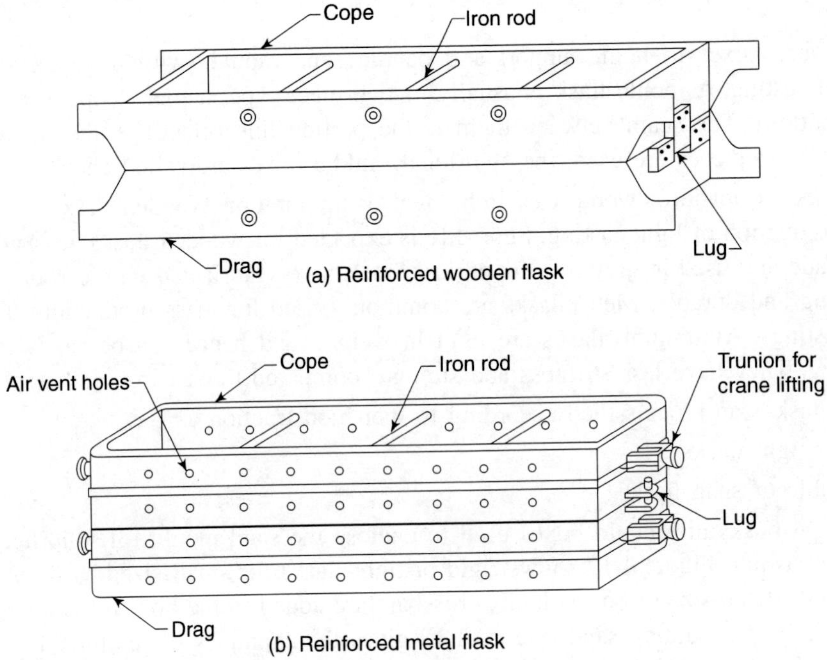

Figure 1.18 Reinforced flasks (a) wood (b) metal

2. **Manufacturing sector** Core manufacturing sector such as steel plant, cement plant, thermal, hydro and nuclear power plants use large number of castings. Some examples are rolls, furnace parts, turbine parts, drive unit gears, machine tool beds and frames, grinding equipment, etc.
3. **Chemical, textile and transport sector** Fertilizer plants, refineries, sugar and chemical plants extensively use cast components. Some examples include valves, pump parts, motor parts, railway rolling stock, etc.

Requirement and value addition of castings in various sectors across the world is shown in Table below.

Application sector	*Requirement*	*Value addition*
Automobile	Very high	Low-medium
Railways and shipping	High	Low
Aerospace and defense	Low	Very high
Sanitary (municipal castings)	Very high	Very low
Machine tool	Medium	Low–medium
Electrical and energy	Medium	Medium
Industrial machinery	Medium	Medium
Construction, farming, mining	Low	High
Biomedical implants	Very low	Very high
Art and luxury products	Very low	Very high

Some of cast components from various industries are shown below.

Casting Applications - Automobile
Wheel Rim
(Al-alloy)
Manifold Pipe
(Al-alloy)
Radiator part
(Cast iron/Brass)
Cylinder Block
(Cast iron)
Differential Casing
(Ductile iron)
Gear Case housing
(Cast iron)
Casting Applications - Railways & Shipping
Rail Engine Frame
(Alloy steel)
Bogie Frame
(Ductile iron)
Railway Plate
(Alloy steel)
Ship Anchor Fitting
(Cast iron)
Ship Fitting
(Ductile iron)
Bollard
(Ductile iron)
Casting Applications - Aerospace
Engine Fuel pump
(Al alloy)
Engine part
(Cu alloy)
Chassis part
(Ti-Al alloy)
Chassis part
(Ti-Al alloy)
Impellor
(Al alloy)
Fuel Pump Mounting
(Mg alloy)
Coupling part
(Mg alloy)
Brake Mounting
(Ti-Al alloy)
Casting Applications - Defence
Tank pump Head
(Cast iron)
Tank Tracks Ice Cleat
(Cast steel)
Fire suppression housing
(Brass)
Cannon
(Bronze)
Cannon Fitting
(Brass)
Tank Fittings
(Al-alloy)
Ammunition Box
(Al-alloy)

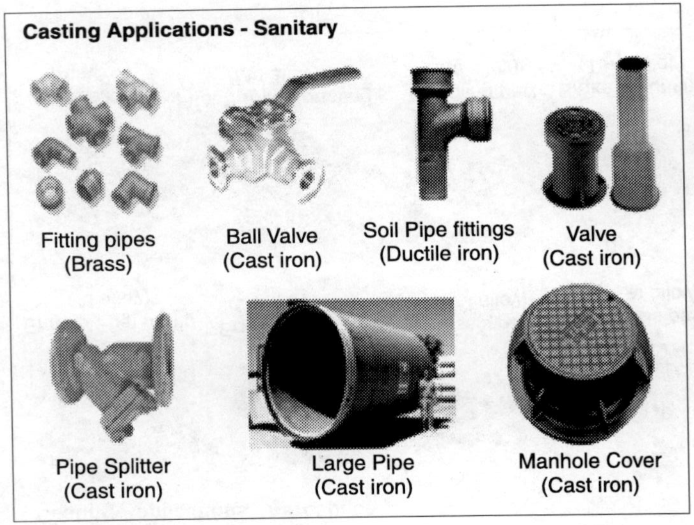
Casting Applications - Sanitary
Fitting pipes (Brass)
Ball Valve (Cast iron)
Soil Pipe fittings (Ductile iron)
Valve (Cast iron)
Pipe Splitter (Cast iron)
Large Pipe (Cast iron)
Manhole Cover (Cast iron)

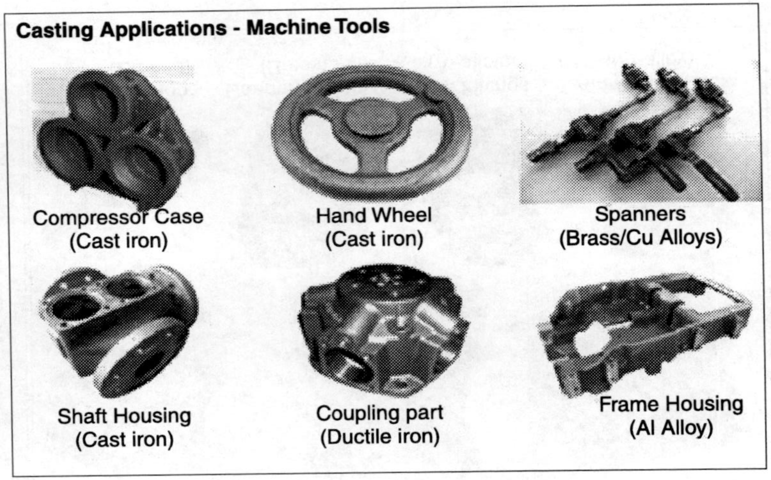
Casting Applications - Machine Tools
Compressor Case (Cast iron)
Hand Wheel (Cast iron)
Spanners (Brass/Cu Alloys)
Shaft Housing (Cast iron)
Coupling part (Ductile iron)
Frame Housing (Al Alloy)

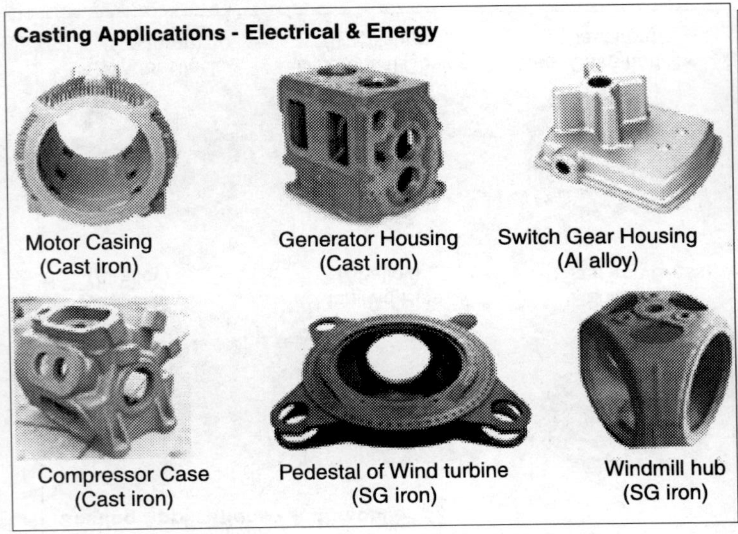
Casting Applications - Electrical & Energy
Motor Casing (Cast iron)
Generator Housing (Cast iron)
Switch Gear Housing (Al alloy)
Compressor Case (Cast iron)
Pedestal of Wind turbine (SG iron)
Windmill hub (SG iron)

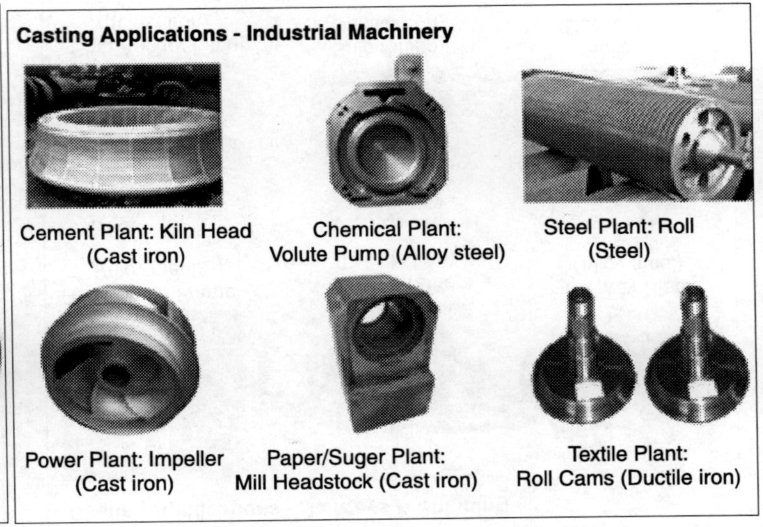
Casting Applications - Industrial Machinery
Cement Plant: Kiln Head (Cast iron)
Chemical Plant: Volute Pump (Alloy steel)
Steel Plant: Roll (Steel)
Power Plant: Impeller (Cast iron)
Paper/Suger Plant: Mill Headstock (Cast iron)
Textile Plant: Roll Cams (Ductile iron)

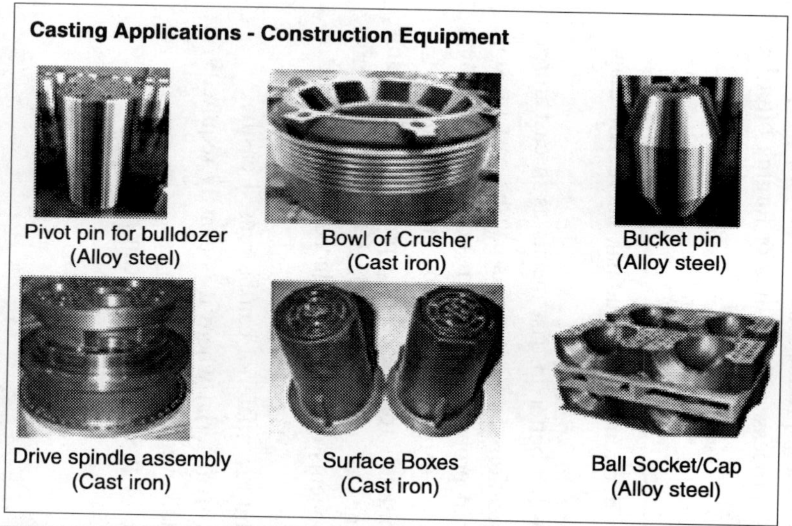
Casting Applications - Construction Equipment
Pivot pin for bulldozer (Alloy steel)
Bowl of Crusher (Cast iron)
Bucket pin (Alloy steel)
Drive spindle assembly (Cast iron)
Surface Boxes (Cast iron)
Ball Socket/Cap (Alloy steel)

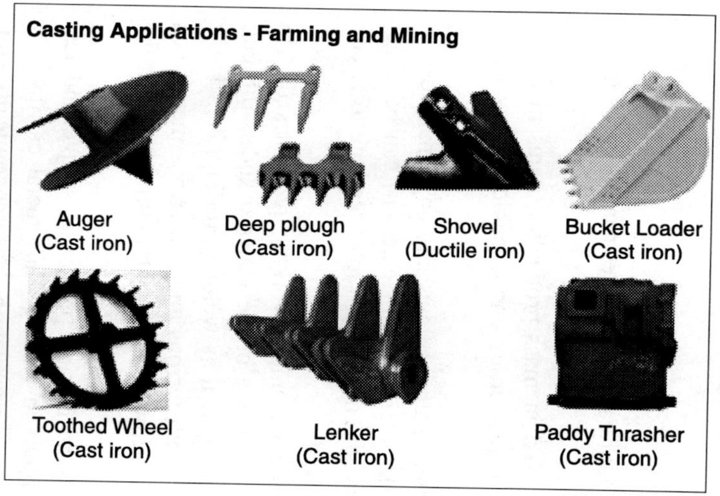
Casting Applications - Farming and Mining
Auger (Cast iron)
Deep plough (Cast iron)
Shovel (Ductile iron)
Bucket Loader (Cast iron)
Toothed Wheel (Cast iron)
Lenker (Cast iron)
Paddy Thrasher (Cast iron)

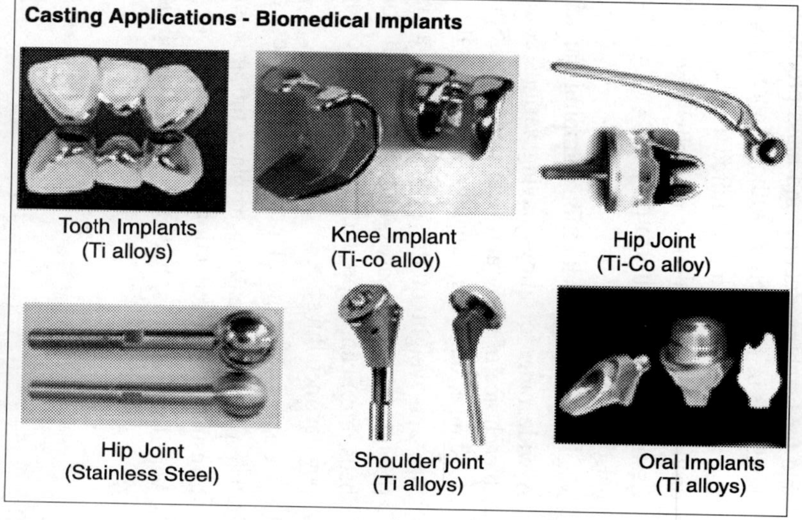
Casting Applications - Biomedical Implants
Tooth Implants (Ti alloys)
Knee Implant (Ti-co alloy)
Hip Joint (Ti-Co alloy)
Hip Joint (Stainless Steel)
Shoulder joint (Ti alloys)
Oral Implants (Ti alloys)

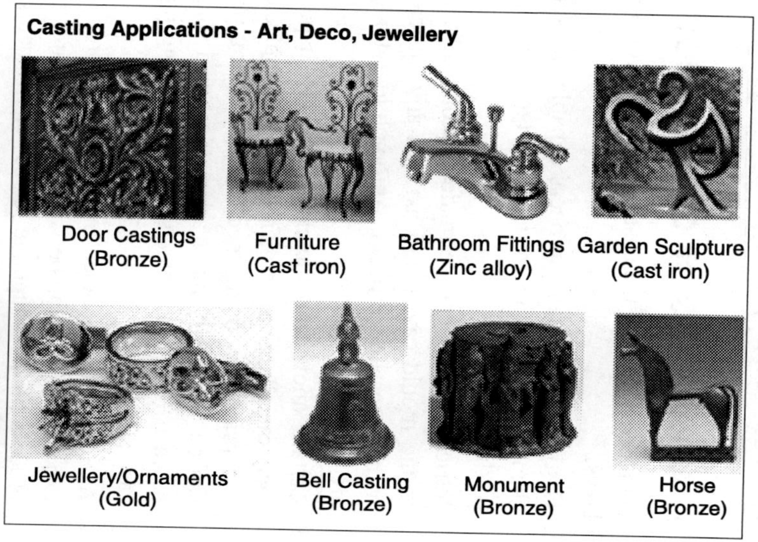
Casting Applications - Art, Deco, Jewellery
Door Castings (Bronze)
Furniture (Cast iron)
Bathroom Fittings (Zinc alloy)
Garden Sculpture (Cast iron)
Jewellery/Ornaments (Gold)
Bell Casting (Bronze)
Monument (Bronze)
Horse (Bronze)

1.10 ADVANTAGES OF CASTING PROCESS

The main advantages of the casting process over other fabrication processes can be summarized as follows:

1. Casting is a versatile process capable of being used in mass production items in large-shaped pieces, with intricate designs and having properties unobtainable by any other methods.
2. Intricate shapes can be made economically, molten material flows into any small section in the mould cavity and as such any intricate shape (internal or external) can be made with the casting process. Automobile cylinder heads, hydraulic valves, carburettors, etc. are a few examples.
3. Certain alloys and metals cannot be shaped by mechanical working processes due to their metallurgical properties. For example, cast iron is a brittle material and is difficult to fabricate by using mechanical working processes such as welding, rolling, etc., but it can be cast easily into various sizes and shapes. Both ferrous and nonferrous metals can be cast. High alloy steels of high melting point to low melting aluminium alloys can be cast.
4. As compared to other manufacturing processes, the casting process is simple and less expensive. The number of operations are minimum in casting, thus metal casting is a process highly adaptable to the requirements of mass production. It is economically suitable for batch production and mass production. Processes like rolling or forging offer less flexibility in production without increasing cost.
5. The necessary tooling required for casting moulds is simple and less expensive. As a result, for the production of a small lot, it is ideal.
6. Casting exhibits uniform properties in all the directions in longitudinal, lateral and diagonal directions.
7. Machinability and vibration damping capacity are better in cast metals as in cast iron.
8. A wide range of alloys having various compositions and properties can be cast by suitable choice of alloy and heat treatment. Special properties like corrosion resistance, heat resistance, damping capacity, high strength, etc., can be obtained by casting process.
9. Size and weight of the product is not a limitation for the casting process. Few grams (jewellery, watch cases, etc.) to several tonnes of casting (machine beds, rolling mill beds, etc.) can be produced. Very large components cannot be produced by forging or by powder metallurgy.
10. In casting, wastage of raw material is less and raw material can be reused easily.
11. It is possible in casting to place the amount of material where it is exactly required. As a result, weight reduction in design can be achieved economically.
12. Very thin sections, because of the flowability of the liquid metal, can be cast.

Summary of advantages of the casting is given below.

Advantages of casting process	
• Versatile and suitable for mass and job production	
• Good dimensional accuracy is possible in some casting processes (tolerance up to 0.1 mm and surface finish 50 microns)	
• It is a one step processes (no joining, etc.)	
• Cheapest method of metal shaping	
Design advantages	
• Size	• Production of prototypes
• Complexity	• Wide range of properties
• Weight saving	• Versatility in casting alloys
Metallurgical advantages	
• Uniform/Isotropic properties	
• Grain size	
• Density	

1.11 LIMITATIONS OF CASTINGS PROCESS

1. Casting is a high energy consuming process. For example, 2000 kWh of power is required to produce a tonne of finished steel casting. Energy costs in the world are growing day by day. Details are given in Table below.

S. No.	Particulars	Energy requirement (kWh/1000 kg casting)
1	Melting of materials	1400
2	Moulding and finishing	130
3	Heat treatment of castings	350
4	Auxiliary services	120
	Total	**2000 kWh/1000 kg casting**

2. Casting is a labour-intensive process compared to other processes and requires safety measures and mechanization.
3. Huge amount of raw material is required to produce a casting. Foundry needs exhaustive buildings, handling systems, large space, and large inventory. For example: To produce 1000 kg of steel casting we need 2200 kg of metallics, 300 kg of refractories, 1200 kg of facing sand, 4000 kg of backing sand and other additives.
4. Dimensional accuracy and surface finish of the castings made by casting processes are the limitations to this technique. Many new casting processes have been developed which can take into consideration the aspects of dimensional accuracy and surface finish. Some

of these processes are die casting process, investment casting process, vacuum-sealed moulding process, and shell moulding process.

5. The cycle time (the time required for making casting starts from the receipt of drawing to final product) for making a component using casting process is high. For example: medium size ferrous casting takes 2-3 months for the first casting. Thus, the entire cycle of order execution for casting can take between 3 months to one year depending on the size, intricacy, composition and quantity to be cast.
6. The working conditions in foundries are quite bad compared to other forming operations like rolling, machining, etc. Workers are exposed to high heat and high dust conditions. The environmental pollution is high in metal casting industries. Many countries like Germany, Switzerland, etc., have strict government regulations.
7. Castings are susceptible to various defects like internal porosity, shrinkage, blow holes, segregation and hot tears. With some materials, it is often difficult to remove defects.
8. Unable to use refractory materials.

1.12 SCOPE OF METAL CASTING INDUSTRY IN INDIA

Metal casting process is the oldest manufacturing process. Metal cast products find their applications in most of the products and almost all automobile products use cast product(s) as their components. It can be said that foundry industry is the 'mother of all industries'.

The growth of automobile industry in India in the past decade has given tremendous boost to foundries. There is a heavy demand for alloy steel casting also from the expanding coal and mineral development sectors. There are around 5100 foundries, both large as well small units, registered in India. Of these, around 3000 units are grey iron foundries, producing about 5.1 million tonnes of grey iron casting. About 300 foundries are in the large sector. Out of total units, 80 per cent are small units, 15 per cent are medium-sized and only 5 per cent are in large sector. In India, the scope of metal casting industry is increasing as the government has made tremendous efforts to improve infrastructure including power generation. The efforts will help metal casting industries, which are power extensive industries, to grow.

The knowledge and application of technology in the area of metal casting will help the industries to excel in all their application areas. The scope of metal casting industry has widened up. It is nowadays not limited to metal products, but the application of cast products also include plastic products, composite, civil and building infrastructure development, bridge construction, etc. The new initiatives and additional scope of foundry industry will require skilled manpower in this field. This will enhance the metal casting industry jobs to huge number of metal casting professionals.

The fast closure of foundry units in advanced countries due to environmental pollution, excess energy requirements and harsh working environment has opened new potential for foundry units in India. In addition to the availability of cheap technical collaboration with buy-back clauses from many advanced western and Japanese foundries, the export potential for quality casting to those countries can be well exploited by modern Indian foundries in the near future.

Present challenges to Indian metal casting industry:

- High cost of technology and related modern equipment.
- Lack of quality raw material like metallurgical coke, pig iron, refractories.
- The cost of energy, which is increasing every time. The shortage of electricity in many regions of India especially during summer season.
- Most Indian foundries use obsolete equipment causing low productivity.
- Industry and taxation law policies have become a barrier in the growth of export business.
- Irregular supply of raw material.
- Environment pollution control measures causes extra expenditure to foundries.
- Lack of trained manpower, especially in melting, moulding and technology at various levels.

Metal casting industries must focus on:

- Quality and not quantity with a spirit of producing right first time and every time
- Waste reduction and improvement of productivity
- Defect prevention and not on defect rectification
- Competition on pricing
- Reduction in lead time
- There should not be any tolerance on defects or delays

Review Questions

1. Classify various manufacturing processes with examples.
2. Explain classification of casting processes.
3. Explain hand tools used in manual casting processes.
4. Write a brief note on the history of casting.
5. Explain the basic steps involved in casting processes.
6. List typical industries which use casting processes.
7. What are the main advantages of casting processes?
8. What are the limitations of casting processes?
9. How do you compare casting processes with machining processes?
10. What are the challenges faced by Indian foundries?

CHAPTER 2

Technology of Pattern Making

2.1 INTRODUCTION

Pattern is a mirror image or replica of the component that has to be manufactured by casting process, with some modifications. Except for various allowances and core prints, a pattern exactly resembles the casting to be produced. The main modifications are the addition of pattern allowances, and the provision of core prints. If the casting is to be hollow, additional patterns called cores are used to create these cavities in the finished product. The quality of the casting produced depends upon the material of the pattern, its design, and construction. A pattern is required even if one object has to be cast. The costs of the pattern and the related equipment are reflected in the cost of the casting. The use of an expensive pattern is justified when the quantity of castings required is substantial. The process of making patterns is called pattern making; the persons who make the patterns are known as pattern makers; and the workshop where the patterns are made is called patterns making shop.

A typical pattern attached with gating and risering system is shown in Figure 2.1.

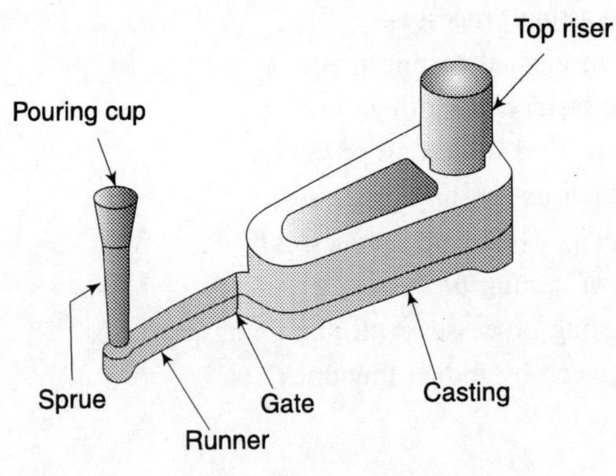

Figure 2.1 A typical pattern with attached gating and risering system

2.2 DIFFERENCE BETWEEN CASTING AND PATTERN

Although pattern is a replica of the casting to be produced, there are few differences between a pattern and a casting.

1. Pattern is usually made of wood. Whereas final casting may be of metal or alloy.
2. The dimensions and shapes of patterns are different from the final casting due to various allowances given on the pattern.
3. Patterns have additional projections called core prints to place cores.
4. Patterns may be made in two or more pieces whereas castings are single products.
5. Patterns may not have all the holes and slots which a casting will have. Holes and slots can be drilled or machined after the metal is casted.

2.3 FUNCTIONS OF A PATTERN

1. A pattern is used to prepare a mould cavity for the purpose of making a casting.
2. For hollow castings, core prints are added to the pattern to support the core in the mould cavity.
3. Gates, riser, and channels are used for feeding molten metal into mould cavity and may become part of the casting. However, in most of the cases they are not an integral part of the pattern.
4. Efforts should be made to provide all allowances in a proper manner on the pattern and the surface of the pattern needs to be made smooth. It results in less casting defects and low cost.
5. A properly constructed pattern minimizes the overall cost of the casting.
6. A pattern helps in establishing a parting line between the cope and drag parts of the mould.
7. A pattern helps in positioning the core properly.

2.4 REQUIREMENTS OF GOOD PATTERN EQUIPMENT

Pattern equipment refers to sets of patterns, supporting plates, coreboxes, sweeps, templates gauges, etc., used in moulding to produce proper mould cavity including assembly of cores. The following are the requirements of good pattern equipment:

1. Dimensional accuracy to get high quality casting.
2. Strength to withstand ramming and abrasion of moulding sand.
3. Rigidity to prevent distortion and warpage through seasonal changes over a long period of use.
4. Good surface for easy removal from mould.
5. Proper colour code to indicate to the moulder, information regarding final casting and precautions during moulding.
6. Should help to increase moulding productivity through suitable design and construction and achieve overall economy in the moulding.

2.5 PATTERN MATERIAL

Some of the materials used for making the patterns are wood, metals and alloys, plastic, plaster of Paris, rubbers, wax, and resins.

Factors to be considered for Selection of Material for Pattern

Before selecting a particular material for pattern, the following factors must be considered:

1. Number of castings to be produced
2. Metal to be cast
3. Mould material used
4. Type of casting process
5. Degree of accuracy and surface finish required
6. Minimum thickness/wall thickness required
7. Method of withdrawal of pattern
8. Size and shape of the casting to be manufactured

Ideal Characteristics of Pattern Materials

The ideal characteristics of pattern material are given as follows:

1. It must be light in weight.
2. It can be worked, shaped and joined easily by gluing, currying and nails.
3. It must be strong, hard and durable.
4. It must have good resistance to wear and abrasion.
5. It must have good resistance to corrosion, and to chemical reactions.
6. It must be dimensionally stable and resistant to variations in temperature and humidity.
7. It must be readily available at low cost.
8. It should be preserved for a long time by applying protective coating on its surface.

2.5.1 Wood

The most commonly used pattern material is wood. Wooden patterns are used when the number of castings to be produced is less. The commonly used woods for making pattern are Teak, Shisham, Kail, Deodar, Sal, Walnut, Mahogany and Pine. They can be classified as hard or soft wood.

1. **Hard wood:** The best quality of pattern wood is obtained from Teak and Mahogany, which are hard, strong, durable and can be finished to a very smooth surface. For larger patterns and medium quantity (say up to 100 components) these are an ideal choice. Teak wood is straight-grained light wood. It is easily workable and wraps easily. Mahogany is a hard wood and durable compared to Pine and Teak. It has uniform grain structure and it is easy to carve and shape. It is costlier than Teak and Pine wood.

2. **Soft wood:** Deodar, Pine, Sal are softer and cheaper than Teak, easily workable but have less life and dimensional stability. For less quality and less critical castings, these are economical.

Seasoning of Wood

Wood is composed of large number of tubular cells. The cell walls of a fibre are held together by organic material. In green condition, the cell cavities and cell walls are full of moisture. In old and dry wood, the cell cavities are empty but cell walls contain upto 30% moisture. The variation in outside atmosphere due to seasonal changes causes changes in this moisture content; if it is below a certain range, it results in distortions/warpage. By natural seasoning over long periods or by artificial seasoning for about a week during which the timber is subjected to hot, moist air in a controlled manner in special heating ovens, the internal moisture of wood can be adjusted to optimum value. This results in dimensional stability, resistance to fungus, higher strength and better working.

Advantages of Wooden Pattern

1. It is readily available and is of light weight.
2. It can be easily shaped and is relatively cheap.
3. It can be preserved easily by applying protective shellac coating.
4. It is easy to finish the pattern using emery paper.
5. It can be joined easily by gluing, currying and nails and intricate shapes can be made easily.
6. It takes less time to make wooden pattern compared to other types such as metallic patterns, etc.

Disadvantages of Wooden Pattern

1. Wood absorbs moisture, which can cause distortion and dimensional changes.
2. It cannot be used for thin cross sections.
3. It wears out quickly as it has less resistance to sand abrasion.
4. It needs preservatives for storing.
5. It cannot be used for a large number of casting impressions.
6. Due to depletion of forests, good hard wood has become quite costly now.
7. Life is short as compared to metallic patterns.

2.5.2 Metallic Patterns

Metallic patterns are preferred when the number of castings required is large to justify their use. They can give good surface finish and withstand high ramming pressures in machine moulding. These patterns are also used in special moulding processes such as shell moulding where the patterns are heated up to 200°C.

The metals commonly used are cast iron, aluminium and copper alloys such as brass and bronze.

The following metals are commonly used for pattern making:

1. **Cast iron:** It is cheaper, stronger and gives good surface finish. It also has good wear resistance against sand abrasion. So it is used in large sizes and long production runs. The high specific gravity of causes handling problems. Cast irons have low impact strength and low reparability.
2. **Brass and bronze:** These are heavier than cast iron and preferred for small castings. They possess good strength and resistance to corrosion and wear. They can be joined easily by soldering. They are preferred for medium to large production range where patterns have to be heated.
3. **Aluminium alloys:** They are light in weight, having low melting point, and easily machinable metals. They have less abrasion resistance against sand abrasion. They are ideal for small production ranges.

Comparative advantages and disadvantages of various pattern materials are shown in Table 2.1.

Table 2.1 Comparison of various pattern materials

Pattern materials	*Advantages*	*Disadvantages*
Aluminium alloys	1. Good machinability 2. Light in weight 3. Good surface finish 4. High corrosion resistance	1. Low strength 2. High cost
Grey Cast Iron	1. Good machinability 2. High strength 3. Low cost 4. Good surface finish 5. Good abrasion resistance	1. Less corrosion resistant 2. Heavy weight 3. It is brittle.
Steel	1. Good surface finish 2. Cheaper than brass 3. High strength/toughness 4. Good repairability/weldability	1. Less corrosion resistant 2. Heavy weight 3. High melting point, more energy is required compared to low melting point alloys.
Brass and Bronze	1. Good surface finish 2. Good machinability 3. Good weldability/solderability 4. High strength 5. High corrosion resistance	1. High cost 2. Heavy weight

(Contd.)

Lead alloys	1. Good machinability	1. High cost 2. High density 3. Low strength
White metals	1. Light in weight 2. Low shrinkage 3. Thin sections can be cast	1. It is a soft material. 2. Abrasion resistance is less.

Advantages of Metallic Pattern

1. It gives good surface finish.
2. Metallic patterns do not absorb moisture and do not warp on storage.
3. Wear rate is less.
4. Accurate dimensions of the castings can be maintained for a large number of casting impressions.
5. It has good durability and strength.
6. It does not change its shape when subjected to moist conditions or is in contact with moulding sand.

Disadvantages of Metallic Pattern

1. Some metal patterns require lots of machining during preparation time. It adds to initial cost of machine, space and cost of the pattern.
2. It may undergo corrosion.
3. Cost is high compared to wood and unsuitable for production of small number of castings.
4. It is heavier compared to wood and difficult to transport/handle.

2.5.3 Plaster of Paris

This material belongs to gypsum family which can be cast and worked with wooden tools.

Advantages

1. It gives good surface finish.
2. It is suitable for casting of intricate shapes.
3. It has high compressive strength.
4. It can be easily worked by wood working tools.
5. No shrinkage allowance is needed for pattern as it expands during solidification.

Disadvantages

1. It is not suitable for mass production.
2. It has low strength.

2.5.4 Wax

These materials are used for investment casting. They are removed by heating and inverting the mould.

Advantages

1. It gives good surface finish.
2. Can be used for thin cross sections.
3. Dimensions can be maintained easily.

Disadvantages

1. Pattern can be used only once, as the pattern is removed by heating.
2. Cost per component is more since every component needs one pattern.

2.5.5 Plastics

A wide variety of plastics are used for pattern making. Different types of plastics like epoxy resins which are chemically hardened as well as polyester resins are used for making patterns. Plastic patterns combine the advantages of wood like less weight, easy machinability and economy with wear resistance and dimensional stability of metals. Plastic patterns are reinforced with fibre glass laminations or even metallic strips. For economy, we can have wooden inserts. Even though quite costly at present in India compared to other developed countries, more use of plastics in pattern equipement is expected in future.

Advantages

1. They are light in weight.
2. They have low shrinkage.
3. They are corrosion resistant, durable and have good strength.
4. They give good surface finish.
5. They have better impact strength than wood.
6. Can be used for thin cross sections.

Disadvantages

1. They require metal reinforcements in thin sections as plastic patterns are fragile.
2. Plastics cannot withstand severe shocks.

2.5.6 Polystyrene

Polystyrene or expanded thermocole is another pattern material which has special property that gasifies on heating. For single quantity casting like prototypes, it is very easy to make a pattern which burns off after pouring the molten material. Polystyrene will give neglible residue. These are used in special mould processes called full mould process, where no hard ramming of mould material is done and the pattern is not removed from the mould for pouring.

Table 2.2 gives the properties of commonly used pattern materials. Table 2.3 gives the pattern material based on expected life.

Table 2.2 The properties of commonly used pattern materials

S. No.	*Properties*	*Wood*	*Mild steel*	*Aluminium alloys*	*Plastics*
1	Machinability	Superior	Good	Good	Good
2	Wear resistance	Poor	Superior	Good	Average
3	Strength	Average	Superior	Good	Good
4	Weight	Superior	Poor	Good	Good
5	Corrosion due to moisture	Superior	Poor	Good	Superior
6	Swelling due to moisture	Poor	Superior	Superior	Superior
7	Reparability	Superior	Good	Poor	Average

Table 2.3 Expected life of different pattern materials

Type of casting	*Number of castings produced before pattern equipment repair*		*Pattern material*
	Pattern	Core	
Small castings (under 600 mm)	2000	2000	Hard wood
	6000	6000	Aluminium/Plastic
	100000	100000	Cast iron
Medium castings (600-1800 mm)	1000	750	Hard wood
	3000	3000	Aluminium/Plastic
Large castings (above 1800 mm)	200	150	Soft wood
	500	500	Hard wood metal reinforced

2.6 TYPES OF PATTERNS

A pattern can be made in many different ways, which are classified into the following:

1. Single-piece pattern or solid pattern
2. Two-piece pattern or split pattern
3. Loose pattern
4. Match plate pattern
5. Cope and drag pattern
6. Follow board pattern
7. Gated pattern
8. Sweep pattern

9. Skeleton pattern
10. Segmental or part pattern
11. Multi-piece pattern

2.6.1 Single-Piece Pattern or Solid Pattern

The single-piece or one-piece pattern is made in one piece without any joints. It is the easiest to fabricate, but can cause some difficulties in making the mould. It is cheapest of all types of patterns. Solid patterns are typically used for geometrically simple parts that are produced in low quantities. This type of pattern is used when pattern withdrawal is easy. The parting line and runner system must be determined separately.

Applications

- Soil tamper
- Stuffing box
- Gland of steam engines

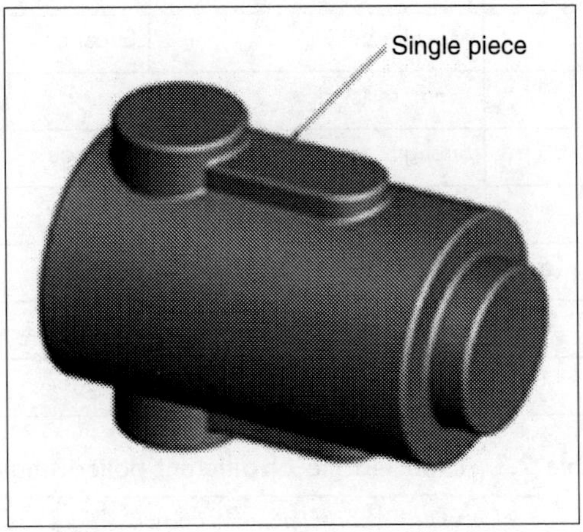

Figure 2.2 Single-piece or solid pattern

2.6.2 Two-Piece Pattern or Split/Pattern

Two-piece patterns are most common in metal casting industries. The pattern is split along the parting surface; the position of split is determined by the shape of the casting. One-half of the pattern is moulded in drag and the other half in cope. The two halves of the pattern must be aligned properly by making use of the dowel pins, which are fitted, to the cope half of the pattern. Split patterns are typically used for parts that are geometrically complex and are produced in moderate quantities.

Applications

- Spindles
- Steam valve bodies
- AK-47 rifle stock

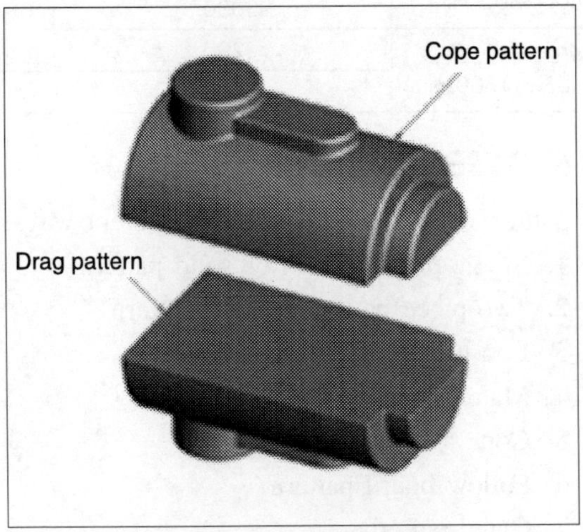

Figure 2.3 Three-piece pattern

2.6.3 Loose Pattern

Sometimes loose pieces need to be added with the pattern. These loose pieces help to facilitate the pattern withdrawal from the mould. This type of pattern is used when the casting has projection, or undercuts. The projections and undercuts may create hindrance in the withdrawal of the pattern from the mould. Hence, loose-piece pattern is used in such cases.

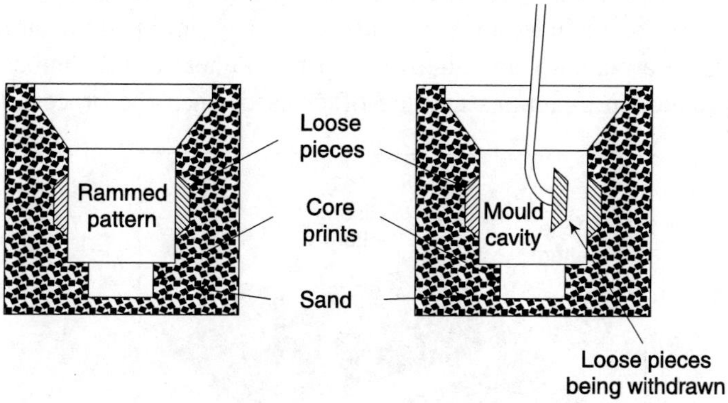

Figure 2.4 Loose-piece pattern

Applications

- Pattern having projections or hanging parts
- Rotor hub
- Axle pin

2.6.4 Match-Plate Pattern

A match-plate pattern is similar to a split pattern, except that each half of the pattern is attached to opposite sides of a single plate. The plate is usually made of wood or metal. This pattern design ensures proper alignment of the mould cavities in the cope and drag and the runner system can be included on the match-plate. Match-plate pattern requires less time to adjust the location of the pattern in the cope and drag portion of pattern is mounted on the opposite of wood or metal plate. Match-plate patterns are used for large production quantities of small size castings and are often used when the process is automated.

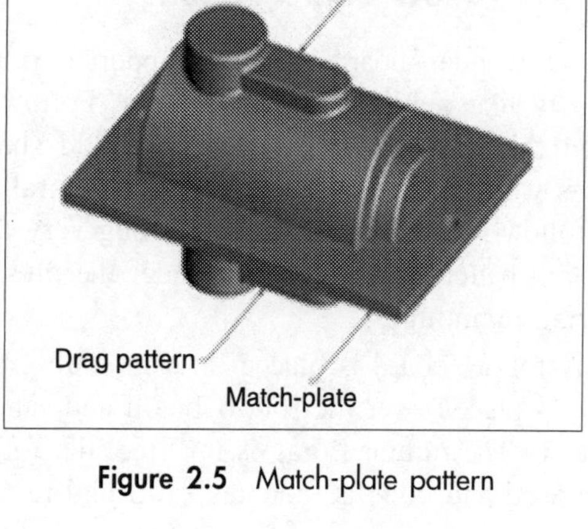

Figure 2.5 Match-plate pattern

Applications

- Piston rings of I.C. engines
- Rotor hub

2.6.5 Cope and Drag Pattern

A cope and drag pattern is similar to a match-plate pattern, except that each half of the pattern is attached to a separate plate and the mould halves are made independently. Just as with a match-plate pattern, the plates ensure proper alignment of the mould cavities in the cope and drag and the runner system can be included on the plates. Cope and drag patterns are often desirable for larger castings, where a match-plate pattern would be too heavy and cumbersome. They are also used for large production quantities and are often used when the process is automated.

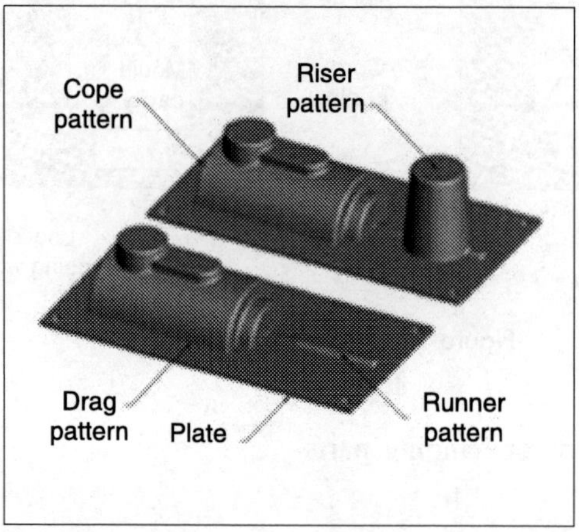

Figure 2.6 Cope and drag pattern

Applications

- Flange pipe
- Water jacket of JCB head
- Hose pipe coupling head

2.6.6 Follow Board Pattern

It is a wooden board used for support during moulding. It acts like a base seat for pattern. Follow boards are used for those patterns which have odd shapes. It supports weak pattern and also acts like natural parting line. A follow board is used for supporting very thin patterns, as thin patterns may collapse under the pressure of sand during ramming.

A follow board is placed on a moulding bed, the pattern is placed over the follow board and sand is rammed over it. The rammed drag is inverted, the follow board is removed and cope is mounted on it and rammed.

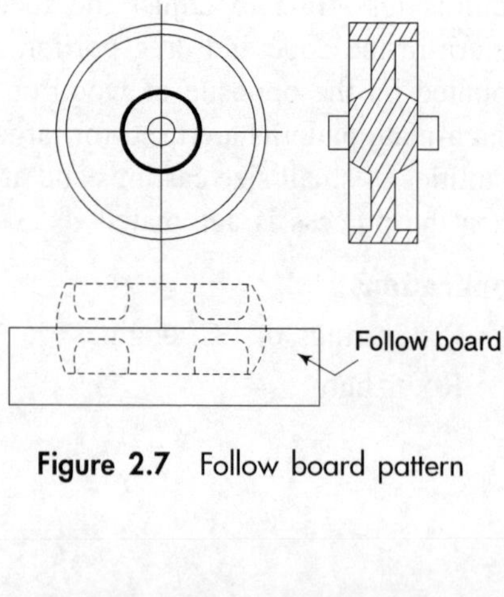

Figure 2.7 Follow board pattern

Application

- It is used for casting master patterns for many applications.

2.6.7 Gated Pattern

Different types of patterns are connected and there is a common sprue which reduces the wastage of metal accumulated in the sprue, runner and riser which, in turn, increases the casting yield. This eliminates the time and inconsistency associated with hand-cutting the gates and runners. Also, since the gating system is designed and fabricated as part of the pattern, the consistency of molten metal flow into the casting and feeding of the casting during solidification is improved. Gated patterns are appropriate for pouring small quantities of castings when quick turn around and low cost are important.

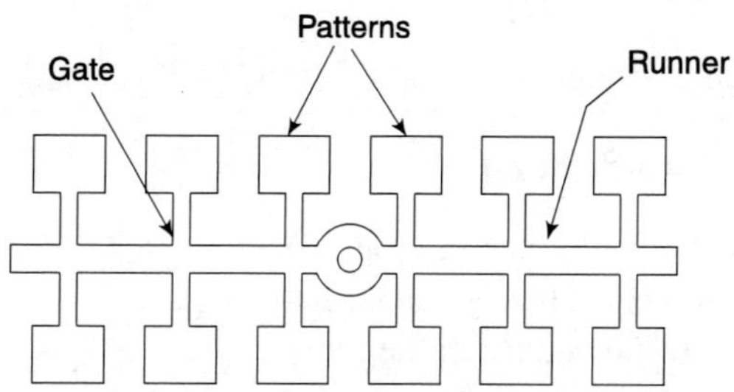

Figure 2.8 Gated pattern

Applications

- Used for small castings such as corner bracket.
- These patterns are used for mass production of small castings.

2.6.8 Sweep Pattern

This is a 2-D pattern used for large castings of circular cross sections and symmetrical shapes.

There is a plate which rotates about a post in this type of pattern. For axi-symmetrical castings such as bells, the full cavity of the casting can be created by sweeping the pattern around the spindle. The pattern is a part of the full shape of the end product.

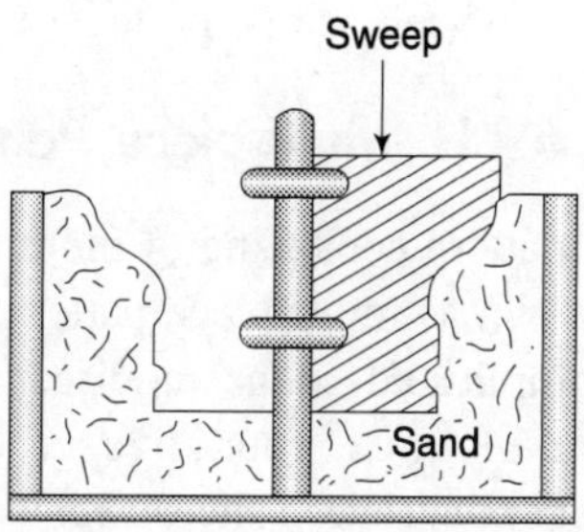

Figure 2.9 Sweep pattern

Applications

- Symmetrical shapes such as wheels, rims
- Large kettles of cast irons
- Bell shapes

2.6.9 Skeleton Pattern

When only a small number of large castings is to be made, it is not economical to make a solid pattern. In such cases however, a skeleton pattern may be used. Just like sweep patterns, these are simple wooden frames that outline the shape of the part to be cast and are also used as guides by the moulder in the hand shaping of the mould. This framework is filled with loam sand and rammed, and surplus sand is removed by strickle board as shown in Figure 2.10. For round shapes, the pattern is made in two halves, which are joined with glue or by means of screw, etc.

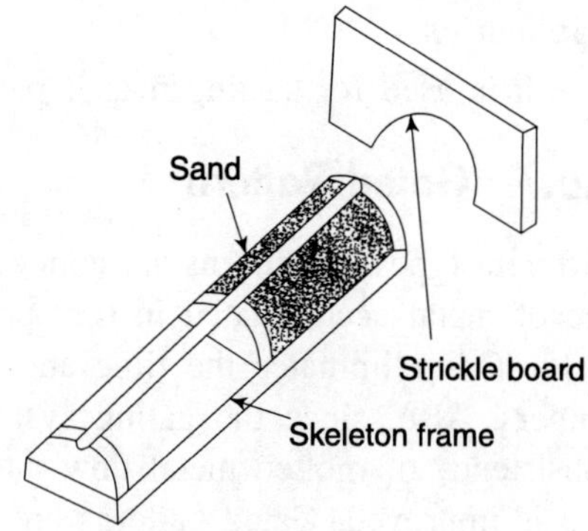

Figure 2.10 Three-piece pattern

Applications

Large castings such as turbines, water pipes, and L-bends

2.6.10 Segmental or Part Pattern

Patterns of this type are used for circular casting. For example, wheel rim, gear blank, etc. The segmental pattern is functionally similar to sweep pattern in the sense, that is, both employ a part of the pattern instead of a complete pattern, for getting the required shape of the mould. It is used for preparing circular castings. In this type it does not revolve continuously like sweep pattern, instead prepares the mould by parts. It completes one portion of the mould and then moves to next position to make the next part of the mould and so on till the mould is completed.

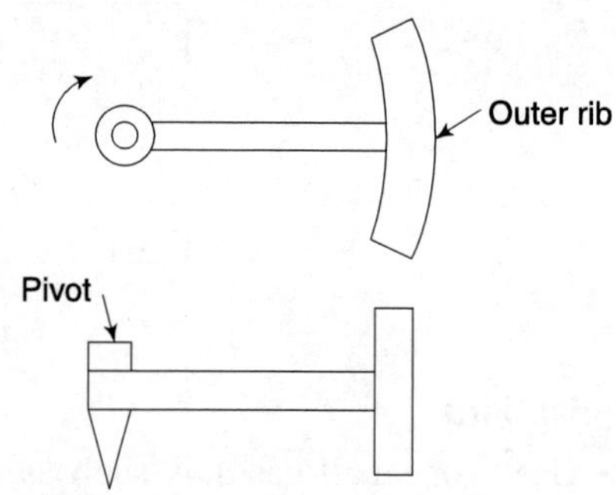

Figure 2.11 Segmental or part pattern

Applications

- Used for circular work like rings, gears, wheels, rims, pulleys, etc.

2.6.11 Multi-piece Pattern

A pattern consisting of more than two pieces is known as multi-piece pattern. Casting of more complicated shape required a pattern consisting of three, four, five or even more parts, depending upon the design of castings. A pattern having three pieces requires three piece flasks and so on.

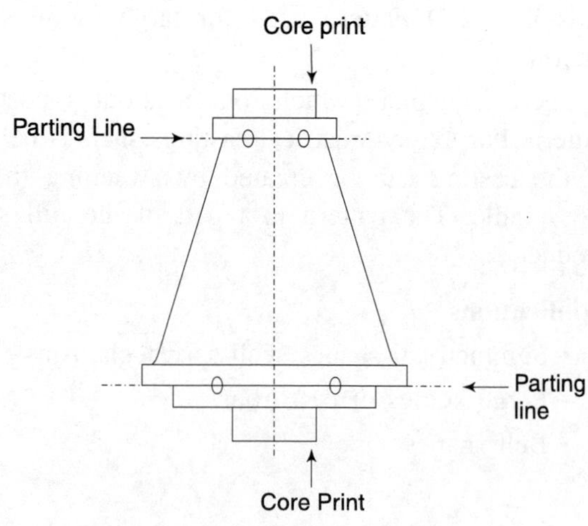

Figure 2.12 Multi-piece pattern

2.7 MASTER PATTERNS

Master pattern is the name given to a pattern having a double contraction or shrinkage allowance. The castings made from the master pattern are used by the patternmaker in making duplicate patterns that will have only a single contraction or shrinkage allowance for the particular metal required of the final casting. An allowance is made not only for the shrinkage of the metal pattern, but also, a second allowance is added for the shrinkage of the casting produced from the metal pattern. For example, a master pattern constructed for an aluminium pattern to be used in producing bronze castings has the shrinkage allowance for aluminium (x per mm) plus the shrinkage allowance for bronze (y per mm) or a total shrinkage allowance of (x + y) per mm. If the final casting is to be machine finished, further allowances must be made for finishing the metal pattern and final casting. The actual metal pattern is made in the foundry, the same as any other casting of similar complexity would be made.

The different types of material that can be used in the construction of master patterns are numerous. These types include wood, plaster, aluminium, brass, bronze, plexiglass, renishape, and more. As comparatively few castings are made from a master pattern, the cost of the materials is held to a minimum so that the pattern can be made as economically as possible. Therefore, for the general purposes of lower cost and of speed in the manufacture of a master pattern, poplar is the most logical material to use.

Master patterns can be used for a variety of manufacturing operations, including sand casting, rubber-plaster moulding, epoxy tooling, hybrid tooling, RTV moulding, and vacuum forming. Depending on the application, nowadays master patterns are made using SLA, and CNC technologies, depending on the specific requirements.

2.8 DESIGN OF PATTERNS

The following factors should be taken into consideration while designing a pattern.

1. The parting line should be so chosen that the smallest portion of the pattern remains in the cope. Careful selection of parting line is required for easy removal of pattern.
2. Critical surfaces which are desired to be cleaner and sound should be placed in drag because the possible defects due to loose sand and inclusion will occur in the cope.
3. As far as possible, full cores should be used instead of jointed half cores. It not only reduces the cost of the cores, but also increases the dimensional accuracy for castings.
4. The use of cores should be minimized by using offset parting wherever possible.
5. Proper allowances must be given wherever necessary.
6. Proper material must always be chosen for pattern making depending upon life, and application.
7. For mass production, the use of several patterns in a mould with a common riser is to be preferred. That is, gated pattern should be used.
8. Wherever there is a sharp corner, a fillet should be provided, and the corners may be rounded up for easy withdrawal of patterns as well as easy flow of the metal in the mould.

9. Abrupt changes in the thickness of the pattern should be avoided by having a gradual change. As far as possible, the wall thickness of various sections should be uniform.
10. As far as possible, the patterns should have a good surface finish, because the surface finish of the casting depends lot on the surface finish of the pattern.

2.8.1 Pattern Allowances

A pattern is always made somewhat larger than the final job to be produced. This excess in dimension is referred as the pattern allowance. Pattern allowance is a vital feature as it affects the dimensional characteristics of the casting.

The liquid metal on solidification undergoes various types of shrinkage. This shrinkage may be because of metal characteristics and also may be because of pattern, gating, risering design. The shrinkage affects the dimensional accuracy needed in the final casting. Hence, in order to get a correct casting, various allowances called pattern allowances need to given on the pattern before making the mould and subsequent pouring in the mould cavity. Similarly, the pattern allowances are also needed to be provided on the core patterns. Selection of accurate allowances greatly helps to reduce costs and avoid rejections.

The allowances usually considered on patterns and coreboxes are:

- Shrinkage/contraction allowance
- Draft/taper allowance
- Machining/finish allowance
- Distortion/camber allowance
- Rapping allowance
- Scaling and grinding allowance

1. Shrinkage allowance

Almost all cast metals shrink or contract volumetrically on cooling. The shrinkage allowance is provided to take care of the contraction of a casting. The total contraction of a casting takes place in three stages:

(i) The contraction of the liquid from the pouring temperature to the freezing temperature.

(ii) The contraction associated with the change of phase from liquid to solid. This is known as **solidifying contraction** or **liquid shrinkage.**

(iii) The contraction of the solid casting from freezing temperature to room temperature. This is known as **solid shrinkage**.

Out of the above three, the first two are taken care of by proper riser and gating systems. Only the last one, that is solid contraction or solid shrinkage is taken care of by the pattern makers by giving a positive shrinkage allowance.

(a) **Liquid shrinkage:** After pouring the liquid metal in a mould, metal starts solidifying. When metal solidifies from liquid state to solidus temperature, there is reduction in volume of the liquid metal. This reduction is known as liquid shrinkage. The reduction in volume is compensated by the riser, which stores liquid metal and the metal in riser remains liquid later than that of metal in cavity. In Figure 2.13, line 2-3 corresponds to liquid shrinkage which is compensated by riser in the mould.

(b) **Solid shrinkage:** Once the casting is solidifies in the mould cavity, the contraction in volume, when the temperature of the casting moves from solidus to room temperature, of the casting occurs. This contraction in volume is taken care of by providing the shrinkage allowances in the pattern. Obviously, the amount of shrinkage allowances depends on the linear coefficient of thermal expansion. The higher the value of this coefficient, the more the value of shrinkage allowances. In Figure 2.13, line 3-4 corresponds to solid shrinkage that is given on pattern.

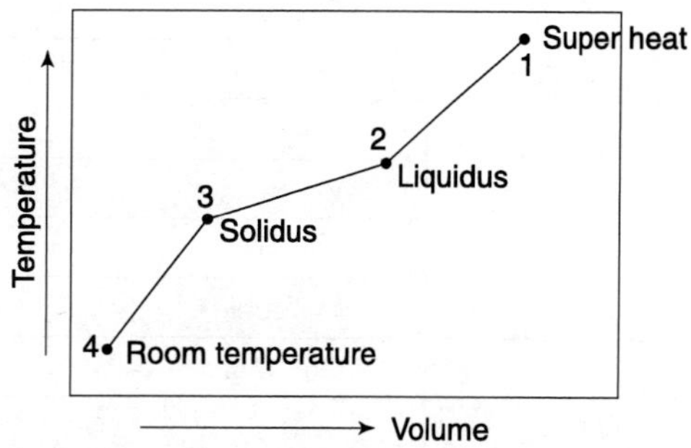

Figure 2.13 Liquid and solid shrinkages

The rate of contraction with temperature is dependent on the material. For example, steel contracts to a higher degree compared to aluminium. To compensate the solid shrinkage, a shrink rule must be used in laying out the measurements for the pattern. Table 2.4 and Table 2.5 give details of the pattern shrinkage or contraction allowance for different alloys.

The shrinkage of metals also depends on:

1. Pouring temperature of the molten metal
2. Design and dimensions of casting
3. Mould material
4. Method of moulding

In foundries, a special scale known as pattern maker's scale is used by pattern makers. This scale is greater in dimensions than ordinary scale and includes pattern allowances due to metal shrinkage. Pattern maker's scale can accommodate four materials, two on each side of the scale.

Table 2.4 Details of pattern contraction allowance for different alloys in percentages

S. No.	*Metal/Alloy*	*Pattern Contraction Allowances %*
1	Aluminium and aluminium alloys	1.3
2	Brass	1.56
3	Bronze	1.80
4	Aluminium bronze	2-2.3

(Contd.)

5	Copper	1.6
6	Cast Iron	
	Grey	0.9 – 1.3
	White	2
	Ductile	0.6-0.8
	Malleable	1 – 1.4
7	Gun Metal	1 – 1.6
8	Lead	2.6
9	Magnesium	2.0
10	Magnesium alloy	1.6
11	Manganese bronze	2.0
12	Nickel	2.0
13	Nickel silicon	1 – 1.6
14	Silicon bronze	1.3 – 1.6
15	Carbon steel	2.0
16	Manganese steel	2.6
17	Chromium steel	2.0
18	Tin	2.0
19	White metal	0.6
20	Zn & Zn alloys	2.6

Table 2.5 Details of pattern contraction allowance for different alloys in terms of absolute values

Materials	*Dimensions*	*Shrinkage Allowances (mm/mm)*
Grey cast iron	Up to 610 mm 610 mm to 1220 mm Over 1220 mm	0.0010 0.009 0.007
Cast steel	Up to 610 mm 610 mm to 1830 mm over 1830 mm	0.021 0.016 0.013
Aluminium	Up to 1220 mm 1220 mm to 1830 mm Over 1830 mm	0.013 0.012 0.010
Magnesium	Up to 1220 mm Over 1220 mm	0.014 0.013

Note: When pattern contains hole, negative allowance has to be given as shown **in Figure 2.14.**

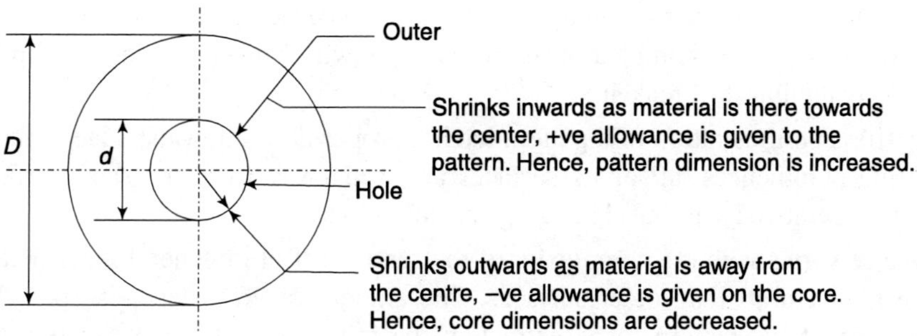

Figure 2.14 Shrinkage allowance

2. Draft allowance or taper allowance

Draft refers to a taper put on the surface parallel to the direction of withdrawal of the pattern from the mould cavity. Draft is provided on all vertical surfaces of the pattern so that it can be removed from the sand without tearing away the sides of the sand mould and without excessive rapping.

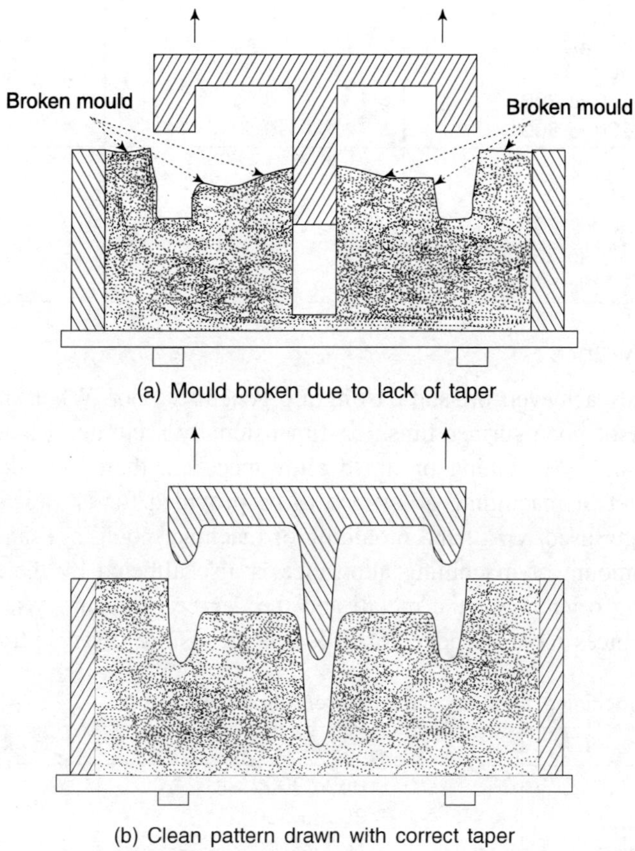

Figure 2.15 (a) Part with no draft allowance and (b) Part with draft allowance

Figure 2.15 (a) shows a pattern having no draft allowance being removed from the pattern. In this case, till the pattern is completely lifted, its sides will remain in contact with the walls of the mould, thus tending to break it.

Figure 2.15 (b) is an illustration of a pattern having proper draft allowance. Here, the moment the pattern lifting commences, all of its surfaces are well away from the sand surface. Thus, the pattern can be removed without damaging the mould cavity.

Draft allowance varies with the complexity of the sand job. But in general inner details of the pattern require higher draft than outer surfaces. The amount of draft depends upon the length of the vertical side of the pattern to be extracted; the intricacy of the pattern; the method of moulding; and pattern material.

Draft allowance for hand moulding = 1.5 – 2.0 mm per 100 mm length for machine moulding = 1 degree. Table 2.6 gives the details of pattern draft allowance for different materials.

Table 2.6 Details of pattern draft allowance for different materials

Pattern material	*Height of the given surface (mm)*	*Draft angle for external surface (degrees)*	*Draft angle for internal surface (degrees)*
Wood	25	3.00	3.00
	25 to 50	1.50	2.50
	50 to 100	1.00	1.50
	100 to 200	0.75	1.00
	200 to 800	0.50	1.00
Metal and plastic	25	1.50	3.00
	25 to 50	1.00	2.00
	50 to 100	0.75	1.00
	100 to 200	0.50	1.00
	200 to 800	0.50	0.75

3. Machining allowance

The finish and accuracy achieved in sand casting are generally poor. When the casting is functionally required to be of good surface finish or dimensionally accurate, it is generally achieved by subsequent machining. Machining or finish allowances are therefore added in the pattern dimension. The amount of machining allowance to be provided for is affected by the method of moulding and casting used, viz., hand moulding or machine moulding, sand casting or metal mould casting. The amount of machining allowance is also affected by the size and shape of the casting; the casting orientation; the metal; and the degree of accuracy and finish required. The machining allowances recommended for different metals are given in Table 2.7.

Table 2.7 Details of machining allowance for different materials

Metals	*Dimensions (mm)*	*Allowances (mm)*
Cast iron	Up to 300	3
	300 to 500	5
	500 to 1000	6.25

(Contd.)

Cast steel	Up to 150 150 to 500 500 to 1000	3 6.25 7.5
Nonferrous	Up to 200 200 to 300 300 to 1000	2.25 3 4

4. Distortion/camber allowance

Sometimes castings get distorted, during solidification, due to their typical shape. For example, if the casting has the form of the letter U, V, T, or L, etc., it will tend to contract at the closed end causing the vertical legs to look slightly inclined. This can be prevented by making the legs of the U, V, T, or L shaped pattern converge slightly (inward) so that the casting after distortion has its sides vertical. Figure 2.16 shows distortion and cambered pattern of I section.

The distortion in casting may occur due to internal stresses. These internal stresses are caused on account of non-uniformity of different sections of the casting and hindered contraction. The following are the main reasons for distortion and warpage:

1. Uneven shrinkage of different parts of casting
2. Poor design of casting
3. Irregular shape of casting
4. Uneven thickness of the different sections

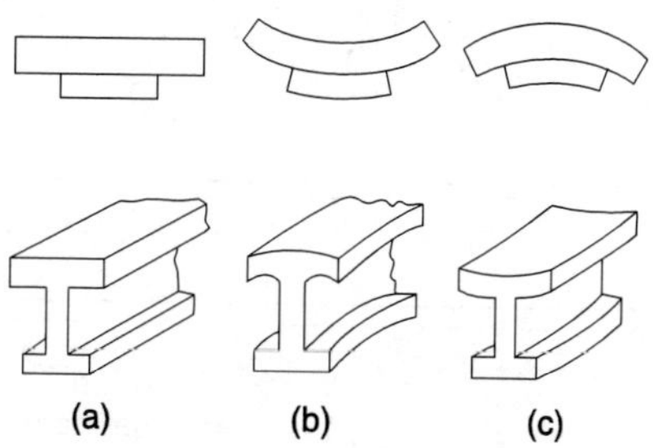

Figure 2.16 (a) Required shape (b) Distorted shape (c) Cambered pattern

Measures taken to prevent the distortion in casting include:

1. Modification of casting design
2. Providing sufficient maching allowance to cover distortion effect
3. Providing suitable allowance on the pattern, called camber or distortion allowance (inverse reflection)

4a. Rapping allowance

Before withdrawal from the sand mould, the pattern is rapped all around the vertical faces to enlarge the mould cavity slightly, which facilitates its removal. Since it enlarges the final casting made, it is desirable that the original pattern dimension be reduced to account for this increase. Thus, rapping allowance is a negative allowance. There is no sure way of quantifying this allowance, since it is highly dependent on the experience of foundry personnel.

5. Scaling and grinding allowance

It is the allowance given to take care of loss of metal due to oxidation/scaling during heat treatment. Also to take care of loss of metal due to grinding while levelling the surface of the casting. Normally the magnitude of this allowance is 0.2 –3 mm.

2.8.2 Pattern Layout

After finding out the pattern allowance, pattern layout drawing must be made preferably to 1:1 scale indicating the given casting drawing of the final machined components. The various allowances are then added in the different views. Figure 2.17 and Figure 2.18 show typical pattern layout drawings.

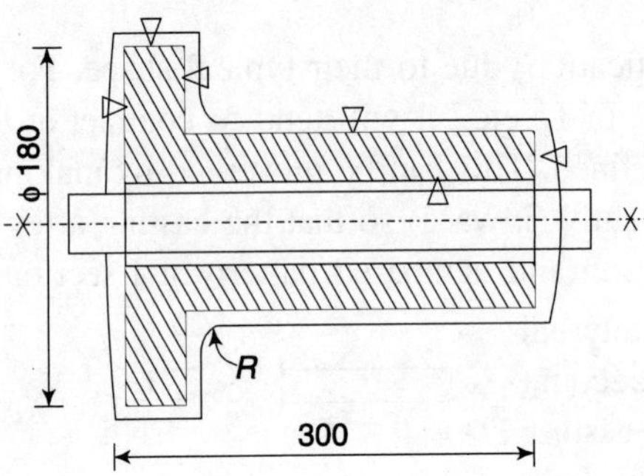

Part Name: Gun metal bush
Material: Gun metal
Type: Wooden split pattern
Contraction allowance: 2%
Machining allowance: 3mm/face
Draft allowance: Positive
Fillet radius: 6 mm

Figure 2.17 Pattern layout for bush

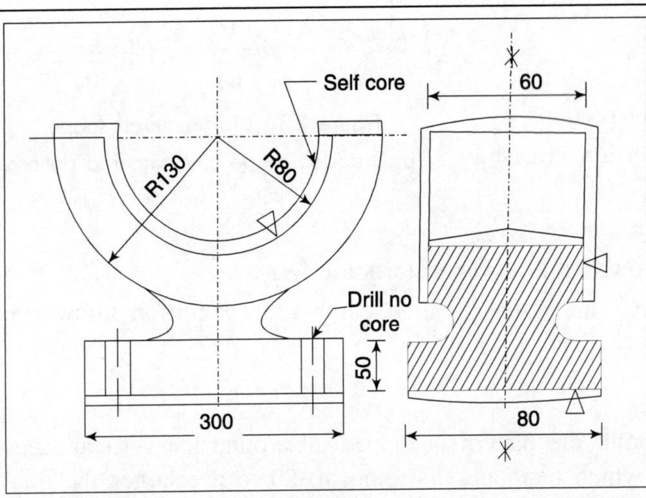

Part Name: Bearing support
Material: Cast iron
Type: Wooden split pattern
Contraction allowance: 1%
Machining allowance: 3mm/face
Draft allowance: Positive

Figure 2.18 Pattern layout for bearing support

The following steps are to be followed for pattern layout:

1. Carefully study the drawing of the casting to be made.
2. Draw two views of the part using pattern maker's scale or otherwise using scale option in CAD software.

3. Add machining allowance as per the specifications.
4. Provide different allowances such as draft allowance, rapping allowance, etc.
5. Locate the position of core prints.
6. Make suitable drawings of cores with allowances, clearly indicating their locations.

Parting line: The layout also clearly shows the parting line used while moulding the pattern that separates the cope and drag mould halves, even when the pattern is not split. Pattern drafts are given according to the parting line. Generally, maximum possible part of the pattern should be taken in the drag to reduce the metallostatic pressure on the cope. As the tendency for defects like sand inclusions, blow holes and shrinkage is raised to the top faces, critical faces of casting are taken to the bottom, leaving non-critical faces to the top, while moulding.

2.8.3 Pattern Construction

Once the layout drawing is done, next step is to construct the pattern. If the pattern is simple, it can be made from a single piece of wood. If the pattern is complicated, then number of wooden pieces, planks, batons or sections of wood are jointed, glued or screwed to form the full pattern. For large patterns, a base frame is constructed from the heavier sections with good joints. Over this frame, wooden batons are fixed forming the outer surfaces; while joining, the directions of grains in wood are considered. The pattern designer decides the size, thickness of joining members, types of joining different pieces and method of finishing to full size, all being indicated by proper drawing with bill of materials.

Properly seasoned and defect-free wood is selected for main frames and joined as per drawing instructions. The individual pieces may be finished by machines or by hand and attached to the main frame. The reinforcing members (called stop-offs) are used if thin and delicate pattern is to be made. Once the body is built it is finished by manual method, using suitable gouges for curved profiles, to the final pattern layout dimensions. The surface of the pattern needs to be finished to a high degree of smoothness by sand paper and putty.

Steps involved in the construction of pattern are:

1. Make a careful study of pattern layout. Locate parting line, number of parts in which the pattern is to be made.
2. Fabricate the different parts of pattern using hand tools and pattern making machines.
3. Join the different parts by nailing, gluing, etc.
4. Inspect the pattern for alignment of different parts and accuracy.
5. Provide suitable fillet in order to avoid sharp corners.
6. Polish the surface of pattern using sand paper and give three coatings of shellac (if it is wooden pattern) to avoid moisture absorption.
7. Paint the pattern as per the pattern colour codes followed in the foundry.

2.8.4 Pattern Colouring

After the patterns are prepared, they are finished by sanding to remove tool marks and other irregularities. After finish, two or three coats of shellac are usually applied to fill up the pores and give a smooth finish. It is essential to obtain a good finish of pattern, since the finish of casting depends on it. For preserving the pattern for a long period, they are spray or brush painted with good quality of enamel paint. Generally, patterns of very large size required for a small number of casting are not painted, whereas the medium or small size patterns are painted in different colors to indicate the position of core prints, stop-off pieces, machined surfaces, etc.

The patterns are normally painted with contrasting colours such that mould maker would understand the functions clearly. The purposes of pattern colouring are to:

1. Identify the core prints, loose pieces, etc.
2. Identify the main body of the pattern and different parts of the pattern.
3. Identify the surfaces to be machined or unmachined.
4. Idicate the type of material to be cast.

Typical colour codes used are given in Table 2.8.

Table 2.8 Colour codes for pattern

Red	Cast surfaces to be machined
Yellow	Seats for core prints
Black	Cast surfaces to be unmachined
No colour	Parting surfaces
Red stripes and yellow base	For loose pieces and seating
Yellow stripes on black background	Core prints for machine openings
Black stripes on yellow background	Supports

2.8.5 Pattern Life

The life of a pattern is the number of good moulds that can be made using a pattern without major repairs. Table 2.9 gives an indication of the life of patterns in relation to the design, material and moulding process adopted.

Table 2.9 Average life of patterns in terms of mould that can be produced per pattern

S. No.	*Pattern Material*	*Pattern Construction/Type*	*Moulding Method*	*No. of Moulds*
1	Soft wood	Loose/Single piece	Hand moulding	Up to 20
		Mounted/Plated pattern	Hand moulding	Up to 50
2	Hard wood	Loose/Single piece	Hand moulding	Up to 50
		Mounted/Plated pattern	Hand moulding	Up to 200
		Mounted/Plated pattern	Machine moulding	Up to 500
3	Aluminium or epoxy resin	Mounted/Plated pattern	Hand moulding	Up to 2000
		Mounted/Plated pattern	Machine moulding	Up to 5000
4	Cast iron/Bronze	Mounted/Plated pattern	Hand moulding	Up to 6000
		Mounted/Plated pattern	Machine moulding	Up to 20000

2.8.6 Core Prints

Whenever a core is to be used to form the internal shape of a casting, it is necessary to provide proper location and support for the core in the mould. Core prints are projects provided on the pattern for this purpose. They are given vertically or horizontally, depending on the position of the core and moulding convenience. Larger taper is given on core prints than on patterns for easy core setting and moulding. Figure 2.19 shows the horizontal and vertical core prints. Table 2.10 gives the size of core prints for a given core size.

Table 2.10 Core print dimensions

S. No.	Diameter or maximum thickness of core (D) (mm)	Size of core prints (P) (mm)
1	Up to 10	15
2	11-19	20
3	20-30	25
4	31-50	30
5	51-80	40
6	81-120	50
7	121-180	60
8	181-250	70
9	251-350	80
10	351-500	90

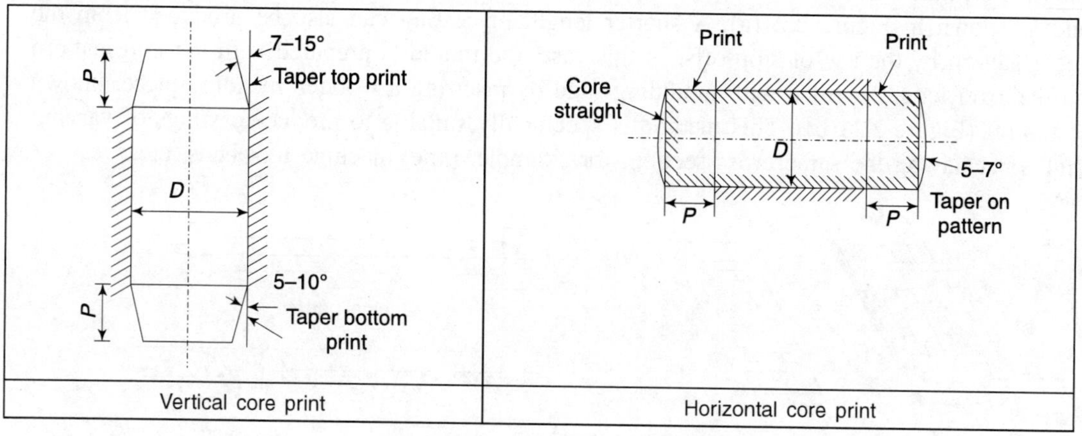

Figure 2.19 Types of core prints

2.8.7 Storing of Pattern

It is important to store the pattern properly to avoid rust, warpage and distortion.

1. Pattern should be stored where temperature and humidity can be controlled, especially in case of precision patterns such as engine block, etc.
2. Pattern should be well codified and numbered so that retrieval and storage becomes easy and minimum time is spent.
3. Pattern should be stored in suitable rack systems.
4. Bin card systems and store ledger should be used for small castings. Automatic storage and retrieval systems (ASRS) should be used for large and precision casting.

2.9 SPECIAL CONSIDERATIONS IN DESIGN OF PATTERNS

2.9.1 Use of Drawback

This is a sand block which is taken away after the removal of cope to facilitate the withdrawal of pattern (Figure 2.20). The drawback is then replaced in its position. While using an old casting as a pattern, a loose piece cannot be used and the use of a core will require the core print to be fitted on the pattern which may not be convenient. In such cases, the use of drawback is most convenient.

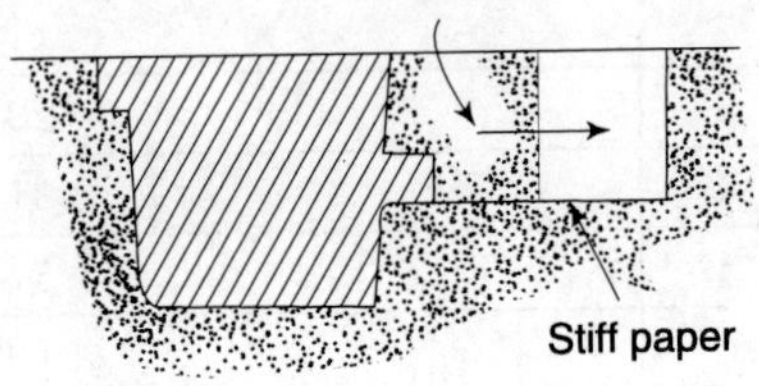

Figure 2.20 Use of drawback

2.9.2 Use of Stop-off

If a casting is thin plate without any ribs, then the pattern will warp and get distorted. In such cases, the use of pieces may be fitted on the pattern to prevent it from distorting. After the pattern is withdrawn, the cavities produced in the mould by the stop-offs are filled with mould sand as shown in Figure 2.21(a). A shorter length of casting can also be produced from the longer pattern by the use of stop-offs. In this case, the mould is prepared with the long pattern and the extra length is filled with moulding sand by inserting a wooden mending piece, known as stop-off (Figure 2.21(b)). This method is specifically suitable to produce castings of varying lengths but having the same cross section, for example, pipe, machine tool beds, etc.

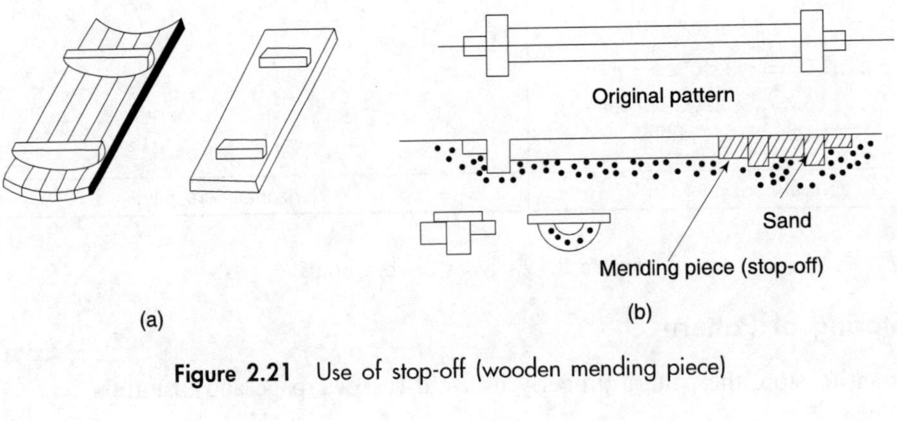

Figure 2.21 Use of stop-off (wooden mending piece)

2.9.3 Use of Fillet

This is the rounding of an internal corner resulting from two intersecting surfaces as shown in Figure 2.22. The shape of this connecting surface is made concave.

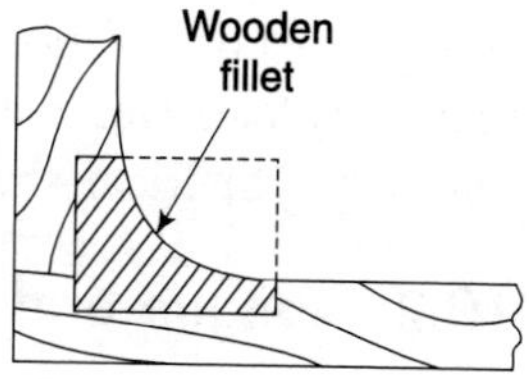

Figure 2.22 Use of fillet

(a) A smooth fillet is made at the sharp corners to provide easy flow of metal into the mould.

(b) Uniform cooling without stress concentration.

(c) More sound casting without defects.

(d) Easy withdrawal of pattern from the mould.

The fillet can be made of wood, wax, leather, plastic or combination of sawdust and glue.

2.10 CONSTRUCTION OF PATTERN AND CORE BOXES

Various types of construction employed for pattern and core boxes include: (a) ring or segmental construction, (b) tongue and groove construction, (c) disc construction, (d) header and stave construction, (e) box construction, and (f) combination of wood and metal.

2.10.1 Ring and Segmental Type

This type of construction consists in building up the whole length in a number of layers or laminates and each layer is made of several segments of wood as shown in Figure 2.23(a).

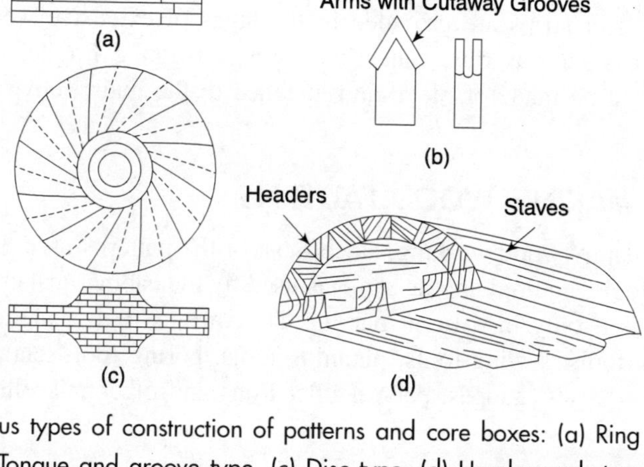

Figure 2.23 Various types of construction of patterns and core boxes: (a) Ring and segmental type, (b) Tongue and groove type, (c) Disc type, (d) Header and stave type

This type of construction offers many advantages such as: (a) requires small wooden pieces, (b) involves least wastage of wood, and (c) results in strong and stable shapes. It is most suited for hollow cylindrical shapes having diameter more than 200 mm.

2.10.2 Tongue and Groove Type

In this case, a groove is cut to accommodate the tongue of the board as shown in Figure 2.23(b). This type of construction is suitable for making the patterns of a circular pulley or fly wheel where the hub and the rim are joined by arms.

2.10.3 Disk Type Construction

In this case also, the laminates are used to build up the desired thickness. Each laminate is formed by joining small pieces of wood laid down in a non-radial manner (Figure 2.23(c)). This type of construction provides structural strength and the pieces remain firm even during machining.

2.10.4 Header and Stave Type Construction

In this case, the ends of hollow pattern are closed by pieces called header on which the lengthwise sections called staves are fixed (Figure 2.23(d)). The headers and staves are glued and shaped finally to the required dimensions. It is used to make large shapes which are too expensive to be made completely solid.

2.10.5 Box Type Construction

This type of construction is useful when the inside of the pattern can be made hollow and the pattern is of large size. The weight of the pattern and the quantity of wood required is reduced.

2.10.6 Wood and Metal Composite Construction

In this case, metal plates or strips are provided in the areas of extensive wear during moulding. The projecting pieces such as ribs, vanes, webs, etc. are quite fragile, if made in wood, and thus, these pieces can be made of metal and attached to the main body of the pattern and core box.

2.11 TOOLS FOR MAKING WOOD PATTERNS

The pattern making is mainly wood working, since most of the patterns used are made in wood. Therefore, most of the tools required are the same as used by the cabinet maker with the addition of some special tools used only in pattern making. The various tools generally employed are measuring and marking tools, sawing tools, planning tools, boring tools, clamping tools, core-box plane, pattern maker's saw, gauges, pattern fillet irons and other miscellaneous tools.

2.11.1 Measuring and Marking Instruments

In marking patterns and coreboxes, the various instruments used for measurements and marking include shrinkage rule, rule, try square, bevel, marking gauge, dividers, outside calipers, inside calipers, vernier calipers, etc. These are shown in Figure 2.24.

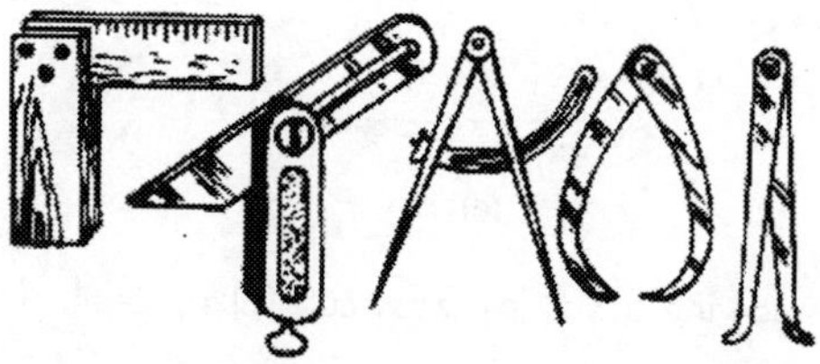

(a) (b) (c) (d) (e)

Figure 2.24 Measuring and marking tools: (a) square, (b) bevel, (c) divider, (d) outside caliper, (e) inside caliper

The shrinkage rule is used to measure the dimensions of workpieces of patterns and core boxes. This rule is made with suitably over sized graduations to take care of allowances for contraction of castings. The try square is used to check right angle and make lines normal to each other on bars and planks.

The marking gauge is used for marking and drawing parallel lines on the plank. The dividers are used for making circles, arcs of circles and for transformation of dimensions. Trammels or beam compass is used to layout the distances and scribe the arcs and circles beyond the capacity of ordinary dividers.

The outside caliper is used to measure outside dimensions or circular shapes and the thickness of pieces. The inside calipers are employed to measure the hole diameters, recesses and the distances between the individual parts of a pattern.

2.11.2 Sawing Tools

The various sawing tools employed in pattern making include coping saw, tenon saw, panel saw, compass saw, back saw, dovetail saw, pattern maker's saw.

Coping saws are employed for cutting curves and irregular shapes from thin shapes of wood. Panel saw can cut in either direction. They have cross cut teeth (Figure 2.25(a)) or rip teeth (Figure 2.25 (b)). Compass saw is used for marking cutouts on the inside of a piece of wood (Figure 2.25(c)).

Back saw refers to the saw having an unsually thin blade reinforced with a rigid steel back. The blade used is hard and tough having a row of fine cross cut teeth filled along the cutting edge. The dovetail saw is extremely thin, has very fine teeth.

Pattern maker's saw is made of a thin steel blade and is designed for the accurate work required in pattern making. Milter saw is an oversized backsaw which is used in conjuntion with miter box.

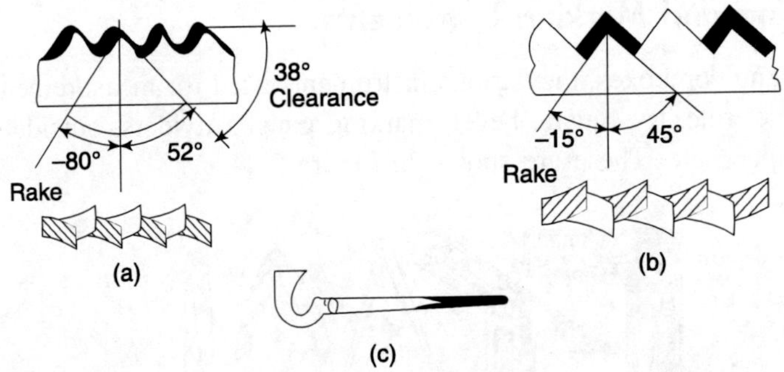

Figure 2.25 Various sawing tools: (a) cross cut, (b) rip teeth, (c) compass saw

2.11.3 Planing Tools

There is a great variety of planing tools used in pattern and core making. The important ones include block plane, rabbet plane, bench plane, router plane, circular plane, core box plane, spokshave, etc.

The block plane is a small, light weight planing tool with an adjustable blade (Figure 2.26). This is made to be worked with one hand. The cutting blade rests on it at an angle of about 20° to the bed of the plane. These planes are particularly suited for small work of accuracy.

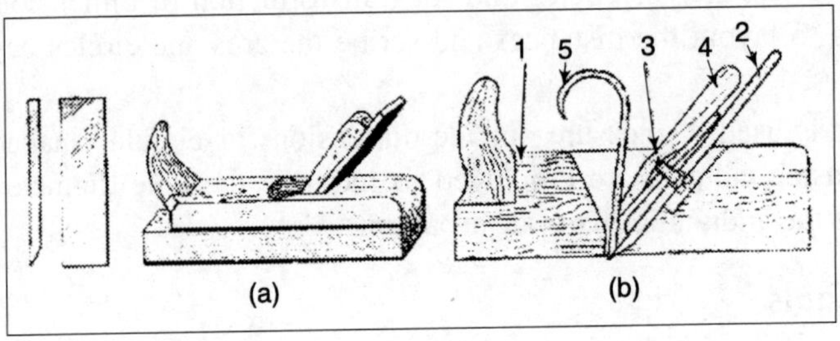

Figure 2.26 Block planes: (a) single iron and (b) double iron (1-block, 2-cutter, 3-curve plane, 4-wedge, 5-shaving)

A rabbet plane is used to plane into corners or against perpendicular surfaces. Bench plane is used for small patterns. A router plane does not resemble other planes which are used to cut the surfaces of the bottom of grooves or other depressions. A circular plane has a flexible steel bottom adjustable for convex or concave surfaces of a wide range of curvatures. These planes are particularly useful in smoothing circular surfaces of patterns. A corebox plane is used to plane a semicircular groove in a pattern. A spoke shave is a small plane with a very short bed. Its name is derived from its first use made for shaping of spokes of wagon wheels. These are available in different types for making inside or outside curves.

2.11.4 Cutting Tools

Varieties of chisels are used for cutting work in pattern making. The important chisels include side chisel, round nose chisel, profile chisel, and ripping chisel.

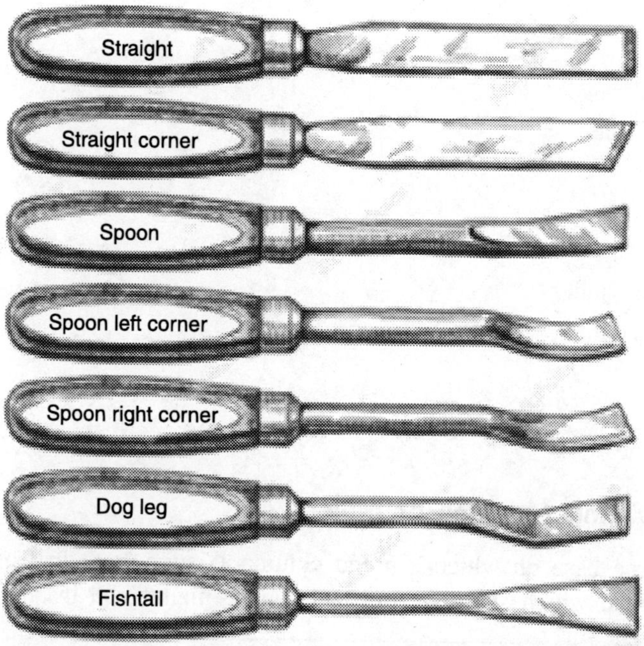

Figure 2.27 Types of cutting tools

2.11.5 Boring Tools

For making round holes in patterns, drill bits are used which are available in a variety of shapes and sizes. The various types of drills are auger bit, twist drill, counter sink, centre bit, expansion bit and multi-spar machine bit as shown in Figure 2.28. Auger bits are held to cut clean holes. A twist drill is used for drilling small holes in wood. A counter sink is used to enlarge hole. Multi-spar is used for larger holes.

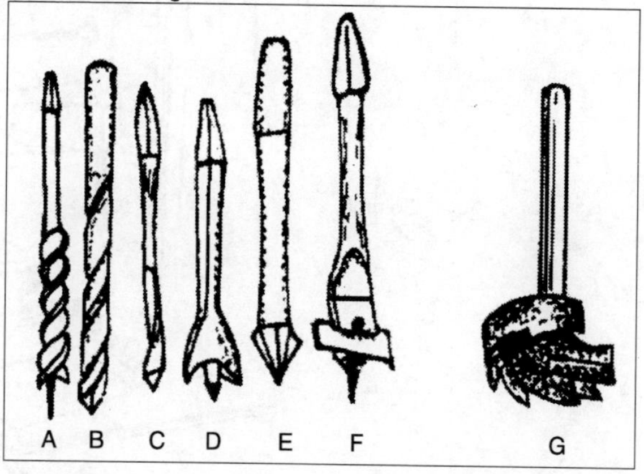

Figure 2.28 Boring tools: A–Auger bit, B–Twist drill, C–Twist drill, D–Centre bit, E–Counter sink, F–Expansion bit, G–Multispar machine bit

2.12 MACHINES FOR WOODEN PATTERN MAKING

Most of the machines employed in wooden pattern making are similar to those used for other jobs in wood working. The machines usually used are:

1. Pattern milling machine or miller
2. Circular saw
3. Band saw
4. Jig saw or scroll saw
5. Wood working lathe
6. Planer
7. Jointer
8. Shaper
9. Disc and bobbin sander
10. Abrasive belt machine
11. Drill press
12. Wood trimmer
13. Grinding machine

2.12.1 Pattern Milling Machine or Miller

This consists of a large base on which a broad column is supported. In front of the base, a table support is provided which slides on rail and can be maintained in any desired position.

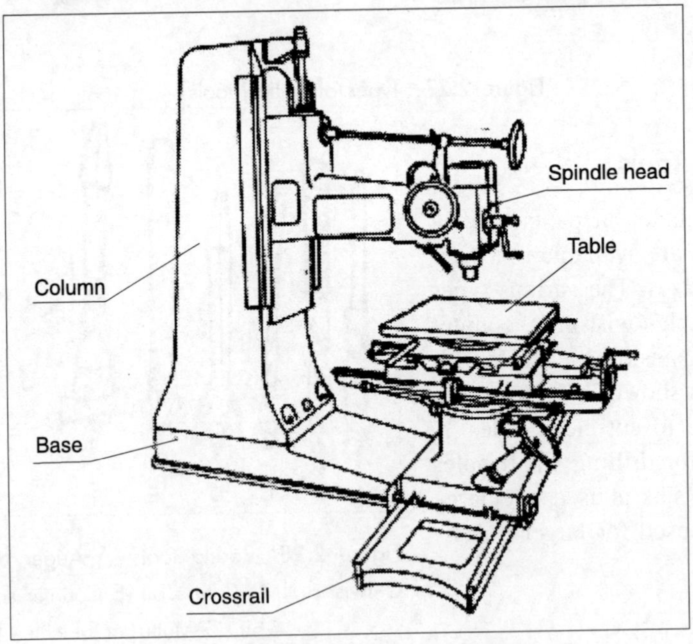

Figure 2.29 Pattern milling machine

The work is mounted on the table and can be moved in horizontal and vertical directions. This machine is equipped with large number of tools, therefore it is capable of performing many operations like boring, drilling, facing, slotting, milling, grooving, shaping, fillet cutting work, cutting radium, forming angular cuts and forming of other shapes.

2.12.2 Circular Saw

It consists of a table, circular cutting blade on bearings and rotating at high speed, driving arrangement for saw blade, cut-off guide required during cross cutting to steer the piece towards the saw blade and a ripping fence acting as a guide during sawing along the grains of wood. Provision is made for tilting the table to provide angular cuts. Cutting speed is 1000 m/s to 3000 m/s.

2.12.3 Band Saw

It consists of metal saw band travelling over the rims of two or more rotating pulleys. It is used to make irregular or curved cuts. It is available in two models, vertical and horizontal.

2.12.4 Jig Saw or Scroll Saw

This is a tiny saw, an ideal machine for cutting small sized work to an intricate and irregular shape. The table of jig saw can also be tilted to carry out angular work. The specific characteristics of this saw is its ability to cut even inside curves by threading the blade through a hole bored into the work. This internal sawing is not practicable in any other wood working machine. The width of the blade is usually 1.5 to 9 mm.

2.12.5 Wood Working Lathe

It is used to make cylindrical or tapered workpieces. It is designed to turn workpieces, both external and internal. In wood turning, lathe work is rotated and shaping is done by tool. It consists of headstock, tailstock, handle slide rest to hold the tool as shown in Figure 2.30.

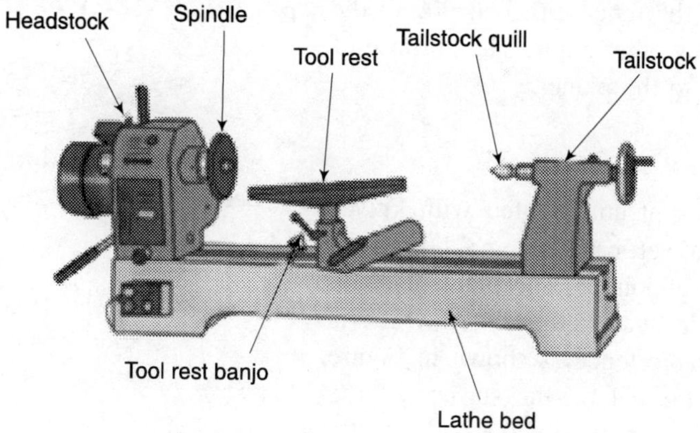

Figure 2.30 Wood working lathe

2.12.6 Wood Working Planer

It is used for planing bars and the planks. The main parts are cutter drum located on the table which is driven by motor. The thickness of shavings cut off the plank can be adjusted by shifting the table by screws (Figure 2.31). The plank is fed by means of feed rolls along the table against a revovling cutter head. The modern planers automatically surface the wood to desired thickness.

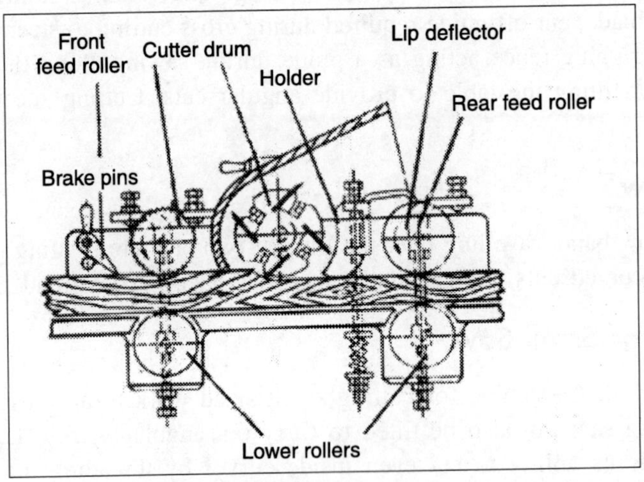

Figure 2.31 Wood working planer

2.12.7 Jointer

It is used for planing straight edges and surfaces of planks avoiding the labour in hand planing. A typical jointer is shown in Figure 2.32. It consists of revolving cutter head, table, adjustable fence guiding the board at a desired angle to the surface of the table. Angular and bevel cuts can be obtained by keeping the fence either at 90° to the table or inclined at any other angle. The purpose of the jointer is similar to the planer.

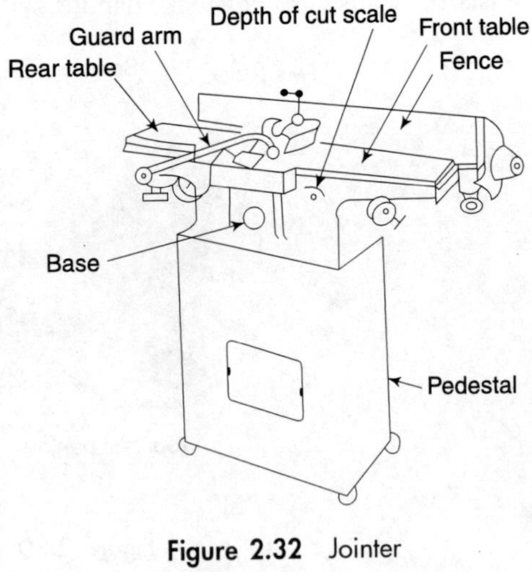

Figure 2.32 Jointer

2.12.8 Shaper

This machine consists of cutter fitted with knives and rotating about a vertical axis and horizontal table similar to that of jointer. The plank in hand is fed along the table against the cutter which is guided by the adjustable fence as shown in Figure 2.33. The shape of the cut on the surface of the wood is the same as the profile of the cutter teeth. The cutter can be designed to produce a variety of shapes.

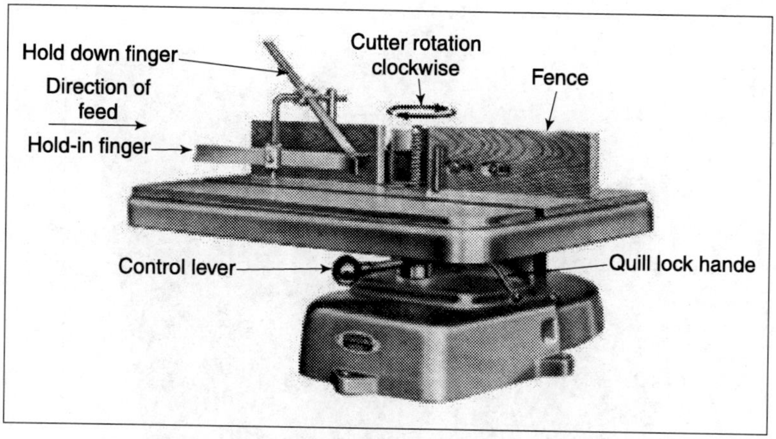

Figure 2.33 Shaper

2.12.9 Disc and Bobbin Sander

This machine consists of a large disk which rotates at high rpm of 300 rpm. A sandpaper of suitable grade is glued and fixed on the outer face of the disc. Table moves up and down through a short stroke of about 50 mm. Bobin sander consists of cylindrical bobbin having a sandpaper around it and is used for curved surfaces.

Disc sander

Bobbin sander

Figure 2.34 Disc and bobbin sander

2.12.10 Abrasive Belt Machine

This machine consists of endless abrasive belt travelling over two drums. This machine is particularly useful in shaping workpieces requiring the use of hand tools. In this machine, the workpiece is supported by an adjustable work table which can be tilted at any desired angle.

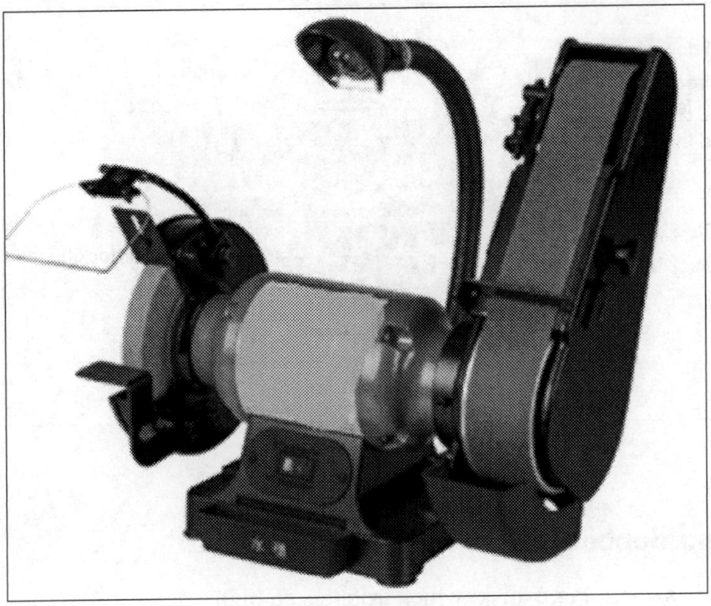

Figure 2.35 Abrasive belt machine

2.12.11 Drill Press

This is an important machine used in pattern shop. In addition to drilling and boring holes, this may be used for mortising, shaping and nuting with special attachments. Drill presses are available in many sizes.

2.12.12 Wood Trimmer

This machine consists of two sharp knives fastened to a movable plate at an angle. The plate can be moved from side to side by hand lever. Although, this is not a motor driven equipment, it is classified as wood working machine. The wood trimmer is used to cut the wood accurately which is not possible by other machines.

2.12.13 Tool Grinding Machine

Grinder is required in a pattern shop for sharpening and shaping of tools and cutters for the efficient use of various machines. The simplest grinding machine is a bench grinder which consists of two grinding wheels fastened to an arbor. One is coarse wheel and other is finer wheel.

Figure 2.36 Bench grinder

Worked Out Examples

Example 2.1 A job shown in Figure 2.37 is to be made of steel by casting process. The mould for this job is made from a wooden pattern. Determine the dimensions of the wooden pattern. Assume machining allowance of 2 mm on each side, shrinkage allowance of 2% and a taper allowance of 1 degree. Allowances given must be cumulative. (First give machining allowance, followed by draft and shrinkage allowances)

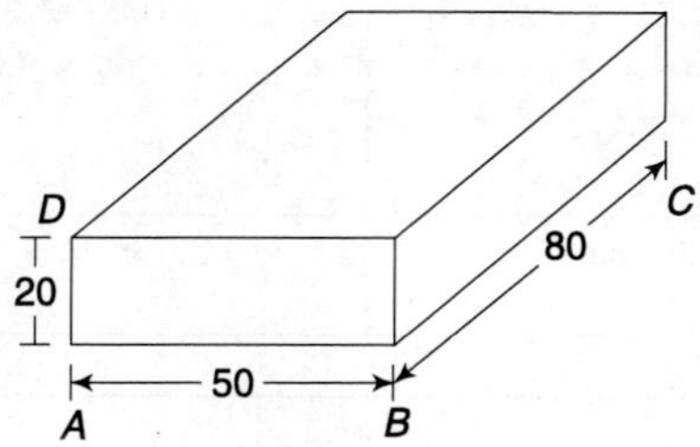

Figure 2.37

Solution

Step 1: Give the machining allowance of 2 mm per side.

Step 2: Give the draft or taper allowance after giving the maching allowance.

Step 3: Give the shrinkage allowance after draft allowance.

Step 1: Give the machining allowance of 2 mm per side.

Step 2: Give the draft or taper allowance after giving the maching allowance.

Taper allowance calculation is shown in the figure.

The draft allowance value x is calculated after giving the machining allowance (Figure 2.38)

$$x = 24 \times \tan 1 = 0.419\,\text{mm}$$

and top surface dimensions are increased to 54.838 × 84.838 mm to provide taper allowance.

Step 3: Give the shrinkage allowance after draft allowance.

Given shrinkage allowance is 2%. Now, the dimensions of pattern are increased by 2% on all sides.

$$\text{Allowance for dimension 54 mm} = 54 + \frac{(54 \times 2)}{100} = 55.08 \cong 55.1 \text{ mm}$$

$$\text{Allowance for dimension 54.838 mm} = 54.838 + \frac{(54.838 \times 2)}{100} = 55.91 \cong 55.9 \text{ mm}$$

Summary of all the dimensions is given in Table 2.11 and final pattern is shown in Figure 2.39.

Table 2.11 Dimensions of pattern

Portion	Machining allowance (mm)	Draft (1°)	Shrinkage allowance (2 %)	Final Pattern Dimension (mm)
AB – 50	50 + 2 + 2 = 54	54 + 0.419 + 0.419 = 54.838 mm (top) 54 mm (bottom)	54.838 + 0.02 × 54.838 = 55.91 (top) 54 + 0.02 × 54 = 55.08 mm (bottom)	55.9 mm 55.1 mm
BC – 80	80 + 2 + 2 = 84	84 + 0.419 + 0.419 = 84.838 mm (top) 84 mm (bottom)	84.838 + 0.02 × 84.838 = 86.53 84 mm + 0.02 × 84 mm = 85.68	86.5 mm 85.7 mm
AD – 20	20 + 2 + 2 = 24	24 + 0 = 24 mm	24 + 0.02 × 24 = 24.48 mm	24.5 mm

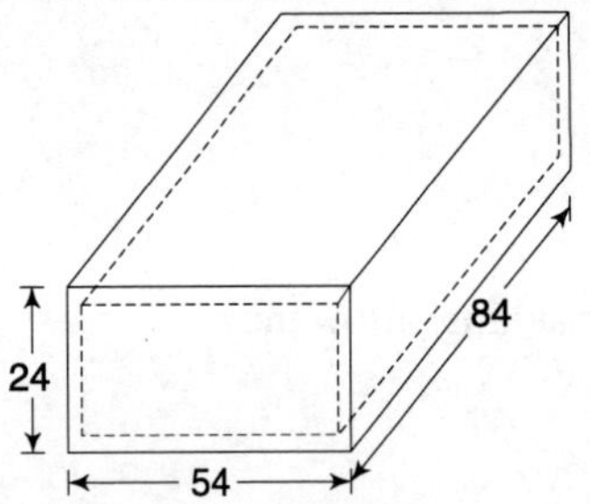

Figure 2.38 Dimensions after giving machining allowance

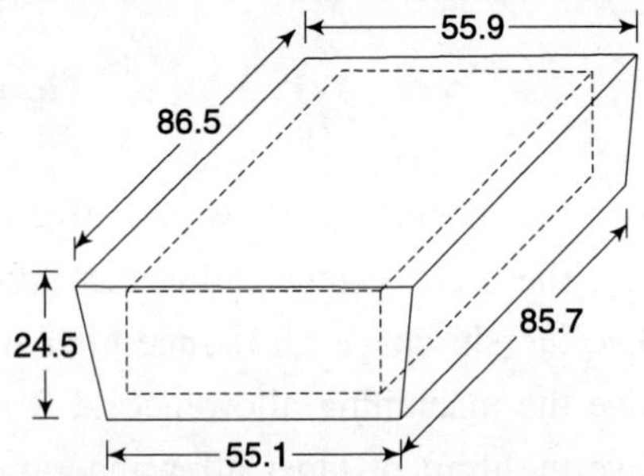

Figure 2.39 Final pattern dimensions

Example 2.2 A job shown here is to be made from steel by casting process. The mould for this job is made from wooden pattern. Determine the dimensions of the wooden pattern assuming machining allowance of 3 mm on each side, shaking allowance of 1 mm on length and width, shrinkage allowance of 3%. Allowances given must be cumulative.

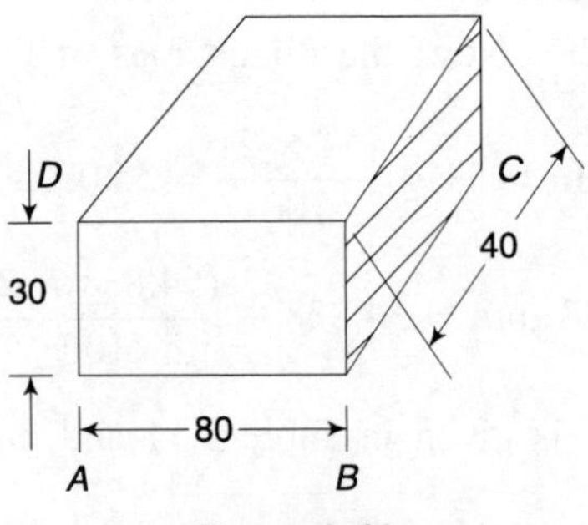

Figure 2.40

Solution

Step 1: Give the machining allowance of 3 mm per side.

Step 2: Give the shrinkage allowance after giving the maching allowance.

Step 3: Give the rapping allowance after machining allowance.

Step 1: Give the machining allowance of 2 mm per side.

3 mm is added to each side of the dimensions as shown in the figure.

Step 2: Give the shrinkage allowance after draft allowance.

Given shrinkage allowance is 3%. Now, the dimensions of pattern are increased by 2% on all sides

$$\text{Allowance for dimension 86 mm} = 86 + \frac{(86 \times 3)}{100} = 88.58 \text{ mm}$$

$$\text{Allowance for dimension 46 mm} = 46 + \frac{(46 \times 3)}{100} = 47.38 \text{ mm}$$

$$\text{Allowance for dimension 36 mm} = 36 + \frac{(36 \times 3)}{100} = 37.08 \text{ mm}$$

Step 3: Give the rapping allowance after machining allowance.

Subtract 1 mm from length and width dimensions. Note that rapping allowance is a negative allowance.

Summary of all the dimensions are given in Table 2.12 and the final pattern is shown in Figure 2.43.

Table 2.12 Dimensions of pattern

Portion	*Machining allowance (mm)*	*Shrinkage allowance (3%)*	*Rapping allowance*	*Final Pattern Dimension (mm)*
AB – 80	80 + 3 + 3 = 86	86 + 0.03 × 86 = 88.58 mm	88.58 – 1	87.58 mm
BC – 40	40 + 3 + 3 = 46	46 + 0.03 × 46 = 47.38 mm	47.38 – 1	46.38 mm
AD – 30	30 + 3 + 3 = 36	36 + 0.03 × 36 = 37.08 mm	37.08	37.08 mm

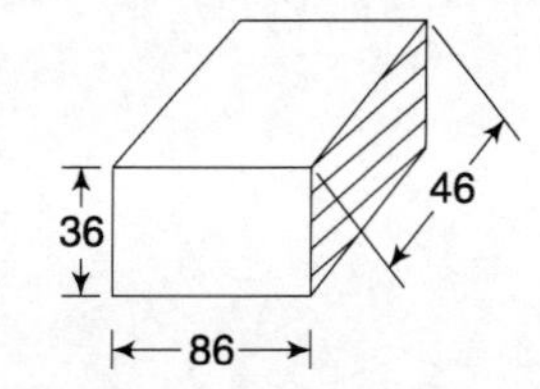

Figure 2.41 Dimensions after giving machining allowance

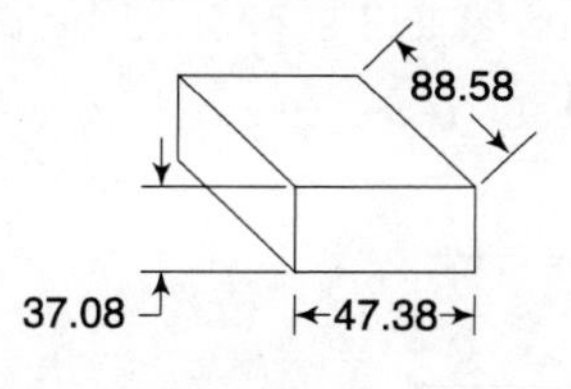

Figure 2.42 Dimensions after giving shrinkage allowance

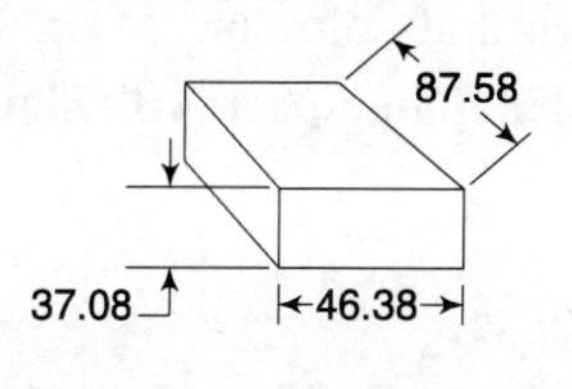

Figure 2.43 Final pattern dimensions

Note: In remaining problems, allowances given are non-cumulative, that is, individual allowances are given and added to get the final pattern dimension. Basic dimensions are considered for the allowance.

For example: If dimension 100 mm is to be given shrinkage allowance of 2% and machining allowance of 3% and rapping allowance of 1% then individual allowances are called for 100 mm and then allowances are added to get the final dimensions. This method is called non-cumulative allowance.

Dimension	Shrinkage allowance	Machining allowance	Rapping allowance	Final dimension
100 mm	100 × 0.02 = 2	100 × 0.03 = 3	100 × 0.01 = 1	100 + 2 + 3 –1 = 104

Example 2.3 Figure 2.44 shows a part is made of cast steel. Allowable shrinkage allowance is 2%. Draft allowance is 2 mm/side, machining allowance of 3 mm to be provided and scale and grinding allowance is 1 mm because the part is to be annealed. Find the final pattern dimensions for a part when it is cast (a) vertically and (b) horizontally. Only grinding is done and no machining is done in the area where core is placed and provide total 2 mm allowance for the core for grinding.

Solution

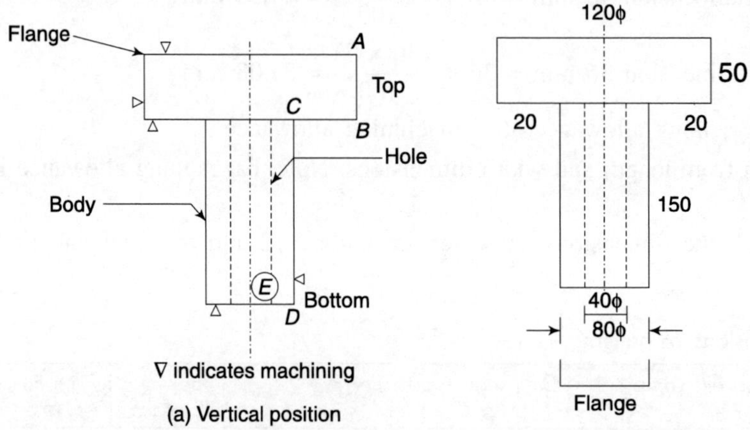

Figure 2.44

Case 1: When the part is cast vertically

Let us use single piece or solid pattern to make this component. All the vertical edges should be given draft allowance. Core is set vertical and supported by chaplets.

Consider flange part AB (Figure 2.45)

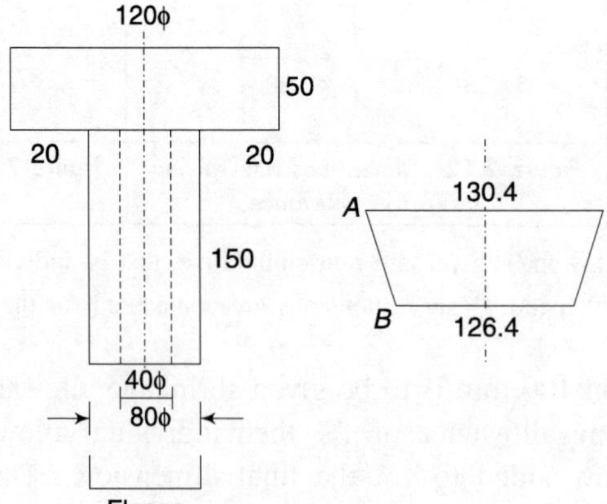

Figure 2.45

Since part AB is vertical, shrinkage, draft, machining and scale grinding allowances are to be given.

Final dimensions of flange (A) = basic dimension + shrinkage of 2% + draft 2 mm/side + machining allowance of 3 mm + scale and grinding allowance

$$= 120 + \left(\frac{2 \times 120}{100}\right) + (2 \times 2) + 3 + 1 = 130.4 \text{ mm}$$

Similarly, dimensions of portion B and vertical AB can be calculated as follows:

Flange (B) = 120 + 2.4 + 4 + 3 + 1 = 126.4 mm

Vertical (AB) Portion = 50 + 1 + 6 + 1 = 58.0 mm

Consider body part CD (Figure 2.46)

Body (C): 80 + 1.6 + 4 + 3 + 1 = ϕ89.6 mm

Body (D): 80 + 1.6 + 3 + 1 = ϕ85.6 mm

Body vertical portion, CD: 150 + 3 + 3 + 1 = 157.0 mm

Core dimensions

Hole: 40 – 0.8 – 1 = 37.2 mm

Length, CD: 150 + 3 + 3 + 1 = 157.0 mm

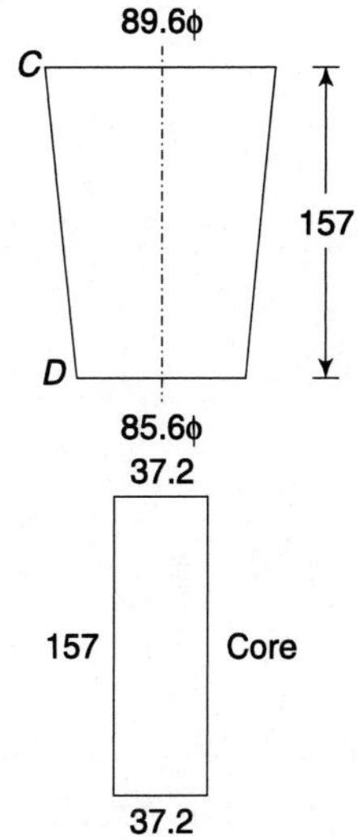

Figure 2.46

Note: The negative allowance is given for the hole. That is on the outer diameter of core. Summary of all the dimensions are given in Table 2.13.

Table 2.13 Dimensions of pattern

Portion	*Shrinkage 2%*	*Draft 2 mm/ side*	*Machining allowance (mm)*	*Scale & grinding allowance*	*Final pattern dimension (mm)*
			Flange		
A – 120 ϕ	2.4	2 × 2 = 4	3	1	130.4
B – 120 ϕ	2.4	-	3	1	126.4
AB – 50	1	-	3 × 2 = 6	1	58.0
			Body		
C – 80 ϕ	1.6	2 × 2 = 4	3	1	89.6
D – 80 ϕ	1.6	-	3	1	85.6
CD – 150	3.0	-	3	1	157.0
			Core		
E – 40 ϕ	–0.8	-	-	–(1 × 1)	38.2
CD – 150	3.0	-	3	1	157.0

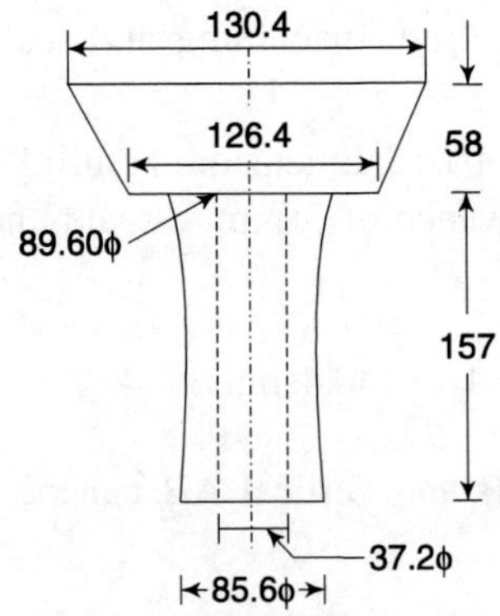

Figure 2.47 Final dimensions of component

Case 2: When the part is cast horizontal

Part is of circular cross section, no draft allowance is required when cast horizontally. Overhang core is to be provided to create hollow space and is supported on core prints and supported by chaplets. Final pattern dimensions are given in Table 2.14.

Table 2.14 Dimensions of pattern

Portion		Shrinkage (2%)	Draft allowance (mm)	Machining allowance (mm)	Scale & grinding allowance	Final pattern dimension (mm)
A	120 ϕ	2.4	-	3	1	126.4
	50	1.0	-	3 × 2 = 6	2 × 1 = 2	59.0
	80	1.6	-	3	1	85.6
B	150	3.0	-	3	1	157
	40 ϕ	−0.8	-	-	(−1 × 2) = −2	37.2
C	150	3.0	-	3	1	157

Example 2.4 Calculate the pattern allowance for the part shown in Figure 2.48. Take shrinkage allowance as 2%, 2 mm per side as draft allowance. Machining allowance 1.5 mm per side. Scaling and grinding allowance as 0.8 mm (total) in all the problems. For cores: Take shrinkage allowance as 1.5%. Draft as 2 mm per side. Machining allowance as 0.5 mm/per side and grinding allowance as 0.8 mm (total).

Solution Part is named as shown in Figure 2.49.

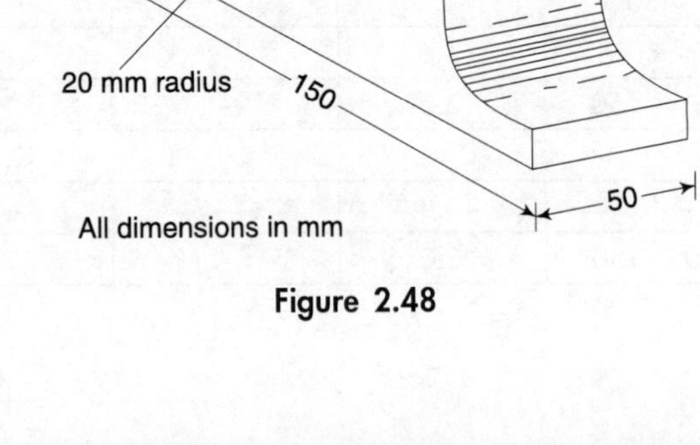

Figure 2.48

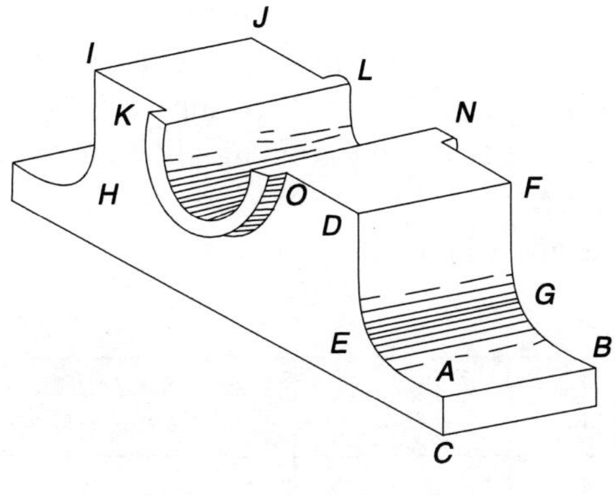

Figure 2.49

It can be cast using single piece patters. The part is to be inverted and cast. Final pattern dimensions are given in Table 2.15. Note **DF = EG = AB = IJ = 50 mm**

Table 2.15 Dimensions of pattern

Portion		Shrinkage allowance	Draft allowance	Machining allowance	Scale & Grinding allowance (mm)	Final dimension (mm)
A	150	3	-	1.5 x 2 = 3	0.8	156.8
C	150	3	2 x 2 = 4	1.5 x 2 = 3	0.8	160.8
AB	50	1	-	1.5 x 2 = 3	0.8	54.8
DI	90	1.8	–2 x 2 = 4	1.5 x 2 = 3	0.8	91.6
EH	90	1.8	2 x 2 = 4	1.5 x 2 = 3	0.8	99.6
KL	60	1.2	-	1.5 x 2 = 3	0.8	65
KO	40ϕ	–0.6	-	–0.5 x 2 = –1	–0.85 x 2 = –1.6	36.8

Example 2.5 Figure 2.50 shows a part is made of high manganese steel. Allowable shrinkage allowance is 2.5%. Draft allowance is 1 mm/side. No machining allowance to be provided. Find the final pattern dimensions for a part when it is cast horizontal. Only grinding is done and no machining is done in the area where core is placed. Assume the parting line passes through mid-plane. Top and bottom portions are shown in Figure 2.50.

The final casting is to be heat-treated (solutionizing and quenching) and therefore 2 mm scale and grinding allowance is to be provided for larger surfaces.

Solution

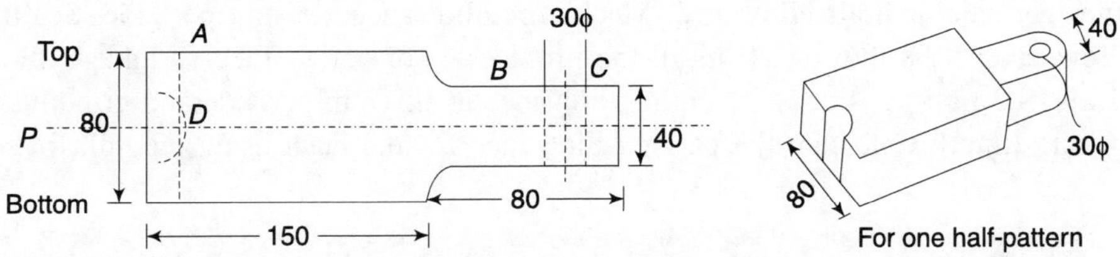

Figure 2.50 Three-piece pattern

Casting is to be made in two pieces with top and bottom portions as shown in Figure 2.50. No machining allowance is given. No special core is required for the portion D. Core is held in position by core prints and chaplets. The final casting is to be heat-treated (solutionizing and quenching) and therefore 2 mm scale and grinding allowance is to be provided for larger surfaces.

Final pattern dimensions are given in Table 2.16.

Table 2.16 Dimensions of pattern

Portion		*Shrinkage 2.5%*	*Draft allowance*	*Scale & Grinding allowance (mm)*	*Final dimension mm*
A	150	3.75	-	2	155.75 Top
		3.75	1	2	156.75 Bottom
	40	1.0	-	2	43
B	80	2.0	-	2	84.00 Top
		2.0	1	2	85.00 Bottom
	20	0.5	-	2	22.5
C	30	−0.75	-	−1	28.25
D	40 ϕ	−1.0	-	−1	38.00
	40	1.0	-	1	42.00 Top
		1.0	1	1	43.0 Bottom

Final dimension of the part is shown in Figure 2.51.

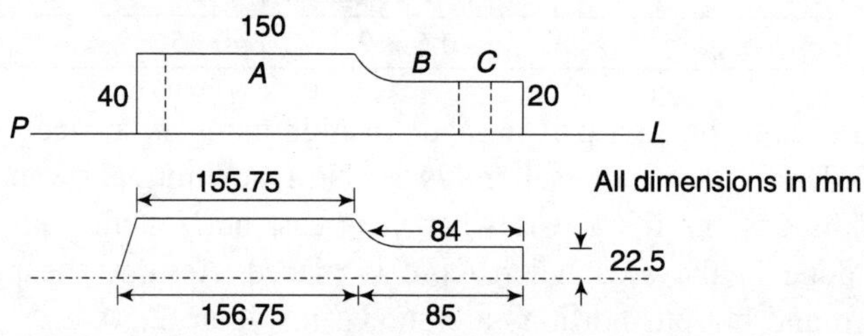

Figure 2.51 Final dimensions of component

Example 2.6 Figure 2.52 shows a bush used in a machine tool. Take shrinkage allowance as 2%, 2 mm per side as draft allowance. Machining allowance 1.5 mm per side. Scaling and grinding allowance as 0.8 mm (total) in all the problems. For cores: Take shrinkage allowance as 1.5%. Draft as 2 mm per side. Machining allowance as 0.5 mm per side and grinding allowance as 0.8 mm (total). Calculate the pattern allowance for the bush using the split pattern.

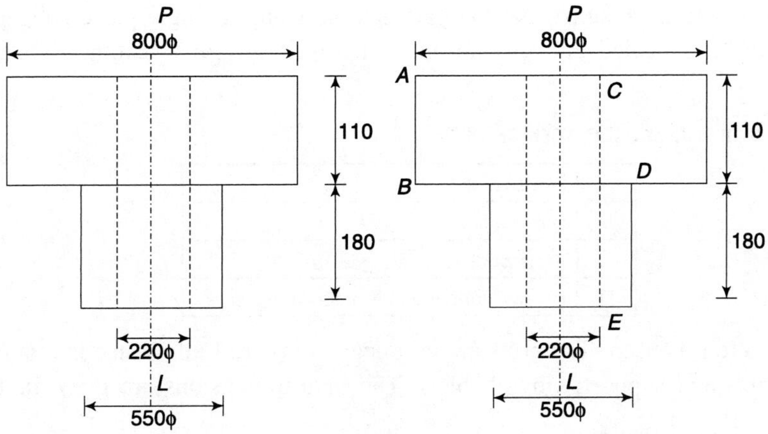

Figure 2.52

Solution Final dimensions of pattern are given in Table 2.17.

Table 2.17 Dimensions of pattern

Portion		Shrinkage (2.5%)	Draft allowance	Machining allowance	Scale & grinding allowance (mm)	Final dimension (mm)
A	400	8	Natural	1.5	0.8	410.3
AB	110	2.2	-	1.5 x 2 = 3	0.8	116
D	275	5.5	Natural	1.5	0.8	282.8
DE	180	3.6	-	1.5	0.8	185.9
C (full core)	220	−3.3	-	−0.5 x 2 = −1	−0.8	215.62
CE	290	5.8	-	1.5 x 2 = 3	0.8	299.6

Example 2.7 Figure 2.53 shows a part is made of grey cast iron. Allowable shrinkage allowance is 1%. Draft allowance is 1 mm/side; machining is done for outer rim of the wheel, on the faces of the hub portion and on either side faces of rim portions. Find the final pattern dimensions for a part when it is cast horizontally. Only grinding is done and no machining is done in

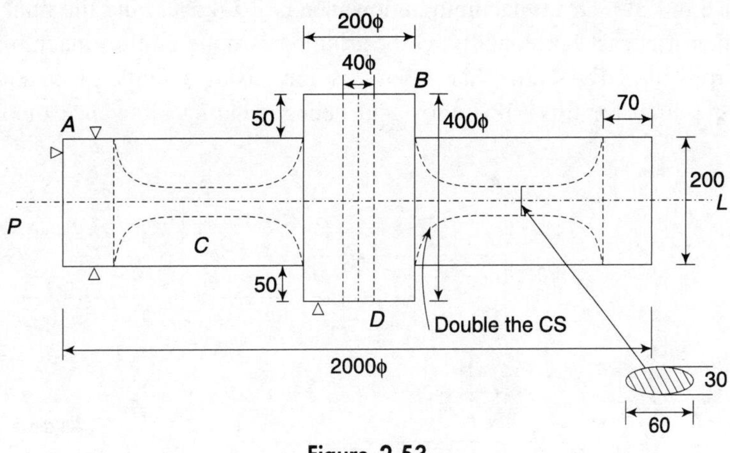

Figure 2.53

area where core is placed. Assume that the part is done using a single piece solid pattern. The final casting is to be annealed and ground and hence grinding and scaling allowances are to be provided.

Solution Let us first mark the various parts

A	Rim portion
B	Hub portion
C	Arms portion
D	Hollow portion to be made using core

Single piece pattern is used. The arms are elliptical so no draft allowance is required. Core is set with core prints and supported by chaplets. The final dimensions are given in Table 2.18.

Table 2.18 Dimensions of pattern

S. No.	Portion		Shrinkage allowance (1%)	Draft allowance (mm)	Machining allowance (mm)	Scaling & grinding allowance (mm)	Final pattern dimension (mm)
1	A	200 ϕ	20	-	10	2	232 Top
			20	2	10	2	234 Bottom
		150	1.5	-	5	2	158.5
2	B	200 ϕ	2	–2	-	1	203 Top
			2	2	-	1	205 Bottom
		200	2	-	5	1	208
3	C	30	0.3	-	-	0.5	30.8
		60	0.6	-	-	0.5	61.1
4	D	40 ϕ	–0.4	-	-	–1	38.6
		400	4	-	10	2	416.00

Example 2.8 Figure 2.54 shows a part is made of Al-Si alloy. Allowable shrinkage allowance is 1.3%. Draft machining allowance is 1.5-4.5%. Find the final pattern dimensions for a part when it is cast horizontally. Only grinding is done and no machining is done in the area where core is placed. Assume that the part is done using a single piece solid pattern. No heat-treatment is required for this Al-Si alloy and hence grinding allowance of 0.2 mm is sufficient.

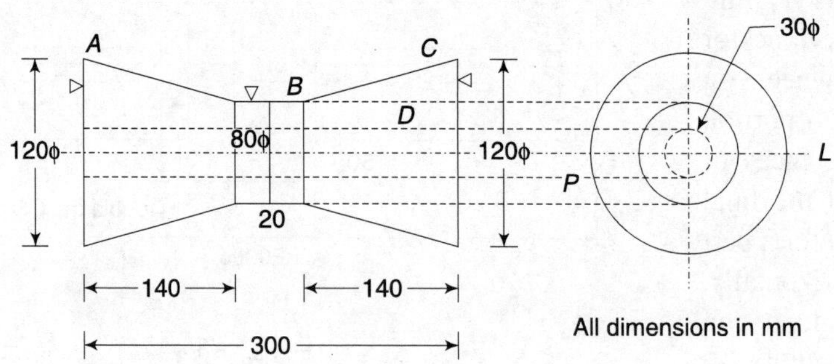

Figure 2.54

Solution The design of the pattern is carried out using the following assumptions.

Since the cross section of the casting is circular and the proposed parting line (PL) is at the mid diameter, draft is naturally present. Hence there is no need of any draft allowance. Core is prepared using corebox for 3300 mm length. It has natural draft. Machining allowance is given on the face 120° and on the OD of 80 (portion B). No machining allowance is given on the 20 mm width portion B. No machining on 30° of the core portion. Since no heat-treatment is required for Al-Si alloys, no scaling allowance is required. Only a nominal grinding allowance of 0.2 mm is given. The core diameter gets reduced due to shrinkage hence negative allowance is given. Machining allowance is given on the length portion 300 mm of core (310.3). Core is horizontal. It is supported on core prints and by using chaplets.

The final dimensions are given in Table 2.19.

Table 2.19 Dimensions of pattern

Portion	Dimension	Shrinkage allowance (mm)	Draft allowance (mm)	Machining allowance (mm)	Scale & grinding allowance (mm)	Final pattern dimension (mm)
A and C	120ϕ	1.56	Natural draft	-	0.2	121.76
	140	1.82	-	3	0.2	145.02
B	80ϕ	1.04	Natural draft	3	0.2	84.24
	20	0.26	-	-	-	20.26
D	30ϕ	−0.39	Core	-	-	29.61
	300	3.9	-	2 × 3 = 6	2 × 0.2 = 0.4	310.3

Example 2.9 Calculate the pattern allowance for Problem 6 (Figure 2.54). Take shrinkage allowance as 2%, 2 mm per side as draft allowance. Machining allowance 1.5 mm per side. Scaling and grinding allowance as 0.8 mm (total) in all the problems. For cores: Take shrinkage allowance as 1.5%. Draft as 2 mm per side. Machining allowance as 0.5 mm per side and grinding allowance as 0.8 mm (total).

Solution Final pattern dimensions are given in Table 2.20.

Table 2.20 Dimensions of pattern

Portion	Dimension	Shrinkage allowance (mm)	Draft allowance (mm)	Machining allowance (mm)	Scale & grinding allowance (mm)	Final patten dimension (mm)
A and C	120ϕ	2.4	Natural draft (circular)	1.5 × 2 = 3	0.8	126.2
	140	2.8	-	1.5	0.8	145.1
B	80 ϕ	1.6	Natural draft	1.5 × 2 = 3	0.8	85.4
	20	0.4	-	-	-	20.4
D	30 ϕ	−0.45	Core	−0.5 × 2 = −1	−0.8	26.75
	300	6	-	1.5 × 2 = 3	0.8	309.8

Example 2.10 Calculate the pattern allowance for the part shown in Figure 2.55. Take shrinkage allowance as 2%, 2 mm per side as draft allowance. Machining allowance 1.5 mm per side. Scaling and grinding allowance as 0.8 mm (total) in all the problems. For cores: Take shrinkage allowance as 1.5%. Draft as 2 mm per side. Machining allowance as 0.5 mm per side and grinding allowance as 0.8 mm (total). (Left as an exercise for students. Hint: See Problem 2.13).

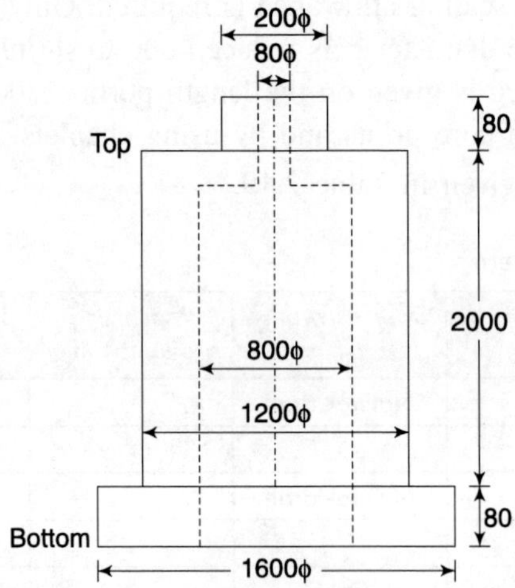

Figure 2.55

Example 2.11 Figure 2.56 shows a part is made of cast iron. Allowable shrinkage allowance is 2%. Part is to be supplied to the customer without machining and heat-treatment. Show pattern and core dimensions separately.

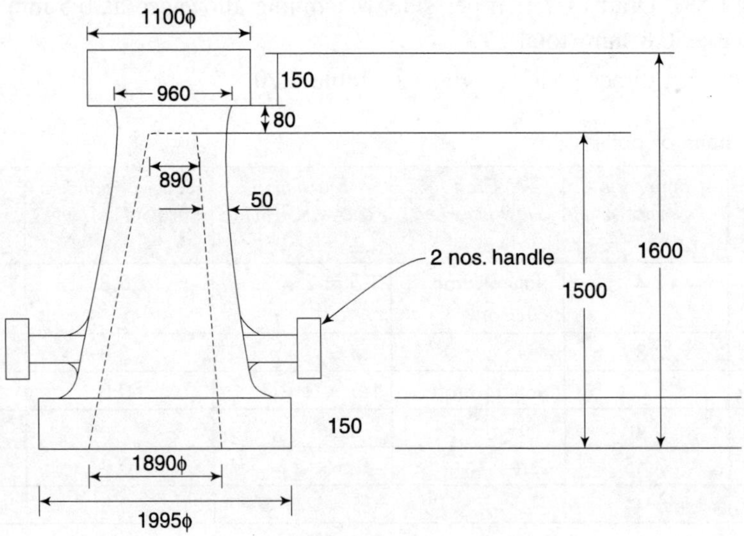

Figure 2.56

The final dimensions are given in Table 2.21.

Table 2.21 Dimensions of pattern

Dimensions (mm)		Shrinkage allowance (mm)	Pattern dimension (mm)	Core dimension (mm)
Total height	1600	32	1632	-
Height	1500	30	1530	-
Bottom flange diameter	1995	39.9	2034.9	-
Top flange diameter	1100	22	1122	-
Flange	150	3	153	-
Core larger diameter	1890	−37.8	-	1852.2
Core smaller diameter	890	−17.8	-	872.2
Core height	1500	30.1	-	1530.1
Wall thickness	50	1	51	-

Example 2.12 The part shown in Figure 2.57 is to be produced by casting process. Take shrinkage allowance as 2%, 2 mm per side as draft allowance. Machining allowance 1.5 mm per side. Scaling and grinding allowance as 0.8 mm (total) in all the problems. For cores: Take shrinkage allowance as 1.5%. Draft as 2 mm per side. Machining allowance as 0.5 mm/per and grinding allowance as 0.8 mm (total). Calculate the pattern allowance considering the parting line is (a) vertical, and (b) horizontal.

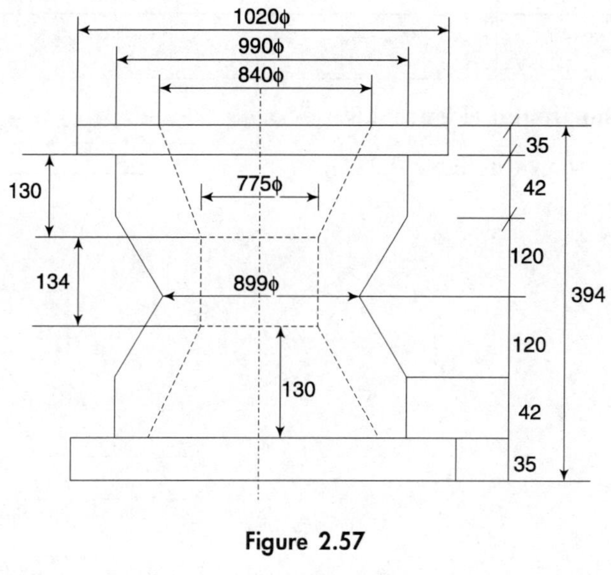

Figure 2.57

Case 1: Parting line vertical (Figure 2.58)

Part is symmetrical in both the axis and parting line shown in the Figure. Final dimensions are given in Table 2.22.

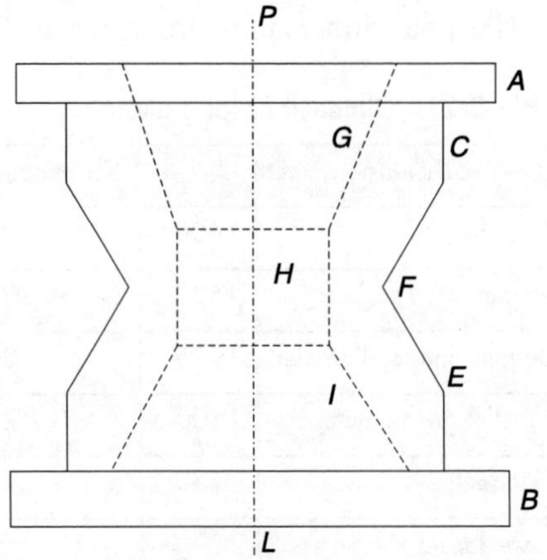

Figure 2.58

Table 2.22 Dimensions of pattern

S. No.	Portion		Shrinkage allowance	Draft allowance (mm)	Machining allowance (mm)	Scaling & grinding allowance (mm)	Final pattern dimension (mm)
1	A/B	510ϕ	10.2	Natural	1.5	0.8	522.5
		35	0.7	-	1.5 × 2 = 3	0.8	39.5
2	C/E	495ϕ	9.9	-	1.5	0.8	507.2
		42	0.84	-	1.5	0.8	45.14
3		120	2.4	-	-	-	122.4
4		420ϕ	−5.04	−0.5 × 2 = −1	-	−0.8 × 2 = −1.6	412.36
		165	3.3	-	1.5	0.8	170.6
5		387.5	−4.65	−0.5 × 2 = −1		−0.8 × 2 = −1.6	380.25

Case 2: Parting line horizontal (Figure 2.59)

Final dimensions are shown in Table 2.23.

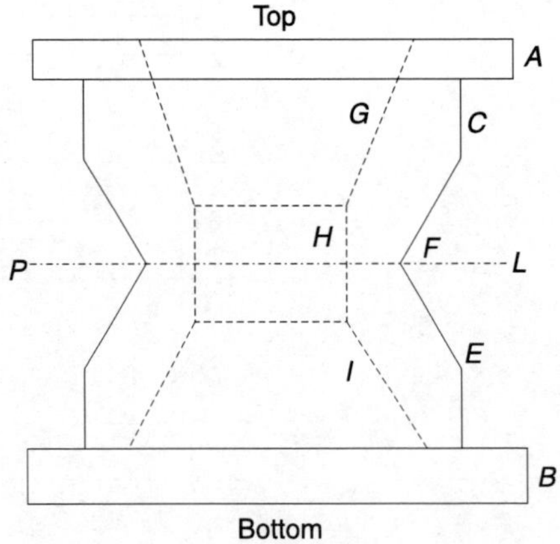

Figure 2.59

Table 2.23 Dimensions of pattern

S. No.	Portion		Shrinkage allowance	Draft allowance (mm)	Machining allowance (mm)	Scaling & grinding allowance (mm)	Final pattern dimension (mm)
1	A/B	1020ϕ	20.4	2 × 2 = 4	1.5	0.8	1046.7
		35	0.7	-	1.5 × 2 = 3	0.8	39.5
		42	0.84	-	1.5	0.8	45.14
2	F	120	2.4	-	1.5	0.8	124.7
		890ϕ	17.98	-	1.5 × 2 = 3	0.8	920.78
3	G/I	840ϕ	−12.6	-	-	−0.8	826.6
		775ϕ	−11.63	-	-	−0.8	762.6
		130	1.95	-	-	0.8	132.75

Example 2.13 The part shown in Figure 2.60 is to be produced by casting process. Take shrinkage allowance as 2%, 2 mm per side as draft allowance. Machining allowance 1.5 mm per side. Scaling and grinding allowance as 0.8 mm (total) in all the problems. For cores: Take shrinkage allowance as 1.5%. Draft as 2 mm per side. Machining allowance as 0.5 mm per side and grinding allowance as 0.8 mm (total). Calculate the pattern allowance considering split pattern with parting line horizontal.

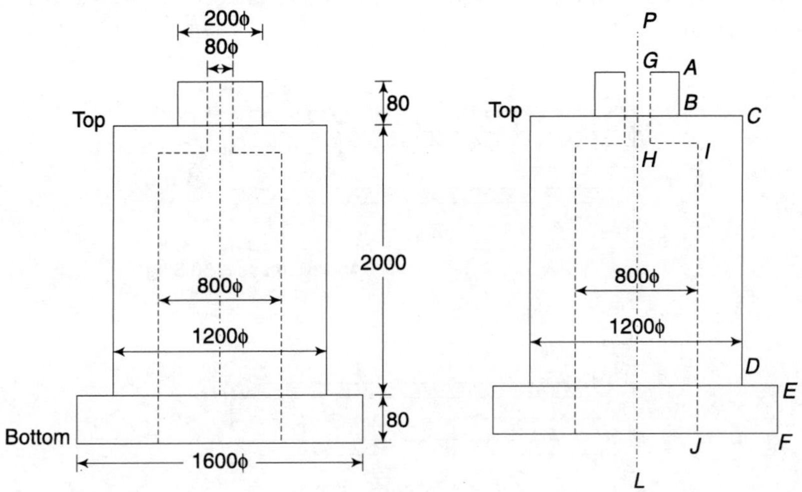

Figure 2.60

Split pattern is used. The final dimensions are given in Table 2.24.

Table 2.24 Dimensions of pattern

S. No.	Portion		Shrinkage allowance	Draft allowance (mm)	Machining allowance (mm)	Scaling & grinding allowance (mm)	Final pattern dimension (mm)
1	A	100 ϕ	2	Natural	1.5	0.8	104.3
2	AB	80	1.6	Natural	1.5 x 2 = 3	0.8	85.4
3	G	40 core	−0.6	-	−0.5	−0.8	38.1
4	GH	150	3	-	1.5 x 2 = 3	0.8	156.8
5	C	600	12	-	1.5	−0.8	614.3
6	CD	2000	49	-	-	−0.8	2040.8
7	I	450 core	−6.75	-	−0.5	0.8	441.95
8	IJ	2010	40.2	-	1.5	0.8	2052.5
9	E	800	16	-	1.5	0.8	818.3
10	EF	80	1.6	-	1.5 x 2 = 3	0.8	85.4

Example 2.14 The part shown in Figure 2.61 is to be produced by casting process. Take shrinkage allowance as 2%, 2 mm per side as draft allowance. Machining allowance 1.5 mm per side. Scaling and grinding allowance as 0.8 mm (total) in all the problems. For cores: Take shrinkage allowance as 1.5%. Draft as 2 mm per side. Machining allowance as 0.5 mm per side and grinding allowance as 0.8 mm (total). Calculate the pattern allowance considering the parting line is horizontal.

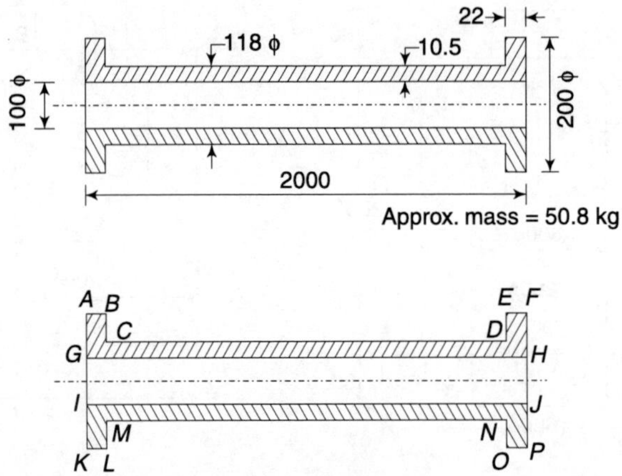

Figure 2.61

Split pattern is used. The final dimensions are given in Table 2.25.

Table 2.25 Dimensions of pattern

S. No.	Portion		Shrinkage allowance	Draft allowance (mm)	Machining allowance (mm)	Scaling & grinding allowance (mm)	Final pattern dimension (mm)
1	AK	200ϕ	4	Natural	1.5	0.8	206.6
2	GI	100ϕ core	−1.5	Natural	−0.5 x 2 = −1	−0.8	96.7
3	AB	22 Top	0.44	-	1.5 x 2 = 3	0.8	26.24
4	GH	22 Bottom	0.44	-	1.5 x 2 = 3	0.8	26.24
5	CD	1956	39.12	-	-	0.8	1995.92
6	CM/DN	118ϕ	2.36	-	1.5	0.8	122.66
7	CQ	10.5	0.21	-	1.5 x 2 = 3	0.8	14.51
8	EF	22 Top	0.44	-	1.5 x 2 = 3	0.8	26.24
9	OP	22 Bottom	0.44	-	1.5 x 2 = 3	0.8	26.24

Example 2.15 The part shown in Figure 2.62 is to be produced by casting process. Take shrinkage allowance as 2%, 2 mm per side as draft allowance. Machining allowance 1.5 mm per side. Scaling and grinding allowance as 0.8 mm (total) in all the problems. For cores: Take shrinkage allowance as 1.5%. Draft as 2 mm per side. Machining allowance as 0.5 mm per side and grinding allowance as 0.8 mm (total). Calculate the pattern allowance. Holes are not to be cast.

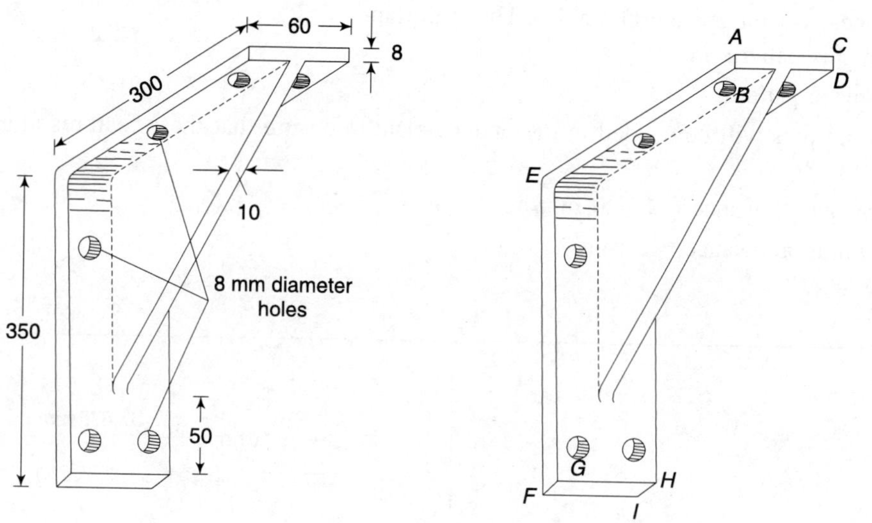

Figure 2.62

Single piece or solid pattern is used. The final dimensions are given in Table 2.26.

Table 2.26 Dimensions of pattern

S. No.	Portion		Shrinkage allowance	Draft allowance (mm)	Machining allowance (mm)	Scaling & grinding allowance (mm)	Final pattern dimension (mm)
1	AB	8	0.16	2	1.5 × 2 = 3	0.8	13.96
2	AE	300	6	-	1.5 × 2 = 3	0.8	309.8
3	EF	350	7	-	1.5 × 2 = 3	0.8	362.8
4	GF/HI	8	0.16	−2	1.5 × 2 = 3	0.8	9.96
5	GH/FI	60	1.2	-	1.5 × 2 = 3	0.8	65
6	AC/BD	60	1.2	-	1.5 × 2 = 3	0.8	65
7	Rib	10	0.2	-	1.5 × 2 = 3	0.8	14

Example 2.16 The blank for the spool shown in the figure is to be sand cast out of A-319, an aluminium casting alloy. Make a sketch of the wooden pattern for this part. Include all necessary allowances for shrinkage and machining.

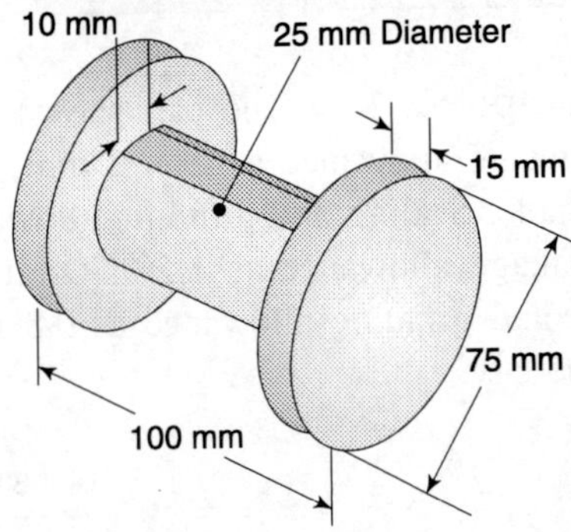

Solution The sketch for a typical greensand casting pattern for the spool is shown. A cross-sectional view is also provided to clearly indicate shrinkage and machining allowances, as well as the draft angles. The important elements of this pattern are:

1. Two-piece pattern.
2. Locating pins will be needed in the pattern plate to ensure that these features align properly.
3. Shrinkage allowance = 10-20 mm/m.
4. Machining allowance = 2 mm.
5. Draft = 3°

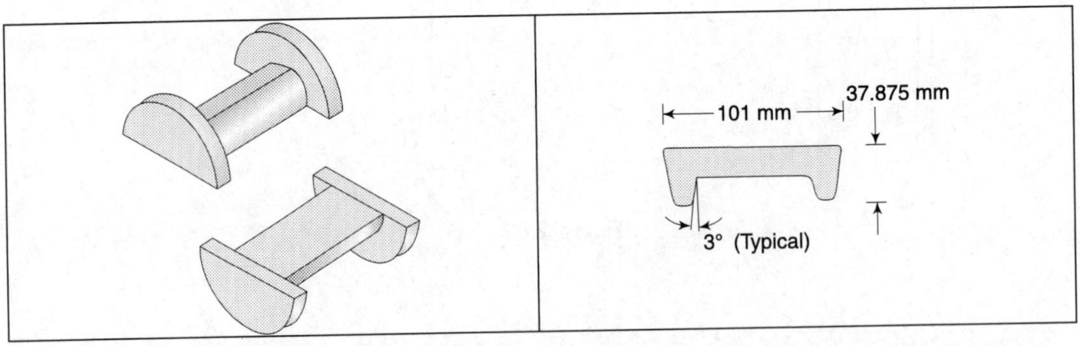

Review Questions

1. Differentiate between casting and pattern.
2. List the ideal characteristics of pattern materials.
3. Explain various pattern materials used in foundry.
4. Write a brief note on colour codes used in pattern shops.
5. Explain various patterns used in casting processes.
6. List typical applications of split and segmented patterns.
7. What are the allowances given on pattern?
8. Explain draft allowance given on pattern with example.
9. How do you compare solid shrinkage with liquid shrinkage?
10. Explain in detail the various tools used in making pattern in pattern shop.

CHAPTER 3

Moulding Materials and Sand Preparations

3.1 INTRODUCTION

The basic function of a mould is to provide a cavity which would receive liquid metal to produce a casting of desired shape and size when the metal solidifies. A core is a solid object which is inserted in the mould cavity so that metal flows around it and upon solidification, the casting is devoid of the metal in the part occupied by the core initially in the mould cavity.

Good castings cannot be made without good moulds. Casting processes are often described by the materials and methods employed in moulding. Most of the castings, approximately 90% of the total production, are cast in bonded sand moulds. Ordinary silica sand mixed with clay and small percentages of additives is made wet with water and used as the mould and core material in the conventional casting process. The average requirement of mould and core material is about 4 to 5 tonnes per production of one tonne of casting. But in extreme cases, this may range from 0.25 tonnes to 10 tonnes depending on the shape and size of the casting. Any mixture of materials, which can be readily moulded and which can produce good castings, can be used as moulding material. Since the ordinary silica sand mixed with proper type of clay has almost all the required desirable properties, at the same time they are available on earth's surface at low cost; this mixture is preferred in foundries.

Main requirements of a mould and core materials are:

1. It must be amenable to the shaping process such as by ramming around patterns and in the coreboxes.
2. The rammed structures must be able to retain their shapes and rigidity once constructed until the metal is cast in the mould and has formed a strong solidified skin.

Before going further into the moulding process, let us differentiate between moulding processes and casting methods.

- Moulding process–methods of making mould and the material used.
- Casting method–moulding process, design of gating and feeding systems, etc.

Moulding methods and materials encompass a wide range of options. The selection of the correct one to use in a specific casting application depends on

1. The alloy,
2. The dimensional tolerances desired,
3. The quality of the casting.
4. The price the customer is willing to pay.

A typical green moulding sand mixture/aggregate should consist of:

- Sand
- Clay
- Moisture
- Additives

Sand is a granular material normally obtainable or artificially produced by disintegrating or crushing of rocks. Sand is an aggregate material essentially consisting of tiny, loose grains, minerals, or rocks which are not larger than 2 mm or smaller than 0.05 mm in diameter. Sand also denotes a class of several minerals—rather than just one mineral—such as silica or quartz. Zircon, olivine, chromite, and ground ceramic minerals, as well as silica are classified as sand when they are in this size range.

In addition to sand, moulding sands may contain from 2 to 50% clay as a binder. The word clay is applied to a particular group of minerals which vary from fireclay (kaolinite) to Western or Southern bentonite (montmorillonite) and a few special clays (halloysite and illite). They are residual or weathered products of various kinds of silicate rocks. Clays have the general structure form of plate or flake in the approximate particle size ranging 10-20 microns in dimension.

Moisture in moulding sand aggregate is as essential as the clay substance itself. The bonding strength of the mould to retain the shape of the casting is developed by the sand-clay-water system. Thus, the purpose of clay will not be served until the required amount of water/moisture is added to it. Water which is present in amounts about 1.5 to 8% activates the clay in the sand, causing the aggregate to develop plasticity and strength.

In addition to these three basic ingredients (sand, clay and water), different types of organic or inorganic materials are added to the moulding sand aggregate as additives in small quantities to impart certain properties such as collapsibility, surface finish, etc. Some of the organic binders include cereals, proteins, pitch, and oils while some inorganic binders are cement, silicates and ester.

3.2 SAND

A foundry moulding mixture passes through four main production stages, namely: preparation and distribution, mould and core production, casting, and cleaning and reclamation. The property requirements of the materials are generally determined by moulding and casting conditions, although the preparation and reclamation stages are also considered in some degree, particularly to the integrated sand systems.

From a general viewpoint, the moulding sand aggregate must be readily mouldable and the mould cavity should retain its shape till the molten metal solidifies to produce a defect-free casting. Certain specific properties have been identified, and testing procedures adapted for their quantitative description.

3.2.1 Desirable Properties of Moulding Sand

Moulding sand is used for making moulds and cores. The moulding sand must be readily mouldable and produce defect-free castings if it is to qualify as a good one. Certain specific properties have been identified and testing procedures adapted for their quantitative description. *Foundry Sand Handbook* sets forth the standard conditions of testing the sand properties. Those properties of most obvious importance include:

1. At the moulding stage

(a) **Flowability** It is the ability of the material to be compacted to a uniform density. The degree of flowability required largely depends upon the method of compaction which may vary from hand ramming with tools to ramming on moulding machines. The material should respond to moulding processes so that it can flow to corners and undercuts of the pattern resulting in almost even ramming effect and thus strength at those points.

(b) **Green strength** It is defined as the ability of the sand to retain the shape of the mould independently without distortion or collapsing when the pattern is withdrawn. Moulding sand that contains moisture is known as greensand.

The green strength and the plasticity of moulding materials should be sufficient to avoid distortion of the mould at the green stage. This property is also known as cohesiveness. Higher the property, better will be the strength of the moulding aggregate. The higher strength is required in the moulding aggregate to retain its shape as the moulding aggregate is subject to metallostatic pressure.

2. During casting

(a) **Thermal stability** It is the ability to retain shape at high temperature. Heat from the casting causes rapid expansion of the sand surface at the metal-mould interface. The mould surface may then crack, buckle, or flake off (scab) unless the moulding sand is relatively stable dimensionally under rapid heating.

(b) **Refractoriness** It is the ability to withstand high temperatures without fusion or other major physical change. High pouring temperatures, such as those for ferrous alloys at 1350 to 1700°C, requires greater refractoriness of the sand. Low-pouring temperature metals, for example, aluminium, at 700°C, do not require a high degree of refractoriness from the sand.

If the fusion temperature of the sand is low, it will fuse and lumps of the sand will be mixed with the metal. It is therefore required that the moulding sand should have higher fusion point to avoid its fusion. The moulding material must be able to withstand the temperature of the pouring liquid metal. This property of the sand is known as refractoriness. It is the ability of the moulding material to resist the temperature of the liquid metal to be poured so that it does not get fused with the metal. The refractoriness of the silica sand is highest (1700°C).

Some refractory materials are given in Table 3.1.

Table 3.1 Properties of some refractory materials

Material	*Melting point*	*Coefficient of linear expansion × 10⁶/°C*
Silica (SiO)	1710	16.2
Alumina (Al_2O_3)	2020	8.0
Magnesia (Mgo)	2800	13.5
Thoria (ThO_2)	3050	9.5
Zirconia (ZrO_2)	2700	6.5
Zircon (ZrO_2), (SiO_2)	2650	4.5
Silicon Carbide (SiC)	2700	3.5
Graphite	4200	-

(c) **Dry strength** Sand should have dry strength to withstand erosive forces and pressure of liquid metal. As the casting is poured, sand adjacent to the hot metal quickly loses its moisture as steam. The moisture is the source of activating the clay and thus helping it to achieve the strength. The dry sand must have strength to resist erosion by the molten metal, and also the metallostatic pressure of the molten metal, or else the mould may enlarge.

(d) **Hot strength** Sand should have hot strength to withstand distortion and deformation during heating at high temperature. In the production of heavy castings, a considerable layer of moulding materials rises to a temperature where the normal or green mechanical properties are no longer the main criterion governing the dimensional stability and resistance to contraction. Depending upon the mass of the casting, the sand must have adequate combination of high temperature properties, including hot strength to withstand mould enlargement due to heating and deformation, crack or breakage due to contraction of the casting. The mould should have sufficient dry strength as well as hot strength to resist erosion, crack propagation and breaking at elevated metal temperatures.

(e) **Thermal conductivity** The material should have sufficient thermal conductivity in order to solidify and cool castings at the maximum rate possible, so that the foundry floor area can have higher utilization.

(f) **Collapsibility** The readiness with which the moulding material with breakdown in knockout and cleaning operations. Heated sand becomes hard and rocklike after casting and is difficult to remove from the casting and may cause the contracting metal to tear or crack.

(g) **Permeability** Ability of the moulding material to allow hot gases to pass through it. When liquid metal is poured into the mould cavity, a large amount of gases are generated. These gases are generated because of the reaction of the moulding material with the liquid metal. The steam in the mould is generated because of the moisture content. In addition, the liquid metal absorbs some gases from the atmosphere and on solidification of metal in the mould, these gases are also released. If these gases, due to reaction, moisture, and atmospheric gases, could not escape from the mould, they may be entrapped inside the casting and can cause various defects in the casting. Hence, the mould must be able to escape these gases or it should be porous. The property of the moulding sand which helps to escape the gases is known as permeability. Higher the permeability of the moulding sand, better it is. Proper venting of the moulding aggregate may also help in escaping the gases which are generated inside the moulding aggregate.

The evaporation of each 1% of moisture from green moulding sand can be shown to generate over 30 times of its own volume of steam; this is paralleled in core sands by gases from volatilisation and decomposition of organic compounds such as core oil and cereals. To provide a path for the escape of gas, permeability of the mould is an essential property, giving protection against surface blows and similar defects.

(h) **Fineness** It is required for the prevention of metal penetration and the production of smooth casting surfaces. The liquid metal is sufficiently fluid at high temperature and has the capacity to flow between the pores of the sand at the metal-mould interface, causing a rough surface. For wetted moulds, the liquid metal penetration will be even higher. Since both permeability and fineness are functions of grain size and distribution, the two properties are in conflict and a compromise is usually necessary. Fineness may be achieved by using fine-grained sands, by continuous grading or by incorporation of filler materials, but all these measures also reduce permeability. An alternative approach is to use a highly permeable moulding material and to obtain surface fineness by the use of mould coatings.

(i) **Finish** Good surface finish of mould cavity produces the same finish to casting. So the grained material is preferred for that property.

(j) **Adhesiveness** Ability of the moulding sand to stick with the walls of moulding boxes. The moulding aggregate should have the property of getting clung to the surface of the moulding boxes. Otherwise, it is difficult to transport the moulding boxes from one place to the other.

3. Storage

(a) Bench life: the ability to retain moulding properties on standing or storage.

(b) Durability: the capacity to withstand repeated cycles of heating and cooling during the casting operations. This will decide the extent of reusability of the moulding sand.

(c) Reusability: it is desired to reduce the expenditure and waste disposal.

Thus, it is evident that the qualities required in a moulding material cannot readily be defined in terms of simple physical properties. For complex aggregate, bulk properties are of greater significance and some of these can be measured directly by simple tests upon sand compacts. Other qualities are represented in specially developed empirical tests designed to reproduce conditions encountered in the foundry. These tests, in conjunction with the direct measurement of more fundamental characteristics such as mechanical grading and chemical composition, provide the basis for the control and development of the moulding material properties.

3.2.2. Base Sand: Characteristics and Types

Granular particles of sand comprise 50 to 95 per cent of the total material in the moulding sand. Most sand moulds and cores are based on silica sand since it is the most readily available and lowest cost moulding material. The base sand may either be clay free, washed, white silica sand or less pure, tan coloured sand containing some small percentage of clay and other materials. Besides silica sand, other sands are often used where special properties are needed. The sand particles used as moulding materials require special characteristics to satisfy the properties and the quality required for moulding the sand aggregate.

(A) Characteristics of Moulding Sand

Apart from considerations of purity and clay content, the average fineness number and particle size distribution are properties of base sand. Some of the important properties of base sands are discussed below.

1. Size and Shape Characteristics

(a) ***Average grain size***

The granular materials are usually classified according to the average size of the grain as follows:

2.00 – 1.00 mm	Very coarse sand
1.00 – 0.50 mm	Coarse sand
0.50 – 0.25 mm	Medium sand
0.25 – 0.10 mm	Fine sand
0.10 – 0.01 mm	Silt
< 0.01 mm	Clay

Grain sizes of foundry sands generally fall within the range of 0.1 to 1.0 mm. foundry men use a parameter known as the AFS fineness number to express the average grain size of sand which is determine by mechanical sieving.

The average grain size is an important parameter to control many properties of the moulding sand aggregate, the most important being the permeability and surface fineness. Coarser sands with greater void space produce moulds with greater permeability and poor surface fineness. On the other hand, finer sand grains produce moulds having higher fineness but poor permeability. Typical variation in permeability of base sand with its grain size is shown in Figure 3.1.

The grain size of base sand also influences the strength of the moulding sand aggregate, an inverse relationship exists between compressive strength and grain size of clay-bonded sands. Figure 3.2 shows the effect of grain size on strength. Finer grains also appeared to be more easily fused (i.e., low refractoriness) than coarser ones. Where maximum refractoriness is required, as in steel moulding sands, the coarser high-purity sands are used.

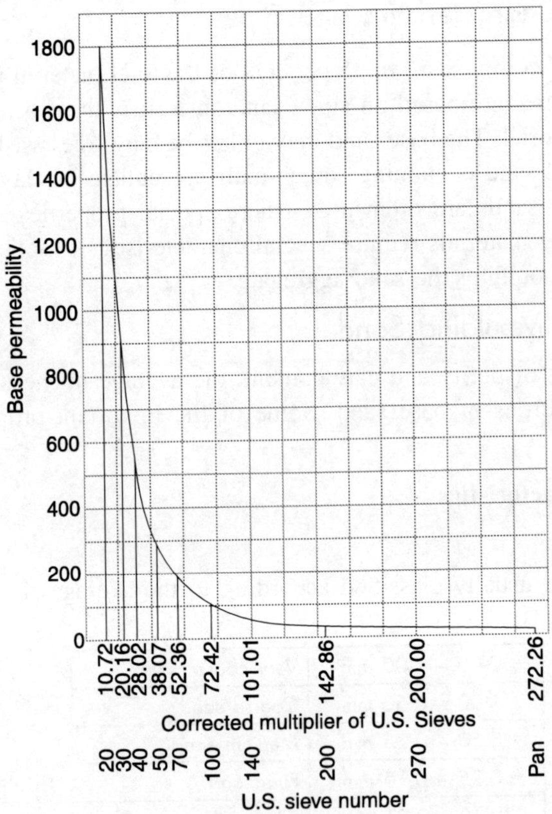

Figure 3.1 Base permeability of silica sand

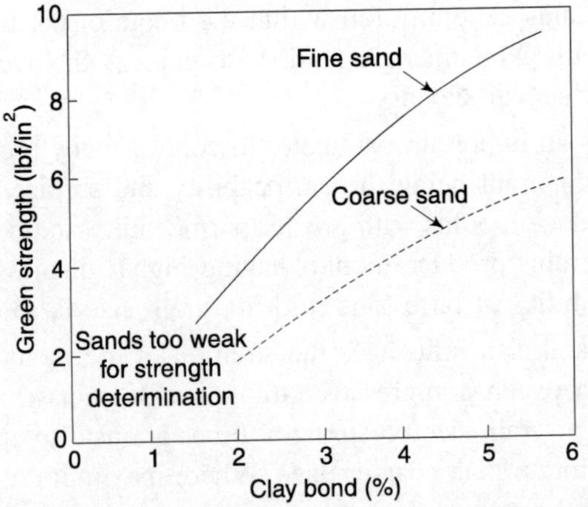

Figure 3.2 Effect of grain size on the strength of clay-bonded sand

(b) *Size distribution*

Besides average grain size, the size distribution of sand is also important. A well-distributed sand grain has high strength but low permeability. Usually sieving is done, like for determining the average grain size, to determine size distribution of a particular sand mixture. From this test, the sand may be categorised by the number of screens over which the bulk fraction is spread as a 2-, 3-, 4-, 5-, etc. screen sand. A screen fraction is arbitrarily defined as one with more than 10 per cent retained on that screen. Thus a 4-screen sand is one where the bulk of the sand is retained on four adjacent screens, each having more than 10% retained on it. The sieve analysis of sand A in Table 3.2 is an example of 5-screen sand while sand B is a 3-screen sand. Although 3-, 5- or 6-screen bulk fraction sands can be used, the 4-screen type seems to be the most versatile over a wider range of conditions.

Table 3.2 Similarity in AFS grain fineness number of two sand samples with different grain size distribution

USA Sieve No.	*Percentage retained*	
	Sand A	*Sand B*
6	0.0	0.0
12	0.0	0.0
20	0.0	0.0
30	1.0	0.0
40	24.0	1.0
50	22.0	24.0
70	16.0	41.0
100	17.0	24.0
140	14.0	7.0
200	4.0	2.0
270	1.7	0.0
Pan	0.3	1.0
Total	100.0	100.0
AFS grain fineness No.	60.0	60.0

(c) *Grain shape*

The shape of sand grain is important with respect to flowability and strength. The shape of individual sand grains may be rounded, angular, sub-angular or compounded (Figure 3.3) depending on the geological origin and transportation history. Rounded sands have the lowest surface area and higher flowability, and need lower binder and moisture to produce moulds with low strength and high permeability. On the other hand, angular sand grains have high surface area and low flowability, and need higher binder and moisture to produce moulds with high strength and low permeability. Compounded sand grains are agglomerated particles of angular or sub-angular grains. They disintegrate at high temperatures and, therefore, are not desirable.

Although the relationship between grain shape and mechanical strength is partly governed by the ramming density attained, a mixture of rounded, angular and sub-angular grains is generally desirable to produce optimum properties.

Average grain shape may be readily observed under microscope or measured by surface area measurement. The average shape of sand grains is quantified by a parameter known as the coefficient of angularity,

$$\text{Coefficient of angularity, } \alpha = \frac{\text{Actual specific surface area }\left(\dfrac{\text{cm}^2}{\text{g}}\right)}{\text{Theoretical specific surface area }\left(\dfrac{\text{cm}^2}{\text{g}}\right)}$$

Here the actual specific surface area is defined as a surface area of all sand grains in a square centimetre area, containing 1g of sand. The theoretical specific surface area, on the other hand, assumes that all the sand grains are spherical.

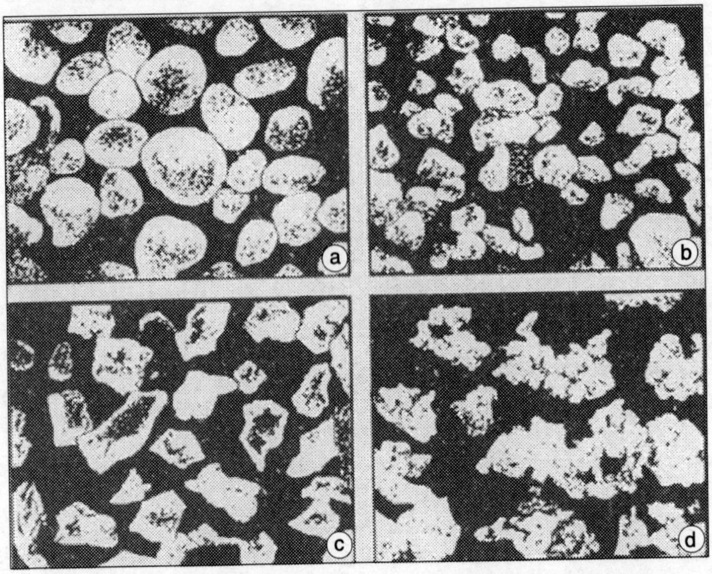

Figure 3.3 Shape of sand grains: (a) Rounded, (b) Sub-angular, (c) Angular, (d) Compounded

Rounded grains	Rounded grains have least contact with one another in a rammed condition and so have higher permeability. Rounded sand grains give poor bonding strength as compared to angular sand grains. Too many smooth and rounded sand grains result in sand wash, sand crack and sand scales. These sands also possess greater flowability.
Sub-angular grains	As compared to rounded grains, sub-angular grains possess better strength and lower permeability. In comparison to angular grains, they possess lower strength and better permeability.

Angular grains	Angular grains have greater bonding strength, less flowability and low permeability than round grain sands. Angular grains have sharp corners and greater contact surface.
Compounded grains	These grains are formed by small grains sticking together. These are least desirable as moulding material due to their tendency to break down at high temperatures. Compounded grains are agglomerated particles of angular or sub-angular sands in moulding or combination of the earlier three shapes.

2. Refractoriness and Thermal Stability

It depends on chemical composition and impurity contents of sand. Most moulding sands are based upon the mineral quartz, which is both geologically abundant and refractory to temperatures approaching 1700°C. As SiO_2 content in sand increases, the refractoriness increases. The purest silica sand containing > 99.8% SiO_2 has the highest refractoriness and thermal stability. The presence of excessive amount of impurities such as iron oxide, feldspar, and limestone lowers the fusion point of sand. So the ideal mixture would be silica sand of high purity with a minimum of binder. For the most severe conditions, as in the case of heavy steel castings, some other sand with high refractoriness, such as zircon sand or olivine sand may be used. For lower melting point alloys, refractoriness can be low and naturally bonded sands, with their much lower content of free silica, can be selected. Typical chemical analyses of high silica sand and natural moulding sand are shown in Table 3.3.

Table 3.3 Chemical analyses of typical foundry sands

	Silica Sand (%)	*Natural Sand (%)*
SiO_2	97.91	78.2
Al_2O_3	1.13	10.1
Fe_2O_3	0.50	10.1
TiO_2	0.04	-
CaO	0.11	2.4
MgO	0.02	1.8
K_2O	0.65	3.1
Na_2O	0.07	0.2
Loss on ignition	0.21	4.1

3. Thermal Expansion

Low thermal expansion is mandatory for moulding sands. However, a major expansion of silica sand occurs in the temperature ranges of 550 to 650°C because of the allotropic transformation of silica. Silica sand is commonly encountered in four different phase transformations and the equilibrium temperature ranges for these modifications are shown in Figure 3.4.

4. Hardness

Adequate hardness of sand is essential so that they do not break during mould preparation, mould handling and pouring of liquid metal.

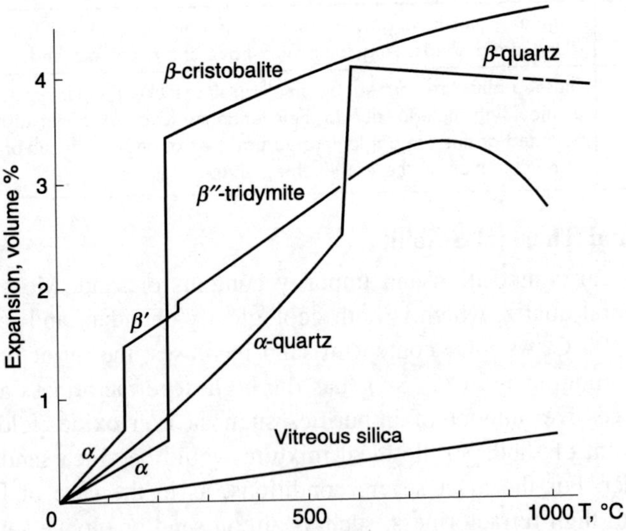

Figure 3.4 Volume change with temperature in silica structures

5. pH Value

Neutral sands having the pH value in the range of 6 to 8 are ideal as moulding sand. Otherwise, they are susceptible to attack by the chemical binders.

B. Classification of Base Sand

The moulding sand mixtures used for sand casting can be divided into two classes:

1. Naturally bonded sand
2. Synthetically bonded sand
    - Silica sand
    - Zircon sand
    - Olivine sand
    - Chromite sand
    - Chamotte sand

Naturally bonded sand is the one containing sufficient clay as mined from the sand pit, so that it can be used directly for moulding after tempering and conditioning. Naturally bonded sands may contain considerable organic matter and from 5% to 20% ordinary clay. It requires 5% to 8% water to develop adequate strength. Such a mixture has less refractoriness than synthetic sands. Hence, are suited for casting low temperature non-ferrous metals and to some extent low melting point cast iron.

Synthetically bonded sand is artificially compounded in the foundry by mixing clay-free silica sand having desired grain size, selected binders which may be organic or inorganic, additives and moisture in proper quantity so as to control the properties of the mould as desired. The composition of moulding sand thus prepared must be carefully controlled for assured, satisfactory and consistent results.

The synthetic sand mould is better than natural sand moulding because:

1. defects in casting due to sand mould expansion can be minimized;
2. more durable than natural sands;
3. easy to handle as it maintains its strength;
4. improved properties that can be controlled;
5. provides better flowability to molten metal; and
6. provides higher permeability.

The application of natural and synthetic sand is given in Table 3.4.

Table 3.4 Applications of natural and synthetic sands

Type of sand	*Application*	*Reasons*
Natural and moulding sand	Light castings	Permeability is not important and only good surface finish may be required.
	Jobbing foundry	Only a few castings are required. Patterns may be of poor quality.
	Mechanised production of castings with a few cores	There is little influx of sand from cores.
	Dry sand moulding	High permeability is not necessary as little gas is evolved. Patching is easy.
Synthetic sand	Heavily cored castings	At shake out, large amount of clay-free core sand enters moulding sand; regular clay addition is possible.
	Mechanised production	Better hardness and uniform properties can be obtained.
	High pressure moulding	Results in good hardness and permeability, easy stripping, high strength and toughness in the mould.

The main ingredients of synthetic moulding sand are:

- Silica grains (SiO_2)
- Binder such as clay
- Moisture to activate the clay and provide plasticity

Silica sand: Most abundantly found either as natural (in river beds) or artificial (crushed quartz). The general chemical nature of silica sand for foundry use is described in Table 3.5.

The principal division between natural and synthetic sands exists in all sections of the industry, but the general trend is towards greater use of synthetic sands, particularly in mechanised foundries. Their chief advantage lies in superior refractoriness and permeability, consistency, and the readiness in which their properties can be controlled, particularly in reclamation systems.

Table 3.5 General chemical composition of silica sand

SiO_2 content	95-96% min.	The higher the silica, the more refractory the sand
Loss on ignition	0.5% max.	Represents organic impurities
Fe_2O_3	0.3% max.	Iron oxide reduces the refractoriness
CaO	0.2% max.	Raises the acid demand value
K_2O_2, Na_2O	0.5% max.	Reduces refractoriness
Acid demand value to pH 4	6 ml max.	High acid demand adversely affects acid catalysed binders

They are not, however, so readily hand worked by the moulder as the naturally bonded sands with their greater latitude of moisture content and somewhat higher green strength. Table 3.6 shows a comparison between the usefulness of natural sand and synthetic sand as foundry sand, while a comparison in properties of different synthetic sands is shown in Table 3.7. Synthetic sands have been most widely used for steel founding, where relatively coarse grained sands are bonded with bentonites and cereals to provide properties such as those shown in Table 3.6 and Table 3.7. Synthetic sand mixture details are given in Table 3.8.

Table 3.6 Comparison of natural sand and synthetic sand

	Natural sand	*Synthetic sand*
Moulding material	Natural sand (sand, clay, organic materials from weeds, trees, bacterium, etc.)	Sand, bentonite, starch, coal.
Sand treatment (use of machine)	Usually not necessary (if poor sand quality, a simple sand treatment will do)	Necessary (sand treatment equipment, mixer, etc.)
Moulding	Easy	Easy (especially when sand treatment is sufficient)
Repair of mould	Easy	Relatively difficult
Shake out	Easy	Easy
Reclamation	Easy (only needs water adjustment)	Easy (although mixing is necessary)
Life of sand	Limited	Not limited
Effects on casting	Mould swelling, sand adherence, gas defects (depending on size of materials).	Similar to natural sand but to a lesser degree (good for rather large size castings).

Table 3.7 Comparison of foundry base sand properties

	Silica	*Olivine*	*Chromite*	*Zircon*
Colour	White light brown	Greenish grey	Black	White brown
Hardness	6.0 – 7.0	6.5 – 7.0	5.5 – 7.0	7.0 – 7.5
Dry bulk density (lb/ft^3)	85 – 100	100 – 125	155 – 165	160 – 185
Specific gravity	2.2 – 2.6	3.2 – 3.6	4.3 – 4.5	4.4 – 4.7
Grain shape	Angular/rounded	Angular	Angular	Angular
Thermal expansion (mm/mm/°C)	0.018	0.0083	0.005	0.003

(Contd.)

Apparent heat transfer	Average	Low	Very low	High
Fusion point, C	1427 – 1760	1538 – 1760	1760 – 1982	2038 – 2204
High temperature reaction	Acidic	Basic	Neutral to basic	Slightly acidic
Wettability with molten metal	Easily	Not generally	Resistant	Resistant
Grain distribution	2 – 5 screens	3 – 4 screens	4 – 5 screens	2 – 3 screens
AFS grain fineness number	25 – 180	40 – 160	50 – 90	95 – 160

Table 3.8 Synthetic sand mixtures used in steel foundry

Properties	*Green moulding aggregate*	*Dry moulding aggregate*
Composition	Medium grade silica sand 5% Bentonite 0.75% Starch	50% Coarse silica sand 50% Medium silica sand 6% Ball clay 3% Bentonite 0.5% Dextrin
Moisture content, %	3.0	6.0
Permeability	170	100
Green compressive strength, kN/m^2	7	8
Dry compressive strength, kN/m^2	0.7	1.7

Advantages of silica sand

Silica sand is mostly used in foundries because:

1. It is a very good refractory material and doesn't fuse or soften even at very high temperatures, i.e., 1650°C, when in contact with molten metal.
2. They can be easily moulded into intricate shapes.
3. They have sufficient porosity or permeability and allow easy escape of gases produced by molten metal and other bonding constituents.
4. They can be used repeatedly for making moulds after addition of some bonding materials.
5. They are cheap and easily available.
6. They are chemically immune to molten metals.

Disadvantage of silica sand

1. It has a high coefficient of thermal expansion.

Types of silica sands

- **Bank sand** Sand from glacial or sedimentary deposits occurring in banks or pits usually containing less than 5 per cent clay and used in synthetic sands and core sands. In the great lakes area, these sands usually contain less than 2 per cent clay and consist of rounded silica grains in the size ranges of 60 to 70, 70 to 80, 80 to 90 and 90 to 100 AFS and over. The sand grains are usually not white because of impurities.

- **Lakes and (or Dune sand)** A sub-angular sand from the great lakes area and especially dune and bottom deposits, substantially free of AFS clay and of 45 to 50 50 to 60, 60 to 70 AFS fineness. Lake sand is usually not white because of impurities.

Zircon sand is the main constituent of igneous rocks and is found in certain metamorphic rocks. It is basically zirconium silicate ($ZrSiO_4$). Zircon ($ZrSiO_4$) is theoretically composed of 67.2% Zirconia (ZrO_2) and 32.8% Silica (SiO_2). However, practical zircon sand usually contains a higher proportion of silica (Table 3.9). Zircon sand has a high specific gravity (4.6) and high thermal conductivity which together cause castings to cool twice faster than silica sand. The chilling effect of zircon sand can be used to produce favourable thermal gradients that promote directional solidification giving sounder castings. The thermal expansion coefficient of zircon is very low (only 1/3 of SiO_2) (Figure 3.5) so that expansion defects can be eliminated. Zircon has higher refractoriness than silica, and moreover it does not react with iron oxide, so sand burn-on defects can be avoided. This produces a better surface finish which reduces cleaning cost and gives a good reproducibility. Zircon sand generally has a fine grading, with AFS number between 140 and 65 (average grain size 115-230 microns); the most frequently used grade is around AFS 100.

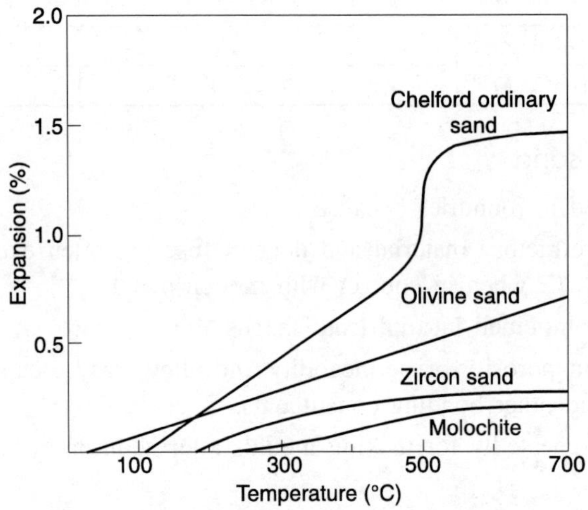

Figure 3.5 Thermal expansions of mould refractories

Zircon is probably the most widely used of the non-silica sands. It is used with chemical binders for high quality steel castings and for critical iron castings such as hydraulic spool valves which contain complex cores, almost totally enclosed by metal, making core removal after casting difficult. Zircon has low acid demand value and can be used with all chemical binder systems. The Cosworth casting process uses the low thermal expansion of zircon sand cores and moulds to cast dimensionally accurate castings. The high cost of zircon sand makes reclamation necessary and thermal reclamation of resin bonded moulds and cores are frequently practised. Because of high cost, instead of making the whole mould, mould washed onto a greensand or silica mould is often used. Some zircon sands contain radioactive minerals which may cause health hazards. The supplier should be contacted to confirm that the level of radioactivity is safe.

Table 3.9 Chemical composition of zircon sand

ZrO_2	TiO_2	SiO_2	Fe_2O_3	Al_2O_3	P_2O_5	Cr_2O_3
65.80	0.26	33.60	0.05	0.25	0.01	0.001

Chromite sand is crushed from the chrome ore whose typical composition is Cr_2O_3 = 44% Fe_2O_3 = 28%, SiO_2 = 2.5%, CaO = 0.5%, and Al_2O_3 + MgO = 25%. The fusion point is about 1800°C. It also requires a very small amount of binder (about 3%). It has high refractoriness and therefore commonly used as facing material in cores and moulds for heavy steel castings. The chromite sand has specific gravity of 4.5. Thermal expansion characteristic of chromite is better than silica and less than zircon sand. It is best suited to austenitic manganese steel castings. Typical composition of chromite sand is shown in Table 3.10.

Table 3.10 Chemical composition of chromite sand

Cr_2O_3	Fe_2O_3	Al_2O_3	SiO_2	MgO	CaO	MnO	TiO_2	V_2O_5
45.3	25.1	14.7	1.6	10.1	0.13	0.26	0.61	0.37

Chromite is a basic sand and has high specific gravity (4.5) and high thermal conductivity which provides a pronounced chilling effect. Thermal expansion is low so expansion defects are unlikely to occur. Chromite sand has a glossy black appearance, and has greater resistance to metal penetration than zircon in spite of its generally coarser grading (typically AFS 70). It has somewhat higher acid demand than other sands, which entails greater additions of acid catalyst when furan resin is used. Apart from this, the sand is compatible with all the usual binder systems. The common binder is bentonite clay (usually 1/2 to 2/3 that of SiO_2). Chromite is generally used for steel casting to provide chilling. It is difficult to reclaim chromite sand since, if it becomes contaminated with silica, its refractoriness is seriously reduced.

Olivine sand contains the minerals, Fosterite (Mg_2SiO_4) and Fayalite (Fe_2SiO_4). The typical composition of olivine sand is shown in Table 3.11. Olivine sand is used mainly for the production of austenitic manganese steel castings (which react with silica and other sands to give serious burn-on defects). It has also been used to avoid the health hazards possible with silica sand. Being a basic sand, olivine has a very high acid demand and is not suitable for use with organic binders (such as furan resins) since all organic binders are acidic in nature. Common binders used with olivine sands are: bentonite, fireclay and sodium silicate. Being a crushed rock, it is highly angular and consequently requires high binder additions. Thermal expansion is regular and quite low.

Table 3.11 Chemical composition of olivine sand

MgO	SiO_2	Fe_2O_3	Al_2O_3
46.0	43.08	8.98	1.02

Chamotte sands are calcined high grade fire clays which are formed by burning approximately at 1450°C and then crushing to required grain size. Typical chemical composition of chamotte sand is shown in Table 3.12.

Table 3.12 Chemical composition of chamotte sand

SiO_2	Al_2O_3	Fe_2O_3	MgO	Na_2O
67.7	29.7	2.1	0.25	Trace

This sand has low thermal expansion property which affects to prevent defects such as scab. But care should be taken when using this sand as it absorbs moisture. The sand is much cheaper than zircon and olivine. Specific characteristics of chamotte sand and negligible affinity to liquid steel make it suitable for steel castings.

Binders used in moulding sands

Sand grains must be held together by some bonding medium. The bonding medium in the moulding material is distributed around the grains so that, as far as possible, the particulate matter is completely coated by the bonding agent. This is necessary because, since a mould involves the use of many sand grains, no matter how the particulate matters are oriented, there will always be an amount of bond available at all points of contact ensuring uniform strength at all parts in the compacted mould. The distribution of the bonding agent around the sand grains is affected by mixing or milling of refractory particles with the bonding medium in sand mills.

- Binders are added to give cohesion to moulding sands.
- Binders provide strength to the moulding sand and enable it to retain its shape as mould cavity.
- Binders should be added in optimum quantity as they reduce refractoriness and permeability. An optimal quantity of binders is needed, as further increases have no effect on properties of foundry sand.

3.3 CLAY

Moulding sands for green and dry sand practice are most commonly bonded with clay, the second most important constituent of the aggregate. In the natural moulding sands the clay occurs in association with the sand grains, whilst the synthetic sands are bonded with selected clays from separate deposits.

Clays are hydrous aluminium silicates obtained as weathered products of silicate rocks and when tempered with water, produce a plastic or semi-plastic mass. Clays have the general structural form of plate or flakes in the approximate particle size range 0.01–1.0 micron in breadth, and plasticity and bond are developed by the addition of water. Net attractive forces are generated between charged hydrated clay particles and between the surfaces of sand grains. The strength of the ionic bond depends on the total surface area of the particles and is strongly influenced by adsorption of exchangeable cations at the free surfaces that modify the balance of local forces between the particles.

On drying, loss of adsorbed water produces shrinkage of the lattice and further strengthening of the bond, so that clay binders are effective in both green and dried condition. Hydration is reversible to temperatures well above the drying range: thus moulds may be dried or cast and the bond can be regenerated by addition of water after each cycle. Heating to progressively higher temperatures, however, removes chemically combined water and causes permanent loss of bonding capacity. The temperature at which this occurs varies with the particular clay but the loss begins at approximately 400°C and is in all cases complete at 700°C. At still higher temperatures, the clays undergo drastic mineralogical changes involving crystallisation of alumina and cristobalite and the formation of mullite.

Although the basic ingredients of a sand mix are only sand, clay and water, other materials are often added in small amounts to moulding and core sands for special purposes. Some of the most important additives include cereals, coal dust, iron oxide, molasses, dextrin, etc.

3.3.1 Clays: Types and Characteristics

According to the American foundry Society (AFS), clay is defined as those particles of sand having a diameter of 20 microns or less and, when suspended in water, fail to settle at a rate of 25 mm per minute. Clay consists of two ingredients: fine silt and true clay. Fine silt is foreign matter of mineral deposit having no bonding power. True clay supplies the necessary bond.

Effective clay is defined as the fraction of the clay present in the total clay of the sand-clay mixture which has effective bonding ability similar to the new clay. It is necessary to maintain a close control on the effective clay present in a sand mix during moulding and particularly in case of closed cycle machine moulding operations.

Based on the composition and structure, the clays used in the foundry as binder can be classified into three groups, i.e., kaolinite, montmorillonite, and illite.

1. Kaolinite or fire clay ($Al_2O_32SiO_2H_2O$)
2. Montmorillonite ($Al_2O_34SiO_22H_2O$)
3. Special clays such as Illite, Halloysite, Attapulgite and Limonite

Kaolinite

It corresponds to the general formula (Al_2O_3 $2SiO_2$ $2H_2O$), and is formed by weathering of feldspar or other aluminous minerals. Kaolinite is the principal constituent of china clay, ball clays and fire clay. Relatively high alumina content of these clays makes them reasonably refractory (softening point, 1600–1700°C), but irreversible dehydration occurs in the temperature range of 400–650°C. The bonding properties of these clays are not high as compared to other clays and hence higher binder content is required, often in the range of 10–20%. Therefore, the use of kaolinite in foundry is very limited.

Montmorillonite

These types of clays can be represented by the basic formula ($Al_2O_34SiO_22H_2O$), but a proportion of Al^{+3} ions are replaced by Mg^{+2} ions in isomorphic substitution. Whenever Mg^{+2} replaces Al^{+3}, there is capacity for adsorption of exchangeable cations such as Na^+ and Ca^{+2} to which the properties of clay are particularly sensitive.

Montmorillonite is the principal mineral constituent of bentonite clays. These clays have a high capacity for water absorption and exceptionally favourable bonding characteristics: strength properties can therefore be derived from addition as low as 3–5%.

Use of bentonites offers the following advantages:

- Bentonite can be circulated in closed systems and the bond is regenerated by adding water.
- Patterns can be stripped easily and thus mould can be made quickly.
- Bentonite resists erosion of moulds.
- Volumetric contraction of bentonites helps in compensating the expansion of silica grains.
- Bentonites retain their capacity for water absorption up to 500–700°C and can thus be regarded as thermally more stable than other clays, with a potentially longer lift in closed systems.

There are basically two types of bentonites. Depending on the kind of substitution metal present, bentonites are of two types, i.e., sodium bentonite known as Western bentonite and calcium bentonite known as Southern bentonite. Sodium bentonites produce better swelling properties, volume increases some 10–20 times; high dry strength which lowers the risk of erosion; better tolerance of variation in water content; low green strength and high resistance to burnout, which reduces clay consumption. The characteristics of calcium bentonites include low swelling, low dry and hot strength, low liquid limit, high plasticity, and high green strength.

Comparative properties of moulding sand with various clays is shown in Table 3.13.

Table 3.13 Comparison of foundry sand properties with various clays

	Sodium bentonite	*Fire clay*	*Calcium bentonite*
Silica sand, AFS GFN 60, weight %	95	88	95
Clay bond weight %	5	12	5
Moisture weight %	2.5	3.0	2.5
Permeability	110	60	108
Green compression strength, kPa	76.53	63.43	97.22
Dry compression strength, kPa	675.70	468.84	372.32

Illite: These clays are produced by the weathering of micas and form the principal source of bond in the natural moulding sands. They do not swell in the same manner as the bentonites but give reasonable strength properties. Irreversible dehydration occurs in the temperature range 500–550°C. Other characteristics of illite clays are fusion point 1380°C, moderate base exchange, and moderate shrinkage to loss of water. Summary of the clays used in foundary is given in Table 3.14

Table 3.14 Summary of the clays

Clays:	**Wood-product binders:*‡**
Fire clay (kaolinite)	Natural resin (e.g., rosin, thermoplastic)
Southern bentonite (calcium montmorillonite)	Sulfite binders (contain lignin, produced in the paper pulp process)
Western bentonite (sodium montmorillonite)	
Secondary mica clays (illite)	Walter-soluble gums, resins, and organic chemicals
Oils:*	**Protein binders (containing nitrogen):**
Vegetables (e.g., linseed oil)	Glue
Marine animal (e.g., whale oil)	Casein
Mineral (used for diluting oils given above)	**Other binders:**
Synthetic resins, thermosetting:*†	Portland cement†
Urea formaldehyde	Pitch (a coal-tar product)*†
Phenol formaldehyde	Molasses (usually applied in water as a spray)
Cereal binders made from corn:*	Cements (e.g., rubber cement)†
Gelatinized starch (made by wet milling, contains starch and gluten)	Sodium silicate (water glass, CO_2 hardening binders)†
	*Harden by baking
Gelatinized corn flour (made by dry-milling hominy)	† Harden at room temperature
Dextrin (made from starch, a water-soluble sugar)	‡Available as either a liquid or a dry power

3.3.2 Properties of Foundry Clays

The basic function of clay as binder is to produce cohesion between the refractory grains in the green or dried state by forming a thin film of coating around each grain. As the bonding layers become continuous and then progressively thick, proportionately less advantage is to be gained from further additions. The general form of relationship between the strength and the binder is shown in Figure 3.6, while typical effect of clay content on relation between mould hardness and green compressive is shown in Figure 3.7.

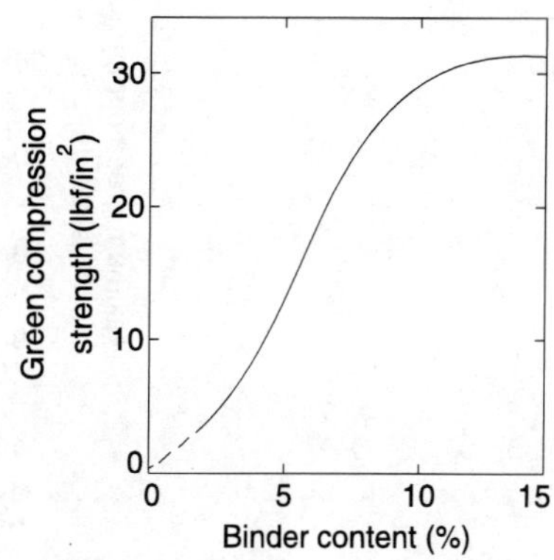

Figure 3.6 Influence of binder content on strength of moulding sand

Generally, Western bentonites are used in sands requiring a higher level of dry compressive strength, in excess of 80 psi, for example. Southern bentonites are used in sands where a lower level of dry compressive strength is acceptable, 40 – 80 psi, for example. Fire clays produce moderate dry strength in the sand. Maximum dry compressive strength over 200 psi can be obtained with mixtures of fire clay and Western bentonite.

Depending on the percentage of clay present, greensand may be classified as clay-saturated or clay-unsaturated mixtures. A clay-saturated greensand is defined as one containing a high

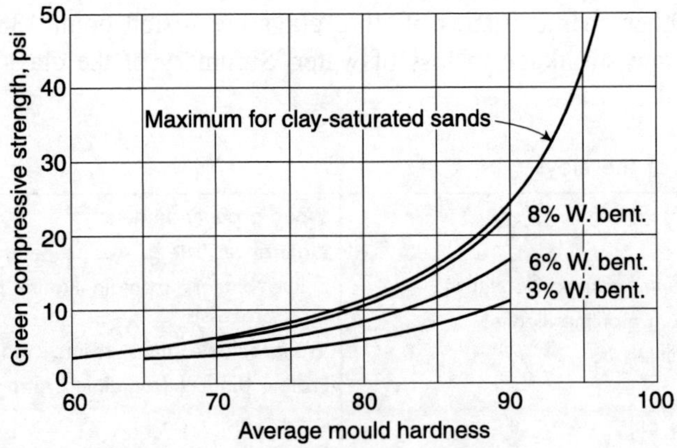

Figure 3.7 Typical effect of clay content on relationship between mould hardness and green compressive strength. The curve marked maximum applies equally well to clay-saturated Southern and Western bentonites and fire clay-bonded sands

enough percentage of clay so that any further increase in clay content will not cause an increase in maximum green compressive strength of the aggregate. This means that, the sand mixture is fully bonded. This definition is depicted graphically in Figure 3.8. The shaded area in the figure represents a variation in maximum strength due to clay purity and source, sieve analysis of base sand, moulding aggregate mixing efficiency and other factors.

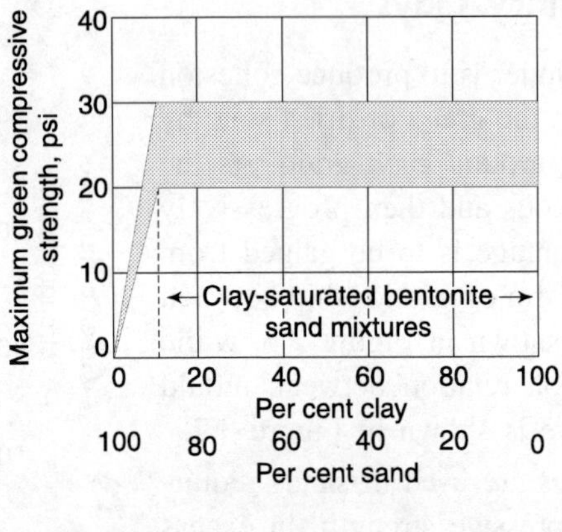

Figure 3.8 Schematic diagram showing the approximate effect of bentonite clays on maximum green compressive strengh of clay sand-water mixtures

The specific percentage of clay required for saturation depends upon purity and type of clay, base sand, and additives. In most cases, however, about 8–12% bentonites or about 20–25% fire clay is sufficient to produce a clay-saturated mixture with the sand of fineness AFS 60–100. Clay-saturated sands are probably the most versatile greensand mixtures for a wide range of casting weights and alloy types. Casting defects due to sand expansion, erosion, etc., are reduced or even eliminated. Since such sands are normally of high strength (14–20 psi green compressive strength), they require adequate ramming (probably over 85 mould hardness) to develop their properties. Clay-unsaturated sand systems, on the other hand, use 4–9% bentonites and 10–15% fire clays. Such sands are used for making lighter castings where expansion defects, erosion, etc., are lesser problems.

3.3.3 Clay and Water

Since the development of bond strength depends upon hydration of the clay, the green strength of a moulding mixture increases with the tempering water content up to an optimum value determined by the proportion of clay. Above this value, additional free water causes the green strength to diminish again as illustrated in Figure 3.9. Dry strength, however, continues to increase to much higher original moisture contents, probably due to improved distribution of the binder and the higher bulk densities attainable. Thus by determination of the optimum water content, the required strength properties can be obtained with minimum use of clay.

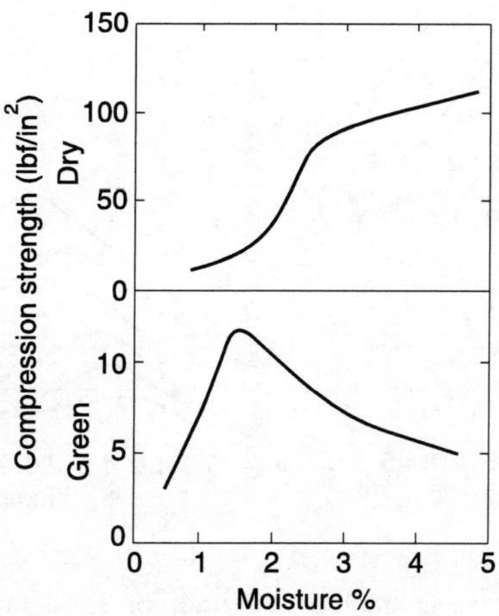

Figure 3.9 Influence of moisture content of green and dry strength of moulding sands (bentonite-bonded silica sand)

Typical combined relationship between clay and water contents and bond strength is illustrated in Figures 3.10 and 3.11 respectively. From these figures it is clear that, for a given clay type and content, there is an optimum water content which forms the greatest number of clay-water-quartz bonds. Too much water causes excessive plasticity and dry strength. Too little water fails to develop adequate strength and plasticity. Control of moisture in the moulding sand so that the best properties are developed is a necessary basis of sand control.

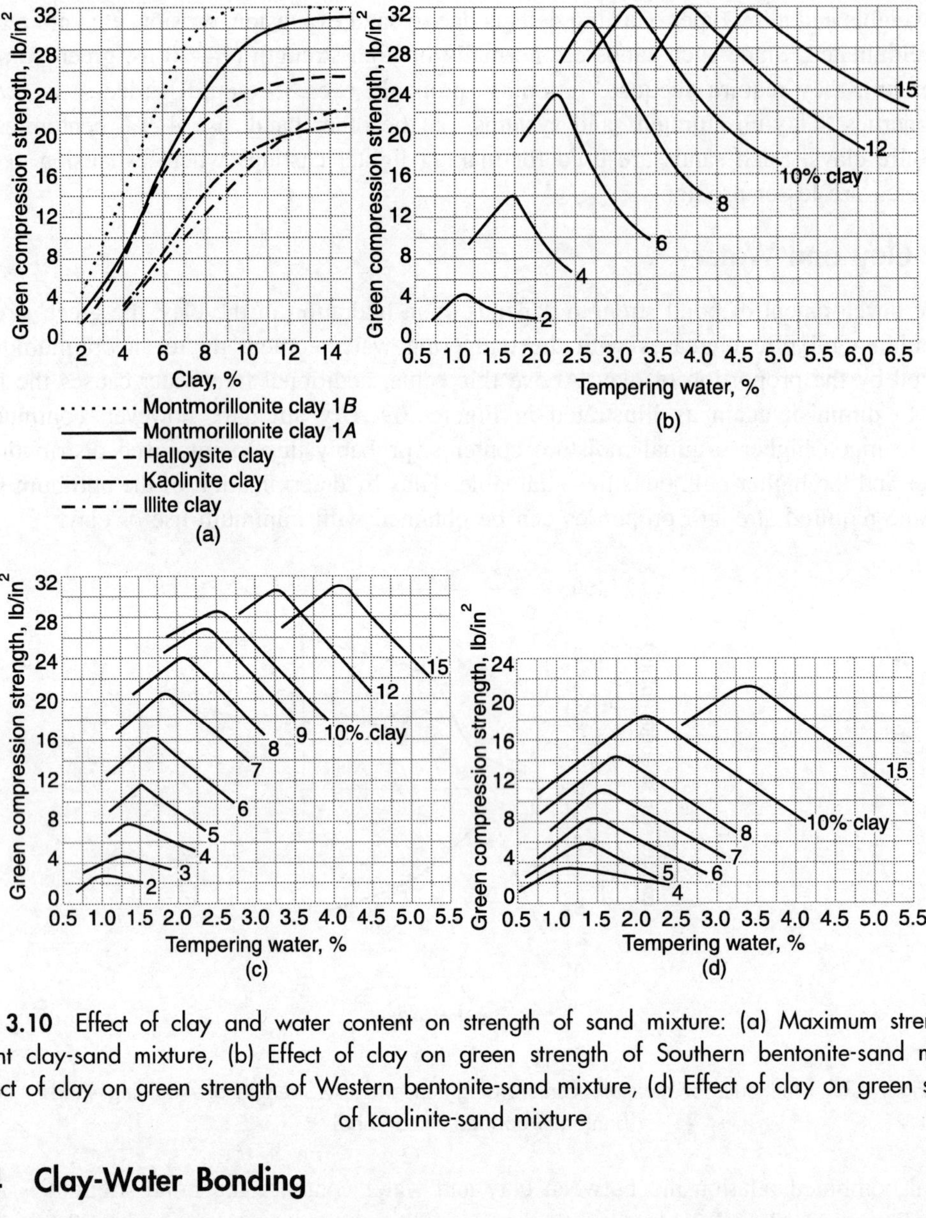

Figure 3.10 Effect of clay and water content on strength of sand mixture: (a) Maximum strength of different clay-sand mixture, (b) Effect of clay on green strength of Southern bentonite-sand mixture, (c) Effect of clay on green strength of Western bentonite-sand mixture, (d) Effect of clay on green strength of kaolinite-sand mixture

3.3.4 Clay-Water Bonding

The science of clays and silica sands, particularly as applied to aggregates used for moulding purposes, has not progressed to the point where we can state clearly what forces are involved in holding particles of clay together. Accordingly, the bonding forces involved may be accounted for by several theories: electrostatic bonding, surface tension forces, and interparticle friction bond.

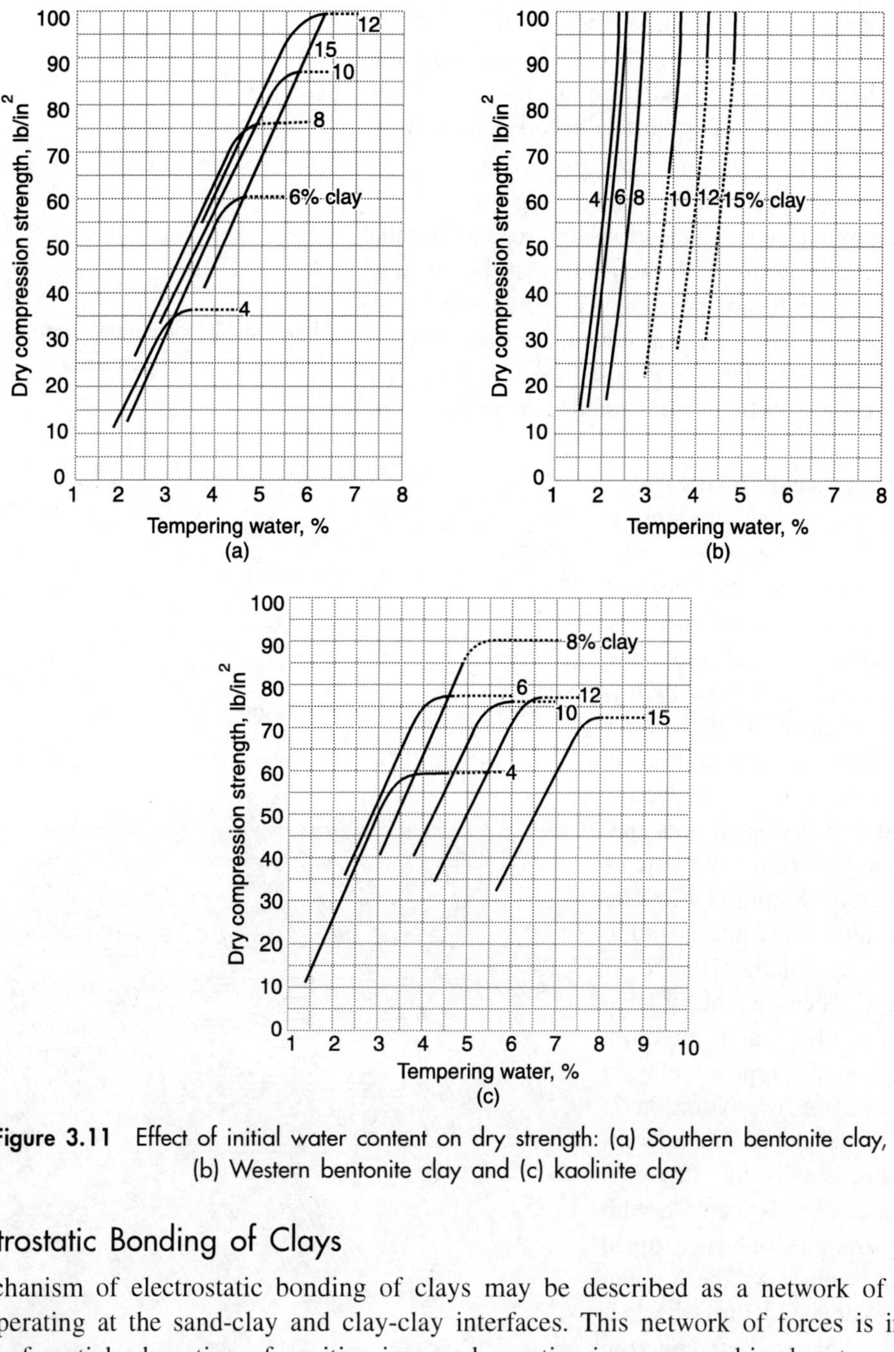

Figure 3.11 Effect of initial water content on dry strength: (a) Southern bentonite clay, (b) Western bentonite clay and (c) kaolinite clay

1. Electrostatic Bonding of Clays

The mechanism of electrostatic bonding of clays may be described as a network of dipolar forces operating at the sand-clay and clay-clay interfaces. This network of forces is initiated by the preferential adsorption of positive ions and negative ions on combined water and clay (hydrated) surfaces.

Dry clay does not provide the necessary bond to hold sand grains firmly together; bond is developed only when clay particles are hydrated. When water is added to dry clay, the negative hydroxyl (OH^-) ions are adsorbed on the nuclei of the clay atoms, owing to unsatisfied valence bonds at the surface of the clay crystal, and form an integral part of the crystal. So the clay-water particle becomes negatively charged.

The positive (H^+) ions in the surrounding water media are attracted by the negative clay ions, but repelled by the nuclei of the clay atoms, with the result that the positive ions take up equilibrium positions. The hydrogen ions and the adsorbed hydroxyl ions about the clay particle comprise a so-called double diffuse layer. A hydrated clay particle or micelle is illustrated schematically in Figure 3.12.

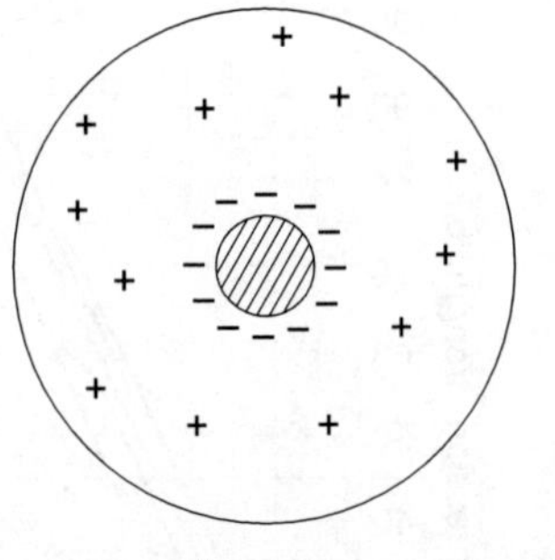

Figure 3.12 Schematic representation of clay micelle

Particles of sand (quartz) also form micelles by the adsorption of hydroxyl and hydrogen ions. When quartz and clay micelles are formed in each other's presence, the hydroxyl ions of the clay micelle exhibit an attraction for the hydrogen ions contained in the quartz micelles. Thus a clay dipole is formed and the result is an electrostatic bond between sand and clay particles and between clay particles as sketched in Figure 3.13. The maximum attractive force is found to be at an optimum distance of separation x. There are many such dipoles in a clay-water medium. Depending on the type of clay, a maximum degree of hydration is necessary to develop a dipole completely. This is why the strengths of clay-bonded sands increase with increasing amounts of water, up to a maximum value. As the amount of water is increased further, water enters the spaces between the dipoles to a distance greater than x, resulting in a decrease in the net intermicellular force (Figure 3.14).

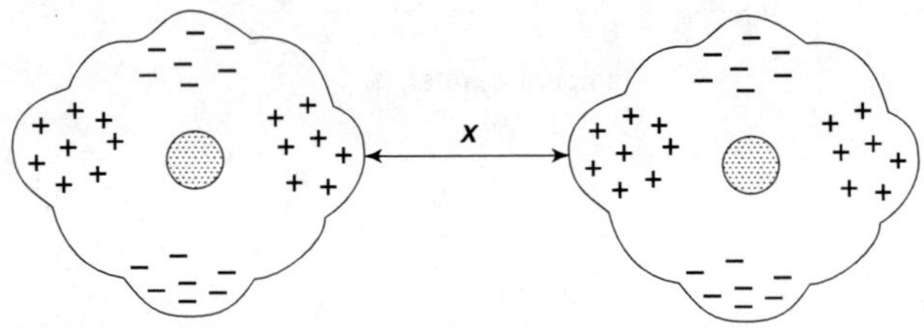

(a)

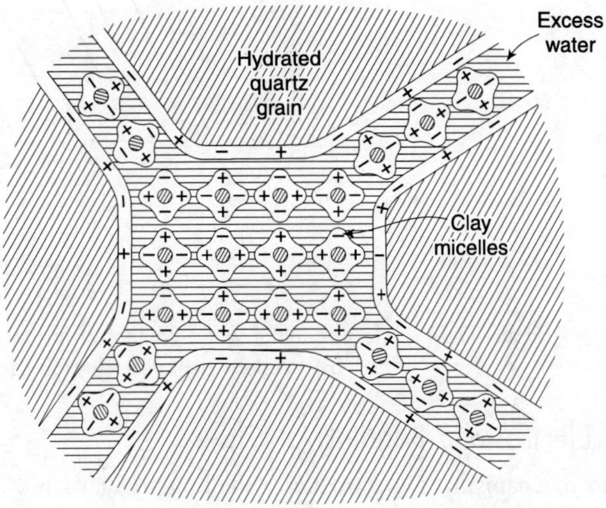

(b)

Figure 3.13 (a) Micellular dipoles, and (b) Disposition of clay and quartz

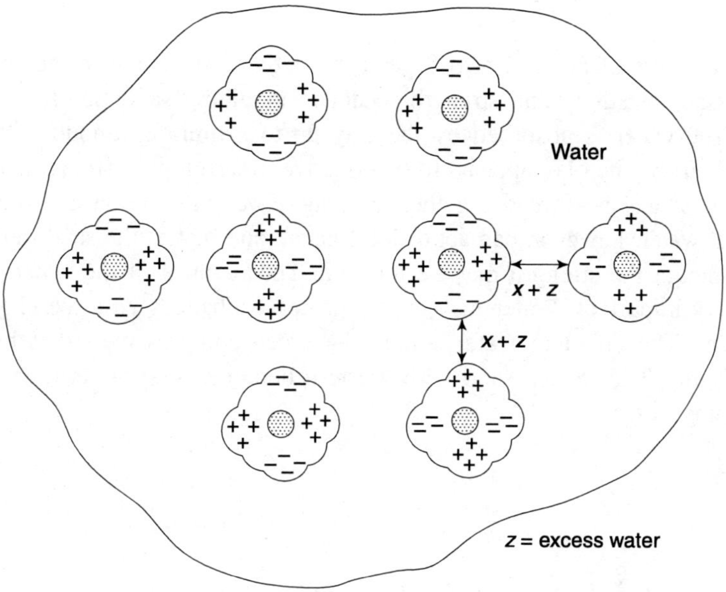

Figure 3.14 Dipole alignment of hydrated clay particles

2. Bonding by Surface Tension Forces

Forces developed by electrostatic interaction between sand-clay particles do not seem strong enough to account for all the strength properties of greensand mixtures, nor in particular for the high resistance to deformation of dry sand aggregates. Another possible source of bond strength is the surface tension of the water surrounding the clay and clay-sand particles, and filling the capillary interstices, particularly the interstices of the clay particles. The bond strength values as high as 880 psi have been obtained, attributable to the surface layers of water acting on a stretched membrane of hydrated clay, forcing the particles together. As the water layer becomes thinner by drying, the forces holding the particles together increase.

3. Bonding due to Interparticle Friction

The geometry of the aggregate can provide another force adding to the strength of the bond between particles. The so-called "block and wedge" theory involves essentially the interparticle friction developed in non-plastic particulate materials under pressure. When moulding sand is rammed inside a flask, most particles are jammed against their neighbours. The resultant inter-particle friction opposes further deformation, and causes a bridging action between long rows of favourably oriented particles and the sides of the flask. The changes in strength properties of sand mixtures due to the use of various sized and shaped sand grains indicate the existence of an interlocking or frictional force.

3.4 WATER

Water, present in amounts of about 1.5 to 8 per cent, activates the clay in the sand, so that it develops the necessary plasticity and strength. Water in moulding sands is often referred to as tempering water. The water is absorbed by the clay up to a limiting amount. Only the water rigidly held (absorbed) by the clay appears to be effective in developing strength. The rigid clay coatings of the grains may be forced together, causing a wedging action and thus developing strength. Additional water, however, can act as a lubricant, and makes the sand more plastic and more mouldable, though the strength may be lowered. Thus control of the water percentage in the aggregate is very important. Water may engage in ion exchanges with the clay if dissolved minerals are present. The effect of water content on green compressive strength of moulding sand is given in Figure 3.15. Silica sand with Western bentonite clay is used as the moulding material for the study.

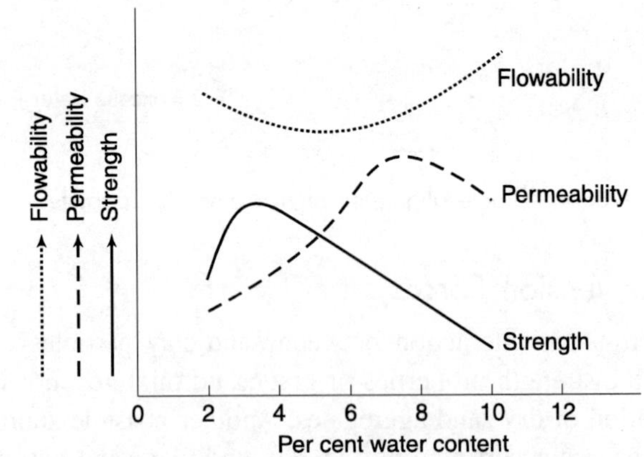

Figure 3.15 Variation of green strength, flowability and permeability with percentage of tempering water

3.5 ADDITIVES

In order to obtain specific characteristics in moulding and core-making sands according to the requirement of molten metal and base sand, suitable additives are mixed during sand preparation. The additives may be of reducing or fibrous nature, or may act as binding agents. These may also help in improving high-temperature plasticity and hot strength, produce anti-metal penetration properties and impart good surface finish to the castings. It is necessary to select the right type and determine the correct proportion of the additive for any given moulding and casting conditions so as to enable the production of flawless castings.

Coal dust It is a finely ground soft coal commonly used in greensand and dry sand moulding for protecting mould surfaces against the action of molten metal and improving surface finish of cast iron castings. When the molten metal comes in contact with mould surfaces containing coal dust, a gaseous envelope is formed which resists the fusion of sand to metal. Use of coal dust increases both green and dry strength, reduces expansion, tendency to scabbing and metal penetration. It, however, tends to reduce the permeability of sand. The sea coal is usually ground to fineness similar to that of the moulding sand in which it is used. Percentages employed in sands are about 2 to 8 per cent. A good quality coal dust suitable for foundry use should have minimum 30% volatile matter, maximum 20% ash, 3% moisture, 1% sulphur and 0.2% phosphorus content. Pitch and fuel oil are also used as reducing agents and have similar effect as coal dust. Typical example is given in Table 3.15.

Table 3.15 Example of sea coal specification

Volatile matter (VCM)	36-40%
Fixed carbon (FC)	55-60%
Ash	3-5%
Ultimate analysis	
H	5.6%
C	80-85%
N	1-3%
O	6-8%
S	0.8% max.
Ash	3-5%
Ash fusion point	1525°C

Iron Oxide Iron oxide powder is used as an additive for both moulding and core making sands to achieve high-temperature plasticity, hot strength and anti-metal penetration characteristics. In core sands, it prevents veining or high-temperature cracking of cores. The use of iron oxide is common in steel foundries both for moulds and cores. In iron foundries, its use is restricted to cores only. Good quality iron oxide should have iron oxide (Fe_2O_3) content not less than 93% and iron content not less than 6%. Its pH value in 10% distilled water solution should not be more than 9%. Its fineness should be 150 mesh BS sieve (106 micron IS sieve). Iron oxide decreases green strength and permeability while improving the hot strength. It reduces collapsibility and makes the shake off of the mould difficult.

Cereals They are used as binder to influence the bonding properties of sand. They are mainly starch and dextrine made from corn. Starch is made by separating the starchy portion from corn by dry or wet milling followed by cooking to gelatinate the starch and finally grinding to required size. Dextrines are produced from corn starch by acidifying with hydrochloric acid and heating in large steam jacketed roasters. This treatment makes the starch soluble in cold water. Sometimes dextrose (crystalline corn sugar) is also used. The cereal binders develop a gelatinous bond with water and are normally employed along with other binders, e.g., in clay bonded moulding sands and in oil and resin bonded core sands. In moulding sands the cereals increase air setting strength, toughness and collapsibility and prevent sand from drying quickly. During pouring, they gasify producing voids between sand grains and allowing their expansion without distortion. An additional advantage is the increased resistance of greensand moulds to friability on air drying. Addition of cereals results in strong moisture retention which delays air drying of sand mixtures and improves bench life. Use of cereal also eliminates scabbing and other expansion defects.

In core sands, cereal binder gives improved green strength and collapsibility with ease of knockout. However, they increase gas evolution and decrease the resistance of dried cores to the absorption of moisture. Dextrin is more effective than starch and it gives the same strength with less moisture. Generally, two types of dextrin are available, yellow and white. Both should have fineness of 100 mesh, BS sieve and moisture not more than 10%. The yellow variety should have a minimum dextrin content of 85% and maximum ash content of 1% whereas the white variety should have minimum dextrin content of 65% and ash content maximum of 0.5%.

Resins They are high melting point gums produced by chemical alterations of rosins (low melting point gums obtainable from pine trees) or produced synthetically. Synthetic resins are of two types, i.e., (a) furans that are urea formaldehyde resins modified with furfuryl alcohol in varying proportions, and (b) phenolics that are phenol formaldehyde modified with urea in varying proportions. In general, furans are more stable even in warm weather, whereas phenolics are cheaper and are resistant to heat and moisture. A large number of resins made in different ways are available suited for particular jobs. Resins are not soluble in water and also do not swell in water. When resins are mixed in the moulding sands, following improvements are obtained:

- Good surface finish of casting.
- Appreciable improvement in dry strength properties.
- Improved frictional properties.
- Decreased toughness and brittleness of green moulding sand.
- Increased flowability of sand and improved dispersion of fines throughout the sand mix and thus, moulds are produced with finer details.
- With the addition of 1.5% neutralised resins, green moulding sand can be stored for longer time.

Molasses It is a commonly used additive both in moulding and core making for iron castings. It is a dark brown viscous liquid containing 60-70% solid sugar obtained as a by-product during sugar refining. It may be used as a substitute of dextrin. It enhances the bench life of sands and imparts high dry strength and collapsibility. Due to high viscosity and wettability, it also increases green strength. Its decomposition at high temperature generates CO_2, which sets up a hardening action of the mould and increases hot compressive strength. On further heating the strength gets decreased, thus making the mould collapsible. However, due to the high hygroscopicity of the mix prepared with molasses, its use is not much favoured for good quality castings.

Linseed Oil (Core Oils) It is the most popular binder for core sand mixes. Linseed oil or other proprietary oils known as core oils, which are made by blending various ingredients such as vegetable oils, mineral oils, animal oils, natural resins and by-product residues from vegetable oil based industries, are used either with cereal binders like dextrin or with dextrin and bentonite. The sand mix develops strength only when the cores prepared from the mix are heated to a temperature of 200-240°C for a specified time that may vary from 1 to 3 hours. The hardening is developed by polymerisation or cross-linking. Overall rate of hardening is controlled by the rate of oxygen supply through sand and thus it emphasises the need of permeability, core venting, core stove ventilation. Reduced dry strength and friability result due to excess baking temperature. A specific characteristic of linseed oil is the resistance of its hardened film to moisture, due to which cores can be stored for prolonged periods without affecting the properties. The cores so prepared have very good baked strength, scratch hardness, permeability and collapsibility and can be stored for a long time. The specific gravity of core oils at 30°C should be about 0.90 and their acid value not more than 10. Standard core oils are of two grades, fast baking and slow baking depending upon the baking time required at 220°C to attain peak tensile strength. Further, depending on the maximum strength developed after baking, core oils are of two types, low strength and high strength.

Sodium Silicate (Water Glass) This is the most common binder used in air-setting or self-hardening processes for moulding and core making. The CO_2 process, ferrosilicon process, cement process, dicalcium silicate process, and others make use of sodium silicate as a binder, along with a solid or gaseous hardener. The variety suitable for the CO_2, process should contain total soluble silica (as SiO_2) 26 to 32%; total alkalinity (as Na_2O) 11 to 13%; relative sp. gr. at 20°C, 1.50 to 1.60; and total invert sugar content, 5 to 10%.

Fibrous Materials These materials are used to improve collapsibility and prevent scabbing and expansion defects. The commonly used materials are wood flour, chaff (dried grass), horsehair or cowhair, sawdust, manure and asbestos. Ground wood flour or other cellulose materials such as cob-flour, cereal hulls, and carbonised cellulose may be added in amounts of 0.5 to 2.0 per cent to moulding sands. They may function to control the expansion of the sand by burning out at elevated temperature. They also can improve collapsibility and flowability of the sand.

Silica Flour Pulverized silica, finer than 200 mesh, is called silica flour. It may be used in amounts up to 35 per cent for the purpose of increasing hot strength of the sand. It also increases the density of the sand for resisting metal penetration.

Table 3.16 gives summary of functions of various additives used in green sand mould aggregate.

Table 3.16 Functions of various additives used in green sand mould aggregate

Purposes	*Additives*
As binder	Cereals, molasses, resin, linseed oil, water glass
Increase collapsibility	Cereals, molasses, fibrous materials
Reduce expansion problems	Cereals, coal dust, firbrous materials
Increase green strength	Coal dust, molasses

(Contd.)

Increase dry strength	Cereals, resin, molasses, coal dust
Increase hot strength	Iron oxide, silica flour
Reduce metal penetration	Coal dust, iron oxide, silica flour
Improve surface finish	Coal dust, resin
Increase flowability	Resin, fibrous materials
Increase bench life	Cereals, resin, molasses

3.6 SAND PREPARATION

One of the important requirements for the preparation of sand is thorough mixing of its various ingredients. This is essential to ensure uniform distribution of the various components in the entire bulk of the sand. The preparation of moulding sand mixture for moulding can be either manual or mechanized. In small foundries and job production shops, the sand handling and conditioning are done manually. Mechanized units are used in large foundries.

Effective moulding of sand preparation usually consists of the following steps:

1. Removal of foreign material, principally fines, metal and hard lumps, as the sand is prepared for reuse.
2. Adequate mixing and tempering. Mulling or other mixing of the sand to distribute the clay and additives should be continued until optimum sand properties are developed. Certain mixtures are much more potent than others. Under-mixed sand may have a non-uniform distribution of ingredients.
3. Aerating, consisting in separating sand grains and in fluffing up of the sand by riddling, screening, or beating the sand, is practised for better moulding results.
4. Control of sand temperature. Cooling of sand is desirable since hot sand over 100°F causes moulding difficulties, e.g., sticking and dropping.

Preparation of sand consists of two processes:

1. Sand tempering
2. Sand conditioning

3.6.1 Sand Tempering

The process by which sufficient moisture is added to the sand is known as sand tempering. For this, water is thrown (3% to 5% for dry sand) over the heap like a spray and then the sand is cut through, a shrovel full at a time, allowing water and air to mix with clay in the sand thus breaking up the lumps. This moistens the clay in the sand and makes it adhesive.

3.6.2 Sand Conditioning

Conditioning is complicated by the fact that the sand is continually being contaminated with core lumps, metal drippings, scale, and wires, as well as being decrepitated by the action of heat. Metal and lump contamination can be reduced by magnetic separation and the screens.

Accumulation of the fine material can in part be controlled by its removal through exhaust systems under the dusty conditions of the shake-out. However, burnt clay and dilution by core sand can be corrected only by additions of clay and other additives to keep the sand properties at the desired level. Green strength and other properties and the combustible material in the sand are lost as the sand is reused unless such additions are made to recover the losses. The clay and other materials added for each cycle of mixing is about 0.15 to 1.00 per cent of the weight of the sand or about 1 to 10 kg of clay per 1000 kg batch, depending on the size of castings being made. The returned sand is of unknown moisture content, and additions for tempering at the proper water percentage must be made.

Detailed procedure to condition the sand

1. The foreign materials, like nails, hard sand lumps and metals from the used sand are removed. For removing the metal pieces, particularly ferrous pieces, sand from the shake-out station is subjected to magnetic separator, which separates the iron pieces, nails, etc., from the used sand. Schematic diagram is shown in Figure 3.16.

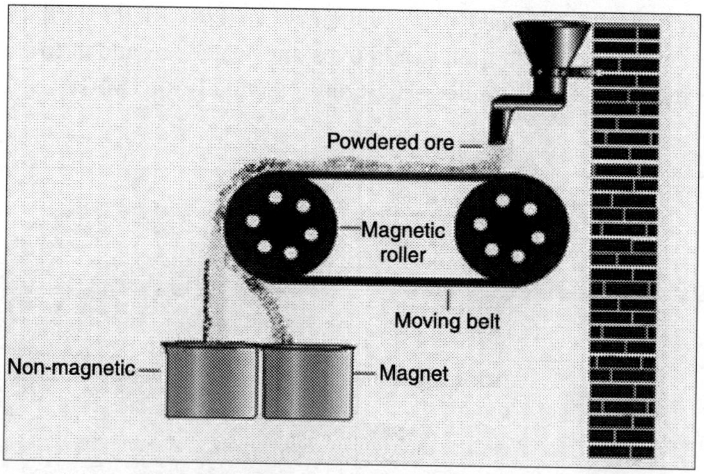

Figure 3.16 Magnetic separator

2. Next, the sand is screened in riddles which separate the hard sand lumps, etc. These riddles may be manual as well as mechanical. Mechanical riddles may be either compressed air operated or electrically operated. Electrically operated riddles are faster and can handle large quantities of sand in a short time.
3. After all the foreign particles are removed the sand is free from the hard lumps, etc., proper amount of pure sand, clay and required additives are added for the loss because of the burnt clay and other materials. As for the moisture content of the returned sand known, it is to be tested and after knowing the moisture the required amount of water is added.
4. There are two methods of adding clay and water to sand. In the first method, water is added to sand followed by clay, while in the other method, clay addition is followed by

water. It has been suggested that the best order of adding ingredients to clay bonded sand is sand with water followed by the binders. In this way, the clay is more quickly and uniformly spread on to all the sand grains. An additional advantage of this mixing order is that less dust is produced during the mulling operation.

5. Now these things are mixed thoroughly in a mixing muller. The main objectives of a mixing muller is to distribute the binders, additives and moisture or water content uniformly all around each sand grain and helps to develop the optimum physical properties by kneading on the sand grains. Inadequate mulling makes the sand mixture weak which can only be compensated by adding more binder.
6. The final step in sand conditioning is the cooling of sand mixture because of the fact that if the moulding sand mixture is hot, it will cause moulding difficulties.

Mixing Muller

Adequate mixing requires proper use of mixers. In mechanized foundry, machines are used for the preparation of sand. After casting, the moulds are broken to take the casting out. Then the sand moulds are crushed and screened using vibratory screens. Metal particles are removed using magnetic separators. Fresh sand, clay, additives and water are added to the sand and are mixed well using a machine called muller. These are essentially of two types: batch type and continuous.

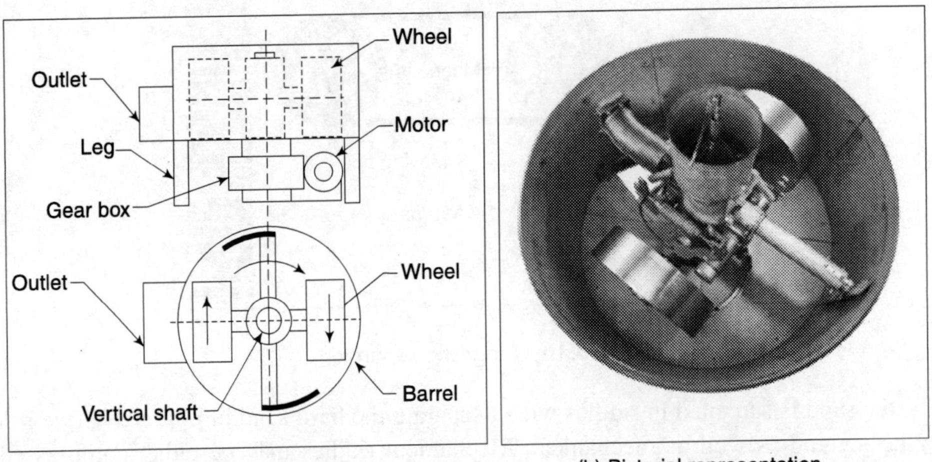

(a) Parts of a muller (b) Pictorial representation

Figure 3.17 Batch-type mixing muller

As shown in Figure 3.17, a batch muller consists of one or two muller wheels and equal number of plough blades, all of them connected to a single driving source. The muller wheels are large and heavy, and continuously roll inside the muller bowl. The plough blades ensure that the sand is continuously agitated. The combined action of both is a sort of kneading action which makes the clay and the moisture uniformly distributed throughout the sand.

A continuous muller consists of two bowls with muller wheels and plough, such that sand, clay and moisture are fed through a hopper into one of the bowls which after getting mulled moves into the second one and then finally out. Thus, well prepared moulding sand is continually available for use. It is generally used for large-scale production.

3.7 MOULD WASH OR MOULD COATINGS OR WASH SPRAY BLACKING

Purely carbonaceous materials such as sea coal, finely powdered graphite or proper compounds are also applied on to mould cavity after the pattern is withdrawn. This is called the mould wash and is done by spraying, swabbing or painting in the form of a wet paste. By this surfacing process, better casting surface finish, greater ease of cleaning castings and less castings results are detected. These are used essentially for the following reasons:

(i) to prevent metal penetration into the sand grains and thus ensure a good casting finish; and

(ii) to avoid mould-metal interaction and prevent sand fusion.

For depositing the mould wash, either water or alcohol can be used as a carrier. But because of the problem of getting the water out of the mould, alcohol is preferred as a carrier. The proprietary washes are available in powder, paste or liquid form. The powder needs to be first prepared and applied whereas the paste and liquid can be applied straightaway. The composition used as mould wash for cast iron and brass contains bentonite clay, cereal powder, graphite powder, water and pinch of benzoate. Liquid base mould are usually used for dry sand mould. For green sand moulds, dusting is done with plumbango powder or graphite powder using moulder's dusting bag. Mould coating powders of different brands are available in the market.

3.8 OTHER TYPES OF SAND

3.8.1 Facing Sand

This sand is used next to the pattern to obtain cleaner and smoother casting surfaces. It must possess high strength and refractoriness because it comes into contact with the molten metal, when the mould is filled.

Generally, sea coal or coal dust is mixed with the system to improve the mouldability and surface finish. The sea coal being carbonaceous, slowly burns due to the heat from the molten metal and give off small amounts of reducing gases. This creates a small gas pressure in the surroundings of the cavity such that molten metal is presented from entering the silica grains or fuse with them. This helps in generating good casting surfaces and also lets the moulding sand peel off from the casting during shake-out.

3.8.2 Backing Sand

This sand is used to back up the facing sand and to fill the rest of the volume of flask. This is normally the reconditioned foundry sand used for removing the bulk of the moulding flask. The moulding flask is filled with backing sand after the pattern is covered with a thin layer of facing sand. Since the casting is not affected to any great extent by the backing sand, it usually contains the burnt facing sand, moulding sand and clay. Close control of the properties is not necessary here. If the clay content is insufficient, a small percentage is added as compensation.

3.8.3 Parting Sand

This is the material which is sprinkled on the pattern and to the parting surface of the mould halves before they are prepared, to prevent the adherence of the moulding sand. This helps in easy withdrawal of the pattern.

Grey iron castings are produced in green and dry condition of used sand moulds. As a measure of economy, a continuous measure of used sand from 50% to 90% is added to moulding sand for grey iron castings. Moreover, 50% fresh materials (sand and clay) and 2% to 3% additives are added to the mixture. The amount of fresh materials and cold dust added depends upon the weight and size of the casting; the heavier the casting and the thicker its walls, the more fresh materials are required. Table 3.17 gives the usual composition and properties of greensand mixtures for light, medium and heavy grey iron castings.

Table 3.17 Composition and properties of greensand mixtures

Constituent or property	*Light castings (less than 200 kg)*	*Medium castings (between 200 to 2000 kg)*	*Heavy castings (more than 2000 kg)*
Sand %	81.5 to 85	89 to 93	88 to 94
Clay %	8 to 9.5	2 to 4	3 to 6
Additive %	Sea coal 7 to 9	Sea coal 4.6 to 6 Wood flour 1	Sea coal 2 to 4 Pitch 1 to 2
Green compressive strength psi	11 to 14	5.5 to 6.5	9 to 10
Permeability No.	80 to 100	90 to 100	120 to 130
Moisture %	3 to 3.7	3 to 3.5	4.5 to 5.5
Mould hardness	70 to 80	70 to 80	80 to 85

3.8.4 Bonding Sand

Sand high in clay content is used to add clay to moulding sand.

3.8.5 Sharp Sand

Sand substantially free of bond. The term has no reference to grain shape. Lake sand is sometimes referred to as sharp sand.

Review Questions

1. What are the main requirements of mould and core material?
2. List the desirable properties of moulding sand.
3. Expalin the characteristics of moulding sand.
4. List the advantages and disadvantages of silica sand.
5. Explain various types of sands used in foundry.
6. Describe the role of binders in preparation of moulding sand.
7. Describe various types of clays used in foundry
8. Explain electrostatic bonding of clays.
9. What are the main additives added in foundry sand?
10. Explain the detailed procedure of sand conditioning.

CHAPTER 4

Testing Sand Properties

4.1 INTRODUCTION

The terminology sand testing is loosely used by foundrymen generally meaning testing of mould and core materials as prepared in the sand plants, but also involving the assessment of certain physical characteristics of sand such as grain size, etc.

The nature of moulding sand can be described by means of the results of standard sand tests which have been adopted by the industry. Complete details of the tests, testing procedure and equipment are set forth in the AFS's *Foundry Sand Handbook*. The tests will only be briefly reviewed here.

The ever-increasing demand for higher accuracy and good surface finish of castings necessitates certainty in the quality of foundry sands. Close examination of moulding sand ensures the following advantages:

1. **Lower cost of materials:** Sand testing often allows the use of less expensive local sands. It also ensures reliable sand mixing and enables a fuller utilization of the inherent properties of moulding sands.
2. **Less number of rejects:** Sand testing on delivery will immediately detect any variation from the standard quality, and adjustment of the sand mixture to specific requirements can be made.
3. **Improved surface finish:** Sand testing allows the choice of sand mixtures to give the desired surface finish.

Thus, sand testing is one of the dominating factors in foundry and pays for itself by obtaining lower per unit cost and increased production resulting from sound castings.

The properties of moulding sand and core sand, like strength, permeability, clay content, moisture, etc., can be determined by means of some tests standardized by American foundry Society (AFS) and Indian Standards (10034-1981, 1918-1966, 1811-1961). Following are the most common routine tests conducted in foundries:

1. Estimation of moisture content in moulding sand.
2. Estimation of clay content in moulding sand.
3. Chemical composition of sand.

4. Grain shape and surface texture of sand.
5. Grain size distribution of sand.
6. Specific surface of sand grains.
7. Water absorption capacity of sand.
8. Refractoriness of sand.
9. Strength of moulding sand through
   - Compression test
   - Shear test
   - Bending test
   - Tensile test
10. Permeability of moulding sand.
11. Flowability of moulding sand.
12. Shatter index of moulding sand.
13. Mould hardness.

4.2 CRITERIA FOR TESTING SAND

4.2.1 Sand Preparation

All sand tests should be performed on samples which will give results that are representative of sand. Recommended procedures for sampling and conditioning of the sand should be used. Moulding sand may be tempered with water and mixed manually or mechanically, preferably by the method similar to that used in the foundry where the hand tests are being performed. Adequate mixing in a laboratory muller requires about 6 to 10 minutes, depending on batch size and muller. After mixing, the sand should be kept in a closed container.

4.2.2 Moisture Content

Moisture is an important element of moulding sand as it affects many properties. This is one of the most frequently performed tests. This test is conducted to find the water content in the moulding sand. Tests are conducted on a sample of the standard sand. The moulding sand should be prepared exactly as is done in the shop on the standard equipment and then carefully enclosed in a closed container to safeguard its moisture content. The moisture content of the moulding sand can be determined in any of the following ways:

Method 1 The moisture content of the sand mixture may be determined by drying a weighted amount of moulding sand *t,* a constant weight between 104-110°C for two hours in a electric furnace and cooling to room temperature in a desiccator and reweighing. The loss in weight gives the amount of moisture which can be expressed as a percentage of the original sand sample.

$$\text{Moisture \%} = \left[\frac{\text{Weight before drying} - \text{Weight after drying}}{\text{Weight before drying}}\right] \times 100$$

Method 2 Alternatively, a moisture teller can also be used for measuring the moisture content. In this, sand is dried by suspending the sample on a dry metallic screen and allowing hot air to flow through the sample. This method of drying completely removes moisture in a matter of minutes compared to two hours in the previous method.

Procedure: The percentage of moisture in a green sand sample can be determined by using moisture teller (Figure 4.1). A weighed representative sample of the moist material is placed in a special pan with fine gauze base and hot air at a thermostatically controlled temperature is blown through it for a predetermined time to remove the moisture. When all the moisture has been removed, the sample pan and contents are rapidly cooled and reweighed. The loss in weight is read directly as percentage moisture on the dial.

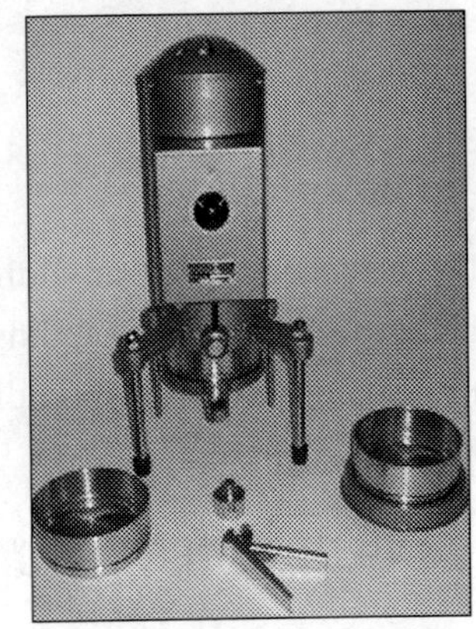

Figure 4.1 Moisture teller

Method 3 Another moisture teller utilises calcium carbide to measure the moisture content. A measured amount of calcium carbide (a little more than that actually required for complete reaction) in a container along with a separate cap consisting of measured quality of moulding sand is kept in the moisture teller. Care has to be taken before closing the apparatus that carbide and sand do not come into contact. The apparatus is then shaken vigorously such that the following reaction takes place:

$$CaC_2 + 2H_2O \rightarrow C_2H_2 + Ca(OH)_2$$

The acetylene (C_2H_2) coming out will be connected to the apparatus that would give directly the amount of acetylene generated which is proportional to moisture present. This instrument is provided with a pressure gauge calibrated to read directly the percentage of moisture present in the moulding sand.

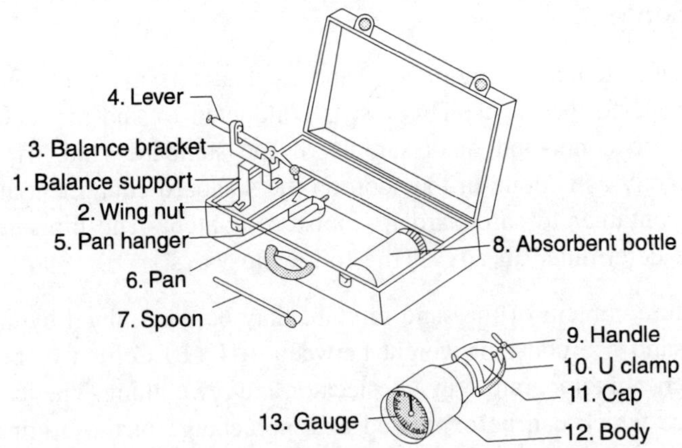

Figure 4.2 Rapid moisture tester

The instrument is portable and needs no power supply. Absorbent compound, i.e., calcium carbide is provided with the instrument in an air-tight bottle.

Procedure: Weigh the sand sample by pre-calibrated balance provided with the instrument. Transfer the same in the cup. Take two spoons of absorbent compound and transfer into the body of the tester. Hold the body horizontally and place the cup in position and tighten the same with the help of clamp and screw. Shake the instrument vigorously to ensure that the above-mentioned reaction occurs. Because of chemical reaction C_2H_2 is generated and the guage pointer moves. Observe the reading when the pointer stops movement. The calibrated scale of the instrument gives the percentage of moisture content in the sample directly.

Some moisture testing instruments are based on the principle that the electrical conductivity of sand varies with the moisture content in it.

4.2.3 Clay Content

The clay content of the moulding sand is determined by dissolving or washing it off the sand.

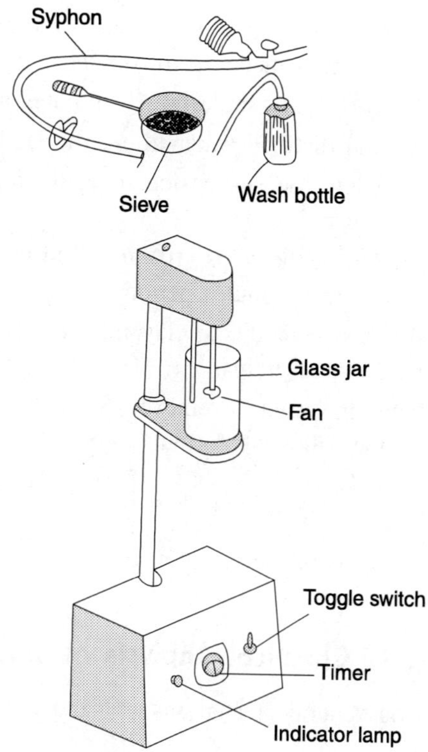

Figure 4.3 Clay content test

The amount of clay is determined by AFS method. According to the AFS, clay in a moulding sand is defined as "particles which fail to settle one inch per minute when suspended in water. These are usually less than 20 microns, or 0.0008 inch in diameter."

The instrument used for determining the clay content is illustrated in Figure 4.3. It consists of a unit having a motorized stirrer, an AFS clay jar, siphon and timer.

Procedure:

1. Take a sample of 50 gm of sand and dry at 105°C to 110°C for an hour.
2. Transfer the sample to AFS clay jar.
3. Add 475 ml of distilled water at room temperature and 25 ml of one per cent solution of caustic soda/sodium hydroxide (NaOH) 25 gm per litre.
4. Stir the contents for 6 minutes by means of a stirrer driven motor.
5. Wash the sand on the stirrer into the jar and fill the jar with distilled water to a height of 6 inches such that the contents are well stirred.
6. Allow the contents to settle for 10 minutes.
7. Siphon off to a depth of exactly 5 inches below the level to which it had been filled, leaving a minimum depth of 1 inch of water in the bottom of the jar.

8. Add distilled water in such a way that the sediments at the bottom are stirred and the height is made up to 6 inches.
9. Allow it to settle for 10 minutes and then siphon off 5 inches of water.
10. Add distilled water filling the jar to 6 inch height stirring the sediments on the bottom.
11. Allow the suspension to settle for exactly 5 minutes and then siphon off 5 inches of water.
12. Repeat the steps (10) and (11) until the water is clear to a depth of 5 inches at the end of the 5 minutes period.

By this method, the clay matter which fails to settle at the rate of 1 inch/min is removed from the sand grains.

Now, the sand is removed from the flask and dried by heating. Clay content is calculated using the following formula:

$$\text{Clay content } \% = \left[\frac{\text{Weight before washing} - \text{Weight after washing}}{\text{Weight before washing}}\right] \times 100$$

Analysis of the clay content of the moulding sands is also performed by the hydrometer and a chemical method.

4.2.4 Chemical Analysis of Sand

Various chemical analyses of sand are carried out to estimate the following in the foundry sand:

- Estimation of $\mathbf{SiO_2}$
- Estimation of $\mathbf{Al_2O_3}$, $\mathbf{Fe_2O_3}$ and $\mathbf{TiO_2}$
- Estimation of $\mathbf{FeO_3}$, $\mathbf{TiO_2}$ and $\mathbf{Al_2O_3}$
- Estimation of **MnO**
- Estimation of **CaO**
- Estimation of **MgO**
- Determination of total alkalis

4.2.5 Sand Grain Size or Sieve Analysis

The purpose of sieve analysis test is to find out the grain size distribution and fineness of the sand. By definition, the AFS grain fineness number is the average grain size and it corresponds to the sieve number whose openings would just pass all the sand grains if all were of the same size. This number is a convenient means of describing the relative fineness of sands, most foundry sands being from about 40 to 220 in average size.

4.2.6 AFS Standard Sand

Sometimes standard sand is described for comparison tests: for example, the effectiveness of the different binders or mixing procedures might be studied. "The AFS standard sand is defined as washed and dried silica sand, AFS fineness number of 50±1, with a 100 per cent through a 40-mesh sieve, 95 per cent through a 50-mesh and remaining on a 70-mesh sieve, and the balance retained on a 100-mesh sieve."

Procedure to determine GFN: The sieve analysis is carried out in an AFS sieve analysis tester (sieve shaker) as shown in Figure 4.4. To find out sand grain size, a sand sample which is devoid of moisture or clay such as the one obtained after the previous testing is to be used. The dried clay-free sand grains are placed on the top sieve of a sieve shaker which contains a series of sieves one up on the other with gradually decreasing mesh sizes. A pan is placed underneath the last sieve. A sample of weighted sand is put on the top sieve of the stack of sieves as per BS410:1986 standard. The mesh sizes are standardized as shown in Table 4.1. The sieves are shaken continuously for a period of 15 minutes. After this shaking operation, the sieves are taken apart and the sand left over each of the sieve is carefully weighed.

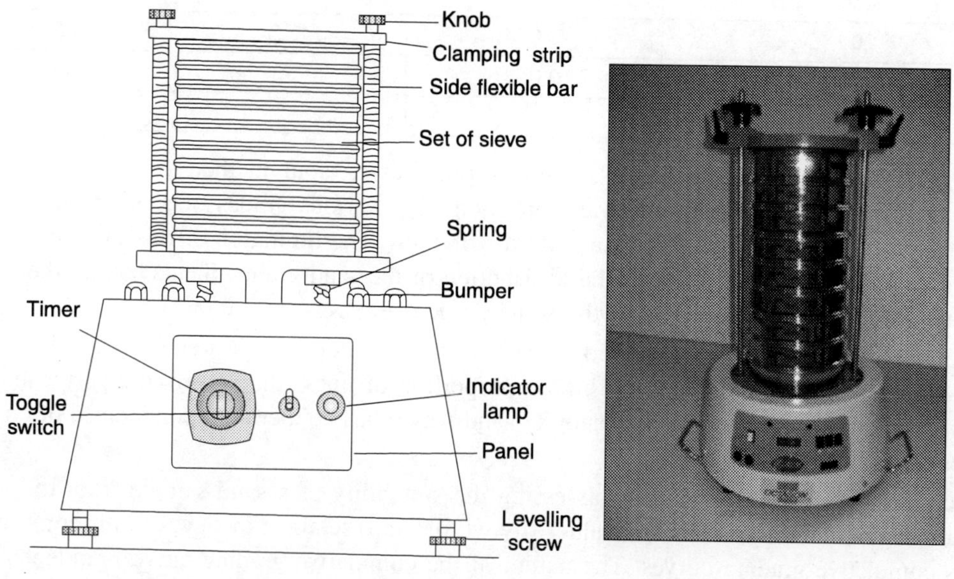

Figure 4.4 Sieve shaker used for finding GFN

To calculate the grain fineness number, each sieve has been given a weightage factor as shown in Table 4.1. The amount retained on each sieve is multiplied by the respective weightage factor, summed up, and then divided by the total mass of the sample, which gives the grain fineness number. The same can be expressed as

$$GFN = \frac{\Sigma M_i f_i}{\Sigma f_i} \tag{4.1}$$

where, M_i = Multiplying factor for the ith sieve

f_i = Amount of sand retained on the ith sieve

Table 4.1 AFS sieve numbers and their Indian equivalents with sizes

US Series Equivalent No. (ASTM)	*Mesh opening (mm)*	*IS Sieve No. microns*	*Multiplying factor*
6	3.327	3.35	3
12	1.651	1.70	5
20	0.833	850	10
30	0.589	600	20
40	0.414	425	30
50	0.295	300	40
70	0.208	210	50
100	0.147	150	70
140	0.104	106	100
200	0.074	75	145
270	0.053	53	200
pan	-	-	300

The determination of grain fineness number provides a rapid method for expressing the average grain size of sand and is of value in comparing various grades of sand from a given deposit or from deposits having similar grain distribution. The limitation of this number is that this number is in no way an index of the distribution of the sand grains. Sands having the same grain fineness number need not be of the same grade.

A sample of sand with higher proportion of the coarse fraction (undesirable for good surface finish) and another sample having too high a proportion of fines (this lowers the permeability) can have the same fineness number as another sand which has neither a preponderance of coarse particles nor of fines.

Hence, a more reliable method for assessing the suitability of a sand sample from the view point of size distribution is to plot % values (by wt. of sand) retained in sieves of different mesh sizes as cumulative grading curves. The nature of the cumulative grading curve is independent of the type of standard sieves used (Figure 4.5).

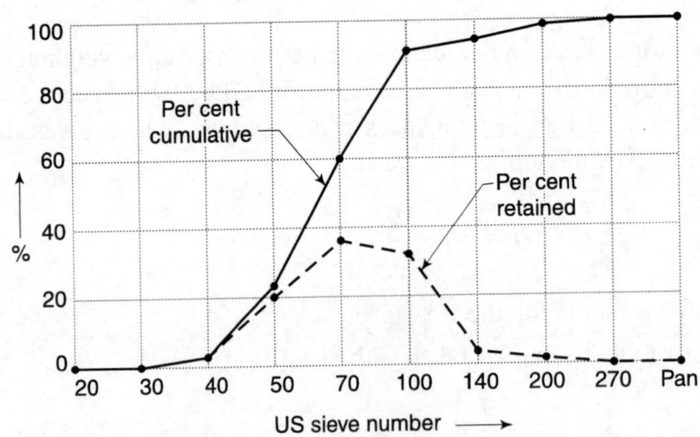

Figure 4.5 Cumulative percentage retained versus sieve number

Cumulative grading curves: A dried 50 gm sample is placed on the top of a series of sieves and shaken for 15 minutes. The portion of the sand grains retained on each sieve is collected and weighed and expressed as percentage of the total sand grade.

Worked Out Examples

Example 4.1 The sieve analysis of sand used in malleable and grey iron foundries for castings which weigh more than 20 kg is carried out. When sand is sieved through the standard sieve, it was found to retain the following amount of sand on respetive sieves. Find the grain fineness number.

Sieve No.	40	50	70	100	140
Retained sample in grams	2.495	13.972	23.952	6.986	2.495

Solution: **Table 4.2** gives the calculations for grain fineness number.

Table 4.2 Sieve analysis

Sieve number	*Multiplication factor (M_i)*	*Retained sample (f_i)*	*Retained percentage (p_i)*	*($M_i \times p_i$)*	*($M_i \times f_i$)*
40	30	2.495	5	150	74.85
50	40	13.972	28	1120	558.88
70	50	23.952	48	2400	1197.6
100	70	6.986	14	980	489.02
140	100	2.495	5	500	249.5
		49.900	**100**	**5150**	**2569.85**

$$\text{GFN} = \frac{5150}{100} = 51.50$$

$$\text{GFN} = \frac{2569.85}{49.90} = 51.50$$

By the above definition, the grain fineness number is the average grain size and corresponds to a sieve number through which all the sand grains would pass through, if they were all of the same size. This is a very convenient way of describing the grain size and its value can be expected between 40 and 220 for those sands used by most of the foundries. Though the sand properties depend on both grain size and grain size distribution, GFN is a very convenient way of finding the sand properties since it takes both into account.

Example 4.2 The sieve analysis of four different sands used in foundry is given below. Find the grain fineness number.

Solution: Table 4.3 gives the sieve analysis for the typical sand which may be used for the purpose mentioned.

Table 4.3 Typical sand sieve analysis

Sieve number	Multiplier (p_i)	Sand 1		Sand 2		Sand 3		Sand 4	
		Retained % (p_i)	$(M_i) \times (p_i)$	Retained % (p_i)	$(M_i) \times (p_i)$	Retained % (p_i)	$(M_i) \times (p_i)$	Retained % (p_i)	$(M_i) \times (p_i)$
30	20	1	20	-	-	-	-	-	-
40	30	33	990	1	30	0.6	18	1.2	36
50	40	44	1760	15	600	1.4	56	6	240
70	50	15	750	32	1600	7	350	7.2	360
100	70	4	280	32	2240	26	1820	14.4	1008
140	100	33	300	15	1500	37	3700	18	1800
200	145	-	-	4	580	22	3190	16.8	2436
270	200	-	-	1	200	4	800	18	3600
Pan	300	-	-	-	-	2	600	18.7	5610
	Total		**4100**		**6750**		**10534**		**15090**
	GFN		**41.00**		**67.50**		**105.34**		**150.90**

Sand 1: Steel casting 50 kg and more

Sand 2: Small steel casting

Sand 3: Light cast iron and copper casting

Sand 4: Aluminium alloy casting

Example 4.3: The sieve analysis of four different sands used in foundry is given below. Find the grain fineness number.

Solution: Table 4.4 gives the sieve analysis for the typical sand which may be used for the purpose mentioned.

Table 4.4 Typical sand sieve analysis

Sieve number	Multiplication factor (M_i)	Retained sample (f_i)	Retained percentage (P_i)	$(M_i) \times (p_i)$	$(M_i) \times (f_i)$
12	5	0.8	1.6	8	4
20	10	0.24	0.48	4.8	2.4
30	20	0.1	0.01	2	2
50	40	1.6	3.2	128	64
70	50	12.8	25.6	1280	640
100	70	15.4	30.8	2156	1078
140	100	10.8	21.6	2160	1080
200	140	2.8	5.6	784	392
270	200	0.5	1	200	100
Pan	300	0.1	0.2	60	30

Total		45.1	90.2	-	
Clay		4.86	9.72	-	
Total		50.00 gm	100%	**6782.8**	**3392.4**

$$GFN = \frac{6782.8}{100} = 67.828$$

$$GFN = \frac{3392.4}{50.00} = 67.848$$

Example 4.4: The sieve analysis of sands used in foundry is given below. Find the grain fineness number. Weight of sample = 50 gm, AFS clay content 5.6 gm, i.e., 11.2%, sand grains 44.4 gm, i.e., 88.8%.

Sieve No.	40	50	70	100	140	200	270	Pan
Retained sample in grams	1.5	0.3	2.5	0.5	8.20	13.00	12.80	5.60

Solution: Table 4.5 gives the sieve analysis for the typical sand which may be used for the purpose mentioned.

Table 4.5 Typical sand seive analysis

Sieve No.	*Opening size in mm*	*Retained sample*		*Multiplier* (M_i)	($M_i \times p_i$)
		Retained sample (f_i)	*Retained percentage* (p_i)		
6	0.327	0.00	0.00	3	0
12	1.651	0.00	0.00	5	0
20	0.833	0.00	0.00	10	0
30	0.589	0.00	0.00	20	0
40	0.414	1.50	3.00	30	90
50	0.295	0.30	0.60	40	24
70	0.208	2.50	5.00	50	250
100	0.147	0.50	1.00	70	70
140	0.104	8.20	16.40	100	1,640
200	0.074	13.00	26.00	140	3,640
270	0.53	12.80	25.60	200	5,120
Pan		5.60	11.20	300	3,360
Total		44.40	88.80		14,194

$$\textit{AFS grain fineness number} = GFN = \frac{\Sigma M_i p_i}{\Sigma p_i} = \frac{14,194}{88.8} = 159.84$$

Example 4.5: The sieve analysis of sands used in foundry is given below. Calculate cumulative percentage retained.

Solution: Solution is shown in Table 4.6.

Table 4.6 Sieve analyis of sand

Sieve No.	*Sieve opening (mm)*	*Wt. of sand retained (gm)*	*% retained*	*Cumulative % retained*
1	1.40	0.00	0.00	0.00
2	1.00	0.00	0.00	0.00
3	0.63	0.12	0.24	0.24
4	0.32	2.14	4.28	4.52
5	0.20	33.40	66.80	71.32
6	0.10	13.02	26.04	97.36
7	0.06	1.02	2.04	99.40
8	Pan	0.30	0.60	100.00
Total		50.00	100.00	

Example 4.6: Table 4.7 gives the sieve analysis of typical two sands as according to standard method of sieve analysis of AFS. One is 56 AFS number 3-screen lake sand and the other 88.1 AFS number 4-screen bank sand, which are blended in the ratio of 60:40 to achieve a 68.7 AFS number 4-screen distribution. Comment on coarse and fine fraction of the sieve analysis.

Table 4.7 Sieve analysis of two typical sands

US Sieve No.	*Lake sand*	*Bank sand*	*Per cent retained 60% lake, 40% bank*
20	0.13	0	0.08
30	0.67	0.03	0.41
40	2.35	0.06	1.43
50	16.02	1.48	10.20
70	45.82	13.30	32.81
100	33.28	43.27	37.28
140	1.23	27.66	11.80
200	0.10	10.9	4.42
270	0.01	1.89	0.76
Pan	0.39	1.42	0.80
AFS No.	56	88.1	68.7

Solution: **The coarse fraction of the sieve analysis** is composed of the total percentage of sand grains retained on the screens coarser than those of the bulk fraction, and in amounts of less than 10 per cent. The total coarse fraction must be limited in amount, usually to less than 4 per cent for a sand of the 4-screen distribution type. This limitation is necessary since an excess of coarse particles contributes to a poor casting surface finish. In addition, coarse particles are easily dislodged from the mould cavity surface and become dirt in the casting.

In the case of a typical 60 to 70 AFS sand of the 4-screen type, the coarse fraction occurs on No. 40 or coarser sieve and must be limited to less than 4 per cent. In a sand of higher AFS fineness number, the coarse fraction might be that on No. 50 or 70 screens.

The fine fraction of the sieve analysis is composed of the total percentage of sand grains retained on screens finer than those of the bulk fraction and in amounts of less than 10 per cent. The total fine fraction must be limited in amount, usually to less than about 5 per cent for a sand of the 4-screen type. This limitation is necessary since an excess of fine particles causes balling to occur during mulling. When water is added, the fines and clay agglomerate to form balls during mulling and thus prevent the clay from being thoroughly disseminated throughout the mass. Although the percentage of fines must be limited to a maximum percentage as stated above, there is also a minimum percentage desired. In the absence of the fines, clay balling can occur. Mulling causes the clay and water to form balls if there are no fines present and again agglomerates of uniformly dispersed ingredients are not formed from mulling. The ability to properly mull the sand and disseminate the ingredients requires this balance of coarse, bulk, and fine sand particles in the base sand.

4.2.7 Specific Surface of Sand Grains

Until now, the methods adopted to determine the grain shape of foundry sands depend almost exclusively on microscopical observation. Hence, the results obtained are prone to the disadvantage of being inaccurate and to personal interpretation.

The average grain shape can be expressed by a factor known as coefficient of angularity which is defined as

$$\textbf{Coefficient of Angularity} = \frac{\text{Actual specific surface area}}{\text{Theoretical specific surface area}}$$

In other words, this angularity may be said as the degree of deviation of the shape of grain from spherical shape.

Actual specific surface area: It is defined as the surface area of all sand grains in sq. cm. contained in one gram of sand.

Theoretical surface specific area: It is defined as the surface area of all sand grains contained in one gram of sand assuming that all the sand grains are spherical.

Since the sphere has a minimum surface for the same weight and since in actual practice all the sand grains are never spherical, therefore the value of the theoretical specific surface area is always smaller than that of the actual specific surface.

The more angular the grains are, the larger is the difference between actual and theoretical specific surface areas.

Determination of actual specific surface area

Hoffmann's method: This method of measuring the actual specific surface of sand is based on the flow of air through a long cylindrical sand bed.

The actual specific surface, S_W, is a function of porosity and permeability of this bed.

$$S_W = \frac{\text{Constant}}{\text{Bulk Density}}\sqrt{\frac{(\text{Porosity})^3}{\text{Permeability}}} \tag{4.2}$$

Basing on a formula of Kozeny, Carman has suggested the following semi-empirical formula

$$S_W = \frac{1}{\delta\,(1-\varepsilon)}\sqrt{\frac{\in^3 Apg}{K\eta QL}} \tag{4.3}$$

δ = Specific weight of material investigated = 2.65 for silica sand

ε = Porosity of the sand bed (pore space volume per volume unit of sand bed)

A = Cross-sectional area of the sand bed (sq. cm.)

p = Pressure difference between the ends of the sand bed (dyne/sq. cm.)

g = Acceleration due to gravity (981 cm/sec.)

K = Constant (= 5, not affected by apparatus and grain size).

η = Dynamic viscosity of air (gm/cm/sec.)

Q = Rate of flow of air through sand bed (c.c./sec.)

L = Length of sand bed (cm)

g & K are constants, therefore simplifying equation (4.3). We get,

$$S_W = \frac{14}{\delta(1-\delta)}\sqrt{\frac{\in^2 Ap}{\eta QL}} \tag{4.4}$$

Now, $$Q = \frac{V}{T} \tag{4.5}$$

V = Total volume of air (c.c.) flowing, the specimen equalling the volume of the length of U-tube between marks M_2 and M_3 (Figure 4.6).

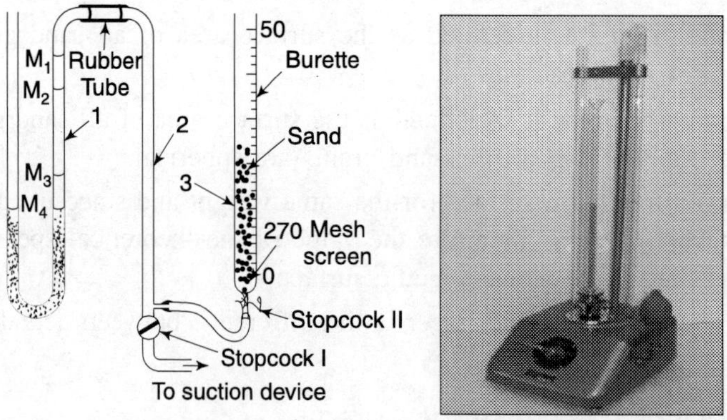

Figure 4.6 Set-up for the measurement of actual surface area of sand

T = Time required by the liquid column for dropping from M_2 to M_3 (seconds).

Therefore,

$$S_W = \frac{14}{\delta(1-\epsilon)}\sqrt{\frac{\epsilon^3 Ap}{V\eta L}} \tag{4.6}$$

The pressure, p, may be expressed for the U-tube method as follows:

$$P = \frac{0.8686 h_1 \gamma}{\log\left(1 + 2\frac{h_1}{h_2}\right)} \tag{4.7}$$

where h_1 = Height between M_3 and M_2 = measuring interval

h_2 = Double height between M_3 and M_4

M_4 = Equilibrium height of the liquid column

γ = Specific weight of the liquid in the U-tube (Kerosene or any suitable mineral oil of low viscosity)

Substituting for p in equation (4.6),

$$S_W = \frac{14}{\delta(1-\epsilon)}\sqrt{\frac{\epsilon^3 A \times 0.8686\, h_1 \gamma}{V\eta L \log\left(1 + 2\frac{h_1}{h_2}\right)}}\sqrt{T} \tag{4.8}$$

or with $\delta(1 - \varepsilon) = D$ = bulk density of the sand bed, the formula may be expressed as follows:

$$S_W = 14 \times \frac{1}{D}\sqrt{\frac{\epsilon^3}{L}}\sqrt{\frac{A \times 0.8686\, h_1 \gamma}{V\eta \log\left(1 + 2\frac{h_1}{h_2}\right)}}\sqrt{T} \tag{4.9}$$

or

$$S_W = X.Y.Z \tag{4.10}$$

where,

$$X = \frac{1}{D}\sqrt{\frac{\epsilon^3}{L}}$$

$$Y = 14\sqrt{\frac{A \times 0.8686 h_1 \gamma}{V\eta \log\left(1 + 2\frac{h_1}{h_2}\right)}}$$

$$Z = \sqrt{T}$$

The value of X is obtained from the volume and the length of the sand bed of known weight yielding also D & ε.

Apparatus: The apparatus is shown in Figure 4.2. Part (1), i.e., the U-tube and part (2) are mounted on one stand whereas part (3), i.e., the burette is clamped in a separate stand. Parts (1) (2) and parts (2) & (3) are connected by means of rubber tubings as shown.

Markings M_1, M_2, M_3 and M_4 are made on the U-tube such that M_1M_2 is approximately equal to 5 cm, M_2M_3 about 25 cm and M_3M_4 also 5 cm. The U-tube contains kerosene oil up to the equilibrium height, M_4. The burette is closed by a 270-mesh screen at the zero mark to support the sand bed.

Procedure

A sample of exactly 100 gm (or 50 gm) of washed and dried sand (clay free) is filled into the burette as uniformly as possible. The burette is slightly tapped by hand or with a pencil until the sand volume cannot be reduced any further. Note the volume of the sand bed.

Now turn the stop cock of the burette to off position. With the help of stop cock I, connect the suction device to the U-tube and raise the liquid level from M_4 to M_1 using the suction device. Now disconnect the suction device from the U-tube by turning off the cock I.

Now turn the stop cock II to on-position. Air will start flowing through the sand bed and the liquid column in the U-tube will begin to drop. Record the time required by the liquid column for dropping from M_3 to M_2 in seconds using a stop watch. From these observations, the actual specific surface area can be calculated using equation (4.9).

Example 4.7 Sample weight = 100 grams

Volume of sand bed = 68.6 cm^3

Flow time T = 42.3 seconds for the apparatus

Used with M_1M_2 = 4 cm

M_2M_3 = 25 cm

M_3M_4 = 4 cm

Inside diameter of U-tube = 1.8 cm

Inside diameter of burette = 1.6 cm

$$\text{Bulk density, } D = \frac{\text{Sample weight}}{\text{Sand bed volume}} = \frac{100}{68.6} = 1.458$$

$$\text{Porosity, } \in = \frac{\delta - D}{\delta}$$

where, δ = specific weight of sand = 2.65

$$\in = \frac{\delta - D}{\delta} = \frac{2.65 - 1.458}{2.65} = 0.45$$

$$\text{Length of sand bed, } L = \frac{\text{Sand bed volume}}{\text{Cross-sectional area of burette}} = \frac{68.6}{\frac{\pi}{4}(1.6)^2} = 34.1 \text{ cm}$$

Substituting these values we get,

$$X = \frac{1}{D}\sqrt{\frac{\in^3}{L}} = \frac{1}{1.458}\sqrt{\frac{0.453}{34.1}} = 3.55 \times 10^{-2}$$

$$Y = 14\sqrt{\frac{A \times 0.8686\, h_1 \gamma}{V\eta \log\left(1 + 2\frac{h_1}{h_2}\right)}}$$

$$Y = 14\sqrt{\frac{\frac{\pi}{4}(1.6)^2 \times 0.8686 \times 25 \times 0.792}{\frac{\pi}{4}(1.8)^2 \times 25\left[\log\left(1 + 2 \times \frac{25}{8}\right)\right] 181 \times 10^{-6}}} = 545$$

$$Z = \sqrt{T} = \sqrt{42.3} = 6.5$$

Now from equation (4.10),

$$S_W = 3.55 \times 10^{-2} \times 545 \times 6.5 = 126 \text{ cm}^2/\text{gm}$$

Determination of theoretical specific surface area

Theory: Assuming all the sand grains to be spherical the theoretical specific surface, S_{th1}, for a particular grain size fraction (i.e., sieve interval), i is given by

$$S_{\text{th1}} = a_i \pi d_i^2 \qquad (4.11)$$

where, a_i = number of sand grains of the particular screen interval (i.e., of the particular grain size fraction)

d_i = the average diameter of the grains

= average diameter of the particular screen interval opening (medium grain size diameter of the grain fraction in cm)

The total theoretical surface is therefore,

$$S_{\text{th}} = \sum a_i \pi d_i^2 \qquad (4.12)$$

If P_i is the weight of particular sieve size fraction in gm

then

$$P_i = a_i \times 1/6 \pi d_i^3 \delta \qquad (4.13)$$

where, δ = specific weight of sand grains = 2.65 for silica

Hence,

$$a_i = 6P_1/\pi d_i^3 \delta \qquad (4.14)$$

$$S_{\text{th}} = \sum a_i \sum d_i^2 \qquad (4.15)$$

$$S_{\text{th}} = \sum \frac{6P_i}{d_i \delta} = \sum \frac{P_i}{d_i} \times 2.26 \qquad (4.16)$$

or theoretical specific surface area per gram of sand,

$$S_{th} = \frac{\sum \frac{P_i}{d_i}}{P} \times 2.26 \tag{4.17}$$

where, P = total weight of sand sample.

Example 4.8 50 gm of dry and clay-free sand sample is placed on top of a series of sieves and shaken for about 15 minutes. After the shaking period, the sand retained on each sieve and the bottom pan is weighed. From this, the theoretical specific surface area can be computed.

Solution:

Table 4.8 Determination of theoretical specific surface area

Sieve No.	*Sieve opening in cm*	*Weight of sand retained in gm (P_i)*	*Average diameter of sand grains in cm (d_i)*	$\left(\frac{P_i}{d_i}\right)$
1.	0.140	0.00	0.1400	0.00
2.	0.100	0.00	0.1200	0.00
3.	0.063	0.12	0.0815	1.47
4.	0.032	2.14	0.0475	45.00
5.	0.020	33.40	0.0260	1285.00
6.	0.010	13.02	0.0150	870.00
7.	0.006	1.02	0.0080	127.50
8.	Pan	0.30	0.0030	100
Total		$\Sigma P_i = P = 50.00$		$\Sigma \left(\frac{P_i}{d_i}\right) = 2428.97$

$$S_{th} = 2.26 \times \Sigma \left(\frac{\frac{P_i}{d_i}}{P}\right) = \frac{(2.26 \times 2428.97)}{50} = 110 \text{ cm}^2/\text{gm}$$

4.2.8 Water Absorption Capacity of Sand

Water absorption capacity is determined in order to reveal internal grain structure.

Apparatus: The apparatus shown in Figure 4.7 consists of a Buckner funnel about 4 inches in diameter and a filtration flask (about 1000 cm^3 in capacity). The filtration flask is connected to a filter pump to enhance filtration.

Method: A hard filter paper is placed in the Buckner funnel and about 200 gm of clay-free sand is placed in

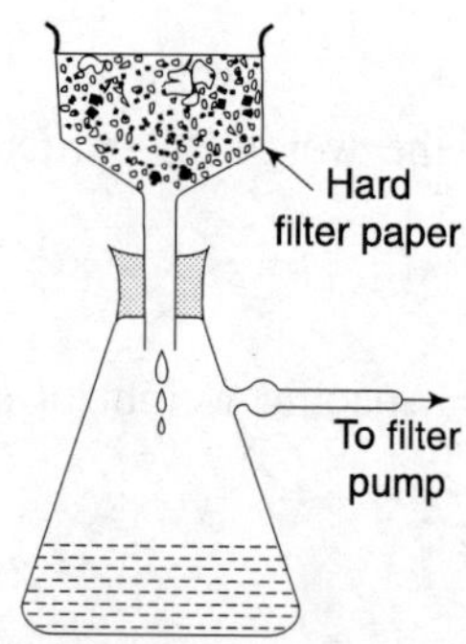

Figure 4.7 Setup for the measurement of water absorption capacity of sand

the Buckner funnel. Now hot water containing some wetting agent (usually alcohol) is poured into the funnel. The sand is then rinsed with ordinary tap water containing some wetting agent. Then the filter pump is connected and the water in the sand above the filter paper is sucked away. When all the water is removed, a fair average sample of 100 gm is taken from the sand in the funnel. This sample is dried in an infra-red dryer, cooled and weighed. This process of drying, cooling and weighing is repeated till constant weight is obtained. The difference in weight (i.e., the retained water content) is an indication of the water absorption capacity of sand.

4.2.9 Refractoriness

The refractoriness is the highest fusion point that is obtainable from sand grains of maximum purity and size. The refractoriness of the moulding sand is judged by firing the AFS standard sand specimens to different temperatures depending upon the type of sand. The fired test pieces are cooled to room temperature and examined under a microscope for surface characteristics or by scratching with a steel needle. If the sand grains remain sharply defined and easily give way to the steel needle, sintering has not yet set in. In the actual experiment, the sand specimen in a porcelain boat is placed into a electric furnace. It is usual practice to start the test from 1000°C and raise the temperature in steps of 100 to 1300°C and in steps of 50 above 1300°C till sintering of the sand grains takes place. At each temperature level, it is kept for at least 3 minutes and then taken out of the oven.

Washed and dried white silica sands of AFS number 30 to 45 are regarded as having highest refractoriness, with fusion point above 1650°C. Impurities such as iron oxide, feldspar, and limestone, which discolour, lower its fusion point. Finer grains appear to be more easily fused than coarser ones where maximum refractoriness is required, as in steel moulding sands, the coarser, high-purity silica sands are used to advantage.

4.2.10 Strength of Moulding Sand

Various types of tests of the strength of moulding sands are performed with a universal-type mechanical testing machine as illustrated in Figure 4.8. Strength in compression, shear, tension and transverse loading may be performed on green sands, dry sands and core sands with this machine and its accessories.

The strength tests conducted are:

1. Compression test (for green, dry or baked specimen)
2. Shear test (for green, dry or baked specimen)
3. Bending test (for dry or baked specimen)
4. Tension test (for dry or baked specimen)
5. Deformation (strain) test

Specimen preparation: Standard specimens of moulding sand are made in identical conditions using a sand rammer. Figure 4.8 shows the schematic diagram of sand rammer. It consists of a calibrated sliding weight for ramming, a little cam, a ramming cam, a plunger and a supporting frame. After preparation of the specimen, first permeability test is conducted and the same specimen can be used for testing strength of the moulding sand.

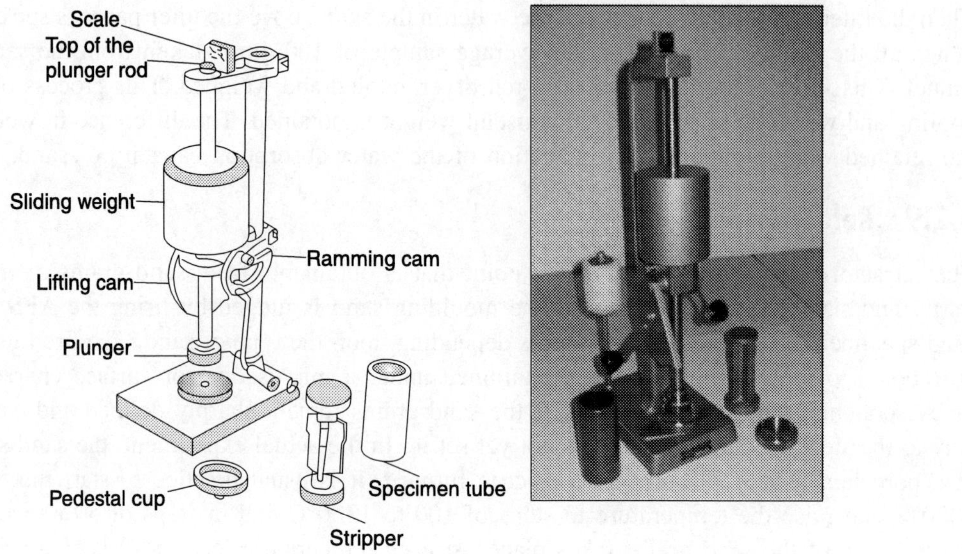

Figure 4.8 Sand rammer

Procedure: Since the permeability and strength of sand depends to a great extent on the degree of ramming, it is necessary that the specimen be prepared under standard conditions. To get reproducible ramming conditions, a laboratory sand rammer is used along with a specimen tube. The measured amount of sand is filled in specimen tube, and a fixed weight of 6.35 to 7.25 kg is allowed to fall on the sand three times from 50.8 + 0.125 mm. The specimen thus produced should have a height of 50.8 + 0.8 mm. To produce this size of specimen usually sand of 145 to 175 gm would be required. The longitudinal specimen for bending test and figure eight specimen for tension test are also made by giving three blows as done for cylindrical specimen. But suitable ramming moulds are to be used for getting the shapes. The specimen sizes for different tests are given in Figure 4.9. Universal testing machine is shown in Figure 4.10.

(a) Green compression test

Green compression test or simply compression test generally refers to stress required to rupture the sand specimen under compressive loading. Green compressive strength is usually simply referred to as the green strength of the sand. It is defined as the maximum load required to break a green mould under axial compression stress. It is important in this and other green strength tests that the tests are performed immediately after the specimen is stripped from the tube. The strength of greensand increases when the sand specimen begins to air-dry. The specimen is kept between two jaws and force is applied slowly to compress the specimen till it breaks. The force applied is measured and the compression strength is calculated. The green strength of the sand is generally in the range of 30 to 160 kPa.

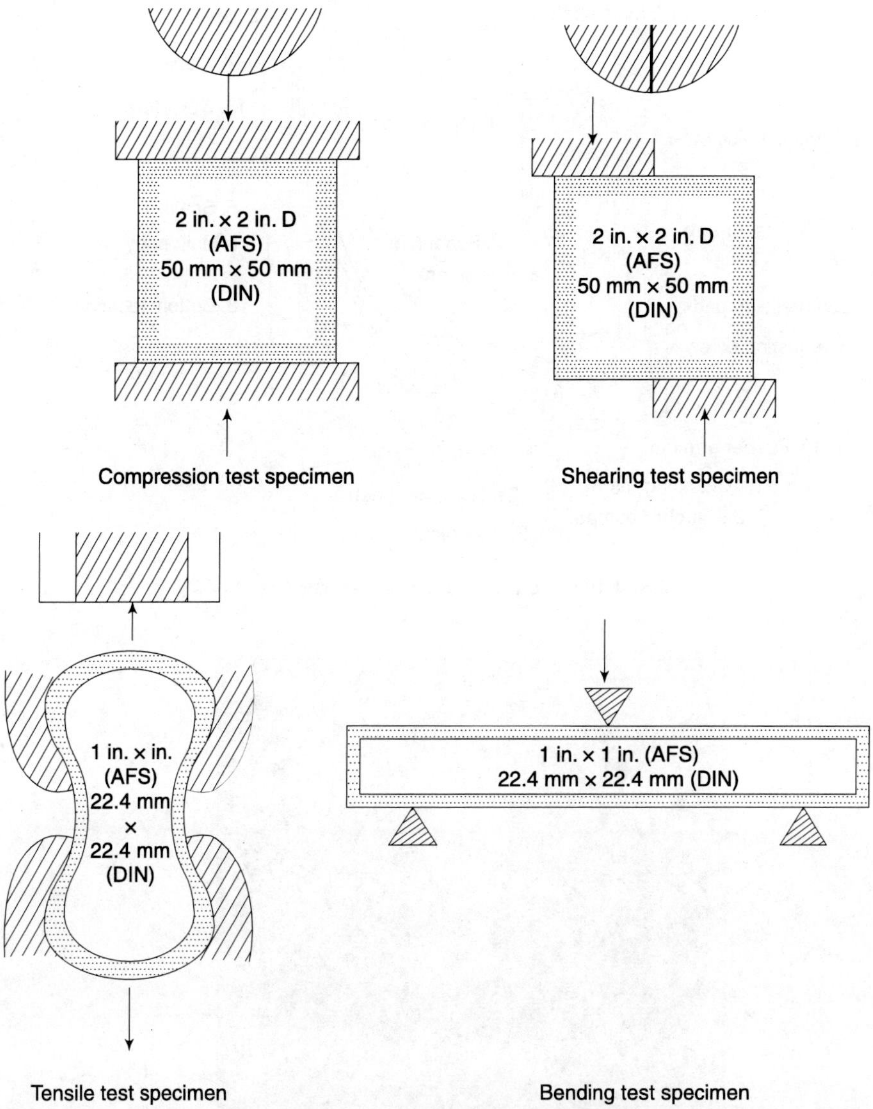

Figure 4.9 Specimens for sand test

Procedure: Rotate the wheel anticlockwise to take the pusher arm (3) back till it stops. Level the instrument with levelling screws so that front edge of pusher plate (20) coincides with 0 on the scale. Ensure that the pendulum is swinging freely. Insert the compression pads (7) in the proper location. When testing green strength, compression pads are inserted at the bottom location (8). For dry strength, the compression pads are inserted at top location (9). Ensure that the magnetic rider is resting against the pusher plate.

Standard compression test specimen is prepared in sand rammer and is held between the compression pads in such a way that the plain surface of the test specimen touches against the pads.

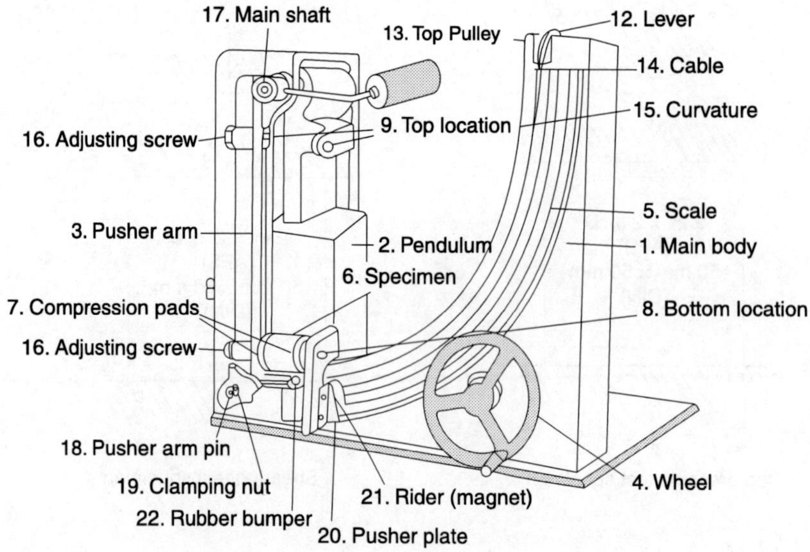

Figure 4.10(a) Universal sand testing machine

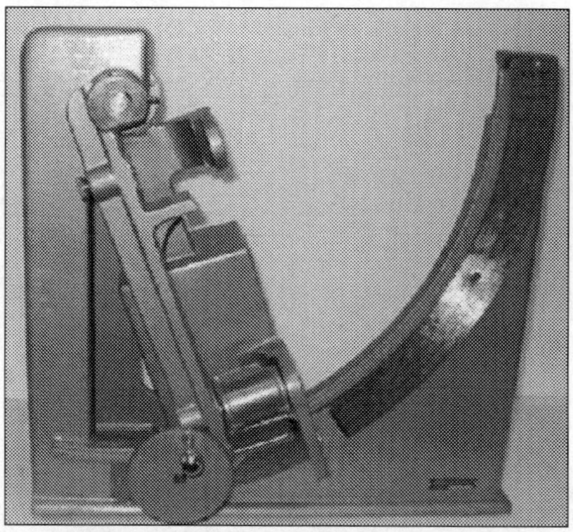

Figure 4.10(b) Universal testing machine

Rotate the wheel in clockwise direction uniformly at the rate so as to obtain the green compression reading of 500 gm/cm^2 on the green scale in 15 seconds. Keep turning the hand wheel until the test specimen breaks. Immediately after the breaking of specimen, the pendulum with the pusher plate drops slightly down whereas the maximum effective compressive strength is indicated on the scale by the lower edge of the magnetic rider. Read the scale according to the location being used. That is for bottom location, read on green compression scale and for top location read on dry compression scale. Take the pusher arm to its orginal position. It is recommended to take three tests on samples prepared from the same batch of sand mix and take the average reading. If there is a gap between the specimen and compression pads, rotate the adjusting screw (16) for removing the gap.

(b) Green shear strength

Green shear strength can be determined with the standard specimen and universal testing machine. This is accomplished by changing the loading surfaces (different adapter) on the testing machine from compression plates to shear plates. The specimen then ruptures in shear along its longitudinal axis when sufficiently loaded. The stress required to shear the specimen along the axis is then represented as the green shear strength. The green strength may vary from 10 to 50 kPa.

Procedure: Remove the compression pads and insert the shear test attachment – greensand at lower location and dry sand at upper location – and place the standard test specimen in between the pads so that plain surface of the test specimen touches against the pads.

Rotate the wheel in clockwise direction uniformly until the specimen breaks. Read the scale according to the location, i.e, upper or lower on the shear scale directly.

Transverse strength test

Remove the shear strength attachment and insert the transverse strength attachment. This attachment has two support arms with locating pins and knife edges. The attachment should be fixed in such a way that the single knife edge is at the centre of the two knife edges of the other support arm as shown in Figure 4.11.

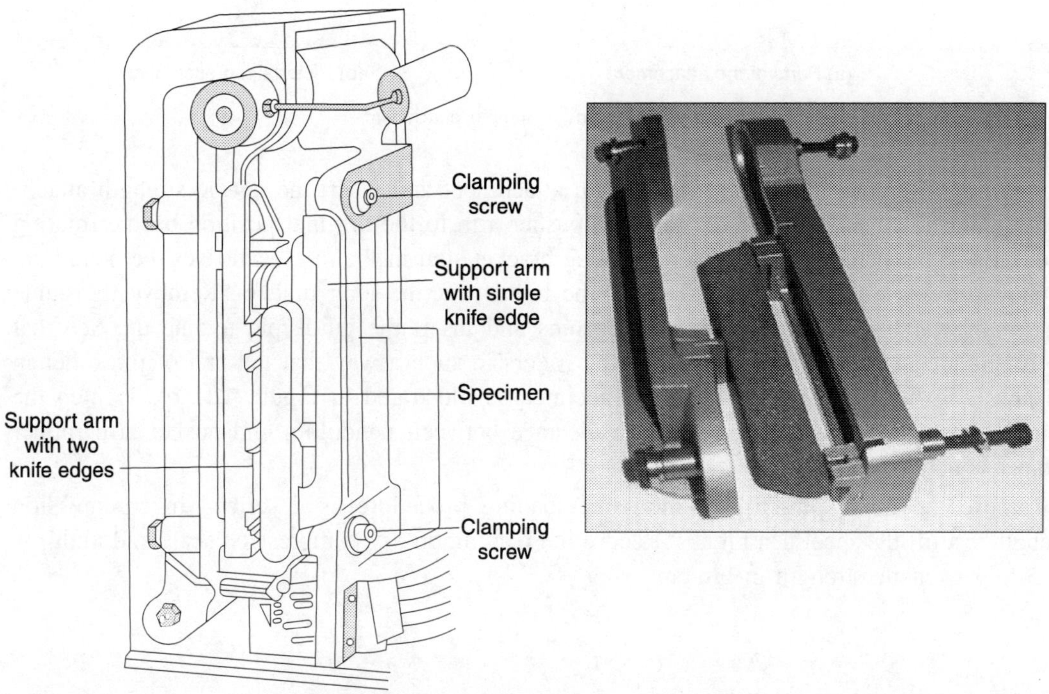

Figure 4.11 Transverse strength attachment

Insert the transverse test specimen between the edges by lifting the pendulum and load the specimen. Follow the same procedure as in compression strength test. When the specimen breaks, record the reading on dry compression scale and multiply by ten to get the transverse strength in gm/cm^2.

(c) Green tensile strength

The tensile strength of green moulding sand using a modified specimen tube for ramming may be determined with the universal sand-strength tester. Green tensile strengths of green moulding sands vary from about 7 to 40 kPa, depending on the type of sand.

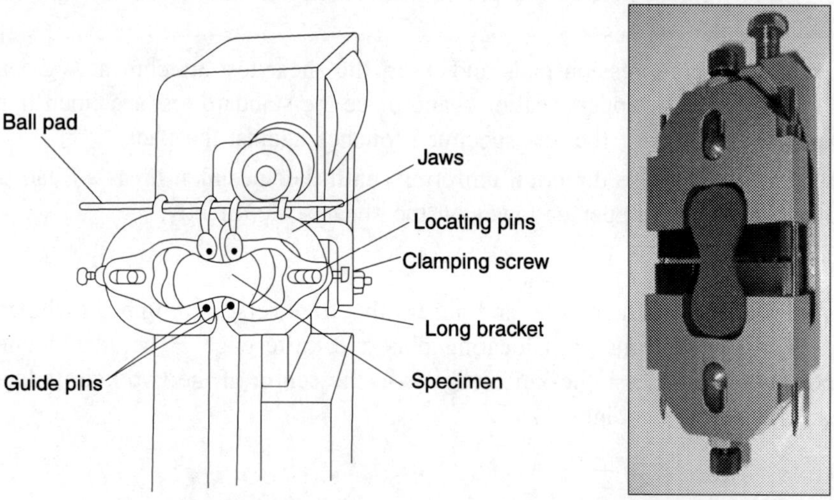

(a) Parts of the attachment (b) Pictorial representation

Figure 4.12 Tensile strength attachment

Procedure: Remove the transverse strength attachment and insert the tensile strength attachment as illustrated in Figure 4.12. Take the pusher arm to the original position by the rotating wheel. Raise the pendulum and insert the long bracket stud in the top location of the pendulum as shown. Release the pendulum. Tighten the bolt to fix the long bracket. Remove the guide pin from jaws. Insert the jaws on locating pins and insert the guide pin to line up. See that the rollers are in proper position. Adjust the rollers in such a way that flat end of the roller in the jaws will flush with the tensile test specimen as illustrated in Figure 4.12(b). Tighten the clamping screws of the jaws, so that the distance between pendulum and pusher arm rubber bumper is about 30 mm.

Remove the guide pin, follow the same loading procedure as described in compression strength test till the specimen breaks. Record the reading on dry compression scale and multiply by 5 to get tensile strength in gm/cm^2.

(d) Dry compression and shear tests

The tests similar to the above can also be carried out with the standard dried between 105° and 110°C for two hours. Since dry compression strength is usually much greater than green strength, higher loads are required on the universal strength machine. Typical dry-compression values on green sands are 140 to 1800 kPa depending on the sand sample.

(e) Deformation (strain) test

Green moulding sand has the ability to deform under load. If the sand deforms too readily under low loads, the mould cavity may not remain true to shape. If the sand has too little ability to deform under load, then it is brittle or lacks toughness. Drawing of patterns then may become difficult. This property may be measured by means of a deformation accessory for the universal strength machine (Figure 4.10). This device measures in inches the amount that the specimen is compressed before it ruptures under compression loading. Deformation of 0.010 to 0.030 in. per in. is observed in many moulding sands. The deformation test is, however, considered a nonstandard test. Sometimes the deformation value is used, together with the green compression strength, to describe toughness of the sand, according to the following formula:

$$STN = D \times GS \times 1000$$

where, STN = Sand toughness number

D = Deformation, 0.001 in.

GS = Green strength

Higher deformation values of the sand mould indicate better capacity to the mould to withstand hydrostatic pressure and volumetric contraction of the casting.

(f) Hot strength

It is important to evaluate the high temperature characteristics of sand. These tests are carried out in an equipment known as dilatometer. The equipment is suitable to carry out hot compression strength, retained strength, expansion and collapsibility, hot gas pressure, hot gas volume, and penetration test. The temperature range of typical dilatometer varies from 0 to 1350°C.

A standard test specimen of 50.8 mm diameter and 50.8 mm height is prepared in a standard sand rammer. The specimen is heated in the dilatometer for a specified testing temperature ranging from 270°C to 1300°C and soaked at that temperature for a fixed period and then subjected to required tests using suitable attachments available with the dilatometer. Common testing temperatures are 250°C, 500°C, 750°C, 1000°C, 1300°C, although others are used.

4.2.11 Permeability

Permeability is a property of foundry sand with respect to how well the sand can vent. In general, permeability is expressed as a number. Permeability number is defined as the volume of air in cm^2 that will pass through a sand specimen of 1 cm^2 cross section and 1 cm height in 1 minute at a pressure of 1 gm/cm^2. For an average mould, the number is about 80 to 130.

The permeability test is conducted for two types of sands:

(a) **Green permeability** is the permeability of the greensand.

(b) **Dry permeability** is the permeability of the moulding sand, dried at 105°C to 110°C to remove the moisture completely.

(a) Factors affecting permeability

Permeability depends on many factors. Some important factors in regulating the degree of permeability are:

1. Grain size
2. Grain shape
3. Grain distribution
4. Type and quantity of bonding material used
5. Density to which sand is rammed
6. Percentage of moisture used for tempering

(b) Testing method

The standard permeability is to measure the time taken by a 2000 cm^3 of air at a pressure typically 980 Pa (10 g/cm^2), to pass through a standard sand specimen confined in a specimen tube. The rate of air passing through a standard specimen under a standard pressure is known as the permeability number.

The standard specimen size is 50.8 mm in diameter and a length of 50.8 mm. The permeability number is obtained by

$$P = \frac{V \times H}{p \times A \times T}$$

where V = Volume of air = 2000 cm^3

H = Height of sand specimen = 5.08 cm

p = Pressure of air gm/cm^2

A = Cross-sectional area of sand specimen = 20.268 cm^2

T = Time in minutes for complete air to pass through

Including the above standard values into the expression, we get

$$P = \frac{501.28}{P \times T}$$

Test procedure: Permeability of moulding sand is determined using the permeability tester. The schematic arrangement of the equipment is shown in Figure 4.13.

This permeability meter enables to read the permeability directly. This apparatus consists of a cylinder (water tank) and another concentric cylinder (also called bell) inside the outer cylinder. The space between the two cylinders is filled with water. A bell having diameter larger than that of the inner cylinder, but smaller than that of the outer cylinder rests on the surface of water.

Cylindrical standard sand specimen, together with the ram tube, is placed on the tapered nose piece. The bell is allowed to sink under its own weight with the help of a multiposition cock. In this way, the air of the streams pass through the nose piece and the permeability is directly recorded from the water manometer. The time for passing the air is noted using a stop watch.

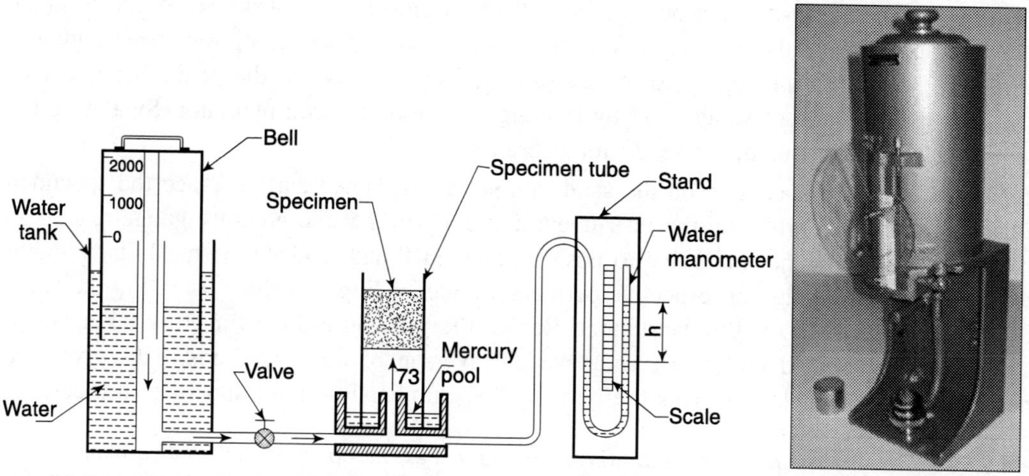

Figure 4.13 Sand permeability tester

Example 4.9 Calculate the permeability number of sand if it takes 1 minute 25 seconds to pass 2000 cm^3 of air at a pressure of 5 gm/cm^2 through the standard sample.

$$p = \text{pressure of air} \frac{\text{g}}{\text{cm}^2} = 5 \frac{\text{g}}{\text{cm}^2}$$

$$T = 1 \text{ min } 25\text{s} = 1 + 25/60 = 1.417 \text{ min}$$

$$\text{Permeability number, } P = \frac{501.28}{5 \times 1.417} = 70.75$$

Electropermeter: This is a handy and simple instrument for direct measurement of the permeability number of greensand, dry sand and baked sand. The instrument is show in Figure 4.14

The instrument consists of a blower, a valve, gauge to read the permeability number directly, orifice, boss with rubber sleeve, mounted on fabricated body and speed regulator. There are three scales:

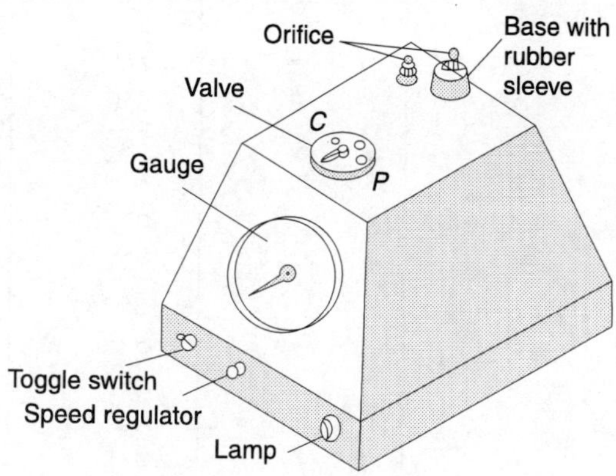

Figure 4.14 Electropermeter

(i) Outer scale to read the permeability number directly with large orifice.

(ii) Middle scale to read the permeability number directly with small orifice.

(iii) Inner scale to read the pressure in mm of water-head.

The valve has two positions: (i) C position to check the set pressure of 100 mm water-head, and (ii) P position to read the permeability number.

Procedure: Place the instrument on a plain platform. Connect the instrument to 220 V single phase AC supply. Ensure that the required proper size orifice is used. Keep the speed regulator to minimum. Switch on the supply to the instrument. Put the valve to the position marked C. Set the gauge to 100 mm of water head by rotating the knob of speed regulator. Switch on the instrument. Now the instrument is ready for operation.

Prepare the standard specimen on the sand rammer as explained earlier. Place the specimen on rubber sleeve boss. Switch on the instrument and confirm that the pressure gauge reads 100 mm of water head. Now put the valve to position marked P and read the permeability number of the test specimen direclty on respective permeability depending on orifice used. Put the valve on C position and switch off the instrument. Repeat the procedure three times by taking fresh test specimen of the sand under test and take the average of three readings. This gives the permeability number. Table 4.9 gives the permeability number for small and large orifices.

Table 4.9 Permeability test pressures and corresponding values, as obtained with orifice plates

Pressure, gm/sq cm	*Permeability*		*Pressure, gm/sq. cm*	*Permeability*	
	Small orifice, 0.5 mm	*Large orifice, 1.5 mm*		*Small orifice, 0.5 mm*	*Large orifice, 1.5 mm*
0.1			3.6	23.4	219
0.2			3.7	22.7	212
0.3			3.8	21.8	205
0.4	....	2450	3.9	21.0	198
0.5	....	2000	4.0	20.0	193
0.6	....	1620	4.1	19.5	185
0.7	....	1350	4.2	19.0	178
0.8	....	1200	4.3	18.4	173
0.9	....	1060	4.4	17.8	167
1.0	....	950	4.5	17.3	163
1.1	....	850	4.6	16.7	156
1.2	....	780	4.7	16.2	151
1.3	....	710	4.8	15.7	146
1.4	....	650	4.9	15.2	142
1.5	....	610	5.0	14.7	138
1.6	....	550	5.1	14.3	134
1.7	....	525	5.2	13.8	128
1.8	....	492	5.3	13.4	126
1.9	....	467	5.4	13.0	122
2.0	49	440	5.5	12.6	119

(Contd.)

Pressure, gm/sq cm	Permeability		Pressure, gm/sq. cm	Premeability	
	Small orifice, 0.5 mm	Large orifice, 1.5 mm		Small orifice, 0.5 mm	Large orifice, 1.5 mm
2.1	47	417	5.6	12.2	115
2.2	44	398	5.7	11.8	112
2.3	42	376	5.8	11.4	108
2.4	40	358	5.9	11.0	105
2.5	38	341	6.0	10.7	102
2.6	36	326	6.1	10.3	99
2.7	34	313	6.2	10.0	96
2.8	33	300	6.3	9.7	93
2.9	31	287	6.4	9.4	90
3.0	30	275	6.5	9.0	88
3.1	29	264	6.6	8.8	85
3.2	28	253	6.7	8.5	82
3.3	27	243	6.8	8.2	80
3.4	25.8	235	6.9	7.9	77
3.5	24.2	226	7.0	7.7	75

4.2.12 Mould Hardness Test

The mould hardness test indicates the resistance of the mould to damage as the metal contacts the mould surface. Mould hardness is measured by the resistance offered by the mould surface to a spring loaded plunger. The mould hardness is tested by means of a hardness tester (Figure 4.15). A spring loaded 1/2 inch diameter steel ball with a mass of 980 is indented into the standard sand specimen prepared. This ball is made to penetrate the sand surface and the penetration of the ball in the mould is indicated on a dial in thousands of an inch. The dial is calibrated to read the hardness directly, i.e., a mould surface which offers no resistance to steel ball would have zero hardness value and a mould which is more rigid and is capable of preventing the steel ball from penetrating would have a hardness value of 100. Usually, an average mould shows an average of 70 to 80.

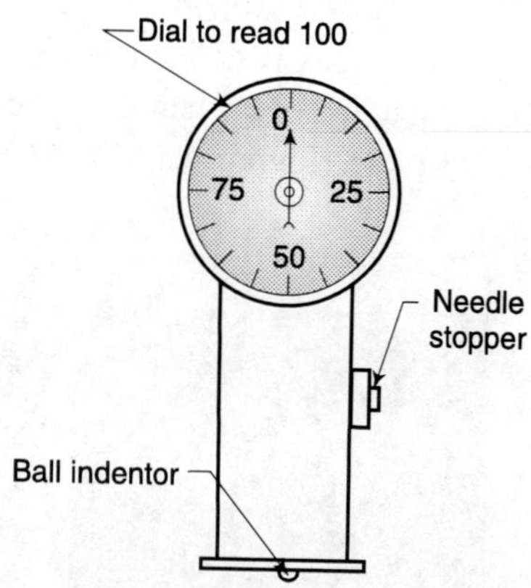

Figure 4.15 Hardness tester

Hardness of the mould surface can also be tested with the help of an "indentation hardness tester". (Figure 4.16). The working of the tester is based on the principle of Brinell hardness testing machine. It consists of an indicator and a spring loaded spherical indenter.

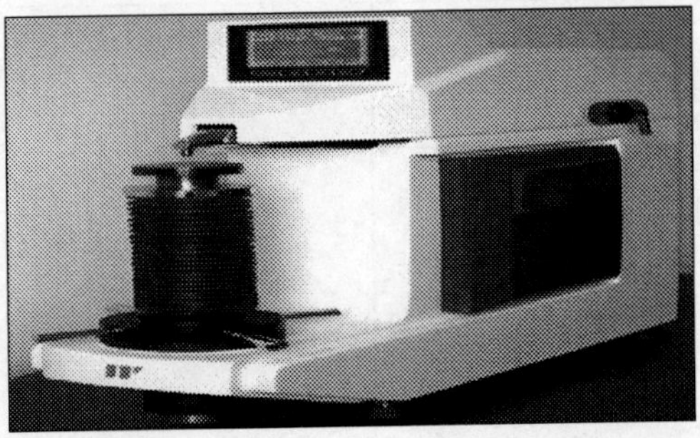

Figure 4.16 Brinell hardness tester

The spherical indenter penetrates the mould surface at the time of testing the depth of penetration w.r.t. the flat reference surface of the tester.

Mould hardness number = $((P)/(D - (D^2 - d^2))$

where,

P – Applied force (N)

D – Diameter of the indenter (mm)

d – Diameter of indentation (mm)

4.2.13 Shatter Index

The shatter index is defined as the proportion by weight of standard specimens which is retained by a particular sieve when the specimens have been shattered by falling from a particular height on solid plate (Figure 4.17).

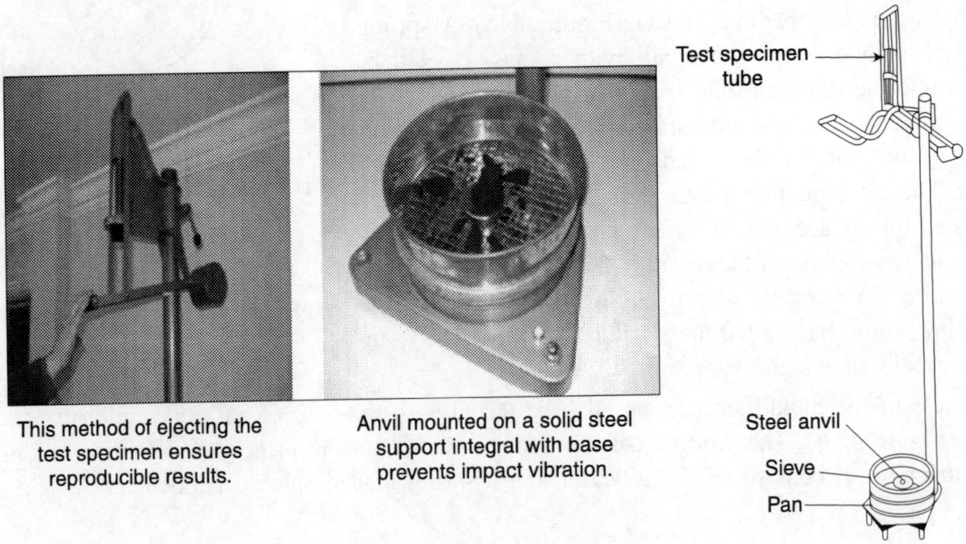

Figure 4.17 Shatter index testing equipment

In this test, the AFS standard specimen is rammed usually by 110 blows and is then allowed to fall on a 12.7 mm (1/2 inch) mesh sieve from a height of 182.88 cm (6 feet). The portion of sand retained on the sieve is measured and is expressed as percentage of the total weight of the specimen and is the measure of the shatter index. This value, when correlated with the compression strength test gives a good indication of the mouldability of a sand, that is, the kind of "lift" which can be obtained when a sand mould is removed from a pattern.

$$\text{Shatter index } \% = \frac{\text{Weight of sand retained on sieve}}{\text{Total weight of AFS specimen}} \times 100$$

Flowability or plasticity is the property of the sand to respond to the moulding process so that when rammed, it will flow all around the pattern and take the desired mould shape. High flowability of sand is desirable for the sand to get compacted to a uniform density and to get good impression of the pattern in the mould.

It is influenced by moisture, permeability, grain shape, fineness, and clay content. The indicator is a specially constructed dial gauge calibrated to read percentage flowability, and may be readily attached to the sand rammer. The stem rests on top of the plunger shaft and the movement between the fourth and fifth drop of the weight is registered as flowability. The further the plunger descends, indicating incomplete flow the lower the percentage of flowability. When the sand rammer is used for ramming test specimens, the indicator is rotated out of position.

Testing Procedure

1. Prepare a metric standard 50 mm diameter test specimen of moulding sand with three drops of the rammer weight.
2. Ram with the fourth drop of the weight and set the indicator into position so that the ball point of the stem comes in contact with the plunger of the rammer. Set the bezel of the indicator to read zero.
3. Ram a fifth time and record the reading on the indicator as percentage flowability. If the plunger does not descend any further, this indicates that the sand has "flowed" as far as possible with the initial four rams; hence the flowability is 100%.

Figure 4.18 Flowability indicator

4.2.14 Miscellaneous Sand Tests

In addition to the sand tests discussed in the preceding sections, a variety of other tests are also used. Among them are:

1. Expansion behaviour and relation to expansion defects
2. Determination of sintering point of base sand

3. Durability. The amount of clay required for rebinding the sand to a constant green strength during repeated cycles of uses for pouring castings can be used to study the life or durability of clays.
4. Measurement of gas evolution in moulding sand: The tendency of the moulding sands to give off gases can be studied by special procedures.
5. Metal penetrations. Metal penetrations have occurred when the molten metal seeps into sand as though it were porous to the liquid. This results in a fused mass of metal and sand which adheres tightly to casting. A test used to study this tendency has been proposed.
6. Loss on ignition. Combustible material in the sand is determined by the per cent loss on ignition of the dry sand.
7. Determination of colloidal properties of clay
8. Determination of flowability of moulding sand
9. Determination of collapsibility

Brief explanation of some of the above commonly used are explained below.

(a) Compactibility test

Compactibility is directly related to the performance of the sand in the moulding operation and detects the degree of temper in the sand mix. The test is run by filling a specimen tube with riddled sand through approximately 10 mm screen mounted at the constant height above the tube. The excess sand is stuck off the tube and the sand is rammed three times. The distance from the top of the tube to the surface of the sand is read as per cent compactibility.

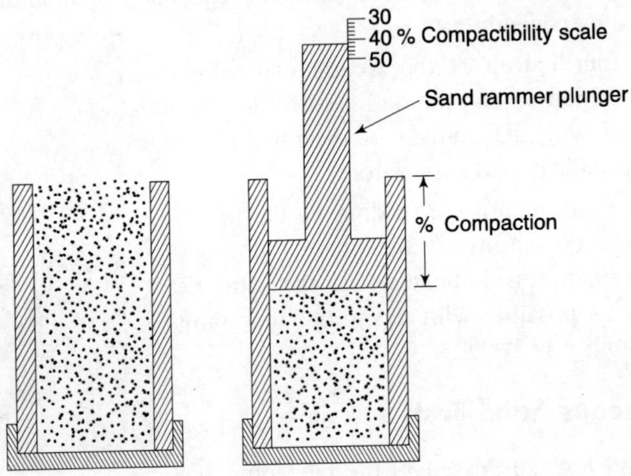

Figure 4.19 Compactibility test

Compactibility is inversely related to bulk density. Moisture is the measure of the water in the sand. A situation when moisture level is insufficient or in excess of what is required to activate the bond must be avoided.

In a mechanised foundry, compactibility of incoming sand may be maintained at a particular target level by continuously adjusting the moisture content in the batch of sand. The cooling of the system sand is very important for optimizing compactibility in green sand moulding. The return sand temperature may be as high as 175°C. The moisture content in the return sand usually ranges between 1.0 and 2.0%. The following parameters have a strong effect on the final sand temperature:

(a) Kinetic energy of hot sand
(b) Retained or added moisture
(c) Air
(d) Intimate contact of the sand mixture with moisture
(e) Time factors that affect cooling

(b) Determination of sintering point of base sand

Sintering point is the lowest temperature at which fusion of the external layer of mould material starts. To measure this property, the sand mixture is heated for three minutes at a constant temperature. After each heating period, the sample is inspected for sintering action.

(c) Determination of colloidal properties of clay

The test is conducted after drying the clay at 110°C. The dried clay of 1gm is taken in a test tube of 15 mm diameter. Distilled water of 15 cc and 0.1 gm of magnesium oxide is added to the test tube and shaked well for 1 minute. The mixture is then allowed to stand still for 24 hours. The ratio of the height of sediments of the total height of the column expressed in percentage gives a measure of colloidal property.

(d) Determination of flowability of moulding sand

This is the property of the moulding sand to be compacted uniformly in a mould cavity with minimum energy. Flowability is generally reduced with increase of clay and moisture. One method of finding flowability is the measurement of deformation of the standard specimen between the fourth and fifth ramming in a standard rammer. The lower the deformation, the better is the flowability. The percentage flowability is calculated as:

$$\text{Percentage flowability} = 100 - 40 \times d$$

where, d is the deformation of the specimen in mm.

The maximum value of d is 2.5 mm, and the corresponding flowability is zero per cent.

(e) Determination of collapsibility

Collapsibility is influenced by the composition of mould mixture, temperature of molten metal and its distribution in the mould. One method of testing collapsibility is dropping a bolt of 20 mm length with conical tail, from a height to penetrate standard specimen after heating to a testing temperature and cooling. The work required to break the specimen is a measure of collapsibility of the mould.

(f) Measurement of gas evolution in moulding sand

The maximum evolution of gases occurs at elevated temperatures due to vaporisation of moisture and burning of organic binders. The total gas generated is measured for a given weight of mixture, during heating to the testing temperature. The amount of gas generated is denoted by the letter G and is expressed in cm^3/gm.

4.2.15 Fluidity Tests

The term fluidity is defined as the ability of the molten metal to continue to flow while it continues to lose temperature and even while it is starting to solidify.

Fluidity, defined in physics as the reciprocal of viscosity, is also defined as the force required to move a surface of unit area at unit velocity past an equivalent parallel surface at unit distance. It is the measure of capacity of a liquid to transmit a dynamic stress in shear.

In terms of casting alloys, the fluidity is defined as the maximum distance (L_f) to which the metal will flow in a standard mould before solidification. The failure of a liquid to fill the mould cavity results not from high viscosity but from premature solidification. Thermal conditions and mode of solidification are the critical factors with respect to cessation of the flow.

Maximum fluidity length versus continuous fluidity length

(a) Maximum fluidity length: In a long mould, the metal will continue to flow until stopped at distance (L_f) by solidification. This is the maximum fluidity length or simply fluidity (L_f) of the metal.

(b) Continuous fluidity length: In a short mould, the migration of the growing choke may reach the end of the channel before the channel is closed. The flowing metal will remelt the choke completely and flow will continue uninterrupted. This is called continuous fluidity length or critical fluidity (L_c) of the metal. Thus, for a channel equal to or smaller in length than (L_c), the flowing capacity or fluidity becomes infinite.

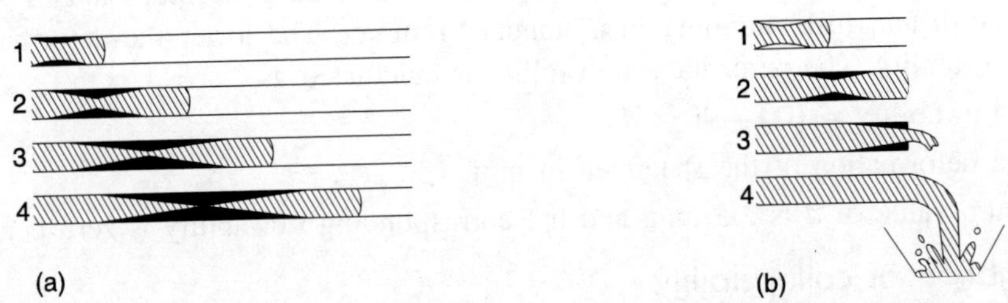

Maximum fluidity length showing the stages of freezing leading to the arrest of flow in a long mould

Continuous flow can occur if the length of the mould does not exceed a critical length

Figure 4.20 Maximum fluidity length versus continuous fluidity length

The casting material's properties which affect the fluidity to a great extent are: viscosity of the melt, heat content of the melt, surface tension, freezing range and specific weight of the liquid metal. Lower the coefficient of viscosity of the molten metal, higher the fluidity. In general, alloys which have a narrow freezing range have higher fluidity compared to the wide freezing range ones.

The mould properties that affect the fluidity are thermal characteristics, permeability and the mould cavity surface. The way in which heat is transferred by the mould from the melt affects the ability of the metal to fill the mould cavity, e.g., in a green sand mould, the fluidity would be lower since more heat is extracted by it than in a dry sand mould.

Measurement of fluidity

Fludity cannot be assessed from individual physical properties. Empirical tests have been devised based on conditions analogus to the casting of metals in the foundry. Measure fluidity as the maximum distance covered by molten metal in standardised system of enclosed channels before cession of flow. Typical test methods are:

- Straight flow channel
- Spiral test
- Standardised fluidity test

Fluidity spiral design is shown in Figure 4.21 and Figure 4.22 shows fluidity spiral for grey cast iron. Figure 4.23 shows fluidity chart for grey cast iron.

The most commonly used testing is the spiral fluidity test. The mould of the same composition that is to be used is prepared with the standard spiral pattern. The casting prepared from the mould is measured for the fluidity which is represented by the length of the spiral solidified. Grey cast iron is most fluid of all the ferrous alloys and its fluidity is affected by the composition factor (CF), given by:

$$CF = \%C + 0.25 \times \%Si + 0.5 \times \%P$$

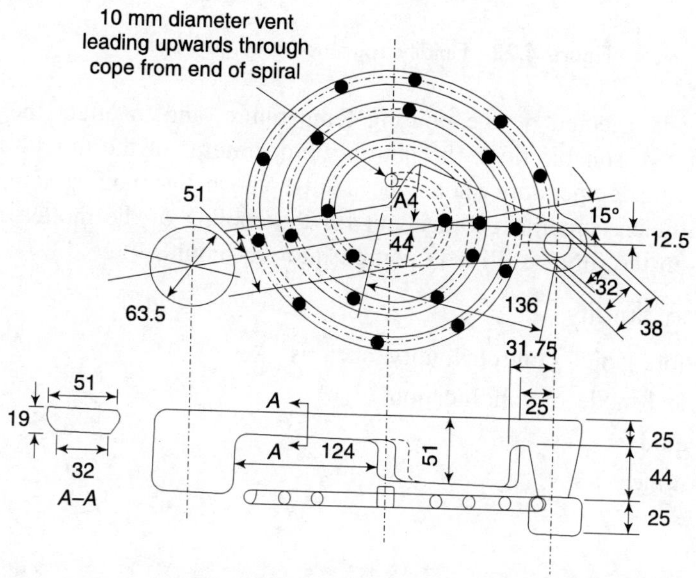

Figure 4.21 Fluidity spiral design

Then the fluidity spiral length in cm can be estimated by means of the following empirical relation:

$$\text{Fluidity, cm} = 37.846 \times \text{CF} + 0.228 \times T - 389.6$$

where, T = pouring temperature, deg C

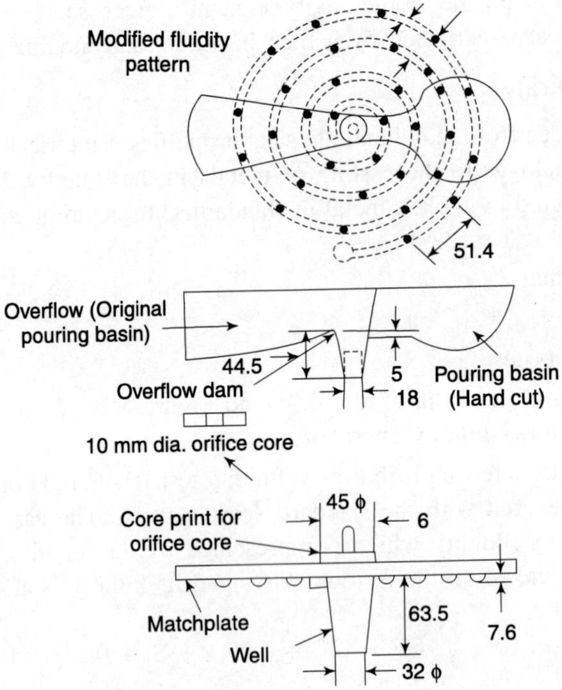

Figure 4.22 Fluidity spiral for gray cast iron

Since the fluidity is affected by the pouring temperature, and to effect the completion of the mould filling in a reasonable time, the pouring temperatures of the mould are accordingly controlled. The pouring temperature should be a little above the melting temperature of the alloy along with sufficient superheat to account for the cooling of the molten metal from the time it is tapped from the furnace till it is poured into the mould.

Factors controlling fluidity

There are many factors which control fluidity, such as

1. Composition and mode of solidification
2. Velocity of liquid
3. Solidification time
    - Modulus

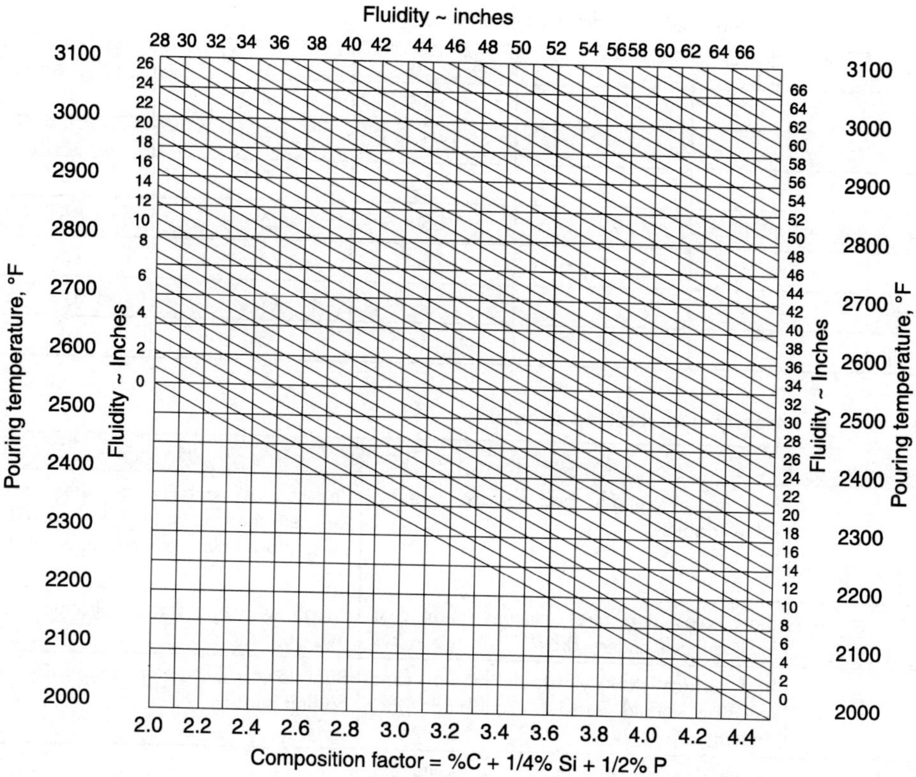

Figure 4.23 Fluidity chart for grey cast iron

- Heat transfer coefficient
- Super heat
- Latent heat
- Mould temperature
- Surface tension

Control property and resulting problems

To have uniform properties of the moulding sand, control is necessary; however, sand properties change from time to time. Therefore, to have a good control of the sand properties, control values are to be determined. Table 4.10 shows control properties and standard control values for steel, cast iron, copper, and aluminium alloys. Control values should not be too low or too high. Table 4.11 shows effect of too low and too high control parameters and associated problems/defects in the casting.

Table 4.10 Control properties and standard control values

Properties	*Steel*	*Cast Iron*	*Copper*	*Aluminium Alloy*
Moisture content	3.2	3.2	4.0	4.0
Permeability	200	150	20	20
GCS, kg/cm^2	0.8	1.0	0.6	0.5
Green surface stability index	92	88	86	85
Flowability	72	75	75	75
Clay content	8	10	12	14
Loss on ignition	3	4	5	6

Table 4.11 Control property and resulting problems

Property	*Too Low*	*Too High*
Moisture	Low green strength, bad surface stability, edge friability	Reduced permeability, low green strength, reduced sand flowability, low mould hardness, blow holes (pin holes), swollen castings
Permeability	Blow holes (pin holes), misruns, head castings, difficult pattern draw	Rough surface, metal penetration
Strength	Bad surface stability, friable mould, sand inclusion, mould errosion, scabbing, poor pattern strip	Bad flowability, under compaction, rough casting surface
Surface stability index	Edge, friability, sand inclusions, mould erosion, scabbing	
Flowability	Soft mould, sand inclusions, swollen castings, bad surface stability mould	
Clay content (active clay)	Low strength, bad surface stability, friable mould, sand inclusions, erosion defect, scabbings	Reduced permeability, knock-out problems, increase of moisture content, cracked casting
Loss on ignition	Poor surface finish, scabbing	Blow or pin-holes

Review Questions

1. What are the common sand tests conducted in foundries?
2. Explain the procedure to find grain fineness number.
3. How do we measure clay content in sand?
4. What is hot strength of a moulding sand?
5. With the help of a neat sketch, explain the permeability test.
6. Explain the working of an electropermeter.
7. Explain mould hardness test.
8. Define shatter index and explain its importance.
9. Explain compactibility test with the help of a neat sketch.
10. Explain the use of fluidity chart with an example.

CHAPTER 5

Types of Sand Moulds

5.1 INTRODUCTION

In order to produce sound castings, moulds are required to have some specific properties, such as:

- It must be strong enough to withstand the temperature and weight of the molten metal.
- It must resist the erosive action of the flowing hot metal.
- It should generate minimum amount of gases as a result of the temperature of the molten metal.
- It should have good venting capacity to allow the generated gases to completely escape from it.

The term moulding process refers to the method of making the mould with the materials used.

Requirements of an ideal mould are:

- Adequate strength and hardeness
- Adequate refractoriness
- Adequate permeability
- Appropriate design of the gating/running system
- Adequate use of feeders, if necessary

A perfect mould is necessary due to the following reasons:

- High temperature of liquid metal
- High pressure of liquid metal and dissolved gases
- Expansion of sand due to high temperature
- Reaction of liquid metal into the mould
- Velocity of liquid metal inside the mould cavity
- Turbulence of liquid metal inside the mould cavity

Ten casting rules as per Professor J Campbell are given as follows:

1. Start with a good quality melt

Immediately prior to casting, the melt shall be prepared, checked, and treated, if necessary, to bring it into conformance with an acceptable minimum standard. However preferably, prepare and use only near-defect-free melt.

2. Avoid turbulent entrainment of the surface film on the liquid

This is the requirement that the liquid metal front (the meniscus) should not go too fast. Maximum meniscus velocity is approximately 0.5 m/s for most liquid metals. This maximum velocity may be raised in constrained running systems or thin section castings. This requirement also implies that the liquid metal must not be allowed to fall more than the critical height corresponding to the height of a sessile drop of the liquid metal.

3. Avoid laminar entrainment of the surface film on the liquid

This is the requirement that no part of the liquid metal front should come to a stop prior to the complete filling of the mould cavity. The advancing liquid metal meniscus must be kept alive (i.e., moving) and therefore free from thickened surface film that may be incorporated into the casting. This is achieved by the liquid front being designed to expand continuously. In practice this means progress only uphill in a continuous uninterrupted upward advance, i.e., in the case of gravity poured casting processes, from the base of the sprue onwards. This implies:

- Only bottom gating is permissible.
- No falling or sliding downhill of liquid metal is allowed.
- No horizontal flow of significant extent.
- No stopping of the advance of the front due to arrest of pouring or waterfall effects, etc.

4. Avoid bubble entrainment

No bubbles of air entrained by the filling system should pass through the liquid metal in the mould cavity. This may be achieved by:

- Properly designed offset step pouring basin; fast back-fill of properly designed sprue; preferred use of stopper; avoidance of the use of wells or other volume-increasing features of filling systems; small volume runner and/or use of ceramic filter close to sprue/runner junction; possible use of bubble traps.
- No interruptions to pouring.

5. Avoid core blows

- No bubbles from the outgassing of cores or moulds should pass through the liquid metal in the mould cavity. Cores to be demonstrated to be of low gas content and/or adequately vented to prevent bubbles from core blows.
- No use of clay-based core or mould repair paste unless demonstrated to be fully dried out.

6. Avoid shrinkage

No feeding uphill in larger section thickness castings because of (i) unreliable pressure gradient, and (ii) complications introduced by convection.

- Demonstrate good feeding design by following all feeding rules, by an approved computer solidification model, and by test castings.
- Control of the level of flash at mould and core joints; (ii) mould coat thickness (if any), and (iii) temperatures of metal and mould.

7. Avoid convection

Assess the freezing time in relation to the time for convection to cause damage. Thin and thick section casting automatically avoids convection problems. For intermediate sections either (i) reduce the problem by avoiding convective loops in the geometry of the casting and rigging, (ii) avoid feeding uphill, or (iii) eliminate convection by roll-over after filling.

8. Reduce segregation

Predict segregation to be within limits of the specification, or agree out-of-specification compositional regions with customer. Avoid channel segregation formation if possible.

9. Reduce residual stress

No quenching into water (cold or hot) following solution treatment of light alloys. (Polymer quenchant or forced air quench may be acceptable if casting stress is shown to be negligible.)

10. Provide location points

All castings to be provided with agreed location points for pickup for dimensional checking and machining.

Classification of Moulds

Moulds used for casting can be broadly divided into the following ways:

(A) Depending on the condition of the mould at the time of pouring

1. Greensand moulds
2. Skin dried moulds
3. Dry sand moulds

(B) Depending on the method used for moulding

1. Bench moulds
2. Hand-made moulds
   (a) Flask mould
   (b) Pit mould
3. Machine-made moulds

5.1.1 Greensand Moulds

Greensand moulding process has a long history. A large production of castings, including the vast majority of those made by machine moulding, is cast in greensand moulds. Greensand is the moulding sand which has been freshly prepared from silica grains, clay and moisture. In a greensand mould, metal is poured immediately and the casting taken out. These are most commonly used and are adapted for rapid production, where the moulding flasks are released quickly. Both ferrous and non-ferrous castings are produced in these moulds.

Moulds for making a tonne of castings may require 4 to 5 tonnes of moulding sand aggregate. The sand-metal ratio may vary from 10:1 to 0.25:1 depending on the type and size of castings and moulding methods employed. In any case, the tonnage of sand which must be handled in a sand casting foundry is large, and its quality must be controlled to make good castings.

It is a popular moulding process for those foundries producing small and medium size castings and may not be able to lose its position due to economic considerations. Schematic diagram of greensand mould is shown in Figure 5.1.

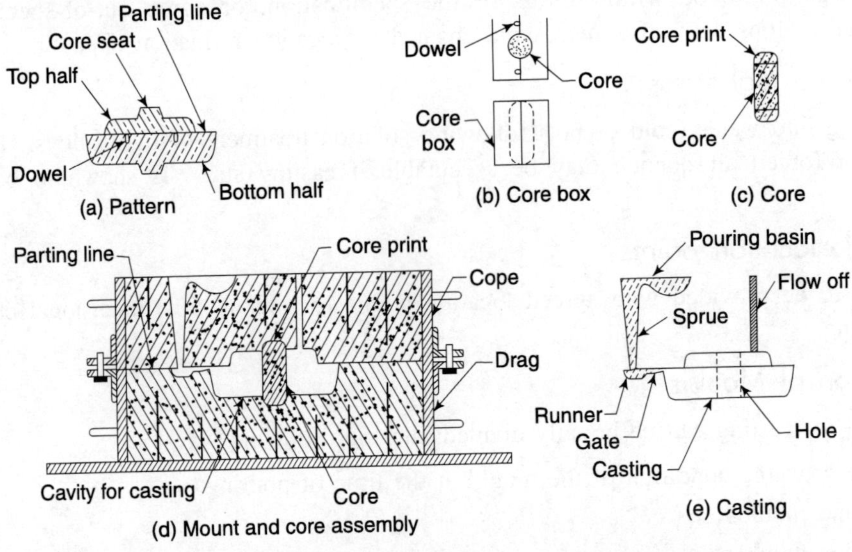

Figure 5.1 Greensand mould

1. **Preparation of the pattern** Most greensand moulding is done with match plate or cope-and-drag patterns. Loose patterns are used when relatively few castings of a type are to be made. In simple hand moulding, the loose pattern is placed on a mould board and is surrounded with a suitable flask.
2. **Making the mould** Moulding requires ramming of sand around the pattern. As the sand is packed, it develops strength and becomes rigid within the flask. Ramming may be done by hand. Both cope and drag are moulded in the same way, but the cope must provide for the sprue. The gating-system parts of the mould cavity are simply channels for the entry of the molten metal.

3. **Core setting** With cope and drag halves of the mould made and the pattern withdrawn, cores are set into the mould cavity to form the internal surfaces of the casting.
4. **Closing and weighting** With cores set, the cope and drag are closed. The cope must usually be weighted down or clamped to the drag to prevent it from floating when the metal is poured.

Steps involved in making a sand mould

1. Initially a suitable size of moulding box for creating suitable wall thickness is selected for a two-piece pattern. Sufficient care should also be taken in such a sense that the moulding box must adjust mould cavity, riser and the gating system (sprue, runner and gates, etc.).
2. Next, place the drag portion of the pattern with the parting surface down on the bottom (ramup) board.
3. The facing sand is then sprinkled carefully all around the pattern so that the pattern does not stick with moulding sand during its withdrawal.
4. The drag is then filled with loose prepared moulding sand and ramming of the moulding sand is done uniformly in the moulding box around the pattern. Fill the moulding sand once again and then perform ramming. Repeat the process three to four times.
5. The excess amount of sand is then removed using strike-off bar to bring moulding sand at the same level of the moulding flask height to complete the drag.
6. The drag is then rolled over and the parting sand is sprinkled over the top of the drag.
7. Now the cope pattern is placed on the drag pattern and alignment is done using dowel pins.
8. Then cope (flask) is placed over the rammed drag and the parting sand is sprinkled all around the cope pattern.
9. Sprue and riser pins are placed in vertical position at suitable locations using support of moulding sand. It will help to form suitable sized cavities for pouring molten metal.
10. Fill the cope with the moulding sand and ram uniformly.
11. Strike off the excess sand from the top of the cope.
12. Remove sprue and riser pins and create vent holes in the cope with a vent wire.
13. Sprinkle parting sand over the top of the cope surface and roll over the cope on the bottom board.
14. Rap and remove both the cope and drag patterns and repair the mould suitably if needed and dressing is applied.
15. The gate is then cut connecting the lower base of sprue basin with runner and then the mould cavity.
16. Bake the mould in case of a dry sand mould.
17. Set the cores in the mould, if needed and close the mould by inverting the cope over the drag.
18. The cope is then clamped with drag and the mould is ready for pouring.

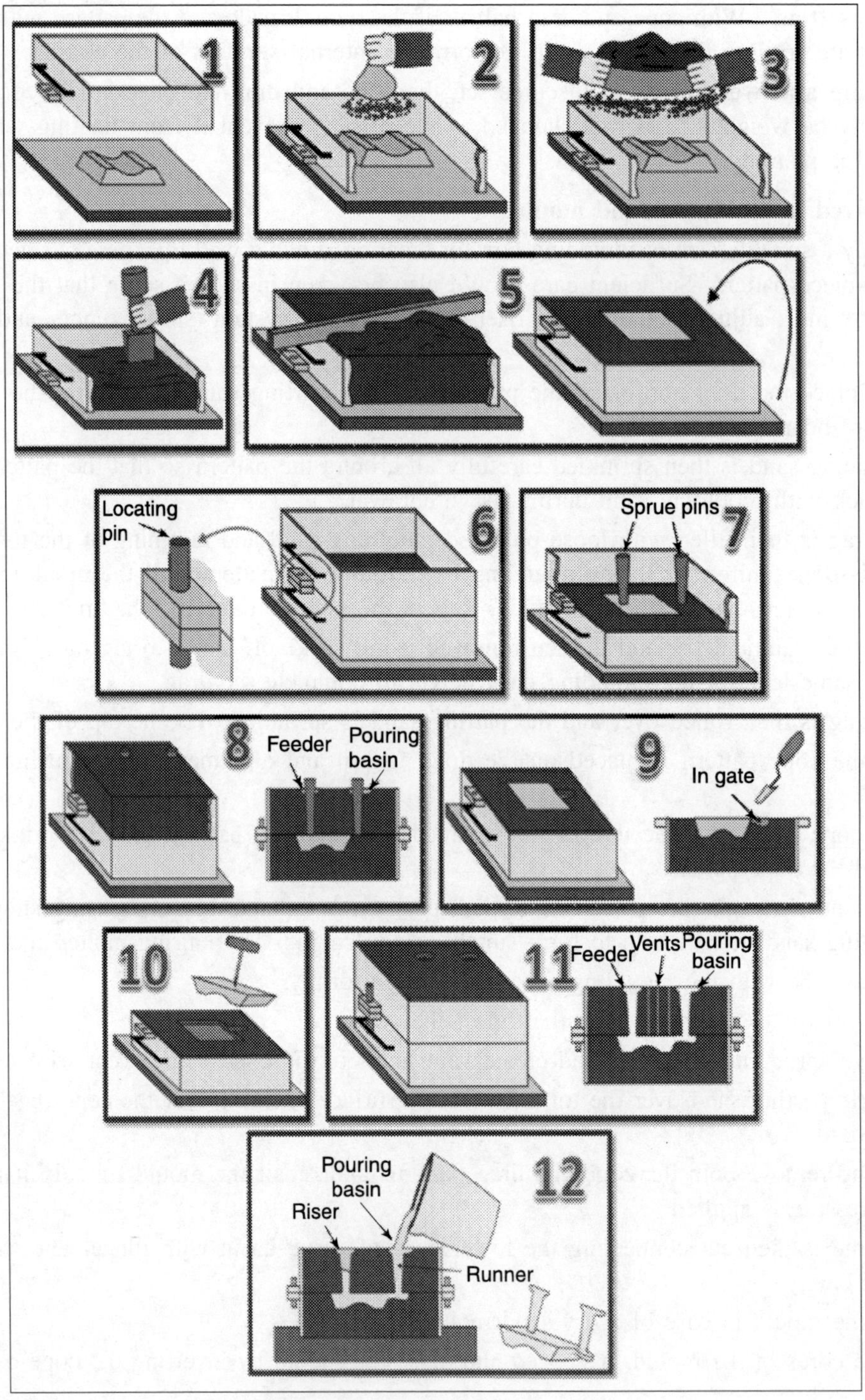

Figure 5.2 Steps in sand casting

Advantages of greensand moulds

- Great flexibility as a production process. Mechanical equipment can be utilized for moulding and its allied operations. Furthermore, greensand can be reused many times by reconditioning it with water, clay, and ether materials. The moulding process can be rapid and repetitive.
- Distortion of mould is less compared to dry sand mould cycle time and space requirement is less.
- Good dimensional accuracy is possible for small sized castings.
- Less danger of hot tearing of castings.
- Least expensive method.
- The sand is readily reconditioned, since there is little dehydration of clay bond.
- Green sand, having low compression strength, offers little resistance to contraction, so the risk of hot tearing is reduced.
- The rapid turnaround of the moulding boxes and the smooth moulding and casting cycle make it suitable for mechanised systems.

Disadvantages of greensand moulds

- Thin, long projections of greensand in a mould cavity are washed away by the molten metal or may not even be mouldable. Cooling fins on air-cooled-engine cylinder blocks and head are an example. Greater strength is then required of the mould.
- Sand casting is more critical than in dry sand moulds.
- The dimensional accuracy and surface finish of greensand castings may not be adequate.
- Dimensional accuracy for large castings is less.
- Large castings require greater mould strength and resistance to erosion than are available in greensands. Erosion of mould is common in large castings.

5.1.2 Dry Sand Moulds

These are the greensand moulds which are completely dried by keeping in an oven between 150 and 350°C for 8-48 hours depending on the binders in the moulding sand. Infrared lamps, torches and hot air are also used as sources for drying. These moulds generally have higher strength and are less likely to be damaged during handling. Better surface finish and dimensional accuracy can be achieved by dry sand moulds. The main disadvantages are the likely distortion of the mould caused during the baking process. Some components made by this method include Engine blocks, large gears, big housings, and construction parts. Both ferrous and non-ferrous metals can be cast in this method.

The advantages of dry sand moulds are:

- Dry sand moulds are generally stronger than greensand moulds and therefore can withstand much metallostatic pressure.
- Good overall dimensional accuracy is possible with these moulds.

- Better surface finish due to application of refractory coatings.
- Pin hole defects can be avoided.

The limitations are:

- Production rate is low, cost is higher.
- Floor space and more flasks are required, which increases the cost.
- Casting defects like hot tears are seen in these types of moulds.

5.1.3 Skin Dried Mould

Though the dry sand mould is preferable for large moulds because of the expense involved, a compromise is achieved by drying only the skin of the mould cavity with which the molten metal comes into contact, instead of the full mould. Skin drying may be performed by torches or electrical heating elements directed at the mould surface. Pouring metal should be completed immediately after the drying process such that moisture from the undried portion would not penetrate the dried skin.

Classification based on methods

5.1.4 Bench Moulding

This type of moulding is preferred for small jobs. The whole moulding operation is carried out on a bench of convenient height. In this process, a minimum of two flasks, namely cope and drag moulding flasks are necessary. But in certain cases, the number of flasks may increase depending upon the number of parting surfaces required.

5.1.5 Handmade Moulds

The traditional methods of making sand moulds are by hand and the process is called hand moulding. Handmade moulds may be in mould boxes or pits in foundries. For small and medium sized casting flasks, moulds are generally used. If the flask mould is made on a work bench, then it is called bench moulding. Other processes, called loam moulding, pit moulding, floor moulding, are also used.

Pit Moulding

Usually, large castings are made in pits instead of drag flasks because of their huge size. In pit moulding, the sand under the pattern is rammed by bedding-in process. The walls and the bottom of the pit are usually reinforced with concrete and a layer of coke is laid on the bottom of the pit to enable easy escape of gas. The coke bed is connected to atmosphere through vent pipes which provide an outlet to the gases. One box is generally required to complete the mould, runner, sprue, pouring basin and gates are cut in it. Typical example is shown in Figure 5.3.

Floor Moulding

This type of moulding is preferred for medium and large size jobs. In this method, only drag portion of moulding flask is used to make the mould and the floor itself is utilized as drag and it is usually performed with dry sand.

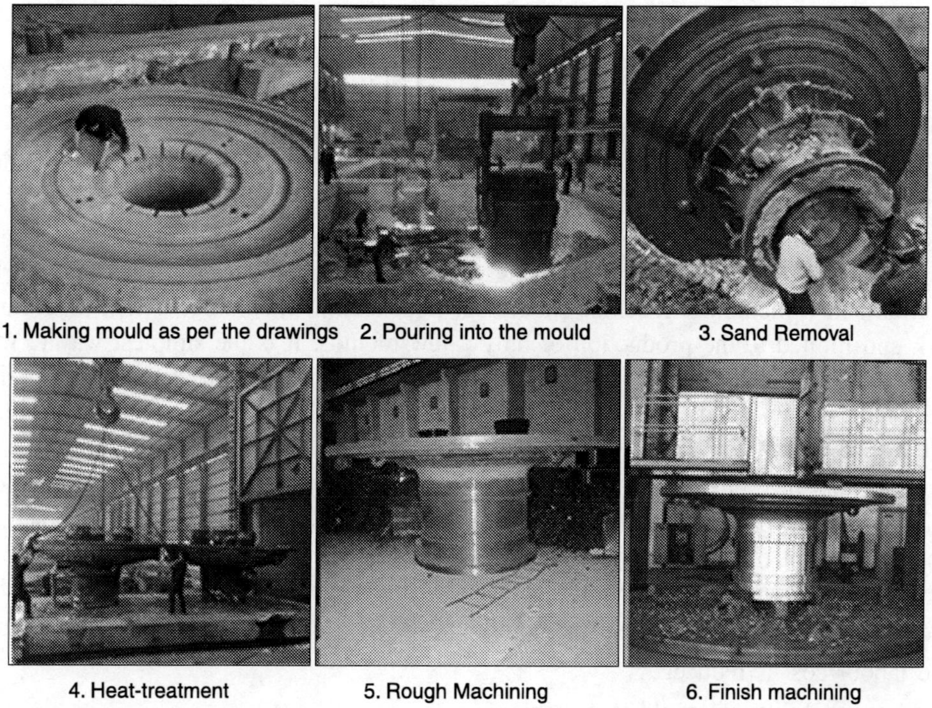

Figure 5.3 Pit moulding of cast steel slag pot

5.1.6 Machine-made Moulds

For mass production of the casting, the general hand moulding technique proves uneconomical and inefficient. The main advantage of machine moulding, besides saving of labour and working time, is the accuracy and uniformity of the castings which can otherwise be only obtained with much time and labour. Or even the cost of machining on the casting can be reduced drastically because it is possible to maintain the tolerances within narrow limits on casting using machine moulding method. Moulding machines thus prepare the moulds at a faster rate and also eliminate the need of employing skilled moulders. The main operations performed by moulding machines are ramming of the moulding sand, rolling over the mould, forming gates, rapping the pattern and its withdrawal. Most of the mould making operations are performed using moulding machines.

5.2 MOULDING MACHINES

Moulding process requires specialised equipment for mould making. Forming the mould is done in one of the following ways:

1. By compacting of the aggregate around the pattern.

2. By free flow of dry aggregate around the pattern. This is well illustrated by shell moulding.
3. By free flow of a slurry or liquid aggregate around the pattern. Plaster and investment moulding illustrate this method.

Variants of the above are also used. The nature of moulding materials determines the mould-forming method and the equipment needed.

Greensand moulding is done by compacting the aggregate around a pattern by ramming, squeezing, jolting, vibration, slinging, blowing or by combination of methods. This work is carried out as bench moulding, machine moulding, floor and pit moulding. Bench moulding is hand work and limited to the production of only a few moulds. It is the simplest way to make one or a few castings of a kind. Floor and pit moulding are suited for the larger casting sizes. By far, the largest tonnage of castings is produced by machine moulding using greensand as the moulding material.

When small and medium sized castings are required in large quantities, machine moulding is generally preferred. In machine moulding, the pattern used is usually of match plate type and is made of metal or plastic. System sand is used for moulding and the majority moulds are poured in green condition. Generally, mechanized foundries use this moulding method.

The advantages of using machines for moulding are:

1. The labour cost is reduced.
2. Good repeatability of mould is possible.
3. Uniform ramming of mould is possible.
4. Good dimensional accuracy of casting is obtained.
5. High production rate is possible.
6. Improves sanitary-hygienic conditions of foundry.
7. Enables flaskless casting (snap or removable flask moulding) for mass production.
8. Reduces casting defects due to better moulding and pattern withdrawal by machine.

Compaction

For typical limits of compaction, bulk density data is to be considered. Minimum bulk density of freshly mixed sand may be as low as 800 kg/mm^3, whereas the maximum after moulding may be as high as 2000 kg/mm^3 or even more.

Even when compacted, the moulding sand mass is composed of approximately 60 to 65 per cent solids and the balance voids. Regardless of the type of bonding clay in the sand, there is a progressive increase in sand bulk density as the amount of work done in compacting the sand increases. The limiting bulk density is achieved when compacting has occurred to the point of sand-grain to sand-grain contact throughout the mass. This will occur at a density of about 1600 kg/mm^3 for typical foundry sands. Application of higher forces will not cause more compaction once the maximum bulk density has been reached. Figure 5.4 shows the variation of mechanical properties for different degrees of compaction in terms of density.

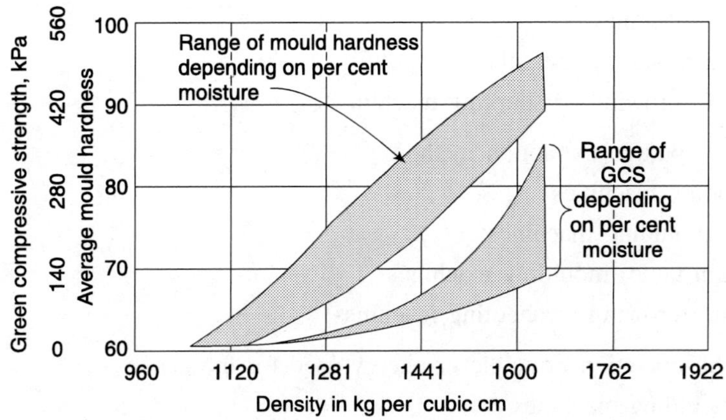

Figure 5.4 Change of mechanical properties due to compaction. Sand used contains mixture of 6% Western bentonite, 2% sea coal, 1% cellulose and the remainder 85 AFS sand

As compaction of the sand occurs, the mechanical properties of the sand change. As the sand density increases, its green compressive strength, green shear strength, tensile strength and mould hardness increase. This means that to raise the bulk density from one level to the next higher level, more work has to be done to overcome the strength of the sand and cause the increased density. Thus, in greensand moulding, there are interacting processes at work as follows:

1. Gross movement of the sand which must occur as sand flows above and adjacent to the pattern and is compacted.
2. Fitting together or packing of sanding grains and expression of the gross voids as the maximum bulk density is approached by compaction.

Increase of mechanical strength of the sand-aggregate as gross movement and packing of the sand grains occur.

The changes sited are the result of work done on the sand by moulding machines. The objective of the moulding machine is to compact the moulding sand to a uniform bulk density of over 90 per cent of maximum, exceeding about 85 mould hardness, producing the sand mechanical properties related thereto.

Moulding machines are classified according to method of sand compaction, the way of extracting pattern from the mould and the kind of energy consumed for the drives.

(A) Classification by sand compaction method used:

1. Squeeze-moulding machines
    - Top squeezing type
    - Bottom squeezing type
    - Diaphragm squeezing type
    - Multiple plunger squeezing type
    - Squeezing and vibrating type
    - Air impulse type

2. Jolt-moulding machines
3. Jolt-squeezing moulding machines
4. Sand slinging (throwing or shooting machines)

(B) Classification by pattern extraction method:
1. Pin-lift moulding machines
2. Pattern-draw moulding machines
3. Roll-over (turn table) moulding machines
4. Sand slinging (throwing or shooting machines)

(C) Classification by operation principle or energy used for the drive:
1. Pneumatic moulding machines
2. Mechanical type moulding machines
3. Hydraulic type moulding machines
4. Electromagnetic type moulding machines

5.2.1 Squeeze Moulding

In the squeeze ramming, a plate slightly smaller than the inside dimensions of the moulding flask is fitted into the flask already filled with the moulding sand. A uniform pressure is applied on the plate, by either moving it down or by moving the flask upwards. The resulting force compacts the sand uniformly. The sand next to the plate rams hardest while the sand below is progressively less hard. Hardness achieved also depends upon the pressure applied on the squeeze plate. This type is suitable for small castings and is generally suitable for shallow flasks. The squeezing may be from the top or bottom.

1. Top Squeeze moulding: In top squeeze moulding, the match plate pattern and the flask are clamped on the machine table and a filling frame is kept over the flask as shown in Figure 5.5. Moulding sand is then filled in both flask and frame. The moulding table with flask and frame

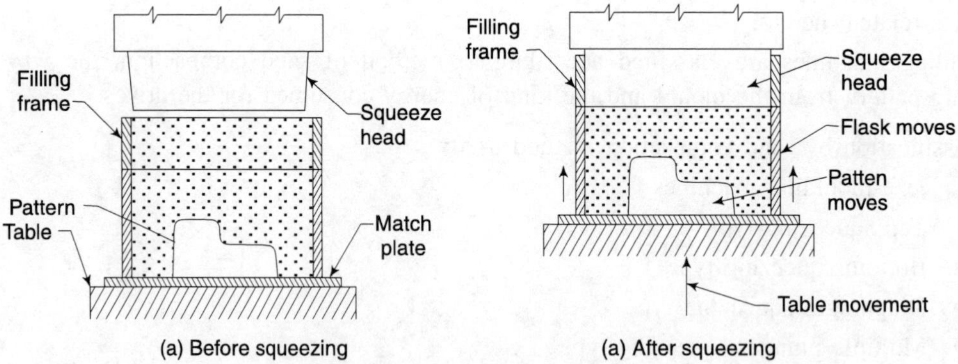

Figure 5.5 Top squeeze moulding

as well as the pattern is then moved upwards so that the stationary squeeze head compacts sand around the pattern. The pattern is then withdrawn and the mould is used for casting.

2. ***Bottom squeeze method*** is almost similar to top squeeze moulding. Here the flask is stationary on the table and the match plate pattern with plunger moves upwards against the flask and thus compacts the sand as shown in Figure 5.6. This method ensures higher compaction of sand over pattern and is technologically more convenient.

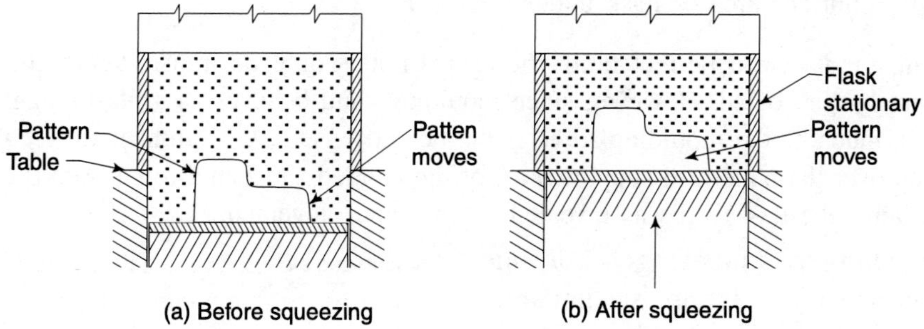

Figure 5.6 Bottom squeeze moulding

Flask with more height is vibrated as well as squeezed to get better results. Low amplitude-high frequency vibration applied on flask diminishes sand friction on the side walls of the flask and results in more efficient and uniform compaction.

3. ***Diaphragm squeeze method*** Compressed air applied on elastic diaphragm can squeeze moulding sand as shown in Figure 5.7. For this type of compaction, the squeeze is fitted with a diaphragm and compressed air at high pressure is applied inside the head. This expands the diaphragm compacting the sand. This system gives more uniform compaction along the flask height over the pattern surface.

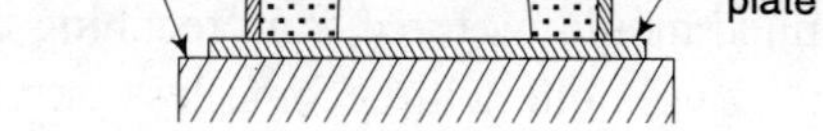

Figure 5.7 Diaphragm squeezing

The maximum squeezing force of a pneumatically operated squeeze type machine is defined by

$$MF = \left(P \times \frac{\pi d_e^2}{4}\right) - W$$

where, MF = Moulding force

P = Squeeze pressure in air cylinder, often assumed to be airline pressure

d_e = Piston diameter of squeeze cylinder

W = Weight of pattern, sand, flask and other accessories on worktable on machine

Hence, the moulding force of a squeeze machine is limited by its piston diameter and the air pressure available, is usually 6-7 bars.

Although MF is relatively constant for a particular machine (and air pressure), the flask size is not. To determine actual moulding pressure (MP) at the flask, the following equation is used:

$$MP = \frac{MF}{A_f}$$

where, MP = Moulding pressure, at the flask surface

MF = Moulding force, applied by squeeze cylinder

A_f = Surface area of flask under force MF

If moulding sand were a perfect fluid, these pressures would be transmitted to the pattern surface uniformly distributed. However, since moulding sand behaves as a plastic aggregate of solid particles, much of the moulding force is inefficiently used. A column of fully compacted sand builds up over the pattern to support most of the applied squeezing force. Additional force is diverted against the flask walls as revealed by the sand movement.

4. Multiple plunger squeezing: Multiple plunger squeezing effect is almost similar to elastic diaphragm squeezing. In Figure 5.8, all the plungers are pressing sand at the same pressure. If the sand resistance is less, it moves more into the mould giving uniform compaction.

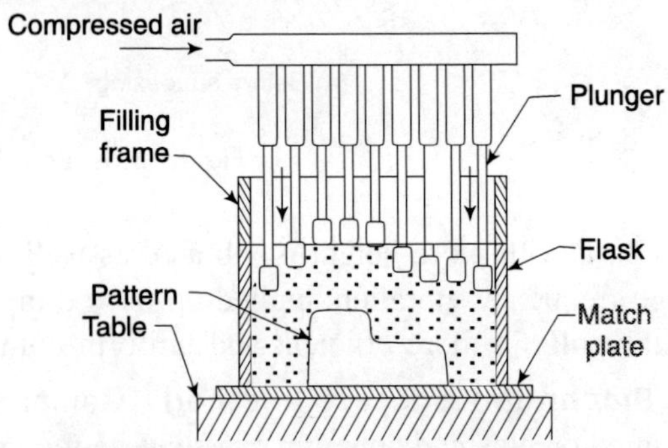

Figure 5.8 Multiple plunger squeezing

Contour squeezing or multiple plunger squeezing is a means of distributing squeeze moulding forces more uniformly in order to obtain a more uniform mould density. Diaphragm squeezing is another means of contour squeezing. Separately actuated feet squeezing small areas also perform contour squeezing. Whatever type of squeezing is used, the objective is to produce a mould usually in excess of 85 mould hardness and 1450 kg/sq. cubic cm density.

Bridging and keying of the sand grains against the flask and each other keeps the total applied moulding force from reaching the pattern surface. Thus, it can be seen that moulding by squeezing alone will become less effective for a given pressure as the depth of the mould half increases. Furthermore, there will be a differential in the degree of packing from the squeeze head to the pattern. Sand density is at a minimum adjacent to the pattern and the hardness of the mould is therefore less than that next to squeeze head. Because of non-uniform pressure distribution, the sand adjacent to the pattern may be non-uniformly rammed. There is a limit to the flask depth that may be properly moulded by squeezing which is dependent on the squeeze machine capacity, pattern contour, moulding sand, etc. To obtain more uniform packing next to pattern, the squeeze method of moulding is used in combination with the jolt method.

5.2.2 Jolt Moulding

In jolt moulding sand is first filled into the flask generally from an overhead hopper and compacting of sand can be attained by jolting the flask with pattern, fixed on the jolting table as shown in Figure 5.9. Using compressed air, the table is lifted from 30 mm to 80 mm. When the piston moves up, the discharge hole is opened and the air escapes dropping the table freely. Compaction takes place due to the gravity force of the flask sand. About 30 to 50 strokes per minute is the frequency of jolting. In jolt moulding, the bottom layers immediately above the pattern surface are well compacted, but the top of the mould remains practically uncompacted.

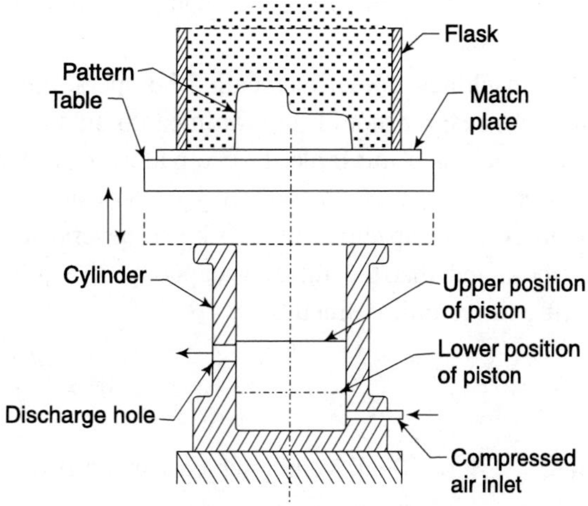

Figure 5.9 Jolt moulding machine

In jolting, packing of the moulding sand is caused by work done by the kinetic energy of the falling sand. The power of jolting results from the conversion of the momentum of the falling sand to work when the sand is instantaneously halted by the jolt table.

$$\text{Jolting power} = \frac{MV}{A}$$

where, M = Sand weight

V = Velocity at instant of jolt = $\sqrt{2gh}$

A = Jolted area

h = Jolt stroke

For a sand weighing 1281 kg/cubic cc. and jolt stroke (h) = 0.75 m

$$\text{Jolting power} = \frac{MV}{A} = \frac{1281 \text{ kg/cubic cc.} \times 3.84 \text{ m}}{1 \text{ cm}^2} = 4913.9 \text{ kg} - \text{m/sec}$$

An important characteristic is evident from the formula. Power for moulding is independent of the flask area and determined mainly by the jolt stroke, a machine characteristic. Of course, the number of times that jolting is done will have a great effect on the degree of sand packing. After about 20 jolts, further jolting causes an asymptotic approach to maximum hardness and density, so that 20 or less is a commonly used number.

In this type of sand packing, the maximum moulding force is applied at the pattern surface. The mould thus is hardest at the pattern surface and parting. Again, if sand were a perfect fluid, it would flow uniformly against the pattern surface under jolting action. Because of bridging and keying, however, non-uniform flow occurs, especially as the vertical depth of the casting increases in deep pockets. Because the sand away from the pattern surface packs less at the pattern surface, it is necessary to finish off the top or back of the mould by squeezing or ramming with a pneumatic rammer. Only the first 5 to 25 mm of sand above the pattern plate is well rammed.

Jolt machines, sometimes called bumpers, are useful in handling many sizes of flasks, especially when larger moulds are rammed up on a moulding floor. The machine shown has a squeeze head as well as a jolt table. Loose pattern or separate copes and drags may be moulded. Heavy flasks are drawn from the pattern by overhead crane. Large core boxes may also be used on jolt machines. Jolt capacities of 250 kg to several tonnes are available in commercial machines. It may be noted that the limiting capacity of a jolt machine is that total weight which it is able to lift and let fall, defined as

$$W_j = \frac{\pi d_j^2}{4} \times p$$

where, W_j = Total weight which can be lifted by jolt cylinder

= Flask weight + pattern weight + jolt-table weight + sand weight

d_j = Diameter of jolt cylinder

p = Air-line pressure

Jolt Squeeze Moulding

Jolt squeeze machines utilize a combination of jolting and squeezing to pack the moulding sand. A combination of the beneficial compaction effects of squeezing and jolting on sand density is then obtained. The drawback of jolt moulding is less compaction of the top layer of the sand in a flask. This can be compensated by squeezing the top layer by a flat head after jolting. Compressed air is used to move the squeeze head downwards. Since the combination of jolt squeeze method results in good and almost even compaction of sand, jolt squeeze moulding machines are commonly used in mechanized and semi-mechanized foundries.

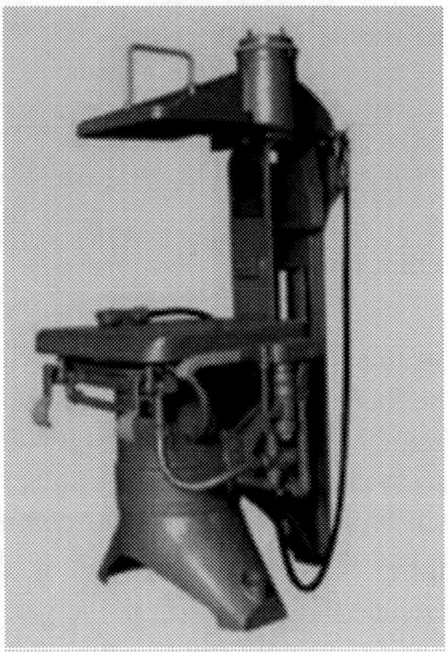

Figure 5.10 Jolt squeeze moulding machine

5.2.3 Sand Slinging

Sand slingers are moulding machines which compact the sand moving at high velocity by impacting the pattern.Sand slinging is done by throwing sand into the flask rapidly with great force. This process develops uniformly high mould hardness. The process is very fast and gives high uniform and ramming. The sand should be uniformly sprayed into the mould by moving the nozzle through the entire area of the flask. Also, it is necessary that the sand be sprayed in layers to achieve uniform hardness. However, the initial cost of the equipment is high compared to other moulding machines described above.

Slingers of several different models are used, including:

1. Stationary slinger
2. Tractor sand slinger
3. Motive slinger
4. Mould blowers

Construction and working: The main part of the sand slinger is a sand shooting head consisting of a rotor and its housing as shown in Figure 15.11. The rotor (impeller) with one or two blades rotates at a speed up to 1500 rpm. A slow speed, 1200 rpm, is used to avoid pattern damage in placing the first sand on the pattern, whereas the faster speed is used for the backup sand. The sand mixture is delivered to this shooter by a conveyor and is thrown to the mould by the rotor blade as a jet. The sand is compacted around the pattern by the kinetic energy and gravity force of the sand jet. Because of the shooting nature of filling the sand, these machines are called sand shooters.

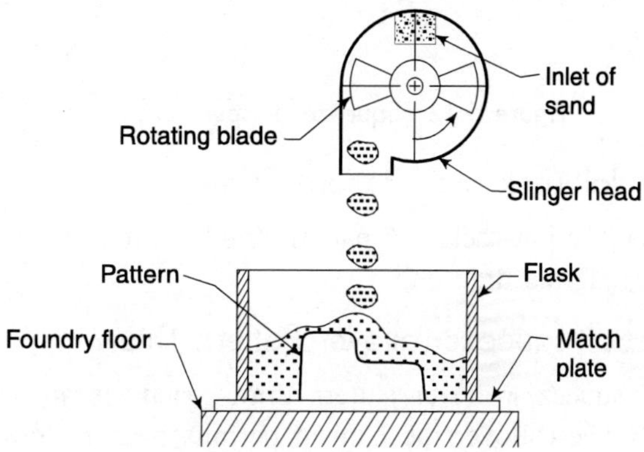

Figure 5.11 Sand slinger head

A sand slinger machine consists of a heavy base, a bin or hooper to carry sand, a bucket elevator to which a number of buckets are attached and a swinging arm which carries a belt conveyor and the sand impeller head. Properly conditioned sand is filled in that bin, which is fed through its bottom to the elevator buckets. These buckets discharge the sand to the belt conveyor which conveys the same to the impeller head. This head can be positioned anywhere on the mould by swinging the arm. The force of sand ejection and striking into the flask is so great that the sand gets packed in the flask uniformly. Sand slingers are generally used for flask moulding of heavy castings in large quantities.

5.2.4 Pattern Extraction or Pattern Stripping Methods

Mechanical pattern stripping is performed by pushing or lifting the flask away from the pattern table, using lift pins. The pattern or its mounting table is vibrated before and during the stripping operation. A slow withdrawal of the pattern during the moments when it leaves the sand is desirable. These steps are incorporated in pin lifts, or lifting bar strippers, which push on the flask, separating the pattern from the mould. Generally, on moulding machines, the mould is lowered away from the mould while the pins or bars hold the flask up.

Slingers of several different models are used, including:

(a) Car-type Ink-squeeze Strippers

- Large size jolt squeeze stripping machines are often used in high production work for medium-sized castings.

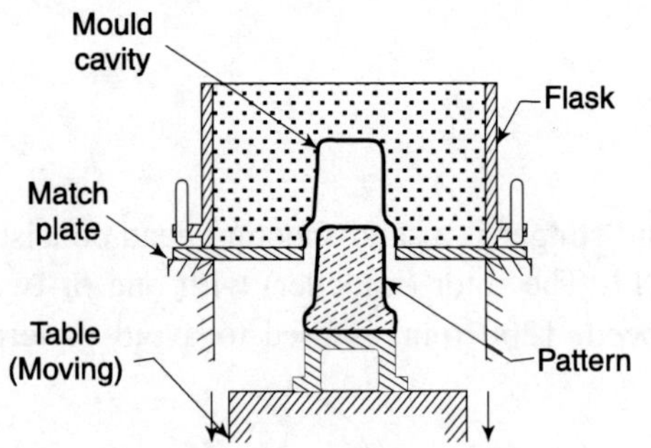

Figure 5.12 Squeeze stripper

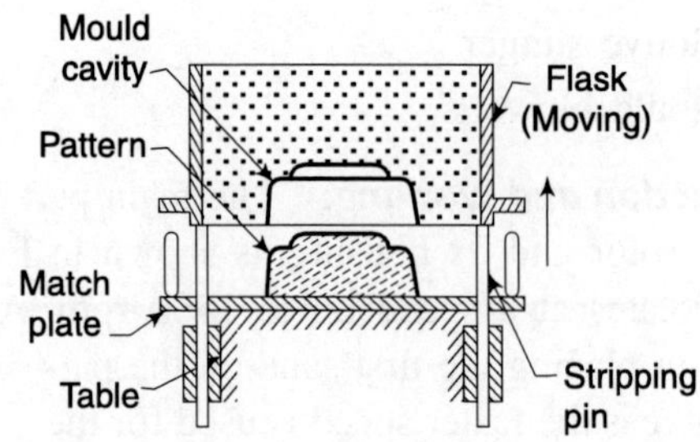

Figure 5.13 Pattern draw

(b) Jolt-rock over Pattern Draw

- The jolt-rock over pattern draw is a machine used for separate cope and drag work, moulding the drag only.

(c) Jolt-squeeze-rollover Pattern Draw

Jolt-squeeze-rollover pattern draw machines are used for the drag in cope and drag moulding. The rollover mechanism permits pattern drawing by lowering the mould away from the pattern.

In machine moulding, the pattern is drawn or stripped from the half mould by various methods. One method is shown in Figure 5.14, where the rammed half mould is raised by stripping pins while the pattern with pattern plate remains on the table. The drawings of pattern through stripping plate are schematically shown. Pattern with pattern plate is lowered, while the moulded flask and the stripping plate remains stationary. This method finds application in stripping patterns with more height.

The pattern drawing principle incorporated in a rollover moulding machine is illustrated in the figure. The mould flask together with pattern and worktable is rotated through 180 degrees about their approximate centre of gravity and then pins lift the table together with pattern out of the mould. The following figure shows the operation of the rock-over pattern draw. The mould flask together with pattern and table is swung over by the arm into the drawing table which is then lowered with the mould away from the pattern.

Closing of the mould

The mould half (drag), after providing metal admitting passages (gating system), finishing and core setting, is closed with the cover (cope). Since some sand falls into the mould during core setting, it may be blown out with a compressed air jet before closing the mould. After closing the mould with cope, the alignment of the two halves is to be checked to get defect-free casting.

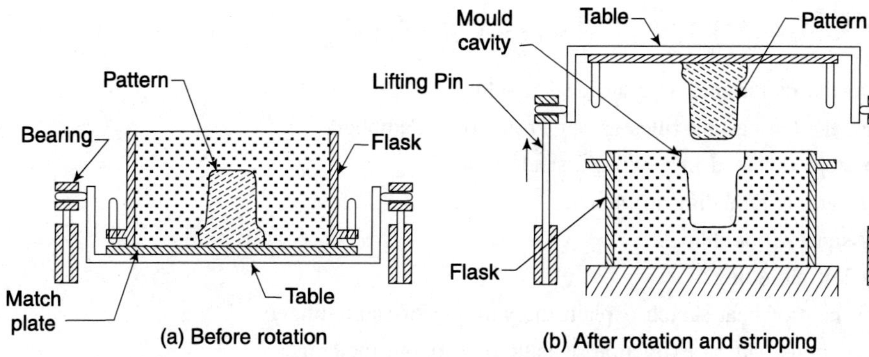

Figure 5.14 Working principle of rollover moulding machine

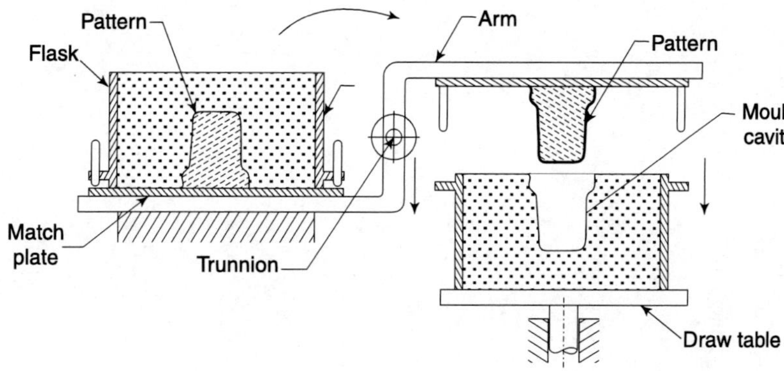

Figure 5.15 Working principle of rock-over moulding machine

The upper box (cope) is prevented from lifting due to pressure of molten metal by clamping it to the lower box (drag). It can be secured in position by using cramps, bolts, wedges or by placing weights as shown in Figure 5.16. The weight or strength of clamping device should be sufficient to withstand the force exerted by the molten metal on the cope during pouring. Now, the mould is ready to fill the molten metal.

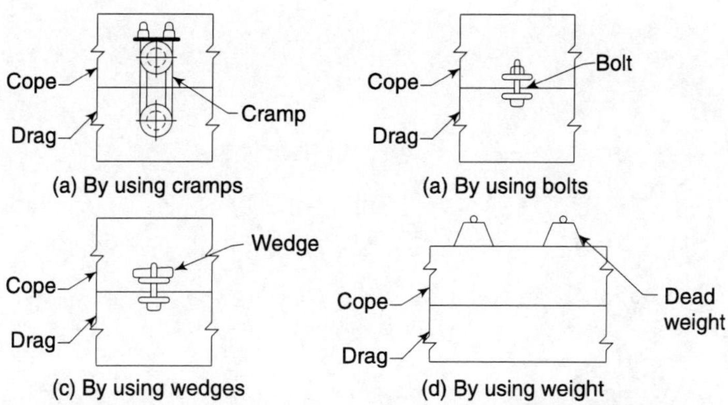

Figure 5.16 Methods of clamping the mould halves

Review Questions

1. What are the requirements of an ideal mould?
2. What are the ten casting rules as per Professor J Campbell?
3. Explain steps involved in making a sand mould.
4. Classify various moulding machines.
5. What is squeeze moulding?
6. What is jolt moulding?
7. With the help of neat sketch explain the working of sand slingers.
8. Explain various pattern extraction or pattern stripping methods.
9. What are the different methods to clamp mould halves?

CHAPTER 6

Cores and Core Making

6.1 INTRODUCTION

A core may be defined as a sand shape, which makes the contour of a casting for which no provision has been made in the pattern for moulding. Cores are made up of sand, metal, plaster or ceramics. Cores are sand shapes which form the inside contour of a casting that is not moulded with the pattern. Core is an obstruction which when positioned in the mould, naturally does not permit the molten metal to fill up the space occupied by the core.

Fitting cores made of suitable materials can form complex internal cavities formed in castings without difficulty. Since cores are usually used to make hollow castings, they are subjected to severe conditions during casting like high temperatures, heavy lifting forces, slim cross section, minimum end supports, etc. Hence, critical design, selection of suitable core materials, and careful making are requisites to the process. Cores are prepared in a separate core making area of a foundry shop.

Essential characteristics of cores:

- Sufficient strength to support itself and to get handled without breaking.
- High permeability to let the mould gases escape through the mould walls.
- Smooth surface to ensure a smooth casting.
- High refractoriness to withstand the action of hot molten metal (metal penetration).
- High collapsibility in order to assess the free contractor of the solidifying metal.
- Ingredients which do not generate mould gases.

6.2 FUNCTIONS OF CORE

The basic function of a sand core is to make internal cavity in a casting. After filling molten metal and solidification of casting, the core material has to be removed from the cavity during fettling and cleaning. This becomes easy, if the core is fully collapsible during or immediately after cooling.

(i) Primary functions

- To provide internal cavity in a casting, so that the component becomes hollow.
- To make a through hole or blind hole in a casting.

(ii) Secondary functions

- It can be used as a part of greensand mould. External surfaces of the component cast are also formed using cores.
- Sometimes, cores are also used to shape those parts of a casting that are very difficult or very costly to produce directly from the pattern. For example: overhangs or backdrafts where it becomes difficult to remove the pattern from the mould.
- Cores may be used to strengthen the mould.
- Cores may be employed to improve the mould surface.
- To make a part of metal feeding system such as pouring cups, strainer in the metal flow line, insert to strengthen the runner, etc.
- To fit as ram-up cores which are coming in the mould to locate other cores, to support chaplets, to hold chills.
- Cores are required to create the recesses, undercuts and interior cavities that are often a part of castings.
- Cores are employed as inserts in moulds to form design features that are otherwise extremely difficult to produce by simple moulding.

6.3 CLASSIFICATION OF CORES

1. Based on the state or condition of core
   - Greensand core
   - Dry sand core
2. Based on the nature of core materials employed
   - Oil bonded cores
   - Resin bonded cores
   - Shell cores
   - Sodium silicate cores
3. Based on the type of core manufacturing process
   - CO_2 process
   - Hot-box process
   - Cold set process
   - Fluid or castable sand process
   - Nishiyama process
   - Furan no-bake system
   - Oil no-bake system
4. Based on the shape and position in the mould:
   - Horizontal core
   - Vertical core

- Hanging core and cover core
- Balanced core
- Wing core or stop-off core
- Ram-up core
- Kiss core

5. Based on the materials used for making cores
   - Sand cores
   - Metal cores.

1. Greensand core

A greensand core is made out of the same sand from which the rest of mould has been made.

2. Dry sand cores

Dry sand cores, unlike greensand cores are not produced as a part of the sand. Dry sand cores are made separately and are independent of that mould. Dry sand core is made up of core sand, which differs very much from the sand out of which the mould is constructed. Dry sand core is made in a core box and it is baked after ramming. Dry sand core is positioned in the mould on core seats formed by core print on the pattern. A dry sand core is inserted in the mould before closing the same.

During casting, cores are surrounded by molten metal. Hence, it should resist erosion, breakage, thermal shock, metal penetration as well as retain its dimensional location for producing good castings. In addition, the removal of cores from casting should be easier. Therefore, the following properties are expected from the mixture used to make core:

- Sufficient strength during core making.
- Enough strength, hardness, etc., during baking of core.
- Dimensional accuracy and strength during storing, handling and assembling.
- Thermal resistance to molten metal and capacity to withstand metallostatic head.
- Ability to pass off gases during pouring and solidification.
- Good collapsibility after casting and cooling so as to remove cores easily.
- High refractoriness, as the cores are subjected to high temperature.

3. Oil bonded cores

Conventional sand cores are produced by mixing silica sand with a small percentage of linseed sand.

4. Resin bonded cores

Phenol resin bonded sand is rammed in a core box. The core is removed from the core box and baked in a core oven at 200 to 250°C to harden the core.

5. Shell cores

Shell cores can be made manually or on machines. The procedure of making shell cores is as follows:

1. The core box is heated to temperature of the order of 200 to 300°C.
2. Sand mixed with about 2 to 5% thermosetting resin of phenolic type is either dumped or blown into the preheated metal core box.
3. Where sand blowing is employed, it is preferred to use resin pre-coated sand to avoid resin segregation.
4. The resin is allowed to melt to the specified thickness.
5. The resin gets cured.
6. Excess sand is dumped and removed.
7. The hardened core is extracted from the core box.
8. Cores thus produced need no further baking.

Advantages of shell cores

- Shell core possesses very smooth surface (3125 micro mm root mean square) and close tolerance (± 0.003 mm/mm).
- Shell core making process can be mechanized and several core making machines are commercially available.
- High permeability is achieved in shell core making.
- Shell cores can be easily stored for future use.
- Shell cores are costly as compared to cores produced by other methods.

6. Sodium silicate – CO_2 cores

This is the most commonly used method for making cores or moulds. Schematic diagram is shown in Figure 6.1. This is a cheap and fast process, therefore it is used extensively.

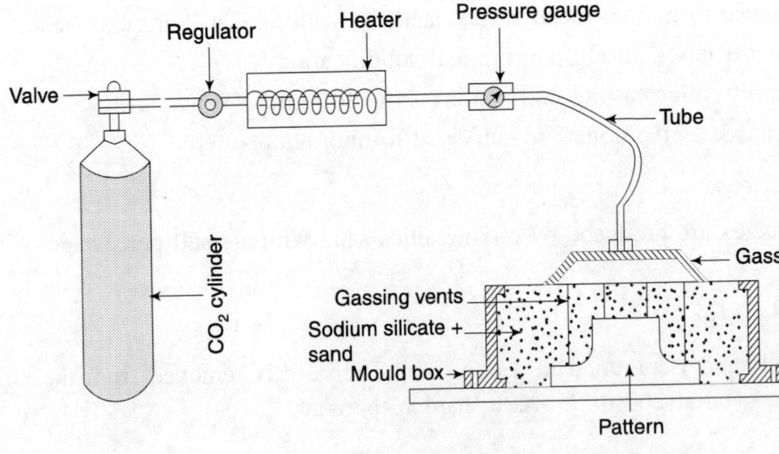

Figure 6.1 Sodium silicate process

This is essentially a quick process of core or mould preparation. The mould is prepared with a mixture of sodium silicate and sand, then treated with carbon dioxide for two to three minutes such that a dry compressive strength of over 1.4 MPa is arrived at. The carbon dioxide is expected to form a weak acid, which hydrolyses the sodium silicate resulting in amorphous silica, which forms the bond. The introduction of carbon dioxide gas starts the reaction by the binder until it becomes solid. The compressive strength of the bond increases with standing time due to dehydration.

Advantages of the process

1. Because of high strength of the bond, core need not be provided with reinforcement.
2. The equipment required is fairly simple because the main requirement is carbon dioxide which is available.
3. No expensive core baking is needed.
4. It does not involve any distortion due to baking and better dimensional accuracies are achieved.

Disadvantages of the process

1. It is an expensive process.
2. Sand mixture does not have good shelf life, therefore it should be used immediately after preparation.
3. The shake-out properties of carbon dioxide moulding are also poor compared to the normal moulding.
4. Moulds and cores get deteriorated from water pick-up if they are kept stored for longer periods before use.
5. They develop non-uniform strength due to the difficulty in achieving uniform gassing.
6. The core sand should be free from impurities; otherwise, the binder consumption would increase.

7. Cold set process

While mixing the core sand an accelerator to the binders, the sand mixture is very flowable and is easily rammed. Curing begins immediately with the addition of accelerators and continues until the core is strong to be removed from the core box. A little heating of the cores hardens it completely. Cold set process is preferred for jobbing production and is employed for making large cores.

8. Castable sand process

A setting or hardening agent such as dicalcium silicate is added to sodium silicate at the time of core sand mixing. The sand mixture possesses high flowability and after being poured in the core box, it chemically hardens after a short interval of time. As compared to the CO_2 process, where it may not be possible to gas the full core uniformly and to obtain uniformly hardened cores, castable sand process produces much better and uniform results. This process is best suited for large jobbing work.

9. Nishiyama process

Nishiyama process uses sodium silicate bonded sand, which is mixed with 2% finely powered ferro silicon. Hardening occurs because of exothermic reaction of silicon with NaOH produced by hydrolysis in the solution of sodium silicate. Cores thus made possess short bench life.

10. Furan no-bake system

The core sand mixture contains washed sand and dried sand with clay content less than 0.5%, furan no-bake resin 2% and activator (phosphoric acid) 40%. The basic reaction between the furan resin and phosphoric acid results in acid dehydration of the resin. The core sand mixture has high flowability and needs reduced rodding (to handle the core). Uniform core hardness, exact core dimension, better fitting cores, lower machining and layout costs, and reduction of oven baking are some of the good characteristics of cores made by this system.

11. Oil no-bake process

The process employs a synthetic oil binder which when mixed with basic sands activated chemically produces cores that can be cured at room temperature. The sand may consist of: Washed and dried sand (500 kg), oil no-bake binders and catalyst (7 kg), oil no-bake crosslinking agent (1.4 kg). In oil no-bake process, the polymerization reaction results in a complete and uniform setting of the complete core sand mass. This process assures better depth of set, fast baking, easier core withdrawal and lower production cost as compared to furan or oil bonding process.

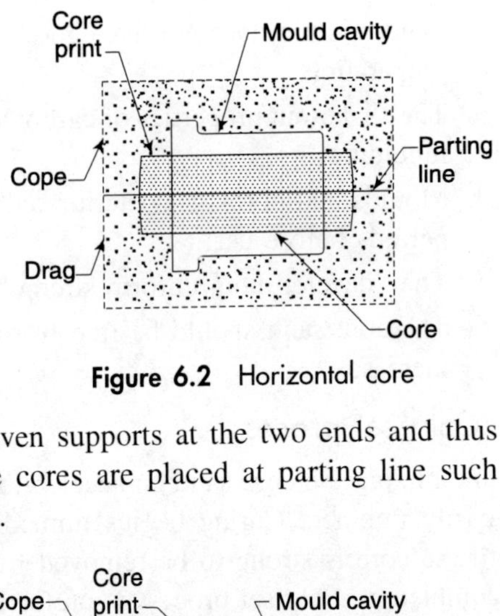

Figure 6.2 Horizontal core

12. Horizontal core

This is the most common type of cores used. It is mostly in cylindrical form and is placed in the mould horizontally as shown in Figure 6.2. These cores are given supports at the two ends and thus they rest on core prints provided by the pattern. These cores are placed at parting line such that one-half lies in the cope and other half in the drag.

13. Vertical core

These cores are similar in shapes to that of the horizontal core but they are placed in a vertical position as shown in Figure 6.3. They are placed both in cope and in drag. Usually, taper is provided on the top as well as at the bottom of cores. Taper on the top is more than the taper at the bottom.

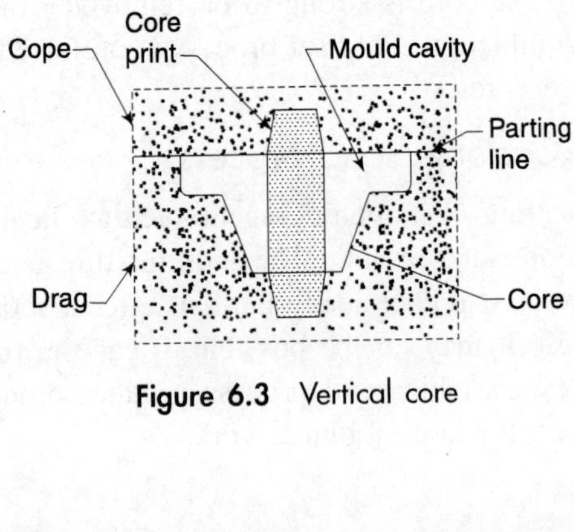

Figure 6.3 Vertical core

14. Hanging core and cover cores

These cores are used when a cored casting is to be completely moulded in the drag by ramming the entire pattern in the drag and the core is required to be suspended from the cope. When the core, while hanging from the cope, does not have any support at the bottom in the drag, it is called a hanging core. But if it has its support in the drag, it is called cover core as shown in Figure 6.4. Usually, these cores are reinforced by wire, etc., for increasing the strength.

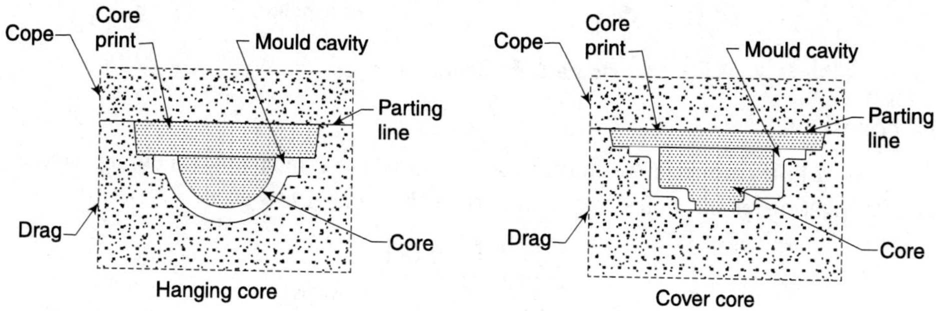

Figure 6.4 Hanging and cover cores

15. Balanced cores

These cores are used when a blind hole along a horizontal axis is required, as shown in Figure 6.5. These cores are supported at only one end and as such, the core prints are made sufficiently long enough to give the proper bearing surfaces. In case the cores are sufficiently long, they may be supported at the other end by means of a chaplet. These chaplets are made of the same material of which the casting is required.

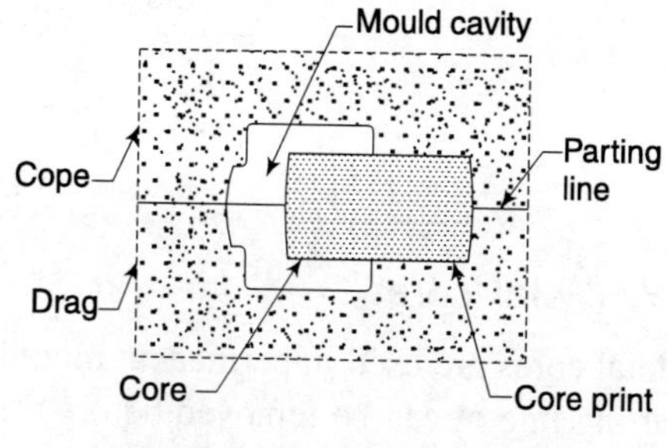

Figure 6.5 Balanced core

16. Wing core or stop off cores

These cores are used when the axis of the desired hole does not coincide with the parting line of the mould, i.e., the core is either required to be placed above or below the parting line as shown in Figure 6.6. The side of the core print is given sufficient amount of taper so that the core can be placed readily in the mould.

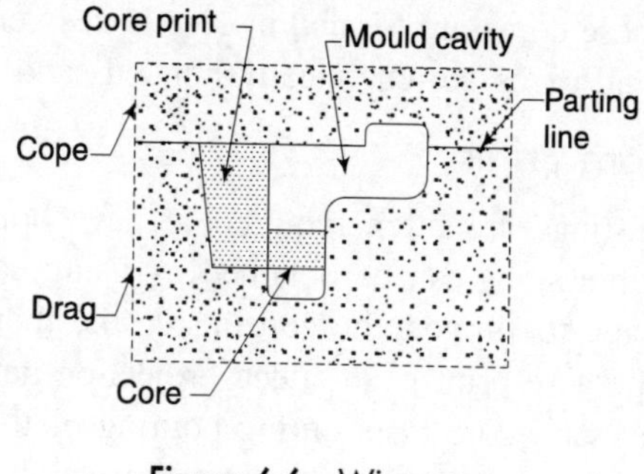

Figure 6.6 Wing core

17. Ram-up cores

These types of cores are set with the pattern in the mould before the mould is rammed. It is used when the cores are located in an inaccessible position or location. It may be used for both interior and exterior portions of a casting.

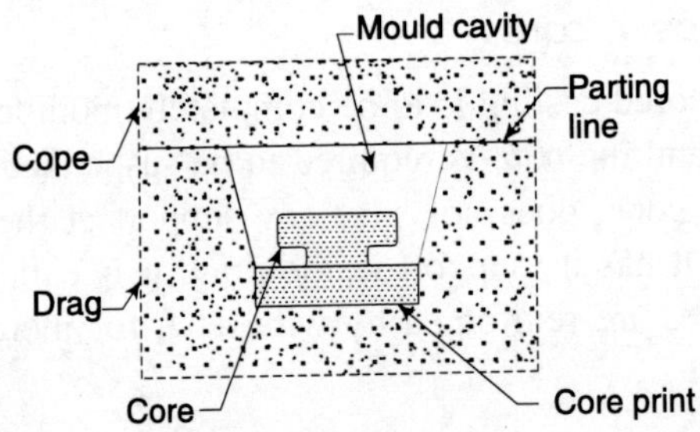

Figure 6.7 Ram-up core

18. Kiss cores

When a number of holes are required in the casting and where dimensional accuracy with regard to the relative location of the holes is not important, kiss cores are used.

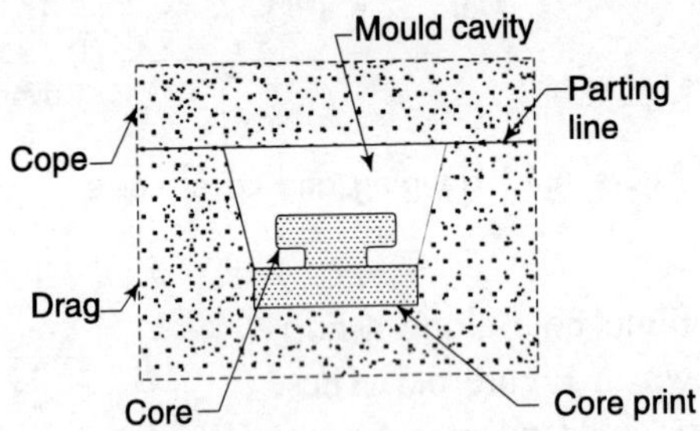

Figure 6.8 Kiss core

19. Metal cores

Metal cores are used in permanent mould casting processes. They should be parallel to the mould parting line or can be removed before the casting is removed from the mould, and shaped so that they are readily freed from the casting. Metal cores are usually made of cast iron or steel.

20. Sand cores

Sand cores are made from materials similar to those used for chemically bonded sand moulds. These cores are formed in core boxes very much similar to patterns boxes used to make mould. Sand cores are chemically bonded sand of complex shapes, and used in all mould types.

Core Prints

Castings are often required to have holes, recesses, etc., of various sizes and shapes. These impressions can be obtained by using cores. So where coring is required, provision should be made to support the core inside the mould cavity. Core prints are used to serve this purpose. The core print is an added projection on the pattern and it forms a seat in the mould on which the sand core rests during pouring of the mould. The core print must be of adequate size and shape so that it can support the weight of the core during the casting operation. Depending upon the requirement a core can be placed horizontally, vertically and can be hanged inside the mould cavity. A typical job, its pattern and the mould cavity with core and core print is shown in Figure 6.9.

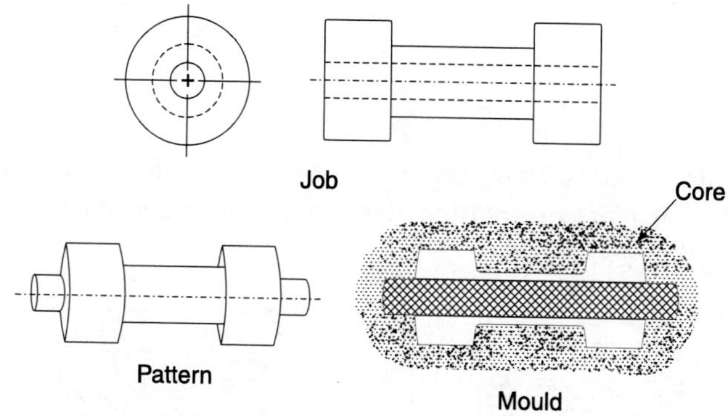

Figure 6.9 A typical job, its pattern and the mould cavity

6.4 TYPES OF CORE PRINTS

The core coming inside a mould cavity has to be supported by the mould wall. An impression in the form of a recess is made in the mould cavity using a projection suitably added on the pattern. This projection on the pattern is referred to as a core print. Thus, a core print is an added projection on the pattern, which forms a seat in the mould for supporting the core extension. The size and shape of the core print should be adequate to support the weight of the core and the lifting forces exerted during the metal filling operation. The common types of core prints can be classified as:

1. Horizontal or parting line core print
2. Vertical or cope and drag core print
3. Hanging or cover core print
4. Balancing core print
5. Wing, stop-off or drop core print
6. Ram-up core print

Corebox

A corebox is basically a pattern for making cores. Coreboxes are employed for ramming cores in them. Coreboxes impart the desired shape to the core sand. Coreboxes range from simple wooden structures to precision metal assemblies, which possess long life under extracting conditions.

Types of coreboxes

(a) Half corebox

(b) Slab or dump corebox

(c) Split corebox
(d) Left- and right-hand corebox
(e) Strickle corebox
(f) Loose piece corebox
(g) Gang corebox

Half corebox

It is used to make cylindrical cores. Half portion of the core is made in the corebox, at one time. Later on, after a number of half core portions have been made and baked, they are cemented together to make full cylindrical cores.

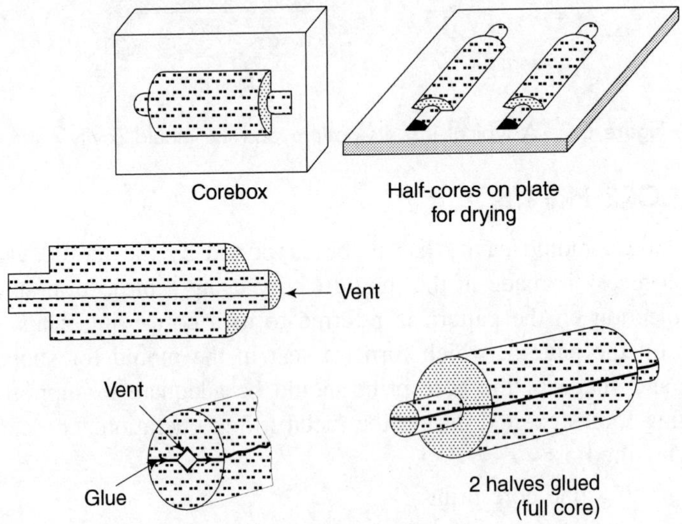

Figure 6.10 Half corebox

Slab or dump corebox

It resembles half corebox as regards to its construction. It is used to make full core at a time. It is used to make rectangular, square, trapezoidal cores. Figure 6.11 show a dump corebox.

Split corebox

Unlike dump and half coreboxes, a split box is in two or three parts. The two portions of the split corebox can be aligned and temporarily joined together by pins of dowels. When joined, the two halves of split corebox depict the full core cavity. For making the core, first of all the two portions of the split corebox are temporarily joined and then sand is rammed. Rammed core can be removed by splitting the corebox again into two or three pieces. Figure 6.12 show a split corebox.

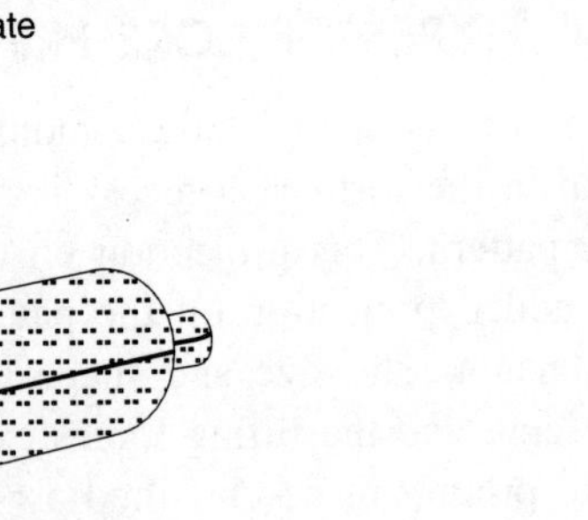

Figure 6.11 Dump corebox

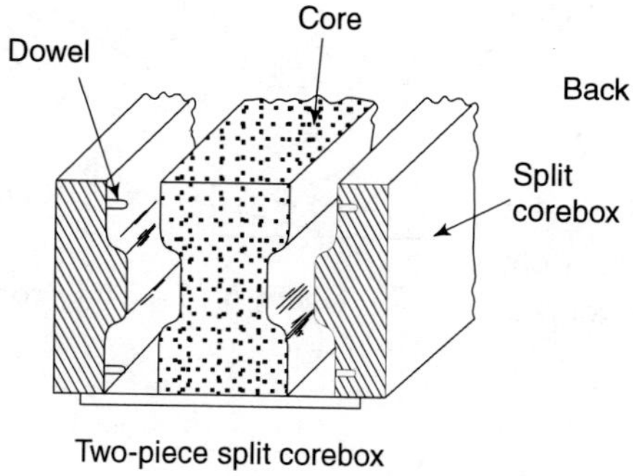

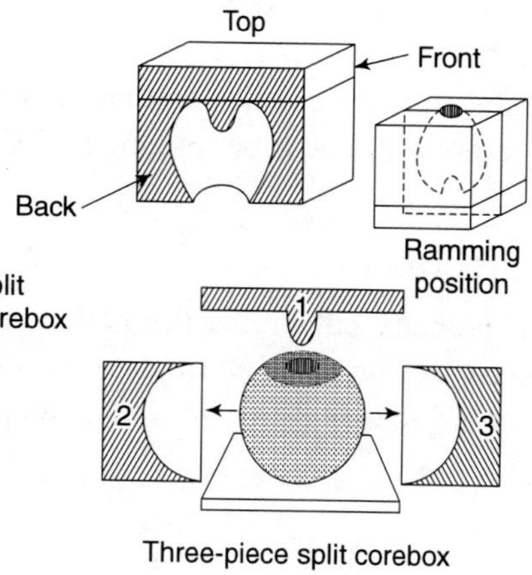

Figure 6.12 Split corebox

Left- and right-hand coreboxes

They are used to make cores for producing pipe bends. Half of the pipe bend core is made in each corebox. Two halves of a pipe bend thus rammed are baked and then cemented together to make the full core. Figure 6.13 shows left- and right-hand coreboxes.

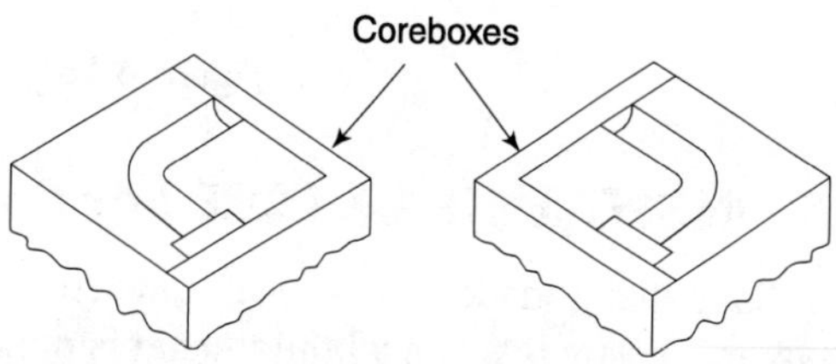

Figure 6.13 Left- and right-hand coreboxes

Strickle coreboxes

Sand is rammed in a dumping corebox. The top portion of the core in the corebox is given a desired shape or contour with the help of a strickle board cut and finished to the desired shape. A strickle board strikes off excess sand not confirming to its shape. A strickle board may be made up of wood. A regular, irregular or unsymmetrical shape which otherwise is difficult to produce on the core, but can be cut on the wooden strickle board can be given to the core. The strickle board method of producing irregular core surfaces costs less as compared to other methods.

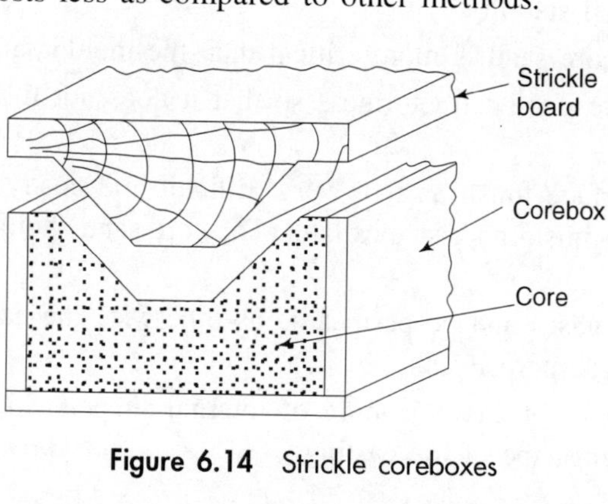

Figure 6.14 Strickle coreboxes

Gang coreboxes

It contains a number of core cavities so that more than one core can be rammed at one time. Figure 6.15 shows gang corebox.

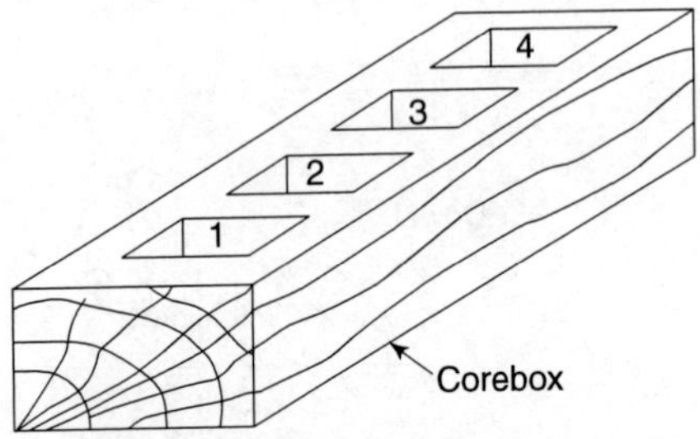

Figure 6.15 Gang coreboxes

Loose piece coreboxes

It resembles half corebox but unlike the half corebox, loose piece corebox can produce two halves of a core, which may be neither identical in size nor in shape.

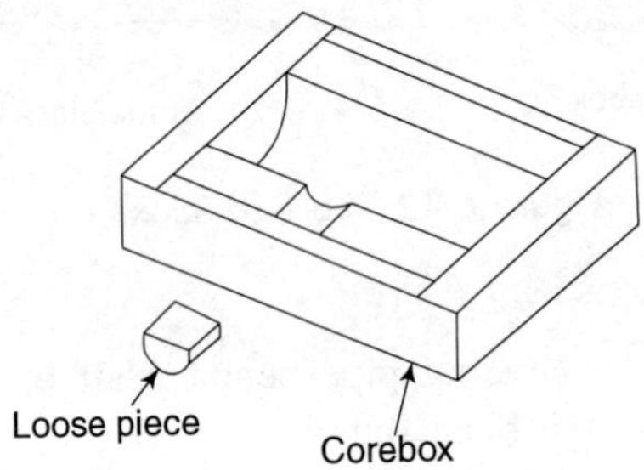

Figure 6.16 Loose piece corebox

6.5 INGREDIENTS OF CORE SANDS

Core is a shape made in core sand placed in a mould to produce an accurate hollow shape or cavity in a casting. When a liquid metal is poured into the mould, it occupies all the empty spaces restricted on the outside by the mould and internally by the core, finally forming a casting.

Sand

To obtain the desirable properties of cores, sand is used as the base material. Silica sand (SiO_2) is generally preferred for nonferrous and grey iron castings. For steel and similar high melting point metals, special sands like Zircon, Olvine, Chamotte, etc., are selected in order to get better refractoriness and thermal stability.

The requirement of the core sand is more critical than the mould sand:

1. Core sand should have good refractoriness so that it does not fuse when surrounded by the liquid metal.
2. Should give good surface finish as it is very difficult and costly in giving complicated casting to machine or finish inner shapes later. The core sand should be of proper fineness accordingly.
3. It should evolve less gases and be permeable as otherwise the casting will have defects like blow holes and gas porosity.
4. It should be collapsible for easy cleaning of internal shape, which is often much more difficult than the outer shape of the casting.

5. It should retain its dimensional accuracy and shape during pouring and casting solidification, possessing proper hot and dry strengths.

Normally, cores are less in volume and more critical than moulds. They are costilier and better sand mixtures are used for core sands compared to moulding sands. The properties tested for core sands are fineness of base sand, green strength, baked strength, permeability, scratch hardness and collapsibility.

Commonly used core sand mixtures are:

1. **Clay bonded sands:** For cheap and simple castings, internal shapes are formed using normal greensands with bentonite (4-5%), dextrin (1-2%) and water (3-4%). For heavier casting, the cores can be made using fire clays (4%) in addition and drying cores in ovens (up to 150°C) to dry away free moisture. The dry sand cores are strong and have less gases compared to green sand.
2. **Oil sands:** Various vegetable oils like linseed oil (up to 15%) can be used to bind silica grain without using clay or water. These do not possess much green strength but develop very good strength over clay bonded sands. Due to excessive cost of oils and requirement of fuel fired ovens and pollution, oil sands are replaced by other sands now.
3. **Resin sands:** Many organic and inorganic resins (urea formaldehyde, phenol formaldehyde) can give very good bonds either in green condition through additional activators and chemicals or by baking to about 220°C. Resin sands give adequate green strength for dimensional stability in green condition, high surface finish and excellent collapsibility. Air-set systems are very popular. Hot box or shell moulding systems sand give highly accurate dimensions as curing takes place within core boxes, before removing from coreboxes.
4. **Sodium silicate sands:** Silica sands with 3-4% sodium silicate can be rammed into the coreboxes and then carbon dioxide is passed through to get hard and cheap cores. The problem of collapsibility and surface finish can be reduced by adding suitable additives and core paints. These cores are used for very large cores as they do not need drying or curing.

6.6 BINDERS AND ADDITIVES

The main purpose of the core binder is to hold the sand grains together.

A good binder will have the following properties:

1. Good strength with high refractoriness
2. Collapse rapidly when metal starts to shrink
3. Will not distort core during baking
4. Maintain strength during storage time
5. Withstand normal handling
6. Disperse properly and evenly throughout the sand mix
7. Should produce a mixture that can be easily formed

Binder added to core mixture can be classified as:

- Inorganic binder
- Organic binder
- Sodium silicate binder for CO_2 processed sand cores

There are many types of binders to mix with core sand. A binder should be selected on the basis of the characteristics that are most suitable for particular use. Some binders require no baking becoming firm at room temperature such as rubber cement, Portland cement and sodium silicate or water glass. In large foundry operations and in some small foundries, sodium silicate is a popular binder as it can be hardened almost instantly by blowing carbon dioxide gas through the mixture. The sodium silicate/CO_2 process hardens through the following reaction:

$$Na_2Si_2O_5{\cdot}H_2O_{(l)} + CO_{2(g)} \rightarrow SiO_{2\,(gel)} + Na_2CO_3{\cdot}H_2O_{\,(glass)}$$

The silica gel that is formed binds individual sand grains together. Oil binders require heating or baking before they develop sufficient strength to withstand the molten metal. Sulphite binders also require heating. The most popular of the sulphite binders is a product of the wood pulp industry. There are many liquid binders made from starches, cereals and sugars.

Inorganic binders can keep considerable strength at elevated temperatures. So they are relatively less collapsible than organic type. Fire clay, bentonite, silica flour and iron oxide are the inorganic binders used in core making.

Organic binders are combustible and are destroyed by heat, therefore they are preferred over inorganic binders. Hence, good collapsibility can be expected. The popular organic binders and additives are as follows:

1. **Core oil:** They are manufactured by blending various ingredients such as vegetable oils, animal oils such as fish oil, petroleum oils, etc. Among the vegetable oils, the more important are linseed oil and soy bean oil. In the mineral oils group, only those oils, which polymerize on baking are preferred. Rosin is used to increase the collapsibility as well as baked strength. It is added about 0.5 to 3% by weight to sand.

   Core oil is in liquid state when it is mixed with the sand but forms a coherent solid film holding the sand grains together when it is baked. Although this drying with certain oils occurs at room temperature and this is why cores made with cores oils are usually baked.
2. **Cereal and water-soluble binders:** About 0.5 to 2% of cereal is added to oil-sand mixture to get more green strength. Water-soluble binders like sugar, molasses, dextrin, etc., are added as binders in addition to core oils to get green and dry strengths. They will reduce the baked strength to some extent.
3. **Pitch:** It's a solid, coal-tar by-product, added about 3% to get better baking strength.
4. **Wood flour:** (1% to 3%) To reduce hot strength and to gain collapsibility.
5. **Thermosetting plastics:** Phenol resins and urea resins are used as thermosetting binders. These are gaining popularity nowadays. They give high strength, collapsibility and they evolve minimum amount of gases. The most common binders under this group are phenol formaldehyde and urea formaldehyde.

6. **Sulphite binders:** Sulphites which are by-products of paper making, are water soluble compounds and used as binders. Baking of the mixture will give good baking properties. They are used along with certain amount of clay.
7. **Dextrin:** This is added to increase the baked strength and collapsibility.
8. **Molasses:** Usually, it is used as a secondary binder to increase the hardness on baking. Molasses is used in the form of molasses water and is sprayed on the core before baking.
9. **Protein binders:** They are used to increase collapsibility.

The simplest core sand mixture commonly used comprises sand, 1% bentonite, 1% core oil, 1% cereal and 2.5 to 6% water.

Sodium silicate about 1.5 to 6% is added to sand as binder and the mixture is hardened by passing CO_2 through the sand mixture after making the core. CO_2 processed cores are widely used nowadays for providing intricate internal cavities. Details about the method are discussed in CO_2 moulding process.

The general composition of the core sand mixture could be core oil (1%) and water (2.5 to 6%). The actual composition to be used depends on the size and shape of the core and the alloy being cast. The properties as controlled by the composition are shown in Figure 6.17. These properties are obtained by baking the standard AFS test samples for 1.5 hours at 175°C.

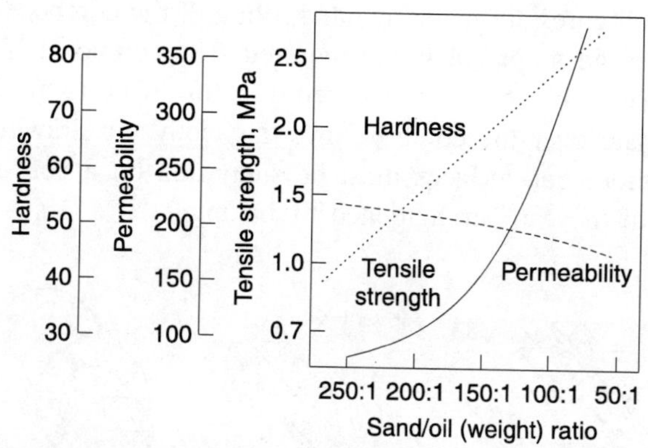

Figure 6.17 Core sand properties for various compositions

6.7 MANUFACTURING OF CORE

Core making basically consists of the following four steps:

1. Core sand preparation
2. Core making
3. Core baking or drying
4. Core finishing and coating

Core sand preparation

Homogeneous and satisfactory mixing of core sand is not possible by hand, and therefore for better and uniform properties, the core sands are generally mixed with the help of any of the following means:

1. Roller miller
2. Core sand mixer
   (i) Vertical revolving arm type
   (ii) Horizontal paddle type

In the case of roller mills, the rolling action of the miller along with the turning over action caused by the ploughs gives a uniform and homogeneous mixing.

Roller mills are suitable for core sands containing cereals binders, whereas the core sand mixer is suitable for all types of core binders. These machines perform the mixing of core sand most thoroughly.

Core making

Core moulding is done manually and with machines. Small cores can be made manually in hand rammed core boxes. Cores on mass scale are rapidly produced on a variety of core making machines.

(i) Manual Process Small cores are made by hand filling of the coreboxes on benches. In this process, mainly a corebox and a core plate are required. The corebox is filled with core sand, rammed, and struck off. Then the core is transferred to a core plate to go for core baking. This is done by placing the plate over the corebox, inverting both and drawing the corebox away from the core. Cores without flat surfaces must be supported on a core drier, which has the shape of the corebox, until they are strengthened by baking.

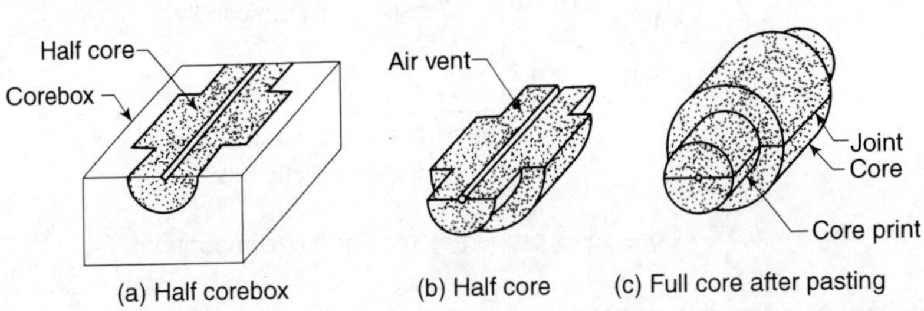

Figure 6.18 Manual process

Bench core making can produce cores of comparatively smaller sizes and in limited numbers. Large size cores, however, have to be handled by two or more men together and with the help of jib or overhead cranes. In the case of large and heavy cores, the moulding may be shifted from the bench to the foundry floor.

(ii) Machine process For large-scale production, core making machines are used. According to the methods employed to compact the core sand, these machines are classified as:

1. Core blowers
2. Squeeze and jolt machines
3. Sand slinger or throwers
4. Core extrusion or die extrusion or stock core machine

Squeeze and jolt machines are rarely used for core moulding. Core blowers and core extrusions are commonly used.

Core blowing

Core blowing is done by filling the corebox cavity with sand, using compressed air. The core sand is suspended in a stream of air and introduced into the cavity at high velocity through the blowing holes. The cavity walls retain the sand and the air is exhausted through various vent holes in the corebox. Core blowing machines are mainly suitable for large quantity parts manufacturing. Since the production rate is high in this machine, the process has become popular in foundries. Figure 6.19 shows the principle of working of core blowing machine.

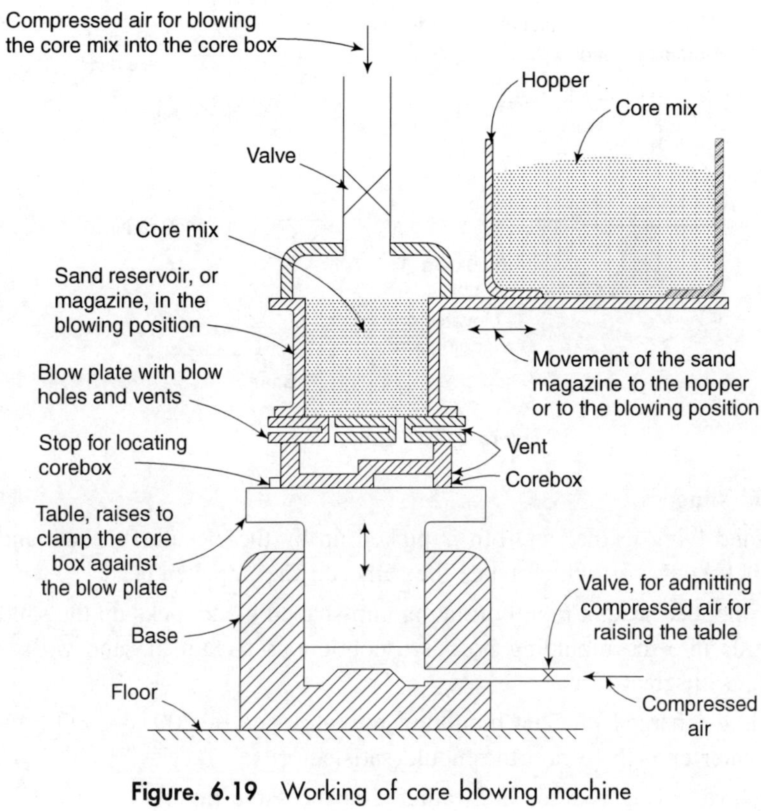

Figure. 6.19 Working of core blowing machine

Operation of core blowing:

1. The sand reservoir is first moved below the hopper, where it is filled with sand.
2. The sand reservoir is then moved to the blowing position. The core box is placed on the table and pressed with the blow plate.
3. The corebox is rapidly filled with sand using a blower at the top with the help of the compressed air at high pressures.
4. The air thus trapped in the corebox is vented out by suitable vents provided in the core box.
5. It is generally understood that because of the blowing operation, larger grains tend to move to the interior of the core and finer grains at the surface, creating a smoother surface.

Sand slinger or throwers

Sand slingers are generally used to make cores of large sizes. It consists of a base, a sand bin, a bucket elevator, a swinging or movable arm, a belt conveyor and the sand impeller.

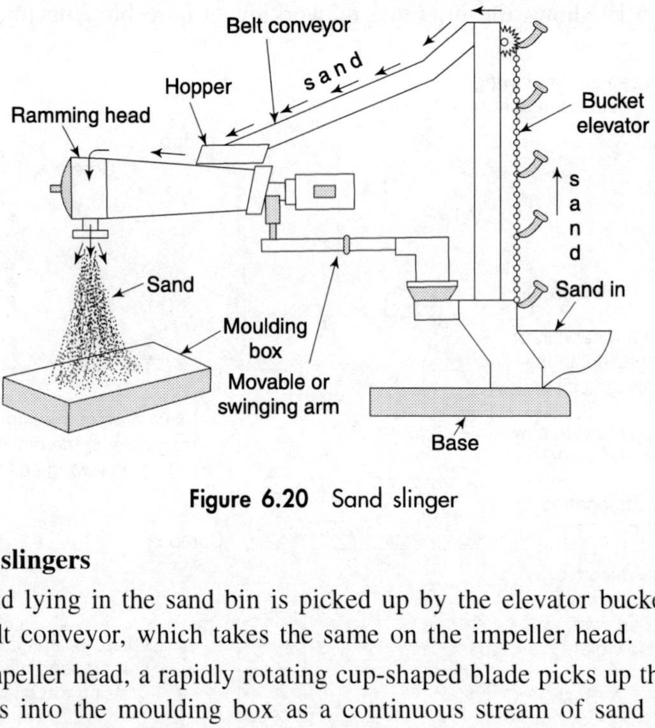

Figure 6.20 Sand slinger

Working of sand slingers

1. Prepared sand lying in the sand bin is picked up by the elevator buckets and is dropped on to the belt conveyor, which takes the same on the impeller head.
2. Inside the impeller head, a rapidly rotating cup-shaped blade picks up the sand and throws it downwards into the moulding box as a continuous stream of sand with machine gun rapidly and with great force.
3. The sand is discharged into the moulding box at a rate of 200 to 2000 kg/minute. This force is greater enough to ram the mould satisfactorily.
4. In the coreboxes, sand is filled and rammed at the same time.

5. The density of the sand that is the result of sand's inertia is uniform throughout the mould.
6. A sand slinger simply packs the sand into the mould and therefore it is often operated together with other systems such as those of rollover and pattern drawing.

Core extrusion machine

Simple cores of regular shapes and uniform cross section can be extruded easily on a core extrusion machine. Cores of square, round, hexagonal and oval sections are produced rapidly on a core extrusion machine. A core extrusion machine has a hopper through which the core sand is fed to a horizontal spiral conveyor. As the spiral conveyor is rotated, it forces core sand through a die of specified shape. Long cores produced can be cut to the desired length.

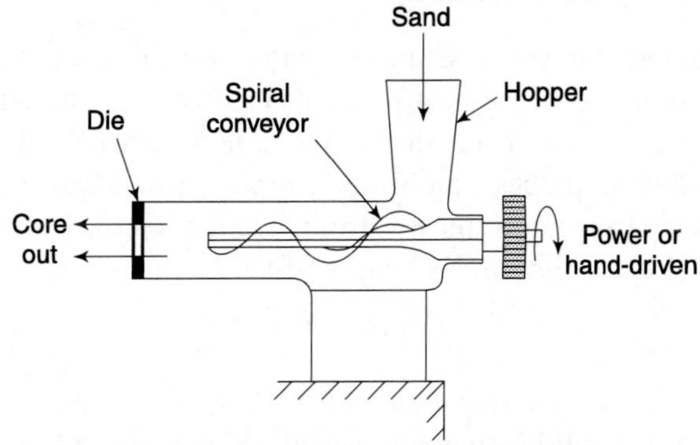

Figure 6.21 Core extrusion machine

Core Venting

Cores surrounded by molten metals are adequately vented to help escape mould gases. Small cores may be vented with pierced holes leading out through core print surfaces. Wax strips may be placed inside the cores to make spaces for mould gases. Large cores can be made hollow and can also be made around pierced metal pipes. If the gas is not vented out then defects like blow holes occur in the casting.

Core Reinforcement

Cores of appreciable length or of fragile design may be reinforced with core wires, rods and arbours. In larger cores, the reinforcement provides an anchoring point for lifting staples used for handling or securing the mould. Reinforcement imparts strength to cores to withstand buoyancy caused by molten metal.

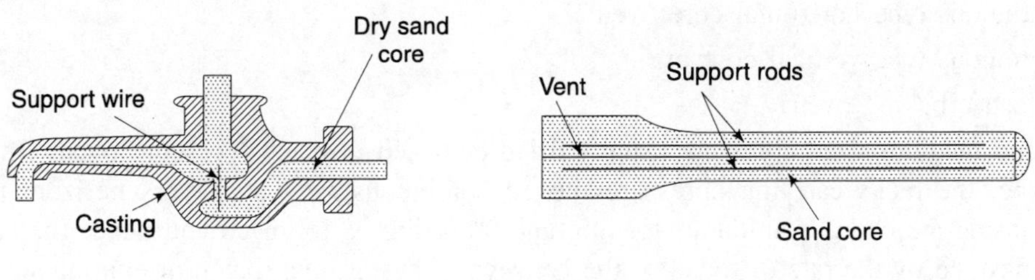

Figure 6.22 Core reinforcement

6.8 CORE DRYING AND BAKING

6.8.1 Core Drying

Drying of cores increases strength and reduces the gas evolution on metal pouring. During drying, the moisture is removed by bringing the temperature of core from 180 to 240°C. The drying process is done inside ovens at a slow rate. Then they are brought to room temperature again slowly in order to avoid cracking and crumbling of surface.

6.8.2 Core Baking

Baking removes moisture and hardens the core binders. Generally, core sand is a poor conductor of heat and hence heat penetrates slowly into the interior sections of the cores. Baking of core is a process of hardening of the core by drying out the binder at a temperature 300° to 350°C. In baking process, moisture is removed first and then during continued heating, the oil binder used changes chemically from liquid to solid by oxygen absorption and polymerization. The total cycle time for baking may be 2 to 6 hours. After baking, a nut-brown colour is obtained for core oil bonded cores. A dark colour is an indication of over-baking, while a light colour shows under-baking.

In a core having thin and thick sections, the thin sections will be over-baked, while thick sections will be optimally baked. Over-baking of cores will result in destroying the binders and hence core will be just a heap of sand. Large cores are baked differently on the surface and in interiors, especially if the oven is too hot. Cores that are not baked fully will create an excess of gas and cause blows in castings.

6.8.3 Core Ovens

Core drying and baking are done inside core ovens. There are many types of core ovens available in the market. The selection depends on the nature, type and size of cores as well as the rate of production required.

Core ovens can be classified as:

1. Batch type core oven
2. Continuous type horizontal core oven
3. Continuous type vertical core oven
4. Dielectric baking oven

Small and medium-sized cores are usually dried or baked in batch type core ovens. In continuous type, the racks carrying cores are moved continuously by conveyors horizontally or vertically inside the oven. Continuous loading and unloading is followed and hence the baking time is controlled by the rate of travel of the conveyor. These ovens (batch or continuous type) are heated by burning gas, oil or coal. They may have separate firing chambers for heating and circulating fans for hot air movement to obtain uniform temperature throughout the oven chamber. Figure 6.23 gives a line sketch of batch type core oven loaded with cores on rack.

Heating is done using gas burners. A blower, fitted on the backside, is used to circulate fresh air inside the oven and to force the moisture to exit along with the heated air through the chimney. The fresh air entry meets the oxygen requirement for gas burning.

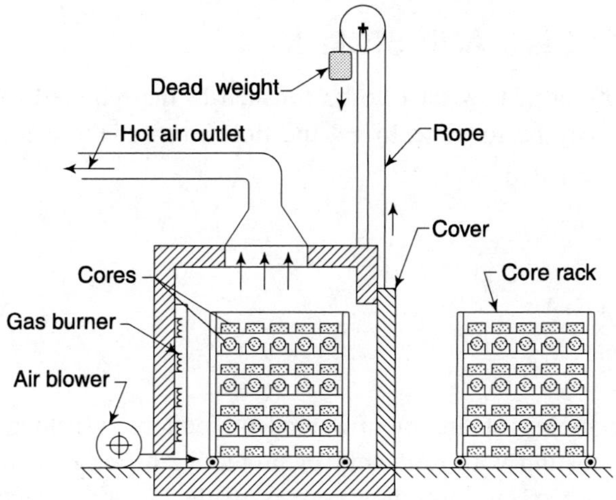

Figure 6.23 Batch-type core oven loaded with core rack

Rapid baking is possible by dielectric heating (Figure 6.24). Induction heating is used for heating materials which are conductors of electricity, like metals, and is done continuously varying magnetic fields. Dielectric heating is done for non-conductors of electricity. In this, alternating electric field is established between the two parallel plates which act as an electric condenser. The material to be heated is placed in between these parallel electrodes. With a high frequency electric current (15 million times/sec) in ON condition, heat is generated into the molecules.

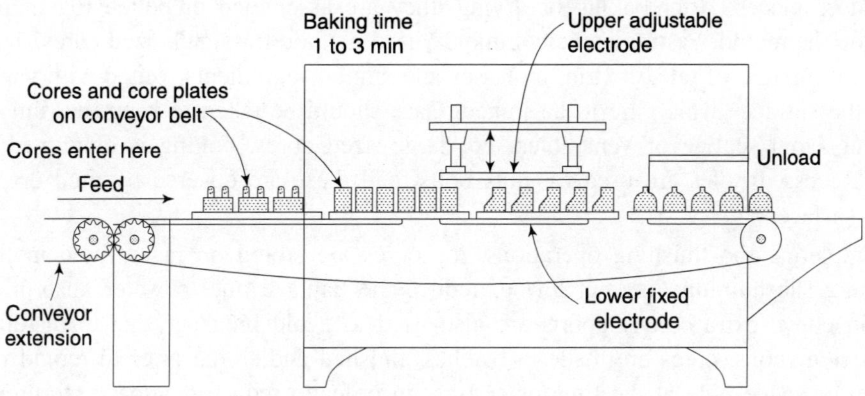

Figure 6.24 Dielectric core baking

In this case, the interior of the cores are heated rapidly as outer surfaces. Thermosetting synthetic resin binders, which cure app. at 250°F and which do not require oxidation are well suited for dielectric heating. Small sized samples can be baked within 30 seconds, while large sections need a few minutes. There is less chance of over-baking or under-baking.

6.9 FINISHING OF CORES AND STORING

After drying or baking of cores, they have to be finished to the required form and size. Sometimes assembly of them may be required to get the final form of the core.

These operations can be listed as:

1. Cleaning
2. Sizing
3. Assembly of core pieces

6.9.1 Cleaning

This process includes trimming of fins or unwanted projections, brushing the sand and other particles on the surface, coating with refractory or protective materials and drying the coating. Finishing process includes venting also. If a core is made into two halves, a vent groove is scratched at the joining surface for the gases to flow into the core prints.

6.9.2 Sizing

Cores after baking may have slight dimensional or shape change. All dimensions are inspected and corrected for critical cases. Templates or gauges are used for fast and accurate contour measurements. Filling, scraping or grinding are the operations employed for sizing the cores.

6.9.3 Assembly of Core Pieces

When a core has intricate shape, large in size or difficult to make as a single piece, it is made in two or more pieces. After baking or drying, they are assembled or bolted together before placing inside the mould. Pasting is the commonly used method for small-sized cores. The paste used may be a mixture of talc, dextrin, molasses and similar ingredients wetted with water. It is applied on the surfaces which are to be joined. Care should be taken to keep the dimensional accuracy and avoid sealing of vent holes. For large-sized cores, bolting is employed to join the pieces. Recessed holes are used for nuts bolts, and these are covered by mud or paste to get smooth surface.

After completing the finishing operations, the cores are stored on racks fitted in the core keeping store. The humidity is kept low to reduce the damage due to water absorption. For intricate thin cores, extra core supports are also used to avoid breakage due to distortion. In mass production shops, cores are made in batches, finished and stored prior to mould making. The moulds are made only at the time of casting, in order to reduce the space requirement.

6.10 MOULD ASSEMBLY

Mould assembly is started by placing the drag in the metal pouring location or platform. Dust and loose sand if any, are blown out using compressed air jet. The cores are then set inside the mould cavity in the required sequence using the core seats and core prints. This process is called core setting.

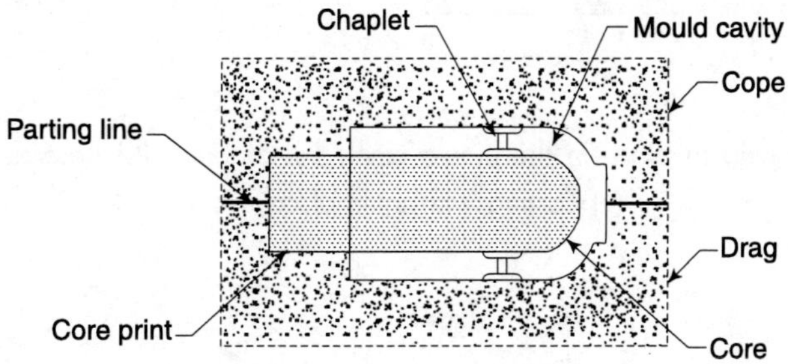

Figure 6.25 Core setting inside the mould

6.11 CORE SETTING

Core setting is the operation of placing cores in moulds. Cores are positioned in the mould by keeping the core prints inside the core seats. They must be securely anchored to the mould otherwise they may be shifted or floated up by the molten metal. While setting a core, first its location relative to the mould cavity is checked and positioned accurately. Then they are bolted to the mould in the case of heavy cores. For small cores, locking at the core prints is sufficient to resist buoyant forces.

6.12 USE OF CHAPLETS

The vertical movement of cores due to buoyancy is usually arrested by supporting the cores using chaplets (gaggers). They are metal forms placed between mould and core surfaces. Chaplets for ferrous alloys are generally made of low-carbon steel and are tin coated or otherwise protected from rusting. When the casting is poured, the chaplet becomes hot and bonds to the casting by diffusion process (Figure 6.26).

Following types of chaplets are used:

1. Double headed chaplets
2. Sheet metal chaplets
3. Cast chaplets
4. Perforated chaplets
5. Radiated chaplets
6. Stem chaplets

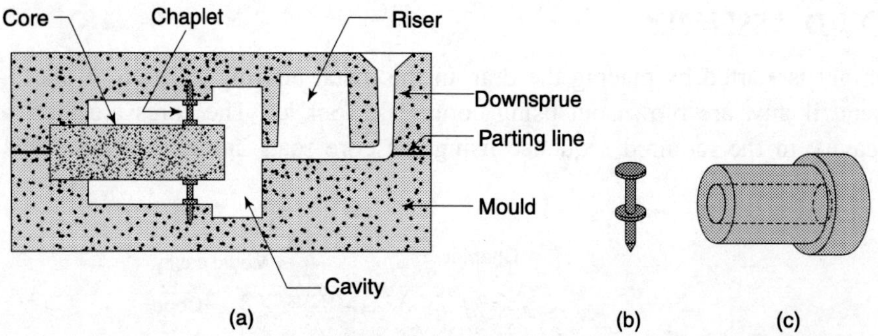

Figure 6.26(A) (a) Core held in place in the mould cavity by chaplets, (b) possible chaplet design, and (c) casting with internal cavity

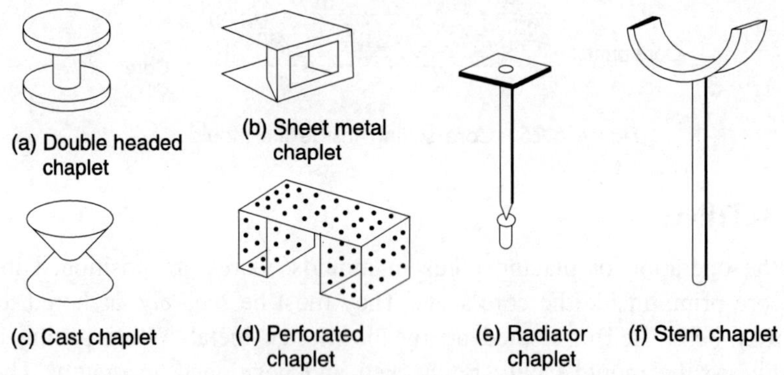

Figure 6.26(B) Forms of chaplets

Internal chills are made of the same metal that is cast. Their surfaces should be clean enough to get good joining to the parent metal. Internal chill is preferred only when the fitting of external chill is very difficult (Figure 6.27).

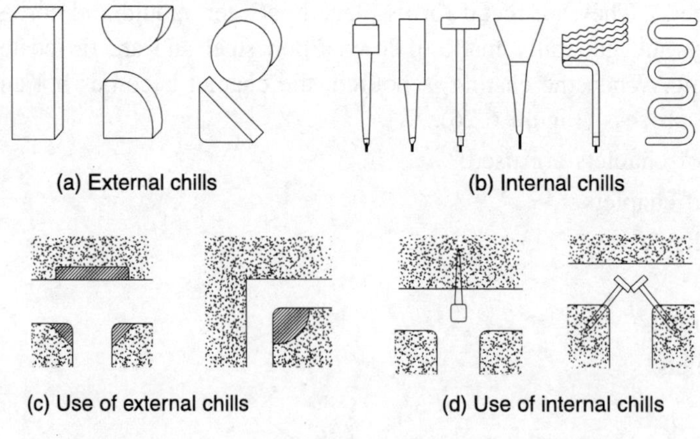

Figure 6.27 Types of chills and their use in moulds

6.13 BUOYANCY FORCES ON CORES

The core prints are provided so that the cores are secured and correctly positioned in the mould cavity. The design of core prints is such as to take care of the weight of the core before pouring and upward metallostatic pressure of the molten metal after pouring. The core prints should also ensure that the core is not shifted during the entry of the metal into mould cavity.

Buoyancy results from the weight of molten metal being displaced by the core, according to Archimedes' principle. The force tending to lift the core is equal to the weight of the displaced liquid less the weight of the core itself.

The main force acting on the core when the metal is poured into the mould cavity is due to buoyancy. The buoyant force can be calculated as the difference in the weight of the liquid metal to that of the core material of the same volume as that of the exposed core. It can be written as

$$P = V(\rho - d)$$

P = Buoyant force, N

V = Volume of the core in the mould cavity, cm^3

ρ = Weight density to the liquid metal, N/cm^3

d = Weight density of the core material = 1.65×10^{-2}, N/cm^3

Table 6.1 lists the weight densities of some of the important foundry materials.

Table 6.1 Densities of foundry materials

S. No.	*Material*	*Density, N/cm^3*
1	Aluminium	0.0265
2	Copper	0.0878
3	Magnesium	0.0171
4	Zinc	0.07
5	Lead	0.1113
6	Carbon steel	0.0771
7	Grey cast iron	0.0686 to 0.0735
8	White cast iron	0.0755
9	Moulding sand	0.0157

The equation would be valid for cases that are similar to the one illustrated in Figures 6.28(a) and 6.28(b) which is more common.

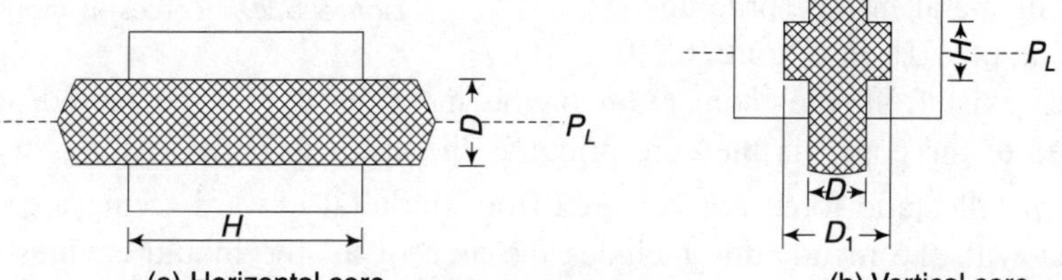

(a) Horizontal core (b) Vertical core

Figure 6.28

For horizontal core,

V = Volume of the core in the mould cavity, cm^3 = $0.25 \times \pi \times D^2 \times H$

For vertical core,

V = Volume of the core in the mould cavity, cm^3 = $0.25 \times \pi\ (D_1^2 - D^2) \times (H\rho - V_d)$

Table 6.2 lists the buoyancy forces experienced by silica sand cores in various liquid metals, expressed as a proportion of the weight of the core.

Table 6.2 Ratio of buoyant force to core weight

S. No.	Liquid metal	Ratio of buoyant force to core weight
1	Aluminium	0.66
2	Brass	4.25
3	Copper	4.50
4	Cast Iron	3.50
5	Steel	3.90

Core print support

Moulding sand (green sand) in a core print will support about 150 kN/m^2. So the core print can support the following load:

Support load (kN) = core print area (m^2) × 150

1 kN = 100 kgf (approximately)

Support load (kgf) = core print area (m^2) × 15000

Opening forces on moulds

Unless a mould is adequately clamped or weighted, the force exerted by the molten metal will open the boxes and cause run-outs. If there are insufficient box bars in a cope box, this same force can cause other problems like distortion and sand lift. It is important therefore to be able to calculate the opening force so that correct weighting or clamping systems can be used.

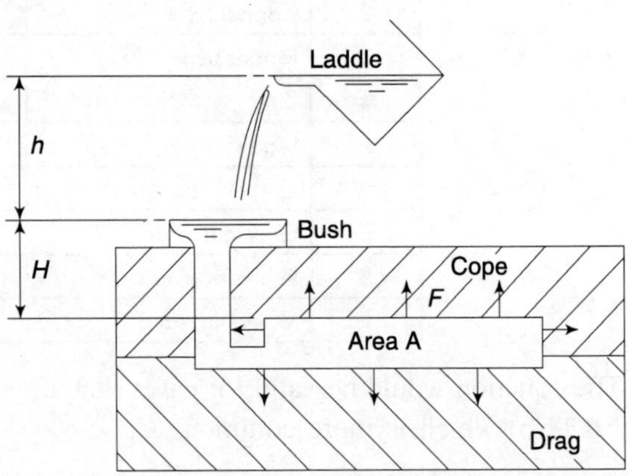

Figure 6.29 Forces in moulds

The major force lifting the cope of the mould is due to the metallostatic pressure of the molten metal. This pressure is due to the height, or head, of metal in the sprue above the top of the mould (H in Figure 6.29). Additional forces exist from the momentum of the metal as it fills the mould and from the forces transmitted to the cope via the core prints as the cores in cored castings try to float.

The opening metallostatic force is calculated from the total upward facing area of the cope mould in contact with the metal (this includes the area of all the mould cavities in the box), the force is

$$F(\text{kgf}) = \frac{A \times H \times d \times 1.5}{1000}$$

where,

A is the upward facing area in cm^2

H (cm) is the height of the top of the sprue above the average height of the upward facing surface

d is density of the molten metal $\left(\frac{\text{g}}{\text{cm}^3}\right)$

1.5 is the safety factor

For ferrous metals, d is 7.5 $\left(\frac{\text{g}}{\text{cm}^3}\right)$

so $F(\text{kgf}) = \frac{A \times H \times 7.5 \times 1.5}{1000} = \frac{11AH}{1000}$

For aluminium alloys, d is 2.7 $\left(\frac{\text{g}}{\text{cm}^3}\right)$

so $F(\text{kgf}) = \frac{A \times H \times 2.7 \times 1.5}{1000} = \frac{4AH}{1000}$

Forces on cores

The core tends to float in the liquid metal and exerts a further upward force. In the case of ferrous alloy, this force is

$$\text{Upward force} = 3.5 \times W(\text{kgf})$$

where W is the weight of the cores in the mould in kg

In aluminium, the floating force can be neglected.

The total resultant on the cope for ferrous metals is

$$\text{Total force} = \frac{11AH}{1000} + 35\text{W (kgf)}$$

Worked Out Examples

Core print design and chaplet

Example 6.1: A core weighing 50 kg has a core print area of 10 cm × 10 cm (the area of the upper, support surface). Find the support force and buoyancy force. Mould is cast in iron.

Solution: The core print area is 10 cm × 10 cm = 0.1 × 0.1 = 0.01 m^2

The print support is

Support (kN) = core print area (m^2) × 150

Support (kN) = 0.01(m^2) × 150 = 1.5 kN = 1500 N

Support (kgf) = core print area (m^2) × 15000 = 0.01 (m^2) × 15000 = 150 kgf

If the mould is cast in iron, the ratio of buoyant force to the core weight is 3.5.

Buoyancy force = 50 × 3.5 = 175 kgf

Therefore, chaplets may be necessary to support the core unless the core print area can be increased.

Example 6.2: Consider a furan resin mould for a large steel valve body casting having an upward facing area of 2500 cm^2 and a sprue height (H) of 30 cm with a core weighing 40 kg. The opening force is

$$\frac{11 \times 2500 \times 30}{1000} + 3.5 \times 40 = 825 + 140 = 965 \text{ kgf}$$

It is easy to see why such large weights are needed to support moulds in jobbing foundries.

Example 6.3: A foundry received an initial order of 10000 pieces of grey iron casting (Figure 6.30) with a possibility of repeat order. It was decided to produce a master pattern made of aluminium from which steel working pattern were to be cast and machined. Using necessary allowances provided, determine the dimensions of the master pattern and its core prints. If the moulding flask allows only a maximum of 15 mm of core print, examine further the possibility of requiring a chaplet and calculate its area, if required.

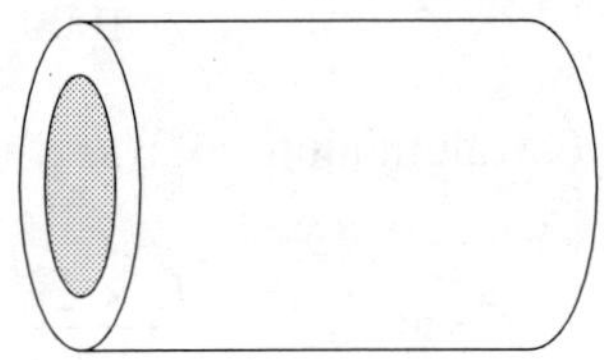

Figure 6.30

Use the following data

Density of grey cast iron	7.2 g/cm^3
Density of steel	7.8 g/cm^3
Density of aluminium	2.7 g/cm^3
Density of core sand	1.6 g/cm^3
Compressive strength of core sand	2.0 kg/cm^2
Compressive strength of chaplets	2.0 kg/cm^2
Factor of safety	5.0
Length of tube (L)	300 mm
Outside diameter	200 mm
Inside diameter	150 mm

Use the following Tables (6.3, 6.4 and 6.5)

Table 6.3 Data for contraction allowances

Contraction rule	*Materials used and place of use*
+8/1000	Cast iron in general, part of thin cast iron
+9/1000	Cast iron products of high shrinkage, part of thin cast steel

+10/1000	Same as above, and aluminium
+12/1000	Aluminiym alloys, bronze, cast steel (thickness 5-7 mm)
+14/1000	High tension brass, cast steel
+16/1000	Cast steel (thickness over 10 mm in general)
+20/1000	Large cast steel
+25/1000	Large cast iron

Table 6.4 Data for general machining allowances

Types of metals and alloys	*Machining allowance (mm)*
Cast irons	
(i) Large size castings (> 1000 mm)	10.0
(ii) Medium size castings (< 150 mm)	3.0
Cast steels	
(i) Large size castings (> 1000 mm)	12.0
(ii) Medium size castings (< 150 mm)	4.3
Non-ferrous materials	
(i) Large size castings (> 1000 mm)	5.0
(ii) Medium size castings (< 150 mm)	1.5

Table 6.5 Data for approximate taper allowances

Height of pattern (mm)	*Shell moulding*	*Sand moulding*		
		Metal		*Wood*
		Machine drawn	*Manual drawn*	*Machine drawn*
Up to 20	0° 45′	1° 30′	3°	3°
20 to 50	0° 30′	1°	1° 30′	1° 30′
:	:	:	:	:
100 to 200	0° 20′	0° 30′	0° 45′	0° 45′

Solution:

Step 1: Calculate the pattern allowance

Here the master pattern will be made of aluminium using which the steel working pattern will be cast. Thus, the master pattern will have double allowances for grey cast iron and steel. Also, since a pattern gives the outside surface of the casting, only the outside diameter, D_o, will be considered while calculating the pattern dimension. The pattern will be a split-pattern type and the cope half of the pattern is shown in Figure 6.31.

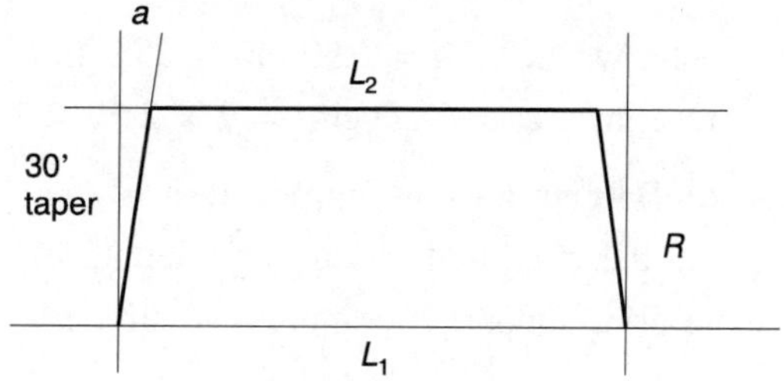

Figure 6.31

Step 2: Calculate the contraction and machining allowances

Contraction allowances = for grey cast iron and steel

$$= \left(\frac{8}{1000}\right) \times 300 + \left(\frac{16}{1000}\right) \times 3000 = 2.4 + 4.8 = 7.2 \text{ mm}$$

Machining allowances = for grey cast iron and steel

$$= 4 + 6 = 10 \text{ mm}$$

Total length of the pattern = L_1 = 300 + 7.2 + 10 = 317.2 mm

Step 3: Calculate the dimension of radius

Contraction allowance = for grey iron + for steel

$$\left(\frac{8}{1000}\right) \times 100 + \left(\frac{16}{1000}\right) \times 1000 = 0.8 + 1.6 = 2.4 \text{ mm}$$

Machining allowances = for grey cast iron and steel

$$= 3.0 + 4.3 = 7.3 \text{ mm}$$

Total radius of the pattern, R = 100 + 2.4 + 7.3 = 109.7 mm

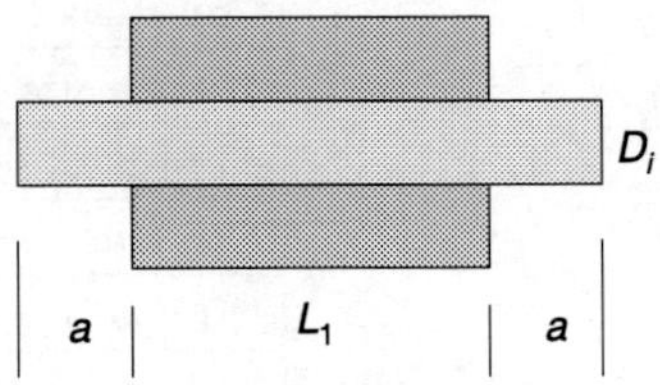

Figure 6.32

Step 4: Calculate the taper allowances

Taper allowance = 30 min = 0.5°

Refer to Figure, $a = R \times \tan 0.5 = 109.7\ (0.0087) = 0.96$ mm

Total length of the pattern at top = $L_2 = L_1 - 2a = 317.2 - 2 \times 0.96 = 315.28$ mm

Step 5: Determine the core dimensions (Figure 6.32)

$$\text{Core buoyancy} = \frac{\pi}{4}\,(D_i^2) \times L \times (\rho_1 - \rho_2)$$

$$= \frac{\pi}{4}\,(15 \text{ cm}^2) \times 31.72 \left(7.2 - 1.65\ \frac{\text{g}}{\text{cm}^3}\right) \times \left(\frac{1}{1000}\right) \text{kg} = 31.11 \text{ kg}$$

Using factor of safety, compressive strength of core print

$$\frac{2.0}{5}\ \frac{\text{kg}}{\text{cm}^2} = \frac{F}{A} = \frac{31.11 \text{ kg}}{2aD_i \text{ cm}^2}$$

Solving we get = a = 2.59 cm = 25.9 mm

Core print area = $2aD_i \text{cm}^2 = 2 \times 2.59 \times 15 = 77.7 \text{ cm}^2$

Step 6: Determine the chaplet area

Using the maximum permissible length of core print (i.e., 15 mm, given data)

Core print support = compressive strength of core × core print area

$$= \frac{2.0}{5}\ \frac{\text{kg}}{\text{cm}^2} \times 2 \times (1.5 \text{ cm}) = 18 \text{ kg}$$

Unsupported load = core buoyancy – core print support

$$= 31.11 - 18 \text{ kg} = 13.11 \text{ kg}$$

Chaplet area required

$$= \text{unsupported load (kg)} \times \text{compressive strength of chaplet} \left(\frac{\text{kg}}{\text{cm}^2}\right)$$

$$= 13.11 \text{ kg} \times 2.0 \frac{\text{kg}}{\text{cm}^2} = 26.22 \text{ cm}^2$$

Example 6.4: A sand core has a volume of 1875 cm^3 and is located inside a sand mould cavity. Determine the buoyancy force tending to lift the core during pouring of molten lead into the mould. Take density of lead 11.3 gm/cm^3 and density of the sand core 1.6 gm/cm^3.

Solution: Density of the sand core is 1.6 gm/cm^3.

Density of lead is 11.3 gm/cm^3.

Weight of the core is 1875(1.6) = 3000 gm = 3.0 kg.

Weight of lead displaced by the core is 1875(11.3) = 21,188 gm = 21.19 kg.

The difference = 21.19 – 3.0 = 18.19 kg.

Given that 1 kg = 9.81 N, the buoyancy force is = 9.81(18.19) = **178.4 N.**

Example 6.5: Find the weight that needs to be kept to compensate for the forces during the pouring in a solid casting of a cast iron pipe of 12.5 cm outside diameter and 10 cm inside diameter with a length of 180 cm. The metal head is to be about 20 cm. While the moulding flask size used for the purpose is 200 × 25 × 20 cm in size. Take the density of core sand as 0.0165 N/cm^3 and liquid metal density to be 0.0771 N/cm^3.

Solution: Volume of the core $= \dfrac{\pi \times 10^2 \times 180}{4} = 14136.75 \text{ cm}^3$

Buoyancy force = $V(\rho - d)$

$$= 14136.75\,(0.0771 - 0.0165) = 856.7 \text{ N}$$

Projected area of the casting = 180 × 12.5 = 2250 cm^2

Upward metallostatic force = 2250 × 0.0771 × 20 = 3469.5 N

Total upward force = 3469.5 + 856.7 = 4326.7 N

Volume of sand in the cope = 200 × 25 × 20 = 100,000 cm^3

Weight of the cope = 100,000 × 0.0165 = 1650 N

Net upward force acting = 4326.7 – 1650 = 2676.7 N

The foundry has to keep a total weight of about 3000 N to compensate for this upward thrust exerted by the core and the metallostatic force.

In the above example, only the static force was considered. However, when the metal is entering the mould cavity at high flow rate, the loss of momentum would exert much higher pressure than that accounted for the static condition. Consequently, more force would be acting on the mould flask than that calculated in the above procedure. It is general practice in the foundries to keep a weight which is about 50% more than the calculated one to account for this dynamic effect.

Example 6.6: A foundry is producing grey iron blocks 300 × 150 × 100 mm. The parting line is at the midpoint of the height of the block so that the mould cavity extends 50 mm into both the cope and drag. The flasks are 500 × 300 × 125 mm, so the combined height of the cope and drag is 250 mm. The mould is poured with molten grey iron, which weighs 0.0597 N/cm^3 and compacted sand weighs 0.0157 N/cm^3. What would be the total lifting force in newtons tending to separate the cope from the drag because of the metallostatic pressure within the mould?

Solution:

Volume of the core = 30 × 15 × 10 = 4500 cm^3

Buoyancy force = $V(\rho - d)$

= 4500 (0.0597 – 0.0157) = 198 N

Projected area of the casting = 30 × 15 = 450 cm^2

Upward metallostatic force = 450 × 0.0597 × 25 = 671.63 N

The total upward force = 198 + 671.63 = 869.63 N

Volume of sand in the cope = 50 × 30 × 12.5 = 18750 cm^3

Weight of the cope = 18750 × 0.0157 = 294.38 N

Net upward force acting = 869.63 – 294.38 = 575.26 N

The foundry has to keep a total weight of about 600 N to compensate for this upward thrust exerted by the core and the metallostatic force.

Example 6.7: A 92% aluminium – 8% copper alloy casting is made in a sand mould using a sand core that weighs 20 kg. Determine the buoyancy force in newtons tending to lift the core during pouring.

Solution Sand density = $1.6 \dfrac{g}{cm^3} = 0.0016\ kg/cm^3$

$$\text{Core volume} = \frac{20}{0.0016} = 12500\ cm^3$$

$$\text{Density of Al – Cu alloys} = 2.8 \frac{g}{cm^3} = 0.00281\ kg/cm^3$$

Weight of displaced Al – Cu alloys = 12500 × 0.00281 = 35.125 kg

Buoyancy force = 35.125 – 20 = 15.125 kg

Buoyancy force in Newtons = 15.125 kg × 9.81 = 148.5 N

Example 6.8: A sand core located inside a mould cavity has a volume of 157 cm^3. It is used in the casting of a cast iron pump housing. Determine the buoyancy force that will tend to lift the core during pouring.

Solution: Sand density = $1.6 \dfrac{g}{cm^3}$ = 0.0016 kg/cm^3

Core weight = 157 × 0.0016 = 0.2512 kg

Density of cast iron = 0.00719 kg/cm^3

Weight of displaced cast iron = 157 × 0.00716 = 1.124 kg

Buoyancy force = 1.124 – 0.2512 = 0.87282 kg

Buoyanct force in Newtons = 0.87292 kg × 9.81 = 8.5633 N

Example 6.9: Chaplets are used to support a sand core inside a sand mould cavity. The design of the chaplets and the manner in which they are placed in the mould cavity surface allows each chaplet to sustain a force of 50N. Several chaplets are located beneath the core to support it before pouring and several other chaplets are placed above the core to resist the buoyancy force during pouring. If the volume of the core is 5500 cm^3 and metal poured is brass. Determine the minimum number of chaplets that should be placed (a) beneath the core and (b) above the core.

Solution: Sand density = $1.6 \dfrac{g}{cm^3}$ = 0.0016 kg/cm^3

Core weight = 5500 × 0.0016 = 8.8 kg

At least 2 chaplets are required beneath to support the weight of the core. Probably 3 or 4 chaplets would be better to achieve the stability.

Density of brass = 0.00867 kg/cm^3

Weight of displaced cast iron = 5500 × 0.00867 = 47.685 kg

Buoyancy force = 47.685 – 8.8 – 38.885 kg

Buoyancy force in Newtons = 38.885 kg × 9.81 = 381.46 N

Nine chaplets are required above the core to resist the buoyancy force.

Example 6.10: A sand core used to form the internal surfaces of steel casting experiences a buoyancy force of 23 kg. The volume of the mould cavity forming the outside surface of the casting is 5000 cm^3. What is the weight of the final casting? Ignore considerations of shrinkage.

Solution: Sand density = $1.6 \dfrac{g}{cm^3}$

Cavity volume = 5000 cm^3

Density of cast iron = $7.82 \dfrac{g}{cm^3}$

Buoyancy force = 7.82 V – 1.6 V = 6.22 V = 23 kg = 23000 g

Displaced volume = $V = \dfrac{23000}{6.22}$ = 3698 cm^3

Volume of the casting = 5000 – 3698 = 1302 cm^3

Weight of the final casting = 1302 × 7.82 = 10,184 = 10.184 kg

Example 6.11: In sand casting, it is important that the cope mould half be held down with sufficient force to keep it from floating when the molten metal is poured in. For the casting shown in Figure 6.33, calculate the minimum amount of the weight necessary to keep the core from floating up as the molten metal is poured in.

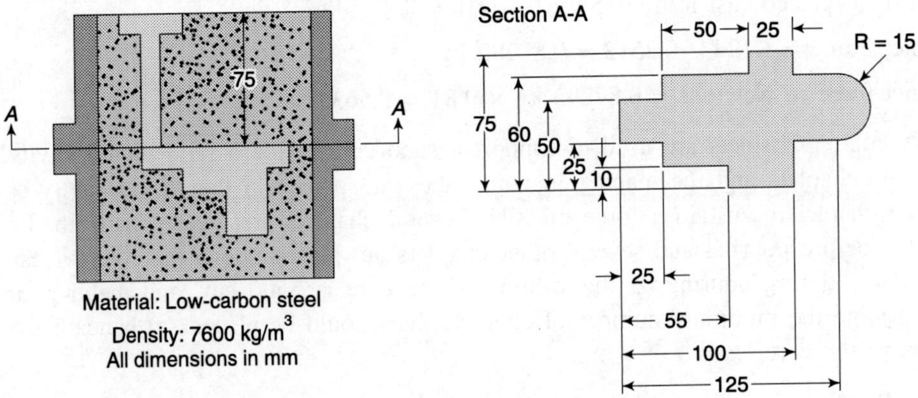

Figure 6.33

Solution: The buoyancy force exerted by the molten metal on the cope is related to the effective height of the metal head above the cope.

The cope mould half must be heavy enough or be weighted sufficiently to keep it from floating when the molten metal is poured into the mould. The buoyancy force on the cope is exerted by metallostatic pressure caused by the metal in the cope above the parting line and can be calculated from the formula,

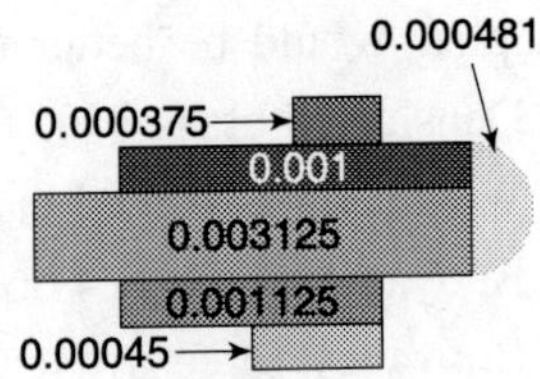

Figure 6.34

F = pressure at parting line × projected area of mould cavity

$$\text{Pressure at parting line} = \rho gh = 7600 \frac{\text{kg}}{\text{m}^3} \times 9.81 \frac{\text{m}}{\text{s}^2} \times 0.075 \text{ m} = 5.592 \text{ kPa.}$$

The projected mould cavity can be calculated from the dimensions given in Figure 6.34 in the problem and is found to be 6553 mm^2 = 0.006553 m^2

Thus, the force is, $F = 0.006553 \times 5.592 \times 1000 = 36.6\text{N}$

Questions from GATE Exams

Example 6.12: In sand casting of hollow part of lead, a cylindrical core of diameter 120 mm and height 180 mm is placed inside the mould cavity. The densities of core material and lead are 1600 kg/m^3 and 11300 kg/m^3 respectively. What is the net force that tends to lift the core while pouring of molten metal? **(GATE 2008-P1)**

Solution: Net buoyancy force = $V g(\rho - d)$

$$V = \text{volume of the casting} = \frac{\pi \times 0.12^2}{4} \times 0.180 = 2.0356 \times 10^{-3}\ \text{m}^3$$

$$\text{Net buoyancy force} = V g(\rho - d) = 2.0356 \times 10^{-3}\text{m}^3 \times 9.81\ \frac{m}{s^2} \times (11300 - 1600)\ \frac{\text{kg}}{\text{m}^3}$$

$$\text{Net buoyancy force} = V g(\rho - d) = 193.71\left(\text{m}^3 \times \frac{\text{m}}{\text{s}^2} \times \frac{\text{kg}}{\text{m}^3}\right) = 193.71\ \frac{\text{kg} - \text{m}}{\text{s}^2} = 193.71\text{N}$$

Example 6.13: A aluminium alloy $\left(\text{density, } 2600\ \frac{\text{kg}}{\text{m}^3}\right)$ casting is to be produced. A cylindrical hole of 100 mm diameter and 100 mm length is made using sand core $\left(\text{density, } 1600\ \frac{\text{kg}}{\text{m}^3}\right)$. What is the net buoyancy force in newton acting on core? **(GATE 2014-Set1)**

Solution: Net buoyancy force = $Vg\ (\rho - d)$

$$V = \text{volume of the casting} = \frac{\pi \times 0.10^2}{4} \times 0.10 = 7.85375 \times 10^{-4}\ \text{m}^3$$

$$\text{Net buoyancy force} = Vg(\rho - d) = 7.85375 \times 10^{-4}\ \text{m}^3 \times 9.81\ \text{m/s}^2 \times (2600 - 1600)\ \text{kg/m}^3$$

$$\text{Net buoyancy force} = Vg(\rho - d) = 7.705\left(\text{m}^3 \times \frac{\text{m}}{\text{s}^2} \times \frac{\text{kg}}{3}\right) = 7.705\ \frac{\text{kg} - \text{m}}{\text{s}^2} = 7.705\text{N}$$

Example 6.14: A flask consisting of cope and drag has the following dimensions: length 300 mm, width 200 mm and total depth 200 mm (cope height 50 mm). A 150 mm × 100 mm × 50 mm cast iron block is to be cast. If the specific weight of the cast iron is 78 kN/m^3 and that of the mould is 16 kN/m^3, find the lifting force on the cope created by buoyant force, weight of sand in the cope, the net force at the liquid sand interface and additional weight to be kept on the cope. **(GATE 1991)**

Solution:

(a) Weight of sand in cope

Weight of sand in the cope = volume of sand in cope × density of sand

Weight of sand in the cope = 0.3 × 0.2 × 0.05 × 16 = 40 kN

(b) Lifting force on the cope created by buoyant force

Weight of the casting = 0.150 × 0.1 × 0.05 × 78 = 58.5 kN

Net buoyancy force = Weight of liquid displaced = weight of casting – weight of cope

Net buoyancy force = 58.5 – 48 = 10.5 kN

The interface force between sand and cope interface is 10.5 kN

(c) Additional weight to be kept on cope

The additional weight to be placed on core = net buoyant force = 10.5 kN.

Review Questions

1. What is the use of a core in moulding processes?
2. What are the functions of a core?
3. Write the classification of a core prints.
4. Explain various core prints used in foundry processes.
5. Explain the various steps in the manufacture of cores.
6. How are ovens classified?
7. With the help of neat sketch, explain various ovens used.
8. Write a brief note on finishing of cores and storing.
9. What are the uses of chaplets?
10. What are the buoyancy forces acting on cores?

Chapter 7

Gating Design

7.1 INTRODUCTION

Pouring of molten metal into a mould is one of the most critical steps in metal casting. The behaviour of the liquid and its subsequent solidification and cooling determine whether the cast shape will be properly formed, internally sound and free from defects. A gating system in a mould is the metal feeding network through which the liquid metal enters a mould and flows to fill the mould cavity. The basic components of a simple gating system in a flask mould are a pouring cup or a pouring basin, a down sprue to carry the liquid metal down to join one or more runners for distributing the metal throughout the mould and the ingates which are the restricted entry points of molten metal from the runner to the mould cavity. A riser is also a feeding unit fitted in proper location and size, so that it acts as a reservoir to fill the undesirable shrinkage cavities formed during solidification of liquid metal. Metals having higher rate of solidification shrinkage are cast with the help of risers.

The success of the pouring operation depends upon:

- Certain qualities of the metal itself, which influence liquid flow. Example, composition, temperature and fluidity. A fluid liquid should have good fluidity and must fill through all the interstices of the mould and good reproduction of design details of casting.
- Properties and design of the mould including the nature of the moulding material and gating technique used to introduce the metal into the mould cavity.

Whilst the metal is in liquid state, the foundryman is also concerned with forces acting on the mould and with volume contraction occurring during cooling to the solidification temperature. A good gating design ensures distribution of the metal in the mould cavity at a proper rate without excessive temperature loss, turbulence, and entrapping gases and slags.

- If the time taken for the liquid metal is long, the solidification may start even before the mould has been completely filled up. To overcome this, use quite a lot of superheat, but then gas solubility may cause a problem.
- If the liquid metal is poured with too high velocity, the mould surface may erode. Thus, an optimum velocity is to be maintained.

According to Walance and Evans, a technically ideal gating system must possess the following attributes:

1. It must introduce metal into the mould cavity with minimum turbulence.
2. It must prevent slag, dross or inclusion in ladle to reach mould cavity.
3. It must avoid aspiration of air and mould gases in metal stream.
4. It must not cause mould or core erosion.
5. It should create favourable temperature gradients in the casting to help directional solidification.
6. It should control the rate of filling of mould cavity avoiding cold shuts due to slow rise as well as erosion due to fast rise.
7. It should be economical to mould and to finish after casting.

7.2 SELECTION CRITERIA

A mould cavity cannot be filled directly as filling of water in a vessel with a bucket. A suitable feeding passage has to be made in the mould so that the heavy liquid metal enters and fills the mould cavity in the desired manner without turbulence. Figure 7.1 shows cross section of a flask mould ready for pouring. The molten metal is poured continuously to the pouring basin so that it flows downwards through the down sprue and enters the runner through sprue base. The runner is connected to the ingates and the molten metal enters the mould cavity through the ingates. Here, the ingates are along the parting plane and the number depends on the size of casting and the design.

While filling the liquid metal, the level of metal gradually rises from the bottom of the mould cavity and the air in the cavity and gases developed due to heat are removed through the 'flow-off' passage provided in the cope. The molten metal comes out through the flow-off hole indicating that the mould filling is complete. Metal pouring is stopped now and the mould with casting is allowed to cool for solidification.

The design of a gating system depends on both the metal and mould compositions.

- For easily oxidizing metals of low melting point such as aluminium, an elaborate gating design is needed to avoid junk (e.g., oxides).
- For cast iron, a short path for the liquid metal is selected to avoid a high pouring temperature.
- For a ceramic mould, the gating design is different from that normally used for a permeable sand mould.

7.2.1 Functions of Gating Systems

The basic requirement of a gating system is to fill the mould cavity with molten metal so that a defect-free casting is obtained at minimum expense. The requirements which are to be satisfied by a gating system can be grouped as primary and secondary functions.

7.2.1.1 Primary Functions of the Gating Systems

1. Economy of size.
2. To provide passage to fill the mould cavity with liquid metal.
3. To fill the mould cavity in a specified time and velocity (usually 0.5 m/s).
4. To fill the molten metal with minimum turbulence.
5. Ease of removal.

7.2.1.2 Secondary Functions of the Gating Systems

1. To provide directional solidification in such a way that the farthest point of mould cavity solidifies first (usually thin section) and the filling point of the mould cavity as the last (usually thick section).
2. To deliver only liquid metal into the mould cavity. No other phases such as slag, dross oxides, sand, air and other gases during the filling process.
3. To feed molten metal with less erosion of the surfaces of the passages and cores. This is important while feeding heavy castings.
4. To remove air and gases trapped inside the mould cavity during the filling process.
5. To establish proper temperature gradient.
6. The elimination of surface turbulence, preferably at an early stage in running system. It should gather fragmented stream due to the long fall through the sprue together again.

Elements of gating systems

1. ***Pouring basin*** The part of gating system is made on or in the top of the mould. The main purpose of the pouring basin is to direct the flow of metal from ladle to the sprue, to help maintain the required rate of liquid metal flow. Sometimes, the metal is directly poured into the top of the sprue, which is made with a funnel-shaped opening. However, better results can be obtained by the use of pouring basin. Pouring basin also reduces the turbulence and vortexing at the sprue entrance. It is placed nearly to the edge of moulding box to balance metal shrinkage or contraction.

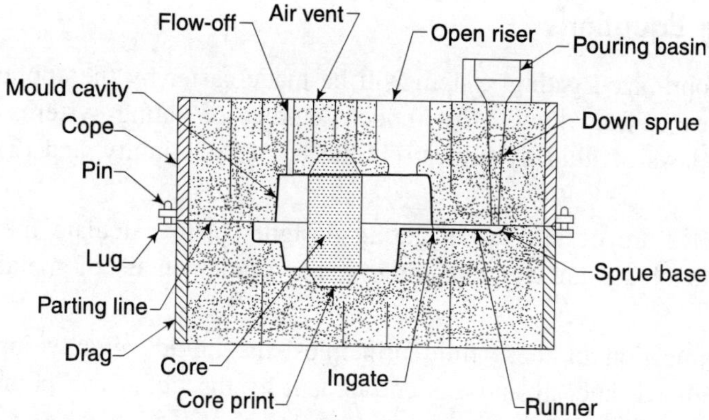

Figure 7.1 Sand mould gating system

2. ***Sprue*** It is a vertical passage that passes through the cope and connects the pouring basin with the runner or gate. As the metal flows down the sprue, its velocity increases. Hence, the section of the sprue should decrease, otherwise the sprue will not remain full of metal with metal leaving the walls of the sprue. This results in aspiration of gases through the surrounding moulding sand. Hence, the sprues are generally tapered downwards and the cross section of a sprue may be square, circular or rectangular.
3. ***Runner*** In large castings the molten metal carried from sprue to gates around the cavity through a passageway, called runner, is generally preferred in drag, sometimes in cope depends on the shape of casting. The runners are of large cross section and often streamlined to slow down and smooth out the flow, and are designed to provide approximately uniform flow rates to the various parts of the mould cavity. It provides to avoid aspiration and turbulence. Runners are commonly made trapezoidal in cross section.
4. ***Gate*** A gate is a passage through which metal flows from the runner to the mould cavity. The location and size of gates are so arranged that they can feed liquid metal to the casting at the rate consistent with the rate of solidification. The gates should break off easily from the casting after solidification. For this, at the junction to the cavity the gates are much reduced in thickness. The best cross section for gates is a trapezoidal one that smoothly passes into a rectangular section at the junction of the cavity.
5. ***Skim bob*** It is an extension along the runner which provides the elimination of ingredient particles in the circular shape portion. Thus, it prevents these impurities from going into the mould cavity and also gives a slow and progressive flow of molten metal into the mould cavity.
6. ***Cavity*** It is a hollow shape or space that is obtained by moulding process. The molten metal enters the mould cavity. The finished product is obtained by mould cavity shape.
7. ***Choke*** It is a part of gating system which has smaller cross-sectional area. It is placed after the skim bob. Its function is to control the flow to lower the flow velocity in the runner and also to minimize sand erosion in the runner. Its another end is connected to gates.

7.2.2 Governing Equations

Proper design of an optimized gating system will be made easier by the application of several fundamental principles of fluid flow. Most modern studies of gating systems are based upon consideration of two laws of fluid dynamics: (1) the law of continuity, and (2) the Bernoulli's theorem.

These two principles are of interest in gating design: (1) to calculate metal velocity and flow rates, and (2) to obtain an understanding of the fundamentals of metal flow in gating systems.

With a proper application of these fluid principles, the liquid velocity inside the running system can be streamlined, and air and gas entrapment by the liquid can be avoided.

The Law of Continuity

For a system with impermeable walls and filled with an incompressible fluid, the rate of flow will be the same at all points in the system.

$$Q = A_1V_1 = A_2V_2$$

where,

Q = Volume flow rate in m^3/s

A = Area of flow passage, m^2

V = Linear velocity of flow, m/s

This law is applicable to both steady and non-steady state flows of incompressible fluids in impermeable walled channels. The permeability of sand moulds can complicate the strict application of this law.

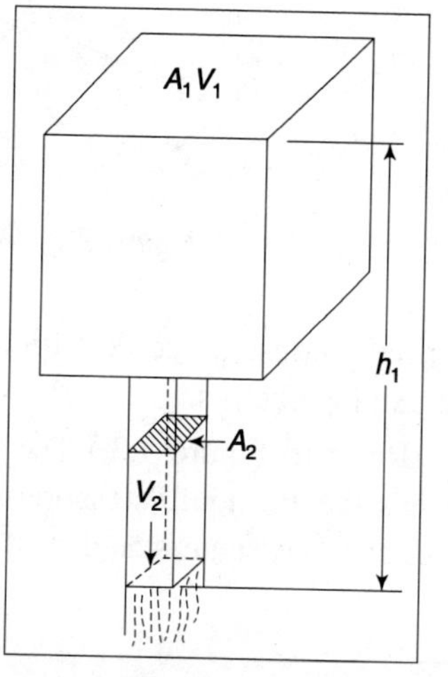

Figure 7.2 Law of continuity

Bernoulli's theorem

The theorem states that, at any point in a full system, the total energy involving the sum of the potential energy, kinetic energy, pressure energy and frictional energy of a flowing liquid is constant.

The equation allows prediction of the effect of the several variables at different points in the gating system, although several conditions inherent in foundry gating system complicate and modify its strict applicaton. For example:

- The equation is for full systems, and at least at the start of pouring, a gating system is empty. This indicates that gating systems should be designed to establish as quickly as possible the flow conditions of a full system.
- The equation assumes an impermeable wall around the flowing metal. In sand foundry practice, the permeability of the mould medium can pose problems, for example, air aspiration in the flowing liquid.
- Additional energy losses due to turbulence and changes in the direction of flow at the bend must be accounted for.
- Heat losses from the liquid metal are not considered, which will set a limit on time over which flow can be maintained. Also, solidifying metals on the walls of the gating system components will alter their design while flow continues.
- The potential energy is maximum at the highest point in the system, that is, the top of the pouring basin.

As metal flows from the basin down the sprue, potential energy changes to kinetic energy as the stream increases in velocity because of gravity. As the sprue fills, a pressure head is developed. Once flow in a filled system is established, the potential and frictional heads become

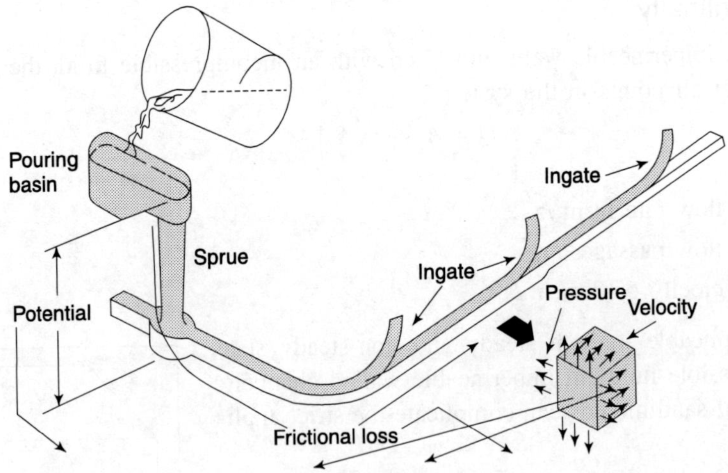

Figure 7.3 Application of Bernoulli's equation to a gating system

virtually constant, so conditions within the gating system are determined by the interplay of the remaining factors.

(Refer to Figure 7.3.) The liquid metal that runs through the various channels in the mould obeys the Bernoulli's theorem which states that the total energy head remains constant at any section. The same stated in the equation form ignoring frictional losses is

$$h + \frac{p}{g\rho_m} + \frac{v^2}{2g} = \text{constant}$$

where,

h = Potential head, m

P = Pressure, P_a

V = Liquid velocity, m/s

ρ_m = Density of liquid metal, kg/m^3

g = Gravitational constant of earth, 9.81 m/s^2

Though quantitatively Bernoulli' theorem may not be applied, it helps to understand qualitatively, the metal flow in the sand mould. As the metal enters the pouring basin, it has the highest potential energy with no kinetic or potential energies. But as the metal moves through the gating system, a loss of energy occurs because of the friction between the molten metal and the mould walls.

Also heat is continuously lost through the mould material though it is not represented in the Bernoulli's equation. This lets the casting solidify.

7.3 CLASSIFICATION OF GATING SYSTEMS

Molten metal can be admitted to a mould cavity in several ways. The gating systems can be broadly grouped into two classes:

1. Based on the location of ingates/entry point of metal to the casting cavity (Figure 7.4)

- Vertical gating/top gating
- Bottom gating
- Horizontal gating

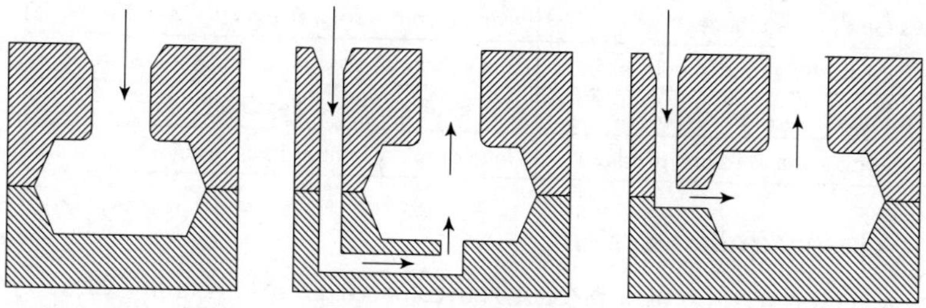

(a) Top pouring system (b) Bottom pouring system (c) Side pouring system

Figure 7.4 Pouring systems

2. Based on the location of ingates/entry point of metal to the casting cavity

- Pressurized or closed gating system
- Unpressurized or open gating system

Vertical gating: The liquid metal is poured vertically to fill the mould with atmospheric pressure at the base. The metal temperature is always high at the top region. Therefore, directional solidification is possible and defect-free castings due to reduction of shrinkage porosity can be obtained. In the top gates, metal enters from the top side and falls to the bottom. This may result in splashing, turbulence, oxidation and damage of mould surfaces. Hence, these gates are not preferred for casting oxidizable alloys and mould cavities having more depths.

Bottom gating: In bottom gating, the liquid metal is filled in the mould from bottom to top. These gates feed metal through the bottom side of the mould cavity so that minimum metal splashing occurs during filling. Reduced erosion and entrapment of gases are advantages of these gates. A bottom gate creates an unfavourable temperature gradient and makes it difficult to achieve directional solidification.

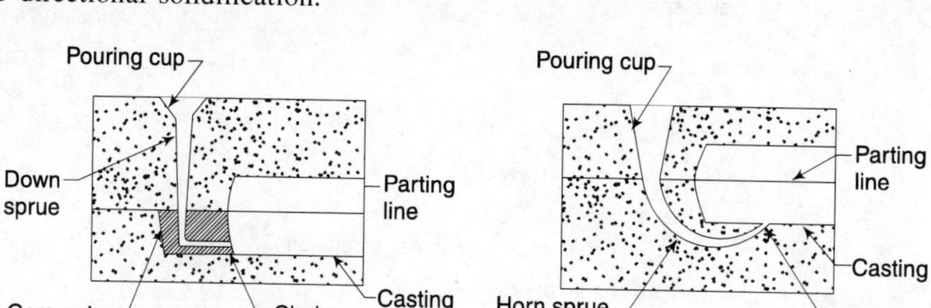

(a) Bottom gate (b) Horn gate

Figure 7.5 Bottom gating

Horizontal gating: In this, additional horizontal portions are introduced for better distribution of the liquid metal with minimum turbulence. This type is normally applied in die casting machines.

Top gating	*Bottom gating*
Faster pouring rate	Filling against gravity, so slow pouring rate
Less fluidity requirement	More fluidity requirement
Severe flashing and mould erosion	Elimination of splashing and mould erosion
Oxidation of liquid metal	Less turbulence and less oxide formation
Favourable temperature gradient	Unfavourable temperature gradient

Pressurized or closed gating system

Figure 7.6 gives a pictorial view of a pressurized (closed) gating system. These types of gates are characterized by gradually decreasing the cross-sectional areas from sprue base to the total area of ingate. Since the total ingate area is less than others, the metal flow is controlled by them. During metal feeding, the whole gating system is filled with molten metal within a few seconds. This is better for separating the slag from molten metal during the flow. At the same time, the metal enters the mould cavity with high velocity like a jet flow. This may lead to metal splashing, capture of air, oxidation of metal and washout of mould walls.

It is to be noted that the down sprue usually has no taper and the sprue base has maximum opening area. At the same time, the total area of ingates is the lowest among the other passages and the ingates are fitted usually at the bottom side of the casting. In Figure 7.2, the critical areas of passage are bottom of sprue a, runner cross section b, and the total of ingate openings $(c_1 + c_2 + c_3)$. A sample of ratio of them of the system is

$$a : b : (c_1 + c_2 + c_3) = 4:3:2$$

Similar ratios keeping the lowest cross section at the ingates are maintained in pressurized (closed) gating system.

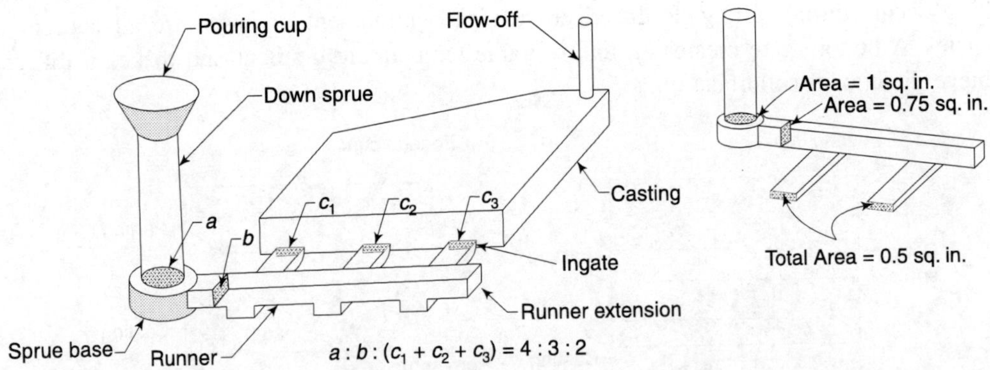

Figure 7.6 Pressurized or closed gating system

Note: Here gate area is the smallest, thus maintaining the back pressure throughout the gating system. This back pressure generates turbulence and thereby minimizes the air aspiration even when straight sprue is used. Not good for light alloys, but good for ferrous castings.

Unpressurized or open gating system

Figure 7.7 shows pictorial view of unpressurized or open gating system designed to cast a slab. In this system, the narrowest opening point is at the bottom of the down sprue. When the liquid metal enters, it fills the tapering down sprue and flows to the runner through the narrowest opening. The flow velocity decreases gradually from the sprue to the runner and enters the mould cavity quietly through the ingates. This entry is without any splash and oxidation, and does not wash out the mould walls and cores. Open gating system keeps the runner and ingates pressure free except the down sprue. Open (unpressurized) gating systems are commonly used for casting steels, aluminium, magnesium and other oxidizable alloys.

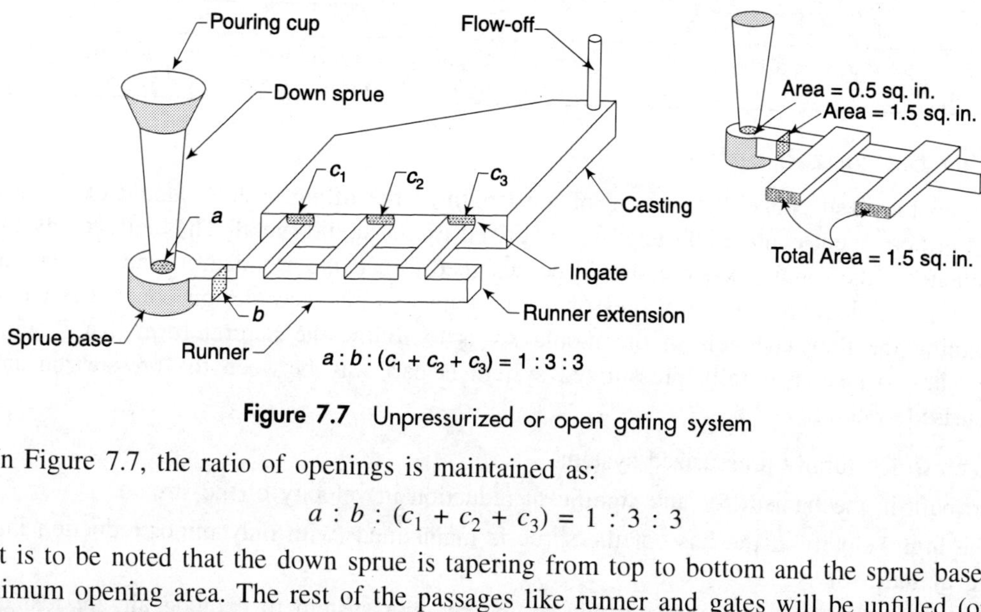

Figure 7.7 Unpressurized or open gating system

In Figure 7.7, the ratio of openings is maintained as:

$$a : b : (c_1 + c_2 + c_3) = 1 : 3 : 3$$

It is to be noted that the down sprue is tapering from top to bottom and the sprue base has minimum opening area. The rest of the passages like runner and gates will be unfilled (open) during feeding of the casting. Gating ratios of different values are used to fulfil different requirements, but the lowest value of the ratio is always given at the bottommost point of the down sprue of the system.

Note: Unpressurised gate system has choke at the bottom of the sprue base, has total runner area and gate areas higher than the sprue area. No pressure is present in the system and hence no turbulence. But chances of air aspiration are possible. Suitable for Al and Mg alloys.

Pressurized systems	*Non-pressurized systems*
Gate is used as choke	Sprue base is used as choke
Turbulence is more	Turbulence is minimised

Velocity increases as liquid flows from sprue to runner to gate.	Velocity decreases as liquid flows from sprue to runner to gate.
Gating system is kept full due to back pressure.	Careful design is required to keep gating system full.
Velocity in each of multiple gates (of equal area) is equal.	Obtaining equal flow in multiple gates is difficult.
Volume of gating systems is smaller producing high casting yield.	High volume of gating system producing low casting yield.
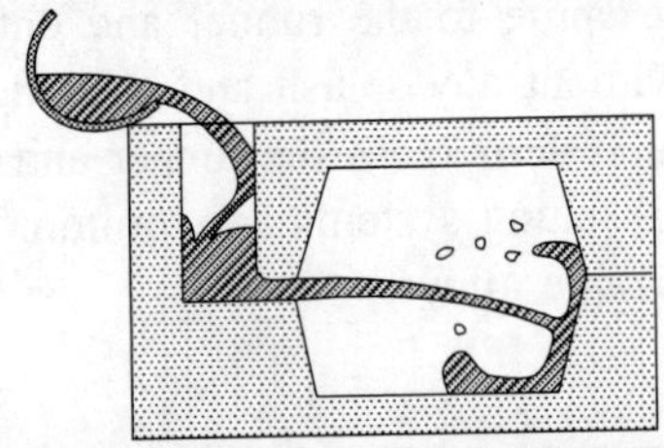	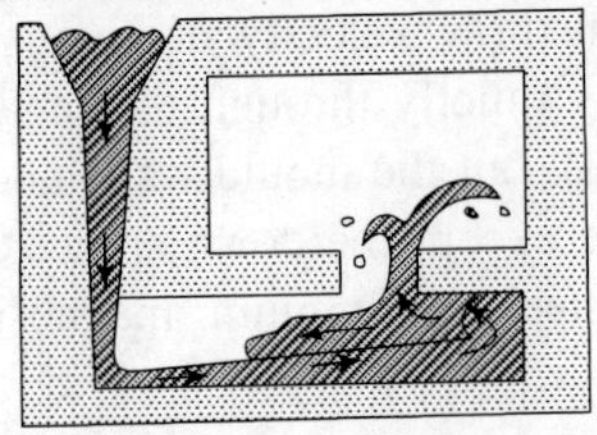
Pressurized system shows the jetting of liquid into the mould cavity.	Unpressurised system shows the fast under jet and the rolling back wave in the oversized runner.

Naturally pressurized system

There is no localized choke. The whole of the length of the filling system should experience its walls in permanent contact and gently pressurized by the liquid metal. Thus, effectively the whole length of the running system should be designed to act like a choke, a kind of continuous choke principle, and uniformly pressurized systems. So, the new design concept is based on designing the flow channels in the mould so as to follow the natural form that flowing metal wishes to take. Naturally pressurized system is part way between the pressurized and unpressurised system.

Problems of naturally pressurized system

- No built-in mechanism for any significant reduction in velocity of the stream.
- The high velocity at the base of the sprue is maintained (with only minor reduction into the mould).
- Thus, the benefits of complete filling of the running system to exclude air are lost on entering the mould cavity.
- The naturally pressurized approach requires completely separate mechanisms to reduce the velocity of the melt through the ingates.

7.3.1 Design of Vertical Gating System

Based on the principles of fluid flow, we can estimate the time taken for the molten metal to fill up the mould. For illustrations, we do this estimation for the following Figures 7.8 and 7.9. The integrated energy balance equation on the basis of per unit mass flow, more commonly known as Bernoulli's equation, will be used.

The Bernoulli's theorem is employed here to calculate the velocity in an ideal fluid system, i.e., in the system where the fluid suffers no energy losses. In real gating systems besides losses due to friction, energy losses occur at all entrances and exits, bends, enlargements and contractions. The exit velocity and flow rates obtained by the above equation would be therefore somewhat higher than those found in actual practice.

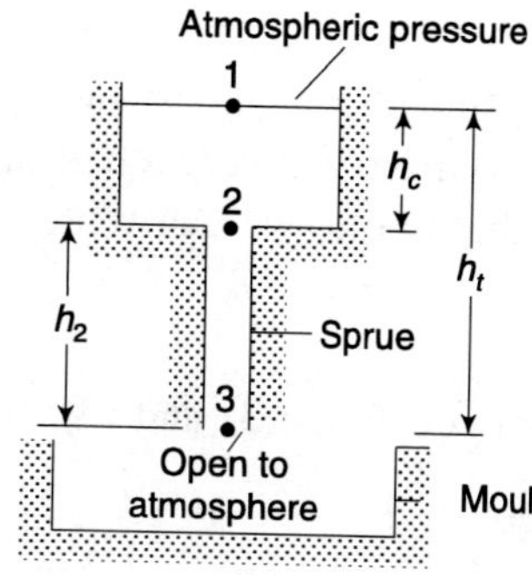

Figure 7.8 Simple vertical gating

In Figure 7.8, it is assumed that the pressure at points 1 and 3 is equal (i.e., $p_1 = p_3$) and that level; is maintained constant. Hence, the velocity at station 1 (v_1) is zero and the frictional losses are neglected. Thus, the energy balance equation between points 1 and 3 gives

$$gh_t = v_3^2/2$$

$$v_3 = \sqrt{(2gh_t)} \quad \text{...(1)}$$

This equation is known as the Torricelli's equation, where g is the acceleration due to gravity and V_3 is the velocity of the liquid metal at the gate, consequently referred to as V_g. So the time taken to fill up the mould,

$$t_f = \frac{Vm}{A_g v_3} \quad \text{...(2)}$$

where A_g is the cross-sectional area of the gate and V_m is the volume of the mould.

7.3.2 Bottom Gating

In Figure 7.2, applying Bernoulli's equation between points 1 and 3, we get

$$gh_t = \frac{p_3}{\rho_m} + \frac{v_3^2}{2} \quad \text{...(3)}$$

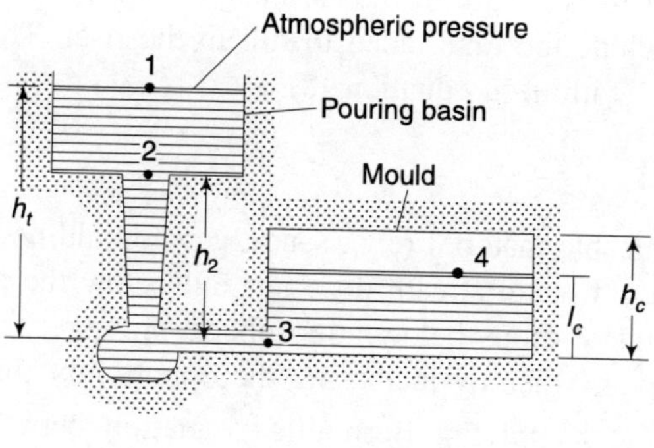

Figure 7.9 Bottom gating

where ρ_m is the density of the liquid metal, p_3 is the gauge pressure at station 3, and h_t is again assumed to be constant. Again applying Bernoulli's equation between points 3 and 4, by assuming that v_4 is very small and all the kinetic energy at station 3 is lost after liquid metal enters the mould, we can write

$$p_3/\rho_m = gh \quad \text{...(4)}$$

From equations **(3)** and **(4)**, the velocity of the liquid metal at the gate we obtain is

$$v_g = v_3 = \sqrt{2g(h_t - h)} \quad \text{...(5)}$$

Equation **(5)** gives the velocity of a jet discharging against a static head h, making the effective head as $(h_t - h)$. Now, for the instant shown, let the metal level in the mould move up through a height dh in a time interval d_t; A_m and A_g are the cross-sectional areas of the mould and the gate, respectively. Then,

$$A_m\, dh = A_g\, v_g\, dt \quad \text{...(6)}$$

Using equations **(5)** and **(6)** we get

$$\frac{1}{\sqrt{2g}} \frac{dh}{\sqrt{(h_t - h)}} = \frac{A_g}{A_m} dt \quad \text{...(7)}$$

At $t = 0$, $h = 0$ and at $t = t_f$ (filling time), $h = h_m$. Integrating equation **(7)** between these limits, we have

$$\frac{1}{\sqrt{2g}} \int_0^{h_m} \frac{dh}{\sqrt{(h_t - h)}} = \frac{A_g}{A_m} \int_0^{t_f} dt$$

$$t_f = \frac{A_m}{A_g} \frac{1}{\sqrt{2g}} 2(\sqrt{h_t} - \sqrt{h_t - h_m}) \quad \text{...(8)}$$

If a riser is used (in order to take care of the shrinkage from the pouring temperature), then pouring time should also include the time taken to fill up the riser. This can be calculated by replacing A_m with A_r and h_m with h_t in equation **(8)**.

7.3.3 Aspiration Effect

For a mould made of a permeable material (e.g., sand), care should be taken to ensure that the pressure anywhere in the liquid metal stream does not fall below the atmospheric pressure. If the pressure in the molten metal drops below atmospheric pressure, then air can be drawn in or aspired into the metal stream, thereby increasing the opportunity for defects in the casting.

Referring to Figure 7.8 and applying Bernoulli's equation between points 2 and 3 we obtain

$$gh_2 + \frac{p_2}{\rho_m} + \frac{v_2^2}{2} = \frac{p_3}{\rho_m} + \frac{v_3^2}{2} \quad \text{...(9)}$$

where p and v refer to the pressure and velocity, respectively, of the liquid metal at stations 2 and 3. If the pressure at point 3 is atmospheric, i.e., $p_3 = 0$, then $p_2 = -\rho_m g h_2$ as $v_2 = v_3$. The gas aspiration will occur at point 2 in the system. Hence, the design in Figure 7.8 is not acceptable. To avoid negative pressure at point 2 (to ensure positive pressure anywhere in the liquid column), the sprue should be tapered, the ideal shape of which can be determined as follows:

Let, in the limiting case, p_2 be equal to zero, when, from equation **(9)**,

$$\frac{v_3^2}{2} = gh_2 + \frac{v_2^2}{2} \qquad ...(10)$$

From the principle of continuity of flow, $A_2 v_2 = A_3 v_3$ where A is the cross-sectional area. Thus,

$$v_2 = \frac{A_3}{A_2} v_3 = R v_3 \qquad ...(11),$$

where, $R = A_3/A_2$. Using equations **(10)** and **(11)**, we obtain

$$\frac{v_3^2}{2g} = h_2 + \frac{R^2 v_3^2}{2g}$$

$$R^2 = 1 - \frac{2gh_2}{v_3^2} \qquad ...(12)$$

Again, $v_3^2 = 2gh_t$ (applying Bernoulli's equation between points 1 and 3, with $p_1 = p_3 = 0$ and $v_1 = 0$). Substituting this in equation **(12)**, we have

$$R^2 = 1 - \frac{h_2}{h_t} = \frac{h_c}{h_t}$$

$$R = \frac{A_3}{A_2} = \sqrt{\frac{h_c}{h_t}} \qquad ...(13)$$

This can be easily seen to be the shape of a freely falling stream when $v_2 = \sqrt{2gh_c}$ and $v_3 = \sqrt{2gh_t}$. Thus, ideally, the sprue profile should be as shown by the solid lines in Figure 7.10 when the pressure throughout the stream is just atmospheric. However, a straight tapered sprue (shown by the dashed lines) is safer (pressure everywhere, except at points 2 and 3, is above atmospheric) and easier to construct. The sprue design in Figure 7.9 is better than that in Figure 7.8. The tapered sprue is designed to conform to the natural form of the flowing stream and therefore reduces turbulence and the possibility of air aspiration. It also tends to fill quickly, establishing the pressure head characteristics of the full flow conditions required by the Bernoulli's equation.

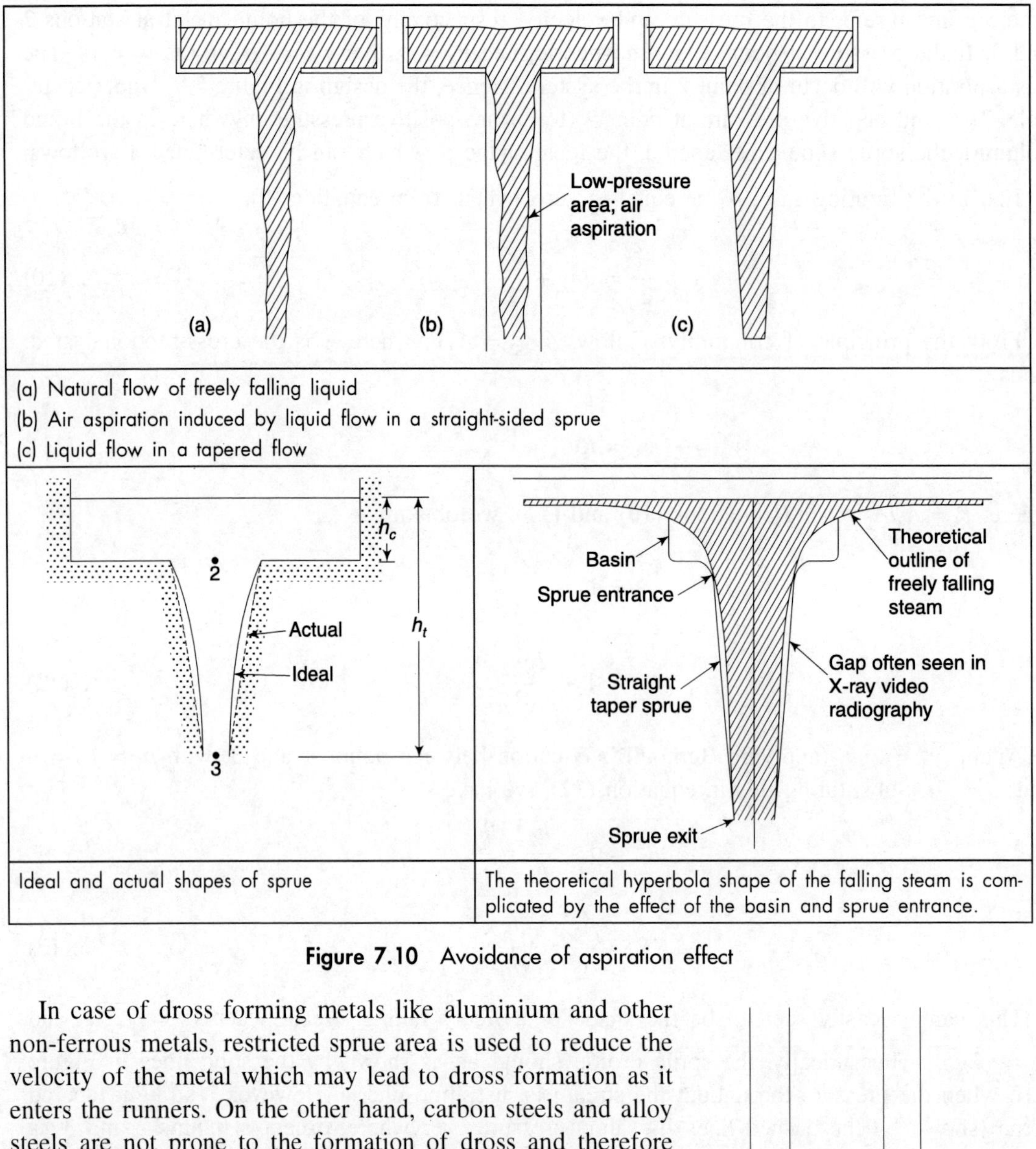

(a) Natural flow of freely falling liquid
(b) Air aspiration induced by liquid flow in a straight-sided sprue
(c) Liquid flow in a tapered flow

Ideal and actual shapes of sprue

The theoretical hyperbola shape of the falling steam is complicated by the effect of the basin and sprue entrance.

Figure 7.10 Avoidance of aspiration effect

In case of dross forming metals like aluminium and other non-ferrous metals, restricted sprue area is used to reduce the velocity of the metal which may lead to dross formation as it enters the runners. On the other hand, carbon steels and alloy steels are not prone to the formation of dross and therefore these precautions are not essential. The pouring temperature of steel is very high and cools rapidly. Therefore, the mould filling must have a fast rate to avoid cooling tendency.

In many high production casting systems, tapered sprue will not be provided. Instead it is compensated by having chokes at the end of sprue or runner as shown in Figure 7.11.

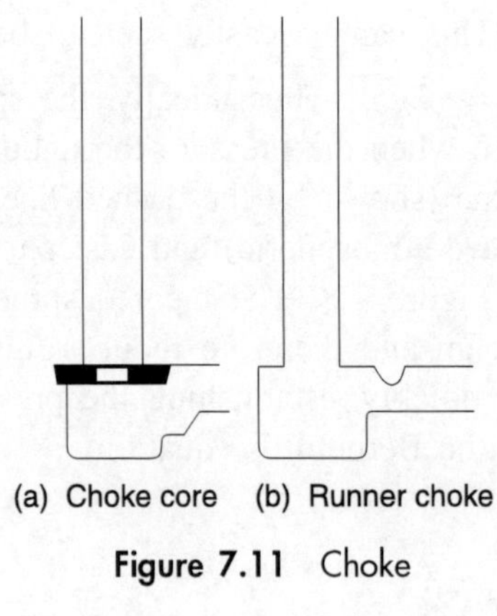

Figure 7.11 Choke

Modes of Liquid Flow

There are two types of metal liquid flows:

1. **Laminar flow:** In laminar flow particles move smoothly, parallel to the direction of flow. The fluid velocity increases from the surface towards the centre of the channel where it reaches the maximum. Laminar flows cause minimum frictional losses.
2. **Turbulent flow:** In turbulent flow, there is irregular movement of liquid particles tumbling in all directions. Other factors which increase the turbulence are:
   (a) Various local resistances such as edges and sharp bends
   (b) Surface area of liquid stream that is exposed to air
   (c) Any suspended second phases like inclusions, slag, gas bubbles, oxide films entrained in the stream.

Reynolds number is used to represent the degree of turbulence. It is defined as

$$R_e = \frac{\rho V d}{\mu} = \frac{Vd}{\nu}$$

V = Mean velocity of liquid metal in m/s

d = Linear dimensions of the channel, m

ρ = Density of liquid metal in kg/m^3

μ = Dynamic or absolute viscosity of liquid in $\frac{\text{N} - \text{s}}{\text{m}^2}$

ν = Kinematic viscosity of liquid metal $= \frac{\mu}{\rho}$

Two critical flow velocities exist. Lower critical velocity below which flow is laminar or viscous. Upper critical velocity above which flow is turbulent.

- Typical critical velocity
  - Al bronze and other highly oxide-prone metals 75 mm/second
  - Al base and Mg base alloys 250 mm/second
  - Copper base and iron base alloys 500 mm/second

Calculations of Reynolds number show that flow of metal in most practical gating systems is turbulent, so a certain amount of turbulence is inevitable and must therefore be tolerated. In practice, design of gating system does not involve the elimination of metal turbulence but rather its reduction to a point where it is not harmful. In general, it is important that the liquid metal flows through the gates at a speed lower than the critical velocity so as to enter the mould cavity smoothly. If the rate of entry is too high, causing the metal to splash then the battle for quality is probably lost. The turbulence inside the mould cavity is the most serious turbulence of all.

7.3.4 Effect of Friction and Velocity Distribution

In the sections 7.2.1 to 7.2.3, we have assumed that the fluid is an incompressible liquid, channel is full and the walls are impermeable. But in practice the sand mould used for casting is permeable; liquid metal in the gating system is always accompanied by mould gases and air and results in friction at mould wall.

Also we have assumed that the velocity of a liquid metal in the sprue and the gate is uniform across the cross section. In fact, the velocity of fluid in contact with any solid surface is zero and is maximum at the axis of the conduit. The velocity distribution within the conduit depends on the shape of the conduit and nature of the flow (i.e., turbulent or laminar).

Further, in our discussion so far, we have also assumed no frictional losses. In real fluids, the frictional losses are always present, especially when there is a sudden contraction in or an enlargement of the flow cross sections. Besides, energy is lost due to resistance of mould wall to the passage of metal (interface friction), internal friction (viscosity) of liquid metal, sudden changes in cross section and sharp changes in direction at bends and junctions. So corrections are to be applied using experimentally determined discharge coefficient, although such losses are not so high.

In the discussion that follows, we shall, in the light of these two factors, i.e., velocity distribution and friction, modify the equation we have already developed.

The non-uniform velocity distribution can be accounted for by modifying the kinetic energy term in the integrated energy balance equation by replacing the (velocity)^2 term $\vec{v}^2/\beta$, where $\vec{v}$ is the average velocity and β is a constant. For a circular conduit, the value of β is equal to 0.5 for laminar flow and approximately equal to 1 for turbulent flow.

The energy loss due to friction in a circular conduit (on the basis of per unit mass) is given by

$$E_{f_1} = 4f\frac{L}{D}\frac{\vec{v}^2}{2} \quad \text{...(14)}$$

where

$\vec{v}$ = Average velocity

D = Diameter of the conduit

L = Length of the conduit

f = Friction factor

The value of f depends on the roughness of the conduit and the nature of the flow. Thus, while using the integrated energy equation between two points, say, 1 and 2, in that order in the direction of flow, E_{f_1} should be added to the energy of station 2.

For a smooth conduit, the value of f is given by the equations

$$f = 16/\text{Re for laminar flow (Re} < 2000) \quad \text{...(15)}$$

$$\frac{1}{\sqrt{f}} = 4\log_{10}(\text{Re}\sqrt{f}) - 0.4 \text{ for turbulent flow (Re} > 2000) \quad \text{...(16)}$$

where Re is Reynolds number. For the range 2100 < Re < 10^5, equation **(16)** can be simplified to the form

$$f = 0.0791\ (\mathrm{Re})^{-1/4} \qquad ...(17)$$

Frictional losses also occur due to a gradual change in the flow direction, e.g., in a 90° bend and similar other fittings. In such a situation, the associated frictional losses are accounted for by using equation **(14)** with an equivalent (*L*/*D*) factor for the bend.

The frictional loss (per unit mass) associated with a sudden enlargement or contraction of a flow area is expressed as

$$E_{f2} = \frac{1}{2}\, e_f\, \vec{v}^2 \qquad ...(18)$$

where $\vec{v}$ is the average velocity of the fluid in the smaller cross section, and e_f is the friction loss factor and it depends on the ratio of flow area and Reynolds number of the flow. For a laminar flow, the length and diameter of the smaller flow cross section have also to be taken into account. The value of e_f depends on whether the flow area is enlarging or contracting in the direction of the flow. The value of e_f for a sharp change in the flow cross sections are shown in Figures 7.12 and 7.13 for sudden expansion and contraction respectively.

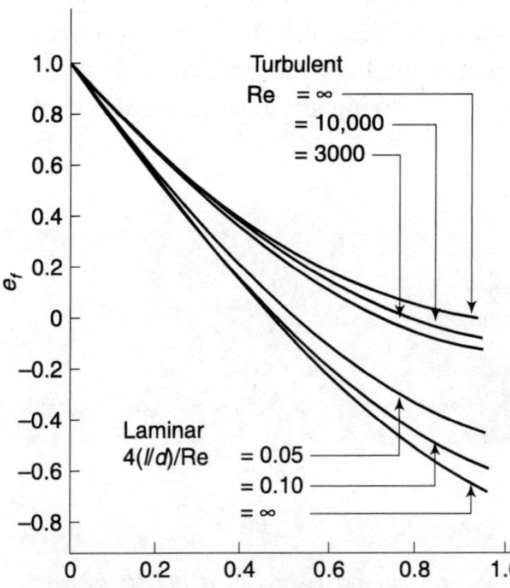

Figures 7.12 Friction loss factor for sudden expansion

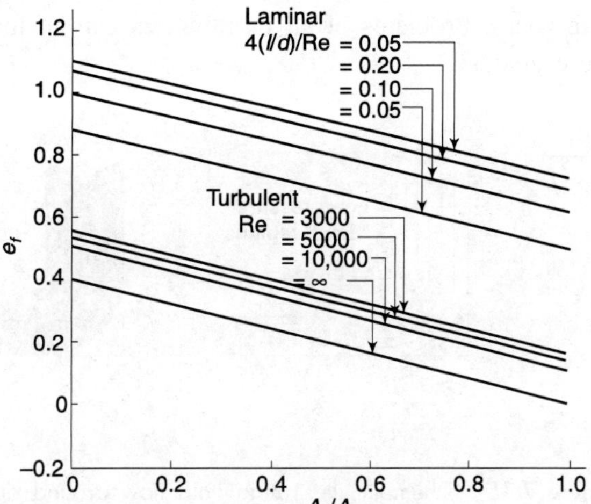

Figures 7.13 Friction loss factor for sudden contraction

In these figures, the Reynolds numbers of the flow have been calculated on the basis of the average flow velocity in the smaller cross sections, and *l* and *d* are length and diameter, respectively of the smaller flow cross sections. The value of e_{fs} for some other types of changes in flow geometry are listed in Figure 7.14 here; e_{fs} refers to the values corresponding to a sharp change of geometry identical initial and final dimensions (i.e., the values of e_{fs} can be obtained from Figures 7.12 and 7.13).

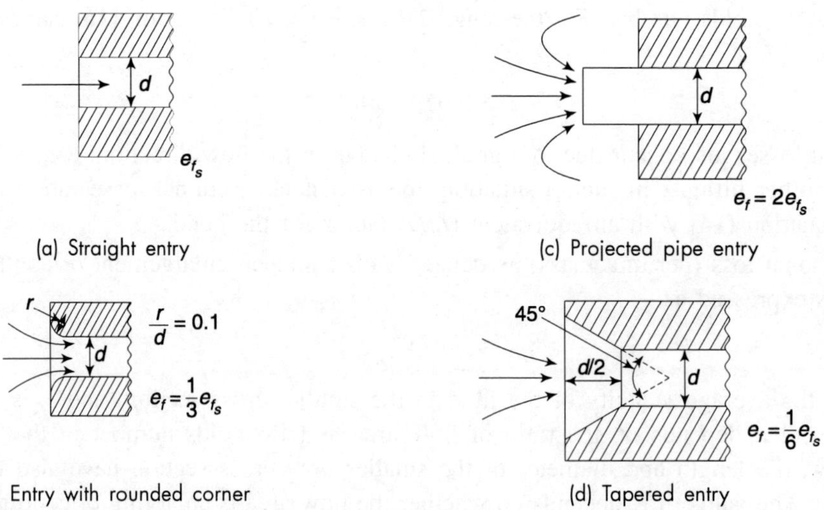

Figure 7.14 Values of e_f for some types of changes in flow geometry

Figure 7.15 illustrates effect of abrupt changes in flow channel cross section. There will be lot of turbulence resulting from sharp corners and metal damages. Steamlined corner minimizes turbulence and metal damage. Sudden changes in the direction of flow can produce low pressure zones. Problems of air entrainment can be minimized by making the change in direction more gradual.

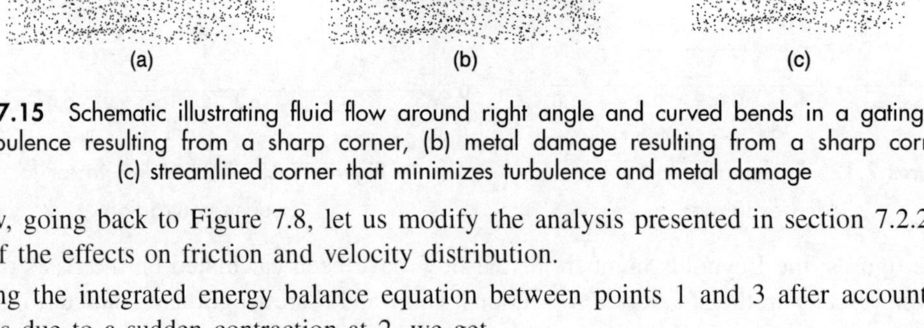

Figure 7.15 Schematic illustrating fluid flow around right angle and curved bends in a gating system: (a) Turbulence resulting from a sharp corner, (b) metal damage resulting from a sharp corner and (c) streamlined corner that minimizes turbulence and metal damage

Now, going back to Figure 7.8, let us modify the analysis presented in section 7.2.2 in the light of the effects on friction and velocity distribution.

Using the integrated energy balance equation between points 1 and 3 after accounting for the loss due to a sudden contraction at 2, we get

$$\frac{p_1}{\rho_m} + 0 + gh_t = \frac{p_3}{\rho_m} + \frac{\vec{v}_3^{\,2}}{2\beta} + E_{f_1} + E_{f_2} \qquad ...(19)$$

where $\vec{v}_3$ is the average fluid velocity in the sprue. With $p_1 = p_3$ and using equations **(14)** and **(18)**, we get

$$2gh_t = \vec{v}_3^2 \left[\frac{1}{\beta} + ef + 4f\frac{l}{d}\right] \qquad ...(20)$$

where d is the diameter of the sprue and l is the length of the sprue (= h_2 in Figure 7.8). Thus, equation **(1)** can be modified as

$$\vec{v}_3 = C_D \sqrt{2gh_t}, \qquad ...(21)$$

where C_D is the discharge coefficient

$$C_D = \left[\frac{1}{\beta} + e_f + 4f\frac{l}{d}\right]^{-1/2} = \frac{1}{\sqrt{\left[\frac{1}{\beta} + e_f + 4f\frac{l}{d}\right]}} \qquad ...(22)$$

In this analysis, we have neglected the fluid velocity (and hence the loss) between points 1 and 2. If the sprue also has a bending or fitting, then E_{f_1} is modified as

$$E_{f_1} = 4f\vec{v}_3^2 \left[\frac{l}{d} + \left(\frac{L}{D}\right)_{eq}\right]$$

The values of $(L/D)_{eq}$ for various types of fittings are listed in standard tables. In such a case, the discharge coefficient C_D is finally given by

$$C_D = \left[\frac{1}{\beta} + e_f + 4_f\frac{1}{d} + \left(\frac{L}{D}\right)_{eq}\right]^{-1/2} \qquad ...(23)$$

7.4 EMPIRICAL RELATIONS USED IN THE DESIGN OF GATING SYSTEMS

In many industries, the various relations we have derived theoretically are not used but established generally by the practice at various foundries and by experimenters. This section deals with design of gating systems using empirical relations.

Sequential steps for design of gating systems:

1. Selection of layout
2. Estimation of optimum pouring time of the casting
3. Calculation of sprue choke area
4. Selection of gating ratio
5. Selection of type of gating and location
6. Calculation of runner, ingate sizes

7.4.1 Selection of a Layout

- Decide first which way up the part is to be cast.

- For two/three-part mould, use casting in cope/middle box, allowing gating at the lowest parts of the casting.
- A heavy section of the casting needs special attention. Orient the heavy part of the casting at the top and plant a feeder there. If this cannot be done, provide chills and/or fins to this section.
- When a general scheme is decided, including the approximate sitting of gates and runners, the provision of feeders (if any) and the location of the sprue, a start can be made on the quantification of the gating system.

7.4.2 Estimation of Optimum Pouring Time of the Casting

One of the objectives as mentioned earlier, for the gating system design, is to fill the mould in the smallest time. The time for complete filling of a mould known as pouring time or filling time. It is an important criterion for design. Too long a pouring time requires a higher pouring temperature and too less a pouring time means turbulent flow mould which makes the casting defect prone. There is thus an optimum pouring time for any given casting.

The pouring time depends on the casting materials, complexity of the casting, section thickness and casting size. The various relations used are not theoretically obtained but established generally by the practice at various foundries and by experimenters. The general considerations for choosing pouring time for grey cast iron may not be much relevant for steels since they lose heat very fast and therefore, the pouring time should be less. For non-ferrous materials, a longer pouring time would be beneficial since they lose heat slowly and also tend to form dross if metal is poured too quickly.

Since the thickness of casting is affected to a great extent, by the ratio of surface area to volume of the casting, it is an important variable in calculating the optimum pouring time in addition to the mass of the casting itself. Normally while considering the mass of the casting, it may not be necessary to consider the mass of the gating system because the gating system is completely filled before metal starts entering the mould cavity. However, if the gating systems are in comparable size to the actual casting, it may be desirable to include them for the calculation.

The following are some standard emprical methods to calculate the filling time for different casting materials.

1. Grey iron casting

Thin walled casting of thickness 2.5-15 mm and weight less than 450 kg

$$t = k\sqrt{W}$$

W = Weight of liquid metal poured in kg

k = Constant taking into account casting wall thickness

Thickness of casting (mm)	2.5 – 3.5	3.5 – 8.0	8.0 – 15
Constant, k	1.63	1.85	2.20

2. Grey cast iron, mass less than 450 kg

Pouring time, t (in seconds) $= K\left(1.41 + \frac{T}{14.59}\right)\sqrt{W}$

where,

$$K = \frac{\text{Fluidity of iron (in inches)}}{40}$$, fluidity can be obtained from Figure 7.16,

T = Average section thickness, mm

W = Mass of the casting, kg

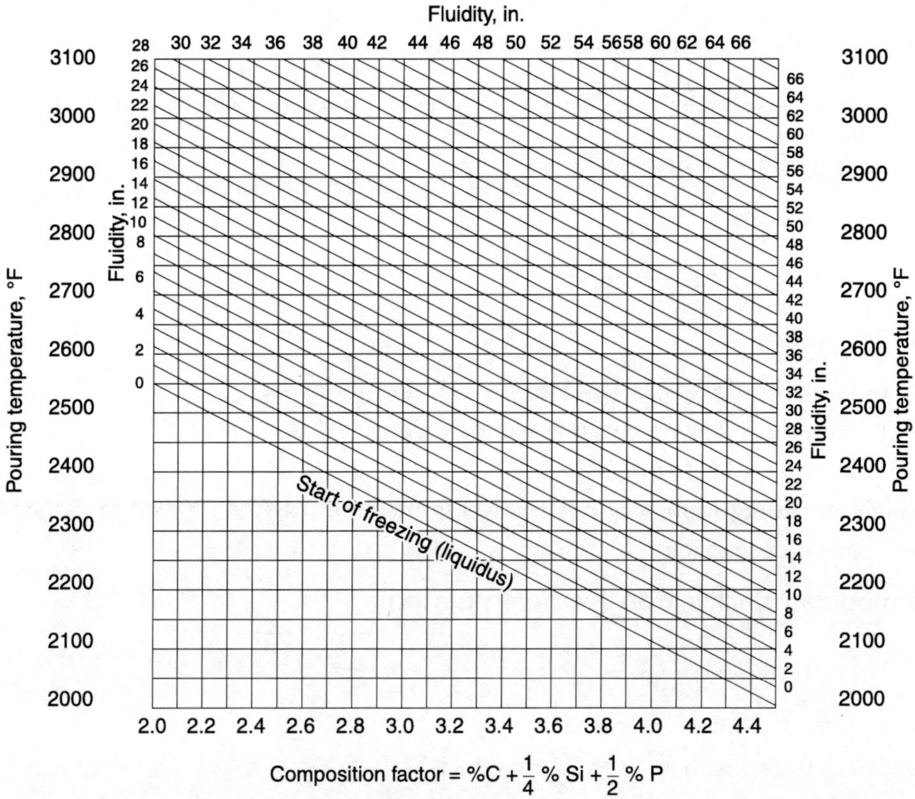

Figure 7.16 Fluidity related to pouring temperature and composition of grey and malleable cast iron

In **FPS** system, the following equation can be used.

Pouring time, t (in seconds) $= K\left(0.95 + \frac{T}{0.853}\right)\sqrt{W}$ for casting weight < 1000 lb

where, $$K = \frac{\text{Fluidity of iron (in inches)}}{40}$$, fluidity can be obtained from chart 1,

T = Average section thickness, inches

W = Mass of the casting, lbs

3. Grey cast iron, mass greater than 450 kg

Pouring time, t (in seconds) $= K\left(1.236 + \dfrac{T}{16.65}\right)\sqrt[3]{W}$

Typical pouring times for cast iron are:

Casting mass	Pouring time in seconds
20 kg	6 to 10
100 kg	15 to 30
100000 kg	60 to 180

In **FPS** system, the following equation can be used.

Pouring time, t (in seconds) $= K\left(0.95 + \dfrac{T}{0.853}\right) W^{\frac{1}{3}}$ for casting weight > 1000 lb

where $K = \dfrac{\text{Fluidity of iron (in inches)}}{40}$, fluidity can be obtained from chart 1,

T = Average section thickness, inches

W = Mass of the casting, lb

4. Steel Castings

Pouring time = t(in seconds) = $(2.4335 - 0.3953 \log W)\sqrt{W}$

In **FPS** system, the following equation can be used

Pouring time = t(in seconds) = $k\sqrt{W}$

W = weight of casting in lb, k = 1.2 (for 100 lb casting), 0.4 (for 100,000 lb casting)

(k values plotted against logW)

5. Shell moulded ductile iron (vertical pouring)

$$\text{Pouring time} = t(\text{in seconds}) = K_1 \sqrt{W}$$

where

K_1 = 2.080 for thinner sections

K_1 = 2.670 for sections 10 to 25 mm thick

K_1 = 2.970 for heavier sections

6. Copper alloy castings

Pouring time = t(in seconds) = $K_2 \sqrt[3]{W}$

K_2 is a constant given by

Top gating	1.30
Bottom gating	1.80
Brass	1.90
Tin bronze	2.80

Another empirical relation is given by

Pouring time = t(in seconds) = $K\ ^3\sqrt{TW}$

T = Average thickness of casting in mm, k = 1.3 to 1.8 (for bottom gated system; lowest value for brass, higher value for tin bronze), k = 1.9 – 2.8 (for top gates systems; lower values for brass, higher values for bronze), W = Weight of the casting, kg.

7. Copper and aluminium alloy castings

Pouring time = t(in seconds) = $k\sqrt{W}$

K = 5 for copper base alloys and k = 8 for Al base alloys. W = Weight of casting in kg.

8. Intricately shaped thin-walled castings of mass up to 450 kg

$$\text{Pouring time} = t(\text{in seconds}) = K_3\,{}^3\sqrt{W'T}$$

where, W' = Mass of the casting with gates and risers, kg

K_3 = A constant as given below

T (mm)	K_3
1.5 to 2.5	1.62
2.5 to 3.5	1.68
3.5 to 8.0	1.85
8.0 to 15.0	2.20

9. For castings above 450 kg and up to 1000 kg

$$\text{Pouring time, } t(\text{in seconds}) = K_4\ ({}^3\sqrt{(W'T)}$$

where, K_4 is a constant given by

T (mm)	K_3
upto 10	1.00
10 to 20	1.35
20 to 40	1.5
above 40	1.7

Typical pouring times for castings whose mass is less than 200 kg and average section thickness of 25 mm are

Grey Cast iron	40 s
Steel	20 s
Brass	15 to 45 s

Pouring rate calculations

1. Determination of the pouring rate (kg/s)

To calculate the optimum pouring rate (in kg/s) for different metals, the following equations can be used.

(a) Ferrous metals and copper base alloys casting

$$R = \frac{W^P}{\left(1.34 + \frac{T}{13.77}\right)}$$

T = Critical casting thickness, mm

W = Weight of the casting, kg

P = Constant

R = Pouring rate in kg/s

Casting weight (kg)	Up to 500	500-5000	5000-15000
Constant (P)	0.5	0.67	0.70

(b) Light metal casting

$$R = b\sqrt{W}$$

W = Weight of the casting, kg

b = Constant

R = Pouring rate in kg/s

Wall thickness (mm)	Below 6 mm	6-12 mm	Above 12 mm
Constant (b)	0.99	0.87	0.47

2. Determination of the adjusted pouring rate (kg/s)

The pouring rate thus obtained from these two equations has to be corrected for metal fluidity (k) and the effect of the friction (f factor) in the gating systems. The adjusted pouring rate (R_a) can be calculated according to the following equation,

$$R_a = \frac{R}{k \times f}$$

The factor f has value of 0.85-0.90 for tapered sprue and 0.70-0.75 for straight sprue.

For all metals other than grey and malleable cast iron, k can be taken as unity. In the cast of the grey and malleable cast iron, the metal fluidity can be calculated from the carbon equivalent (Figure 7.16).

Typical pouring rates used for different casting alloys are given in Table 11.1. The calculated values from any of the above formulae may be checked from the same table.

Table 7.1 Typical pouring rates for sand castings

Metal	Pouring rate in kg/s for castings of mass, kg			
	Up to 10	10 – 50	50 – 100	100 – 500
Pig iron	1.1	1.5 – 2.0	3.0 – 4.0	3.5 – 6.0
Steel	1.2 – 1.4	1.9 – 2.5	4.0 – 5.0	4.5 – 7.0
Aluminium alloys	0.25 – 0.3	0.5 – 0.7	1.0 – 1.3	1.2 – 2.0

7.4.3 Calulation of Sprue Choke Area

Having calculated the optimum pouring time, it is now required to establish the main control area which meters the metal flow into the mould cavity so that the mould is completely filled within the calculated pouring time. This is controlling the choke area. Normally, the choke area happens to be at the bottom of the sprue and hence, the first element to be designed in the gating system is the sprue size and its proportions. The main advantage in having sprue bottom as the choke area is that proper flow characteristics are established early in the mould.

(a) Choke area calculations (Figure 7.17)

The choke area can be calculated using Bernoulli's equation as

$$A = \left(\frac{W}{\rho t C\sqrt{2gH}}\right)$$

where, A = Choke area, mm^2
W = Casting mass, kg
t = Pouring time, seconds
ρ = Mass density of the molten metal, kg/mm^3
g = Acceleration due to gravity, mm/s^2
H = Effective metallostatic head (sprue height), mm
C = Efficiency factor or discharge factor

The effective sprue height H, of a mould depends on the casting dimensions and the type of gating used. The effective sprue heads can be calculated using the following relation:

Top gate, $H = h$

Bottom gate, $H = h - \left(\frac{c}{2}\right)$

Parting gate × 3 $H = h - \left(\frac{p^2}{2c}\right)$

where, h = Height of sprue
p = Height of mould cavity in cope
c = Total height of mould cavity

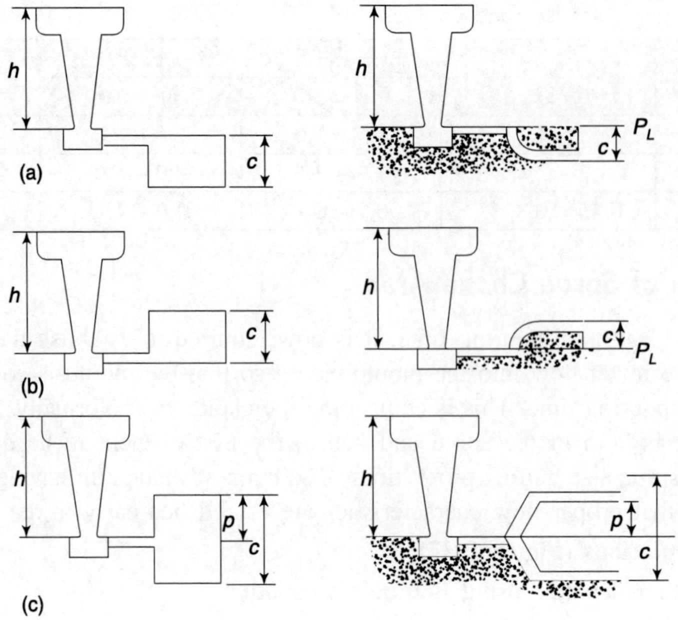

Figure 7.17 Choke area calculations

The efficiency coefficient (discharge factor) of the gating system depends on the various sections that are normally used in a gating system. The elements of a gating system should normally be circular in cross section since they have lower surface area to volume ratio which would reduce losses and have less friction. Moreover, streamlining the various gating elements would greatly increase the volumetric efficiency of the gating system and allow for smaller size gates and runners which would increase the casting yield.

Whenever a runner changes direction or joins with another runner or gate, there is some loss in the metal head, all of which when taken properly into consideration would give the overall efficiency of the gating system. Hence,

$$C = \left[\frac{1}{\sqrt{(1 + k_1 \left(\frac{A^2}{A_1^2}\right) + k_2 \left(\frac{A^2}{A_2^2}\right) + \ldots}} \right]$$

where k_1, k_2, ... are loss coefficients occurring at changes in the direction or area as given in Table 7.2.

A_1, A_2, ... are areas downstream from changes.

A is the choke area.

Table 7.2 Values of loss coefficients for various gate elements

Gate elements	*Sharp*	*Round*
Sprue entry from pouring cup	0.75	0.20
Bend of sprue into runner	2.00	1.00
Right angle bend in runner		
Square cross section	2.00	1.50
Round cross section	1.50	1.00
Junction at right angles to runners	4.0 to 6.0	–
Junction with 25% or more area reduction from runner into ingates	2.00	0.50
Runner choke when choke area approximately (1/3)rd runner area, plus bend of sprue into runner	13.00	–
Losses from wall friction		
Round channel loss $= \dfrac{0.02L}{D}$		
Square channel loss $= \dfrac{0.06L}{D}$		
Rectangular channel loss $= 0.07L\left(\dfrac{A+B}{2AB}\right)$		
where, L = Length; D = Diameter of round or side of square; A = One side of rectangle; B = Other side of rectangle		

Though, this is the most rigorous way of calculating the efficiency factor, it may not be necessary to go to this length all the time. Average values of the efficiency factor are provided for typical gating systems in Table 7.3 which may be used for calculating the gating.

Though it is preferable to have the choke in the sprue, it may sometimes be convenient to mould a straight sprue in which case, the choke is provided in the runners. The efficiency factors for such systems are also provided in Table 7.3.

For aluminium castings, a metal flow rate of 4.044 g/min for 1 sq. mm of sprue area in an unpressurised gating is found to be sufficient for achieving sound castings.

Table 7.3 Efficiency coefficients, C for various types of gating systems

Type of system	*Tapered choked sprue*	*Straight sprue runner choke*
Single runner	0.90	0.73
Two runners with multiple ingates, no bends in runners	0.90	0.73
Two runners with multiple ingates, 90° bends in runners	0.85	0.70

(b) Sprue design calculations avoiding aspiration effects

As explained earlier, the sprues should be tapered down to take into account the gain in velocity of the metal as it flows down reducing the aspiration. The exact tapering can be obtained by the equation of continuity. Denoting the top and choke sections of the sprue by the subscripts t and c respectively we can get,

$$A_t V_t = A_c V_c$$

$$A_t = \frac{A_c V_c}{V_t}$$

Since the velocities are proportional to the square of the potential heads, as can be derived from Bernoulli's equation,

$$A_t = A_c \left(\sqrt{\frac{h_c}{h_t}} \right)$$

The square root suggests that the profile of the sprue should be parabolic if exactly done as per the above equation. But making a parabolic sprue is too inconvenient and therefore a straight taper is preferable. It has been found that a straight tapered sprue is able to effectively reduce the air aspiration as well as increase the flow rate compared to a parallel sprue.

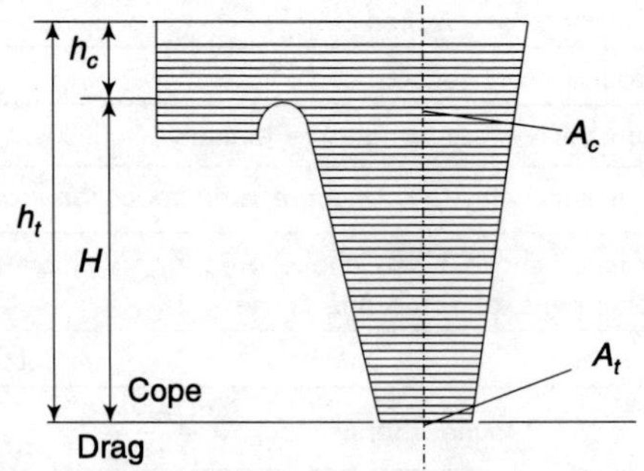

Figure 7.18 Sprue design

In order to arrive at the dimensions of the sprue at the top and subsequent taper, one has to consider the head of the metal in the pouring basin as shown in Fig. 7.18. Metal at the entry of the sprue would be moving with a velocity of $\sqrt{2gh}$ and hence

$$A_t = A_c \left(\sqrt{\frac{h_t}{h_c}} \right)$$

where, H = Actual sprue height

and $h_t = h_c + H$

Table 7.4 shows the theoretical values of area ratios of top and choke portions of the sprue based on sprue height and metal head in the pouring basin.

Table 7.4 Theoretical ratios of sprue top and choke areas based on pouring basin depth

Sprue height, mm	*Depth in pouring basin, mm*				
	50	100	150	200	250
50	1.414	1.225	1.155	1.118	1.095
100	1.732	1.414	1.291	1.225	1.183
150	2.000	1.581	1.414	1.323	1.265
200	2.236	1.732	1.528	1.414	1.342
250	2.450	1.871	1.633	1.500	1.414
375	2.915	2.179	1.871	1.696	1.581
500	3.317	2.450	2.082	1.871	1.732
600	3.742	2.739	2.309	2.062	1.897

Though these ratios are theoretically correct, often it is not possible to control exactly the amount of the head in the pouring basin while pouring. Hence, it is a general practice to neglect the effect of the pouring basin head and proportion of the sprue top solely based on the sprue height alone.

(c) Other gating elements

(i) *Pouring basin* The main function of a pouring basin is to reduce the momentum of the liquid flowing into the mould by settling the first into it. In order that the metal enters the sprue without any turbulence it is necessary that the pouring basin be deep enough, and also the entrance to the sprue of smooth radius of at least 25 mm. Experience shows that the pouring basin depth of 2.5 times the sprue entrance diameter is enough for smooth metal flow and to prevent vortex formation, as shown in Figure 7.19.

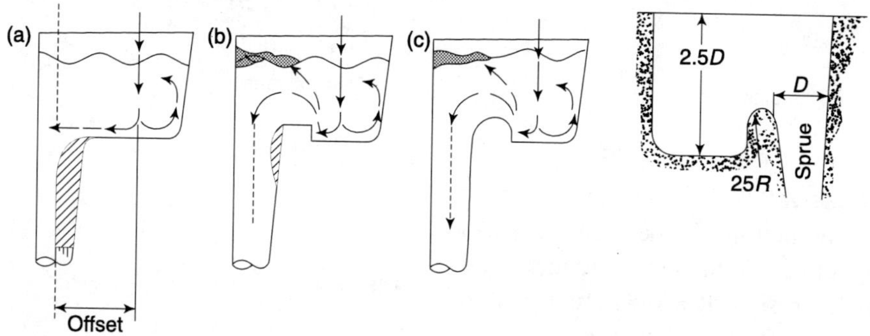

Figure 7.19 Offset pouring basins: (a) without step (not used), (b) sharp step (not used), (c) residual step (used commonly)

The offset blind end brings the vertical downward velocity to a stop. It avoids the incoming liquid which goes straight down the sprue with unchecked velocity and taking with it unwanted components such as air and dross, etc. The step (or weir) (a few mm to 20 mm high) is essential to eliminate the fast horizontal components of flow over the top of the sprue. It also reverses the downward velocity to make an upward flow, giving some opportunity for lighther phases such as slag and bubbles to separate prior to entering the sprue. The provison of a generous radius over the top of the step smoothing the entrance into the sprue, further aids the smooth, laminar flow of metals.

In order that vortex is not formed during pouring, it is necessary that the pouring basin be kept full. Furthur provision should be made in the pouring basin so that constant conditions of flow are established. This can be achieved by using a delay screen or a strainer core, as in Figure 7.20. A delay screen is a small piece of perforated thin tin sheet placed in the pouring basin at the top of the down sprue. This actually melts because of the heat from the metal and in the process delays the entrance of the metal into the sprue thus filling the pouring basin fully. This ensures a constant flow of metal as also excludes slag and dirt since only metal from below is allowed to go into the sprue. A similar effect is also achieved by a strainer core which is a ceramic coated screen with many small holes. The strainer restricts the flow of metal into the sprue and thus helps in quick filling of the pouring basin.

The metal should be poured steadily into the pouring basin keeping the lip of the ladle as close as possible. Pouring basins are most desirable with castings in alloys which form troublesome oxide skins (aluminium, aluminium bronze, etc.).

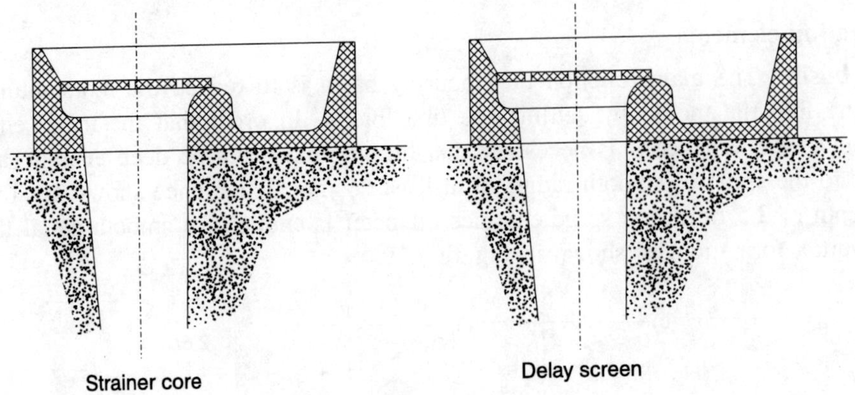

Figure 7.20 Pouring basin designs

(ii) *Sprue Basewell* The provision of sprue base well at the bottom of the sprue helps in reducing the velocity of the incoming metal and also the mould erosion. It avoids initial splash and suppresses the formation of vena contracta. Reasonable proportions for a sprue base well are presented in Figure 7.21. A general guideline could be that the sprue well area should be five times that of the sprue choke area and the well depth should be approximately equal to that of the runner. For a narrow and deep runner, the well diameter should be 2.5 times the width of the runner in a two runner system, and twice its width in one runner system.

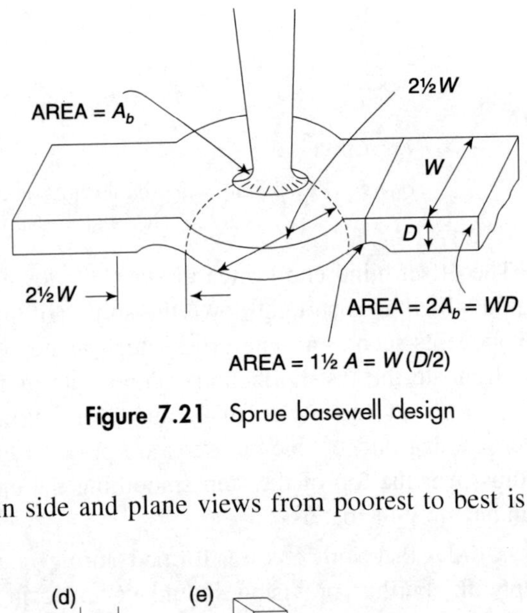

Figure 7.21 Sprue basewell design

A wide variety of the sprue/runner junctions in side and plane views from poorest to best is shown in Figure 7.22.

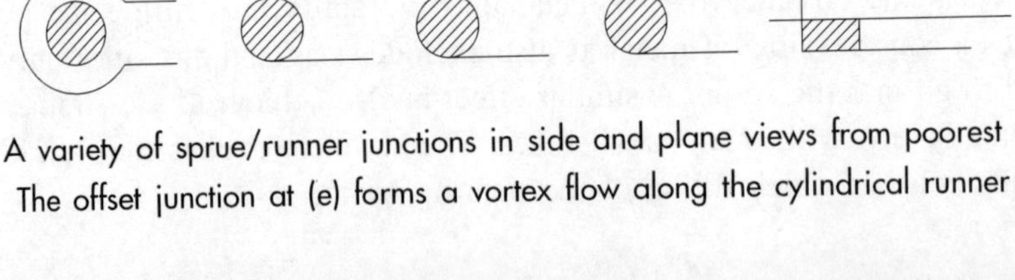

Figure 7.22 A variety of sprue/runner junctions in side and plane views from poorest (a) to best (d). The offset junction at (e) forms a vortex flow along the cylindrical runner

The optimized size of a well is double the diameter of the sprue exit and double the depth of the runner as shown in Figure 7.23.

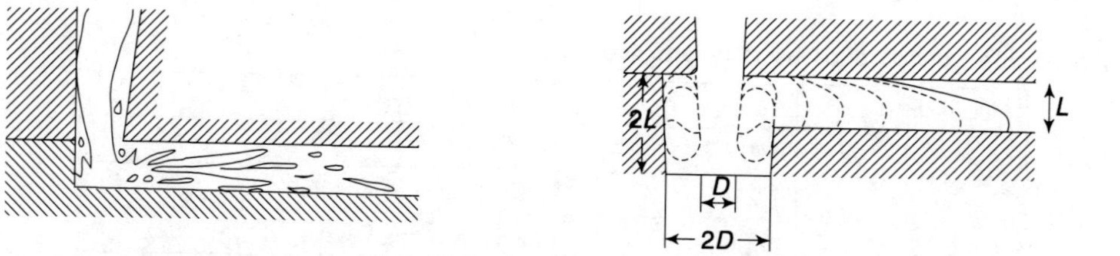

Sprue well avoids initial splash and suppresses formation of vena contracta.

The optimized size of a well is double the diameter of the sprue exit and double the depth of the runner.

Figure 7.23 Optimised size of a well

Do not place the gate at the base of the sprue so that the high velocity of the falling stream is redirected straight into the mould (Figure 7.24).

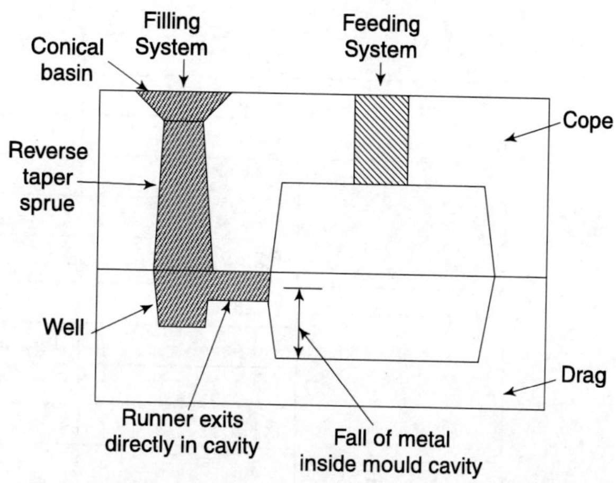

Figure 7.24. Position of sprue and basewell cause high velocity of stream

(iii) *Design guidelines for Pouring Bush/Basin*

- Decide on the type of pouring basin used.
- Conical pouring cup (can be used for small casting). For hand ladle filling (50 mm above the entrance to the sprue), monograms can be used. When poured directly from furnace (usually from a height), monogram cannot be used.
- Pouring bush/basin (for larger castings).

7.4.4 Selection of Gating Ratio

Many authors of previous gating studies have determined the ratios for gating designs as a useful tool. The selection of gating ratio depends on the characteristics of the metal being cast like fluidity, tendency of dross or slag formation, pouring temperature, shrinkage behaviour, etc. It also depends on the material characteristics like resistance to erosion, scabbing tendency and type of moulds like greensand, dry sand, CO_2, no-bake cold box, hot box, etc. The type of gating system may be pressurized or non-pressurized depending on the above parameters which finally control the rate of flow in the different parts of the mould.

The gating ratio refers to the proportion of the cross-sectional areas between the sprue, runner and ingates. These ratios are made to determine the cross sections between several portions of the gating system. The standard convention takes the ratio of sprue area:runner area: ingate area, where the sprue choke is the smallest cross-sectional area in the sprue. Table 7.5 gives gating ratios used in practice for various materials.

Table 7.5 Some gating ratios used in practice

Material	Ratio	Remarks
Aluminium	1:2:1	Pressurized system
	1:1.2:2	
	1:2:4	Dross forming materials
	1:3:3	Non-pressurized system
	1:4:4	Dross forming materials
	1:6:6	
Aluminium bronze	1:2.88:4.8	
Brass	1:1:1	
	1:1:3	
	1.6:1.3:1	
Copper	2:8:1	
	3:9:1	
Ductile iron	1.15:1.1:1	
	1.25:1.13:1	
	1.33:2.67:1	
Grey cast iron	1:1.3:1.1	
	1:4:4	Used in some foundries
	1.4:1.2:1	
	2:1.5:1	
	2:3:1	
	4:3:1	
	1:1:1	In case of single gating system
	1:1.41:1	In case of double gating system
	1:1.73:1	In case of three gating systems
	1:2:1	In case of four gating systems
Magnesium	1:2:2	
	1:4:4	
Malleable iron	1:2:9.5	
	1.5:1:2.5	
	2:1:4.9	

Steels	1:1:7	
	1:2:1	
	1:2:1.5	
	1:2:2	Non pressurized system
	1:3:3	
	1.6:1.3:1	
	1:1:1	Fin gated
SG iron	4:8:3	For intricate casting (pressurised systems)
	10:9:8	Dry sand mould
	1:2:2	Shell mould, vertical pouring
	1.2:1:2	Reverse choke

7.4.5 Selection of Type of Gating and Location

The selection of the gating and its location depends on various factors as given below.

1. Casting configuration
2. Method of moulding
3. Casting wall thickness
4. Weight of metal entering mould
5. Mould material

7.4.6 Calculation of Ingate Sizes and Runners Design

Ingate sizes design: The ingate can be considered a weir with no reduction in cross section of the steam at the gate. Then the rate of flow of molten metal through the gates depends on the free height of the metal in the runner and the gate area and the velocity with which metal is flowing in the runner. The free height, h can be calculated as:

$$h = 1.6 \sqrt[3]{\frac{Q^2}{gb^2} + \frac{V^2}{2g}} \text{ (mm)}$$

Q = Metal flow rate, mm^3/s

b = Gate width, mm

V = Metal velocity in runner, mm/s

g = Acceleration due to gravity, mm/s^2

Having obtained the head of the metal, the height of the gate h, is given by

$h_1 = h - 5$ mm

Gates higher than this will not fill completely and those lower than this will increase the velocity of the steam entering it.

The ingates are generally made wider compared to depth up to a ratio of 4. This facilitates in the severing of gating from the casting after solidification. It may sometimes be preferable to reduce the actual connection between the ingate and the casting by means of a neck down, wash burn or dry sand core so that the removal of the gating is simplified.

The following points should be kept in mind while choosing the positioning of the ingates:

1. Ingate should not be located near a protruding part of the mould to avoid the striking of vertical mould walls by the molten metal stream.
2. Ingates should preferably be placed along the longitudinal axis of the mould wall.
3. Ingates should not be placed near a core print or a chill.
4. Ingate cross-sectional area should preferably be smaller than the smallest thickness of the casting so that the ingates solidify first and isolate the castings gating system. This would reduce the possibility of air aspiration through the gating system in case of metal shrinkage.

Design of runner

The runner is that part of the filling system that acts to distribute the melt horizontally around the mould, reaching distant parts of the mould cavity quickly to reduce heat loss problems. The runner is usually necessarily horizontal following the normal mould joint in conventional horizontally parted moulds. For vertically jointed moulds, or investment moulds where there is little geometrical constraint, the runner can be inclined uphill. Runner should be arranged under the casting, so that the runner is connected to the mould cavity by vertical gates. This will fill the runner completely prior to rising through the gates and into the mould cavity.

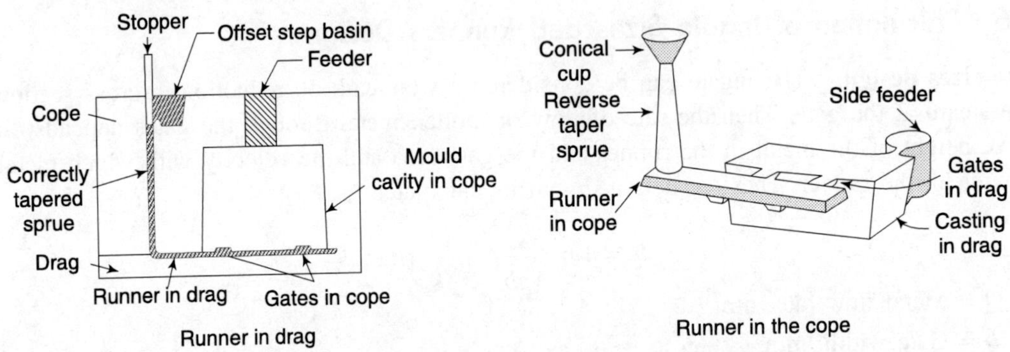

Figure 7.25 Runner design

Moulding runner in cope: This is particularly used in iron and steel foundries to minimize the danger of run-outs and separation of metal and slag. The gates will inevitably start to fill and allow metal into the mould cavity before the runner is full. The first metal and its load of slag enters the gates immediately before filling the runner and thus prior to the chance of trapping slag against the upper surface of the runner. In short, the runner in the cope results in the violation of the fundamental no fall criterion. The runner in the cope is not recommended for any type of casting including grey cast iron. Table 7.6 shows Webster's (1964) findings on runner to sprue exit area ratio on the filling efficiency.

Table 7.6 Runner to sprue exit area ratios by Webster's findings

Runner/Sprue exit area ratio	*Webster's findings*
1.0	High metal velocity
2.0	Optimum metal velocity, runner fills rapidly and excludes air bubbles reasonably efficiently
3.0	Starts to be difficult
4.0	Simply wasteful

Recent video X-ray radiographic studies have made it clear that even the expansion of the area of flow by a factor of 2 cannot prevent a serious amount of surface turbulence. The best that can be achieved without damage is by a 20% increase in area of the runner. Any greater expansion of the runner will cause the runner to be completely filled and so permit conditions for damage.

The expanding, rectangular runner:

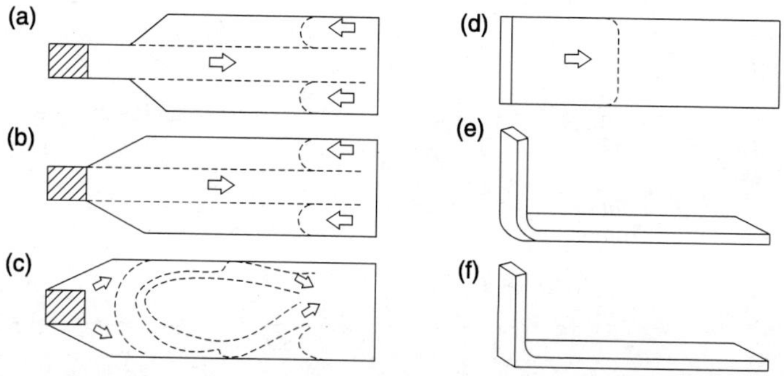

Figure 7.26 Plane view of a square section sprue connected to shallow rectangular runner showing attemps to expand the runner (a and b) that fail completely. Attempt (c) is better but flow ricochets off the walls generating a central starved, low pressure region. (d) A slot sprue and slot runner produce a uniform flow distribution in the runner shown in (e) (recommended) and (f) (probably acceptable)

Tapered runner

Refer to Figure 7.27. When the runner has two or more gates, the momentum of the flowing liquid causes the furthest gate, number 3 to be favoured. The rapid flow past the opening of gate 1 will create a reduced pressure region in the adjacent gate at this point, drawing liquid out of the casting! The flow may be either in or out of gate 2, but at such a reduced amount as to probably be negligible. For the present case, it would have been best to have only gate 3. Where more than one gate is attached to the runner, the runner needs to be reduced in cross section as each gate is passed. Smooth, straight taper

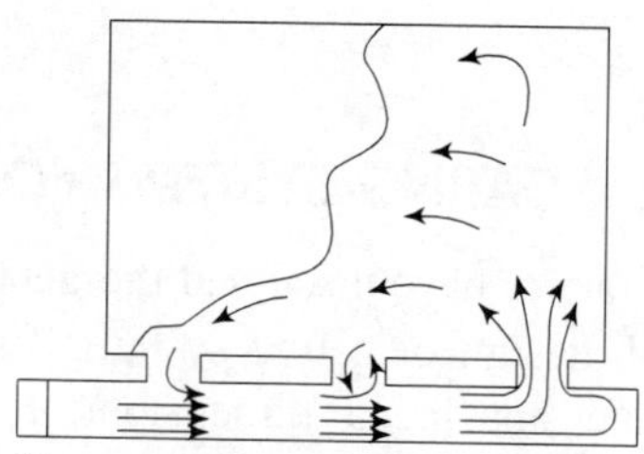

Figure 7.27 Unbalanced delivery of melt into the mould as a result of an incorrect runner design

geometry does a reasonable job of distributing the flow evenly (Figures 7.28 and 7.29).

Use of filters to reduce flow velocity

One of the most effective devices to reduce the speed of flow in the runner is the use of a filter. The close spacing of the walls of its capillaries ensures a high degree of viscous drag. Flow rate can often be reduced by a factor of 4 or 5. Various types of filters are shown in Figure 7.30.

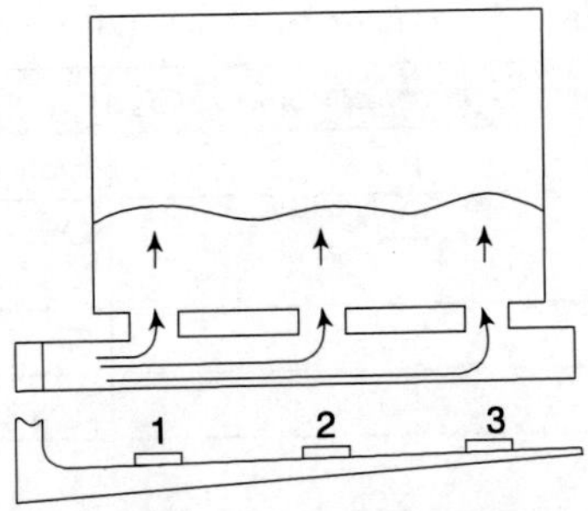

Figure 7.28 A tolerable balanced design using tapered runner

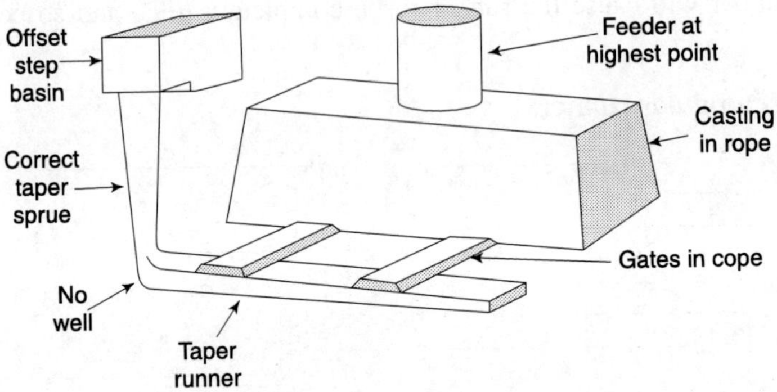

Figure 7.29 Taper runner

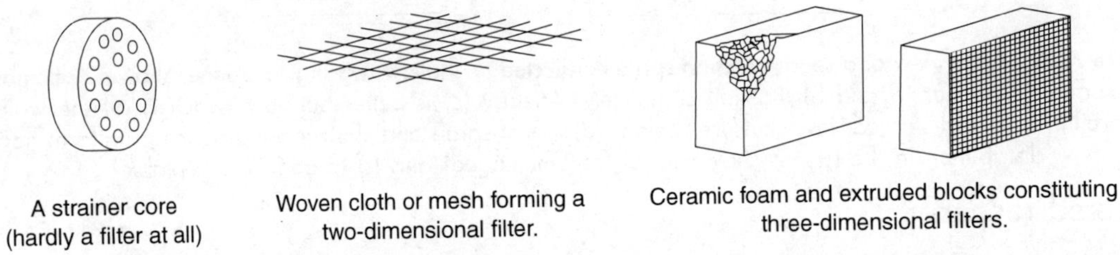

Figure 7.30 Use of filters

7.5 GATING SYSTEMS FOR VARIOUS MATERIALS

So far we have discussed the general guidelines for design of gating systems. This section deals with design of gating systems for specific materials. Design includes calculation of the sprue, runner and ingate dimensions. Following commonly used materials are considered:

1. Gating system for cast iron casting
2. Gating system for SG iron castings
3. Gating systems for steel castings
4. Gating systems for nonferrous alloys.

7.5.1 Gating System for Cast Iron Casting

Cast iron casting being normally of intricate design combined with section sensitivity and chilling tendency of liquid metal pose maximum problems in gating systems design.

(a) Pouring time

According to Czikel,

(i) Pouring time = t(in seconds) = $\dfrac{M}{E}$

where, M = Modulus of casting = $\dfrac{\text{casting volume}}{\text{casting surface area}}$ (cm)

E = Constant = 0.015 to 0.035

For machine tool casting, use E = 0.025

(ii) According to Nielson,

Pouring time = t(in seconds) = $0.32\ (t_w \times G)^{0.4}$

where, t_w = Wall thickness (mm)

G = Casting weight in kg.

(b) Sprue exit area

$$\text{Sprue exit area (cm}^2) = \frac{22.6\ W}{c \times \rho \times t \times \sqrt{h}}$$

where,

W = Casting weight (kg)

c = Discharge coefficient = 0.8 – 1.2 (intricate to simple)

ρ = Specific gravity of metal

t = Pouring time (seconds)

h = Pressure head (cm)

For short tapered sprue, diameter at entrance = 1.414 × exit diameter

For long tapered sprue, diameter at entrance = 1.732 × exit diameter

Sprue well base diameter = 2.24 × D

(c) Gating ratio

Sprue exit area: total runner area: total ingate area

Expressed as 1:k:1

where,

k = 1 for single ingate

k = 1.41 for two ingates

k = 1.73 for three ingates

k = 2 for four or more ingates

1:4:4 is also used in some foundries

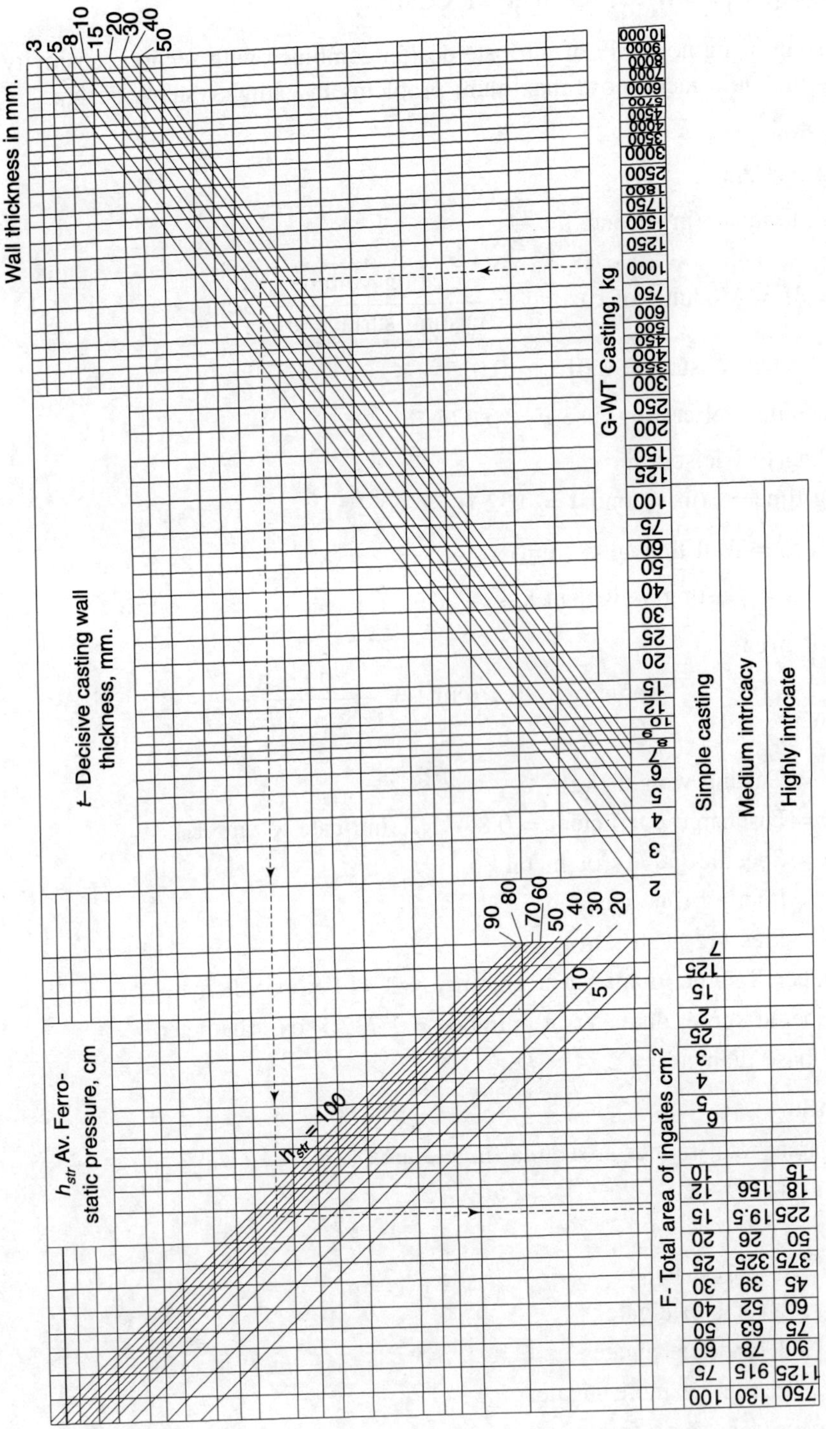

Figure 7.31 Sobolev's nomogram

Sobolev's nomogram (Figure 7.31) is referred from the practical data of many reputed foundries of East European cast iron foundries and accepted by most foundries. The casting weight, decisive wall thickness (t) in mm, average ferrostatic head in cm and the class of casting intricacy are variable parameters to be considered.

For example, for a casting of 1000 kg (A), a vertical line to intersect wall thickness 20 mm draw line (B). From there, draw a horizontal line to intersect 60 cm ferrostatic pressure head line at (C). Draw a vertical line from (C) to the table row medium intricacy class (D) and read the value 19.5 sq. cm directly as the total ingate area. Each gate size is decided as per the casting shape by choosing the number of ingates.

The value obtained from this method is highly practical and is used to confirm/approximate the total area calculated from other theoretical methods.

For grey iron castings which are not so complicated and whose weight is lower than 100 kg per piece, it is recommended to use the data in Fundator's table (Table 7.7) which gives the total area of cross section and number of ingates. The area given is the lowest permissible and applies to green sand moulds only.

Table 7.7 Fundator's table

Weight of casting	*Number of ingates depending on wall thickness (in mm)*					*Cross-sectional area of ingate (in sq. cm)*
	3 – 5	5 – 8	8 – 10	10 – 15	15 – 20	
Upto 0.5	1	1	1	1	1	0.3 – 0.4
0.5 – 1.0	1	1	1	1	1	0.4 – 0.5
1.0 – 3.0	1	1	1	1	1	0.5 – 0.7
3.0 – 5.0	2	2	1	1	1	0.6 – 0.8
5.0 – 10	3	3	2	2	2	0.6 – 0.8
10 – 15	*	3	2	2	2	0.7 – 0.8
15 – 20	*	4	4	3	3	0.8 – 0.9
20 – 30	*	4	4	3	3	0.8 – 1.1
30 – 40	*	5	4	3	3	0.9 – 1.2
40 – 60	*	4 – 5	4	3 – 4	3	1.0 – 1.5
60 – 100	*	5 – 6	5	4 – 5	4	1.0 – 1.5

7.5.2 Gating System for SG Iron Casting

In SG iron, magnesium is added in the molten state. Magnesium has a higher affinity towards oxygen and hence metal when comes in contact with atmospheric oxygen, produces slag. The slag and non-metallic particles like magnesium oxide, sulphide and silicate are filtered out before pouring SG iron. Figure 7.32 shows the use of ceramic foam filters used for making crankshaft of SG iron.

(a) **Pouring time**

According to R. W. White,

Pouring time = t (in seconds) = $0.97 \sqrt{\text{Pouring weight (kg)}}$

(b) **Sprue exit area**

$$\text{Sprue exit area (cm}^2) = \frac{\text{Pouring weight (kg) W}}{1.1 \times \text{pouring time (sec.)}}$$

(c) **Gating ratio**

Karsay suggested non-pressurized system for large and intricate casting, where turbulent flow is detrimental for good quality.

Pressurized systems are used for less intricate casting.

Sprue exit area: total runner area: total ingate area

Expressed as 4:8:3

The minimum gate thickness should not be less than 3 mm width of ingate = four times the thickness

$$\text{Sprue area} = \frac{4}{3} \times \text{choke area of ingates}$$

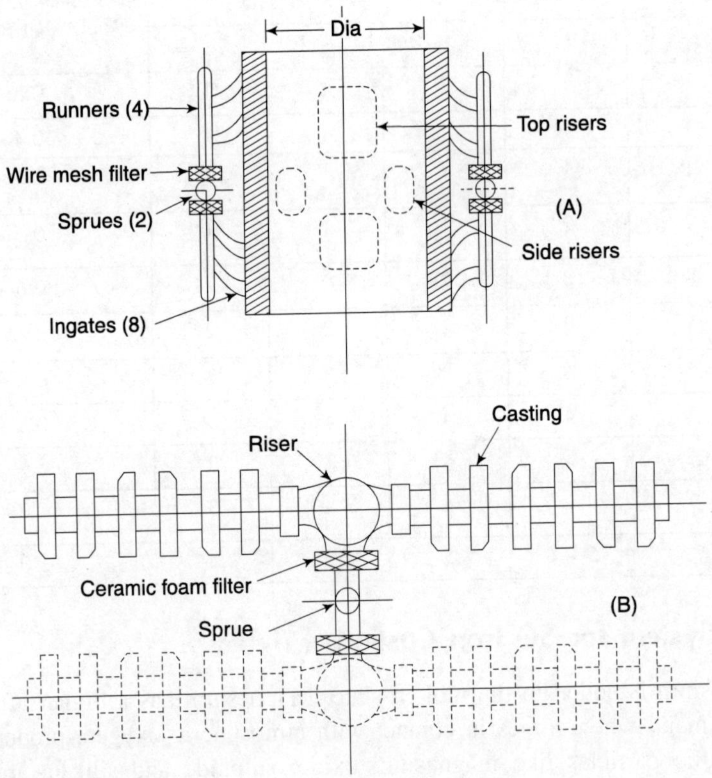

Figure 7.32 Use of ceramic foam filters in gating systems for crankshaft

7.5.3 Gating System for Steel Casting

Design of gating systems for steel casting is less critical compared to gating systems for cast metals and alloys. The following are the important considerations:

1. Due to high temperature gradient (melting temperature of iron is 1500°C and mould material is at room temperature), mould shall experience thermal shock.
2. Erosion of the mould metal due to heavy density of steel.
3. Fast filling is required to avoid laps, cold shuts, etc., in intricate castings.
4. Location of ingate should promote directional solidification but not hinder as steel is prone to hot tearing tendency.

Some of the common practical solutions for the above problems are given below:

1. Use refractory gating, sprue and sleeves in place of normal sand to minimize sand erosion, sand wash and entrapment of impurities.
2. Use bottom, side and step gating for deep castings to avoid large fall of metal into the mould.
3. Location of the ingates must be such that heat concentration is avoided.
4. Multiple sprues for large casting for fast filling of liquid metal in the mould cavity.

(a) Pouring time

$$\text{Pouring time} = t \text{ (in seconds)} = K^3\sqrt{t_w W}$$

t_w = Wall thickness, mm
W = Weight of the casting, kg
K = Time coefficent varying from 1.2 to 2.6

(b) Sprue exit area

$$\text{Sprue exit area (cm}^2) = \frac{W \text{ (kg)}}{0.31 \times \text{pouring time (sec.)} \times \mu \times \sqrt{h}}$$

t_w = Wall thickness, mm
W = Weight of the casting, kg
μ = Coefficeint of mould resistance to flow of metal
h = Average ferrostatic pressure, cm

(c) Gating ratio

Normally non-pressurized system is used for steel casting.

Sprue exit area: total runner area: total ingate area

Expressed as 1:1.5:2 or 1:2:2

In steel foundries, the following diameter sprues made out of refractory sleeves are used.

Small castings up to 10 kg	$\phi 30$
For castings 10-300 kg	$\phi 40$
For castings 300-2000 kg	$\phi 50$, $\phi 60$, $\phi 80$ or larger

7.5.4. Gating System Design for Nonferrous Alloys

Some of the important considerations are:

- To avoid the production of dross and its passage into casting, the system should be capable of preventing dross slag present in the ladle form reaching the mould cavity.
- To avoid the entrapment of air or mould gases in the metal system.
- Dross inclusions in the casting can be prevented by avoiding turbulent flow of metal from the pouring crucible to the casting cavity and by removing dross from the molten metal before it enters the casting cavity. Perforated metal screens are used in gating system in magnesium foundries for removal of dross.

Steps in design

Taking simple plate casting as an example for outlining the steps involved in methoding:

- Design the pouring basin which will contain enough metal to keep the gating system full. Pouring basin can have projections to trap slag before entering the sprue.
- Decide the pouring time approximately, i.e., about 50 seconds for castings weighing more than 100 kg, and of the order of 20 to 30 seconds for the smaller castings weighing about 5 kg and more. The pouring time selected shall be in such a way that it avoids cold-shuts.
- Estimate the total weight of metal to be poured into the casting. For a good radiographic quality casting, the thumb is that about 2.5 to 3 times the casting weight are to be poured.

(a) Metal flow rate

$$\text{Metal flow rate} = \frac{\text{Weight of metal to be poured}}{\text{Pouring time}}$$

Sprue diameter is to be designed to ensure this flow of metal.

$$A = \frac{G \times 1000}{\rho \times c \times t \times \sqrt{2gh}}$$

A = Choke area (sprue area at bottom), cm.

G = Casting weight + gating + riser, kg.

ρ = Density of molten metal, gm/cc.

h = Effective height of metal head.

c = Discharge coefficient.

(combined c of both gating and moulding can be taken as 0.4 for thin-walled casting and 0.7 for thick-walled casting).

t = Pouring time in seconds.

(b) Feed metal criteria

$$k_m = G/t$$

k_m = 0.3 Gkg/sec. (for aluminium alloy).

k_m = 0.4 Gkg/sec. (for magnesium alloy).

If these conditions are satisfied, then feeding is adequate.

- Normally, for a casting weighing more than 35 kg more than one sprue is to be provided. Sprue height is determined by mould box height and diameter of sprue at the top is determined by the draft taper requirement. Area at the top of the sprue is normally three times that of the bottom; a round sprue is easier for moulding. Rectangular sprues are preferred for magnesium alloys.
- Sprue well is an enlargement at the bottom of the sprue and is provided to reduce the turbulence that occurs in changing the direction of flow from vertical to horizontal sprue base; can be either cylindrical or rectangular.

  Depth of sprue well = 1.5 to 2 times runner depth.

  Length of sprue well = 4 × (sprue diameter).
- Metal screens are to be placed in sprue well. Wire mesh is provided to trap solid inclusions in the metal. Screens with 20 to 50% openings are to be used and are to be given coating to avoid inclusions.
- Metal is distributed around the mould cavity by horizontal runners starting at the sprue base and entering at a location as low as possible. Runners are extended beyond the last ingate to act as traps to catch dirt and oxide formed in the gating system. Runners' cross-sectional area is calculated as two to four times the sprue choke area. Individual runner area is obtained by dividing this value by the number of runners connected to the sprue. Runner shapes are normally trapezoidal for smooth flow. All dimensions of runner's cross section are calculated based on the sprue cross-sectional area.

(c) Ingate design

Ingate area is four times the space choke area. Individual ingate areas are calculated by dividing the total area by the number of ingates. Space between ingates should be roughly two times the gate width.

- Light metal alloys like aluminium bronzes and high tensile brass require bottom gates or (reversed) horn gates.
- High pressurized system is not recommended for these alloys.
- Cast iron, phosphor-bronze and gun metal are gated by a variety of methods including top gating methods.

- Alloys susceptible to oxidation and to formation of strong surface films need quiet and progressive mould filling with least division of the liquid metal stream.

(d) Use of chills

Chills are metallic inserts, usually made of C.I. or steel, rammed into the sand mould to increase the heat absorption locally to achieve 'directional solidification'. This reduces thermal stresses from being created due to uneven cooling especially in aluminium, copper and steel castings. Chills help in extending feeding distance of risers and reduce the number of risers required. In cast iron, chilling is used to achieve localized fine-grained and harder areas wherever required. Chills help to avoid defects caused by overheating at entry points of ingates in complicated castings.

Worked Out Examples

Example 7.1: Two gating designs for a mould of 50 cm × 25 cm × 15 cm are shown in Figure 7.33. The cross-sectional area of gate is 5 cm^2. Determine the filling time for both the designs.

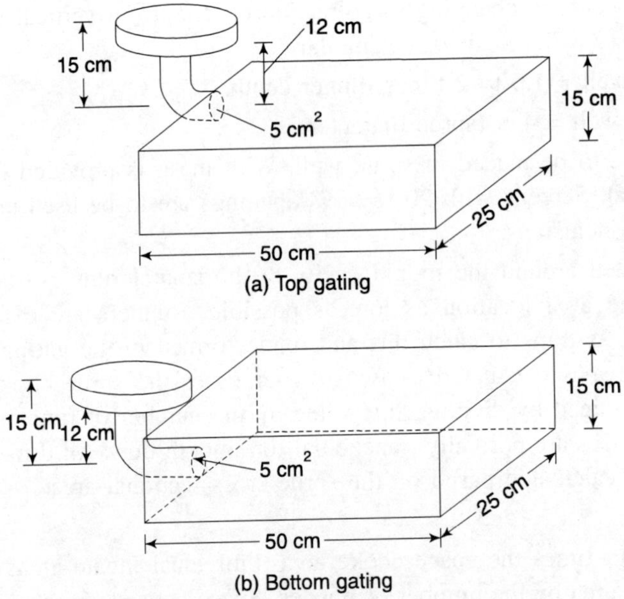

Figure 7.33 Top and bottom gating designs

Solution: For Figure 7.32a, since h_t = 15 cm, from equation **(1)**, we have

$$v_3 = \sqrt{2 \times 981 \times 15} \text{ cm/sec} = 171.6 \text{ cm/sec.}$$

The volume of the mould is $V = 50 \times 25 \times 15$ cm^3 and the cross-sectional area of the gate is $V = 50 \times 25 \times 15$ cm^2. So, from equation **(2)**, we get

$$t_f = \frac{50 \times 25 \times 15}{5 \times 171.6} \text{ sec} = 21.86 \text{ sec.}$$

For Figure 7.33b, $h_t = 15$ cm, $h_m = 15$ cm, $A_m = 50 \times 25$ cm^2 and $A_g = 5$ cm^2. Using equation **(8)**, we have,

$$t_f = \frac{50 \times 25}{5} \frac{\sqrt{2}}{\sqrt{9.81}} \sqrt{15} \text{ sec} = 43.71 \text{ sec.}$$

It should be noted that in Figure 7.33b the time taken is double of that in Figure 7.33a. We can easily verify that this will always be so if $h_m = h_t$.

Example 7.2: Figure 7.34 shows a mould along with the riser for casting a plate 20 cm × 20 cm × 5 cm. Determine the area A_g such that the mould and riser get filled up within 10 seconds after the down sprue has been filled. It should be noted that $A_3 > A_g$ since below the down sprue, a flat gate is attached to the casting. Neglect the frictional and orifice effects.

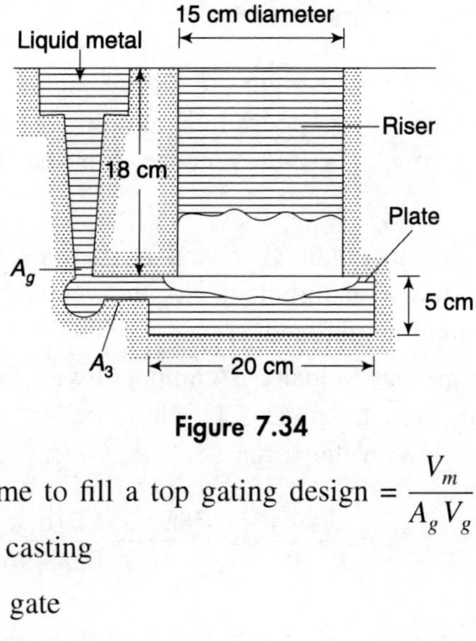

Figure 7.34

Solution: We know that time to fill a top gating design $= \dfrac{V_m}{A_g V_g}$

where V_m = Volume of casting

V_g = Velocity at gate

Also, $V_g = \sqrt{2gh_t}$

$$t_1 = \frac{V_m}{A_g \sqrt{2gh_t}} \quad \text{...(1)}$$

$$t_1 = \frac{0.2 \times 0.2 \times 0.05}{(A_g \times \sqrt{2 \times 9.81 \times 0.18}} = \frac{1.065 \times 10^{-3}}{A_g} \quad \text{...(2)}$$

For a bottom gating design,

$$\text{Time to fill} = \frac{A_m}{A_g} \sqrt{\frac{2}{g}} \times [\sqrt{h_t} - \sqrt{(h_t - h_m]} \quad \text{...(3)}$$

where, A_m = Area of cross section of molten metal casting
h_t = Height of pouring
h_m = Height of metal casting from bottom gate

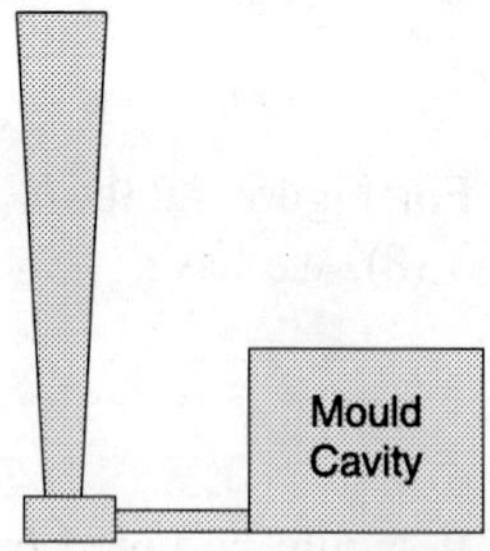

Thus,
$$t_2 = \frac{\frac{\pi}{4} \times .15^2}{A_g} \sqrt{\frac{2}{g}} \times [\sqrt{0.18} - \sqrt{0}]$$

$$t_2 = \frac{3.387 \times 10^{-3}}{A_g} \quad ...(4)$$

Also it is given that $T = t_1 + t_2$; where T = 10 seconds

$$10 = \frac{1.065 \times 10^{-3}}{Ag} + \frac{3.387 \times 10^{-3}}{A_g}$$

$$A_g = 4.45 \times 10^{-4} \text{m}^2 = 4.45 \text{ cm}^2$$

Example 7.3: Solve Example 7.1 (for Figure 7.32) by including the friction and velocity distribution effects. The liquid being poured is molten Fe with the properties, ρ_m = 2700 kg/m^3, kinematic viscosity, η = 0.00496 kg/m-sec. For the 90° turn at the end of the sprue, $(L/D)_{eq}$ = 25.

Solution: It should be noted that to obtain β, f, and e_f (before calculating C_D), we should determine whether the flow is laminar or turbulent. So, the velocity must be determined first. This calls for an iterative procedure as outlined.

As a first approximation, the average velocity is computed with the assumption that C_D = 1. With C_D = 1, $\vec{v}_3$ = 1.716 m/sec [see Example 7.1]. Now, $(\pi/4)d^2$ = 5 cm^2 or d = 0.252 m. Hence, Reynolds number of the flow in the sprue is

$$Re = \frac{\rho_m \vec{v}_3 d}{\eta} = \frac{7800 \times 1.716 \times 0.0252}{0.00496} \approx 68{,}000$$

So, the flow is turbulent. Thus,

β = 1,

$e_f \approx$ 0.45 [from Figure 7.13 with $A_2/A_1 \approx 0$, where A_2 and A_1 are the cross-sectional areas of the sprue and the pouring basin, respectively.]

$$f = \frac{0.0791}{(68000)^{1/4}} = 0.0049 \text{ [using equation (17)].}$$

Now, using equation **(23)** with l = 0.12 m, we have

$$C_D = \left[1 + 0.45 + 4 \times 0.0049\left(\frac{0.12}{0.0252} + 25\right)\right]^{-1/2} = 0.7$$

From equation **(21)**,

$$v_3 = C_D \sqrt{2gh_t} = 0.7 \times 1.716 \text{ m/sec} = 1.20 \text{ m/sec.}$$

We recalculate C_D on the basis of this value of $\vec{v}_3$ until it converges.

For the second approximation,

$$R_e = \frac{1.2}{1.716} \times 68{,}000 = 47{,}500$$

When

$$\beta = 1,$$

$$e_f \approx 0.46,$$

$$f = \frac{0.0791}{(47{,}500)^{1/4}} = 0.0054$$

$$C_D = \left[1 + 0.46 + 4 \times 0.0054 \left(\frac{0.12}{0.0252} + 25\right)\right]^{-1/2} = 0.69 \quad (\text{starting value of } C_D = 0.7)$$

So, $\vec{v}_3 = 0.69 \times 1.716 \text{ m/sec} = 1.18 \text{ m/sec.}$

Thus, the value of $\vec{v}_3$ has converged with sufficient accuracy. Now using equation **(2)**, we find the time taken to fill up the mould is

$$t_f = \frac{V}{A_g \vec{v}_3} = \frac{50 \times 25 \times 15}{5 \times 118} = 31.7 \text{ sec.}$$

This time is approximately 43% higher than that obtained in Example 7.1 (neglecting all the losses).

Example 7.4: Figure 7.35 shows a ladle having an internal diameter of 1 m with a capacity height of 1.2 m. It has a 45° tapered nozzle to a 75 mm exit diameter.

(i) Calculate the time required to empty the ladle if it is filled with an Al-Si alloy at 704°C.

(ii) Estimate the discharge rate in kg/sec (a) initially and (b) when the ladle is 75% empty.

Given $\eta_m = 2700$ kg/m^3 $\eta = 0.00273$ kg/m-sec.

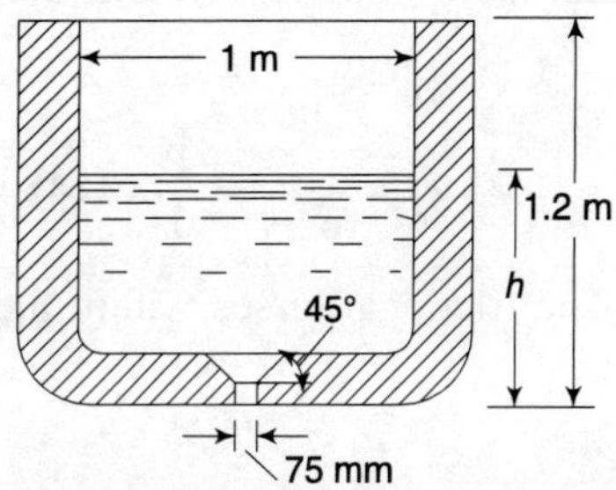

Figure 7.35 Description of nozzle

Solution: Let at any instant the level of liquid metal be h when the average velocity of the liquid metal through the nozzle is

$$\vec{v}_n = C_D \sqrt{2gh}$$

The mass flow discharge rate is

$$\dot{m} = \rho_m A_n \vec{v}_n = \rho_m A_n C_D \sqrt{2gh} \qquad \textbf{(a)}$$

where A_n is the nozzle area. Again, $\dot{m} = -\rho_m A_l dh/dt$, where A_l is the ladle cross-sectional area.

Thus,

$$-\frac{dh}{dt} = \frac{A_n}{A_l} C_D \sqrt{2gh} \quad \textbf{(b)}$$

This is, of course, under the assumption that C_D remains almost constant for all values of h. Now, the time taken (t_f) for the level to come down from h_i to h_f can be obtained by integrating equation **(b)** as

$$-\int_{h_i}^{hf} \frac{dh}{\sqrt{h}} = \frac{A_n}{A} C_D \sqrt{2g} \int_0^{t_f} dt$$

$$2(\sqrt{h_i} - \sqrt{h_f}) = \frac{A_n}{A} C_D \sqrt{2g}\, t_f$$

$$t_f = \sqrt{\frac{2}{g}} \frac{A_l}{A_n} \frac{1}{C_D} (\sqrt{h_i} - \sqrt{h_f}) \quad \textbf{(c)}$$

For the given data, let us first examine how C_D varies with h to justify the assumption that C_D remains constant.

First, let us take $h = h_i = 1.2$ m. As a first approximation, we take $C_D = 1$ when

$$\vec{v}_n = \sqrt{2gh} = 4.85 \text{ m/s},$$

$$Re = \frac{2700 \times 4.85 \times 0.075}{0.00273} = 3.6 \times 10^5$$

So, the flow is turbulent and consequently

$$\beta = 1,$$

$$e_f = \frac{1}{6} e_{f_s} = \frac{1}{6} \times 0.45 = 0.075 \qquad \text{(From Fig. 7.14 with } A_2/A_1 \approx 0)$$

The frictional losses within the ladle are negligible. Hence,

$$C_D = (1 + 0.075)^{-1/2} = 0.96$$

As this value of C_D is very near unity, there is no need to iterate further.

Second, let us consider the case when when $h \approx 0$, $\vec{v}_n \approx 0$, $Re \to 0$, i.e., the flow is laminar.

So, $\beta = \frac{1}{2}$ and $e_f \approx 0.82$. Thus,

$$C_D = (2 + 0.82)^{-1/2} \approx 0.6.$$

Third, let $h = 5$ cm $= 0.05$ m and assume that $C_D = 1$. Then,

$$v_n = \sqrt{2 \times 9.81 \times 0.5} \approx 1 \text{ m/sec.}$$

Thus, $$Re = \frac{2700 \times 1 \times 0.075}{0.00273} = 74175$$ (i.e., the flow is turbulent).

So,

$$\beta \approx 1$$

$$e_f = \frac{1}{6}\, e_{f_s} = \frac{1}{6} \times 0.45 = 0.075$$

(Again from Figure 7.14 and Figure 7.13 with $A_2/A_1 \approx 0$),

i.e., $C_D \approx 0.96$.

So, C_D remains equal to 0.96 for almost the entire period (except when $h \to 0$), justifying our assumption. We are now in a position to calculate the time and discharge rate.

(i) Here, $$A_l = \frac{\pi}{4} \times 1\text{m}^2 = 0.785 \text{ m}^2$$

$$A_n = \frac{\pi}{4} \times (0.075)^2 \text{ m}^2 \quad [\text{with } h_f = 0 \text{ from equation } \textbf{(c)}]$$

$$t_f = \sqrt{\frac{2}{9.81}} \times \frac{0.785}{4.4 \times 10^{-3}} \times \frac{1}{0.96}\sqrt{1.2} \text{ sec} = 91.9 \text{ seconds}$$

(ii) The initial discharge rate, from equation **(a)** with $h = 1.2$ m, is given as

$$\dot{m} = 2700 \times 4.4 \times 10^{-3} \times 0.96\sqrt{2 \times 9.81 \times 1.2} \text{ kg/sec}$$

$$= 55.34 \text{ kg/sec}$$

The discharge rate when the ladle is 75% empty, i.e., with $h = 0.25 \times 1.2$ m, is obtained from equation **(a)** as $\dot{m} = 27.67$ kg/sec.

Example 7.5: A ladle containing molten steel is placed in a pressurized chamber (Figure 7.36). The air pressure in the chamber is increased to 0.3 MPa in order to force the molten metal into the mould through the pouring tube. Estimate the time taken to fill up the mould. Also account for the frictional losses in the tube and the losses due to sudden contraction and expansion. Initially, half of the tube length is submerged into the melt. Steel density is 7600 kg/m^3 and kinematic viscosity is 0.00595 kg/m-s.

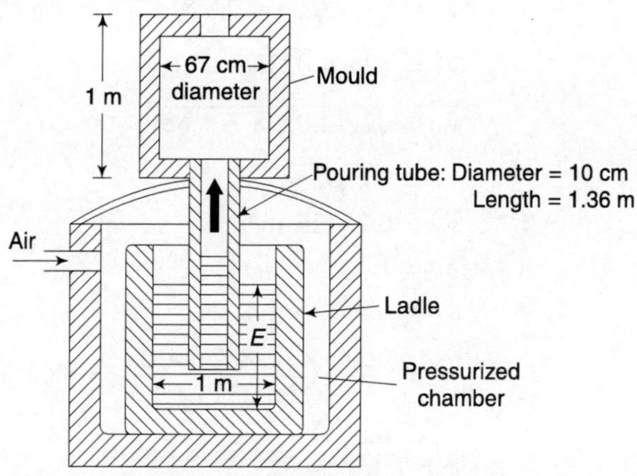

Figure 7.36

Solution: Without losses

$$\frac{P_1}{\rho g} + \frac{v_1^2}{2g} = \frac{P_2}{\rho g} + \frac{v_2^2}{2g}$$

$$\frac{v_1^2}{2g} = \frac{0.3 \times 10^6}{7600 \times 9.81} + \frac{v_2^2}{2g}$$

$$A_2 v_2 = A_1 v_1$$

$$\frac{\pi}{4} \times 0.1^2 \times v_2 = \frac{\pi}{4} \times 0.67^2 \times v_1$$

$$\frac{v_2}{v_1} = 44.89$$

$$-\frac{v_2^2}{44.89^2 \times 2g} + \frac{v_2^2}{2g} + \frac{0.3 \times 10^6}{7600 \times 9.81} = x$$

$$0.995 \times \frac{v_3^2}{2g} = (-x + 4.0238)$$

$$\int_{0.68}^{1.68} \frac{-\left(\frac{dx}{dt}\right)^2}{\sqrt{4.0238 - x}} \times \sqrt{\frac{0.995}{2g}} = \int dt$$

$$0.1125^2 \times [-\sqrt{4.0238 - x})]_{0.68}^{1.68} \times 2 = t_f$$

$$0.1125^2 \times 1.8286 - 1.530947 \times 100 \times 2 = t_f$$

$$t_f = 6.69 \text{ seconds}$$

Now taking losses,

$$\frac{P_1}{\rho g} + \frac{v_1^2}{2g} + h_1 = \frac{P_2}{\rho g} + \frac{v_2^2}{2g}$$

$$v_2 = \sqrt{-2g(h_t - 4.0238)}$$

$$v_2 = \sqrt{2g(4.0238 - h_t)}$$

$$v_2 = \sqrt{2g(4.0238 - 1.68)}$$

$$v_2 = \sqrt{45.985}$$

$$v_2 = 6.78125 \text{ m/s}$$

We assume

$$c_d = 1$$

$$v_3 = 6.78125 \text{ m/s}$$

$$R_e = \frac{\rho v d}{\mu} = \frac{7600 \times 6.78125 \times 0.1}{0.00595} = 8.66176 \times 10^5$$

$$f = \frac{0.0791}{Re^{0.25}} = 2.59283 \times 10^{-3}$$

$$\left(\frac{l}{D}\right)_{eq} = \left(\frac{1.36}{0.1}\right)$$

$$c_{f2} = 0.3 \text{ (Assumption)}$$

$$\frac{A_2}{A_1} = 0.2$$

$$c_d = \left(1 + 0.3 + 4 \times 2.59283 \times 10^{-3} \times \frac{1.36}{0.1}\right)^{-0.5}$$

$$c_d = (1 + 0.3 + 0.141049952)^{-0.5}$$

$$c_d = (1.44)^{-0.5} = 0.8330296$$

$$R_e = 8.66176 \times 10^5 \times 0.8330296 = 721550$$

$$f = \frac{0.0791}{R_e^{0.25}} = 2.71399 \times 10^{-3}$$

$$c_{f2} = 0.3 \text{ (Assumption)}$$

$$c_d = \left(1 + 0.3 + 4 \times 2.71399 \times 10^{-3} \times \frac{1.36}{0.1}\right)^{-0.5}$$

$$c_d = (1.44764)^{-0.5} = 0.8311$$

$$v_3 = 0.8311311399 \times 6.78125 \times 4.936 \text{ m/s}$$

$$t_f = \frac{V_3}{A_g v_3} = \frac{1 \times \frac{\pi}{4} \times 0.67^2}{\frac{\pi}{4} \times 0.1^2 \times 4.936}$$

$$t_f = 10.02 \text{ seconds (with friction losses)}$$

Example 7.6: Figure 7.37 shows the casting of a seamless metal pipe in a cooled mould. A pressure pouring technique is used to fill the mould quickly, as depicted in the figure. Determine the expression for estimating the time taken to fill up only the mould. Assume that the level (h_1) in the ladle does not change. You may neglect the frictional losses but not the entrance losses.

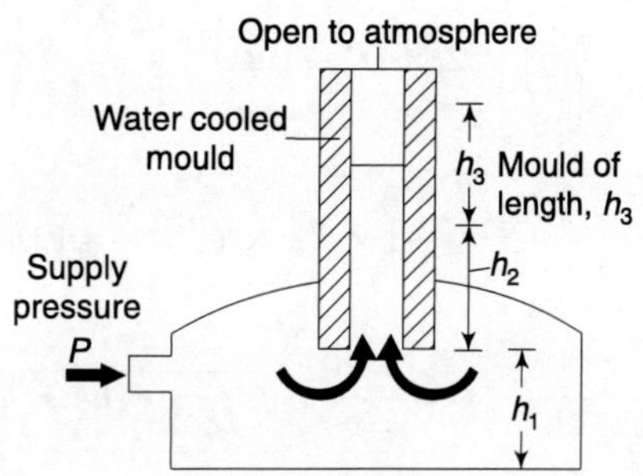

Figure 7.37

Solution: As we know that in Figure 7.37, the supply pressure is P and mould is open to atmosphere.

Since the mould is projected entry

$$e_f = 2e_{f_s}$$

Now consider two points 1 and 2 as shown in figure, at a distance x from each other.

Applying Bernoulli's equation between the two points

$$\frac{P_1}{\rho g} + h_t + \frac{v_1^2}{2g} = \frac{P_2}{\rho g} + (h_t + x) + \frac{v_2^2}{2g} + e_f \frac{\vec{v}^2}{g}$$

$$\frac{P}{\rho g} = x + \frac{v_2^2}{2g\beta} + \frac{\rho_f \vec{v}^2}{g}$$

$$\frac{P - \rho g x}{\rho} = \frac{v_2^2}{2\beta} + e_f v_2^2$$

$$\frac{2(P - \rho g x)}{\rho} = v_2^2\left(\frac{1}{2\beta} + e_f\right)$$

Consider the flow to be turbulent

$$\frac{2(P - \rho g x)}{\rho}\,\frac{1}{(0.5 + 2e_{f_s})} = v_2^2$$

$$v_2 = \left(\frac{2(0.5 + 2e_{f_s})^{-1}}{\rho} \times (P - \rho g x)\right)^{0.5}$$

$$\frac{dx}{dt} = v_2 - \left(\frac{2(0.5 + 2e_{f_s})^{-1})}{\rho} \times (P - \rho g x)\right)^{0.5}$$

$$\int_{h_2}^{h_2+h_3} \frac{dx}{(P - \rho g x)^{0.5}} = \left(\frac{2(0.5 + 2e_{f_s})^{-1}}{\rho}\right)^{0.5} \times t_f$$

$$\left[\left(\frac{-1}{\rho g}\right) \times 2 \times (P - \rho g x)^{0.5}\right]_{h_2}^{h_2+h_3} = \left(\frac{2(0.5 + 2e_{f_s})^{-1}}{\rho}\right)^{0.5} \times t_f$$

$$\left(\frac{-1}{\rho g}\right) \times \sqrt{2} \times (\sqrt{(P - \rho g(h_2 + h_3)} - \sqrt{(P - \rho g h_2)}) = \left(\frac{2(0.5 + 2e_{f_s})^{-1}}{\rho}\right)^{0.5} \times t_f$$

$$t_f = \frac{-\sqrt{P}}{\rho g}(\sqrt{P - \rho g(h_2 + h_3)} - \sqrt{(P - \rho g h_2)}\,\sqrt{(2 \times (0.5 + 2e_{f_s})}$$

$$t_f = \sqrt{\frac{1}{\rho g}}\left(\sqrt{P - \rho g h_2} - \sqrt{P - \rho g(h_2 + h_3}\right)\sqrt{(2 \times (0.5 + 2e_{f_s})}$$

$$t_f = \left(\sqrt{\frac{P}{\rho g} - h_2} - \sqrt{\frac{P}{\rho g} - (h_2 + h_3)}\right)\sqrt{(2 \times (0.5 + 2e_{f_s})}$$

Example 7.7: Calculate the optimum pouring time for a casting whose mass is 20 kg and having an average section thickness of 15 mm. The materials of the casting are grey cast iron and steel. Take the fluidity of iron as 700 mm (28 inches).

Grey cast iron

Pouring time, $t = K\left(1.41 + \dfrac{T}{14.59}\right)\sqrt{W}$

$$= \left(\frac{28}{40}\right)\left(1.41 + \frac{15}{14.59}\right)\sqrt{20} = 7.632 \text{ seconds}$$

Steel

Pouring time, $t = (2.4335 - 0.3953 \log W)\sqrt{W}$

Pouring time, $t = (2.4335 - 0.3953 \log 20)\sqrt{20} = 8.5825$ seconds

Example 7.8: Calculate the optimum pouring time for a casting whose mass is 100 kg and a thickness, 25 mm. Fluidity of iron is 32 inches. Calculate both for cast iron and steel.

Grey cast iron

Pouring time, $t = K\left(1.41 + \dfrac{T}{14.59}\right)\sqrt{W}$

$$= \left(\frac{32}{40}\right)\left(1.41 + \frac{25}{14.59}\right)\sqrt{100} = 24.988 \text{ seconds}$$

Steel

Pouring time, $t = (2.4335\text{-}0.3953 \log W)\sqrt{W}$

Pouring time, $t = (2.4335\text{-}0.3953 \log 100)\sqrt{100} = 16.429$ seconds

Example 7.9: For the casting shown in Figure 7.38, which is to be made in cast iron, calculate the choke area.

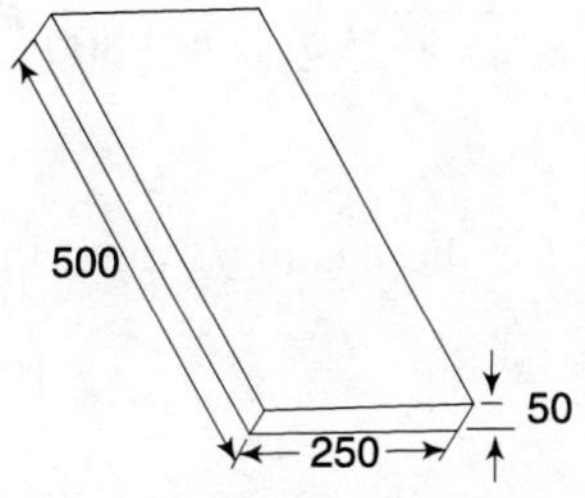

(a) Casting

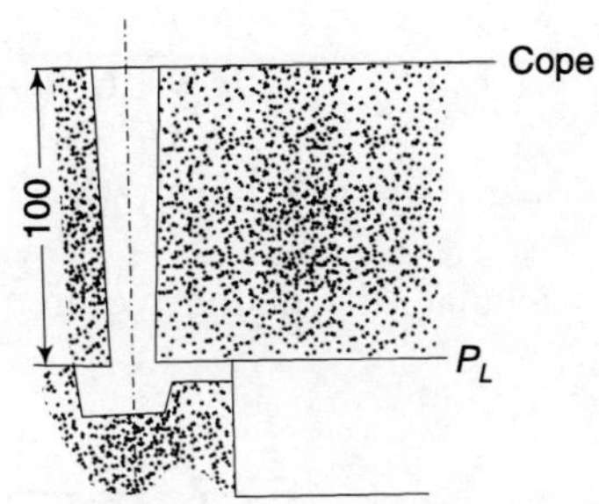

(b) Moulding arrangement

Figure 7.38

Volume of the casting $= 500 \times 250 \times 50$

$= 6.25 \times 10^6 \text{ mm}^3$

Weight of the casting $= 7.86 \times 10^{-6} \times 6.25 \times 10^6$

$= 49.125$ kg

Assume a carbon equivalent of 4.0 and a pouring temperature of 1300°C, so that fluidity obtained from Figure 4.22 is 22 inches.

$$\text{Pouring time, } t = \left(\frac{22}{40}\right)\left[1.41 + \left(\frac{50}{14.59}\right)\right]\sqrt{49.125} = 18.647 \text{ seconds} \cong 19 \text{ seconds}$$

Assuming a top gating with 100 mm cope height,

Effective sprue height = 100 mm

C, efficiency factor from Table 7.3 = 0.85

$$\text{Choke area, } A = \left(\frac{49.125}{(6.09 \times 10^6 \times 19 \times 0.85 \times \sqrt{2 \times 9800 \times 100}}\right)$$

$$= 376.587 \text{ mm}^2$$

Example 7.10: Design the ingate dimensions for pouring a 15 kg casting in 10 s with the runner having a cross section of 625 mm^2 and two ingates of 25 mm width each.

Assuming equal flow through gates,

$$\text{Flow rate} = \frac{15}{2 \times 10} = 0.75 \text{ kg/s}$$

Assuming the metal to have a density of 6.9 g/cm^3

$$Q = \frac{0.75 \times 1000}{6.9 \times 10^3} = 0.109 \times 10^6 \text{ mm}^3\text{/s}$$

Let the runner gate junction be the same as 625 mm^2.

$$\text{Velocity of metal, } V = \frac{0.109 \times 10^6}{625} = 174.4 \text{ mm}$$

$$\text{Free height,} \quad h = 1.6 \sqrt[3]{\frac{0.109^2 \times 10^{12}}{9800 \times 25 \times 25} + \frac{174.4^2}{2 \times 9800}} = 21.506 \text{ mm}$$

Hence, gate height = 21.506 – 5 = 17 mm

Example 7.11: Calculate the gating requirements for the casting shown in Figure 7.39 to be cast in C30 steel.

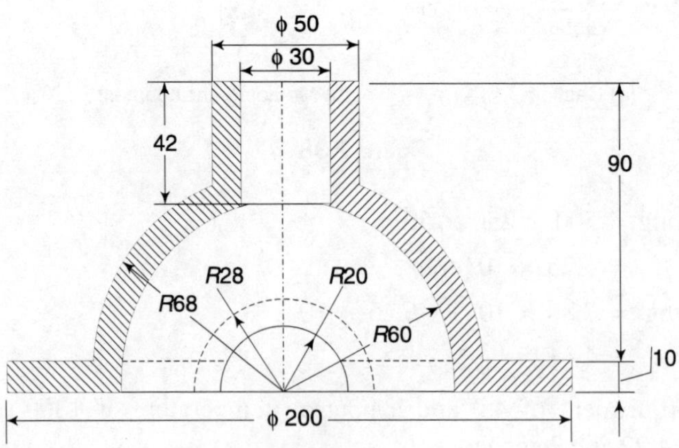

Figure 7.39

Step 1: Calculation of casting weight

To calculate weight, we shall find the volume of the casting,

$$\text{Hemisphere} = \frac{4}{3} \times \frac{\pi}{2} \times (68^3 - 60^3) = 206156 \text{ mm}^3$$

$$\text{Flange} = \frac{\pi}{4} \times (200^2 - 120^2) = 201062 \text{ mm}^3$$

$$\text{Top cylinder} = \frac{\pi}{4} \times (58^2 - 30^2) \times 42 = 81279 \text{ mm}^3$$

$$\text{Total volume} = 206156 + 201062 + 81279 = 488497 \text{ mm}^3$$

$$\text{Mass of the casting} = 488497 \times 7.86 \times 10^{-6} = 3.84 \text{ kg}$$

There are two castings in the mould and their weight = 3.84 kg × 2 = 7.68 kg

Assuming a yield of 0.60

$$\text{Weight of metal poured} = \frac{7.68 \text{ kg}}{0.60} = 12.8 \text{ kg}$$

$$\text{Pouring time} = (2.4335 - 0.3953 \times \log 12.8) \times 12.8^{0.5} = 7.14 \text{ seconds}$$

$$\text{Pouring rate} = \left(\frac{12.8}{7.14}\right) = 1.793 \text{ kg/s}$$

Choosing a parting gate and two castings in the mould, the casting arrangement is shown in Figure 7.40. Note that the effective sprue height is 150 mm.

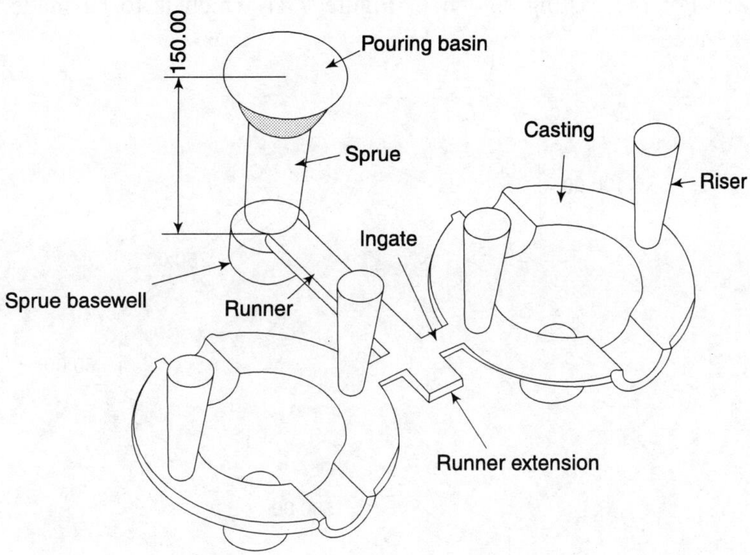

Figure 7.40 Casting arrangement for component

Step 2: Calculation of sprue exit area

Effective sprue height = 150 mm

Assume that this is a non-pressurised gating system that is one choke is located in sprue with one runner and two ingates. For this combination, the efficiency coefficient of the gating system from Table 7.2 is obtained as 0.73.

$$\text{Choke area} = \frac{12.8}{7.12 \times 0.73 \times 7.7 \times 10^{-6} \times \sqrt{2 \times 9800 \times 150}} = 186.53 \text{ mm}^3$$

$$\text{Choke diameter} = 15.41 \text{ mm} \cong 16 \text{ mm}$$

Step 3: Gating ratio

Assuming a gating ratio of 1:2:2

Step 4: Runner sizing

$$\text{Runner area} = 2 \times \pi \times \frac{16^2}{4} = 402.12 \text{ mm}^2$$

Assuming two runners and a rectangular cross section with 2:1 ratio, we get the runner proportions as 16 × 25.

Step 5: Ingate sizing and yield calculations

Assume two ingates so that metal through each gate = 96.14 mm/s

$$\text{Each gate area} = 0.5 \times 402.12 \text{ mm}^2 = 201.06 \text{ mm}^2$$

Assuming an aspect ratio of 4, the size of the gate is 728 mm.

Example 7.12: For the casting shown in Figure 7.41 which is to be made in cast iron, calculate the choke area.

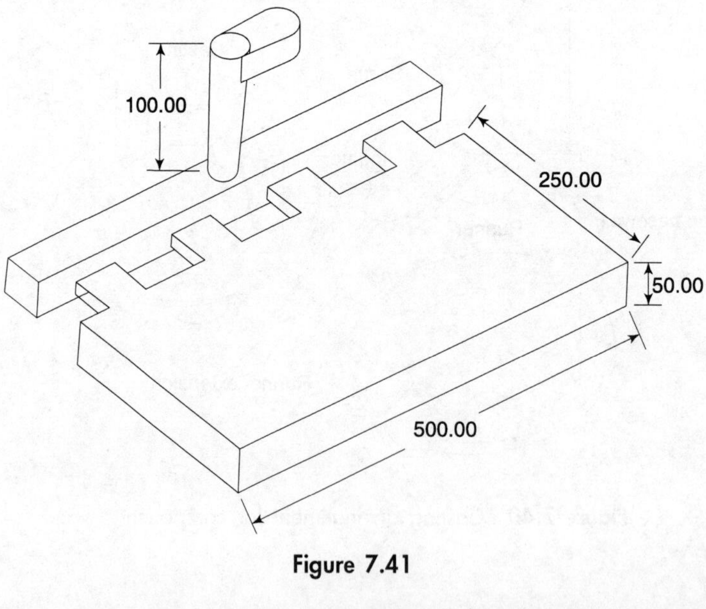

Figure 7.41

Volume of the casting = 500 × 250 × 50 = 6.25×10^6 mm^3

Weight of the casting = $7.86 \times 10^{-6} \times 6.25 \times 10^6$ mm^3 = 49.125 kg

Assuming a composition factor of 4 and pouring temperature of 1300°C, the fluidity is 22 inches.

$$\text{Pouring time, } t = \frac{22}{40}\left(1.41 + \frac{50}{14.59}\right) \times 49.125^{0.5} = 19 \text{ seconds}$$

Calculate the effective sprue height, assuming top gating system with 100 mm cope height.
Effective sprue height = 100 mm

Select the efficiency fator, C = 0.73 for pressurised gating system consisting of two runners.

Assuming the density of the metal as 6.90×10^{-6} kg/mm^3

$$\text{Choke area} = \frac{49.125}{6.90 \times 10^{-6} \times 19 \times 0.73 \times \sqrt{2 \times 9800 \times 100}} = 366.648 \text{ mm}^3$$

In a pressurised gating system, the choke area is located in ingates, with four ingates; the ingate area of each is 90 mm^2. Assuming rectangular cross section for ingates, the ingate dimensions would be 15 × 6 mm.

Example 7.13: Design the gating requirements for the casting shown in Figure 7.42 to be cast in aluminium. Use the following data:

Material	:	Aluminium
Moulding process	:	Greensand
Pouring	:	Hand ladle
Density	:	2400 kg/cubic metre

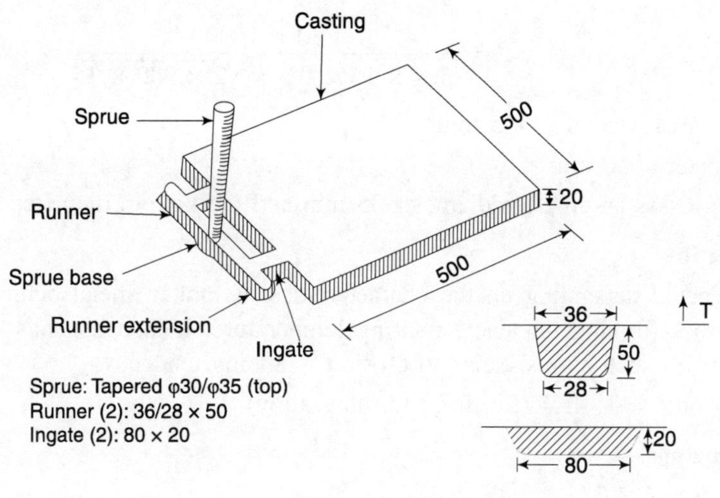

Figure 7.42

Solution:

Step 1: Calculation of casting weight

Casting weight = 0.5 × 0.5 × 0.02 × 2400 = 12 kg

Total weight of pour = 12 × 2.5 (assume) = 30 kg (including runners and risers)

Head of pouring = 150 mm (height of the ladle from parting line/cope)

Discharge coefficient = 0.5 for medium thick wall

Assumed time of pouring: 15 seconds (up to 20 seconds for small castings)

Step 2: Calculation of sprue exit area

The choke area can be calculated using Bernoulli's equation as

$$A = \left(\frac{W}{(\rho t C\sqrt{2gH}}\right)$$

where, A = Choke area, mm^2 (cm^2)

W = Casting mass, kg

t = Pouring time, seconds

ρ = Mass density of the molten metal, $\frac{\text{kg}}{\text{mm}^3}\left(\frac{\text{gm}}{\text{cm}^3}\right)$

g = Acceleration due to gravity, $\frac{\text{mm}}{\text{s}^2}\left(\frac{\text{cm}}{\text{s}^2}\right)$

H = Effective metallostatic head (sprue height), mm (cm)

C = Efficiency factor or discharge factor

$$\text{Sprue base area, } A = \frac{W \times 1000}{\rho t C\sqrt{2gH}}$$

$$A = \frac{30 \times 1000}{2.4 \times 15 \times 0.5 \times \sqrt{2 \times 980 \times 15}} = 7.75 \text{ cm}^2$$

Minimum sprue area needed = 775 mm^2.

Calculated diameter = 30 mm

But considering losses assume tap diameter 35 mm and the bottom diameter 30 mm.

Step 3: Gating ratio

Gating ratio is selected depending on the characteristics of molten metal being cast such as fluidity, slag or dross forming tendency, pouring temperature. It also depends on the mould material characteristics such as resistance to erosion, scabbing tendency.

Assume, gating ratio = 1: 4: 4 (for dross forming alloy)

Step 4: Runner sizing

Total runner area = 4 × 775 = 3100 mm^2

Assume two runners each of 1550 mm^2 cross-sectional area

Assuming inner as trapezoidal section of depth 50 mm, parallel arms 36 mm and 28 mm respectively.

Step 5: Ingate sizing

Considering two ingates each of cross section area 1550 mm^2

Assuming rectangular section, w: b = 4: 1, (this is to avoid junction hot spot and thin sections.)

Therefore, ingate section = 80 mm × 20 mm.

Example 7.14: Design the gating system for the part shown in Figure 7.43. Material is cast iron (C = 3.6% and Si = 1.6%). Density of cast iron can be taken as 7000 kg/m^3. Greensand is used as the moulding sand. Pouring temperature is 1343°C (2450°F). No feeder is to be used. Unpressurised gating with bottom pouring is recommended. Gating ratio to use – 1:2:4

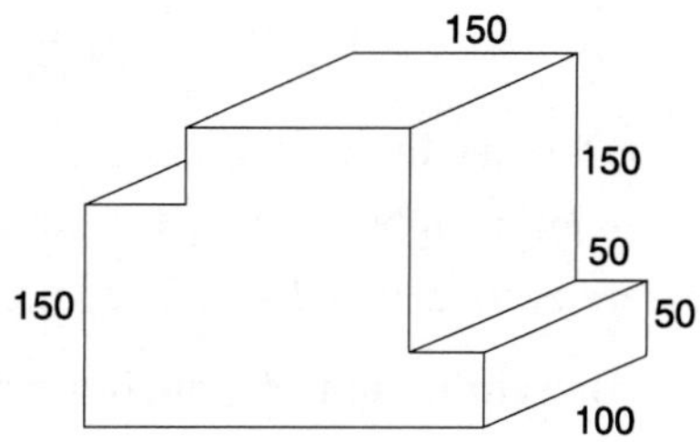

(All dimensions are in mm)

Figure 7.43

Step 1: Selection of a Layout

- Decide first which way up the part is to be cast.
- For two/three-part mould, use casting in cope/middle box, allowing gating at the lowest parts of the casting.
- A heavy section of the casting needs special attention. Orient the heavy part of the casting at the top and plant a feeder there. If this cannot be done, provide chills and/or fins to this section.
- When a general scheme is decided, including the approximate sitting of gates and runners, the provision of feeders (if any) and the location of the sprue, a start can be made on the quantification of the gating system.

Using the above guidelines, the layout for the part is prepared as shown in Figure 7.44.

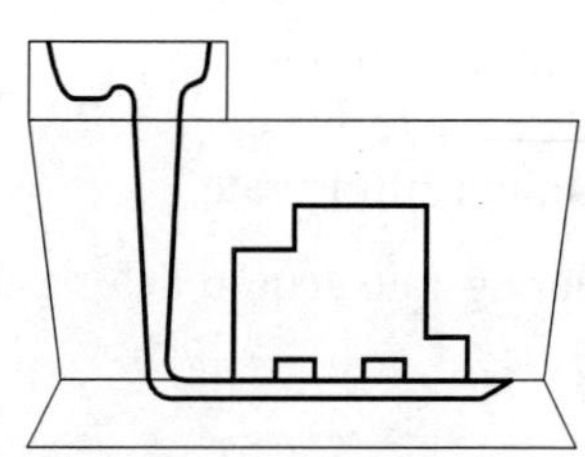

Figure 7.44 Layout of the part

Step 2: Design of Pouring Bush/Basin

- Decide on the type of pouring basin used.
- Conical pouring cup (can be used for small casting). For hand ladle filling (50 mm above the entrance to the sprue) monograms can be used. When poured directly from furnace (usually from a height) monogram cannot be used.
- Pouring bush/basin (for larger casting)

Based on the above guidelines, offset pouring basin is used.

Step 3: Design of Down-Runner or Sprue

3.1 Calculate weight and volume of liquid metal

- Calculate the weight of liquid metal to be poured (includes the casting itself and the rigging, which include the feeder and the gating systems).

- Initially, an extra 20 to 60% of casting weight depending on the size of the casting can be used as the rigging weight. An approximate estimate is nearly always good enough, although a revised value can always be used in a subsequent iteration of the rigging design calculations to obtain an accurate value for the weight of the rigging.

Casting volume,

$$V_c = (75 \times 150 + 150 \times 200 + 50 \times 50) \times 100 = 4375000 \text{ mm}^3 = 4.375 \times 10^{-3} \text{ m}^3$$

Casting weight,

$$W_c = V_c \times \rho_c = 4.375 \times 10^{-3} \text{ m}^3 \times 7000 \frac{\text{kg}}{\text{m}^3} = 30.625 \text{ kg}$$

For this medium-sized casting, let us select the rigging weight = 40% casting weight

Rigging weight = 40% of casting weight = 12.25 kg

Therefore, total liquid metal requirement = weight of casting + weight of rigging = 42.875 kg

3.2 Determination of pouring rate

- Determine/decide on total filling time of casting.
- Pouring time < solidification time of thinnest section of casting.
- Find solidification time from modulus and the Chvorinov's rule.
- Experience/trial-and-error are needed.
- Find the average/initial filling rate (kg/s or m^3/s) in the gating system.

Average filling rate, $$\frac{\text{kg}}{\text{s}} = \frac{\text{Weight of casting (kg)}}{\text{Filling time}}$$

Initial filling rate, $$\frac{\text{kg}}{\text{s}} = 1.5 \times \text{Average filling rate}$$

Pouring rate required for cast iron,

$$R = \frac{W^P}{\left(1.34 + \dfrac{T}{13.77}\right)}$$

T = Critical casting thickness = 50 mm

W = Weight of the casting = 42.875 kg

P = Constant = 0.5

R = Pouring rate in kg/s

$$R = \frac{42.875^{0.5}}{\left(1.34 + \dfrac{50}{13.77}\right)} = 1.32 \text{ kg/second}$$

Now the adjusted pouring rate,

$$R_a = \frac{R}{(k \times f)}$$

Composition of iron given, C = 3.6% and Si = 1.6%

Composition factor, %C + $\frac{\%Si}{4}$ = 3.6 + 1.6/4 = 4%

Pouring temperature = 2450°F

Refer to chart 1, for CF = 4 and T = 2450°F, we get fluidity = 27

Now k can be calculated as $k = \frac{\text{Fluidity}}{40} = \frac{27}{40} = 0.675$

Assume f = 0.90

$$R_a = \frac{R}{(k \times f)} = \frac{1.32 \text{ kg/sec}}{(0.675 \times 0.90)} = 2.17 \text{ kg/sec}$$

$$\text{The filling time, } \frac{W}{R_a} = \frac{42.875}{2.17} = 19.76 \text{ sec}$$

3.3 Select the velocity of flow

The selected velocity of flow must be lower or equal to the critical velocity of liquid metal, which is as follows:

Al-bronze (75 mm/s)

Al-base and Mg-base alloys (250 mm/s)

Cu-base and Fe-base alloys (500 mm/s)

3.4 Calculation of the effective metal head of the casting

$$\text{Effective metal head, } h_p = H - 0.5\left(\frac{h_2^2}{h_1}\right)$$

h_p = Metal head pressure at the sprue base

H = Height of sprue

h_1 = Height of the casting above parting line or height of casting in cope

h_2 = Height of casting

$h_1 = h_2$ = 200 mm

H = Casting height + extra height on top of casting + height of pouring basin

H = 200 + 50 + 50 = 300 mm

$$\text{Effective metal head, } h_p = H - 0.5\left(\frac{h_2^2}{h_1}\right) = 300 - 0.5\left(\frac{200^2}{200}\right) = 200 \text{ mm}$$

Coefficient of discharge = 0.8

$$G = 9.81 \text{ m/s}^2$$

3.5 Calculation of choke area

Smallest area in gating system that regulates the rate of liquid flow throughtout the gating system is choke,

$$A = \frac{W}{\rho tc\sqrt{2gh_p}} = \frac{W}{\rho tV}$$

A = Choke area, W = Casting weight including all elements

ρ = Density of the molten material = 7000 kg/m^3

t = Pouring time = 19.76 seconds

C = Discharge coefficient = 0.8

g = Acceleration due to gravity = 9.81 m/s^2

h_p = Effective height of metal head = 200 mm

V = Liquid metal velocity, m/s

If unpressurised gating is used, the sprue exit area acts as choke. On the other hand, in the pressurised system, gate will act as choke.

$$A = \frac{W}{\rho tc\sqrt{2gh_p}} = \frac{42.875}{7000 \times 19.76 \times 0.8\sqrt{2 \times 9.81 \times 0.2}} = 1.96 \times 10^{-4}\ \text{m}^2 = 196\ \text{mm}^2$$

3.6 Calculate the sprue dimensions

Sprue exit area = choke area

Calculate the sprue top area

$$A_1 = A_2\sqrt{\frac{h_2}{h_1}}$$

A_1 = Sprue top area

A_2 = Sprue bottom area

h_1 = Distance between ladle and sprue top

h_2 = Distance between ladle and sprue bottom

Since unpressurised system is used, spure base area = choke area

Sprue base area = 196 mm^2 = 15.80 mm diameter ≅ 16 mm diameter

$$A_1 = A_2\sqrt{\frac{h_2}{h_1}} = 196\sqrt{\frac{300}{50}} = 576.1\ \text{mm}^2 = 27.08\ \text{mm diameter} \cong 28\ \text{mm diameter}$$

4. Design of runner and ingates

4.1 Calculation of runner dimensions

- The runner will have dimensions as specified by the gating ratio. For example, for a gating ratio of 1:2:4 (unpressurised system), runner area = 2 × sprue exit area.
- The cross section of the runner is usually rectangular and thickness of runner should be close to the height of sessile drop of the corresponding liquid metal (16 mg for Mg-base, 13 mm for Al-base, 8 for Cu and Fe-base alloys) to avoid turbulence (even if the liquid velocity exceeds the critical velocity).
- Take a simple 90° turn from sprue exit into the runner so streamline the sprue exit and runner junction. The inside radius of this turn should be at least approximately 1 or 2 times the thickness of the runner.

In our problem, gating ratio is 1:2:4.

Runner area = 2 × sprue exit area

Runner area = 2 × 196 mm^2 = 392 mm^2 = 15 × 26 mm rectangle

4.2 Calculation of gate dimensions

- Use the gating ratio to determine the total gate area.
- Gate area = 2 × runner area.
- For pressurised system, it will act as choke.
- For multiple gates, divide the total gate area by the number of gates to determine the individual gate area.
- Usually, the thickness of gate equals the thickness of the runner. Length of gate is suitably designed to access the mould cavity.
- Avoid hot spot in the gate junction

$$\text{Gate modulus} = \frac{1}{2} \times \text{local casting modulus}$$

$$\text{Gate thickness} = \frac{1}{2} \times \text{local casting thickness}$$

For the current design,

Gate area = 2 × runner area = 2 × 392 mm^2 = 784 mm^2

Using two gates, area of individual gate = 392 mm^2 = 15 × 26 mm rectangle

5. Determination of sprue dimensions using nomogram

Calculation of sprue, runner and ingate dimensions of the gating design can be made with the use of nomogram (Figure 7.45). Once the average fill rate and sprue height have been decided, the sprue bottom area, gate area and sprue top area can be found from such nomogram. Assumptions made in the preparation of the nomograms are:

1. Unpressurised gating systems
2. Intial pouring rate = 1.5 × average pouring rate
3. $C = 0.8$ (sprue loss) and $k = 0.5$ (basin loss)
4. Critical gate velocity: 250 mm/s for Al-based alloys, 500 mm/s for Cu- and Fe-based alloys
5. 20% safety factor over size of the sprue top area
6. $h_1 = 50$ mm (height of liquid at the pouring basin)
7. Density data: 2500 kg/m^3 for Al-base alloys, 7000 kg/m^3 for Cu-, Fe-, Ni- and Co-based alloys.

Averge flow rate = 2.17 kg/s

Initial flow rate = 1.5 × average flow rate = 1.5 × 2.17 = 3.26 kg/s

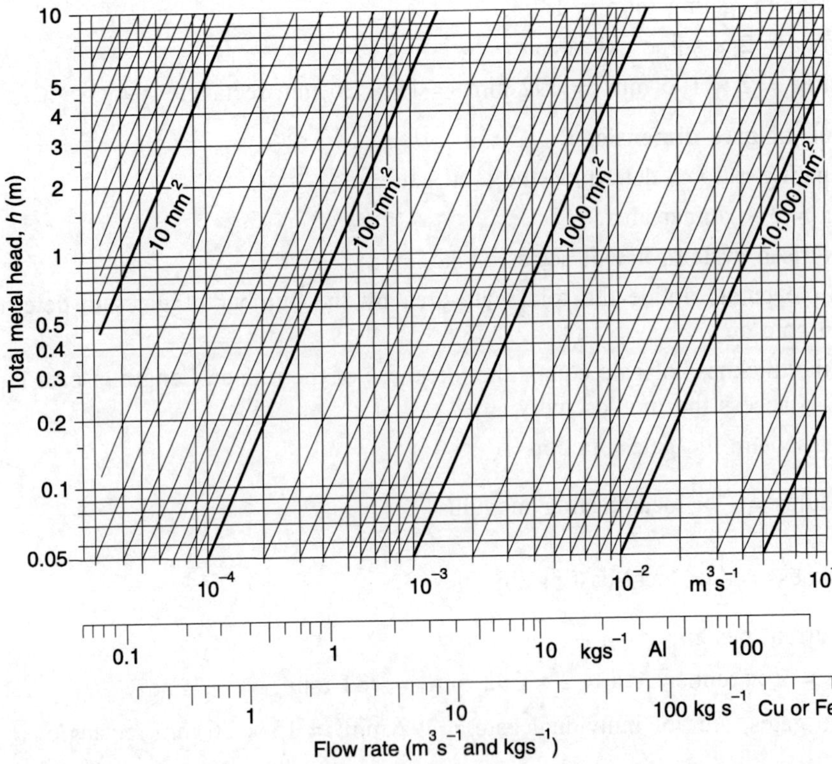

Figure 7.45 Nomogram giving approximate sprue area for light and dense material as a function of initial flow rate and head height

Sprue top area (for h = 50 mm) = 510 mm^2 = 26 mm diameter

Sprue base area (for h = 300 mm) = 210 mm^2 = 17 mm diameter

Example 7.15: The down sprue leading into the runner of a certain mould has length of 175 mm. The cross-sectional area of the base at the sprue is 400 mm^2. The mould cavity has a volume of 0.001 mm^2. Determine (a) the velocity of the molten metal flow through the base of the down sprue, (b) the volume flow rate, and (c) the time required to fill the mould.

Solution:

$$\text{Velocity, } v = \sqrt{2gh} = \sqrt{2 \times 9810 \times 175} = 1852.97 \frac{\text{mm}}{\text{s}} = 1.85297 \frac{\text{m}}{\text{s}}$$

$$\text{Volume flow rate} = \text{Velocity} \times \text{area} = 1852.97 \times 400 = 741188.23 \frac{\text{mm}^3}{\text{s}}$$

$$\text{Time to fill the cavity} = \frac{\text{Volume of mould cavity}}{\text{Volume flow rate of metal}} = \frac{1000000}{741188.23} = 1.3492 \text{ s} \cong 1.35 \text{ s}$$

Example 7.16: The flow rate of liquid metal into the down sprue of mould = 1 litre/sec. The cross-sectional area at the top of sprue is 800 mm^2 and its length is 175 mm. What area should be used at the base of the sprue to avoid aspiration of the molten metal?

Solution:

Velocity, $v = \sqrt{2gh} = \sqrt{2 \times 9810 \times 175} = 1852.97$ mm/s = 1.853 m/s

Volume flow rate = 1.0 litre per sec = 1000 cc per sec = 1000000 mm^3/s

Assuming volumetric continuity,

$$\text{Area at the base} = \frac{\text{Volume flow rate}}{\text{Velocity}} = \frac{1000000}{1853} = 539.6 \text{ mm}^2 \cong 540 \text{ mm}^2$$

Example 7.17: Molten metal can be poured into the pouring cup of a sand mould at a steady rate of 1000 cm^3/s. The molten metal overflows the pouring cup and flows into the down sprue. The cross section of the sprue is round, with the diameter at the top being 3.4 cm. If the sprue length is 25 cm long. Determine the proper diameter at its base so as to maintain the same volume flow rate.

$$\text{Velocity, } v = \sqrt{2gh} = \sqrt{2 \times 9810 \times 250} = 2214.72 \frac{\text{mm}}{\text{s}} = 2.214 \frac{\text{m}}{\text{s}}$$

Assuming volumetric continuity,

$$\text{Area at the base} = \frac{\text{Volume flow rate}}{\text{Velocity}} = \frac{1000000}{2214.72} = 451.523 \text{ mm}^2 \cong 452 \text{ mm}^2$$

$$\text{Area at the base} = \frac{\pi d^2}{4} = 452 \text{ mm}^2$$

Solving, we get diameter at the base = 23.977 mm ≅ 24 mm

Example 7.18: While pouring into a sand mould, the molten metal can be poured into the down sprue at a constant flow rate during the time it takes to fill the mould. At the end of pouring, the sprue is filled and there is negligible metal in the pouring cup. The down sprue is 150 mm long. Its cross-sectional area at the top = 520 mm^2 and at the base = 400 mm^2. The cross-sectional area of the runner = 400 mm^2 and it is 200 mm long before leading into the mould cavity whose volume = 10^6 mm^3. The volume of the riser located along the runner near the mould cavity = 410000 mm^3. It takes three seconds to fill the entire mould including cavity, riser, runner and sprue. This is more than the calculated theoretical time required indicating a loss of velocity due to friction in the sprue and runner. Find (a) theoretical velocity and flow rate at the base of the down sprue, (b) the total volume of the mould, (c) the actual velocity and flow rate at the base of the sprue, and (d) the loss of head in the gating system due to friction.

Solution:

(a) Theoretical flow rate and flow rate at the base of down sprue

$$\text{Velocity, } v = \sqrt{2gh} = \sqrt{2 \times 9810 \times 150} = 1715.52 \frac{\text{mm}}{\text{s}} \cong 1.715 \text{ m/s}$$

$$\text{Volume flow rate} = \text{Velocity} \times \text{area} = 1715.52 \times 400 = 686206.97 \frac{\text{mm}^3}{\text{s}}$$

(b) The total volume of the mould

Total volume of the mould is given by

Volume of mould cavity + volume of riser + averge volume of down sprue + volume of runner

$$= 10^6 + 410000 + 0.5(520 + 400) \times 150 + 400 \times 200 = 1559000 \text{ mm}^3$$

(c) Actual flow rate

$$\text{Actual flow rate} = \frac{1559000}{3} = 519666.7 \ \frac{\text{mm}^3}{\text{s}}$$

$$\text{Actual velocity} = \frac{1559000}{400 \text{ mm}^2} = 1299.17 \ \frac{\text{mm}}{\text{s}}$$

(d) Loss of head due to friction

$$\text{Velocity, } v = \sqrt{2gh} = \sqrt{2 \times 9810 \times h} = 1299.17$$

Solving, we get actual head, h = 86.07 mm

$$\text{Head loss} = 150 - 86.07 \text{ mm} = 63.93 \text{ mm}$$

Example 7.19: Assume that the top of a round sprue has a diameter of 100 mm and is at a height of 150 mm from the runner. Plot the profile of the sprue diameter as a function of its height. Assume that the sprue has a diameter of 25 mm at the bottom.

Solution: The difficulty is that the reference location for height measurements is not known. Often chokes or wells are used to control flow, but this problem will be solved assuming that proper flow is to be attained by considering hydrodynamics in the design of the sprue.

Let,

d_1 = Diameter at the top of the sprue = 100 mm

d_2 = Diameter at the bottom of the sprue = 75 mm

h_1 = Height of the sprue from the top

h_2 = Height of the sprue from the bottom

We can write the relation to avoid aspiration effect as

$$\frac{d_1^2}{d_2^2} = \sqrt{\frac{h_2}{h_1}}$$

The equation can be represented as

$$d = C(h)^{-0.25}$$

where C is a constant and to be found out from boundary condition.

Assume the boundary condition,

$$h = h_o \text{ at diameter, } d = d_1 = 100 \text{ mm}$$

where h_o is the height at the top of sprue to the reference location.

$$h = h_o + 300 \text{ mm at diameter, } d = d_2 = 25 \text{ mm}$$

First boundary condition yields

$$d_1 = C(h_1)^{-0.25}$$

$$100 = C(h_o)^{-0.25}$$

$$C = 100(h_o)^{0.25} \quad ...(a)$$

Second boundary condition yields

$$25 = C(h_o + 300)^{-0.25}$$

Using

$$C = 100(h_o)^{0.25}$$

$$25 = C(h_o + 300)^{-0.25}$$

$$25 = 100(h_o)^{0.25}\,(h_o + 300)^{-0.25} \quad ...(b)$$

Solving (a) and (b), we get $C \cong 104.1$

$$h_o = 1.176 \text{ mm}$$

$$d = 104.1\,(h + 1.176)^{-0.25}$$

Note that $h = 0$ is the location of the top of the sprue, $h = 300$ is the bottom of the sprue, and that the sprue is axisymmetric. The profile can be plotted as shown in Figure 7.46.

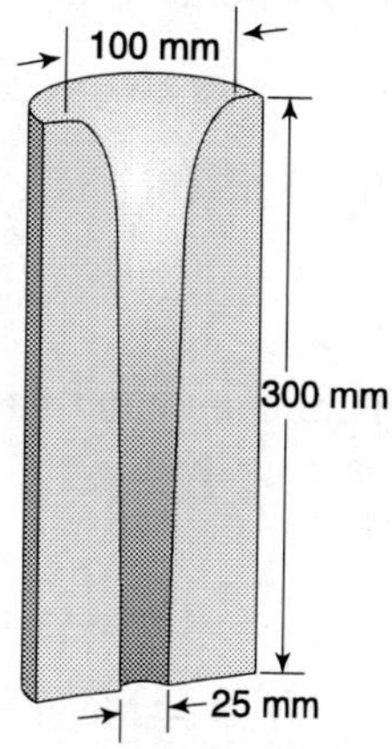

Figure 7.46

Example 7.20: A sprue is 200 mm long and has a diameter of 125 mm at the top, where the metal is poured. If a flow rate of 60,000 mm^3/s is to be achieved, what should be the diameter of the bottom of the sprue?

Solution: Assuming the flow is frictionless, the velocity of the molten metal at the bottom of the sprue is given by

$$\text{Velocity, } v = \sqrt{2gh} = \sqrt{2 \times 9810 \times 200} = 1980.901 \frac{\text{mm}}{\text{s}} \cong 1.98 \text{ m/s}$$

$$\text{Volume flow rate} = \text{Velocity} \times \text{Area} = 1715.52 \times 400 = 686206.97 \frac{\text{mm}^3}{\text{s}}$$

$$\text{Area} = \frac{\text{Volume flow rate}}{\text{Velocity}} = \frac{60000}{1980.901} = 30.289 \text{ mm}^2 \cong 30.3 \text{ mm}^2$$

$$\text{Area at the base} = \frac{\pi d^2}{4} = 30.3 \text{ mm}^2$$

Solving, we get diameter at the base = 6.210 mm ≅ 6.3 mm

Example 7.21: Aluminium is poured into a sand mould. The metal level in the pouring basin is 250 mm above the metal level in the mould, and the runner is circular with a 10 mm diameter. What is the velocity and rate of the flow of the metal into the mould? Is the flow turbulent or laminar?

Solution: Assume that there is no frictional loss in sprue and pressure changes are negligible; we can write

$$h_1 + \frac{v_1^2}{2g} = h_2 + \frac{v_2^2}{2g}$$

where the subscript 1 indicates the top of the sprue and 2 the bottom. If we assume that the velocity at the top of the sprue is very low (as would occur with the normal case of a pouring basin on top of the sprue with a large cross-sectional area), then $v_1 = 0$. Therefore, the velocity at the bottom of the sprue is

$$v_2 = \sqrt{2g(h_1 - h_2)} = \sqrt{2 \times 9810 \times 250} = 2215 \ \frac{\text{mm}}{\text{s}} = 2.215 \text{ m/s}$$

If the opening is 10 mm diameter, the flow rate is

$$Q = 2.215 \times \frac{\pi \times 0.01^2}{4} = 1.740 \times 10^{-4} \ \frac{\text{m}^3}{\text{s}}.$$

Pure aluminium has a density of 2700 kg/m^3 and a viscosity is 0.0015 Pa-s. The Reynolds number is then,

$$R_e = \frac{\rho V d}{\mu} = \frac{Vd}{\nu}$$

V = Mean velocity of liquid metal in $\frac{\text{m}}{\text{s}}$ = 2.215 m/s

d = Linear dimensions of the channel, 10 mm = 10×10^{-3} m

ρ = Density of liquid metal in $\frac{\text{kg}}{\text{m}^3} = 2700 \ \frac{\text{kg}}{\text{m}^3}$

μ = Dynamic or absoulte viscosity of liquid in $\frac{\text{N} - \text{s}}{\text{m}^2} = 0.0015$

ν = Kinematic viscosity of liquid metal = $\frac{\mu}{\rho}$

$$R_e = \frac{\rho V d}{\mu} = \frac{2700 \times 2.215 \times 10 \times 10^{-3}}{0.0015} = 39870$$

The flow is turbulent.

Example 7.22: The volume flow rate of metal into a mould is 0.2 m^3/min. The top of the sprue has a diameter of 20 mm, and its length is 200 mm. What diameter should be specified at the bottom of the sprue to prevent aspiration? What is the resultant velocity and Reynolds number at the bottom of the sprue if the metal being cast is aluminium with a viscosity of 0.004 $\frac{\text{N-s}}{\text{m}^2}$?

Solution: The metal flow rate is $Q = 0.02 \ \frac{\text{m}^3}{\text{min}} = 3.334 \times 10^{-4} \ \frac{\text{m}^3}{\text{sec}}$. Let us use the subscripts 1 and 2 for the top and bottom of the sprue respectively.

Diameter of sprue at top, d_1 = 20 mm = 0.02 m

$$\text{Area of sprue at top} = \frac{\pi d_1^2}{4} = \frac{\pi \times 0.02^2}{4} = 3.14 \times 10^{-4} \text{ m}^2$$

$$v_1 = \frac{\text{Volume flow rate}}{\text{Area}} = \frac{3.334 \times 10^{-4}}{3.14 \times 10^{-4}} = 1.062 \text{ m/s}$$

Neglecting the frictional losses and recognizing that the pressure at the top and bottom of the sprue is atmospheric. We can write,

$$h_1 + \frac{v_1^2}{2g} = h_2 + \frac{v_2^2}{2g}$$

Substituting the values we get,

$$0.2 \text{ m} + \frac{1.062^2 \left(\frac{\text{m}^2}{\text{s}^2}\right)}{2 \times 9.81 \left(\frac{\text{m}}{\text{s}^2}\right)} = 0 + \frac{v_2^2}{2 \times 9.81 \left(\frac{\text{m}}{\text{s}^2}\right)}$$

Solving we get,

$$v_2 = 2.23 \text{ m/s}$$

To avoid aspiration effect, the sprue opening should be the same as that required by flow continuity equation.

$$Q = A_2 v_2 = 3.334 \times 10^{-4} \frac{\text{m}^3}{\text{sec}} = \text{A}_2 \times 2.23$$

Solving we get, $A_2 = 1.49 \times 10^{-4} \text{ m}^2$

Diameter, $d_2 = 14 \text{ mm}.$

Now we can calculate the Reynolds number,

$$Re = \frac{\rho V d}{\mu} = \frac{Vd}{\nu}$$

V = Mean velocity of liquid metal in $\frac{\text{m}}{\text{s}} = 2.23$ m/s

d = Linear dimensions of the channel, 14 mm = 14×10^{-3} m

ρ = Density of liquid metal in $\frac{\text{kg}}{\text{m}^3} = 2700 \frac{\text{kg}}{\text{m}^3}$

μ = Dynamic or absoulte viscosity of liquid in $\frac{\text{N-s}}{\text{m}^2} = 0.004$

ν = Kinematic viscosity of liquid metal = $\frac{\mu}{\rho}$

$$R_e = \frac{\rho V d}{\rho} = \frac{2700 \times 2.23 \times 14 \times 10^{-3}}{0.004} = 21073.5$$

The flow is turbulent.

Example 7.23: A sprue is 300 mm long and has a diameter of 125 mm at the top. The molten metal level in the pouring basin (which is much larger than the top of the sprue) is taken to be 75 mm from the top of the sprue for design purposes. If a flow rate of 655483 mm^3/sec is to be achieved. What should be diameter at the bottom of the sprue? Will the sprue aspirate? Explain.

Solution: Assuming the flow is frictionless, the velocity of the molten metal at the bottom of the sprue is given by

$$v = \sqrt{2gh} = \sqrt{2 \times 9810 \times 300} = 2426.11 \text{ mm/s}$$

Volume flow rate = Velocity × area = 2426.11 mm/s × area = 655483 mm^3/sec

Solving, we get the area = 270.17 mm^2

For a circular runner, the diameter would be 270.17 mm^2 = $\dfrac{\pi d^2}{4}$

Solving we get the area, d = 18.54 mm

Condition to prevent the aspiration is given by

Let,

d_1 = Diameter at the top of the sprue = 125 mm

d_2 = Diameter at the bottom of the sprue

h_1 = Height of the sprue from the top of the pouring basin = 75 mm

h_2 = Height of the sprue from the bottom = 375 mm

We can write the relation to avoid aspiration effect as

$$\frac{d_1^2}{d_2^2} = \sqrt{\frac{h_2}{h_1}}$$

$$\frac{125^2}{d_2^2} = \sqrt{\frac{375}{75}}$$

Solving, we get d_2 = 83.59 mm > 18.54 mm

Thus, sprue will not aspirate.

Example 7.24: Pure aluminium is poured into a sand mould. The metal level in the pouring basin is 250 mm above the metal level in the mould and runner is circular with a 10 mm diameter. (a) What are the velocity and rate of flow of the metal into the mould? (b) Is the flow turbulent or laminar? (c) What runner diameter is needed to ensure a Reynolds number of 2000? d) How long will a 327741 mm^3 casting take to fill with such a runner? (e) How long would it take for the sprue to feed a casting with a square cross section of 50 mm per side and a height of 100 mm? Assume the sprue is frictionless.

Solution:

(a) Velocity and flow rate of metal into the mould

Assume that there is no frictional loss in sprue and pressure changes are negligible, we can write

$$h_1 + \frac{v_1^2}{2g} = h_2 + \frac{v_2^2}{2g}$$

where subscript 1 indicates the top of the sprue and 2 the bottom. If we assume that the velocity at the top of the sprue is very low (as would occur with the normal case of a pouring basin on top of the sprue with a large cross-sectional area), then $v_1 = 0$. Therefore, the velocity at the bottom of the sprue is

$$v_2 = \sqrt{2g(h_1 - h_2)} = \sqrt{2 \times 9810 \times 250} = 2214.72 \frac{\text{mm}}{\text{s}}$$

For a circular runner, the area would be $A_2 = \frac{\pi d^2}{4} = \frac{\pi \times 10^2}{4} = 78.54 \text{ mm}^2$

Flow rate, $Q = A_2 v_2 = 78.54 \text{ mm}^2 \times 2214.72 \frac{\text{mm}}{\text{s}} = 173938.57 \text{ mm}^3/\text{sec}$

(b) Is the flow turbulent or laminar?

Pure aluminium has a density of 2700 kg/m^3 and a viscosity is 0.0015 Pa-s at 700°C. The Reynolds number is then,

$$R_e = \frac{\rho V d}{\mu} = \frac{Vd}{\nu}$$

V = Mean velocity of liquid metal in $\frac{\text{m}}{\text{s}}$ = 2.21472 m/s

d = Linear dimensions of the channel, 10 mm = 10×10^{-3} m

ρ = Density of liquid metal in $\frac{\text{kg}}{\text{m}^3} = 2700 \frac{\text{kg}}{\text{m}^3}$

μ = Dynamic or absoulte viscosity of liquid in $\frac{\text{N-s}}{\text{m}^2}$ = 0.0015

ν = Kinematic viscosity of liquid metal = $\frac{\mu}{\rho}$

$$R_e = \frac{\rho V d}{\mu} = \frac{2700 \times 2.21472 \times 10 \times 10^{-3}}{0.0015} = 40346.49 \cong 40347$$

The flow is turbulent. The velocity and/or diameter are to be decreased to make the flow laminar.

(c) What runner diameter is needed to ensure a Reynolds number of 2000?

Reynolds number of 2000 can be achieved by reducing the channel diameter. So that

$$R_e = \frac{\rho V d}{\mu} = \frac{2700 \times 2.21472 \times d}{0.0015} = 2000$$

Solving, we get $d = 5.017 \times 10^{-4}$ mm = 0.5016 mm

For this diameter, the initial flow rate would be

$$Q = A_2 v_2 = \frac{\pi \times 0.5^2}{4} \times 2214.72 \frac{\text{mm}}{\text{s}} = 434.84 \text{ mm}^3/\text{sec}$$

(d) How long will a 327741 mm^3 casting take to fill with such a runner?

$$\text{Time taken to fill cavity of 327741 mm}^3 = \frac{\text{Volume of cavity}}{\text{Flow rate}}$$

$$= \frac{327741}{434.84} = 753.7 \text{ seconds}$$

Time taken to fill the cavity is 753.7 seconds (about 12 minutes) and only if the initial flow rate could be maintained, which is generally not the case. Such a long filling time is unacceptable. Since it is likely that metal will solidify in the runner and thus not fill the mould completely. Also, with such a small runner, additional mechanisms need to be considered like surface tension and friction would severely reduce the velocity in the Reynolds number calculation above. This is generally the case with castings: to design a sprue and runner system that maintains laminar flow in the fluid would result in excessively long fill times.

(e) If the sprue cross section is square.

Volume of the casting = Area of square cross section × height = 2500 × 100 = 250000 mm^3 with a constant cross-sectional area of 2500 mm^2.

The velocity will change as the mould fills, because the pouring basin height above the molten metal will decrease. The velocity will vary according to:

$$v = \sqrt{2gh}$$

$$\text{Flow rate, } Q = Av = \frac{\pi \times d^2}{4} \times \sqrt{2gh}$$

Mould cavity fills at a rate of $\dfrac{Q}{2500 \text{ mm}^2}$

$$\frac{dh}{dt} = \frac{Q}{A} = -\frac{\dfrac{\pi \times d^2}{4} \times \sqrt{2gh}}{A}$$

Minus is added so that h refers to the height difference between the metal level in the mould and the runner, which decreases with respect to time. Separating the variables,

$$\frac{dh}{\sqrt{h}} = \frac{Q}{A} = -\frac{\pi d^2 \sqrt{2g}}{4A}\, dt$$

Integrating we get,

$$(2\sqrt{h})_{0.050}^{0.300} = -\frac{\pi d^2 \sqrt{2g}}{4A} \{t\}_0^t$$

Using $d = 0.01$ m and $A = 2500$ mm^2 we get

$t = 4.66$ seconds.

Example 7.25: A mould has a down sprue whose length is 20 cm and cross-sectional area of the base of the down sprue is 1 cm^2. The down sprue feeds a horizontal runner leading into the mould cavity of volume of 1000 cm^3. What is the time required to fill the mould cavity? **(GATE 2005)**

Solution: Length of sprue = 20 cm

Cross-sectional area of base of sprue = 1 cm^2

Volume of the casting = 100 cm^3

$$\text{Velocity, } v = \sqrt{2gh} = \sqrt{2 \times 9810 \times 200} = 1980.91 \frac{\text{mm}}{\text{s}} = 198.091 \frac{\text{cm}}{\text{s}}$$

$$\text{Time required for filling} = \frac{\text{Volume}}{\text{Cross-sectional area} \times \text{velocity}} = \frac{1000 \text{ cm}^3}{198.091 \frac{\text{cm}}{\text{s}}} = 5.05 \text{ seconds.}$$

Example 7.26: In a sand casting operation, the total liquid head is maintained constant such that it is equal to mould height. The time taken to fill the mould with a top gate is t_a. If the same mould is filled with the bottom gate as t_b. Ignore the time required to fill the runner and frictional effects. Assume atmospheric pressure at the top molten metal surfaces. What is the relation between t_a and t_b? **(GATE 2006)**

Solution:

$$\text{Time required for filling} = \frac{\text{Volume}}{\text{Cross-sectional area} \times \text{velocity}}$$

$$t_a = \frac{\text{Volume}}{A_c \times v_{max}}$$

$$t_b = \frac{\text{volume}}{A_c \times v_{avg}}$$

$$\frac{t_a}{t_b} = \frac{\text{volume}}{A_c \times v_{max}} \times \frac{A_c \times V_{avg}}{\text{Volume}} = \frac{v_{avg}}{v_{max}}$$

But $$V_{avg} = \frac{v_{max} + v_{min}}{2}$$

Assume $$v_{min} = 0$$

$$V_{avg} = \frac{v_{max}}{2}$$

$$\frac{t_a}{t_b} = \frac{\frac{v_{max}}{2}}{v_{max}}$$

$$\frac{t_a}{t_b} = \frac{1}{2}$$

This gives $t_b = 2\ t_a$

Example 7.27: A 200 mm long down sprue has an area of cross section of 650 mm^2 where the pouring basin meets the down sprue (that is at the beginning of the down sprue). A constant head of molten metal is maintained by the pouring basin. The molten metal flow rate is 6.5×10^3 mm^3/s considering the end of the down sprue to be open to atmospheric and acceleration due to gravity of 10^4 mm/s^2. What is the area of down sprue in mm^2 at its end without aspiration effect? **(GATE 2007)**

Solution: Height of sprue, $h_2 = 200$ mm

$$A_2 = 650 \text{ mm}^2$$

$$\text{Flow rate, } Q = 6.5 \times 10^3 \text{ mm}^3/\text{s}$$

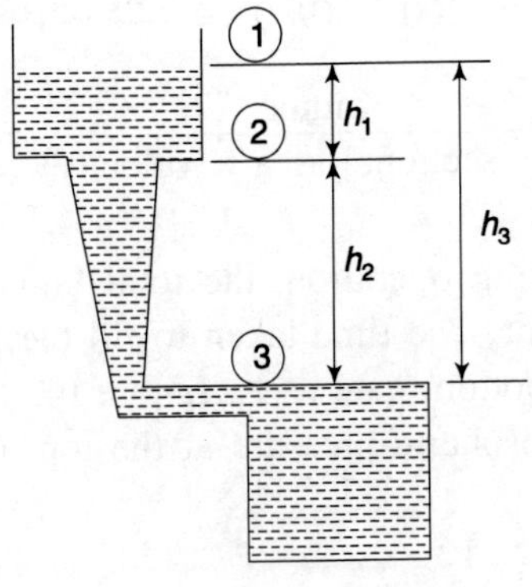

Figure 7.47

Acceleration due to gravity, $g = 10^4$ mm/s^2

$$\text{Velocity, } v_2 = \frac{\text{Volume}}{\text{Cross-sectional area}} = \frac{6.5 \times 10^3 \text{ mm}^3/\text{s}}{650 \text{ mm}^2} = 1000 \text{ mm/s}$$

h_1 = 50 mm (Height of molten metal in the pouring basin, above reference)

h_3 = 50 mm + 200 mm = 250 mm

Flow rate, $Q = A_2 v_2 = A_3 v_3$

$$6.5 \times 10^3 \text{ mm}^3/\text{s} = A_3 \sqrt{2 \times 10^4 \times 250}$$

Solving we get, $A_3 = 290.7$ mm^2

Example 7.28: The height of the down sprue is 175 mm and its cross-sectional area at the base is 200 mm^2. The cross-sectional area of the horizontal runner is also 200 mm^2. Assuming no losses, find the time to fill the mould cavity of volume 1000000 mm^3. Use g = 10 m/s^2.

(GATE 2002-P2)

Solution: Length of sprue = 175 mm

Cross-sectional area of base of sprue = 200 mm^2

Volume of the casting = 1000000 mm^3

$$g = 10 \text{ m/s}^2 = 10000 \text{ mm/s}^2$$

$$\text{Velocity, } v = \sqrt{2gh} = \sqrt{2 \times 10000 \times 175} = 1870.83 \frac{\text{mm}}{\text{s}} = 18.7083 \frac{\text{cm}}{\text{s}}$$

$$\text{Time required for filling} = \frac{\text{Volume}}{\text{Cross-sectional area} \times \text{Velocity}}$$

$$\text{Time required for filling} = \frac{1000000 \text{ mm}^3}{200 \text{ mm}^2 \times 1870.83 \frac{\text{mm}}{\text{s}}} = 2.67 \text{ seconds}$$

Example 7.29: In a sand casting process, a sprue of 10 mm base diameter and 250 mm height leads to a runner which fills a cubical cavity of 100 mm side. What is the volume flow area in mm³/s and what is mould filling time in seconds? **(GATE 2007-P1)**

Solution: Length of sprue = 250 mm

$$\text{Cross-sectional area of base of sprue} = \frac{\pi \times d^2}{4} = \frac{\pi \times 10^2}{4} = 78.5375 \text{ mm}^2 = 78.54 \text{ mm}^2$$

$$\text{Volume of the casting} = 100 \times 100 \times 100 = 10^6 \text{ mm}^3$$

$$g = 9810 \text{ mm/s}^2$$

$$\text{Velocity, } v = \sqrt{2gh} = \sqrt{2 \times 9810 \times 250} = 2214.7233 \frac{\text{mm}}{\text{s}} \cong 2215 \frac{\text{mm}}{\text{s}}$$

$$\text{Flow rate, } Q = \text{Area} \times \text{Velocity} = 78.54 \text{ mm}^2 \times 2215 \frac{\text{mm}}{\text{s}} = 173966 \text{ mm}^3/\text{sec}$$

$$\text{Time required for filling} = \frac{\text{Volume}}{\text{Cross-sectional area} \times \text{velocity}}$$

$$\text{Time required for filling} = \frac{1000000 \text{ mm}^3}{173966 \text{ mm}^3/\text{sec}} = 5.748 \text{ sec} \cong 5.75 \text{ sec}$$

Example 7.30: A mould having dimensions 100 × 90 × 20 (all in mm) is filled with molten metal through a gate with a height, h and cross-sectional area, A. The mould filling time is t_1. The height now is quadrupled and cross section is halved. What is the corresponding filling time and ratio of t_2/t_1? **(GATE 2012-P1)**

Solution:

$$\text{Time required for filling} = \frac{\text{Volume}}{\text{Cross-sectional area} \times \text{Velocity}}$$

$$t_1 = \frac{V}{A \times \text{Velocity}}$$

$$\text{Velocity, } v = \sqrt{2gh}$$

$$t_1 = \frac{V}{A \times \sqrt{2gh}} \quad \text{...(a)}$$

$$t_2 = \frac{V}{\frac{A}{2} \times \sqrt{2g(4h)}} = \frac{V}{A \times \sqrt{2gh}} \quad \text{...(b)}$$

$$\frac{t_1}{t_2} = \frac{\dfrac{V}{A \times \sqrt{2gh}}}{\dfrac{V}{A \times \sqrt{2gh}}} = 1$$

Example 7.31: In metal casting process, molten copper alloys is poured into a sand mould. The level of molten metal in the pouring basin is at the height of 300 mm from the runner having a diameter of 10 mm. If the density and melting temperature of the molten copper are 9000 kg/m^3 and 1000 cc respectively. What is the rate of flow of molten metal into the mould neglecting the friction and other losses in cm^3/s? **(GATE 2015-P1)**

Solution:

Length of sprue = 300 mm

$$\text{Cross-sectional area of base of sprue} = \frac{\pi \times d^2}{4} = \frac{\pi \times 10^2}{4} = 78.5375 \text{ mm}^2 = 0.7854 \text{ cm}^2$$

Volume of the casting = 1000 cm^3

$$g = 9810 \text{ mm/s}^2$$

$$\text{Velocity, } v = \sqrt{2gh} = \sqrt{2 \times 9810 \times 300} = 2426.11 \frac{\text{mm}}{\text{s}} \cong 242.61 \frac{\text{cm}}{\text{s}}$$

$$\text{Flow rate, } Q = \text{Area} \times \text{Velocity} = 0.7854 \text{ cm}^2 \times 242.61 \frac{\text{cm}}{\text{s}} = 190.546 \text{ cm}^3/\text{sec}$$

Example 7.32: A down sprue of 180 mm length has a diameter of 20 mm at its top end. The liquid metal in the pouring cup is maintained up to 60 mm height. What should be the theoretical diameter of down sprue at its lower end to ensure that there is no aspiration effect? **(GATE 1987)**

Solution:

Method 1

Height of sprue, h_2 = 180 mm

$$\text{Area of top of sprue} = \frac{\pi \times d_2^2}{4} = \frac{\pi \times 20^2}{4} = 314.15 \text{ mm}^2$$

$$A_2 = 314.15 \text{ mm}^2$$

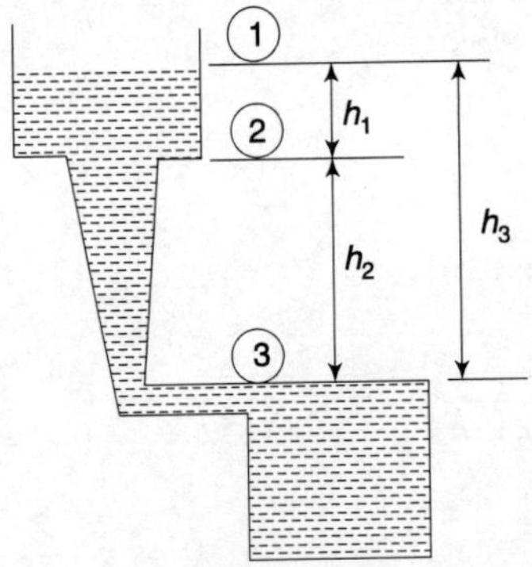

Figure 7.48

h_1 = 60 mm (height of molten metal in the pouring basin, above reference)

$$\text{Velocity, } v_2 = \sqrt{2 \times g \times h_1} = \sqrt{2 \times 9810 \times 60} = 1084.988 \frac{\text{mm}}{\text{sec}} \cong 1085 \text{ mm/sec}$$

$$h_3 = 60 \text{ mm} + 180 \text{ mm} = 240 \text{ mm}$$

Velocity, $v_3 = \sqrt{2 \times g \times h_3} = \sqrt{2 \times 9810 \times 240} = 2169.977 \frac{\text{mm}}{\text{sec}} \cong 2170$ mm/sec

$$\text{Flow rate, } Q = A_2 v_2 = A_3 v_3$$

$$314.15 \text{ mm}^2 \times 1085 \text{ mm/sec} = A_3 \times 2170 \text{ mm/sec}$$

$$\text{Solving we get, } A_3 = 157.08 \text{ mm}^2$$

$$\text{Area of bottom of sprue} = \frac{\pi \times d_3^2}{4} = 157.08 \text{ mm}^2$$

$$\text{Solving we get, } d_3 = 14.14 \text{ mm}$$

Method 2

Let,

d_2 = Diameter at the top of the sprue = 20 mm

d_3 = Diameter at the bottom of the sprue

h_1 = Height of the sprue from the top of the pouring basin = 60 mm

h_2 = Height of the sprue from the bottom of the pouring basin = 180 mm

h_3 = Total height of the sprue = 180 + 60 = 240 mm

We can write the relation to avoid aspiration effect as

$$\frac{d_2^2}{d_3^2} = \sqrt{\frac{h_1}{h_3}}$$

$$\frac{20^2}{d_3^2} = \sqrt{\frac{60}{240}}$$

Solving we get, $d_3 = 14.14$ mm

Example 7.33: In a particular mould design, the down sprue has an area of cross section of 6.45 cm^2 where the pouring basin leads into the sprue. The sprue is 20 cm long. The required metal flow rate at the top section of the sprue is 820 cm^3/s. Determine the pouring height necessary above the spure top. Also determine the area of cross section of the sprue at its bottom to avoid the aspiration of liquid molten metal. **(GATE 1999)**

Solution:

Method 1

d_2 = diameter at the top of the sprue

$$\text{Area of top of sprue} = \frac{\pi \times d_2^2}{4} = 6.45 \text{ cm}^2$$

Solving we get,

d_2 = 2.87 cm

d_3 = Diameter at the bottom of the sprue

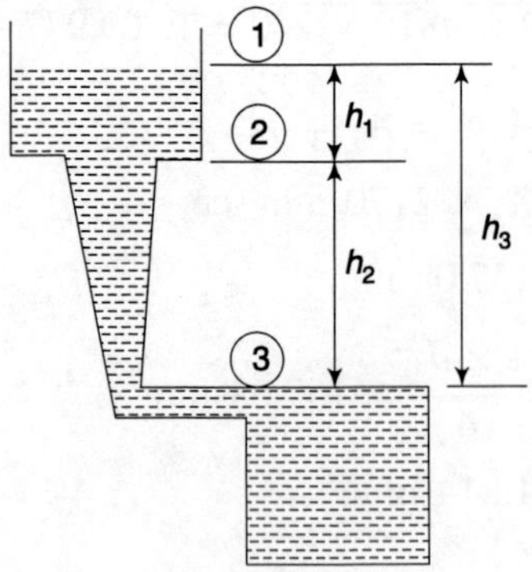

Figure 7.49

$$\text{Velocity, } v_2 = \frac{\text{Volume}}{\text{Cross-sectional area}} = \frac{820 \text{ cm}^3/\text{s}}{6.45 \text{ cm}^2} = 127.13 \text{ cm/s}$$

h_1 = Height of molten metal in the pouring basin

$$\text{Velocity, } v_2 = \sqrt{2 \times g \times h_1} = 127.13 \frac{\text{cm}}{\text{s}}$$

Solving we get, $h_1 = 8.24$ cm

$$h_1 = 8.24 \text{ cm}$$

$$h_3 = 8.24 \text{ cm} + 20 \text{ cm} = 28.24 \text{ cm}$$

$$\text{Flow rate, } Q = A_2 v_2 = A_3 v_3$$

$$820 \text{ cm}^3/\text{s} = A_3 \sqrt{2 \times 981 \times 28.24}$$

Solving we get, $A_3 = 3.484 \text{ cm}^2$

$$\text{Area of bottom of sprue} = \frac{\pi \times d_3^2}{4} = 3.484 \text{ cm}^2$$

Solving we get, $d_3 = 2.106 \text{ cm} \cong 2.11 \text{ cm}$

Method 2

h_1 = Height of the sprue from the top of the pouring basin = 8.24

h_2 = Height of the sprue from the bottom from pouring basin = 20 mm

h_3 = Total height of the sprue = 28.24 cm

We can write the relation to avoid aspiration effect as

$$\frac{d_2^2}{d_3^2} = \sqrt{\frac{h_1}{h_3}}$$

$$\frac{2.87^2}{d_3^2} = \sqrt{\frac{8.24}{28.24}}$$

Solving we get, $d_3 = 2.11$ cm

Example 7.34: A rectangular block of steel cross section 100 mm × 150 mm and 250 mm height is to be cast without any riser. The block is moulded entirely in the drag of greensand flask and is top gated. The cope of the flask is 200 mm height and the height of the metal during pouring is 100 mm above the cope level. A tapered sprue is used and gating ratio is 1:4:2. Assume no energy losses in the systems. The time taken is seconds to fill the casting cavity, if the tapered sprue with 500 mm^2 exit area is used. **(GATE 1991-P1)**

Solution: Volume of the casting = 100 mm × 150 mm × 250 mm = 3750000 mm^3

Block is entirely in drag and top gate is used.

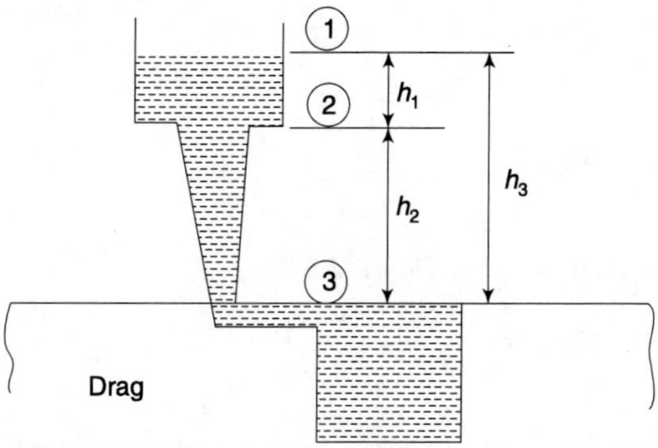

Figure 7.50

h_2 = Height of the cope = 200 mm

h_1 = Pouring height = 100 mm

h_3 = Total height = 300 mm

Area of bottom of sprue, A_3 = 500 mm^2 = 5 cm^2

Velocity, $v_3 = \sqrt{2 \times g \times h_3} = \sqrt{2 \times 981 \times 30} = 242.61 \ \frac{\text{cm}}{\text{s}}$

Flow rate, $Q = A_3 v_3$

$Q = 5 \text{ cm}^2 \times 242.61 \ \frac{\text{cm}}{\text{s}} = 1213.053 \ \frac{\text{cm}^3}{\text{s}} = 1213.053 \times 1000 \ \frac{\text{mm}^3}{\text{s}}$

$$\text{Time required for filling} = \frac{\text{Volume}}{\text{Cross-sectional area} \times \text{Velocity}} = \text{Volume flow rate}$$

$$\text{Time required for filling} = \frac{3750000 \text{ mm}^3}{1213.053 \times 1000 \text{ mm}^3/\text{s}} = 3.0913 \text{ sec} \cong 3.1 \text{ seconds}$$

Example 7.35: A cylinder of 150 mm diameter and 200 mm height is to be cast without any riser. The cylinder is moulded entirely in the drag of a green sand flask and top gated. The cope of the flask is 200 mm height and height of metal during pouring is 50 mm above the cope. A tapered sprue is used and gating ratio is 1:1.5:2. What is the time taken in seconds to fill the casting cavity neglecting the energy losses if the ingate area is 400 mm^2?

(GATE 1994-P1)

Solution: Volume of the casting = $\frac{\pi \times 150^2}{4} \times 200$ mm = 3534291 mm^3

Block is entirely in drag and top gate is used.

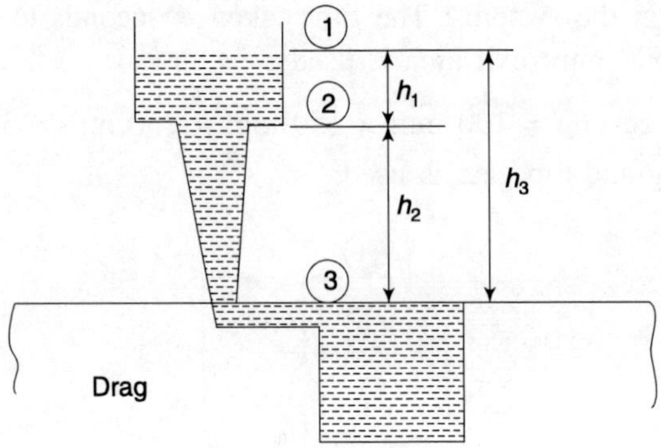

Figure 7.51

h_2 = Height of the cope = 200 mm

h_1 = Pouring height = 50 mm

h_3 = Total height = 250 mm

Gating ratio = 1:1.5:2

Gating ratio = area of sprue: area of runner: area of ingate

Area of ingate is given as 400 mm^2

Therefore, area of sprue = 200 mm^2

Area of bottom of sprue, A_3 = 400 mm^2/2 = 2 cm^2

Velocity, $v_3 = \sqrt{2 \times g \times h_3} = \sqrt{2 \times 981 \times 25} = 221.47 \frac{\text{cm}}{\text{s}}$

Flow rate, $Q = A_3 v_3$

$Q = 2 \text{ cm}^2 \times 221.47 \frac{\text{cm}}{\text{s}} = 442.95 \frac{\text{cm}^3}{\text{s}} = 442.95 \times 1000 \frac{\text{mm}^3}{\text{s}}$

Time required for filling = $\frac{\text{Volume}}{\text{Cross-sectional area} \times \text{Velocity}}$ = Volume flow rate

Time required for filling = $\frac{3534291}{442.95 \times 1000}$ = 7.979 sec ≅ 8 seconds

Review Questions

1. What are the primary functions of the gating systems?
2. What are the secondary functions of the gating systems?
3. What are the classifications of gating systems?
4. Differentiate between top gating and bottom gating systems.
5. Differentiate between pressurised and non-pressurised gating systems.
6. Explain steps involved in sprue design.
7. How do you calculate mould filling time?
8. How can we avoid aspiration effect?
9. What is gating ratio?
10. Explain various steps involved in the design of runner and ingates.

CHAPTER 8

Risers and Its Design

8.1 INTRODUCTION

The casting process is essentially a phase transformation of metals from liquid to solid state in a preshaped cavity. This phase-transformation is accompanied by some volumetric shrinkage in most of the cast metals. If casting does not have a source of more molten metal to feed it as it shrinks, it results in casting defects. The purpose of a riser is to compensate that extra molten metal which is going to shrink. Risers are considered a part of the gating system, since they are feeding castings. The volumetric shrinkage of metals during solidification is noted as 6.6% for pure aluminium and 4.92% for copper. Steel has 2.5 to 4% shrinkage, whereas for grey cast iron the shrinkage is 1.9% to negative, depending on graphitization. Grey cast iron needs less feeding than other metals as the amount of shrinkage is compensated by the process of graphitization so, cast irons generally do not require risers for feeding. Steel and most of the non-ferrous alloys have wide solidification range which requires sufficient feed metal to obtain sound casting.

8.2 GENERAL SHRINKAGE BEHAVIOUR

Molten metal occupies considerably more volume than the solidified casting, causing a number of problems. Liquid metals experience three different contractions during solidification (Figure 8.1):

1. Liquid contraction is not troublesome
2. Solidification contraction causes feeding problems
3. Solid contraction causes hot tearing/cracking.

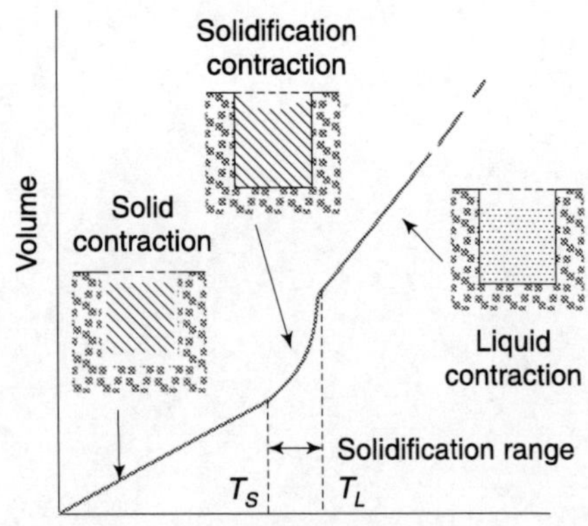

Figure 8.1 Schematic illustration of three shrinkage regimes: in the liquid, during freezing, and in the solid.

Once pouring is completed, the original supply of liquid metal is removed. Thus, some source of molten metal must be provided to feed the shrinkage that occurs as the metal cools in the mould.

In some cases, especially with small castings, the gates and runners are designed so that they are able to feed the shrinkage. Sometimes, gates are also used to scatter shrinkage, distributing it so that shrinkage at any one point is slight and the soundness of the casting is not appreciably affected. Sometimes, heavy sections of a casting feed light or thin sections. In most cases, however, we must use reservoirs called feeders/risers to supply molten metal to the casting after pouring. Properly used, the feeder will feed shrinkage even while solidification is going on. In other words, it will feed both liquid shrinkage and solidification shrinkage.

Liquid contracts on freezing because of rearrangement of atoms from open "randomly-packed" structure to a regular "densely-packed" structure. FCC and HCP solids contract more during solidification. FCC/HCP materials contract 3.2-7.2%. BCC materials contract by 2.0-3.2%. Table 8.1 shows typical values of contraction.

Table 8.1 Values of contraction

Metal	*Crystal structure*	*Volume change %*
Al	FCC	7.14%
Cu	FCC	5.30%
Mg	HCP	4.10%
Zn	HCP	4.08%
Fe	BCC	3.16%
Li	BCC	2.74%
Si	Diam	−2.90%
Bi	Rhom	−3.32%

Based on the major contracts in solidification behaviour, alloys can be classified as:

1. **Alloy freezing with marked skin formation** In this direction, solidification is favoured. Extremely sound with large cavity is possible.
2. **Alloys showing extensive pasty zones during freezing** In this direction, solidification is not favoured unless heavy chill or large temperature gradient is used. Casting may have widespread gas and shrinkage porosity.
3. **Grey cast irons** which show an expansion stage during freezing. It undergoes two successive stages of freezing. Freezing of austenite dendrites as in alloy and freezing of eutectic with precipitation of graphite. Thus, expansion offsets the contraction. Summary is given in Table 8.2.

Table 8.2 Solidification behaviour

Short freezing range	*Long freezing range*	*Expansion range*
Low carbon steel	Medium and high carbon steels	Grey cast iron
Brasses	Nickel-based alloys	Ductile cast iron
Aluminium bronzes	Phosphor bronzes	
High tensile brasses	Gun metals	
Aluminium	Magnesium alloys	
Copper	Complex aluminium alloys	

8.3 BASIC REQUIREMENTS OF A RISER SYSTEM

- Feeder should be thermally adequate; it should freeze after the casting section has solidified.
- The riser should be volumetrically adequate. It should compensate for the required amount of feed metal.
- The riser should have adequate feeding range. The feeding system should adequately cover the entire casting – the feeding ranges of the risers should cover all parts of the casting to eliminate centre-line shrinkage and ensure soundness.

8.4 BASIC WORKING OF A RISER

Casting of a slab with a riser fitted at one end is shown in Figure 8.2(a). Once the molten metal is poured, solidification starts from the skin of the mould. The free end of the slab has more contact surface area to remove heat from the metal, so solidification is fast and completed first at this end. The molten metal available near the end compensates the shrinkage cavity, resulting in good solid metal at the end. This behaviour of free end is called end effect. After the initial solidification at the skin of the mould cavity, the solidification process progresses gradually towards the centre-line of the slab. This is referred to as progressive solidification. As solidification progresses, metal shrinkage continues and the liquid metal flows by gravity from the riser to the cavities for filling. This results in a depression of top skin of the riser, which is called pipe of the riser. In this figure, the intermediate stage of solidification is shown by a thin line. The solidification completes gradually at the free end, and it moves to the riser side draining the molten metal available in the riser. This movement of gradual completion of solidification from one end to the molten metal available area is called directional solidification. Due to the presence of riser, which has thicker cylindrical cross section and elevated position, the liquid metal can flow into the centre-line portion of the slab. This transfers the shrinkage cavity to the riser. Step-by-step process of liquid shrinkage is shown in Figure 8.3.

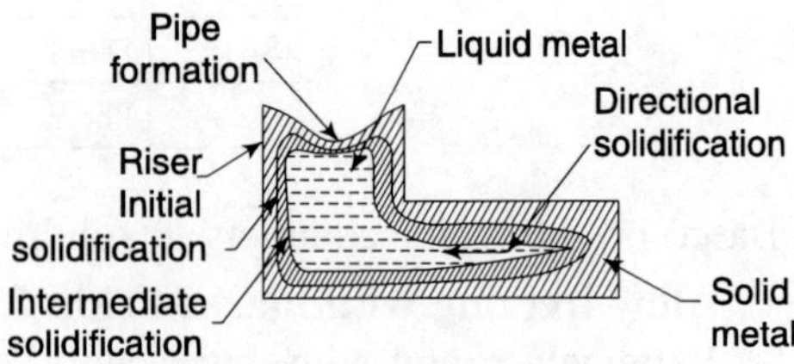

Figure 8.2 Working of a riser

8.5 FUNCTIONS OF A RISER

1. The main function of a riser is to supply the molten metal to the solidifying casting and eliminate shrinkage cavity.
2. It should allow easy escape of air and gases.
3. It should provide necessary temperature gradient to promote directional solidification.
4. It should solidify after the casting has solidified.
5. It should promote minimum wastage of metal.
6. Solidification time of riser should be greater than the casting.

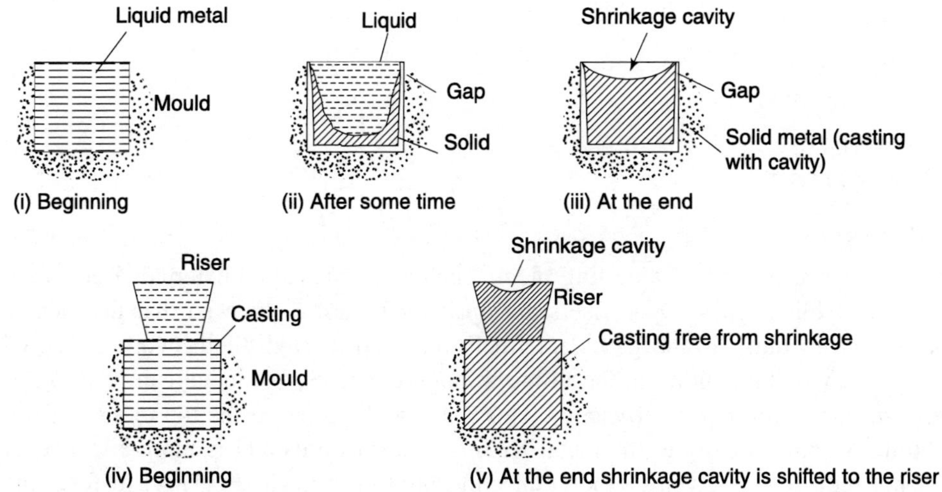

Figure 8.3 Liquid shrinkage compensation by riser

8.6 TYPES OF RISERS

Risers are generally classified as top risers, side risers and blind risers. Top risers are placed on the top of the hot spot whereas the side feeders are placed at the side of the hot spot usually at the parting line. Top feeders are most effective because they are easy to mould, easy to fettle and act efficiently by gravitational forces. They also help to exhaust mould gases and indicate filling of mould cavity with metal forming complete casting. Whenever possible, top risers are to be used.

Side risers are placed on casting having hot spots not accessible for direct top risers. They help in reducing excess metal of padding and reduce felting cost.

Blind risers are used to feed localized hot spots which are below the parting line and not accessible for top as well as side risers. Sometimes, when a casting shape has a profile with different heights, the higher spots are fed by open risers and lower spots by blind risers. Blind risers reduce the requirement of feed metal which would otherwise need excessive height of open risers. Atmospheric pressure can be made to act on liquid metal contained in a blind riser by providing William or Washburn core to penetrate solidifying surface skin. Blind risers are extensively used for intricate casting like valves, etc.

Detailed classification of risers is as follows:

(A) External risers

1. Open-top riser
2. Open-side riser
3. Blind-top riser
4. Blind-side riser

(B) Internal risers
1. Open-top riser
2. Open-side riser
3. Blind-top riser
4. Blind-side riser

External risers

If a riser is located outside of a casting block, such a riser is called external riser. These risers can be open or blind types. Open risers are open to the atmosphere so that they are fitted in the cope of the mould. Air and gases are freely removed through them, while filling of metal in the mould cavity. In addition to the filling through the gates, hot metal can be added through an open riser for increasing its efficiency. This riser can be fitted to the top or side of a casting. If the fitting is above a casting, the metal feeding is downwards and is more effective. A small dimensional reduction is given to the connecting point as shown (Fig. 8.4) to form the neck. This will not reduce the efficiency, at the same time the riser can be easily removed without damaging the shape of casting during fettling.

For a side riser, the bottom of the cylinder is converted to hemispherical form for minimum surface area. The neck portion is connected to the casting as shown in Figure 8.4 (b) and 8.4(d). Side risers are slightly less efficient than top risers in feeding of metal.

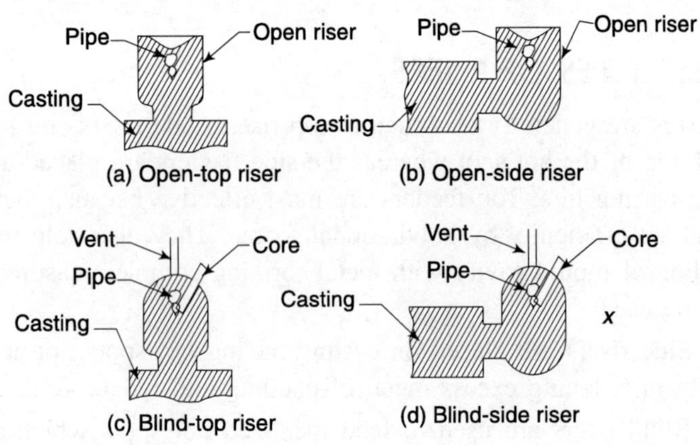

Figure. 8.4 External open and blind risers

If the riser feeding location is at too low a level from top of the mould, blind risers (not open to atmosphere) are used. Here, the top and bottom ends of the cylinder are converted to hemispherical form in order to reduce the surface area. The feeding direction may be downwards or to the side, and they are called top risers or side risers respectively (Figures 8.4(c) and (d)). An air vent is provided from the riser cavity to the top of the mould to remove air and gases during filling of metal. A sand core or a pipe is fitted extra to the centre of the riser in order to admit atmospheric pressure for the flow of molten metal from the riser to the casting.

Internal risers

If a riser is surrounded by or inside of a casting block or frame, such a riser is called internal riser. Figure 8.5 shows internal risers of open and blind types. The feeding of the riser may be from the top or side as in the external risers. If the riser is surrounded by hot metal of the casting, the working efficiency of the riser is improved and the size can be proportionally reduced. But the chance to provide an internal riser is rare in common casting shapes.

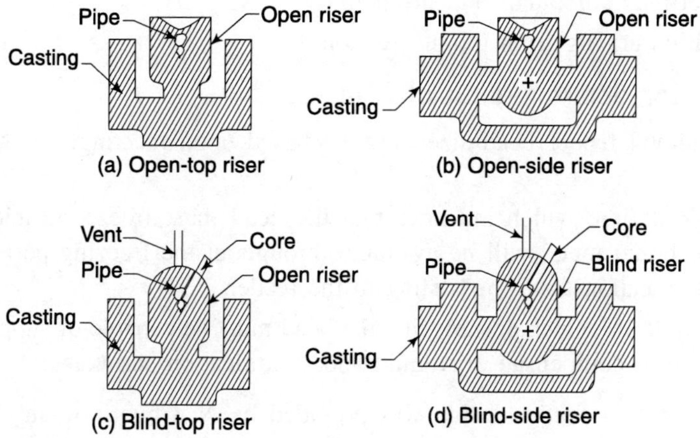

Figure 8.5 Internal open and blind risers

Advantages of blind risers:

1. Bottom gating through the side riser provides proper temperature gradients; the last metal to enter, and the hottest, is at the bottom of the casting and in the riser which provides conditions for good feeding; also the sand in these regions is preheated.
2. Blind risers can be placed at any position in the mould.
3. The cylindrical riser with a hemispherical dome approaches a minimum ratio of surface area to volume and thus slows down cooling.
4. Blind risers can be connected to the flat side of a casting, (in Fig. 8.6), and can be removed more easily than if connected to a contoured surface, as top, open risers usually are.

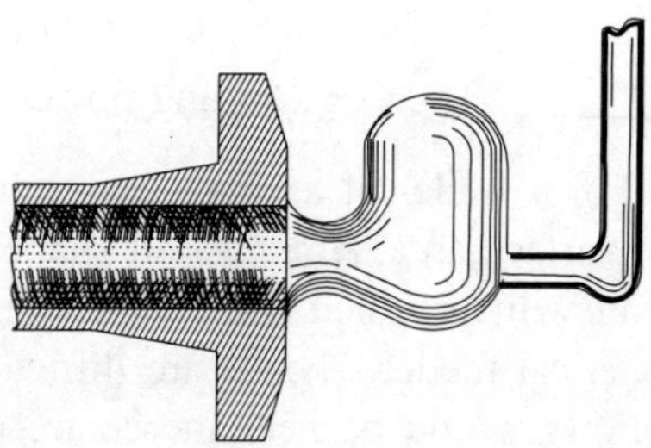

Figure 8.6 Blind risers

8.7 DETERMINATION OF OPTIMUM RISER SIZE

Risers are designed to compensate the solidification shrinkage of a casting and make it free from shrinkage porosity. Riser design depends on the following factors:

1. Riser shape
2. Riser size as a function of casting shape
3. Location of risers
4. Grouping of castings

5. Riser connection to the castings
6. Use of chills
7. Use of exothermic compounds and insulators
8. Special condition arising from joining sections

Various methods for determination of riser size

Methods for determining feeder head size must be based upon meeting the separate requirements.

- Freezing time criterion, which indicates that the head must freeze sufficiently slowly to ensure that the liquid metal will be available throughout the freezing period, so enabling directional solidification from the casting to the feeder.
- Feed volume criterion, which ensures that the head must be capable of supplying sufficient volume of liquid to compensate for liquid and solidification shrinkage.

1. The theoretical basis for feeder design was provided by N. Chvorinov, in 1940, through a series of experiments. He stated that the freezing time of a casting is related to its geometry.

$$t = \frac{K^2 \times V^2}{SA^2}$$

where,

t = Freezing time of casting
V = Volume of casting
SA = Surface area of casting
K = Chilling power of the mould

For a feeder of a given size to have a maximum solidification time, it must have the smallest surface area. Amongst all shapes, a sphere has the smallest surface area. A spherical feeder head will remain liquid for the longest period of time amongst all shapes of unit volume. A spherical feeder's heads are difficult to mould and cause feeding problems. The last metal to solidify would be near the centre of the sphere, where it will not be used to feed a casting. Practicalities dictate the use of cylinders for most feeders, although hemispherical feeders or a hemispherical base of cylindrical feeders are used. Height of the riser should not be less than one half times its diameter and the maximum height should be no more than one and a half times its diameter. This is because of V/A ratio. Dimensions outside those limits would make the V/A ratio lower than it is within them.

Besides maximizing freezing time, other factors influencing the shape of a feeder head include:

- The timing of the demand for feed metal, affecting the shape of the shrinkage cavity in the head.
- The permissible area of junction with the casting: this should be as small as possible to minimize fettling costs.

2. Heuver, in 1943, introduced the 'inscribed circle method' to identify hot spots requiring feeding in complicated casting sections.

A circle (representing a sphere) is inscribed in the critical cross section. Then a number of circles are drawn, gradually increasing in size towards the riser by 10 to 20% in diameter till the bottom of riser.

Riser size provided is between 1.5 and 3.0 times the diameters of the biggest circle as shown in Figure 8.7. The outer points of the circles are joined by a smooth curve to provide "pad", assuring a path for riser metal to reach the "hot spot" efficiently. Even though not very scientific, this method is widely used even now in the industry because of its simplicity and effectiveness.

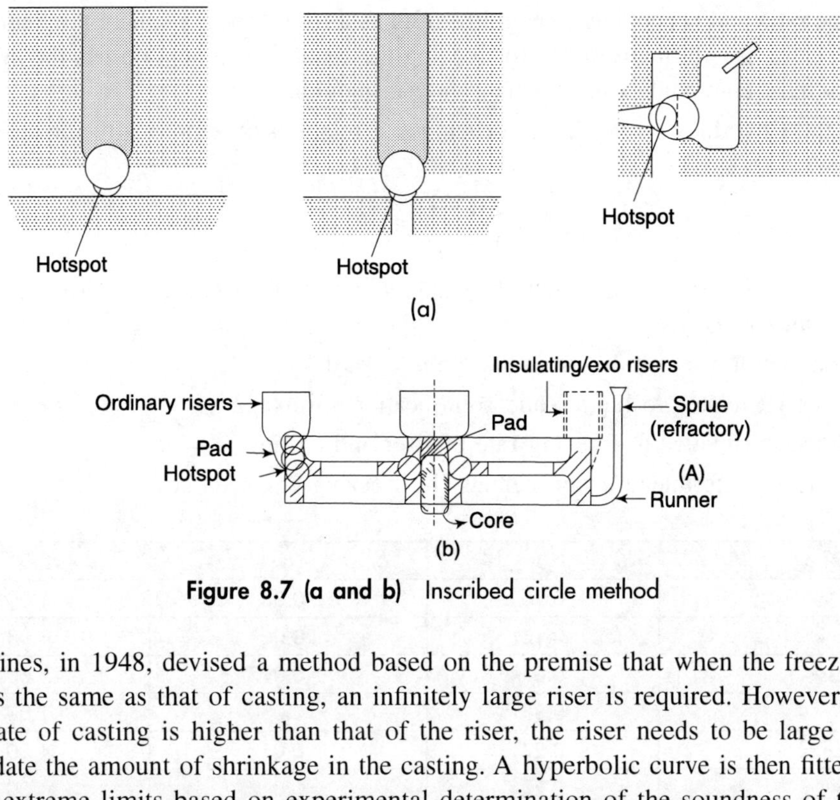

Figure 8.7 (a and b) Inscribed circle method

3. J.B. Caines, in 1948, devised a method based on the premise that when the freezing rate of the riser is the same as that of casting, an infinitely large riser is required. However, when the freezing rate of casting is higher than that of the riser, the riser needs to be large enough to accommodate the amount of shrinkage in the casting. A hyperbolic curve is then fitted between these two extreme limits based on experimental determination of the soundness of the casting made with different risers.

Since solidification of the casting occurs by losing the heat from the surfaces and the amount of heat is given by the volume of the casting. The cooling characteristics of a casting can be represented by the surface area to the volume ratio. Since the riser is also similar to the casting in its solidification behaviour, the riser characteristic can also be specified by the surface area to the volume ratio.

If this ratio of the casting is higher, then it is exposed to cool faster. Chvorinov has shown that the solidification time is proportional to the square of the ratio of volume to surface area of the casting. The constant of proportionality, called mould constant, depends on pouring temperature, casting and mould thermal characteristics.

$$t = K\left(\frac{V}{SA}\right)^2$$

The 'freezing ratio', X of a mould is defined as the ratio of cooling characteristic of casting to the riser.

$$X = \frac{SA_{casting}/Volume_{casting}}{SA_{riser}/Volume_{riser}}$$

In order to be able to feed the casting, the riser should solidify last and hence its freezing ratio should be greater than the unity. It may be argued that the sphere has the lowest surface area to volume ratio and hence that it should be used as a riser. But in a sphere the hottest metal being at the centre, it is difficult to use for feeding the casting. The next best is the cylindrical type which is most commonly used for their ease in moulding.

Based on the Chvorinov's rule, Caine developed a relationship empirically for the freezing ratio as follows:

$$X = \frac{a}{Y - b} - c$$

X = Freezing ratio = (surface area of casting/volume of the casting)/(surface area of riser/volume of the riser)

Y = Volume ratio = volume of riser/volume of casting

b = Constant related to liquid-solid solidification shrinkage

c = Constant, relative to freezing rate of riser and casting

a = Constant determined experimentally by freezing characteristics

	a	b	c
Steel	0.10	0.03	1.00
Aluminium	0.10	0.06	1.08
Cast iron, brass	0.04	0.017	1.00
Grey cast iron	0.33	0.030	1.00
Aluminium bronze	0.24	0.017	1.00
Silicon bronze	0.24	0.017	1.00

The following equations are used for calculating the risering requirements for aluminium alloys LM4 (Cu 2-4%; Si 4-6%) and LM11 (Cu 4-5%).

When height to diameter of the riser equals 1,

$$\text{LM4}: X = \frac{33.45}{23.98 - Y}$$

$$\text{LM11}: X = \frac{17.11}{12.58 - Y}$$

When height to diameter ratio of riser equals 1.5,

$$\text{LM4}: X = \frac{32.09}{23.98 - Y}$$

$$\text{LM11}: X = \frac{16.75}{12.58 - Y}$$

For nodular iron in CO_2 moulds, the following equation may be used

$$X = \frac{39.44}{27.78 - Y}$$

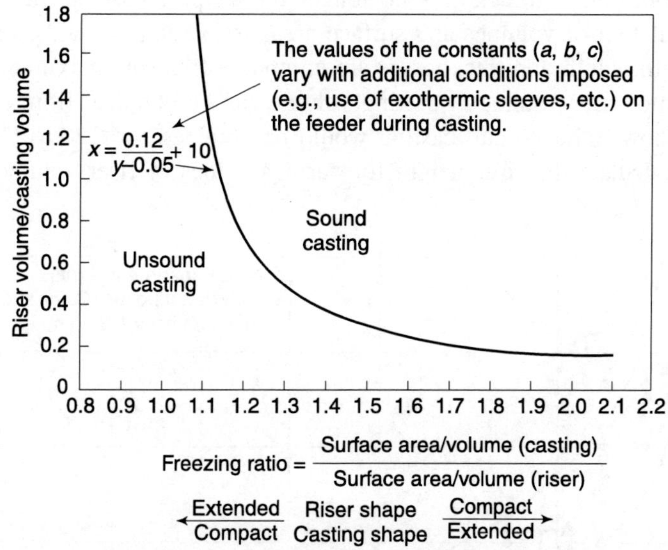

Figure 8.8 Caine's equation

The above equation when plotted will be as shown in Figure 8.8. The line shows the locus of the points that separate the sound castings and castings with shrinkage in steel castings. It is clear from Figure 8.8 that for identical freezing rates of feeder and casting, feeder volume requirement becomes infinite. When casting freezes increasingly rapidly relative to the feeder, the feeder volume requirement decreases to minimum, which is represented by the shrinkage requirement alone.

Disadvantages of Caine's method

(i) Requires trial-and-error calculation to obtain desired feeder size.

(ii) Nature of shape of shrinkage cavity generated in feeder affects feeder size.

4. Myskowski, Bishop, Pellinc and Johnson Naval Research Laboratories, US, (1951-1956), have modified Caine's method to include the 'Shape-Factor' which is given by (L+W/T) the length, width and thickness of the casting zone to be fed. This is also known as the Naval Research Laboratory (NRL) method. Curves have been drawn for shape factor against volume ratio if casting to riser is as shown in Figure 8.9. The curve is only true for values of height:

diameter ratio between 0.5 and 1.0. J.F. Wallace in 1960 developed these further and provided curves to directly find riser dimensions. Correction factors for riser-type, joining of thick and thin section, etc., were given. This method is essentially a simplification of Caine's method that defines a shape factor to replace the freezing ratio.

$$\text{Shape Factor (s)} = \frac{\text{Length + Width}}{\text{Thickness}}$$

With increasing values of S (thin casting), the feeder head diminishes towards the limiting level at which the controlling influence is no longer the shape factor but the volume of feed metal required. Calculation of volumes and surface areas is too complicated compared to shape factor where the length, width and thicknesses are computed from the maximum dimensions of casting section. Ratio of riser volume to casting volume can be obtained from the graph shown in Figure 8.9. This shows when sound casting would be obtained for C20 to C50 steel. Figure 8.9 can be used to calculate the riser height for various values of riser volumes ranging from 0 to 6500 cm^3.

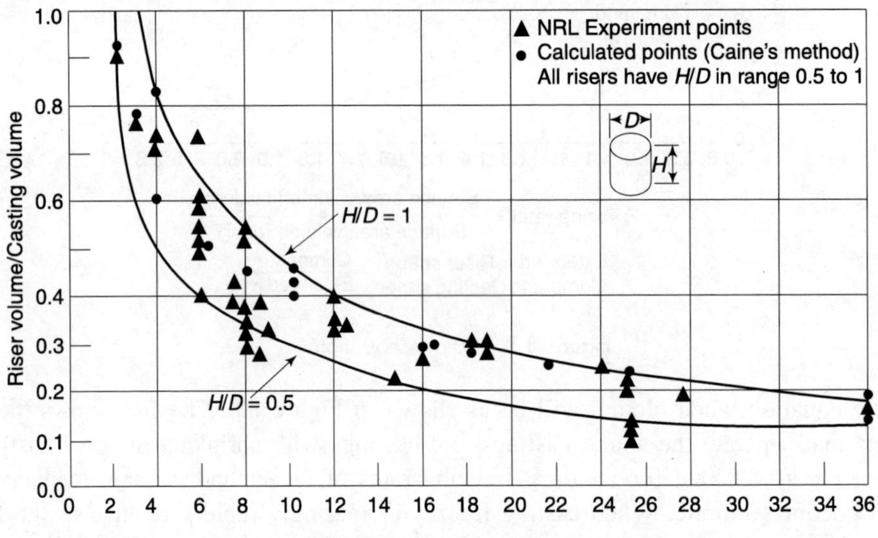

Figure 8.9 NRL method

The values obtained by NRL method are generally conservative in nature and correspond to Caine's method for most of the simple geometries.

The risers should be located at the heaviest sections as they themselves act as feeders for the thin sections. When smaller sections are connected to thicker sections, the risers should have larger volume to cater to this appendage. But if this branch is thin, no change in riser volume is required. Some examples where such branches are to be accounted for are presented in Figures 8.10 (a), (b) and (c).

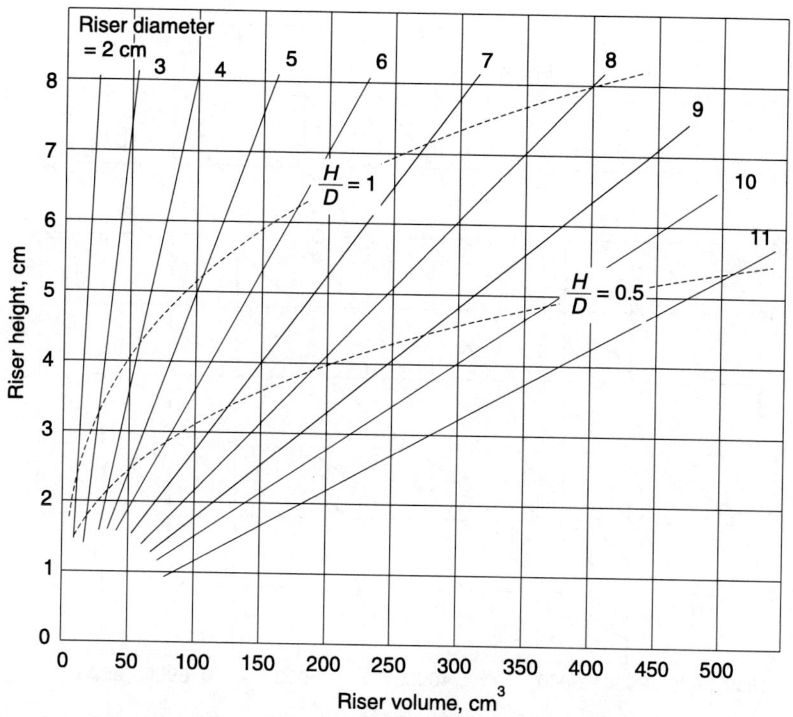

Figure 8.10(a) Chart for riser dimension using NRL method

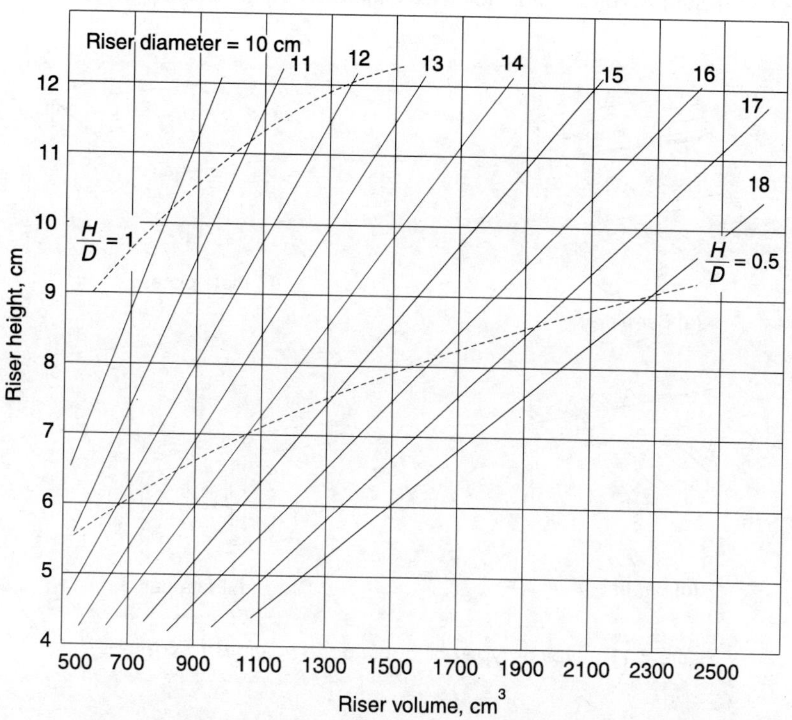

Figure 8.10(b) Chart for riser dimension using NRL method

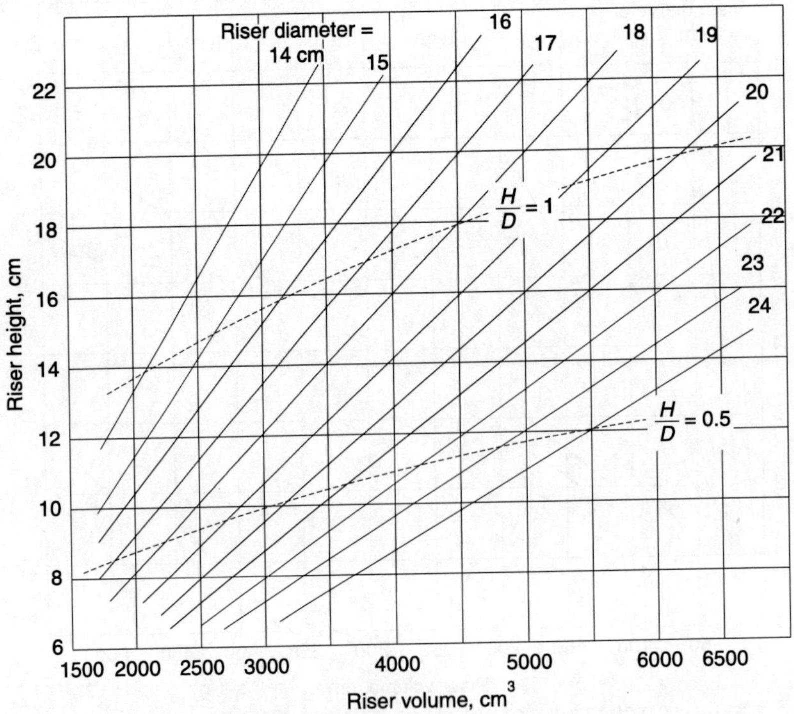

Figure 8.10(c) Chart for riser dimension using NRL method

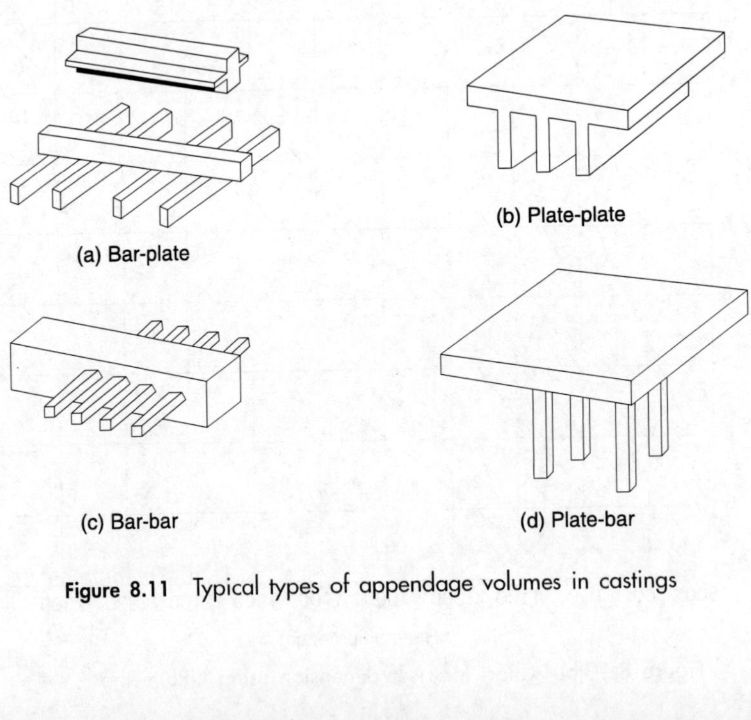

Figure 8.11 Typical types of appendage volumes in castings

The increase in riser volume due to addition of branches (parasitic volumes) can be estimated from Figure 8.12.

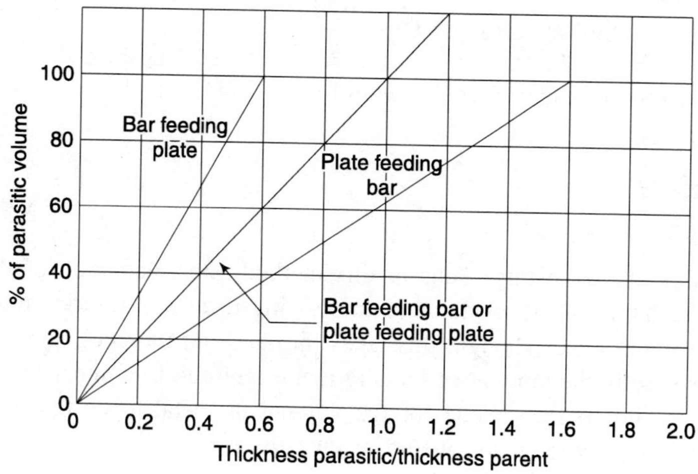

Figure 8.12 Appendage volume calculation chart

The other shapes that are of interest are the hollow cylindrical shapes. In these, since the heat removal is restricted, a correction factor (k) needs to be applied to get effective plate thickness. If T is the true wall thickness, then the correction factor is given as in Table 8.3.

Table 8.3 Correction factor for hollow cylindrical shapes

Core diameter	0.5 T	T	2T	4T
Correction factor, k	1.17	1.14	1.02	1.00

$$\text{Shape factor} = \frac{L + W}{kT}$$

5. Modulus method by R. Wlodawer. R. Wlodawer in 1966, introduced the 'modulus' concept, which has the most extensive application in foundries today from Chvorinov's equation,

$$t = \frac{V^2}{K^2 \times SA^2}$$

Wlodawer termed the ratio [volume of casting/cooling surface area] as 'modulus' (M_r).

It has been empirically established that if the modulus of the riser exceeds the modulus of the casting by a factor of 1.2, the feeding during solidification would be satisfactory.

The modulus is the inverse of the cooling characteristic (surface area/volume) as defined earlier. In steel castings, it is generally preferable to choose a riser with a height to diameter ratio of 1.

$$\text{Volume} = \frac{\pi \times D^3}{4}$$

where, D = diameter of riser

The bottom end of the riser is in contact with the casting and thus does not contribute to the calculation of surface area.

$$\text{Surface area} = \pi D^2 + \frac{\pi D^2}{4}$$

The modulus of such a cylinder riser, M_r would be

$$M_r = 0.2\ D$$

Since, $M_r = 1.2\ M_c$

$$D = 6\ M_c$$

where, M_r is modulus of the casting. Thus in this method, the calculation of the riser size is simplified to the calculation of the modulus of the casting itself and no trial-and-error solution as is given in the previous case. Though, this takes into account the cooling effect of the riser, it does not consider exactly the amount of feeding metal required to compensate for the shrinkage of the casting. If allowance is made for the volume of metal to be fed to counteract the contraction of the casting, the equation would change to

$$D^3 - 5.46\ M_c D^2 - 0.05093\ V_c = 0$$

where, V_c = Volume of casting.

The above is valid when the height to diameter ratio of the riser is one. When the third term in the equation relating to feed volume is neglected, the previous simplified equation would be arrived at.

With 'chunky' castings, e.g., cubes, the volume component may be negligible, but for those 'rangy' castings, similar to plate like, the influence of volume component becomes increasingly significant. It is sometimes useful to have a parameter called 'ranginess factor', R to define the casting type. It may be defined as,

$$R = \frac{\text{Modulus of a cube of same volume as casting}}{\text{Modulus of casting}}$$

In Table 8.4, the second column shows the moduli of simple geometric shapes, and the third column shows the calculation of modulus of complex shapes.

Table 8.4 Modulus formulae for some common shapes

(a)	Cube	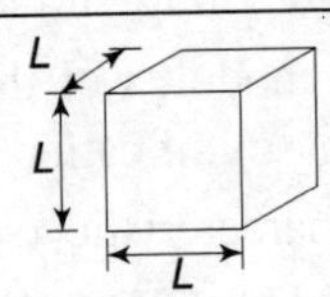 Side = L	$\frac{L}{6}$
(b)	Cylinder	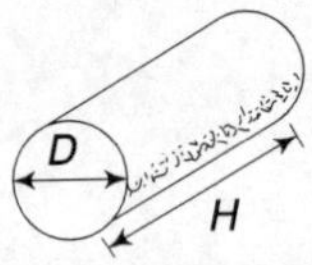 Diameter = D, Height = H	$\frac{DH}{2D + 4H}$ *Note:* If $H = D$, the modulus is $\frac{D}{6}$

(c)	Disc	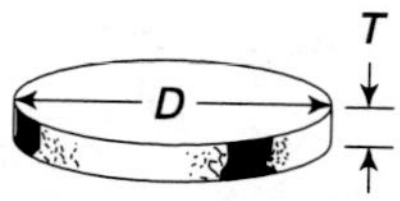	Diameter = D Thickness = T	$\frac{DT}{2D + 4T}$
(d)	Bar or plate	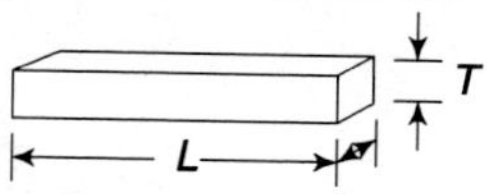	Length = L Width = W Thickness = T	$\frac{TWL}{2(TW + WL + LT)}$
(e)	Endless cylinder (ends terminated by another part of casting)	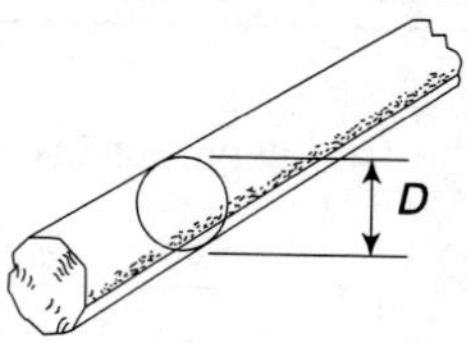	Diameter = D	$\frac{D}{4}$ *Note:* Because radial heat flow is faster than that from a flat surface, calculated moduli for endless cylinders may be reduced by multiplying by 0.85.
(f)	Endless plate (teminated on all sides by another part of casting)	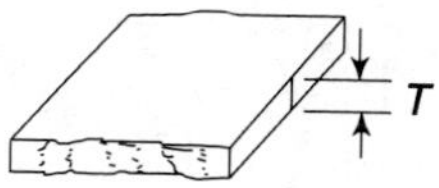	Thickness = T	$\frac{T}{2}$
(g)	Endless bar (ends terminated by another part of casting)	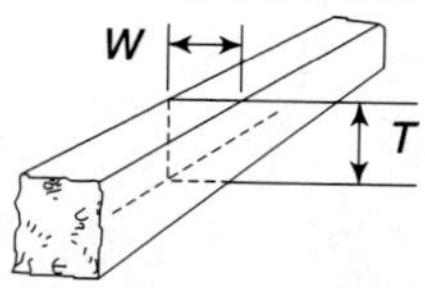	Thickness = T Width = W	$\frac{TW}{2(W + T)}$
(h)	Endless hollow cylinder	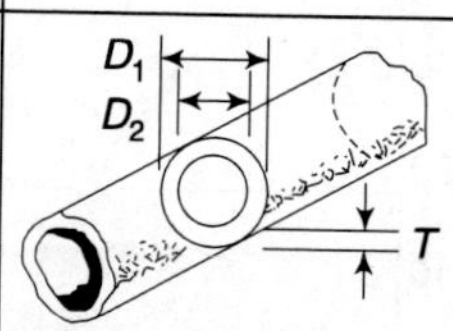	OD = D_1 Dia. core = D_2 Wall thickness = T	$\frac{D_1 - D_2}{4} = \frac{T}{2}$
(i)	Annulus	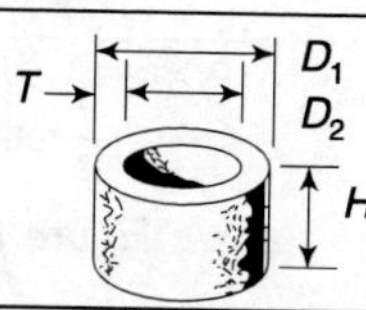	OD = D_1 Dia. core = D_2	$\frac{(D_1 - D_2)}{2(D_1 - D_2) + H} = \frac{TH}{2(T + H)}$

• More complicated shapes should be broken down into simple shapes.	• For more complex shapes, a general formula of the following can be used: $M = \frac{\text{cross-sectional area}}{\text{perimeter}}$
• Moduli of the individual simple shapes should be determined.	**Example:** For simple rectangular shape, $M = \frac{a.b}{2(a + b)}$
• The section having the highest modulus should be considered sensitive for porosity formation.	• If any of the sections contains directly non-cooling surface, its dimension (c) should be excluded from the perimeter

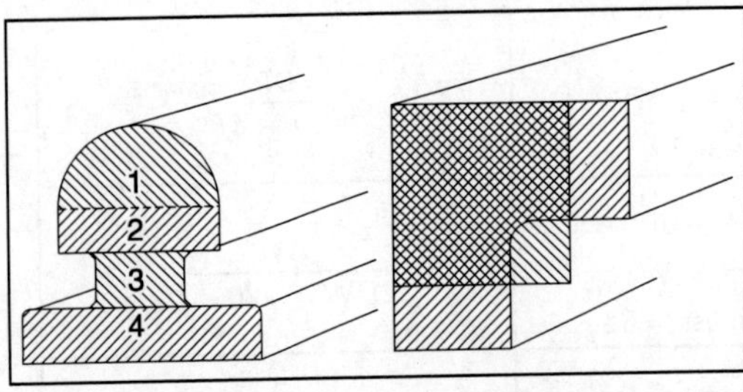

Example: For simple rectangular section,

$$M = \frac{a.b}{2(a + b) - c}$$

The dimensions of any feeder head must satisfy the two separate criteria: (1) modulus (representing freezing time), and (2) volume feed capacity. Figure 8.13 illustrates the factors determining the feeder head volume. Theoretically feeder head requires for plates of various proportions (spherical head, ignoring non-cooling interface). Curve A shows the minimum head volume percentage to satisfy freezing time criterion (based on the relation $M_r = 1.2\ M_c$). Curve B shows minimum head volume percentage to satisfy volume feed demand criterion (based on 3% specific shrinkage and 20% metal utilization). Curve C shows composite feeding curve.

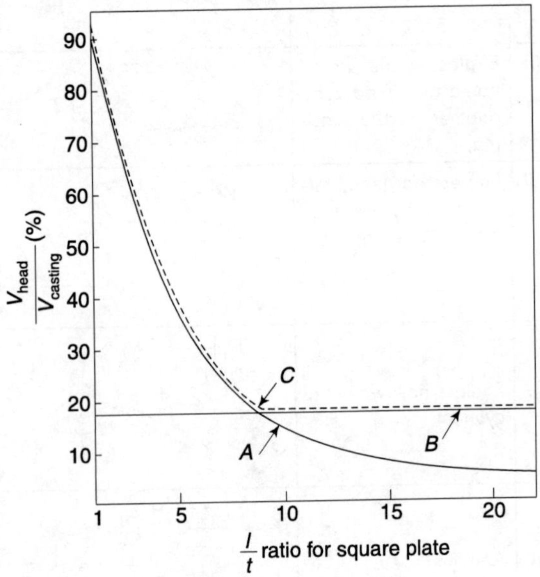

Figure 8.13

The ranges of casting dimensions over which freezing time and volume feed capacity respectively control feeder head size depend upon the specific shrinkage of the alloy.

6. Adams and Taylor, in 1993, incorporated solidification shrinkage into Chvorinov's equation and obtained the most efficient riser from the freezing point of view, the one in which solidification ceases simultaneously with solidification in the casting.

$$\frac{A_r}{A_c} = (1 - \beta)\ \frac{V_r}{V_c}$$

where β is the coefficient related to solidification shrinkage and mould dilation.

7. Ruddle, in 1976, modified the above to get the equation

$$D^3 - K_1 M_c D^2 - K_2 V_2 = 0$$

where D = Diameter of the riser

K_1, K_2 = Coefficients depending on total volumetric contraction, surface and top insulation factors or riser, H/D ratio of riser and a safety factor for criticality of casting

M_c = Casting modulus

V_c = Casting volume

This is highly accurate and provides the best theoretical/practical base for risering design as it takes into account both volume effect and modulus effect of casting and other important parameters in modern feeding systems application.

Ruddle in, 1982, developed a computer programme using the above equation for feeding system design giving economic data in the print-out.

8. Geometric programming method by R.C. Greese (1971)

These are based on the solution of generalized polynomials from Adam Taylor's risering equation with heat-transfer data and modulus factors.

9. Computer solutions are obtained based on the problem of minimizing of riser volume subject to the constraint $t_r > t_c$ (solidification time of riser and casting).

In a specific case, we get $H = 6 \times \dfrac{q_r}{q_c} \times \dfrac{V_c}{SA_c}$

where q_r, q_c are the solidification constants for riser and casting and H is the height of the riser.

10. Computer programming technique – Corbett (1983): In extensive research work by FOSECO, the team developed a computer program for the feeding system based on Heuver's inscribed circle and shape factor method, modifying it for different latest feeding aids like insulating and exothermic risers. The program has facilities for graphics, and cost calculation in addition to feeder design. Exothermic sleeves are designed to provide effective feeding of metal by extending the solidification time of risers used on steel, SG iron, and grey iron castings. Exothermic sleeve compositions comprise (a) an oxidizable metal where the oxidizable metal comprises aluminium dross, and (b) an oxidizing agent capable of generating an exothermic reaction. Details of insulating sleeve sizes are given in Table 8.5.

Table 8.5 Insulating sleeve size selection

S. No.	Sleeve Size		Effective sleeve modulus (mm)	Casting modulus	Equlavalent sand riser dimensions		Estimated weight of metal contained in sand riser (kg)	Estimated weight of metal contained in sleeve riser (kg)
	ID (mm)	Ht (mm)			ID (mm)	Ht (mm)		
1.	150	150	37.5	31.2	195	162.5	34.1	15.5
2.	150	200	40.2	33.5	195	227.5	47.5	21.7
3.	150	250	41.7	34.8	195	292.5	61.4	27.9
4.	180	200	46.5	38.8	234	227.5	68.7	31.3
5.	180	250	48.7	40.6	234	292.5	88.4	40.2
6.	180	300	50.3	41.9	234	357.5	108.0	49.2
7.	200	200	50.6	42.1	260	227.5	84.9	38.6
8.	200	250	53.2	44.3	260	292.5	109.1	49.7
9.	200	300	55.0	45.8	260	357.5	133.4	60.7
10.	250	300	66.2	55.2	325	357.5	208.4	94.9

Contd.

11.	250	350	68.1	56.8	325	422.5	246.3	112.1
12.	250	400	69.6	58.0	325	487.5	284.2	129.4
13.	300	350	78.5	65.4	390	403	338.3	154.0
14.	300	400	80.7	67.2	390	468	392.8	178.8
15.	300	450	82.4	68.7	390	533	447.4	203.6
16.	350	350	88.7	73.9	455	403	460.4	209.6
17.	350	400	91.5	76.3	455	468	534.7	243.2
18.	350	450	93.7	78.1	455	533	608.9	277.2
19.	350	500	95.6	79.6	455	598	683.2	311.0
20.	400	350	98.3	81.9	520	403	601.4	273.7
21.	400	400	101.7	84.8	520	468	698.4	317.9
22.	400	450	104.5	87.1	520	533	795.4	362.0
23.	400	500	106.8	89.0	520	598	892.3	406.1
24.	400	550	108.7	90.6	520	663	989.4	450.3
25.	450	400	105.7	88.1	562.5	425.0	742.1	380.0
26.	450	450	109.1	90.9	562.5	487.5	851.2	435.8
27.	450	500	112	93.3	562.5	550.0	960.0	491.7
28.	450	550	114.4	95.3	562.5	612.5	1,069.5	547.6
29.	450	600	116.4	97.0	562.5	675.0	1,178.6	603.4
30.	500	450	118.3	98.6	625	487.5	1,050.9	538.1
31.	500	500	121.7	101.4	625	550.0	1,185.6	607.0
32.	500	550	124.5	103.7	625	612.5	1,320.4	676.0
33.	500	600	126.9	105.7	625	675.0	1,455.1	745.0
34.	500	650	128.9	107.4	625	737.5	1,590.0	814.1
35.	500	700	130.7	108.9	625	800.0	1,724.5	882.9

Risering requirement for cast iron

During solidification, grey cast iron has different rates of volumetric changes depending on the carbon and silicon compositions in the final melt. Wallace and Evans have calculated the riser sizes required based on the surface area and volume of the casting for high strength grey cast iron (3% C; 1.5% Si; 0.05% P; 0.10% S; 0.8% Mn) being a representative composition. Risering curves for grey cast iron in green and dry sand moulds are presented in Figures 8.14 and 8.15.

The soft grey cast iron may not require any risering, but if we have a doubt then a small shrink bob of 2 to 4% of casting volume may be provided at the necessary point. Whenever more than one riser are to be used for any casting, then casting is divided into a number of parts and the risering requirements are to be calculated separately for each part. The way the riser is connected to the casting is important. The connection called neck should solidify first so that shrinkage cavity is located in the riser itself. The removal of riser should be made sim-

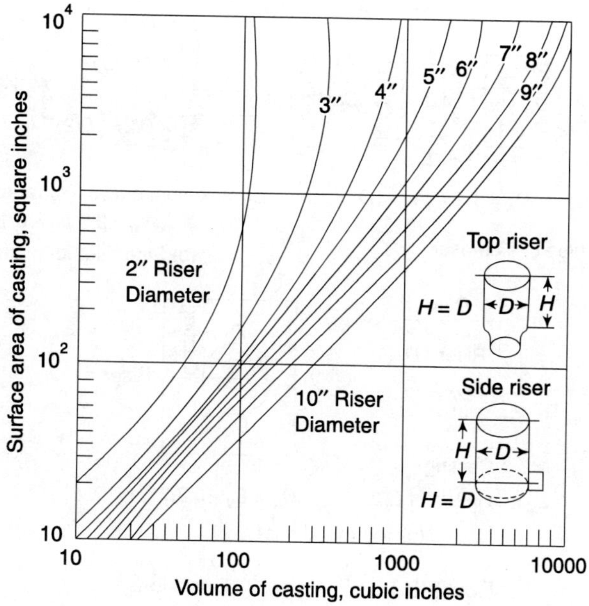

Figure 8.14 Risering curves for grey CI (greensand mould)

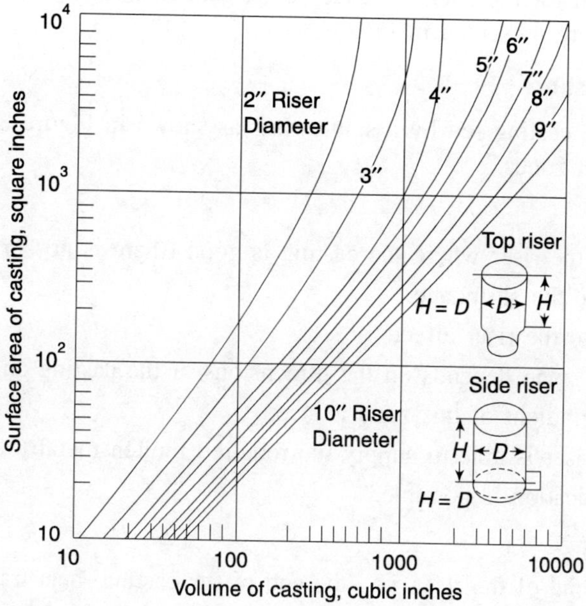

Figure 8.15 Risering curves for grey CI (dry sand mould)

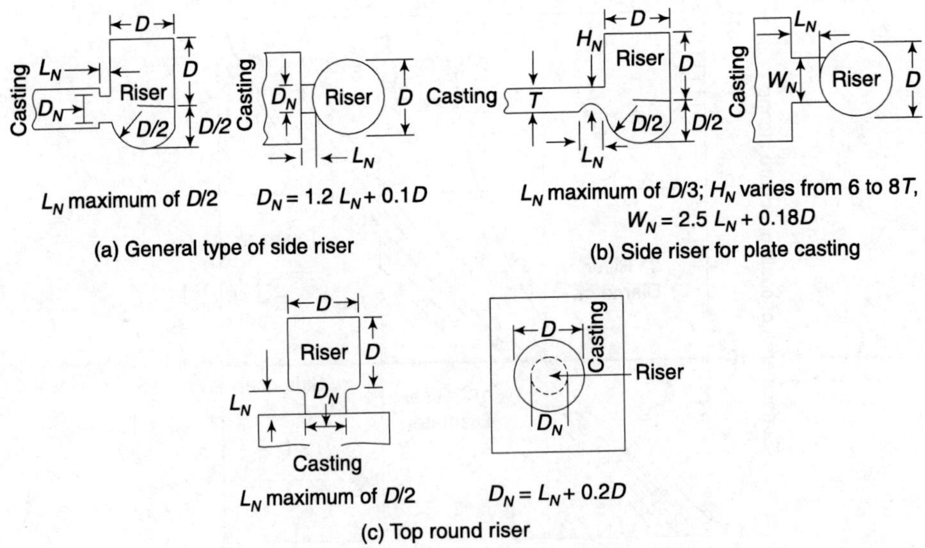

Figure 8.16 Riser neck dimensions

pler by making the connection smaller. But if the connection is too small, then it may solidify before casting and thus defeats the function of casting. It is a normal practice to provide a thin ceramic or core material for the neck to reduce cross-sectional area. The neck dimensions for various types of risers are presented in Figure 8.16.

Distinct zones in casting

Consider a simple plate casting fed by a single riser as shown in Figure 8.17. It is found that there exist three distinct zones.

Zone A:

- This zone is near the riser where the casting is good (there is no shrinkage in the zone).
- The riser will have the shrinkage.
- This zone is due to the riser effect.
- The length of this zone depends on the dimensions of the casting and riser.
- The temperature gradient of this zone is steep.
- In this zone, the riser is able to supply the required molten metal to the casting.
- Directional solidification is possible.

Zone C:

- Near the farthest end of the riser, i.e., the end of the casting, heat transfer is faster as the mould is able to extract heat from the ends. Therefore, the molten metal starts solidifying faster and steep temperature gradients are established.

- Directional solidification is possible.
- Molten metal starts solidifying from the end moving towards the riser. This establishes end effect and the casting zone will be strong.

Zone B:

- This zone lies between the riser and the end of the casting. The molten metal lying in this zone does not have steep temperature gradients. No directional solidification occurs.
- There will be intermixing of liquids and solids forming a pasty or mushy zone.
- There will be locking of dendrites making the flow of molten metal difficult. The riser is unable to force the molten metal here, hence feeding is difficult.
- Invariably centre line shrinkage occurs and the zone will be unsound.
- The length of this unsound zone depends on the total length of the casting. Shorter the length, lesser will be this unsound zone.
- By providing a chill in the casting or by having another riser will eliminate the defect or unsound zone.
- By providing taper/padding along the length of the casting, unsound zone can be eliminated.
- The temperature profile along the length of casting from riser to end of the casting is shown in Figure 1.14.
- It should be noted that steep temperature gradient which establishes directional solidification has to be promoted to provide a sound casting.

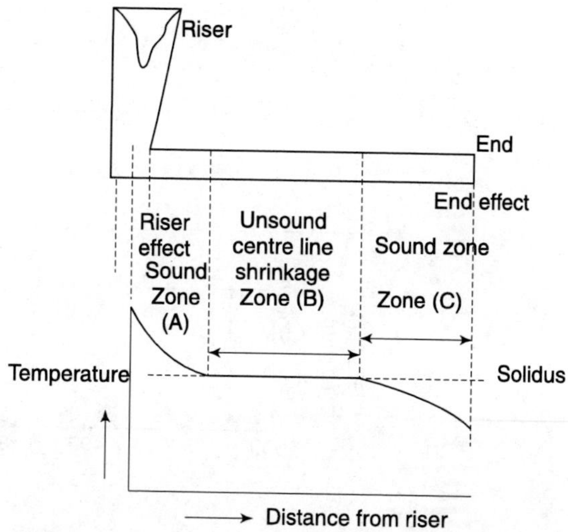

Figure 8.17 Feeding of plate casting by a riser

8.8 FEEDING DISTANCES

While calculating the risering dimensions, it was assumed that riser would be able to feed, whatever the length of the casting may be. But if the casting is too long, the riser would not be able to feed the entire length of casting. Feeder heads are normally placed in direct contact with the heavier sections of a casting, since this enables directional solidification to be maintained throughout freezing. In a complex casting, the shape is divided into a number of natural zones of feeding, each centred on a heavy section separated from the remainder of the casting by more constrained numbers. Each zone is then fed by a separately calculated feeder. In normal conditions, there will be a limit to how far feed liquid can be provided along a flow path. Up to this distance from the feeder, the casting will be sound. Beyond this distance, the casting will exhibit porosity. For many extended castings, feeding range is the limiting factor as appreciable temperature gradients are needed for feeding and these are difficult to include in parallel sections over long distances.

Examples to obtain castings free from dispersed porosity:

- Steels: 0.2-0.4°C gradients/cm for plate and 1.5-2.6°C/cm for bars.
- For nonferrous alloys: 5.5°C gradients/cm and 13°C gradients/cm (long freezing range alloys).

Figure 8.18 shows the temperature distribution in solidifying steel bar: steep temperature gradients against end walls and shallow gradients adjacent to feeder heads. Isothermal plateau in intermediate regions can be observed from the data. Feeding can only take place over very short distances in such plateau zones and centre line shrinkage in remaining areas.

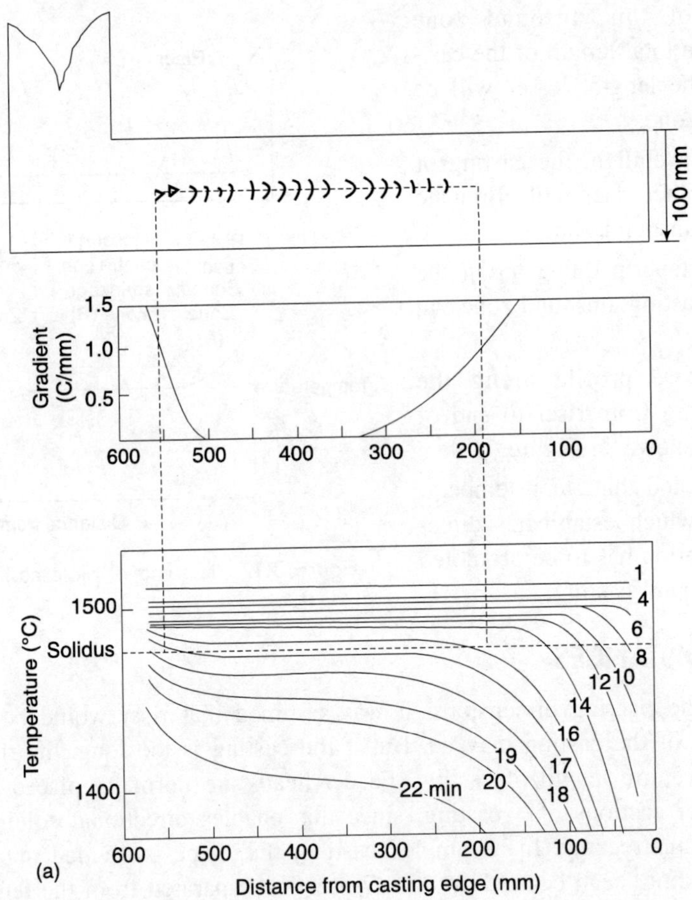

Figure 8.18 Temperature distributions in solidifying steel bar

The total castings could be classified as bars, plates or spherical and cubical sections. In cubical and spherical sections, feeding would not be a problem. Naval Research Laboratory after extensive tests on steel castings has arrived at a number of rules for these feeding distances. Figures 8.19 and 8.20 show feeding distances for bars, plates and plates of varying thickness.

- When a long bar or plate is cast without feeder, a certain length of the casting from each end of the plate or bar is sound. This results from directional solidification that developed at the ends because of faster cooling rate. This is called **end effect (EE)**.
- When a long bar or plate is cast using a feeder at the centre of the casting, a certain distance from feeder (in any direction) the bar or plate is sound. This is called **feeder effect (FE)**.
- The use of chills provides a powerful influence for extending the feeding range of heads when placed at the intermediate position of feeders. The spacing between the feeder heads in this case is more than doubled, which also greatly increases the casting yield.

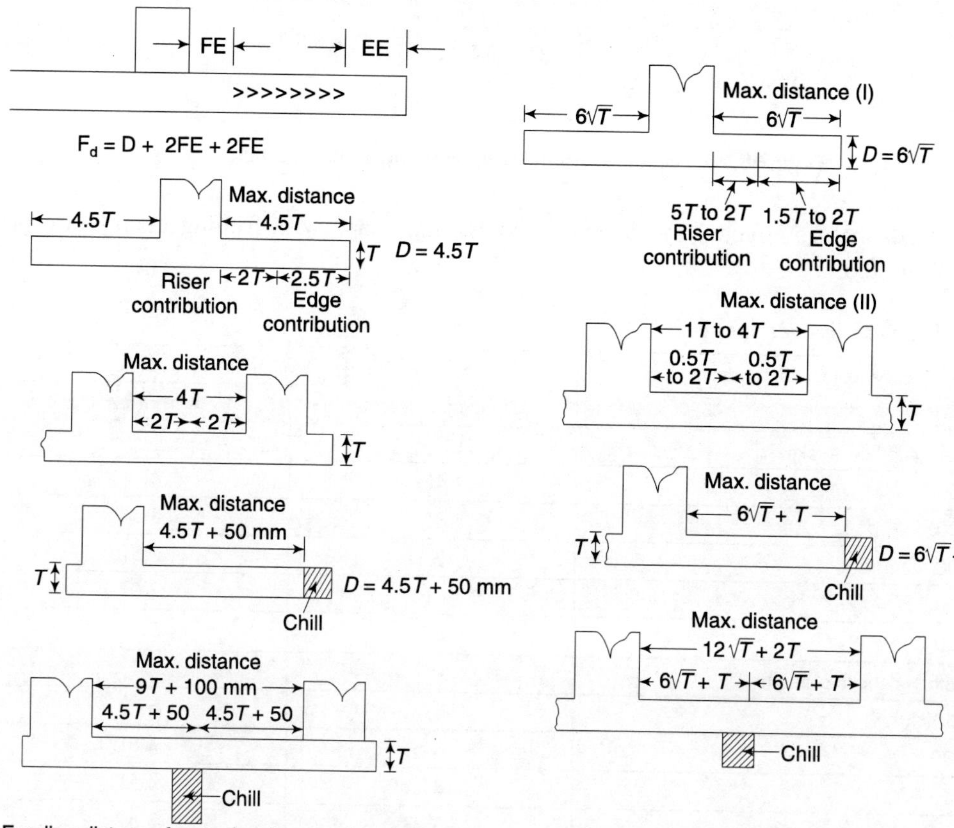

Feeding distance for steel plates cast in greensand

Feeding distance relationships for bars

Figure 8.19 Feeding distances for bars

Figure 8.20 shows feeding distances for varying plate thickness. When the plates joined at their edges to form stepped members (where the conjunction of two plates differ in thickness by a factor exceeding 1.4), the feeding distance in the thinner or parasite plate is increased, when compared with that when the latter is cast alone.

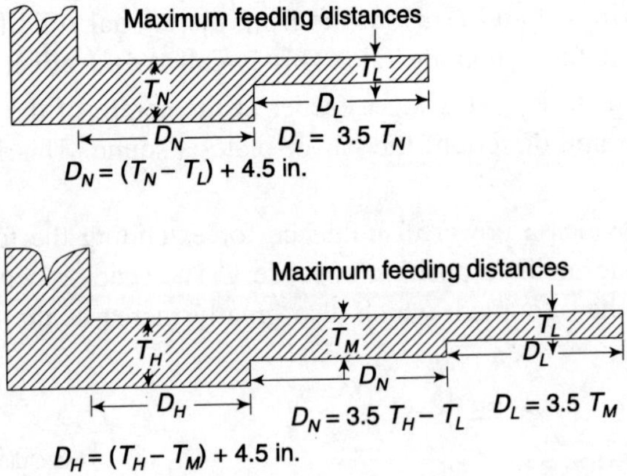

Figure 8.20 Feeding distances for varying plate thickness

Feeding distances as affected by section thickness and shape with feeding method are presented in Table 8.6.

Table 8.6 Feeding distance for risers

Section thickness, T (mm)	*Feeding distance in multiples of thickness, T*	
	Horizontal feeding	*Vertical feeding*
Square bars (width = thickness)		
Up to 50	4T	4T
100	3T	3.5T
200	2.1T	2.5T
300	T	2.0T
>300	6T	6T
Plates (width = 2 x thickness)		
Up to 50	6T	6T
100	4.5T	4.8T
200	4T	4.5T
300	3.3T	3.5T
>300	10T – 405 mm	16T – 350 mm
Plates (width > 3 x thickness)		
Up to 100	4.5T	4.5T
200	3.5T	3.5T
>300	2.8T	11.8T – 130 mm

Feeding distance increases with increasing freezing range. For comparison of feeding range in different alloys, Flinn introduced the useful concept of a centre line resistance factor (CRF).

$$\text{CRF} = \frac{\text{Time during which crystals are present at the centre line}}{\text{Total solidification time}} \times 100$$

CRF is a measure of the periods during which metal flow for feeding through a central channel is successively free and obstructed by growing crystals.

- For short freezing range alloys, CRF is low and have decreased feeding distances. The solidified skin grows rapidly and blocks the channel.
- For long freezing range alloys, CRF is high, and have increased feeding distance. The outer solid layer is far from being solid during much of the period of solidification and connection of liquid through this dendritic pasty zone will allow feeding to continue.

Table 8.7 shows typical centre line resistance factors for various alloys in sand moulds.

Table 8.7 Centre line resistance factors of typical alloys cast in sand moulds

Alloy	*Centre line resistance factor*
Copper (99.8%)	<1
Lead (99.0%)	17
60-40 Brass	26
18.8 steel (0.2% c)	35
0.6% steel	54
Monel	64
88/10/2 gun metal	95
Al 4% copper	96

8.9 CHILLS AND PADS

A chill is a metal piece added nearby the casting surface or kept inside the casting cavity so that more heat is absorbed from the molten metal quickly in order to get directional solidification in that portion of the casting. A chill is also used to eliminate hot pockets which may bring shrinkage cavity or porosity. Chills normally provide a steeper temperature gradient so that directional solidification as required in the casting be obtained. Chills are metallic objects having a higher heat absorbing capability than the sand mould. There are two types of chills, external chills and internal chills.

8.9.1 External Chills

External chills are metal-inserts kept at proper places outside of the mould cavity surface so that, heat removal rate is increased by freezing at those points. External chills are used for steel, cast iron and copper. These inserts may be of standard shapes or with special form to suit the mould cavity shape. Figure 8.21(a) shows forms of external chills. They are fitted as inserts, after coating with silica flour or similar refractory materials. Figure 8.21(c) shows examples of external chills fitted at junctions. Providing a chill at the edge may not normally have the desired effect as the temperature gradient is steeper at the end of the casting since heat is removed from all sides as shown in Figure 8.22. However if it is placed between two risers, it would have maximum effect. In those castings which have ribs or bosses, the joint is a source of hot spot. This hot spot could be eliminated by using a chill of appropriate size at the end of the rib or the appendage. To calculate the proportions of the chill, the results by Myskowski are used as shown in Figure 8.23.

8.9.2 Internal Chills

The internal chills are placed inside the mould cavity where external chill cannot be provided. These chills remove heat from the surroundings and become a part of the casting when it solidifies. Cleanliness of internal chills is far more important because they are surrounded on all sides by the molten metal. Figure 8.21(b) shows forms of internal chills. They are usually fitted at the hot spots as shown in Figure 8.21(d) to result in directional solidification.

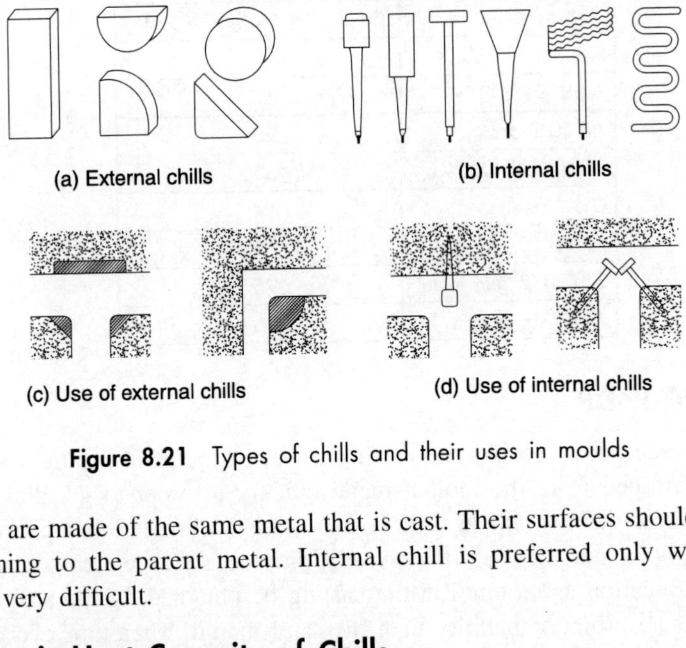

Figure 8.21 Types of chills and their uses in moulds

Internal chills are made of the same metal that is cast. Their surfaces should be clean enough to get good joining to the parent metal. Internal chill is preferred only when the fitting of external chill is very difficult.

8.9.3 Volumetric Heat Capacity of Chills

- Chills are used in the sand mould to extract heat at a faster rate corresponding to a thicker section. The intention is to ensure that solidification of casting progresses towards the riser and no hot spots are created resulting in shrinkage.
- Chills promote directional solidification. They act as heat sinks. They are made of cast iron, steel, copper, etc., but the use of steel and cast iron are very common.

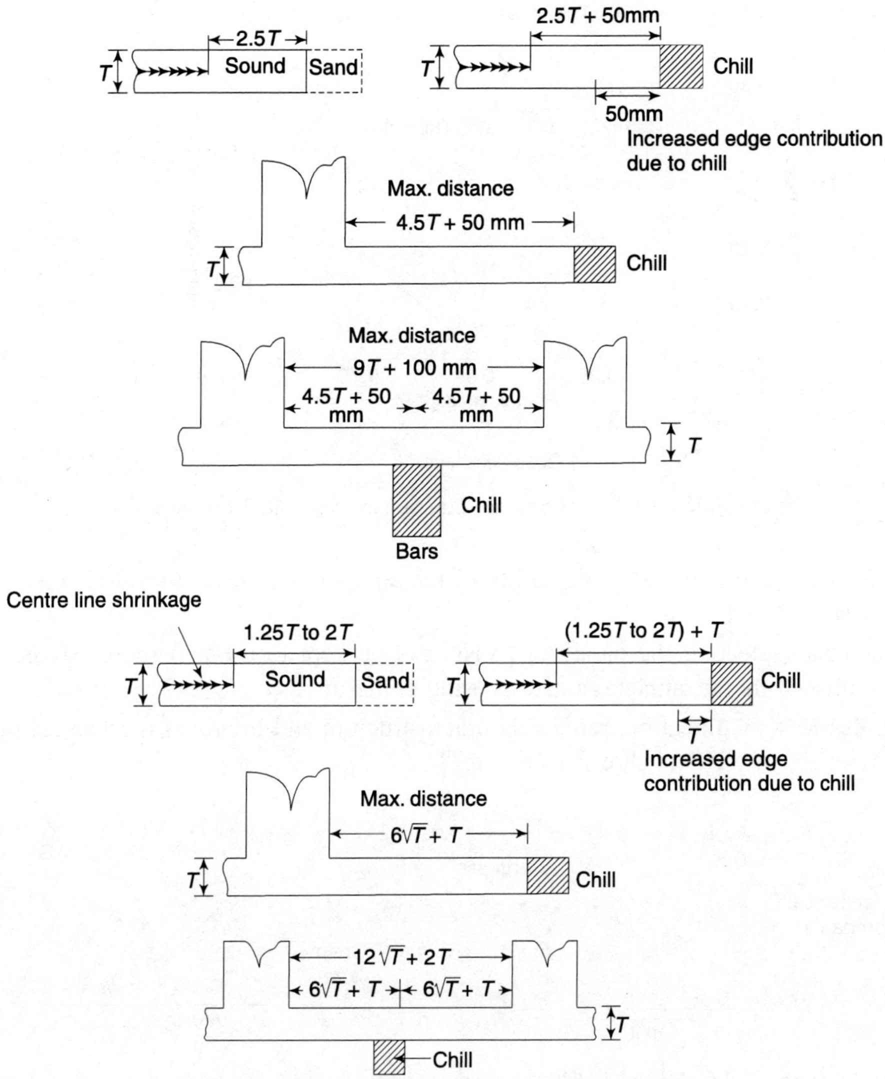

Figure 8.22 Effect of chills on feeding distance

- They are incorporated in the sand during moulding and are collected at the shake out and reused. The important features of the chills are volume of the chill, density of the chill and the specific heat of the chill.
- A parameter, volumetric heat capacity (VHC) is used to describe the chill performance. **VHC** = volume of the chill (V) × density of the chill material (e) × specific heat of chill material (Cp).

$$\mathbf{VHC = V.e.Cp}$$

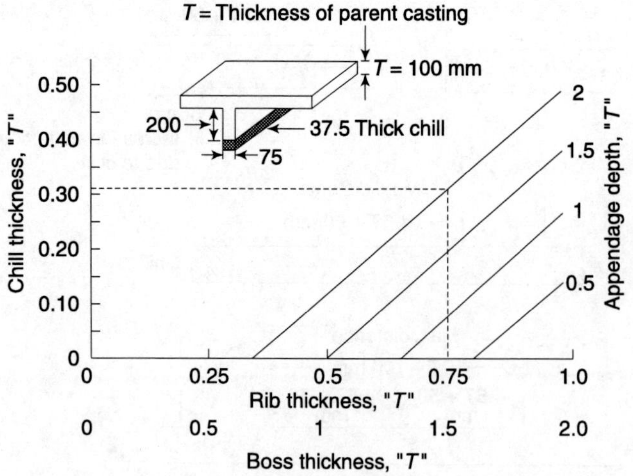

Figure 8.23 Chill proportions required in plate steel castings

- As VHC of the chill increases, the ability of the chill to extra heat from the molten metal also increases.
- By judiciously selecting the parameter, VHC can be kept to the maximum. Studies have well established the advantage gained in using chills in casting.
- VHC influences solidification, refines the microstructure and improves mechanical properties. The effect of VHC is shown in Figure 8.24.

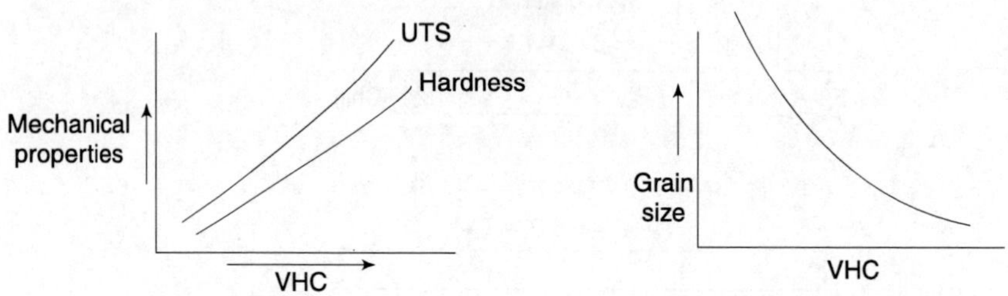

Figure 8.24 Effect of VHC on mechanical properties and grain size

- Another parameter, modulus of chill which takes into account the ratio of volume of the chill to the surface area of the chill, is also used to assess the chill efficiency.

$$\text{Modulus of chill} = \frac{\text{Volume of chill}}{\text{Surface area of chill}} = \frac{V}{SA}$$

- For a given casting as $\left(\frac{V}{SA}\right)$ increases, heat extraction will be faster, grain structure is refined and mechanical properties are enhanced. The effect of V/SA of chill on mechanical properties and grain size is shown in Figure 8.25.

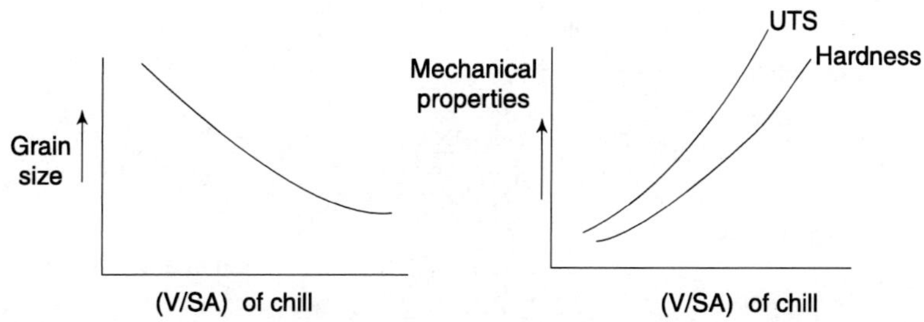

Figure 8.25 Effect of (V/SA) of chill on mechanical properties and grain size

The use of chills, primarily to extend feeding distance, also exerts a profound influence on feeding requirement. By creating an earlier demand for the feed metal, the freezing time needed in the feeder head is shortened. Thus, the reduction in the effective modulus of a wholly or partially chilled casting enables feeder head dimensions to be reduced towards the level at which volume feed capacity governs the size of head required. Figure 8.26 shows the influence of chilling on head volume required to satisfy freezing time criterion (spherical head : plates of various proportions).

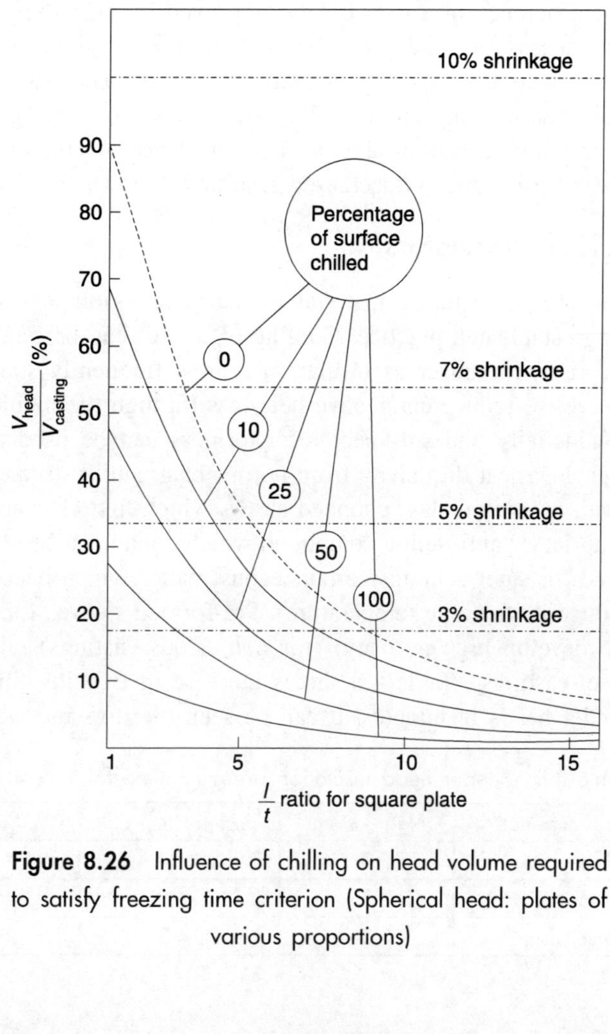

Figure 8.26 Influence of chilling on head volume required to satisfy freezing time criterion (Spherical head: plates of various proportions)

8.9.4 Pads

Pads are extra fittings or thickness of metal given on castings to get definite directional solidification towards the location of the riser. Figure 8.27 shows one example of the use of a pad on a casting. The projection on the right side of the casting has uniform thickness for certain length and at the end, there is a thick portion by design. In order to get directional solidification from the right extreme end, an external chill can be fitted at the end and a tapering pad at the middle portion as shown. Here, the chill fastens the freezing at the end, and the pad directs the solidification towards the riser by increasing the thickness in a tapered form. If the presence of a pad is making problem in the functioning of the component, that has to be removed by machining after the casting process.

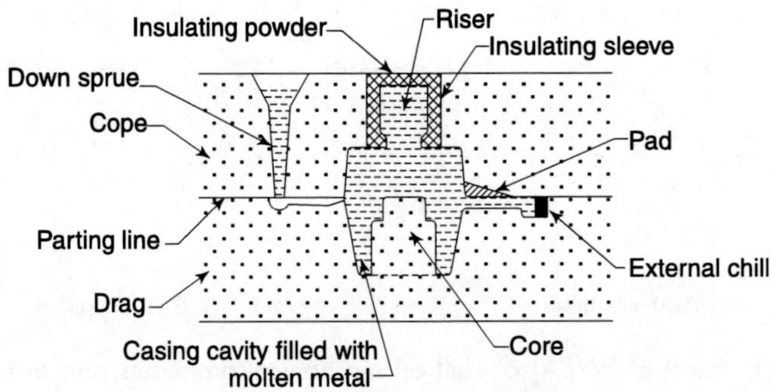

Figure 8.27 Use of chill, pad and insulator

8.10 USE OF INSULATORS AND EXOTHERMIC COMPOUNDS

The efficiency of a riser can be improved by extending the solidification time in the riser. It is thus necessary to keep the metal in the riser in liquid form for as long a period as required so that it would feed the casting till it solidifies. The aids used for this purpose are called feeding aids. Thermal insulation of the riser portion or adding of heat by exothermic compounds are two common feeding aids used in industries. As the efficiency increases, the volume of metal wasted for a riser is decreased resulting in lower cost of production of the casting.

8.10.1 Insulators

The use of insulting material to retard radiation losses from the exposed metal surface is a long established practice. Cooling of a riser can be retarded by insulating the top and sides of the riser, moreover as insulation is now frequently sought by lining the heads with moulded sleeves to reduce conductive heat loss through the mould. Any material which has low thermal conductivity and sufficient refractoriness can be used as insulator. The materials used derive their low heat diffusivity from porous or granular structures. Top covering materials include dry sand, powdered slag, chopped straw (which first char and then burn away to leave a bulky ash, proprietary anti-piping compound which contains a certain amount of exothermic material; those used for steel contains carbonaceous matter which becomes absorbed by the metal and locally reduce its freezing temperature). Pre-formed sleeves for lining includes foamed gypsum plaster (to develop high porosity) (for nonferrous castings), diatomaceous silica and vermiculite (for steel castings). Table 8.8 shows increase in freezing times for 100 mm × 100 mm cylindrical feeder heads in minutes. It can be seen that top and wall insulation give better results.

Table 8.8 Feeder head insulation: freezing time for 4 x 4 inches in cylindrical feeder heads (min)

Material	*No treatment*	*Top insulation only*	*Wall insulation only*	*Both top and wall insulation*
Steel	5.0	13.4	7.5	43.0
Copper	8.2	14.0	15.1	45.0
Aluminium	12.3	14.3	31.1	45.6

8.10.2 Exothermic Compounds

To maintain metal in molten form inside the riser till the solidification period of a casting, extra heat may be added to the riser by the use of exothermic compounds. Adding of small quantities of these in powder form inside the riser, they react with the molten metal and produce additional heat. They are added in two ways:

- Material is **added on top of feeder** to control feeding. Composition must be compatible with that of the casting.
- Material is **mixed with bonding material and water** to mould sleeve for lining feeder head. The added substance helps delaying exothermic reaction and extends the period during which heat is generated. This diminishes the danger of contamination, since the exothermic reaction is confined to the mould wall. The period during which heat is generated is controlled.

Aluminium powder and iron oxide form a rigorous exothermic reaction resulting in strong heat input. Oxidants and oxidizers, together with other additions, may also be applied to get heat input. Electric arc on the top of riser can also be employed for adding heat for large castings. Other exothermic materials that can be used are graphite or charcoal powder, rice hulls and thermite mixtures. The carbonaceous materials called 'anti-piping compounds', besides being cheap, provide the insulation and get oxidized slowly. But thermite mixtures get oxidized very quickly and release heat over a very small period and subsequently, the effect would be that of only the residue acting as an insulator. As such these are not generally used. For steel casting, an insulating shield on the top of an open riser is very effective since it reduces considerable amount of heat loss by radiation. Loose cellulose material such as rice husk is used as top insulation.

Thermit reactions are given by

$$2Al + Fe_2O_3 = 2Fe + Al_2O_3 + \text{Heat } (\Delta H_{298} = -853 \text{ kJ})$$

$$8Al + 3Fe_2O_3 = 9Fe + 4Al_2O_3 + \text{Heat } (\Delta H_{298} = -3347 \text{ kJ})$$

Figure 8.28 shows that use of the insulating compound and exothermic mixture in the feeder reduces the piping tendency and decreases the amount of metal required in the feeder. Net effect of using these artificial means is to reduce the size of the feeder. This also means a higher casting yield. Figure 8.29 shows the metal utilization of feeders of various forms moulded in sand. The (a) cylindrical and (b) hemispherical heads have been treated with normal feeding compounds, (c) efficiency of the reverse taper heads depends on detailed geometry and (d) exothermic sleeve. Metal utilization for ductile iron plates happens with (e) cylindrical sand feeder, (f) insulating feeder and (g) cruciform exothermic feeder. Figure 8.29 shows the chart for selection of exothermic feeder head dimensions.

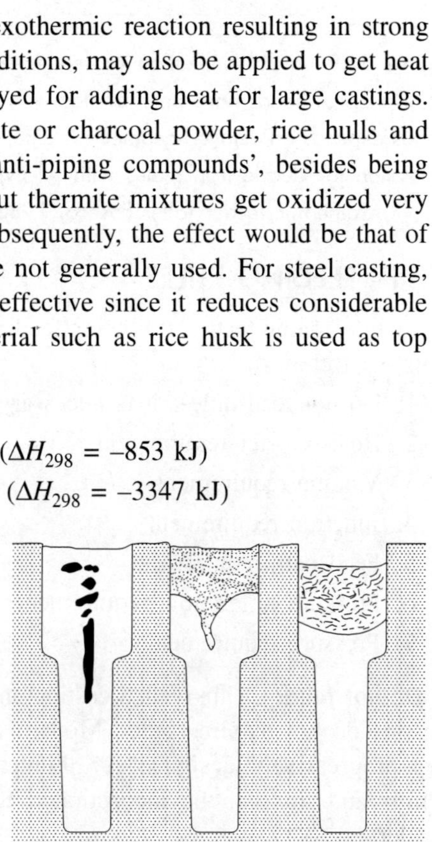

Figure 8.28 Use of an insulating compound or exothermic mixture in the feeder reduces the piping tendency and decreases the amount of metal required in the feeder

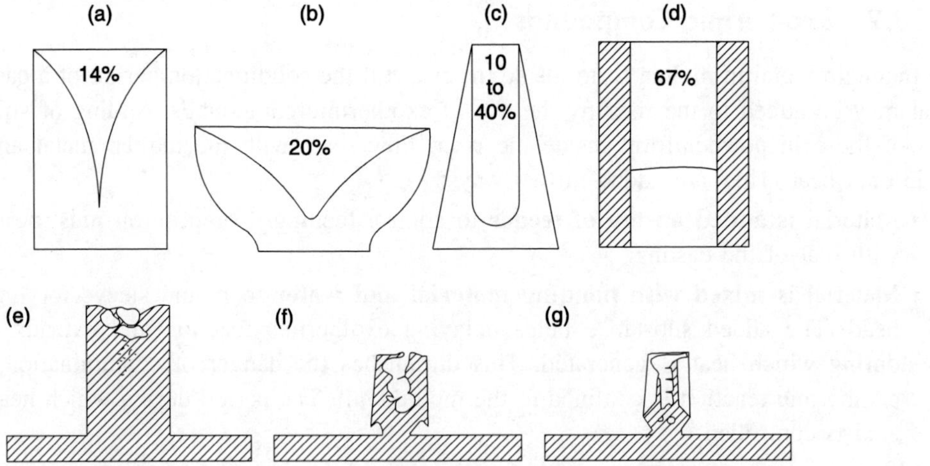

Figure 8.29 Metal utilization of feeders of various forms moulded in sand. (a) cylindrical and (b) hemispherical heads have been treated with normal feeding compounds; (c) efficiency of the reverse taper heads depends on detailed geometry (Heine, 1982, 1983); and (d) exothermic sleeve (Beeley, 1972). Metal utilization for ductile iron plates with (e) cylindrical sand feeder; (f) insulating feeder; and (g) cruciform exothermic feeder (after Foseco, 1988)

8.11 FEEDING RULES

Before designing feeder, following feeding rules must be followed:

1. Do not feed unless it is necessary
2. Heat transfer requirement
3. Volume requirement
4. Junction requirement
5. Feed path requirement
6. Pressure differential requirement
7. Pressure requirement

1. Do not feed: Thin walled casting and small casting do not require feeder. Modern castings are progressively designed for thinner walls. Chills and fins can be incorporated to avoid feeder.

2. Heat transfer requirement: Feeder should solidify last. Modulus of the feeder must be at least 1.2 times the modulus of the casting ($M_r = 1.2\ M_c$). Modulus of the feeder can be increased by use of an insulating or exothermic

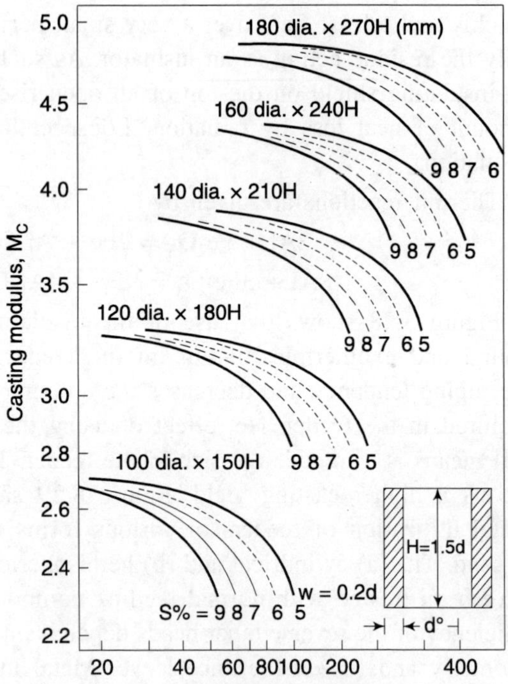

Figure 8.30 Chart for selection of exothermic feeder head dimensions

sleeve or by applying insulating or exothermic powder on top of the feeder. For example, a cylindrical feeder in an insulating material is only 0.63 D in diameter compared with diameter D in sand and therefore, this insulated feeder has only 40% of the volume of the sand feeder.

3. Mass flow requirement: Feeder must contain sufficient liquid to meet the volume contraction requirement of the casting. Although we may have provided a feeder of such a size that it would theoretically contain liquid until after the casting is solid, in fact it may still be too small to deliver the volume of feed liquid that the casting demands.

Normal feeders are relatively inefficient in the amount of feed metal that they are able to provide. This is because they themselves freez at the same time as the casting depletes the liquid reserves of the reservoir.

$$\eta V_r = \alpha(V_c + V_r)$$

$$V_r = \left[\frac{\alpha}{\eta - \alpha}\right](V_c)$$

where V_r = Volume of riser

V_c = Volume of casting

η = Feeder efficiency

α = Thermal coefficient of expansion

For aluminium alloy casting,

$\alpha = 7\%$, $V_r = V_c$ and maximum yield is 50%

For steel alloy (FCC) castings,

$\alpha = 4\%$, $V_r = 0.4\ V_c$

For steel alloy (BCC) castings,

$\alpha = 3\%$, $V_r = 0.27\ V_c$

Therefore, compared to aluminium alloys, the smaller solidification shrinkage of ferrous metal reduces the volume requirement of the riser considerably.

For graphite cast irons, the feed volume reduces approximately to zero in the region of 3.6 to 4.0% carbon equivalent (CE). A feeder may still be required, because of the difference in timing between the feed demand and the graphite expansion.

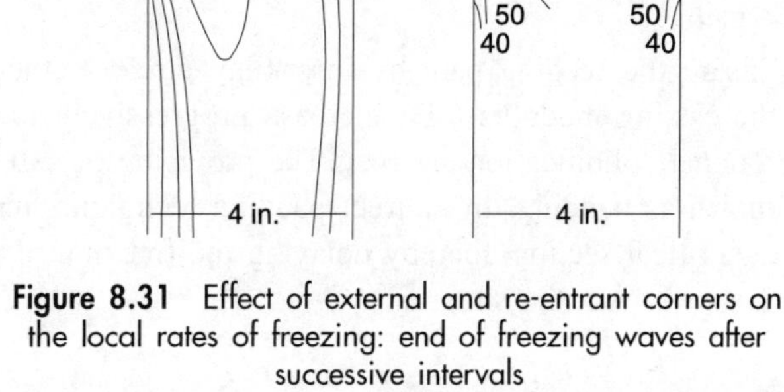

Figure 8.31 Effect of external and re-entrant corners on the local rates of freezing: end of freezing waves after successive intervals

4. Junction requirement: The junction between the casting and the riser should not create a hot spot. Feeders and ingates should not be planted on the casting so as to create a T or L junction. They are best added as extensions to a section, as an elongation to a wall or plate, effectively moving the junction off the casting. Figure 8.31 shows the effect of external and re-entrant corners on the local rates of freezing: end of freezing waves after successive intervals.

If there is no alternative to the placing of feeder directly on casting, avoid hot spot in the middle of junction. For *T* junction, keep (M_r = 1.2 M_c) and for *L* junction keep M_r = 1.33 M_c. Reduce the junction effect by design modification and chills. Figure 8.32 show some design modifications and use of chills to avoid junction effect.

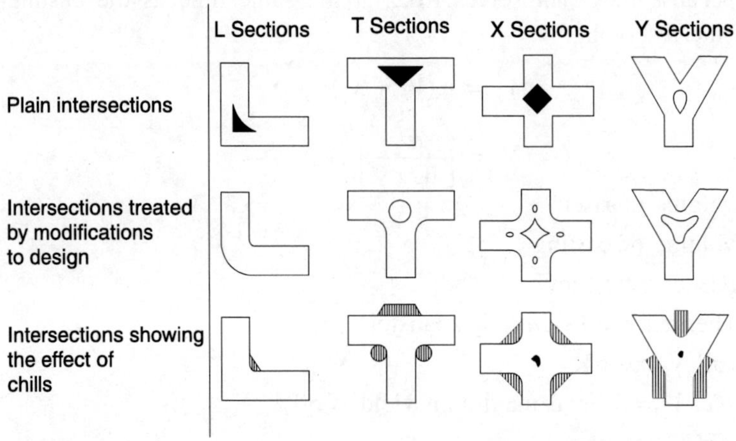

Figure 8.32 Ways to avoid the junction effect

5. Path feed requirement: There must be a feed path to allow feed metal to reach those regions that require it. The highest modulus regions in casting are either potential regions for shrinkage porosity if left unfed or may be feeding paths. The following are the ways to ensure feed path:

- Directional solidification towards feeder.
- Minimum temperature gradient: when the temperature gradient at the solidus falls to below critical value (0.1-2 K/mm) then porosity is observed even in well degassed material.
- Feeding distance: there will be a limit to how far feed liquid can be provided along a flow path.

Ensure the feeding path by inscribing circle method. If the circle diameter (2D drawings) or the casting modulus (3D) increase progressively towards the feeder, then the conditions of directional solidification are met. The provision of extra metal on the casting to thicker section is known as padding. In indirect padding technique, molten metal is fed to a position adjacent to the critical section, thereby delaying the freezing of neighbouring area of the casting. Figure 8.33 shows both the situations.

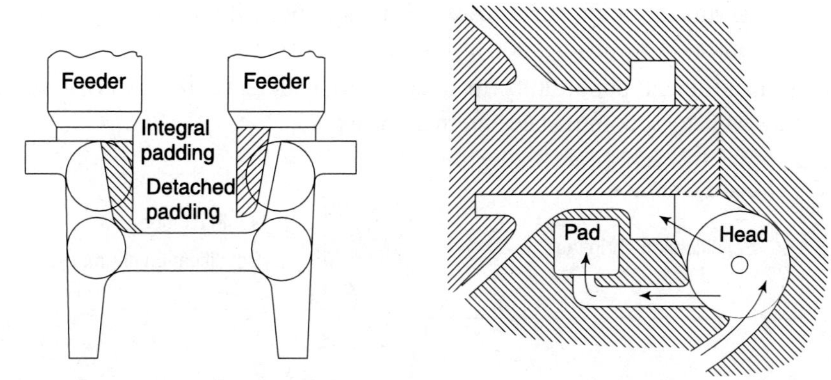

Use of Heuvers' circle to determine the amount of attached padding described by Beeley (1972)

Use of indirect padding described by Daybell (1953)

Figure 8.33 Feed path using Heuvers' circle and indirect padding method

6. Pressure gradient requirement: There must be sufficient pressure differential to cause the feed material to flow in the correct direction. Place feeders high to feed downhill and place gates low to fill uphill. If the pressure gradient needed to cause the liquid to flow along the path is not available, then feed liquid will not flow to where it is needed. A positive pressure gradient from the outside to inside of the casting will help to ensure that the feed material (either solid or liquid) travels along the flow path into those parts of the casting experiencing shrinkage.

The driving forces for liquid flow are:

1. Positive pressure such as atmospheric pressure and/or the pressure due to the hydrostatic head of the metal in the feeder.
2. Reduced or even negative pressure generated within poorly fed regions of some castings.

7. Pressure requirement: There must be sufficient pressure at all points in the casting to maintain the dimensional accuracy of the casting and to suppress the formation and growth of cavities. This is a hydrostatic requirement relating to the suppression of porosity contrast with the previous pressure gradient requirement that relates to the hydrodynamic requirement of flow.

Figure 8.34 shows the comparison between the external size and internal shrinkage porosity in a casting as result of (a) moderate pressure in the liquid and adequately rigid mould; (b) too much pressure and/or weak mould.

A fall in internal pressure may cause a variety of problems:

1. Liquid may be sucked from the cast surface.
2. The internal pressure may fall just sufficiently to unfurl, but not fully open the population of biofilms.
3. The internal pressure may fall sufficiently to open biofilms so that a distribution of fine and dispersed micro porosity will appear.

4. The internal shrinkage may cause macro shrinkage porosity, especially if there are large biofilms present as a result of poor filling of the casting.
5. If there is insufficient opportunity to open internal defects, the external surface of the casting may sink to accommodate internal shrinkage.

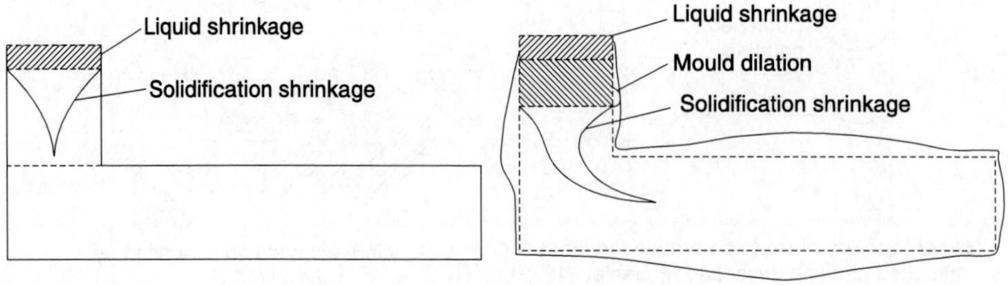

Figure 8.34 Comparison between the external size and internal shrinkage porosity in a casting as a result of (a) moderate pressure in the liquid and adequately rigid mould; (b) too much pressure and/or weak mould

Worked Out Examples

Example 8.1: Calculate the size of the cylindrical riser (height and diameter equal) necessary to feed a steel slab casting 25 × 25 × 5 cm with a side riser, casting is poured horizontally into the mould.

Volume of casting = 25 × 25 × 5 = 3125 cm^2

Surface area of casting = 2 × 25 × 25 + 4 × 25 × 5 = 1750 cm^2

$$\text{Volume of the riser} = \frac{\pi \times D^3}{4}$$

where D is the riser diameter.

$$\text{Surface area of the riser} = \pi D^2 + \frac{\pi \times D^2}{4} = 1.25\,\pi D^2 \ (11)$$

$$\text{Freezing ratio, } X = \frac{1750/3125}{1.25\pi D^2/0.25\,\pi D^3} = 0.122\,D$$

$$Y = \frac{\text{Volume of riser}}{\text{Volume of casting}} = \frac{0.25\,\pi D^3}{3125} = 0.000251\,D^3$$

Substituting this in Caine's equation for steels

$$0.112D = \frac{0.10}{0.000251\,D^3 - 0.03} + 1$$

On simplification, we get $D^4 - 8.9286D^3 - 119.52\,D = 2490$

By trial-and-error, we get $D = 11.44 \text{ cm} \approx 12 \text{ cm}$

Example 8.2: Calculate the riser size, for Example 8.1, using the modulus method.

Since it is a slab of 25 × 25 × 5 cm, it can be considered a long bar with cross section 25 × 5.

$$\text{Modulus of casting, } M_C = \frac{25 \times 5}{2(25+5)} = \frac{125}{60} = 2.08 \text{ cm}$$

$$\text{The riser diameter, } D = 6 \times M_C = 6 \times \frac{125}{60} = 12.5 \text{ cm}$$

Example 8.3: Calculate the diameter of the cylindrical riser for size of a bar of width (a), depth (b) and unit length. Use the modulus method.

Given

Width = a, depth = b, length = unit length = l

Volume, $V = a \times b \times l = a \times b \times 1 = ab$

Cooling surface area = $2a + 2b$

Assuming that there is no heat dissipated from faces (ab) across the section.

$$\text{Modulus of casting, } M_c = \frac{V}{SA} = \frac{ab}{2(a+b)}$$

Similarly, for a cylindrical riser, feeding the bar, assuming

H = Height

D = Diameter of the riser, and when $H = D$,

$$\text{Volume of riser, } V_r = \frac{\pi \times D^2}{4} \times D = \frac{\pi \times D^3}{4}$$

Cooling surface area (neglecting bottom face)

$$\text{Surface area, } SA = \pi \times D \times D + \frac{\pi \times D \times D}{4} = \frac{5 \times \pi \times D^2}{4}$$

$$\text{Therefore, modulus of riser, } M_r = \frac{V_r}{SA_r} = \frac{D}{5}$$

Therefore, diameter of cylindrical riser = $M_r \times 5$

For effective feeding, as riser also loses heat during solidification along with casting, we have to increase its modulus by 20%.

Modulus of riser = $1.2\ M_C$

Therefore, diameter of a cylindrical riser = 1.2 × 5 × modulus of casting

$$= 6 \times \frac{ab}{2(a+b)}$$

Example 8.4: Calculate the riser dimensions for a slab of 25 × 25 × 5 cm. Use NRL method.

$$\text{Shape Factor} = \frac{25+25}{5} = 10$$

From Figure 8.9

Riser volume, $V_r = 0.47 \times$ casting volume

$$= 0.47 \times 25 \times 25 \times 5 = 1468.75 \text{ cm}^3$$

For a cylindrical riser of equal diameter and height,

$$V_r = 0.25 \times \pi D^3$$

$$D = \sqrt[3]{\frac{4 \times 1468.75}{\pi}} = \sqrt[3]{1870} = 12.32 \text{ cm}$$

Example 8.5: A cylindrical riser must be designed for a sand-casting mould. The casting itself is a steel rectangular plate with dimensions 7.5 cm × 12.5 cm × 2.0 cm. Previous observations have indicated that the total solidification time for this casting is 1.6 minutes. The cylinder for the riser will have a diameter-to-height ratio, 1.0. Determine the dimensions of the riser so that its solidification time is two minutes.

First determine the *V/A* ratio for the plate.

Volume, $V = 7.5 \times 12.5 \times 2.0 = 187.5 \text{ cm}^3$

Surface area, $A = 2(7.5 \times 12.5 + 7.5 \times 2.0 + 12.5 \times 2.0) = 267.5 \text{ cm}^2$

Solidification time = 1.6 minutes, thus we can determine the mould constant, using a value of $n = 2$ in the equation.

$$t = K\left(\frac{V}{SA}\right)^n = K\left(\frac{V}{SA}\right)^2$$

$$K = \frac{t}{(V/SA)^2} = \frac{1.6}{(187.5/267.5)^2} = 3.26 \text{ min/cm}^2$$

Next we must design the riser so that its total solidification time is 2 minutes, using the same value of mould constant. The volume of the riser is given by

$$V = \frac{\pi D^2 h}{4}$$

and the surface area is given by

$$SA = \pi D h + \frac{2\pi D^2}{4}$$

Since we are using a *D/H* ratio = 1.0, then $D = H$. Substituting D for H in the volume and area formulas, we get

$$V = \frac{\pi D^3}{4}$$

$$A = \pi D^2 + \frac{2\pi D^2}{4} = 1.5\ \pi D^2$$

Thus, the *V/A* ratio = *D*/6. Using this ratio in Chvorinov's equation, we have

$$t = 2.0 = 3.26\left(\frac{D}{6}\right)^2 = 0.09056\ D^2$$

$$D^2 = \frac{2.0}{0.09056} = 22.086 \text{ cm}^2$$

$$D = 4.7 \text{ cm}$$

Since $H = D$; then $H = 4.7$ cm also.

The riser represents waste metal that will be separated from the cast part and remelted to make subsequent castings. It is desirable for the volume of metal in the riser to be minimum. Since the geometry of the riser is normally selected to maximize the V/A ratio, this tends to reduce the riser volume as much as possible.

Example 8.6: Consider a cube of side L for a casting to be fed by a cylindrical riser of height, H = 1.5 diameter. Assume top riser and neglect surface area of common surface.

$$\text{Volume}_{\text{casting}} = L \times L \times L = L^3 \quad ...(1)$$

$$\text{Surface area}_{\text{casting}} = L \times L \times L = L^3 \quad ...(2)$$

$$\left(\frac{V}{SA}\right)_{\text{casting}} = \frac{L^3}{5L^2} \quad ...(3)$$

$$(\text{Volume})_{\text{riser}} = \frac{\pi}{4} D^2H = \frac{\pi}{4} D^2 (1.5D) = \frac{1.5}{4} \pi D^3 \quad (\text{Since } H = 1.5) \quad ...(4)$$

$$(\text{Surface area})_{\text{riser}} = \pi dh + \frac{\pi}{4} D^2$$

$$= \pi d(1.5D) + \frac{\pi}{4} D^2$$

$$= 1.5\, \pi D^2 + \frac{\pi}{4} D^2$$

$$= \frac{(6 + 1)}{4} \pi D^2 = \frac{7}{4} \pi D^2 \quad ...(5)$$

From (4) and (5),

$$\left(\frac{V}{SA}\right)_{\text{Riser}} = \frac{1.5\, \pi D^3}{4} \times \frac{4}{7\pi D^2} = \frac{1.5D}{7} \quad ...(6)$$

For proper risering $\left(\frac{V}{SA}\right)_{\text{Riser}} \geq \left(\frac{V}{SA}\right)_{\text{casting}}$

$$\frac{1.5D}{7} \geq \frac{L}{5}$$

$$D \geq \left(\frac{7L}{5 \times 1.5}\right)$$

Therefore, $D \geq 0.933\, L$

Hence, $H \geq 1.5\, (0.933L) = 1.4\, L$

Example 8.7: Calculate the riser dimensions for the casting shown in Figure 8.35.

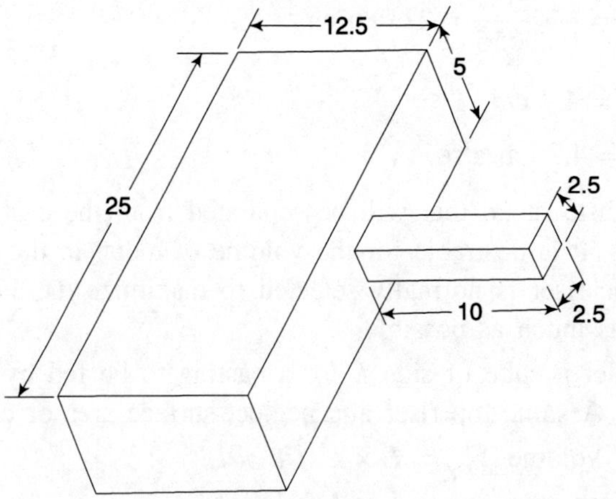

Figure 8.35

First neglect the branch shown and calculate the shape factor for the main plate

$$\text{Shape factor} = \frac{25 + 12.5}{5} = 7.5$$

Volume of the casting, V_c = 25 × 25 × 5 = 3125 cm^3

Volume of the riser, V_r = 0.575 × 3125 = 1796.875 cm^3

The branch volume = 2.5 × 2.5 × 10 = 62.5 cm^3

This is a plate feeding the bar with a thickness ratio of 0.5. Hence, from Figure 8.12, we can get the parasitic volume as 30%

Hence, the riser volume = 0.30 × 62.5 + 1796.875 = 33691.406 cm^3

Riser diameter, D = 10.531 cm

Example 8.8: Calculate the riser diameter for an annular cylinder of outside diameter 30 cm, inside diameter 10 cm and height 30 cm.

Core diameter = 10 cm

Plate thickness = 10 cm

Correction factor = 1.14 from Table 8.2

$$\text{Shape factor} = \frac{20\pi + 30}{1.14 \times 10} = 8.143$$

Casting volume = 0.25 × π × (30^2 – 20^2) × 30 = 18849 cm^3

Riser volume = 0.48 × 18849 = 9047.52 cm^3

Riser diameter = 22.585 cm.

Example 8.9: Calculate the feeding arrangement for the bearing housing shown in Figure 8.36.

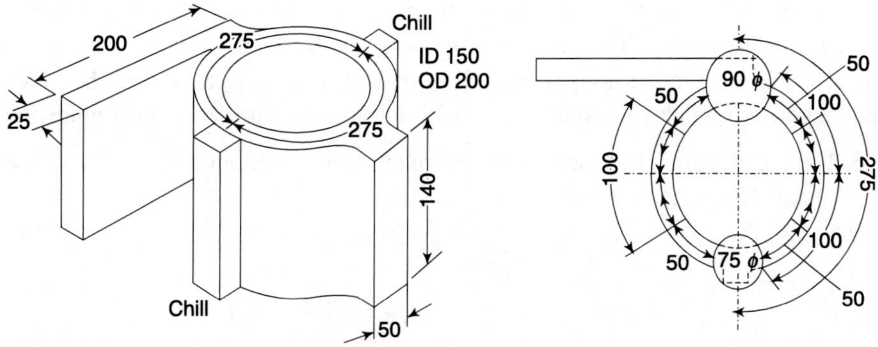

Figure 8.36

The bearing housing essentially consists of a base plate and cylindrical section which can also be unrolled and considered a plate. The length of the plate is given by

$$L = 0.5 \times \pi \times (200 + 150) = 550 \text{ mm}$$

The length of the half cylinder is 275 mm, as shown in the figure. Since the distance is too long to be fed by a single riser, it was decided to place two risers as shown at the two points of heavy sections.

For the half cylinder,

$$\text{Shape factor} = \frac{275 + 140}{25} = 16.6$$

From Figure 8.9, riser volume = 0.25 × casting volume

$$\text{Casting volume} = (27.5 + 5) \times 140 \times 2.5 = 1137.5 \text{ cm}^3$$

Since the thickness of the parentage and appendage are same, the total volume can be directly obtained without any modification as shown above.

$$\text{Riser volume} = 0.25 \times 1137.5 = 284.375 \text{ cm}^3$$

The above can be achieved with a riser diameter of 7.5 cm and 6.5 cm long.

Similarly, the second portion,

$$\text{Volume of casting} = (27.5 + 20) \times 14 \times 2.5 = 1662.5 \text{ cm}^3$$

Since the shape factor is same as the previous one,

$$\text{Volume of the riser} = 0.25 \times 1662.5 = 415.625 \text{ cm}^3$$

This can be achieved with a riser of 9 cm diameter and 6.6 cm long.

Since each of these will be able to feed only up to a distance of 5 cm on either side, it is necessary to augment the feeding distance by providing chills at the centre points between two risers, this increases the feeding distance to $(4.5T + 5) = 4.5 \times 2.5 + 5 = 16.25$ cm which is more than 0.5 × 27.5 cm. But one problem still persists because feeding length along the height is only 4.5 × 2.5 = 11.25 cm whereas actual length is more at 14. Hence the bottom portion should be provided with a ring chill. However, these can be omitted, since the risers were actually attached at a point which is longer than 2.5 cm and therefore, they may be able to take care of extra length. A test casting should be made to actually check the presupposition.

Example 8.10: Calculate the risering requirements for the mounting base casting shown in Figure 8.37 using NRL method.

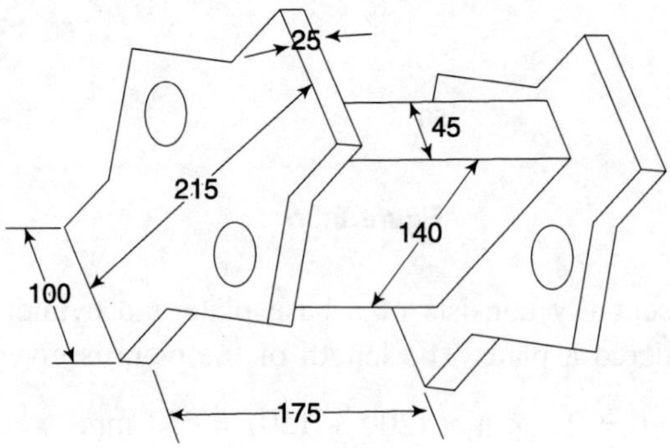

Figure 8.37

Considering the central heavy portions, it was decided to have a single feeder centrally placed. Feeding range = $4T$ = 4 × 45 = 180 mm.

It can easily feed the entire height (140 mm)

The shape factor for the main section (45 × 140 × 175) is 0.5

$$\text{Shape factor} = \frac{140 + 175}{45} = 7$$

From Figure 8.9, riser volume = 0.5 × casting volume

Next, considering the appendages,

Volume of parent section = 140 × 175 × 45 = 1102500 mm^3.

Neglecting the triangular portion and the hole,

The appendage volume = 2 × 100 × 215 × 25 = 1075000 mm^3

$$\frac{\text{Appendage thickness}}{\text{Parent thickness}} = \frac{25}{45} = 0.555$$

From the chart in Figure 8.12 (for plate-plate)

Effective appendage volume = 0.555 × 1075000 = 596625 mm^3

Equivalent casting volume = 1102500 + 596625 = 1699125 mm^3

Riser volume = 0.5 × 1699125 = 849563 mm^3

From this volume and from the charts in Figure 8.10, we obtain the riser sizes as 11.5 cm in diameter and 8.5 cm in the height.

Example 8.11: Calculate the riser dimensions for the following cast steel gear wheel rim shown in Figure 8.38. Take height of the rim (*a*) = 220 mm, width of the rim (*b*) = 100 mm, non-cooling web thickness (*C*) = 30 mm, outside diameter = 1000 mm, and inside diameter = 800 mm.

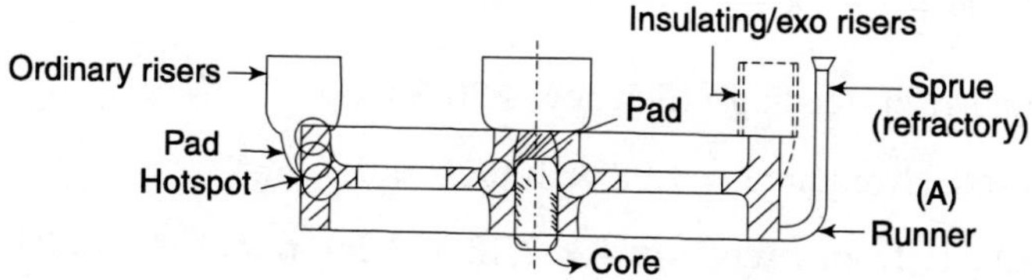

Figure 8.38 Cast steel gear wheel

Divide the casting into suitable zones for feeding depending on geometry: This being a uniform symmetrical section no division needed.

(a) Calculate the modulus of each zone.

$$\text{Modulus of casting, } M_c = \frac{a \times b}{2(a + b) - c}$$

where, a = Height of rim (220) $\quad OD$ = 1000 mm

b = Width of rim (100) $\quad ID$ = 800 mm

c = Non-cooling web thickness (30)

$$M_c = \frac{220 \times 100}{(2 \times 320) - 30} = 36.1 \text{ mm}$$

(b) Riser modulus, $M_r = 1.2 \times M_c = 1.2 \times 36.1 = 43.3$ mm

(c) Riser size

If ordinary sand riser diameter = $5 \times M_r = 5 \times 43.3 = 217$

Say, diameter = 220 mm

Height = 220 mm

If insulating/sleeve riser (Refer Table 8.3)

Sleeve diameter = 180 mm

Height = 200 mm

(d) Number of risers

For critical castings assuming a circumferential coverage of 50% by risers,

Total coverage by risers = $\pi \times$ mean diameter $\times 0.5 = \pi \times 900 \times 0.5 = 1414$ mm

$$\text{Number of risers} = \frac{1414}{220} = 6.42$$

Approximately say six risers.

Check for feed metal required

Weight of casting = $\pi/4 \times (100 - 64) \times 1 \times 7.8$ kg = 220 kg

Feed metal volume = Casting volume × % volume shrinkage

$$\text{Feed metal volume} = \frac{220}{72} \times \frac{6.5}{100} = 1.98 \text{ dm}^3$$

For ordinary sand riser–feeding efficiency – 20%

Volume of feed-metal required = $1.98 \times 100/20 = 9.9$ dm^3

(e) Feed metal available from risers = $\pi/4 \times 2.22 \times 2.2 \times 6 \times 0.2 = 10.03$ dm^3

As this is found adequate, finally feeding system consists of 6 ordinary risers of diameter, 220 mm and height, 220 mm or six insulating risers of diameter, 180 mm and height, 200 mm.

$$\text{Weight of six ordinary risers} = 6 \times \frac{\pi}{4} \times 2.23 \times 7.2 = 360$$

$$\text{Yield with ordinary risers} = \frac{220}{220 + 360} = 38\%$$

$$\text{Weight of six insulating risers} = 6 \times \frac{\pi}{4} \times 1.82 \times 2 \times 7.2 = 220 \text{ kg}$$

$$\text{Yield with insulating risers} = \frac{220}{220 + 360} = 50\%$$

Example 8.12: Calculate the feeding distance for: (a) plate casting with single riser without chill, (b) plate casting with single riser with chill, (c) plate casting with two risers without chill and (d) plate casting with two risers with chill.

Solution: For a plate casting, $FD = 4.5\ T$; single riser, T = Thickness of the casting

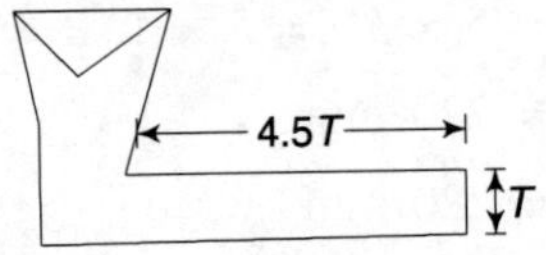

Fig 8.39 Plate casting with a single riser

Plate casting rising with a chill, $FD = 4.5T + 2$ inches

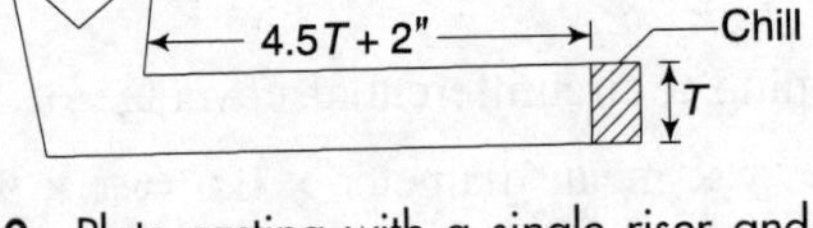

Fig 8.40 Plate casting with a single riser and a chill

Plate casting with two risers, $FD = 4T$

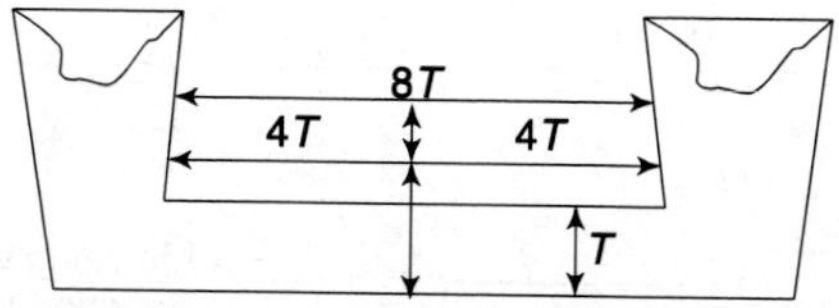

Fig 8.41 Plate casting with two risers

Plate casting with two risers and chill, $FD = 9T + 4$ inches

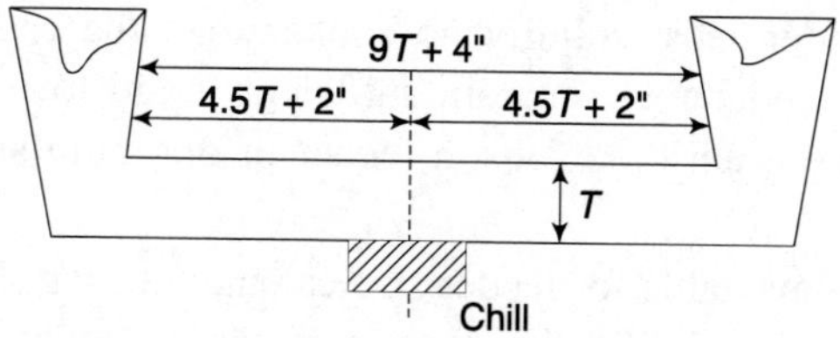

Figure 8.42 Plate casting with two risers and a chill

Example 8.13: Calculate the number of risers for the following example shown in Figure 8.43. Take feeding distance as $4T$. Thickness of the casting (T) = 30 mm, diameter of the casting (D) = 100 mm, diameter of the riser (d) = 15 mm.

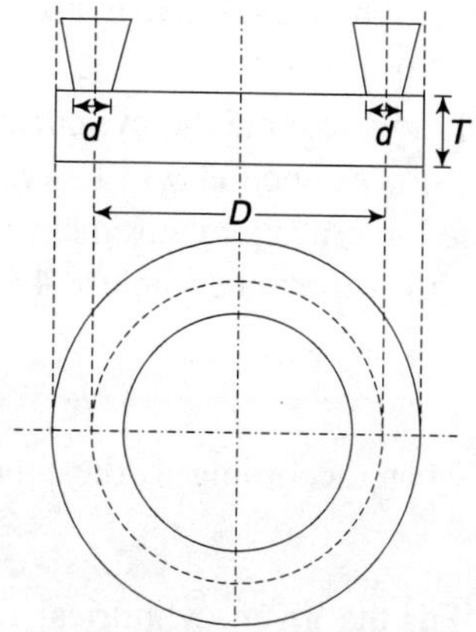

Figure 8.43 Calculation of number of risers required

Solution:

n = Number of risers

d = Diameter of riser

T = Thickness of casting

D = Mean diameter of circle of circular casting

Circumference, $C = \pi D$

Feeding distance $= n \times d = n \times 4T$

$$\text{Hence number of risers, } n = \left[\frac{C}{4T + d}\right] = \left[\frac{\pi D}{4T + d}\right] = \left[\frac{\pi \times 100}{4 \times 35 + 20}\right] = 1.9625 \cong 2 \text{ risers}$$

Example 8.14: Design a suitable feeding system for the plate casting shown in Figure 8.44 (dimensions are in mm).

Solution:

Step1: Calculate the volume of the casting

$$V_C = 1000 \times 200 \times 20 = 4000000 \text{ mm}^3 = 4000 \text{ cm}^3 = 4000 \text{ cc}$$

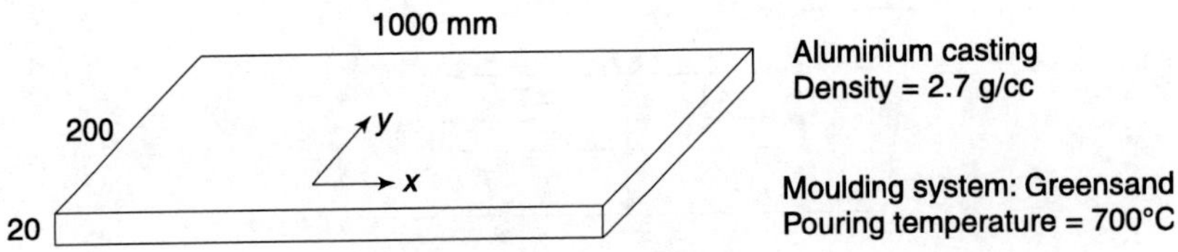

Figure 8.44 Plate casting

Step 2: Calculate the dimensions of the feeder based on heat transfer criteria

As a general rule, the feeder must solidify no sooner than the casting has solidified (heat transfer criterion). Also, the feeder must contain sufficient liquid to meet the volume contraction requirement (feed volume criterion). Thus, a feeder of optimum size must satisfy both the these requirements.

There are many techniques available for feeder size calculation. We shall use modulus method here. According to this method, solidification time depends on square of modulus of casting. Sphere, having the smallest modulus per unit volume, is therefore the ideal shape for a feeder. But a spherical feeder is practically too difficult to mould and consequently, cylindrical feeders are mostly used.

Let us assume the cylindrical feeder with H = 1.5 D will be placed on the top of the casting. Thus, a T Junction will be created between the casting and the feeder. Now let us first consider junction criteria requirement of the feeding design. Consider the casting as an infinite plate, its modulus from the Table 8.4 is given by

$$M_c = \frac{\text{Thickness of plate}}{2} = \frac{20}{2} = 10 \text{ mm} = 1\text{cm}$$

Then, according to the junction criterion, the modulus of the feeder

$$M_F = 2\ M_c = 2 \times 1 = 2 \text{ cm}$$

For the given cylindrical feeder (H = 1.5 D) with base-uncooled (i.e., not in direct contact with the liquid metal) its modulus,

$$M_F = \frac{3D}{14} = 2 \text{ cm}$$

D = 9.33 cm

H = 1.5 × 9.33 cm = 14 cm

Step 3: Calculate the volume of the feeder

The volume of a single feeder, D = 9.33 cm and height, H = 14 cm is

$$V_F = \frac{\pi \times D^2}{4} \times H = \frac{(\pi \times 9.33^2)}{4} \times 14 = 957 \text{ cm}^3 = 957 \text{ cc}$$

Step 4: Calculate the feeding distance and number of the feeders

In normal conditions, there is a limit to how far a feeder can feed along a flow path. Up to this distance from feeder, the casting will be sound. Beyond this distance, the casting will exhibit porosity.

Now we have to determine the number of feeders of this size and shape to be required by the given casting. To determine this, we need to calculate the feeding distance of this feeder. Let us consider the following four options of using the feeder on the top of the casting (Figure 8.45).

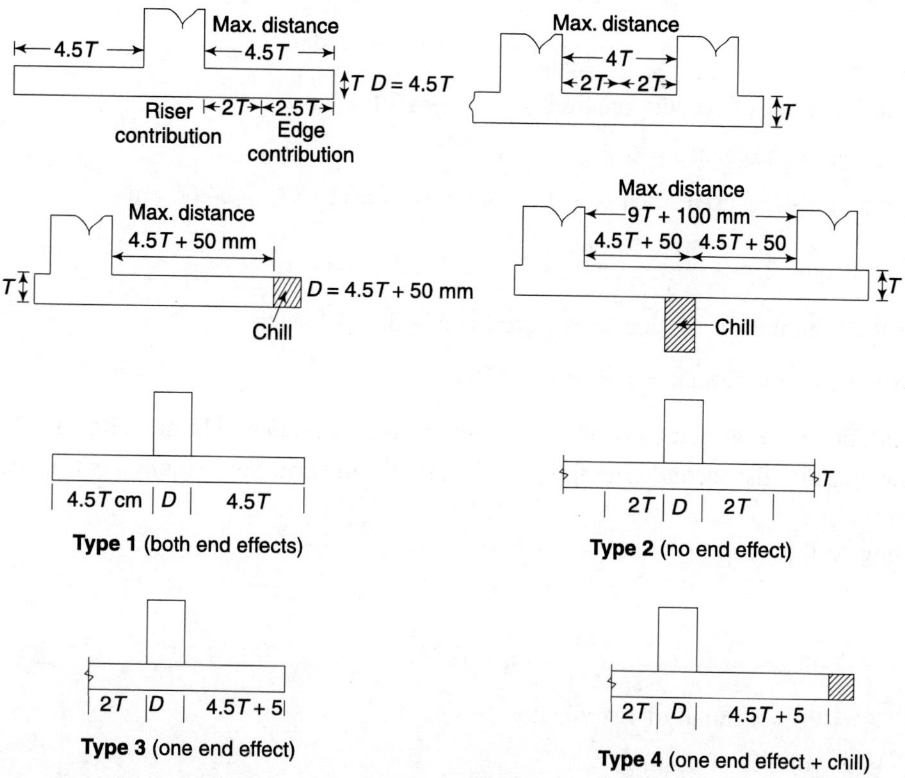

Figure 8.45

Then, the total feeding distance for each situation is as follows:

For type 1, $F_d = D + 9T = 9.33 + 9 \times 2 = 27.33$ cm

For type 2, $F_d = D + 4T = 9.33 + 4 \times 2 = 17.33$ cm

For type 3, $F_d = D + 6.5T = 9.33 + 6.5 \times 2 = 22.33$ cm

For type 4, $F_d = D + 6.5T = 9.33 + 6.5 \times 2 + 5 \text{ cm} = 27.33$ cm

Thus, if a feeder is placed at the centre (Type 1), it will feed the whole casting along the Y direction. Along the X direction, a multiple number of feeders is required (Figure 8.46).

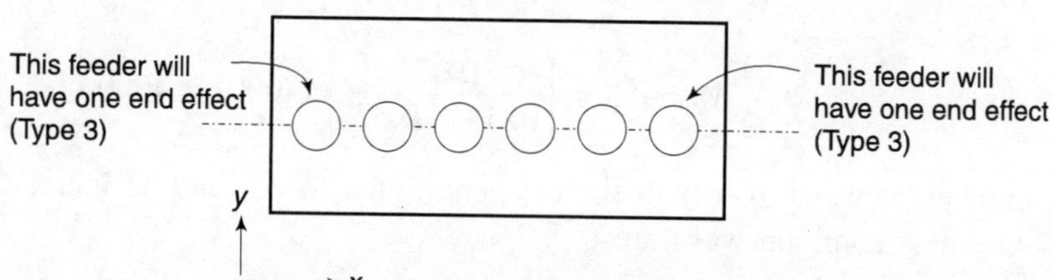

Figure 8.46

The two extreme end feeders (Type 3) will cover a distance of 2 (22.33) = 44.66 cm

Distance remaining to feed = 100 – 44.66 = 55.34 cm

Within this distance, the feeders to be placed will have no end effect (Type 2)

Thus, number of type 2 feeders required within this distance of 55.34 cm

$$= \frac{55.34}{17.33} \text{ cm} = 3.19 \cong 4 \text{ feeders}$$

Thus, total number of feeders required = 2 + 4 = 6 feeders

Total volume of the feeder = 6 × 957 = 5742 cc

If two end chills are used, then 2 × Type 4, F_d = 2 × (27.33) = 54.66 cm

$$\text{Numbers of type 2} = \frac{(100 - 54.66)}{17.33} = 2.62 \text{ cm} \cong 3 \text{ feeders}$$

So, the total number of the feeder required = 2 + 3 = 5 feeders

Total volume of the feeder = 5 × 957 = 4785 cc

Step 5: Calculate the dimensions of the feeder based on feeder volume criteria

Let us now consider the second criterion, that is, the volume criterion, requirement of the feeding design.

According to this rule,

$$\eta V_r = \alpha(V_c + V_r)$$

$$V_r = \left[\frac{\alpha}{\eta - \alpha}\right](V_c)$$

where, $V_r = V_F$ = Volume of riser/feeder

V_c = Volume of casting

η = Feeder efficiency

α = Thermal coefficient of expansion

For aluminium alloy casting,

$$\alpha = 7\%, = 0.07 \text{ and } \eta = 0.14$$

$$V_r = \left[\frac{\alpha}{\eta - \alpha}\right](V_c)$$

$$V_r = V_F = \left[\frac{\alpha}{\eta - \alpha}\right](V_c)$$

$$V_r = V_F = \left[\frac{0.07}{(0.14 - 0.07)}\right](4000) = 4000 \text{ cc}$$

There are a number of ways by which the efficiency of a feeder can be increased (Figure 8.47). Some of the most common ways are:

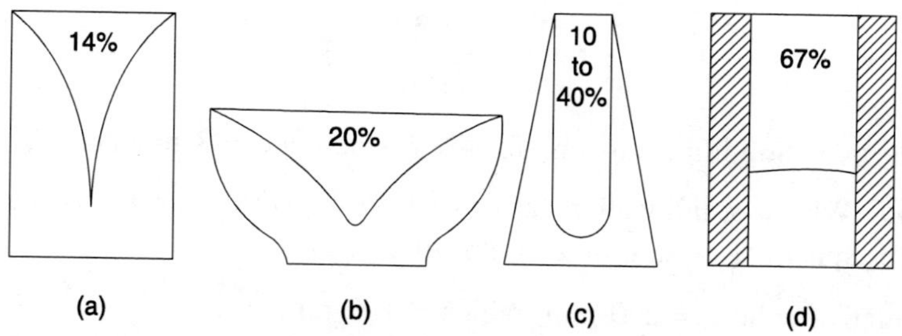

Figure 8.47 Metal utilisation of feeder of various forms moulded in sand, (a) cylindrical and (b) hemispherical head have been treated with normal feeding compounds; (c) the efficiency of the reverse tapered heads depends on detailed geometry, and (d) shows an exothermic sleeve.

- By using feeder head with a higher feeding efficiency
- Feeder insulation
- Use of exothermic materials.

If an insulating sleeve is used, the efficiency of feeder is 67%. Then,

$$V_F = \frac{(0.07 \times 4000)}{(0.67 - 0.07)} = 467 \text{ cc}$$

Comparing these two criteria, the volume calculated by junction criterion is greater than that calculated by volume criterion. Thus, the former will apply.

Hence, the volume of feeder will be 5742 CC, and its dimensions

$D = 9.33$ cm and

$H = 14.0$ cm

Example 8.15: Two cubical castings of the same metal and sizes of 2 cm side and 4 cm side are moulded in greensand. If the smaller casting solidifies in 2 minutes. What is the expected time of solidification of larger casting? **(GATE 1989)**

Solution: We know that solidification time is proportional to modulus (M)

$$\text{Modulus, } M = \frac{\text{Volume}}{\text{Surface area}} = \frac{a^3}{6a^2} = \frac{a}{6}$$

$$M_1 \text{ for 2 cm side } = \frac{a}{6} = \frac{2}{6} = \frac{1}{3}$$

$$M_2 \text{ for 4 cm side } = \frac{a}{6} = \frac{4}{6} = \frac{2}{3}$$

t_{s1} = Time of solidification for 2 cm side cube = 2 minutes

t_{s2} = Time of solidification for 4 cm side cube

$$\frac{2}{t_{s2}} = \frac{M_1^2}{M_2^2} = \frac{(1/3)^2}{(2/3)^2} = \frac{1}{4}$$

Solving we get, t_{s2} = Solidification time for 4 cm side cube = 8 minutes.

Example 8.16: With a solidification factor of 0.97×10^6 s/m^2, what is the solidification time in seconds for spherical casting of 200 mm diameter? **(GATE 2003)**

Solution: Diameter of sphere = 200 mm, radius = 100 mm = 0.1 m

We know that solidification time is proportional to modulus (M)

$$\text{Modulus, } M = \frac{\text{Volume}}{\text{Surface area}}$$

$$\text{Volume of sphere} = \frac{4\pi r^3}{3}$$

$$\text{Surface area of sphere} = 4\pi r^2$$

$$\text{Modulus, } M = \frac{\text{Volume}}{\text{Surface area}} = \frac{r}{3}$$

$$t_s \propto M^2$$

$$t_s = (\text{constant})\, M^2$$

$$t_s = (0.97 \times 10^6)\left(\frac{s}{m^2}\right) \times \left(\frac{0.1}{3}\right)^2 = 1077.78 \text{ seconds} \cong 1078 \text{ seconds}$$

Example 8.17: Volume of a cube of side "l" and volume of spheres of radius r are equal. Both the cube and spheres are solid and made of same material. They are being cast. What is the ratio of the solidification time of cube to that of sphere? **(GATE 2007)**

Solution: We know that solidification time is proportional to modulus (M)

$$\text{Modulus, } M = \frac{\text{Volume}}{\text{Surface area}}$$

$$M_1 \text{ for cube} = \frac{V}{6l^2}$$

$$M_2 \text{ for sphere} = \frac{V}{4\pi r^2}$$

t_{s1} = Time of solidification for cube

t_{s2} = Time of solidification for sphere

$$\frac{t_{s1}}{t_{s2}} = \frac{M_1^2}{M_2^2} = \frac{\left(\dfrac{V}{6l^2}\right)^2}{\left(\dfrac{V}{4\pi r^2}\right)^2} = \frac{(4\pi r^2)^2}{(6l^2)^2} = \frac{(4\pi)^2 r^4}{(6)^2 l^4} = 4.386 \times \left(\frac{r}{l}\right)^4$$

Example 8.18: For the case of moulding, it is decided to replace a spherical riser of diameter 100 mm by a cylindrical riser. Determine the size of cylindrical riser that will have identical solidification time. Assume blind risering and ratio of the height to radius of cylinder is 2.

(GATE 2010)

Solution: We know that solidification time is proportional to modulus (M)

$$\frac{\text{Height}}{\text{Radius}} = \frac{h}{r} = 2$$

$$\text{Height} = 2 \times \text{radius } (r) = \text{diameter } (d)$$

Since the spherical riser is replaced by the cylindrical riser, solidification time for both must be same.

$$\text{Modulus, } M = \frac{\text{Volume}}{\text{Surface area}}$$

$$\text{Surface area of the cylinder} = \pi dh + 2 \times \frac{\pi d^2}{4} = \frac{3\pi d^2}{2}$$

$$\text{Volume of the cylinder} = \frac{\pi d^2 h}{4} = \frac{\pi d^3}{4}$$

$$M_1 \text{ for cylinder} = \frac{\frac{\pi d^3}{4}}{\frac{3\pi d^2}{2}} = \frac{\pi d^3}{4} \times \frac{2}{3\pi d^2} = \frac{d}{6}$$

$$M_2 \text{ for sphere} = \frac{\frac{4\pi r^3}{3}}{4\pi r^2} = \frac{4\pi r^3}{3} \times \frac{1}{4\pi r^2} = \frac{r}{3} = \frac{d}{6}$$

t_{s1} = Time of solidification for cube

t_{s2} = Time of solidification for sphere

$$\frac{t_{s1}}{t_{s2}} = \frac{M_1^2}{M_2^2} = \frac{\left(\frac{d}{6}\right)^2_{\text{cylinder}}}{\left(\frac{d}{6}\right)^2_{\text{sphere}}} = 1$$

Therefore, diameter of cylinder = diameter of sphere = 100 mm

Example 8.19: A 10 mm thick steel bar is to be horizontally cast with two correctly spaced top risers of adequate feeding capacity. Assuming end effect without chill, what should be the theoretical length of the bar?

(GATE 2002-P1)

Solution: Length of the casting plate with given two risers without considering the wall effect and without the use of chills is at least $13t$, therefore,

Theoretical length of the bar = $13t$ = 130 mm

Example 8.20: A casting of size 100 mm × 100 mm × 50 mm is required. Assume the volume shrinkage of casting as 2.6%. If the height of the riser is 80 mm and riser volume desired is four times the shrinkage in casting. What is the appropriate riser diameter in mm?

(GATE 2003)

Volume of the casting = 100 mm × 100 mm × 50 mm = 5×10^5 mm^3

Surface area of the casting = 100 × 100 × 2 + 100 × 50 × 4 = 40000 mm^3

Volume shrinkage of casting is 2.6%, therefore amount of shrinkage is

$$\text{Shrinkage volume} = \frac{2.6}{100} \times 5 \times 10^5 \text{ mm}^3 = 13 \times 10^3 \text{ mm}^3$$

It is given that riser volume is four times the shrinkage volume

Riser volume = $4 \times 13 \times 10^3$ mm^3 = 52×10^3 mm^3

Height, h = 80 mm

$$\text{Riser volume} = \frac{\pi d^2 h}{4} = 52 \times 10^3 \text{ mm}^3$$

Solving we get, d = 28.769 mm ≅ 29 mm.

Additional check also to be made. Riser must be the last to solidify, applying this condition we get,

t_{s1} = Time of solidification for riser

t_{s2} = Time of solidification for casting

$$\text{Surface area of the cylinder} = \pi dh + 2 \times \frac{\pi d^2}{4} = 80\pi d + \frac{\pi d^2}{2}$$

$$\text{Volume of the cylinder} = \frac{\pi d^2 \times 80}{4} = 20\pi d^2$$

$$M_1 \text{ for casting} = \frac{\text{Volume}}{\text{Surface area}} = \frac{5 \times 10^5}{40000} = 12.5$$

$$M_2 \text{ for riser} = \frac{\text{Volume}}{\text{Surface area}} = \frac{20\pi d^2}{80\pi d + \frac{\pi d^2}{2}} = \frac{d}{4 + \frac{d}{40}} = \frac{d}{\frac{160 + d}{40}} = \frac{40d}{160 + d}$$

$$t_{s1} \geq t_{s2}$$

$$M_1 \geq M_2$$

$$\frac{40d}{160 + d} \geq 12.5$$

Solving we get, d = 72.72 mm

Example 8.21: What is the shape factor for a casting in the form of an annular cylinder of outside diameter 30 cm, inside diameter 20 cm and height 30 cm. Take correction factor = 1.

(GATE 2005-P1)

Solution:

$$\text{Shape factor} = \frac{\text{Length + width}}{\text{Thickness}}$$

$$\text{Shape factor} = \frac{\text{Height of cylinder} + \pi \times \text{averarge diameter}}{\text{Thickness of shell}}$$

$$\text{Shape factor} = \frac{30 + \pi \times \dfrac{30 + 20}{2}}{\dfrac{30 - 20}{2}} = 21.7$$

Example 8.22: A cast steel slab of dimensions 30 × 20 × 5 cm is poured horizontally using a side riser. The riser is cylindrical with diameter and height, both equal to D. What is the freezing ratio of the mould? **(GATE 2005-P1)**

Solution: Volume of the casting = 30 × 20 × 5 = 3000 cm^3

Surface area of the casting = 30 × 20 × 2 + 20 × 5 × 2 + 30 × 5 × 2 = 1700 cm^3

$$M_1 \text{ for casting} = \frac{3000}{1700} = 1.764$$

$$\text{Surface area of the cylinder} = \pi dh + 2 \times \frac{\pi d^2}{4} = \frac{3\pi d^2}{2}$$

$$\text{Volume of the cylinder} = \frac{\pi d^2 h}{4} = \frac{\pi d^3}{4}$$

$$M_2 \text{ for riser} = \frac{\dfrac{\pi d^3}{4}}{\dfrac{3\pi d^2}{2}} = \frac{\pi d^3}{4} \times \frac{2}{3\pi d^2} = \frac{d}{6}$$

$$\text{Freezing ratio} = \frac{M_2}{M_1} = \frac{\dfrac{d}{6}}{1.764} = 0.09448\text{d}$$

Example 8.23: A casting of size 400 mm × 200 mm × 140 mm solidifies in 20 minutes. What is the solidification time for a casting 400 mm × 200 mm × 35 mm under similar casting conditions? **(GATE 2006-P1)**

Solution: Volume of the casting, V_1 = 400 mm × 200 mm × 140 mm = 11.2 × 10^6 mm^3

Surface area of the castings,

$$S_1 = 400 \times 200 \times 2 + 200 \times 140 \times 2 + 140 \times 400 \times 2 = 328000 \text{ mm}^2$$

Volume of the casting, V_2 = 400 mm × 200 mm × 35 mm = 2.8 × 10^6 mm^3

Surface area of the casting,

$$S_2 = 400 \times 200 \times 2 + 200 \times 35 \times 2 + 35 \times 400 \times 2 = 202000 \text{ mm}^2$$

$$M_1 \text{ for casting } 1 = \frac{11.2 \times 10^6 \text{ mm}^3}{328000 \text{ mm}^2} = 34.146$$

$$M_2 \text{ for casting } 2 = \frac{2.8 \times 10^6 \text{ mm}^3}{202000 \text{ mm}^2} = 13.861$$

t_{s1} = Time of solidification for casting 1

t_{s2} = Time of solidification for casting 2

$$\frac{20}{t_{s2}} = \frac{M_1^2}{M_2^2} = \frac{(34.146)_1^2}{(13.861)_2^2}$$

Solving we get, t_{s2} = 3.29 min

Example 8.24: A solid cylinder of diameter D and height equal to D and a solid cube of side L are being sand cast by using the same material. Assuming there is no super heat in both cases, what is the ratio of solidification time of the cylinder to solidification time of cube? **(GATE 2009-P1)**

Solution: For cylinder, diameter = height = D

For cube, side length = L

$$\text{Modulus for cube, } M_1 = \frac{\text{Volume}}{\text{Surface area}} = \frac{L^3}{6L^2} = \frac{L}{6}$$

$$\text{Surface area of the cylinder} = \pi dh + 2 \times \frac{\pi d^2}{4} = \frac{3\pi d^2}{2}$$

$$\text{Volume of the cylinder} = \frac{\pi d^2 h}{4} = \frac{\pi d^3}{4}$$

$$M_2 \text{ for cylinder} = \frac{\frac{\pi d^3}{4}}{\frac{3\pi d^2}{2}} = \frac{\pi d^3}{4} \times \frac{2}{3\pi d^2} = \frac{d}{6}$$

t_{s1} = Time of solidification for cylinder

t_{s2} = Time of solidification for cube

The ratio of solidification of cylinder to cube is given by

$$\frac{t_{s1}}{t_{s2}} = \frac{M_1^2}{M_2^2} = \frac{\left(\frac{d}{6}\right)^2_{\text{cylinder}}}{\left(\frac{L}{6}\right)^2_{\text{cube}}} = \left(\frac{d}{L}\right)^2$$

Example 8.25: In a sand casting process, a sphere and a cylinder of equal volumes are separately cast from the same molten metal under identical conditions. The height and diameters of the cylinder are equal. What is the ratio of the solidification time of the sphere to that of the cylinder? **(GATE 2011-P1)**

Solution: Volume of the sphere = Volume of the cylinder

$$\frac{4\pi r_s^3}{3} = 2\pi r_c^3$$

where r_c and r_s are radius of cylinder and sphere respectively.

$$\frac{4\pi r_s^3}{3} = 2\pi r_c^3$$

$$r_s = 1.1446\ r_c$$

$$M_2 \text{ for cylinder} = \frac{\frac{\pi d^3}{4}}{\frac{3\pi d^2}{2}} = \frac{\pi d^3}{4} \times \frac{2}{3\pi d^2} = \frac{d}{6} = \frac{r_c}{3}$$

$$M_1 \text{ for sphere} = \frac{\frac{4\pi r^3}{3}}{4\pi r^2} = \frac{4\pi r^3}{3} \times \frac{1}{4\pi r^2} = \frac{r}{3} = \frac{d}{6} = \frac{r_s}{3} = \frac{1.1446 \times r_c}{3} = 0.38151 r_c$$

t_{s1} = Time of solidification for sphere

t_{s2} = Time of solidification for cylinder

$$\frac{ts_1}{ts_2} = \frac{M_1^2}{M_2^2} = \frac{(0.38151 r_c)^2_{\text{sphere}}}{\left(\frac{r_c}{3}\right)^2_{\text{cylinder}}} = 1.3098 \cong 1.31$$

Example 8.26: A cylindrical blind riser with diameter '*d*' and height '*h*' is placed on the top of mould cavity in a closed type sand mould as shown in figure below. If the riser is of constant volume, then for what ratio of *h*:*d*, the rate of solidification in the riser is least?

(GATE 2014 ME-Set 3)

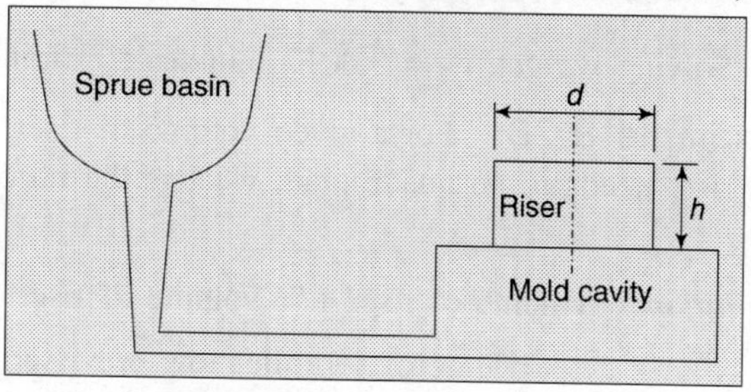

Solution: The given casting is a top riser, hence for top riser the optimum relationship between diameter and height of the riser is $h = d/2$ or $d = 2h$.

$$\text{Volume of the cylinder, } V = \frac{\pi d^2 h}{4}$$

$$h = \frac{4V}{\pi d^2}$$

$$\text{Surface area of the riser exposed for heat transfer, } A = \pi dh + \frac{\pi d^2}{4}$$

$$A = \pi dh + \frac{\pi d^2}{4} = \pi d\left(\frac{4V}{\pi d^2}\right) + \frac{\pi d^2}{4}$$

$$A = \frac{4V}{d} + \frac{\pi d^2}{4}$$

For A to be minimum, $\frac{\partial A}{\partial d} = 0 = -\frac{4V}{d^2} + \frac{\pi \times 2d}{4}$

This gives $d = 2h$

Thus, if the riser is of constant volume, then for ratio of $h{:}d = 1{:}2$ the rate of solidification in the riser is least.

Example 8.27: A cylindrical riser of 6 cm diameter and 6 cm height has to be designed for a sand casting mould for producing a steel rectangular plate casing of 7 cm × 10 cm × 2 cm dimensions, having the solidification time of 1.36 minutes. Determine the total solidification time in minutes for the riser. **(GATE 2014-Set 4)**

Solution: Volume of the casting, $V_c = 7 \text{ cm} \times 10 \text{ cm} \times 2 \text{ cm} = 140 \text{ cm}^3$

Surface area of the casting, $S_c = 7 \times 10 \times 2 + 10 \times 2 \times 2 + 2 \times 7 \times 2 = 208 \text{ cm}^2$

$$M_1 \text{ for rectangular casting} = \frac{140}{208} = 0.6731$$

$$\text{Surface area of the cylinder} = \pi dh + 2 \times \frac{\pi d^2}{4} = \frac{3\pi d^2}{2}$$

$$\text{Volume of the cylinder} = \frac{\pi d^2 h}{4} = \frac{\pi d^3}{4}$$

$$M_2 \text{ for riser} = \frac{\frac{\pi d^3}{4}}{\frac{3\pi d^2}{2}} = \frac{\pi d^3}{4} \times \frac{2}{3\pi d^2} = \frac{d}{6} = \frac{6}{6} = 1$$

t_{s1} = Time of solidification for rectangular casting = 1.36

t_{s2} = Time of solidification for riser

$$\frac{1.36}{t_{s2}} = \frac{M_1^2}{M_2^2} = \frac{(0.6731)_1^2}{(1)_2^2}$$

Solving we get, $t_{s2} = 3.0017 \text{ minutes} \cong 3 \text{ minutes}$

Example 8.28: For a given volume of riser, if the solidification time of the molten metal in riser needs to be quadrupled, what is the ratio of surface area of the riser? **(GATE 2014-P1)**

Solution: Volume of riser 1 = Volume of riser 2 = V

t_{s1} = Time of solidification for riser 1

t_{s2} = Time of solidification for riser 2

$t_{s2} = 4t_{s1}$

$$\text{Modulus for riser 1, } M_1 = \frac{\text{Volume}}{\text{Surface area}} = \frac{V_1}{A_1} = \frac{V}{A_1}$$

$$\text{Modulus for riser 2, } M_2 = \frac{\text{Volume}}{\text{Surface area}} = \frac{V_2}{A_2} = \frac{V}{A_2}$$

$$\frac{t_{s1}}{t_{s2}} = \frac{M_1^2}{M_2^2}$$

$$\frac{t_{s1}}{4t_{s1}} = \frac{\left(\frac{V}{A_1}\right)_1^2}{\left(\frac{V}{A_2}\right)_2^2}$$

which gives $$A_1 = 2A_2$$

Example 8.29: Solidification time of the casting is proportional to $\left(\frac{V}{A}\right)^2$, where V is the volume of the casting and A is the total casting surface area losing heat. Two cubes of same material and size are cast using sand casting process. The top face of one of the cubes is completely insulated. What is the ratio of solidification time for the cube with the top face insulated to that of the other cube?

(GATE 2015-set 1)

Volume of riser 1 = Volume of riser 2 = V

$$t_s \propto \left(\frac{V}{A}\right)^2$$

Let cube 1 have all its faces exposed to heat transfer

Let L be the length of the cube

Surface area, $A_1 = 6L^2$

Cube 2 has one face out of its six faces insulated

Surface area, $A_2 = 5L^2$

t_{s1} = Time of solidification for cube 1

t_{s2} = Time of solidification for cube 2

$$\text{Modulus for cube 1, } M_1 = \frac{\text{Volume}}{\text{Surface area}} = \frac{V_1}{A_1} = \frac{V}{6L^2}$$

$$\text{Modulus for cube 2, } M_2 = \frac{\text{Volume}}{\text{Surface area}} = \frac{V_2}{A_2} = \frac{V}{5L^2}$$

$$\frac{t_{s1}}{ts_2} = \frac{M_1^2}{M_2^2}$$

$$\frac{t_{s1}}{t_{s2}} = \frac{\left(\frac{V}{6L^2}\right)_1^2}{\left(\frac{V}{5L_2}\right)_2^2}$$

which gives $$t_{s2} = 1.44\ t_{s1}$$

Example 8.30: A cube and a sphere is made of cast iron, having volume of 1000 cm^3. They were cast under identical conditions. The time taken for solidification of cube was four seconds. What is the time of solidification for sphere?

(GATE 2015-Set 2)

Solution: Volume of cube = Volume of sphere = 1000 cm^3

Time taken for solidification of cube = 4 seconds

Volume of the sphere = Volume of the cube

$$\frac{4\pi r_s^3}{3} = L^3$$

where r_s and L are radius of sphere and length of cube respectively.

$$\frac{4\pi r_s^3}{3} = L^3$$

$$r_s = 0.6204\ L$$

$$M_2 \text{ for cube } = \frac{L^3}{6L^2} = \frac{L}{6}$$

$$M_1 \text{ for sphere } = \frac{\frac{4\pi r^3}{3}}{4\pi r^2} = \frac{4\pi r^3}{3} \times \frac{1}{4\pi r^2} = \frac{r}{3} = \frac{d}{6} = \frac{r_s}{3} = \frac{0.6204L}{3}$$

t_{s1} = Time of solidification for sphere

t_{s2} = Time of solidification for cube

$$\frac{t_{s1}}{t_{s2}} = \frac{M_1^2}{M_2^2} = \frac{\left(\frac{0.6204L}{3}\right)^2_{\text{sphere}}}{\left(\frac{L}{6}\right)^2_{\text{cube}}} = \frac{0.04276}{0.02777} = 1.53936 \cong 1.54$$

$$\frac{t_{s1}}{4} = 1.54$$

which gives $\quad t_{s2} = 1.54 \times 4 = 6.16$ seconds

Example 8.31: The dimensions of a cylindrical side riser (height = diameter) for a 25 cm × 15 cm × 5 cm steel casting are to be determined. For the tabulated shape factor given below, what is the diameter of riser in cm?

Shape factor	2	4	6	8	10	12
Riser volume/casting volume	1	0.7	0.55	0.5	0.4	0.35

Solution: Side riser with height = Diameter

$$\text{Shape factor} = \frac{\text{Length + width}}{\text{Thickness}}$$

$$\text{Shape factor} = \frac{25 + 15}{5} = 8$$

Referring to the table for shape factor = 8, we get

$$\frac{\text{Riser volume}}{\text{Casting volume}} = 0.5$$

Riser volume = 0.5 × casting volume

Riser volume = 0.5 × (25 × 15 × 5) = 937.5 cm^3

Riser volume = $\frac{\pi d^2 h}{4} = \frac{\pi d^3}{4}$ = 937.5 cm^3

Solving we get, diameter = height = 10.6 cm

Example 8.32: What is the ratio of solidification time of a cylindrical casting (height = radius) to that of cubic casting of side two times the height of cylindrical casting?

(GATE 2015-Set 3)

Solution: t_{s1} = Time of solidification for sphere

t_{s2} = Time of solidification for cylinder

$L = 2h = 2r$

$$\text{Surface area of the cylinder} = \pi dh + 2 \times \frac{\pi d^2}{4} = \pi d\frac{d}{2} + 2 \times \frac{\pi d^2}{4} = \pi d^2$$

$$\text{Volume of the cylinder} = \frac{\pi d^2 h}{4} = \frac{\pi d^2 \frac{d}{2}}{4} = \frac{\pi d^3}{8}$$

$$\text{Modulus for cylinder, } M_1 = \frac{\frac{\pi d^3}{8}}{\pi d^2} = \frac{\pi d^3}{8} \times \frac{1}{\pi d^2} = \frac{d}{8} = \frac{r}{4}$$

$$\text{Modulus for cube, } M_2 = \frac{\text{Volume}}{\text{Surface area}} = \frac{L^3}{6L^2} = \frac{L}{6} = \frac{2r}{6} = \frac{r}{3}$$

t_{s1} = Time of solidification for cylinder

t_{s2} = Time of solidification for cube

$$\frac{t_{s1}}{ts_2} = \frac{M_1^2}{M_2^2} = \frac{\left(\frac{r}{4}\right)^2_{\text{cylinder}}}{\left(\frac{r}{3}\right)^2_{\text{cube}}} = 0.5625$$

$$\frac{t_{s1}}{t_{s2}} = \frac{\text{Time of solidification for cylinder}}{\text{Time of solidification for cube}} = 0.5625$$

which gives t_{s2} = 1.54 × 4 = 6.16 seconds

Example 8.33: A cylindrical job with diameter 200 mm and height 100 mm is to be cast using the modulus method of riser design. Assume that the bottom surface of cylindrical riser does not contribute to the cooling surface. If the diameter of the riser is equal to its height, what is the height of the riser in mm? **(GATE 2016-Set 1)**

Solution: According to the modulus method,

$$M_{\text{riser}} = 1.2\ M_{\text{casting}}$$

$$\text{Surface area of the riser} = \pi dh + \frac{\pi d^2}{4} = \frac{5\pi d^2}{4}$$

$$\text{Volume of the riser} = \frac{\pi d^2 h}{4} = \frac{\pi d^3}{4}$$

$$\text{Modulus for riser} = \frac{\dfrac{\pi d^3}{4}}{\dfrac{3\pi d^2}{2}} = \frac{\pi d^3}{4} \times \frac{4}{5\pi d^2} = \frac{d}{5}$$

$$\text{Modulus for casting} = \frac{\dfrac{\pi \times 200^2 \times 100}{4}}{\pi \times 200 \times 100 + 2 \times \dfrac{\pi \times 200^2}{4}} = \frac{3141500}{62830 + 31415 \times 2} = 25$$

$$M_{\text{riser}} = 1.2\, M_{\text{casting}}$$

$$\frac{d}{5} = 1.2 \times 25$$

Solving, $d = 150$ mm

Example 8.34: Two cast iron blocks *P* and *Q*, each of 500 mm length, have the same cross-sectional area. Block *P* has rectangular cross section of 100 mm × 200 mm. Block *Q* is of square cross section. Both *P* and *Q* were cast under same conditions with all their surfaces enclosed within the mould. The solidification time of a casting is proportional to squares of ratio of its volume to its surface area. What is the ratio of solidification of block *P* to that of block *Q*? **(GATE 2016-P1)**

Solution: Length of the block is same = 500 mm

Cross-sectional area is same

$$L^2 = 100 \times 200$$

$$L = 141.42 \text{ mm}$$

Block *P* has rectangular cross section.

Volume of the casting, $V_p = 500 \times 200 \times 100 = 10000000 \text{ mm}^3$

Surface area of the casting,

$$S_p = 100 \times 200 \times 2 + 200 \times 500 \times 2 + 2 \times 500 \times 100 = 340000 \text{ mm}^2$$

$$M_p \text{ for rectangular casting} = \frac{10000000}{340000} = 29.4117$$

Block *Q* has square cross section.

Volume of the casting, $V_q = 141.42 \times 141.42 \times 500 = 9999808.2 \text{ mm}^3$

Surface area of the square casting,

$$S_q = 141.42 \times 500 \times 4 + 141.42 \times 141.42 \times 2 = 322839.23 \text{ mm}^2$$

$$M_q \text{ for square casting} = \frac{9999808.2}{322839.23} = 30.9746$$

t_p = Time of solidification for block P

t_q = Time of solidification for block Q

$$\frac{t_p}{t_q} = \frac{M_p^2}{M_q^2} = \frac{(29.4117)_p^2}{((30.9746)_q^2} = 0.9016$$

Example 8.35: Calculate the ratio of the solidification times of two steel cylindrical risers of sizes 30 cm in diameter and 60 cm in height, and 60 cm in diameter by 30 cm in height subjected to identical conditions of cooling. **(GATE 1992)**

Solution:

Riser 1

$$\text{Surface area of the riser 1} = \pi dh + 2\frac{\pi d^2}{4} = \pi \times 30 \times 60 + 2\,\frac{\pi \times 30^2}{4} = 7068.375 \text{ cm}^2$$

$$\text{Volume of the riser 1} = \frac{\pi d^2 h}{4} = \frac{\pi \times 30^2 \times 60}{4} = 42410.25 \text{ cm}^3$$

$$\text{Modulus for riser 1} = \frac{42410.25}{7068.375} = 6$$

Riser 2

$$\text{Surface area of the riser 2} = \pi dh + 2\,\frac{\pi d^2}{4} = \pi \times 60 \times 30 + 2\,\frac{\pi \times 60^2}{4} = 11309.4 \text{ cm}^2$$

$$\text{Volume of the riser 2} = \frac{\pi d^2 h}{4} = \frac{\pi \times 60^2 \times 30}{4} = 84820.5 \text{ cm}^3$$

$$\text{Modulus for riser 2} = \frac{84820.5}{11309.4} = 7.5$$

t_1 = Time of solidification for riser 1

t_2 = Time of solidification for riser 2

$$\frac{t_1}{t_2} = \frac{M_1^2}{M_2^2} = \frac{(6)^2}{(7.5)^2} = 0.64$$

Example 8.36: An aluminium cube of 10 cm side has be cast along with a cylindrical riser of height equal to its diameter. The riser is not insulated on any surface. If the volume shrinkage of aluminium during solidification is 6%, calculate the shrinkage volume of cube on solidification and minimum size of the riser so that it can provide the shrinkage volume. **(GATE 1993)**

Solution: Volume of the casting = 10 cm × 10 cm × 10 cm = 1000 cm^3

Surface area of the casting = $6 \times 10^2 = 600 \text{ cm}^2$

Volume shrinkage of casting is 6%, therefore amount of shrinkage is

$$\text{Shrinkage volume} = \frac{6}{100} \times 1000 \text{ cm}^3 = 60 \text{ cm}^3$$

Let riser volume = three times the shrinkage volume

Riser volume = 3 × 60 = 180 cm^3

Height = h = diameter = d

$$\text{Riser volume} = \frac{\pi d^2 h}{4} = \frac{\pi d^3}{4} = 180 \text{ cm}^3$$

Solving we get, d = 6.2 cm

Additional check also to be made. Riser must be the last to solidify, applying this condition we get,

t_{s1} = Time of solidification for riser

t_{s2} = Time of solidification for casting

$$\text{Surface area of the riser} = \pi dh + 2 \times \frac{\pi d^2}{4} = \frac{3\pi d^2}{2}$$

$$\text{Volume of the riser} = \frac{\pi d^2 \times h}{4} = \frac{\pi d^3}{4}$$

$$M_1 \text{ for riser} = \frac{\text{Volume}}{\text{Surface area}} = \frac{\frac{\pi d^3}{4}}{\frac{3\pi d^2}{2}} = \frac{d}{6}$$

$$M_2 \text{ for casting} = \frac{\text{Volume}}{\text{Surface area}} = \frac{1000}{600} = \frac{10}{6}$$

$$t_{s1} \geq t_{s2}$$

$$M_1 \geq M_2$$

$$\frac{d}{6} \geq \frac{10}{6}$$

Solving we get, d = 10 cm

Hence, diameter = height of riser = 10 cm.

Example 8.37: Two castings of the same metal have the same surface area. One casting is in the form of sphere and other a cube. What is the ratio of solidification time for the sphere to cube? **(GATE 1998)**

Solution: Volume of sphere = volume of cube

$$\frac{4\pi r_s^3}{3} = L^3$$

where r_s and L are radius of sphere and length of cube respectively.

$$r_s = 0.6204\ L$$

$$M_2 \text{ for cube } = \frac{L^3}{6L^2} = \frac{L}{6}$$

$$M_1 \text{ for sphere } = \frac{\frac{4\pi r^3}{3}}{4\pi r^2} = \frac{4\pi r^3}{3} \times \frac{1}{4\pi r^2} = \frac{r}{3} = \frac{d}{6} = \frac{r_s}{3} = \frac{0.6204L}{3}$$

t_{s1} = Time of solidification for sphere

t_{s2} = Time of solidification for cube

$$\frac{t_{s1}}{t_{s2}} = \frac{M_1^2}{M_2^2} = \frac{\left(\frac{0.6204L}{3}\right)^2_{\text{sphere}}}{\left(\frac{L}{6}\right)^2_{\text{cube}}} = \frac{0.04276}{0.02777} = 1.53936 \cong 1.54$$

Example 8.38: Two castings, a cube and a slab of the same material solidify under identical conditions. The volume of the casting is equal but the slab dimensions are in the ratio of 1:2:4. Find the ratio of the solidification times of the cube to that of slab. **(GATE 2000)**

Solution: Volume of cube casting = Volume of slab

Volume of slab, $V_s = 1 \times 2 \times 4 = 8$ units

Surface area of slab, $A_s = 1 \times 2 \times 2 + 2 \times 4 \times 2 + 4 \times 1 \times 2 = 28$ units

Volume of the cube, $V_c = L^3$

$$L^3 = 8$$

$$L = 2 \text{ units}$$

$$M_1 \text{ for cube } = \frac{L}{6} = \frac{1}{3}$$

$$M_2 \text{ for slab } = \frac{\text{Volume}}{\text{Surface area}} = \frac{8}{28}$$

t_{s1} = Time of solidification for cube

t_{s2} = Time of solidification for slab

$$\frac{ts_1}{ts_2} = \frac{M_1^2}{M_2^2} = \frac{\left(\frac{1}{3}\right)^2}{\left(\frac{8}{28}\right)^2} = 1.36$$

Example 8.39: A steel plate 40 cm × 30 cm × 0.3 cm is to be cast. The volume shrinkage of steel during solidification is 3%. A cylindrical side riser with diameter, 4 cm and height, 4 cm is used. The riser volume should be at least three times that dictated by shrinkage considerations. Is the riser volume sufficient, if not what should be the riser size? **(GATE 1990)**

Solution: Volume of the casting = 40 cm × 30 cm × 0.3 cm = 360 cm^3

Surface area of the casting = 40 × 30 × 2 + 30 × 0.3 × 2 + 0.3 × 40 × 2 = 2442 cm^2

Volume shrinkage of casting is 6%, therefore amount of shrinkage is

$$\text{Shrinkage volume} = \frac{3}{100} \times 360 \text{ cm}^3 = 10.8 \text{ cm}^3$$

Let riser volume = three times the shrinkage volume

Riser volume = 3 × 10.8 = 32.4 cm^3

Height = h = diameter = d

$$\text{Riser volume} = \frac{\pi d^2 h}{4} = \frac{\pi d^3}{4} = 32.4 \text{ cm}^3$$

Solving we get, d = 4.5 cm

Additional check also to be made. Riser must be the last to solidify, applying this condition we get,

t_{s1} = Time of solidification for riser

t_{s2} = Time of solidification for casting

$$\text{Surface area of the riser} = \pi dh + 2 \times \frac{\pi d^2}{4} = \frac{3\pi d^2}{2}$$

$$\text{Volume of the riser} = \frac{\pi d^2 \times h}{4} = \frac{\pi d^3}{4}$$

$$M_1 \text{ for riser} = \frac{\text{Volume}}{\text{Surface area}} = \frac{\frac{\pi d^3}{4}}{\frac{3\pi d^2}{2}} = \frac{d}{6} = \frac{4}{6}$$

$$M_2 \text{ for casting} = \frac{\text{Volume}}{\text{Surface area}} = \frac{360}{2442} = 0.1474$$

$$t_{s1} \geq t_{s2}$$

$$M_1 \geq M_2$$

$$\frac{4}{6} \geq 0.1474$$

This condition is also satisfied.

Example 8.40: If the freezing ratio equals to (casting area/casting volume)/(riser area/riser volume) and volumetric ratio equals to riser volume/casting volume, for a disc casting of 20 cm diameter and 10 cm thick are 1.4 and 0.8 respectively. What are the corresponding values for a square disc of 20 cm side and 10 cm thickness. Same cylindrical riser with height to diameter ratio unity is used in both cases. **(GATE 1992 –P1)**

Solution:

	Circular disc		Square disc
Diameter (cm)	20	Side (cm)	20
Height (cm)	10	Thickness (cm)	10
Freezing ratio	1.4		
Volumetric ratio	0.8		

Same cylindrical riser (diameter = height) is used in both cases.

Circular disc

$$\text{Surface area of circular disc} = \pi dh + 2\frac{\pi d^2}{4} = \pi \times 20 \times 10 + 2\,\frac{\pi \times 20^2}{4} = 1256.6 \text{ cm}^2$$

$$\text{Volume of the circular disc} = \frac{\pi d^2 h}{4} = \frac{\pi \times 20^2 \times 10}{4} = 3141.5 \text{ cm}^3$$

$$\text{Freezing ratio} = \frac{\left(\dfrac{\text{Area}}{\text{Volume}}\right)_{\text{circular}}}{\left(\dfrac{\text{Area}}{\text{Volume}}\right)_{\text{riser}}} = \frac{\left(\dfrac{1256.6}{3141.5}\right)_{\text{circular}}}{\left(\dfrac{\text{Area}}{\text{Volume}}\right)_{\text{riser}}} = 1.4$$

Riser with $d = h$

$$\text{Surface area of the riser} = \pi dh + 2 \times \frac{\pi d^2}{4} = \frac{3\pi d^2}{2}$$

$$\text{Volume of the riser} = \frac{\pi d^2 \times h}{4} = \frac{\pi d^3}{4}$$

$$\frac{\text{Surface area}}{\text{Volume}} = \frac{\left(\dfrac{3\pi d^2}{2}\right)}{\left(\dfrac{\pi d^3}{4}\right)} = \frac{6}{d}$$

$$\text{Freezing ratio}_{\text{circular}} = \frac{\left(\dfrac{\text{Area}}{\text{Volume}}\right)_{\text{circular}}}{\left(\dfrac{\text{Area}}{\text{Volume}}\right)_{\text{riser}}} = \frac{\left(\dfrac{1256.6}{3141.5}\right)_{\text{circular}}}{\left(\dfrac{6}{d}\right)_{\text{riser}}} = 1.4$$

$$\frac{\left(\dfrac{1256.6}{3141.5}\right)_{\text{circular}}}{\left(\dfrac{6}{d}\right)_{\text{riser}}} = 1.4$$

Solving we get, $d = 21$ cm

Square disc

$$\text{Surface area of square disc} = 20 \times 20 \times 2 + 20 \times 10 \times 4 = 1600 \text{ cm}^2$$

Volume of the square disc = 20 × 20 × 10 = 4000 cm^3

$$\text{Freezing ratio}_{\text{square}} = \frac{\left(\frac{\text{Area}}{\text{Volume}}\right)_{\text{square}}}{\left(\frac{\text{Area}}{\text{Volume}}\right)_{\text{riser}}} = \frac{\left(\frac{1600}{4000}\right)_{\text{square}}}{(0.28571)_{\text{riser}}} = 1.4$$

$$\text{Volumetric ratio} = \frac{(\text{Volume})_{\text{riser}}}{(\text{Volume})_{\text{casting}}} = \frac{V_r}{V_c}$$

$$\text{Surface area of the riser} = \pi dh + 2 \times \frac{\pi d^2}{4} = \frac{3\pi d^2}{2} = \frac{3 \times \pi \times 21^2}{2} = 2078.102 \text{ cm}^2$$

$$\text{Volume of the riser} = \frac{\pi d^2 \times h}{4} = \frac{\pi d^3}{4} = 7273.36 \text{ cm}^3$$

$$\text{Volumetric ratio} = \frac{(\text{Volume})_{\text{riser}}}{(\text{Volume})_{\text{casting}}} = \frac{V_r}{V_c} = \frac{7273.36}{4000} = 1.81$$

Example 8.41: A casting of size 200 mm × 100 mm × 70 mm solidifies in ten minutes. Estimate the solidification time for 200 mm × 100 mm × 10 mm casting under similar conditions.

(GATE 1993-P1)

Solution:

Block 1

Volume of the casting, V_1 = 200 × 100 × 70 = 14 × 10^5 mm^3

Surface area of the casting,

$$S_1 = 200 \times 100 \times 2 + 100 \times 70 \times 2 + 70 \times 200 \times 2 = 82000 \text{ mm}^2$$

$$M_1 \text{ for rectangular casting} = \frac{14 \times 10^5}{82000} = 17.073$$

Block 2

Volume of the casting, V_2 = 200 × 100 × 10 = 2 × 10^5 mm^3

Surface area of the square casting,

$$S_2 = 200 \times 100 \times 2 + 100 \times 10 \times 2 + 10 \times 200 \times 2 = 46000 \text{ mm}^2$$

$$M_2 \text{ for square casting} = \frac{2 \times 10^5}{46000} = 4.3478$$

t_1 = Time of solidification for block 1

t_2 = Time of solidification for block 2

$$\frac{10}{t_2} = \frac{M_1^2}{M_2^2} = \frac{(17.073)^2}{(4.3478)^2} = 15.4196$$

Solving we get, $t_2 = 0.6485$ minute $\cong 0.65$ minute

Example 8.42: Three castings are of the same material, volume and cast under similar conditions. One is sphere, another is cube and the third is cylinder with its length being equal to its diameter. **(GATE 1995-P)**

(a) What is the ratio of solidification times of sphere to the cube?

(b) What is the ratio of solidification times of sphere to the cylinder?

Solution: Volume of sphere = Volume of cube

$$\frac{4\pi r_s^3}{3} = L^3$$

where r_s and L are radius of sphere and length of cube respectively.

$$r_s = 0.6204\ L$$

$$M_2 \text{ for cube } = \frac{L^3}{6L^2} = \frac{L}{6}$$

$$M_1 \text{ for sphere } = \frac{\frac{4\pi r^3}{3}}{4\pi r^2} = \frac{4\pi r^3}{3} \times \frac{1}{4\pi r^2} = \frac{r}{3} = \frac{d}{6} = \frac{r_s}{3} = \frac{0.6204L}{3}$$

ts_1 = Time of solidification for sphere

ts_2 = Time of solidification for cube

$$\frac{t_{s1}}{t_{s2}} = \frac{M_1^2}{M_2^2} = \frac{\left(\frac{0.6204L}{3}\right)^2_{\text{sphere}}}{\left(\frac{L}{6}\right)^2_{\text{cube}}} = \frac{0.04276}{0.02777} = 1.53936 \cong 1.54$$

Volume of the sphere = Volume of the cylinder

$$\frac{4\pi r_s^3}{3} = 2\pi r_c^3$$

where r_c and n_s are radius of cylinder and sphere respectively.

$$r_s = 1.1446\ r_c$$

$$M_2 \text{ for cylinder } = \frac{\left(\frac{\pi d^3}{4}\right)}{\frac{3\pi d^2}{2}} = \frac{\pi d^3}{4} \times \frac{2}{3\pi d^2} = \frac{d}{6} = \frac{r_c}{3}$$

$$M_1 \text{ for sphere } = \frac{\frac{4\pi r^3}{3}}{4\pi r^2} = \frac{4\pi r^3}{3} \times \frac{1}{4\pi r^2} = \frac{r}{3} = \frac{d}{6} = \frac{r_s}{3} = \frac{1.1446 \times r_c}{3} = 0.38151 r_c$$

t_{s1} = Time of solidification for sphere

t_{s2} = Time of solidification for cylinder

$$\frac{t_{s1}}{t_{s2}} = \frac{M_1^2}{M_2^2} = \frac{(0.38151 r_c)^2_{\text{sphere}}}{\left(\frac{r_c}{3}\right)^2_{\text{cylinder}}} = 1.3098 \cong 1.31$$

Example 8.43: In the casting of steel under certain mould conditions, the mould constant in Chvorinov's rule is known to be 4 min/cm^2, based on previous experience. The casting is a flat plate whose length = 30 cm, width = 10 cm, and thickness = 20 mm. Determine how long it will take for the casting to solidify.

Solution: Volume, $V = 30 \times 10 \times 2 = 600$ cm^3

Surface area, $A = 2(30 \times 10 + 30 \times 2 + 10 \times 2) = 760$ cm^2

According to Chvorinov's rule,

$$\text{Modulus, } M = \frac{\text{Volume}}{\text{Surface area}} = \frac{600}{760} = 0.7894$$

$$t_s \propto M^2$$

$$t_s = (\text{constant})\, M^2$$

$$t_s = (4 \text{ min/cm}^2) \times (0.7894)^2 = 2.4930 \cong 2.5 \text{ minutes}$$

Example 8.44: A disk-shaped part is to be cast out of aluminium. The diameter of the disk = 500 mm and its thickness = 20 mm. If the mould constant = 2 sec/mm^2 in Chvorinov's rule, how long will it take for the casting to solidify?

Solution:

$$\text{Surface area of circular disc} = \pi dh + 2\frac{\pi d^2}{4} = \pi \times 500 \times 20 + 2\frac{\pi \times 500^2}{4} = 424102 \text{ mm}^2$$

$$\text{Volume of the circular disc} = \frac{\pi d^2 h}{4} = \frac{\pi \times 500^2 \times 20}{4} = 3926875 \text{ mm}^3$$

$$\text{Modulus, } M = \frac{\text{Volume}}{\text{Surface area}} = \frac{3926875}{424102} = 9.2593$$

$$t_s \propto M^2$$

$$t_s = (\text{constant})\, M^2$$

$$t_s = (2 \text{ sec/mm}^2) \times (9.2593)^2 = 171.46 \cong 171.5 \text{ seconds} \cong 2.86 \text{ minutes}$$

Example 8.45: In casting experiments performed using a certain alloy and type of sand mould; it took 155 seconds for a cube-shaped casting to solidify. The cube was 50 mm on a side. (a) Determine the value of the mould constant in Chvorinov's rule. (b) If the same alloy and mould type were used, find the total solidification time for a cylindrical casting in which the diameter = 30 mm and length = 50 mm.

Solution: Volume, $V = 50^3 = 125{,}000 \text{ mm}^3$

Area, $A = 6 \times 50^2 = 15{,}000 \text{ mm}^2$

$$\text{Modulus, } M = \frac{\text{Volume}}{\text{Surface area}} = \frac{125{,}000}{15{,}000} = 8.3333$$

$t_s \propto M^2$

$t_s = (\text{constant})\, M^2$

$155 = (K) \times (8.3333)^2$

Solving we get, $k = 2.232 \text{ sec/mm}^2$

$$\text{Surface area} = \pi dh + 2\,\frac{\pi d^2}{4} = \pi \times 30 \times 50 + 2\,\frac{\pi \times 30^2}{4} = 6125.925 \text{ mm}^2$$

$$\text{Volume} = \frac{\pi d^2 h}{4} = \frac{\pi \times 30^2 \times 50}{4} = 35341.875 \text{ mm}^3$$

$$\text{Modulus, } M = \frac{\text{Volume}}{\text{Surface area}} = \frac{35341.875}{6125.925} = 5.769$$

$t_s \propto M^2$

$t_s = (\text{constant})\, M^2$

$t_s = (2.232) \times (5.769)^2 = 74.289 \cong 74.3 \text{ seconds} \cong 1.24 \text{ minutes}$

Example 8.46: A steel casting has a cylindrical geometry with 100 mm diameter and weighs 9 kg. This casting takes 6 minutes to completely solidify. Another cylindrical casting with the same diameter-to-length ratio weighs 5.4 kg. This casting is made of the same steel, and the same conditions of mould and pouring were used. Determine: (a) the mould constant in Chvorinov's rule, (b) the dimensions, and (c) the total solidification time of the lighter casting. The density of steel = 7850 kg/m^3.

Solution:

(a) For steel, $\rho = 7850 \text{ kg/m}^3$

$$\text{Weight, } W = \rho V,\ V = \frac{W}{\rho} = \frac{9}{7850} = 1.1464 \times 10^{-3} \text{ m}^3 = 1146496.82 \text{ mm}^3$$

$$\text{Volume} = \frac{\pi d^2 h}{4} = \frac{\pi \times 100^2 \times h}{4} = 1146496.82 \text{ mm}^3$$

Solving we get, $h = 145.98 \text{ mm} \cong 146 \text{ mm}$

$$\text{Surface area} = \pi dh + 2\,\frac{\pi d^2}{4} = \pi \times 100 \times 146 + 2\,\frac{\pi \times 100^2}{4} = 61573.4 \text{ mm}^2$$

$$\text{Modulus, } M = \frac{\text{Volume}}{\text{Surface area}} = \frac{1146496.82}{61573.4} = 18.62$$

$$t_s \propto M^2$$

$$t_s = (\text{constant})\ M^2$$

$$6 \times 60 = (K) \times (18.62)^2$$

Solving we get, $k = 1.0383$ sec/mm^2

(b) Find dimensions of smaller cylindrical casting with same D/h ratio and weight = 5.4 kg

Weight, $W = \rho V$, $V = \dfrac{W}{\rho} = \dfrac{5.4}{7850} = 6.8789 \times 10^{-3}\ \text{m}^3 = 687898.1\ \text{mm}^3$

Ratio $= \dfrac{d}{h} = \dfrac{100}{146} = 0.685$ or $h = 1.46d$

Volume $= \dfrac{\pi d^2 h}{4} = \dfrac{\pi \times d^2 \times 1.46d}{4} = 687898.1$

Solving we get, $d = 84.34$ mm and $h = 1.46 \times 84.34 = 123.13$ mm

Solidification time

Surface area $= \pi dh + 2\dfrac{\pi d^2}{4} = \pi \times 84.34 \times 123.13 + 2\dfrac{\pi \times 84.34^2}{4} = 43796.91\ \text{mm}^2$

Modulus, $M = \dfrac{\text{Volume}}{\text{Surface area}} = \dfrac{687898.1}{43796.91} = 15.70$

$$t_s \propto M^2$$

$$t_s = (\text{constant})\ M^2$$

$$t_s = (1.0383) \times (15.70)^2$$

Solving we get, $t_s = 256.144$ sec = 4.27 minutes

Example 8.47: The total solidification times of three casting shapes are to be compared: (1) a sphere with diameter 10 cm, (2) a cylinder with diameter and length both 10 cm, and (3) a cube with each side 10 cm. The same casting alloy is used in the three cases. (a) Determine the relative solidification times for each geometry. (b) Based on the results of part (a), which geometric element would make the best riser? (c) If the mould constant is 3.5 min/cm^2 in Chvorinov's rule, compute the total solidification time for each casting.

Solution: For ease of computation, make the substitution 10 cm = 1 decimetre (1 dm)

Part (a)

(1) Sphere

Volume, $V = \dfrac{\pi D^3}{6} = \dfrac{\pi 1^3}{6} = \dfrac{\pi 1^3}{6}\ \text{dm}^3$

Area, $A = \pi \times D^2 = \pi 1^2 = \pi\ \text{dm}^2$

Modulus, $M = \dfrac{\text{Volume}}{\text{Surface area}} = \dfrac{\frac{\pi}{6}}{\pi} = 0.1667$

$$t_s \propto M^2$$

$$t_s = (\text{constant})\ M^2$$

$$t_{s\text{-sphere}} = (K) \times (0.1667)^2$$

(2) Cylinder

$$\text{Volume, } V = \frac{\pi D^2 H}{4} = \frac{\pi \times 1^2 \times 1}{4} = \frac{\pi}{4}\ \text{dm}^3$$

$$\text{Area, } A = 2 \times \frac{\pi \times D^2}{4} + \pi D L = \frac{\pi \times 1^2}{2} + \pi \times 1 \times 1 = 1.5\ \pi\ \text{dm}^2$$

$$\text{Modulus, } M = \frac{\text{Volume}}{\text{Surface area}} = \frac{\frac{\pi}{4}}{1.5\pi} = 0.1667$$

$$t_s \propto M^2$$

$$t_s = (\text{constant})\ M^2$$

$$t_{s\text{-cylinder}} = (K) \times (0.1667)^2$$

(3) Cube

$$\text{Volume, } V = L^3 = 1\ \text{dm}^3$$

$$\text{Area, } A = 6L^2 = 6\ \text{dm}^2$$

$$\text{Modulus, } M = \frac{\text{Volume}}{\text{Surface area}} = \frac{1}{6} = 0.1667$$

$$t_s \propto M^2$$

$$t_s = (\text{constant})\ M^2$$

$$t_{s\text{-cube}} = (K) \times (0.1667)^2$$

Part (b)

All three shapes are equivalent as risers.

Part (c)

If $K = 3.5$ min/cm^2 = 350 min/dm^2

$t_{s\text{-cube}} = (350) \times (0.1667)^2 = 9.723$ minutes. Note, however, that the volumes of three geometries are different. Volume of sphere = 524 cm^3, volume of cylinder = 784.5 cm^3 and volume of cube = 1000 cm^3. Accordingly, we might revise our answer to **Part (b)** and choose the sphere on the basis that it wastes less metal than other shapes.

Example 8.48: The total solidification times of three casting shapes are to be compared: (1) a sphere, (2) a cylinder, in which the length-to-diameter ratio = 1.0, and (3) a cube. For all three geometries, the volume = 1000 cm^3. The same casting alloy is used in the three cases. (a) Determine the relative solidification times for each geometry. (b) Based on the results of part (a), which geometric element would make the best riser? (c) If the mould constant = 3.5 min/cm^2 in Chvorinov's rule, compute the total solidification time for each casting.

Solution: For ease of computation, make the substitution 10 cm = 1 decimeter (1 dm)

Part (a)

(1) Sphere

Sphere volume, $V = \dfrac{\pi D^3}{6} = 1.0$ dm^3, solving $D = 1.241$ dm

Sphere area, $A = \pi \times D^2 = \pi \times 1.241^2 = 4.836$ dm^2

$$\text{Modulus, } M = \frac{\text{Volume}}{\text{Surface area}} = \frac{1}{4.836} = 0.2067$$

$$t_s \propto M^2$$

$$t_s = (\text{constant})\ M^2$$

$$t_{\text{s-sphere}} = (K) \times (0.2067)^2 = 0.0428\ K$$

(2) Cylinder

Cylinder volume, $V = \dfrac{\pi D^2 H}{4} = 1$ dm^3, solving we get $D = H = 1.084$ dm

Cylinder area, $A = 2 \times \dfrac{\pi \times D^2}{4} + \pi DL = \dfrac{\pi \times 1.084^2}{2} + \pi \times 1.084 \times 1.084 = 5.536$ dm^2

$$\text{Modulus, } M = \frac{\text{Volume}}{\text{Surface area}} = \frac{1}{5.536} = 0.1806 \text{ dm}$$

$$t_s \propto M^2$$

$$t_s = (\text{constant})\ M^2$$

$$t_{\text{s-cylinder}} = (K) \times (0.1806)^2 = 0.0326\ \text{K}$$

(3) Cube

Volume, $V = L^3 = 1$ dm^3, solving we get $L = 1$ dm

Area, $A = 6L^2 = 6$ dm^2

$$\text{Modulus, } M = \frac{\text{Volume}}{\text{Surface area}} = \frac{1}{6} = 0.1667$$

$$t_s \propto M^2$$

$$t_s = (\text{constant})\ M^2$$

$$t_{\text{s-cube}} = (K) \times (0.1667)^2 = 0.02778\ \text{K}$$

Part (b)

Sphere would be the best riser, since V/A ratio is greatest.

Part (c)

If $K = 3.5\ \dfrac{\text{min}}{\text{cm}^2} = 350$ min/dm^2

$$t_{s\text{-cube}} = 0.02778 \times 350 = 9.72 \quad \text{minutes}$$

$$t_{s\text{-sphere}} = 0.0428 \times 350 = 14.98 \text{ minutes}$$

$$t_{s\text{-cylinder}} = 0.0326 \times 350 = 11.41 \text{ minutes}$$

Example 8.49: A cylindrical riser is to be used for a sand-casting mould. For a given cylinder volume, determine the diameter-to-length ratio that will maximize the time to solidify.

Solution: To maximize t_s, the V/A ratio must be maximized.

Cylinder volume, $V = \dfrac{\pi D^2 L}{4}$, solving we get $L = \dfrac{4V}{\pi D^2}$ ...(a)

Cylinder area, $A = 2 \times \dfrac{\pi \times D^2}{4} + \pi DL$ ...(b)

Using value of L from (a)

$$\text{Cylinder area, } A = 2 \times \frac{\pi \times D^2}{4} + \pi D\left(\frac{4V}{\pi D^2}\right) = \frac{\pi \times D^2}{2} + \left(\frac{4V}{D}\right)$$

Differentiate the area equation with respect to D:

$$\frac{dA}{dD} = \pi \times D - \left(\frac{4V}{D^2}\right) = 0$$

$$D = \left(\frac{4V}{\pi}\right)^{\frac{1}{3}} \qquad \text{...(c)}$$

From the previous expression for L, substituting in the equation for D that we have developed,

$$L = \frac{4V}{\pi D^2} = \frac{4V}{\pi\left(\left(\frac{4V}{\pi}\right)^{\frac{1}{3}}\right)^2} = \left(\frac{4V}{\pi}\right)^{\frac{1}{3}}$$

Thus, optimal values are

$$L = D = \left(\frac{4V}{\pi}\right)^{\frac{1}{3}}$$

And therefore, the optimal ratio $\dfrac{D}{L} = 1$

Example 8.50: A riser in the shape of a sphere is to be designed for a sand casting mould. The casting is a rectangular plate, with length = 200 mm, width = 100 mm, and thickness = 18 mm. If the total solidification time of the casting itself is known to be 3.5 minutes, determine the diameter of the riser so that it will take 25% longer for the riser to solidify.

Solution: Casting volume, $V = LWt = 200(100)(18) = 360{,}000 \text{ mm}^3$

Casting area, $A = 2(200 \times 100 + 200 \times 18 + 100 \times 18) = 50{,}800 \text{ mm}^2$

$$\text{Modulus, } M = \frac{\text{Volume}}{\text{Surface area}} = \frac{360{,}000}{50{,}800} = 7.0866$$

$t_s \propto M^2$

$t_s = (\text{constant})\ M^2$

$t_s = (K)\ 7.0866^2$

$$K = \frac{3.5}{7.0866^2} = 0.0697 \text{ min/mm}^2$$

$$\text{Sphere volume, } V = \frac{\pi D^3}{6} = 0.5236\ D^3$$

$$\text{Sphere area, } A = \pi \times D^2 = 3.1416\ D^2$$

$$\text{Modulus, } M = \frac{\text{Volume}}{\text{Surface area}} = \frac{0.5236\ D^3}{3.1416\ D^2} = 0.1667\ D$$

$t_s \propto M^2$

$t_s = (\text{constant})\ M^2$

$t_s = 1.25 \times 3.5 = 4.375$ minutes

$$4.375 \text{ minutes} = (0.0697) \times (0.1667D)^2$$

$$D = 47.5 \text{ mm}$$

Example 8.51: A cylinder with a diameter-to-height ratio of 1 solidifies in four minutes in a sand casting operation. What is the solidification time if the cylinder height is tripled? What is the time if the diameter is tripled?

Solution: Cylinder volume, $V = \dfrac{\pi D^2 L}{4}$

$$\text{Cylinder area, } A = 2 \times \frac{\pi \times D^2}{4} + \pi DL$$

$$\text{Modulus, } M = \frac{\text{Volume}}{\text{Surface area}} = \frac{\dfrac{\pi D^2 L}{4}}{2 \times \dfrac{\pi \times D^2}{4} + \pi DL} = \left(\frac{DL}{2D + 4L}\right)$$

$$t_s \propto M^2$$

$t_s = (\text{constant})\ M^2$

$t_s = (K)\left(\dfrac{DL}{2D + 4L}\right)^2 = 4$ minutes

Solving we get, $K = \dfrac{24}{D}$

If the height is tripled, then we can use $D_2 = D$ and $L_2 = 3L$ to obtain the solidification time for the new condition as:

$$t_s = (K)\left(\frac{D_2\ L_2}{2D_2 + 4L_2}\right)^2$$

$$t_s = \left(\frac{24}{D}\right)\left(\frac{D \times (3L)}{2D + 4(3L)}\right)^2$$

$$t_s = \left(\frac{24}{D}\right)\left(\frac{3D}{2\frac{D}{L} + 4(3)}\right)^2$$

Solving we get, t_s = 1 minute

If the diameter is tripled, so that $D_3 = D$ and $L_3 = 3L$

$$t_s = (K)\left(\frac{D_3L_3}{2D_3 + 4L_3}\right)^2$$

$$t_s = \left(\frac{24}{D}\right)\left(\frac{3D \times (L)}{2 \times 3D + 4(L)}\right)^2$$

$$t_s = \left(\frac{24}{D}\right)\left(\frac{3}{6\frac{D}{L} + 4}\right)^2$$

Solving we get, t_s = 2.16 minutes

Example 8.52: The optimum shape of a riser is spherical to ensure that it cools more slowly than the casting it feeds. Spherically shaped risers, however, are difficult to cast. (a) Sketch the shape of a blind riser that is easy to mould, but also has the smallest possible surface area-to-volume ratio. (b) Compare the solidification time of the riser in part (a) to that of a riser shaped like a right circular cylinder. Assume that the volume of each riser is the same, and that for each the height is equal to the diameter.

Solution:

(a) A sketch of a blind riser that is easy to cast is shown in Figure 8.48, consisting of a cylindrical and hemispherical portion.

(b) Note that the height of the cylindrical portion is equal to its radius (so that the total height of the riser is equal to its diameter). The volume,

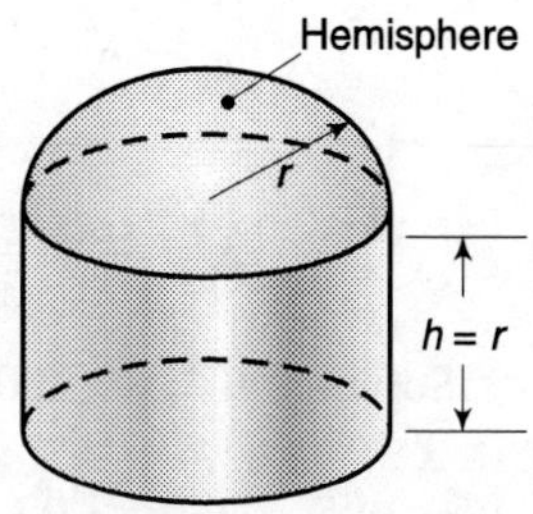

Figure 8.48

$$V = \pi r^2 h + \frac{1}{2}\left(\frac{4\pi r^3}{3}\right) = \frac{5\pi r^3}{3} \quad \text{...(a)}$$

Letting V be unity, we have

$$r = \left(\frac{3}{5\pi}\right)^{\frac{1}{3}} \quad \text{...(b)}$$

The surface area of the riser,

$$A = 2\pi rh + \pi r^2 + \frac{1}{2}(4\pi r^2) = 5\pi r^2 \quad \text{...(c)}$$

Using Eq. (b) into Eq. (c) we get,

$$A = 5\pi r^2 = 5\pi\left\{\left(\frac{3}{5\pi}\right)^{\frac{1}{3}}\right\}^2 = 5.21$$

$$\text{Modulus, } M = \frac{\text{Volume}}{\text{Surface area}} = \frac{1}{5.21}$$

$$t_s \propto M^2$$

$$t_s = (\text{constant})\ M^2$$

$$t_s = (K)\left(\frac{1}{5.21}\right)^2 = 0.0368\ K$$

We know that the solidification time for a cylinder with a height equal to its diameter is 0.033C. This blind riser will cool a little slower but not much so, and is easiest to cast.

Example 8.53: A sphere with a diameter of 75 mm solidifies in three minutes in a sand-casting operation. Using the same equipment and metal, estimate the time required to produce the part shown in Figure 8.49.

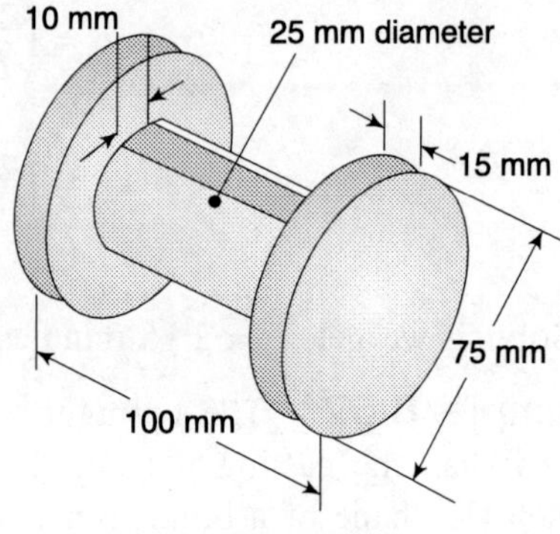

Figure 8.49

Solution: For the sphere, note that $V = \frac{4\pi r^3}{3}$ and $A = 4\pi r^2$. From Chvorinov's rule,

$$\text{Modulus, } M = \frac{\text{Volume}}{\text{Surface area}} = \frac{\frac{4\pi r^3}{3}}{4\pi r^2} = \frac{r}{3}$$

$$t_s \propto M^2$$

$$t_s = (\text{constant})\ M^2$$

$$t_s = (K)\left(\frac{r}{3}\right)^2 = K\left(\frac{r^2}{9}\right)$$

$$3 = (K)\left(\frac{r}{3}\right)^2 = K\left(\frac{\left(\frac{0.075}{2}\right)^2}{9}\right)$$

Solving, $K = 19{,}200\ \text{min/m}^2$

$V = 2\left(\frac{\pi}{4}(0.075)^2\right)0.015 + \frac{\pi}{4}(0.025^2)(0.070) = 1.669 \times 10^{(-4)}\ \text{m}^3$. To calculate the surface area, the outside discs will be considered separately from the inner cylinder. For the outside discs, the area for each disc,

$$A_0 = \pi dl + 2\left(\frac{\pi d^2}{4}\right) = \pi \times 0.075 \times 0.015 + 2\left(\frac{\pi \times 0.075^2}{4}\right) = 0.01237\ \text{m}^2$$

The exposed exterior of the inner cylinder,

$$A_i = \pi dl = \pi \times 0.025 \times 0.070 = 0.00550\ \text{m}^2$$

The total surface area is then (note that the faces of the inner cylinder need to be subtracted from total),

$$A = 2A_0 + A_i - 2\left(\frac{\pi}{4}\right) \times (0.025)^2 = 0.02926\ \text{m}^2$$

From Chvorinov's rule

$$t_s = (K)\left(\frac{V}{A}\right)^2 = 19{,}200\left(\frac{1.669 \times 10^{-4}}{0.02926}\right)^2 = 0.625 \text{ minute}$$

Example 8.54: The constant C in Chvorinov's rule is given as 3 s/mm^2 and is used to produce a cylindrical casting with a diameter of 75 mm and a height of 125 mm. Estimate the time for the casting to fully solidify. The mould can be broken safely when the solidified shell is at least 20 mm. Assuming the cylinder cools evenly, how much time must transpire after pouring the molten metal before the mould can be broken?

Solution: Note that for the cylinder

$$A = 2\left(\frac{\pi d^2}{4}\right) + \pi dh$$

$$A = 2\left(\frac{\pi \times 75^2}{4}\right) + \pi \times 75 \times 125 = 38{,}290 \text{ mm}^2$$

$$V = \left(\frac{\pi}{4}(\text{d})^2\right)h = \frac{\pi}{4}(75^2)(125) = 5.522 \times 10^5 \text{ mm}^3$$

From Chvorinov's rule,

$$t_s = (K)\left(\frac{V}{A}\right)^2 = 3\left(\frac{5.522 \times 10^5}{38{,}290}\right)^2 = 624 \text{ seconds or just over 10 minutes to solidify}$$

The second part of the problem is far more difficult, and different answers can be obtained depending on the method of analysis. The solution is not as straightforward as it may seem initially. For example, one could say that the 20 mm wall is 53.3% of the thickness, so that 0.533(624) = 333 seconds is needed. However, this would not be sufficient because an annular section at an outside radius has more material than one closer to the centre. It is thus reasonable and conservative to consider the time required for the remaining cylinder to solidify. Using h = 85 mm and d = 35 mm, the solidification time is found to be 21.8 seconds. Therefore, one still has to wait for 602 seconds before the mould can be broken.

Example 8.55: A jeweller wishes to produce twenty gold rings in one investment-casting operation. The wax parts are attached to a wax central sprue with a 20 mm diameter. The rings are located in four rows, each 15 mm from the other on the sprue. The rings require a 3 mm diameter, 12 mm long runner to the sprue. Estimate the weight of gold needed to completely fill the rings, runners and sprues. Assume a typical ring has a 25 mm outer diameter, 19 mm inner diameter and 5 mm width. The specific gravity of gold is 19.3.

Solution: The volume of each ring is

$$V = \left(\frac{\pi}{4}(0.025)^2\right) - \left(\frac{\pi}{4}(0.019)^2\right) = 1.036 \times 10^{-6} \text{ m}^3$$

The total volume for 20 rings is 2.0735 × 10^{-5} m^3. There are twenty runners to the sprue, so this volume component is

$$V = 20\left(\frac{\pi}{4}(d)^2\right)L = 20\left(\frac{\pi}{4}(0.003)^2\right) \times 0.012 = 1.696 \times 10^{-6} \text{ m}^3$$

The central sprue has a length that needs four rows, so that the total length is 45 mm. The central sprue volume is

$$V = \left(\frac{\pi}{4}(d)^2\right) L = \left(\frac{\pi}{4}(0.020)^2\right) \times 0.045 = 3.534 \times 10^{-6} \text{ m}^3$$

The total volume is then 2.5965×10^{-5} m^3, not including the metal in the pouring basin, if any. The specific gravity of gold is given as 19.3, thus its density is 19,300 kg/m^3. Therefore, the jeweller needs 0.50 kg of gold.

Example 8.56: Plot the solidification time of a cylinder with a volume of 15,000 mm^3 as a function of its diameter-to-length ratio. Assume that the solidification time for a 25 mm diameter sphere is three minutes.

Solution: Note that for a sphere

$$A = 4\pi r^2$$

$$V = \left(\frac{4\pi \times r^3}{3}\right)$$

Using $r = 12.5$ mm

$$V = 8181 \text{ mm}^3$$

$$A = 1963 \text{ mm}^2$$

$$\text{Modulus, } M = \frac{\text{Volume}}{\text{Surface area}} = \frac{8181}{1963}$$

$$t_s \propto M^2$$

$$t_s = (\text{constant})\, M^2$$

$$t_s = (K)\left(\frac{8181}{1963}\right)^2 = 3 \text{ minutes}$$

$$K = 0.1727 \text{ min/mm}^2$$

This value of the constant in Chvorinov's rule will be used for the analysis of the cylinder. For a cylinder,

$$A = \frac{\pi d^2}{2} + \pi dh$$

$$V = \frac{\pi d^2}{2} h$$

For a cylinder with a volume of 15000 mm^3, the relationship between the diameter and height will therefore be

$$d = \sqrt{\frac{15000 \times 4}{\pi h}}$$

The solidification time is given by Chvorinov's rule as

$$t_s \propto M^2$$

$$t_s = (\text{constant})\, M^2$$

$$t_s = (0.1727)\left(\frac{\frac{\pi d^2}{2}h}{\frac{\pi d^2}{2} + \pi dh}\right)^2$$

This is solved numerically. Plot is shown in Figure 8.50.

```
figure
V=15000;
ds=25;
rs=ds/2;
As=4*pi*rs^2;
Vs=4*rs^3*pi/3;
ts=3;
C=ts*As*As/Vs/Vs;
hmin=5;
hmax=50;
dh=(hmax-hmin)/100;
h=hmin : dh : hmax;
d=(V.*4./pi./h).^(0.5);
Ac=pi.*d.^2./2.+pi.*d.*h;
Vc=pi.*d.*d.*h./4.;
t=C.*(Vc./Ac).^2;
d_over_h=d./h;
plot(d_over_h,t)
```

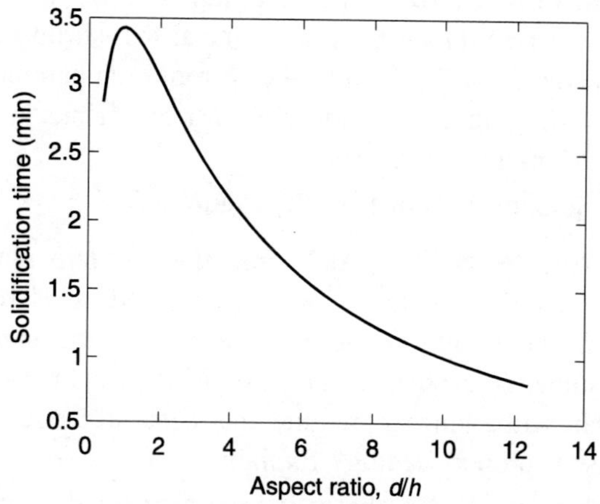

Figure 8.50

Example 8.57: A cylindrical casting has a diameter of 300 mm and length of 1 m. Another casting of the same metal is rectangular in cross section, with a width-to-thickness ratio of 3, and has the same length and cross-sectional area as the round casting. Both pieces are cast under the same conditions. The cylindrical casting solidifies in three minutes. How long does the rectangular cross section casting take to solidify?

Solution: The castings have the same length and cross-sectional area (thus, the same volume) and the same casting conditions, hence the same K value. The total surface area of the round casting, with r = 150 mm, is

$$A_{\text{round}} = 2\pi rl + 2\pi r^2$$

$$A_{\text{round}} = 2\pi \times 0.150 \times 1 + 2\pi \times (0.150)^2 = 1.084 \text{ m}^2$$

The cross-sectional area of the round casting is $\pi r^2 = \pi \times 0.150^2 = 0.07068 \text{ m}^2$. The rectangular cross section has sides x and $3x$, so that

$$0.07068 = 3x^2, \text{ solving we get } x = 0.153 \text{ m}$$

Hence, the perimeter of the rectangular casting with the same cross-sectional area and axes ratio of 3 is 1.228 m. The total surface area is

$$A_{\text{rect}} = 2 \times 0.07068 + 1.228 \times 1 = 1.369 \text{ m}^2$$

According to Chvorinov's rule, cooling time for a constant volume is inversely proportional to surface area squared. Therefore,

$$\frac{t_{\text{rect}}}{t_{\text{round}}} = \left(\frac{1.084}{1.369}\right)^2 = 0.6265$$

Since it took three minutes for the cylindrical casting to solidify, the rectangular casting will take 0.6265 × 3 = 1.88 minutes.

Example 8.58: Sketch a graph of specific volume versus temperature for a metal that shrinks as it cools from the liquid state to room temperature. On the graph, mark the area where shrinkage is compensated for by risers.

Solution is shown in the Figure 8.51.

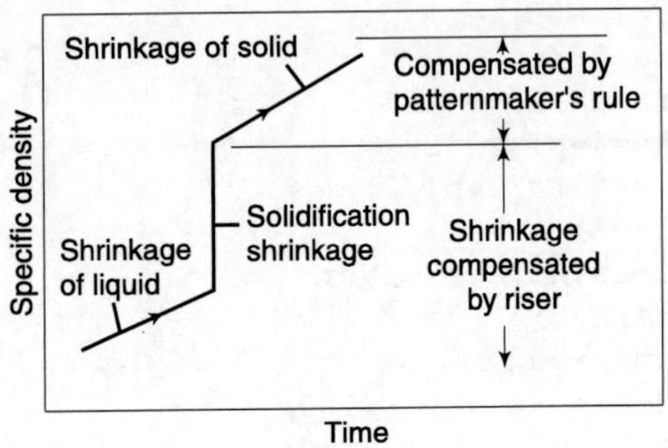

Figure 8.51

Example 8.59: A 75 mm thick square plate and a right circular cylinder with a radius of 100 mm and height of 50 mm each have the same volume. If each is to be cast using a cylindrical riser, will each part require the same size riser to ensure proper feeding? Explain.

Solution: Recall that it is important for the riser to solidify after the casting has solidified. A casting that solidifies rapidly would most likely require a smaller riser than one which solidifies over a longer period of time. Let's now calculate the relative solidification times, using Chvorinov's rule. For the cylindrical part, we have

$$V_{\text{cylinder}} = \pi r^2 h = \pi \times 0.1^2 \times 0.050 = 0.00157 \text{ m}^3$$

The surface area of the cylinder,

$$A_{\text{cylinder}} = 2\pi r^2 + 2\pi rh = 2 \times \pi \times 0.1^2 + 2 \times \pi \times 0.1 \times 0.05 = 0.0942 \text{ m}^2$$

$$\text{Modulus, } M = \frac{\text{Volume}}{\text{Surface area}} = \frac{0.00157}{0.0942}$$

$$t_s \propto M^2$$

$$t_s = (\text{constant})\, M^2$$

$$t_{\text{s-cylinder}} = (K)\left(\frac{0.00157}{0.0942}\right)^2 = 2.78 \times 10^{-4}\, K$$

For a square plate with sides L and height $h = 0.075$ m, and the same volume as the cylinder, we have,

$$V_{\text{plate}} = 0.00157 \text{ m}^3 = L^2 h = L^2 \times 0.075$$

Solving we get, $L = 0.144$ m

$$A_{\text{cylinder}} = 2L^2 + 4Lh = 2 \times 0.144^2 + 4 \times 0.144 \times 0.075 = 0.0847 \text{ m}^2$$

$$\text{Modulus, } M = \frac{\text{Volume}}{\text{Surface area}} = \frac{0.00157}{0.0847}$$

$$t_s \propto M^2$$

$$t_s = (\text{constant})\ M^2$$

$$t_{\text{s-plate}} = (K)\left(\frac{0.00157}{(0.0847}\right)^2 = 3.43 \times 10^{-4}\ K$$

Therefore, the cylindrical casting will take longer to solidify and will thus require a larger riser.

Review Questions

1. Differentiate between riser and gating systems.
2. What are the main functions of riser systems?
3. What are the basic requirements of riser systems?
4. Explain the basic working of a riser.
5. Write Caine's equation.
6. What is shape factor and explain its significance.
7. Explain the usefulness of Chvorinov's rule.
8. Explain distinct zone in casting.
9. Explain the heat transfer criteria in casting.
10. Write a note on use of insulators and exothermic compounds.

CHAPTER 9

Cooling and Solidification of Castings

9.1 INTRODUCTION

Many properties of castings are directly related to their final structures developed after the casting operation. The final structure of a casting, however, depends on its solidification history. A foundry man is always interested in producing sound castings free from defects as porosity, shrinkage, crack, segregation, etc. To produce a casting free from such defects, it is necessary to understand the phenomena of the solidification of castings and the various variables causing the defects during solidification. This chapter covers the fundamentals of the solidification process and various other factors which should not be ignored for proper feeding of the castings.

9.2 SOLIDIFICATION OF METALS

Solidification involves the transformation of the molten metal back into the solid state. Transition from liquid to solid takes place by **nucleation** and then by **growth**. Nucleation is a process of aggregation of clusters of atoms representing the first appearance of the new phase. Kinetic energies of several atoms of molten metal reaches a low enough value so that they form a short-range order, called a **nucleus**. The nucleation process of solids is influenced to some extent by the structure and properties of the liquid. Growth means getting bigger as more heat is extracted, this newly formed nucleus continues to grow at the expense of liquid phase. The driving force for solidification is proportional to the undercooling provided that the latent heat and the entropy of fusion do not vary much with temperature. A nucleation event occurs when one of these clusters continues to gain more atoms than it loses. The solidification process differs depending on whether the metal is a pure element or an alloy.

Solidification of Pure Metals A pure metal solidifies at a constant temperature equal to its freezing point, which is the same as its melting point. The melting points of pure metals are well-known and documented. For example: aluminium solidifies at 660°C, iron solidifies at 537°C and tungsten solidifies at 3410°C.

The process occurs over time as shown in the plot in Figure 9.1, called a cooling curve. The actual freezing takes time, called the *local solidification time* in casting. Local solidification time is the time between the beginning of freezing and completion of freezing. In this time, the molten metal heat of fusion is delivered into the mould. The *total solidification time* is the time taken between pouring and complete solidification.

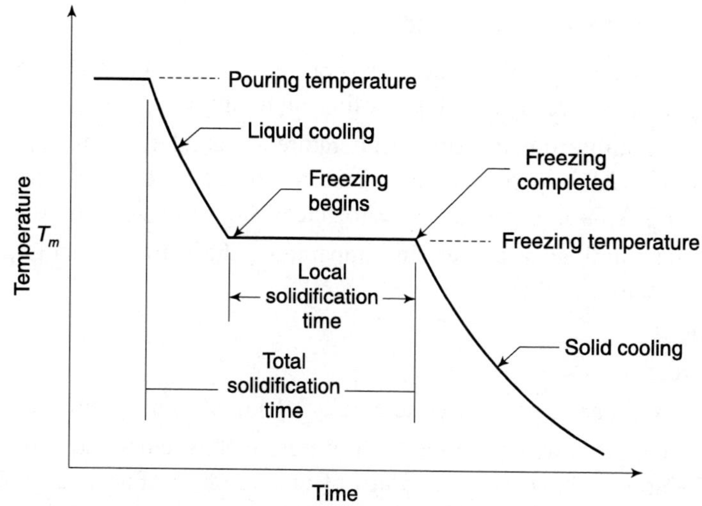

Figure 9.1 Cooling curve for a pure metal during casting

After the casting has completely solidified, cooling continues at a rate indicated by the downward slope of the cooling curve. Because of the chilling action of the mould wall, a thin skin of solid metal is initially formed at the interface immediately after pouring. Thickness of the skin increases to form a shell around the molten metal as solidification progresses inwards towards the centre of the cavity. The rate at which freezing proceeds depends on heat transfer into the mould, as well as the thermal properties of the metal. It is of interest to examine the metallic grain formation and growth during this solidification process. The metal which forms the initial skin has been rapidly cooled by the extraction of heat through the mould wall. This cooling action causes the grains in the skin to be fine and randomly oriented. As cooling continues, further grain formation and growth occur in a direction away from the heat transfer. Since the heat transfer takes place through the skin and mould wall, the grains grow inwardly as needles or spines of solid metal. As these spines enlarge, lateral branches form, and as these branches grow, further branches form at right angles to the first branches. This type of grain growth is referred to as **dendritic growth,** and it occurs not only in the freezing of pure metals but alloys as well. These tree-like structures are gradually filled in during freezing, as additional metal is continually deposited onto the dendrites until complete solidification has occurred. The grains resulting from this dendritic growth take on a preferred orientation, tending to be coarse, columnar grains aligned towards the centre of the casting. The resulting grain formation is illustrated in Figure 9.2.

Figure 9.2 Characteristic grain structure in a casting of a pure metal, showing randomly oriented grains of small size near the mould wall, and large columnar grains oriented towards the centre of the casting

Concept of undercooling in pure metals

In most cases, transition during solidification is not easy. Solidification cannot begin at the equilibrium freezing temperature, T_m, during cooling of liquid.

Undercooling, Δt = equilibrium freezing temperature – actual freezing temperature undercooling occurs because metal experiences reluctance to convert into a solid despite cooling to a temperature below T_m. This is especially true for clean liquid in a clean container.

Although pure metal solidifies at a constant temperature, the nature of cooling curve of metals depends on many factors like

- cooling conditions
- latent heat of freezing
- temperature, specific heat, and heat conductivity of liquid and mould material.

Effect of undercooling is shown in Figure 9.3. For pure metals and some alloys, undercooling can be very large. Table 9.1 shows typical values of undercooling. For amorphous solids (e.g., glass), the liquid is so sluggish that undercooling and a sharp freezing point is never observed as shown in Figure 9.4.

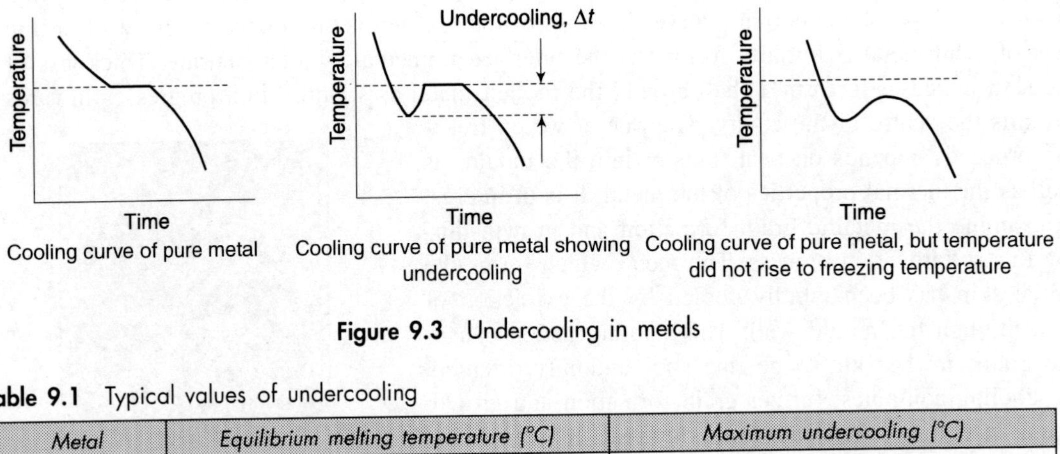

Figure 9.3 Undercooling in metals

Table 9.1 Typical values of undercooling

Metal	*Equilibrium melting temperature (°C)*	*Maximum undercooling (°C)*
Lead	327.6	80
Bismuth	271	90
Tin	232.1	118
Copper	1083.6	236
Iron	1536	295
Nickel	1453	319

Solidification of alloys: In most alloys, solidification will not occur at a particular single temperature. It happens at a temperature range. This range depends on the alloy composition. The exact range depends on the alloy system and the particular composition.

Solidification begins when the temperature drops below the liquidus temperature and is completed when it reaches the solidus temperature. Within this temperature range, the alloys are in a mushy or pasty state consisting of columnar dendrites and liquid metal between the dendrite arms. Dendrites have three-dimensional arms and branches (secondary arms), which eventually interlock. The width of the mushy zone is described in terms of a temperature difference, known as the freezing range. Freezing zone = Liquidus temperature – solidus temperature. Pure metals have zero freezing range (solidification front moves as a plane without forming a mushy zone). For alloys, short freezing range ≥ temperature difference of less than 50°C. Long freezing range ≥ temperature difference of less than 110°C. Ferrous castings generally have narrow mushy zones. Aluminium and magnesium alloys have wide mushy zones.

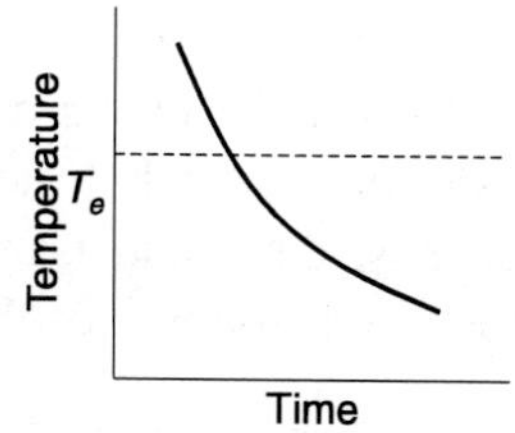

Figure 9.4 Cooling curve of an amorphous metal

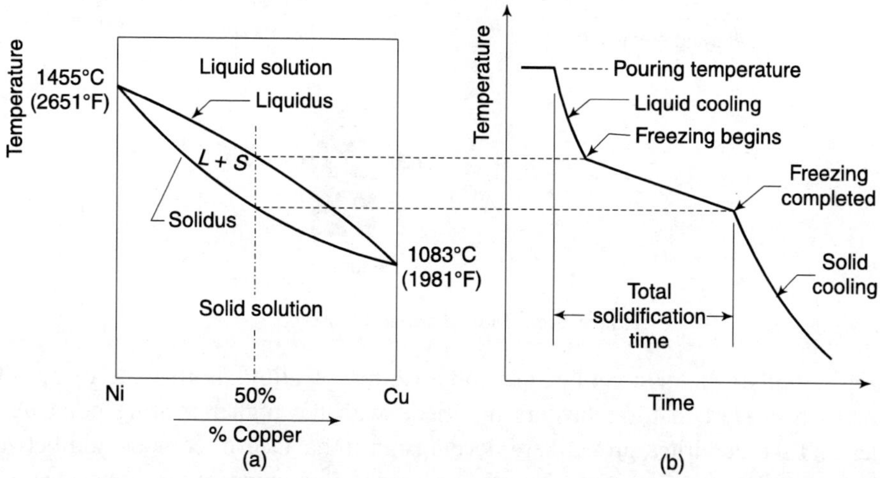

Figure 9.5 (a) Phase diagram for a copper–nickel alloy system and (b) associated cooling curve for a 50% Ni–50%Cu composition during casting

Mushy zone formation in alloys: Solidification of an alloy can be explained with reference to Figure 9.5, which shows the phase diagram for a particular alloy system and the cooling curve for a given composition. As temperature drops, freezing begins at the temperature indicated by the **liquidus** and is completed when the **solidus** is reached. The start of freezing is similar to that of the pure metal. A thin skin is formed at the mould wall due to the large temperature gradient at this surface. Freezing then progresses as before through the formation of dendrites that grow away from the walls. However, owing to the temperature spread between the liquidus and solidus, the nature of the dendritic growth is such that an advancing zone is formed in which both liquid and solid metals coexist. The solid portions are the dendrite structures that have formed sufficiently to trap small islands of liquid metal in the matrix. This solid–liquid region has a soft consistency that has motivated its name as the mushy zone. Depending on the conditions of freezing, the **mushy zone** (Figure 9.6) can be relatively narrow, or it can exist throughout most of the casting. The latter condition is promoted by factors such as slow heat transfer out of the hot metal and a wide difference between liquidus and solidus temperatures. Gradually, the liquid islands in the dendrite matrix solidify as the temperature of the casting drops to the solidus for the given alloy composition.

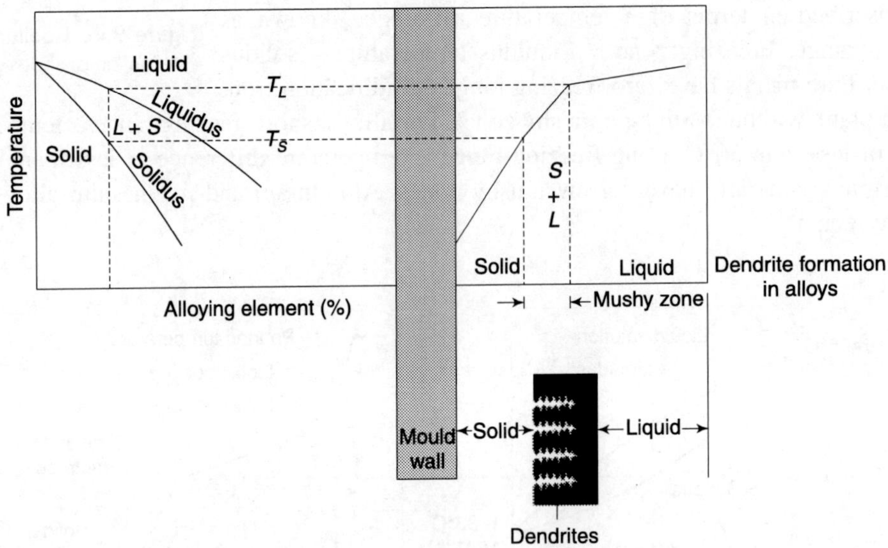

Figure 9.6 Mushy zones in alloys

Segregation: Another factor complicating solidification of alloys is that the composition of the dendrites as they start to form favours the metal with the higher melting point. As freezing continues and the dendrites grow, there develops an imbalance in composition between the metal that has solidified and the remaining molten metal. This composition imbalance is finally manifested in the completed casting in the form of segregation of the elements. The segregation is of two types, microscopic and macroscopic.

Microscopic segregation: At the microscopic level, the chemical composition varies throughout each individual grain. This is due to the fact that the beginning spine of each dendrite has a higher proportion of one of the elements in the alloy. As the dendrite grows in its local vicinity, it must expand using the remaining liquid metal that has been partially depleted from the first component. Finally, the last metal to freeze in each grain is that which has been trapped by the branches of the dendrite, and its composition is even further out of balance. Thus, we have a variation in chemical composition within single grains of the casting. Mechanical properties of castings are sensitive to micro-segregation. They can usually be removed by homogenisation treatment. Examples of microscopic segregation are cellular, dendritic, and grain boundary segregation.

Figure 9.7 Characteristic grain structures in an alloy casting, showing segregation of alloying components in the center of casting

Macroscopic segregation: At the macroscopic level, the chemical composition varies throughout the entire casting. Since the regions of the casting that freeze first (at the outside near the mould walls) are richer in one component than the other, the remaining molten alloy is deprived of that component by the time freezing occurs at the interior. Thus, there is a general segregation through the cross section of the casting, sometimes called **ingot segregation**, as illustrated in Figure 9.7. Macro-segregation refers to zonal segregation of the solute in different parts of casting occurring over a long distance ranging from 1 cm to 1 m and, therefore, cannot be removed. They result either from changes that occur in the liquid before the solidification front has proceeded or as the result of fluid motion in the mushy zone behind the solidification front. Examples include: normal or positive segregation (or coring), V-segregation, inverse V-segregation, negative segregation, and surface segregation.

Eutectic Alloys: Eutectic alloys constitute an exception to the general process by which alloys solidify. A eutectic alloy is a particular composition in an alloy system for which the solidus and liquidus are at the same temperature. Hence, solidification occurs at a constant temperature rather than over a temperature range. The effect can be seen in the phase diagram of the lead–tin system. Pure lead has a melting point of 327°C while pure tin melts at 232°C. Although most lead–tin alloys exhibit the typical solidus–liquidus temperature range, the particular composition of 61.9% tin and 38.1% lead has a melting (freezing) point of 183°C. This composition is the eutectic composition of the lead–tin alloy system, and 183°C is its eutectic temperature. Lead–tin alloys are not commonly used in casting, but Pb–Sn compositions near the eutectic are used for electrical soldering, where the low melting point is an advantage. Examples of eutectic alloys encountered in casting include aluminium–silicon (11.6% Si) and cast iron (4.3% C).

9.2.1 Shrinkage

Our discussion of solidification has neglected the impact of shrinkage that occurs during cooling and freezing. Shrinkage occurs in three steps:

(1) liquid contraction during cooling prior to solidification;

(2) contraction during the phase change from liquid to solid, called solidification shrinkage; and

(3) thermal contraction of the solidified casting during cooling to room temperature.

The three steps can be explained with reference to a cylindrical casting made in an open mould, as shown in Figure 9.8. The molten metal immediately after pouring is shown in part (0) of the series. Contraction of the liquid metal during cooling from pouring temperature to freezing temperature causes the height of the liquid to be reduced from its starting level as in (1) of the figure. The amount of this liquid contraction is usually around 0.5%. Solidification shrinkage, seen in part (2), has two effects. First, contraction causes a further reduction in the height of the casting. Second, the amount of liquid metal available to feed the top centre portion of the casting becomes restricted. This is usually the last region to freeze, and the absence of metal creates a void in the casting at this location. This shrinkage cavity is called a pipe by foundry men.

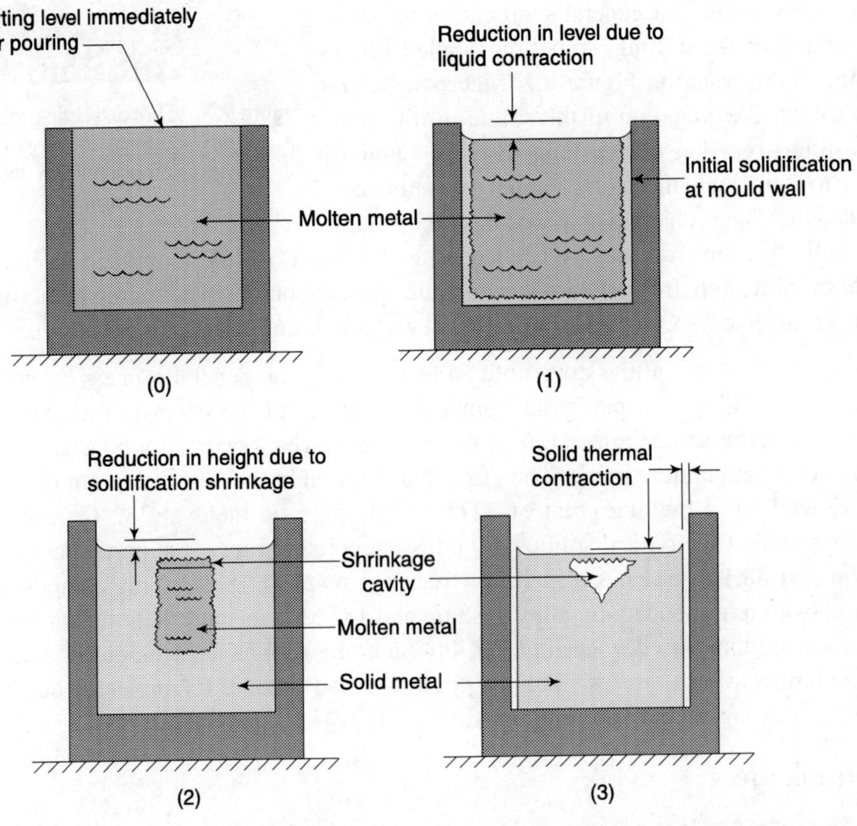

Figure 9.8 Shrinkage of a cylindrical casting during solidification and cooling: (0) starting level of molten metal immediately after pouring; (1) reduction in level caused by liquid contraction during cooling; (2) reduction in height and formation of shrinkage cavity caused by solidification shrinkage; and (3) further reduction in height and diameter due to thermal contraction during cooling of the solid metal. For clarity, dimensional reductions are exaggerated in our sketches

Once solidified, the casting experiences further contraction in height and diameter while cooling, as in (3) of Figure 9.8. This shrinkage is determined by the solid metal's coefficient of thermal expansion, which in this case is applied in reverse to determine contraction. Solidification shrinkage occurs in nearly all metals because the solid phase has a higher density than the liquid phase. The phase transformation that accompanies solidification causes reduction in the volume per unit weight of metal. The exception is cast iron containing high carbon content, whose solidification during the final stages of freezing is complicated by a period of graphitization, which results in expansion that tends to counteract the volumetric decrease associated with the phase change.

Compensation for solidification shrinkage is achieved in several ways depending on the casting operation. In sand casting, liquid metal is supplied to the cavity by means of risers. In die casting, the molten metal is applied under pressure. Pattern makers account for thermal contraction by making the mould cavities oversized. The amount by which the mould must be made larger relative to the final casting size is called the pattern shrinkage allowance. Although the shrinkage is volumetric, the dimensions of the casting are expressed linearly, so the allowances must be applied accordingly. Special "shrink rules" with slightly elongated scales are used to make the patterns and moulds larger than the desired casting by the appropriate amount.

There are few methods by which damages due to shrinkage can be minimized. They are called directional solidification methods.

Method 1

Providing risers It is desirable for the regions of the casting far away from the liquid metal supply to freeze first and for solidification to progress from these remote regions towards the location of riser. In this way, molten metal will continually be available from the risers to prevent shrinkage voids during freezing. For example, the regions of the cast with lower V/A ratios should be placed far away from the riser location. Solidification will start from these locations and it will progress towards the riser location where bulkier sections of the cast are present. Hence, the bulkier sections will continually receive molten metal from the risers till freezing.

Method 2

Providing chills Chills can be provided at appropriate locations in order to have rapid solidification at those points. Internal and external chills can be provided. Various refractory mould materials—sand, investment and plaster—are all poor absorbers of heat. The various chill materials have high order of magnitude of chilling powers compared to the refractory mould materials.

External chills They are metal pieces, placed adjacent to the pattern and subsequently packing the sand around it to make the rest of the mould in the normal way. These chills can extract heat from the molten metal more rapidly than the surrounding sand in order to promote localized solidification. They are mainly used in sections of the casting that are difficult to supply with molten metal. Figure 9.9 shows the effect of external chill. If the chill is not used in this case then there will be defects as shown in Figure 9.9(b).

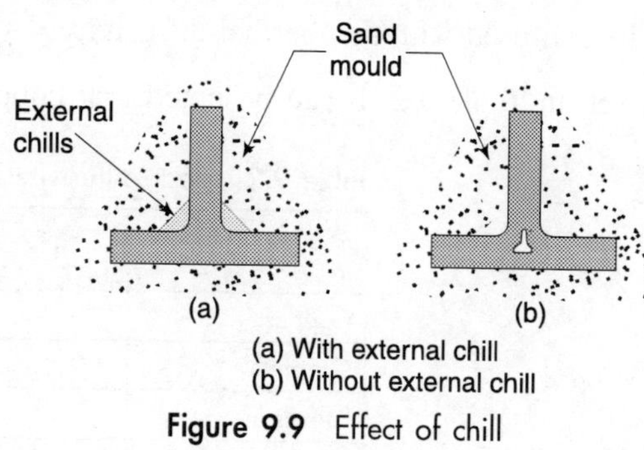

Figure 9.9 Effect of chill

Internal chills are metal pieces that are deliberately put inside the mould cavity to cause localised cooling and become integral part of the casting. The internal chill should have a chemical composition similar to the metal being poured, so that it can be made out of same cast metal.

To have an effective bonding with the liquid metal and to eliminate causing porosity problems, the metal piece must be cleaned thoroughly (devoid of any oxide film and moisture).

Figure 9.10 shows the freezing time of a plate 225 mm × 150 mm × 50 mm in Al_5Si_3Cu alloy at various distances from the chilled end. It can be observed that as the heat capacity increases for a given distance from the chill, the solidification time decreases.

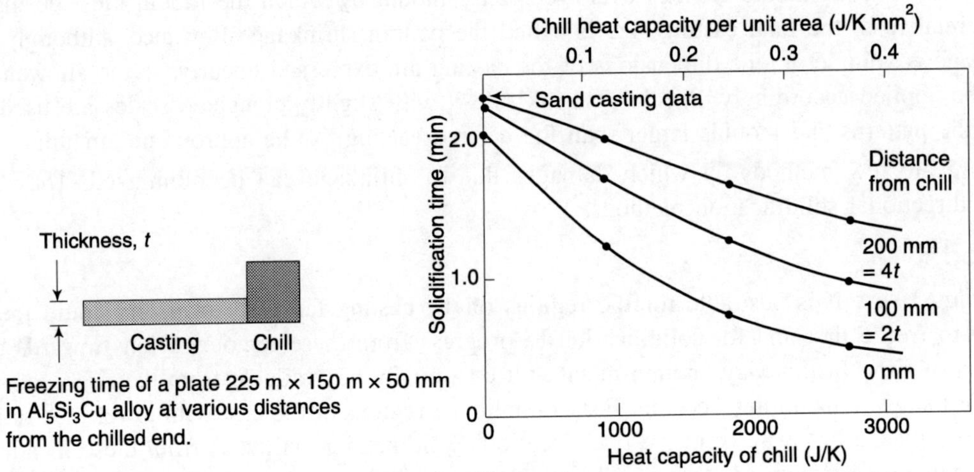

Figure 9.10 Freezing time for plate

The ability of the metal to be a chill depends on its capacity of absorbing heat, known as heat diffusivity.

$$\text{Heat diffusivity} = \sqrt{K\rho c_p}$$

where

K = thermal conductivity, ρ = density, c_p = heat capacity of chill material. This term should not be confused with the thermal diffusivity = $\frac{K}{\rho c_p}$. Heat diffusivity for some common materials is given in Table 9.2. It can be noted that copper has the highest chilling power.

Table 9.2 Heat diffusivity for some common materials

Material	*Heat diffusivity ($Jm^{-2}\ K^{-1}\ s^{(1/2)}$)*
Copper	37000
Graphite	22136
Pure Iron	16186
Sand	1015
Investment	671
Plaster	566

A full chilling power of the material can only be developed if the material is infinitely thick. Otherwise, the piece of metal becomes saturated with the heat and after some time it does not absorb any heat. The amount of heat a chill can actually absorb can be defined by the term volumetric heat capacity.

$$\text{Volumetric heat capacity} = \rho c_p V$$

where,

ρ = density, c_p = heat capacity of chill material and V is the volume of the chill.

Figure 9.11 shows the relative diffusivities (ability to diffuse heat away in a large chill) and heat capacities (ability to absorb heat if relatively small) of chill material.

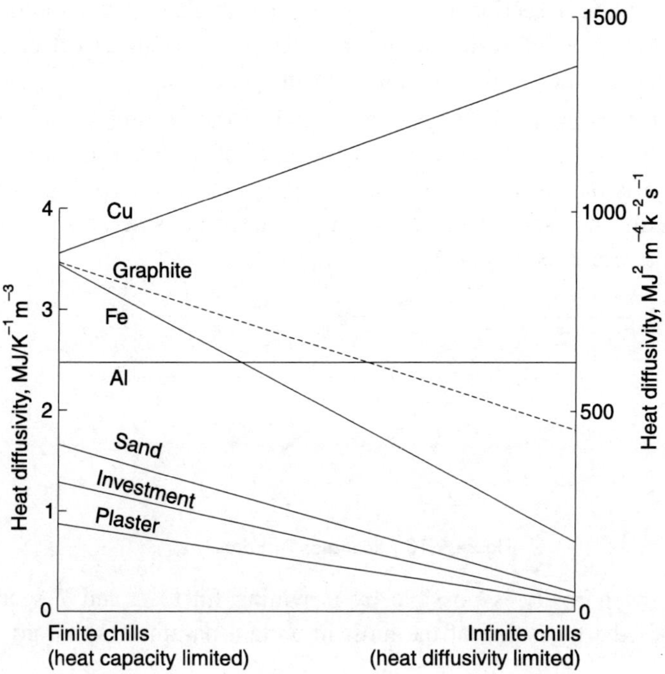

Figure 9.11 Relative diffusivities of chill mateial

Method 3

Providing fins: Pins are thin projections of solids used for localised cooling. Before we look specifically at fins on castings, it is worth spending some time to consider the concepts involved in junctions of all types between different cast sections. Figure 9.12 shows a complete range of T-junctions between walls of different relative thickness. When the wall forming the upright of the T is thin, it acts as a cooling fin, chilling the junction and the adjacent wall (the top cross of the T) of the casting.

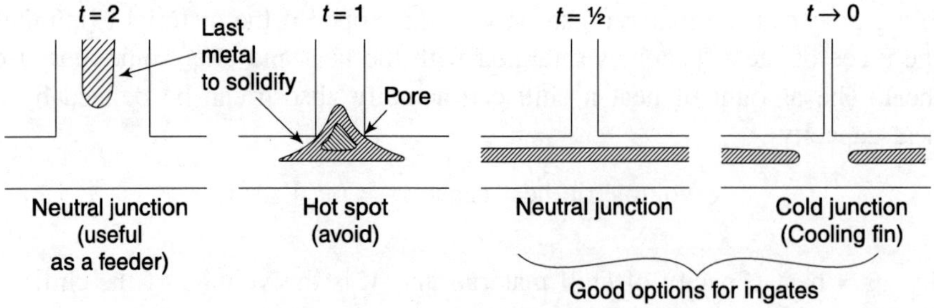

Figure 9.12 Range of T-junctions

When the upright of the T-section has increased to a thickness of half the casting section thickness, then the junction is close to thermal balance, the cooling effect of the fin balancing the hot-spot effect of the concentration of metal in the junction.

By the time the upright of the *T* has become equal to the casting section, the junction is a hot spot. When finally the section thickness of the upright of the *T* is twice the casting section, then the junction is balanced once again, with the casting now acting as the mild chill to counter the effect of the hot spot at the junction. We have considered these junctions merely in the form of the intersections of plates.

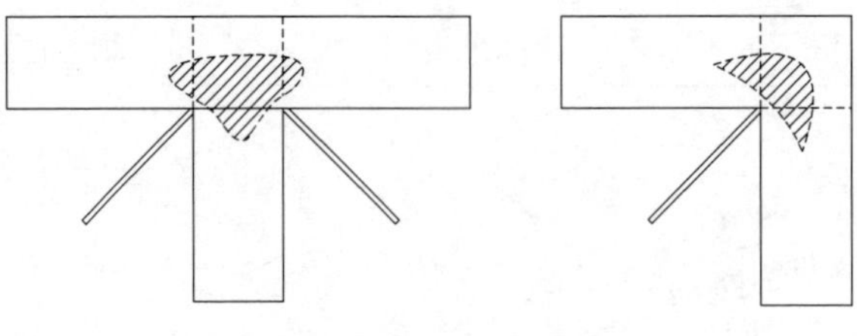

Figure 9.13 Reduction in porosity

The reduction in the porosity is possible by providing fin to *L* and *T* section as shown in Figure 9.13. Length of the fin is one of the most important parameters that governs rate of heat transfer.

The advantages of fins in casting are as follows:

1. The fin is always provided on the casting, because it is an integral part of the tooling. Thus, unlike a chill, the placing of it cannot be forgotten.
2. It is always exactly in the correct place. It cannot be wrongly sited before the making of the mould (the incorrect positioning of chill is possible. Although the location of the chill is normally carefully painted on the pattern, the application of the first coat of mould release agent usually does an effective job in elimination of all traces of this).
3. It cannot be displaced or lifted during the making of the mould, if the chill lifts slightly during the filling of the sand, then displacement of chill results in sand inclusion defects around the chill.
4. An increase in productivity has been reported as a result of using the fin.

5. Fins can be easily cut off. Chills require substantial dressing, especially if the chill has *V* grooves or has been misplaced during moulding.
6. The fin does not cause scrap casting because of condensation of moisture and other volatiles. Condensation of moisture using chills can cause blow defects.
7. The fin does not require to be retrieved from the sand system.
8. The fin does not wear out. Old chills become rounded over time and thermal conductivity of the chill may change due to oxidation (say, oxidation of graphite flakes in the iron).
9. Sometimes it is possible to solve a localised feeding problem. The typical example is the isolated boss in the centre of the plate by chilling with a fin instead of providing a local supply of the feed metal. In this case, the fin is cheaper than the feeder.

9.3 RATE OF SOLIDIFICATION

A reservoir of liquid metal, called riser, is used to compensate for the shrinkage that takes place from the pouring temperature up to solidification. In this respect, grey cast iron is an interesting exception where solidification occurs in two stages. The shrinkage associated with the first stage may well be compensated by the expansion that takes place during the second stage, and as stated, a riser may not be necessary. To ensure that the riser does not solidify before the casting, we should have an idea of the time taken by the casting, to solidify. Moreover, the placement (location) of the riser can be judiciously chosen if an estimate of the time taken by the casting to solidify up to a certain distance from the mould face is available.

The heat rejected by the liquid metal is dissipated through the mould wall. The heat, released as a result of cooling and solidification of the liquid metal, passes through different layers. The temperature distribution in these layers, at any instant, is schematically shown in Figure 9.14 The thermal resistances which govern the entire solidification process are those of the liquid, the solidified metal, the metal-mould interface, the mould, and the ambient air. These five different regions are indicated by the numbers 1 to 5 in the figure. The solidification process is quite complicated especially when complex geometry, freezing of alloys, or temperature dependence of thermal properties is considered. In what follows, we shall discuss the solidification of pure metals in some cases of practical interest. In doing so, we shall depend on the situation, make simplifying assumptions to neglect the thermal resistance of one or more of the regions shown in the figure.

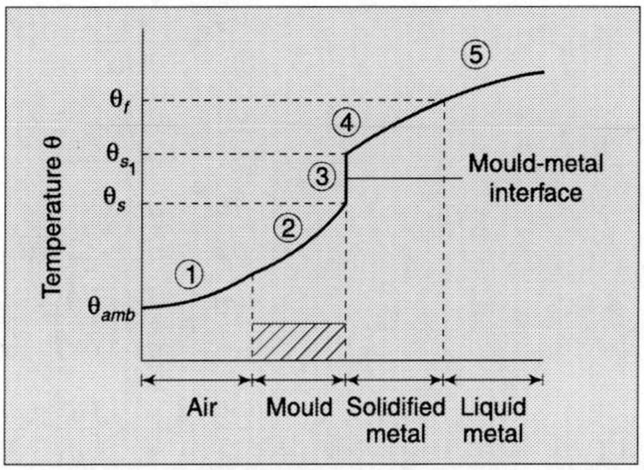

Figure 9.14 Temperature distribution in different layers

9.3.1 Solidification of a Large Casting in an Insulating Mould

During the solidification of a large casting in an insulating mould, like the one used in the sand or investment casting, almost the entire thermal resistance is offered by the mould. Hence, the analysis we give computes the freezing time by considering only the thermal resistance of region 2 (Figure 9.14).

Consider a mould face *AB* as shown in Figure 9.15. The large mould, initially at a temperature of 60°C, is assumed to be extended up to infinity in the *x*-direction. At time $t = 0$, the liquid metal at temperature θ_p is poured into the mould. We also assume that the metal just in contact with the mould face solidifies instantaneously. In other words, the temperature of the mould face is raised to θ_f (freezing temperature of the metal) at $t = 0$ and is maintained at that value till the completion of solidification. The temperature distribution within the mould at a subsequent time *t* (assuming one-dimensional heat conduction in the *x*-direction) for such a case is given by[1]

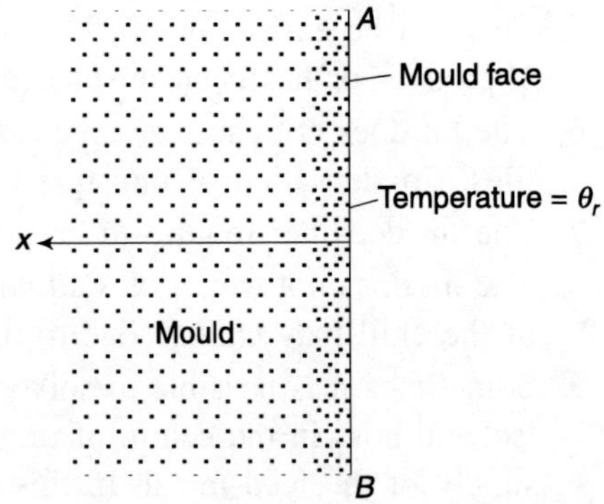

Figure 9.15 Mould face and coordinate systems

Solution: Given

$$\frac{\partial^2\theta}{\partial x^2} = \frac{1}{\alpha}\frac{\partial\theta}{\partial t} \quad ...(1)$$

Let

$$\zeta = \frac{x}{2\sqrt{\alpha t}}$$

Also,

$$\frac{\partial\zeta}{\partial x} = \frac{1}{2\sqrt{\alpha t}} \quad ...(2)$$

$$\frac{\partial\theta}{\partial x} = \frac{\partial\theta}{\partial\zeta} \times \frac{\partial\zeta}{\partial x}$$

$$\frac{\partial\theta}{\partial x} = \frac{\partial\theta}{\partial\zeta} \times \frac{1}{2\sqrt{\alpha t}} \quad ...(3)$$

$$\frac{\partial^2\theta}{\partial x^2} = \frac{\partial}{\partial x}\left(\frac{\partial\theta}{\partial x}\right) = \frac{\partial}{\partial x}\left(\frac{\zeta}{x} \times \frac{\partial\theta}{\partial\zeta}\right)$$

$$\frac{\partial^2\theta}{\partial x^2} = \frac{\zeta}{x} \times \left(\frac{\partial}{\partial x}\left(\frac{\partial\theta}{\partial\zeta}\right)\right)$$

(since $\frac{\zeta}{x}$ is a pure function of time)

$$\frac{\partial^2\theta}{\partial x^2} = \frac{\zeta}{x}\left(\frac{\partial}{\partial\zeta}\frac{\partial\theta}{\partial\zeta}\frac{\partial\zeta}{\partial x}\right)$$

$$\frac{\partial^2\theta}{\partial x^2} = \left(\frac{\zeta}{x}\right)^2 \times \frac{\partial^2\theta}{\partial^2\zeta} \text{ since } \frac{\partial\zeta}{\partial x} = \frac{\zeta}{x} \quad ...(4)$$

[1] Carslaw, H.S. and Jaeger, J.C., Conduction of Heat in Solids, Clarendon Press, Oxford, 1959.

With respect to time,

$$\frac{\partial \theta}{\partial t} = \frac{\partial \theta}{\partial \zeta} \times \frac{\partial \zeta}{\partial t}$$

$$\frac{\partial \theta}{\partial t} = \frac{\partial \theta}{\partial \zeta} \times \frac{x}{2\sqrt{\alpha}} \times \frac{-1}{2} \times t^{-\frac{3}{2}}$$

$$\frac{\partial \theta}{\partial t} = \frac{\partial \theta}{\partial \zeta} \times \frac{x}{2\sqrt{\alpha t}} \times \frac{-1}{2} \times t^{-1}$$

$$\frac{\partial \theta}{\partial t} = -\frac{\zeta}{2t} \times \frac{\partial \theta}{\partial \zeta} \qquad ...(5)$$

Substituting equations (4) and (5) in equation (1)

$$\left(\frac{\zeta}{x}\right)^2 \frac{\partial^2 \theta}{\partial \zeta^2} = \frac{1}{\alpha} \times \frac{-\zeta}{2t} \times \frac{\partial \theta}{\partial \zeta}$$

$$\frac{1}{4\alpha t} \times \frac{\partial^2 \theta}{\partial \zeta^2} = \frac{-\zeta}{2\alpha t} \times \frac{\partial \theta}{\partial \zeta}$$

$$\frac{\partial^2 \theta}{\partial \zeta^2} + 2\zeta \frac{\partial \theta}{\partial \zeta} = 0$$

$$e^{\zeta^2} \frac{\partial^2 \theta}{\partial \zeta^2} + 2\zeta e^{\zeta^2} \frac{\partial \theta}{\partial \zeta} = 0$$

$$\frac{\partial}{\partial \zeta}\left(e^{\zeta^2} \frac{\partial \theta}{\partial \zeta}\right) = 0 \qquad ...(6)$$

Integrating both sides,

$$e^{\zeta^2} \frac{\partial \theta}{\partial \zeta} = c_1$$

$$\frac{\partial \theta}{\partial \zeta} = c_1 \times e^{-\zeta^2}$$

$$\theta = c_1 \int_0^{\zeta} e^{-\zeta^2} d\zeta + c_2 \qquad ...(7)$$

Using boundary conditions,

$$\text{at } x = 0,\ \zeta = 0,\ \theta = \theta_f \qquad ...(8)$$

$$\text{at } x = \infty,\ \zeta = \infty,\ \theta = \theta_o \qquad ...(9)$$

Using equation (8) we get,

$$\theta_f = 0 + c_2$$

$$c_2 = \theta_f \qquad ...(10)$$

From equation (9)

$$\theta_o = c_1 \int_0^{\infty} e^{-\zeta^2} d\zeta + \theta_f$$

$$\theta_o - \theta_f = c_1 \times \frac{\sqrt{\pi}}{2}, \text{ since } \int_0^\infty e^{-x^2}\, dx = \frac{\sqrt{\pi}}{2}$$

$$c_1 = \frac{2}{\sqrt{\pi}} (\theta_o - \theta_f) \qquad ...(11)$$

Substituting equations (10) and (11) in equation (7), we get,

$$\theta = \frac{2}{\sqrt{\pi}} (\theta_o - \theta_f) \int_0^\zeta e^{-\zeta^2}\, d\zeta + \theta_f + \theta_o - \theta_o$$

$$\theta = \frac{2}{\sqrt{\pi}} \times \theta_o + (\theta_f - \theta_o) \times \left(1 - \frac{2}{\sqrt{\pi}} \int_0^\zeta e^{-\zeta^2}\, d\zeta\right)$$

$$\theta = \theta_o - (\theta_f - \theta_o)\,(1 - \operatorname{erf}(\zeta)) \qquad ...(12)$$

At the mould surface,

Rate of heat flow is given by

$$Q = -kA \left.\frac{\partial \theta}{\partial x}\right|_{x=0}$$

$$\left.\frac{\partial \theta}{\partial x}\right|_{x=0} = (\theta_f - \theta_o) \times \left(-\frac{2}{\sqrt{\pi}}\, e^{-\zeta^2} \frac{\partial \zeta}{\partial x}\right)$$

$$\left.\frac{\partial \theta}{\partial x}\right|_{x=0} = (\theta_f - \theta_o) \times \left(-\frac{2}{\sqrt{\pi}}\, e^{\zeta^2} \frac{1}{2\sqrt{\alpha t}}\right)$$

Also, $x = 0$, $\zeta = 0$

$$\left.\frac{\partial \theta}{\partial x}\right|_{x=0} = (\theta_f - \theta_o) \times \left(-\frac{1}{\sqrt{\pi \alpha t}}\right)$$

Hence,

$$Q = -kA \frac{(\theta_f - \theta_o)}{\sqrt{\pi \alpha t}} \qquad ...(13)$$

Heat rejected from the metal,

$$Q = \rho_m V(L + C_m(\theta_p - \theta_o)) \qquad ...(14)$$

where

ρ_m is the density of the metal

L is the latent heat

C_m is the specific heat capacity

θ_p is the pouring temperature

From equation (13), we know that

$$Q = \int_0^{t_s} \frac{kA(\theta_f - \theta_o)}{\sqrt{\theta \alpha t}}\, dt \qquad ...(15)$$

Equating equations (14) and (15)

$$\frac{kA(\theta_f - \theta_o)}{\sqrt{\pi \alpha}} \times 2\sqrt{t_s} = \rho_m V(L + C_p\,(\theta_p - \theta_f)$$

$$t_s = \left(\frac{V}{A}\right)^2 \times \left(\frac{\rho_m((L + c_p(\theta_p - \theta_f)) \times \sqrt{\pi\alpha}}{2kA\,(\theta_f - \theta_o)}\right)^2$$

The temperature distribution within the mould at a subsequent time, t is given by[2]

$$\theta_X(T) = \theta_0 + (\theta_F - \theta_0)\left[1 - \text{erf}\,\frac{X}{2\sqrt{\alpha t}}\right] \qquad ...(16)$$

erf(z) is called the error function. This function is very useful for heat Conduction problems in semi-infinite media. It represents the convergent series

$$\text{erf}(z) = \frac{2}{\sqrt{\pi}}\left(z - \frac{z^3}{3.1!} + \frac{z^5}{5.2!} - \frac{z^7}{7.3!} + ...\right)$$

or the integral

$$\text{erf}(z) = \frac{2}{\sqrt{\pi}}\int_0^x e^{-x^2}\,dx \qquad ...(17)$$

The values of this function for various values of its argument z are given in mathematical tables and handbooks. We shall use the result

$$\frac{d}{dz}\,\text{erf}\,(z) = \frac{2}{\sqrt{\pi}}\,e^{-z^2} \qquad (18)$$

in this chapter.

Where $\theta_x(t)$ is the temperature at a distance x from the mould face at an instant t and α is thermal diffusivity of the mould material and is equal to $k/(\rho c)$ with k = conductivity, ρ = density, and c = specific heat of the mould material. Now, the rate of heat flow through the mould face of any instant t is given by

$$Q = kA\frac{\partial \theta x}{\partial x} \text{ at } x = 0 \qquad ...(19)$$

where, A is the cross-sectional area of the mould-metal interface (approximately the surface area of the casting). Using equations (16) and (19), we obtain

$$Q = \frac{kA(\theta_f - \theta_0)}{\sqrt{\pi\alpha t}} \qquad ...(20)$$

Thus, the total quantity of heat flow across the mould face up to a certain time t_0,

$$Q_{t_0} = \int_0^{t_0} Q dt = \frac{kA(\theta_f - \theta_0)}{\sqrt{\pi\alpha}}\sqrt{t_0} \qquad ...(21)$$

Next, let us calculate the heat the liquid metal rejects in order to solidify. If the liquid metal has a latent heat L, a specific heat c_m, and a density ρ_m, then the heat rejected is given by

$$\theta_r = \rho_m\,V[L + C_m\,(\theta_p - \theta_f)] \qquad ...(22)$$

[2] Carslaw, H.S. and Jaeger, J.C., Conduction of Heat in Solids, Clarendon Press, Oxford, 1959.

or

$$\gamma = \left\{ \frac{(\rho_m \sqrt{\pi\alpha}\, [L + c_m (\theta_p - \theta_f)]}{2k(\theta_f - \theta_0)} \right\} \qquad ...(23)$$

It should be noted that the foregoing analysis assumes a plane metal-mould interface *AB*, not usually encountered in engineering practice. Often, we are required to find out the freezing time of complex contours. For such contours, all we need to do is observe (without any precise calculations) the following basic features to know whether the analysis we have given underestimates or overestimates the actual freezing time. To observe these features, we consider three types of metal-mould interfaces (see Figure 9.16) namely, (i) convex, (ii) plane (used in our analysis), and (iii) concave.

In Figure 9.16a, the heat flow is more divergent, and consequently the rate is somewhat more than that in Figure 9.16b. Thus, the freezing time in such a case is overestimated by the foregoing analysis. Similarly, in Figure 9.16c, the heat flow is more convergent, and consequently the rate is somewhat less than that in Figure 9.16b. So, the freezing time in such a case is underestimated by the analysis we have given.

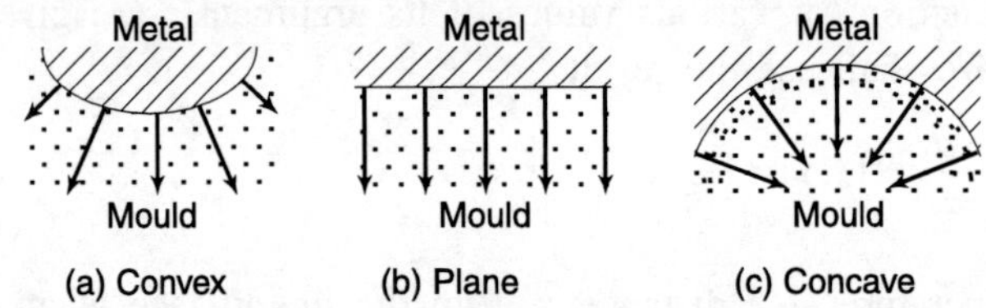

Figure 9.16 Heat flow across various types of metal mould interfaces

The quantitative results of the effect of the mould-casting interface on the freezing time can be obtained for some basic shapes. Before we give these results, we define two non-dimensional parameters, namely

$$\beta = \frac{(V/A)}{\sqrt{\alpha t_s}} \qquad ...(24)$$

$$\lambda = \frac{(\theta_f - \theta_0)}{\rho_m L'} \rho_c \qquad ...(25)$$

with

$$L' = L + C_m (\theta_p - \theta_f) \qquad ...(26)$$

In terms of these parameters, equation (23) can be rewritten as

$$\beta = \lambda \frac{2}{\sqrt{\pi}} \qquad ...(27)$$

A similar result for an infinitely long cylinder is

$$\beta = \lambda \left(\frac{2}{\sqrt{\pi}} + \frac{1}{4\beta} \right) \qquad ...(28a)$$

and for the sphere is

$$\beta = \lambda\left(\frac{2}{\sqrt{\pi}} + \frac{1}{3\beta}\right) \qquad ...(28b)$$

9.3.2 Solidification with Predominant Interface Resistance

In some common casting processes, the heat flow is controlled significantly by the thermal resistance of the mould-metal interface (indicated as region 3 in Figure 9.14). These processes include permanent mould casting and die casting. The condition of no contact resistance exists only when the mould-metal contact is so intimate that a perfect wetting occurs, i.e., the casting gets soldered to the mould face. In this section, we shall consider the solidification process assuming that the thermal resistance at the interface is of overriding importance. In such a case, the temperature distribution, assuming no superheat, is as shown in Figures 9.17 and 9.18. We are considering again a problem of one-dimensional heat flow. Here, the rate of heat flow through the interface is

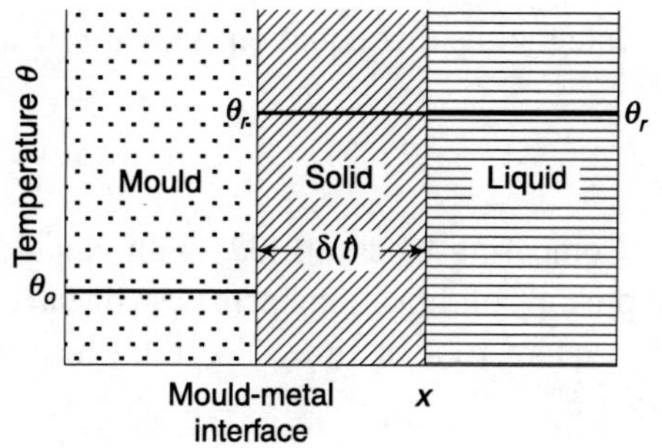

Figure 9.17 Temperature distribution in mould and solidified and liquid metal

$$Q = h_f\,(\theta_f - \theta_0)A \qquad ...(29)$$

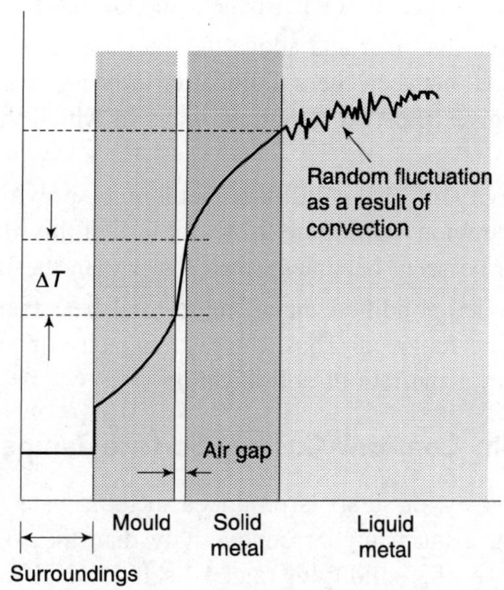

Figure 9.18 Interface resistance

where h_f is the film heat transfer coefficient of the interface and A is the surface area of the interface. If the solidification front at this instant is at a distance

$$\text{Rate of heat released} = \rho_m AL\frac{d\delta}{dt} \qquad ...(30)$$

Therefore, equating the right-hand sides of (29) and (30), we have

$$\frac{d\delta}{dt} = \frac{h_f(\theta_f - \theta_0)}{\rho_m L} \qquad ...(31)$$

Integrating this equation with $\delta = 0$ at $t = 0$, we get

$$\delta(t) = \frac{h_f\,(\theta_f - \theta_0)}{\rho_m L} \qquad ...(32)$$

Comparing equation (32) with equation (c) of Example 9.1, we see that, in the former, the depth of solidification increases linearly with time whereas in the latter, it is proportional to the square root of time. The total heat rejected by the casting is given by equation (32) with $\theta_p = \theta_f$.

The heat flow through the interface during the period of solidification t_s is $h_f\,(\theta_f - \theta_0)\,At_s$. So, t_s is obtained by equating this expression with equation (32), after substituting $\theta_p = \theta_f$, as

$$t_s = \frac{\rho_m L}{h_f(\theta_f - \theta_0)} \cdot \frac{V}{A} \qquad ...(33)$$

Equation (33) is helpful in estimating the solidification time of small, thin-section parts cast in a heavy metal mould as used in a die or permanent mould casting.

It may be noted at this stage that over and above the interface resistance, we have discussed, there are significant differences between the solidification process in a sand mould and that in a chill or metal mould. We give here two important ways in which the latter differs from the former:

(i) The thermal conductivity of the solidified metal may provide considerable thermal resistance, as shown by region 4 of Figure 9.14. Because of this, the surface temperature of the casting (θ_s), as can be seen, becomes much lower than the freezing temperature θ_f.

(ii) Because of the sub-cooled solidified metal, more total heat than that considered in the earlier sections has to be removed. Thus, the heat capacity of the solidifying metal also plays an important role in the rate of solidification.

9.3.3 Solidification with Constant Casting Surface Temperature

If large, slab-shaped casting (say, of steel) is produced in thin, water cooled mould made out of metal (say copper) having a much higher conductivity than the solidified casting, then the thermal resistance provided by the solidifying metal itself is significant. In such a case, the predominant thermal resistance is offered by region 4 (see Fig. 9.14). Neglecting the thermal resistances of all the other regions, the temperature distribution at any instant takes the shape shown in Figure 9.19. Here the mould metals interface (or the casting surface) temperature θ_s can be assumed to remain constant at its initial value θ_0, and θ_f indicates the freezing temperature of the metal and this is also taken as the pouring temperature. At any instant t, $\delta\,(t)$ indicates the depth of solidification. The process can be idealized, without much error, as a one-dimensional one. Hence, the solidification time t_s is obtained from $\delta(t_s) = h/2$, where h is the thickness of the slab being cast. The temperature profile within the range $0 < x < \delta(t)$ is given by

$$\frac{\theta - \theta_s}{\theta_\infty - \theta_s} = \text{erf}\left(\frac{x}{\sqrt{\alpha_s t}}\right) \qquad ...(34)$$

where α_s is the thermal diffusivity of the solidified metal and θ_∞ is a constant of integration. At $x = \delta(t)$, $\theta = \theta_f$; substituting these in equation (34), we get

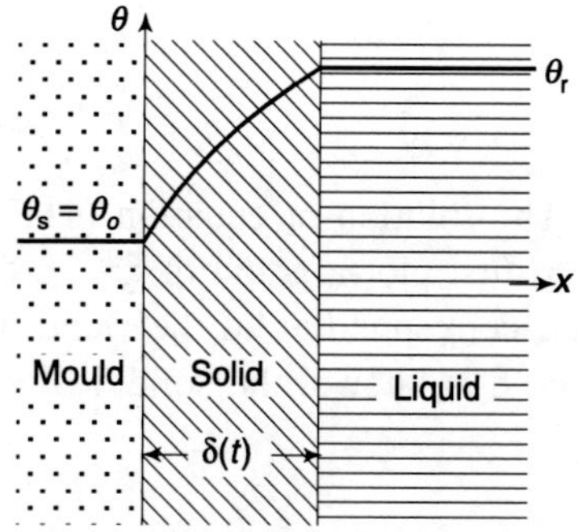

Figure 9.19 Surface temperature distribution with constant casting

$$\text{erf}\left[\frac{\delta(t)}{(2\sqrt{\alpha_s t}}\right] = \frac{\theta_f - \theta_s}{\theta_\infty - \theta_s} = \text{constant} = \lambda(\text{say}) \qquad ...(35)$$

$$\delta(t) = 2\xi\sqrt{\alpha_s t} \qquad ...(36)$$

where erf $(\xi) = \lambda$. Once again, we find the depth of solidification varies as the square root of time. Now, the constant ξ can be determined as follows. Considering the rate of energy flow balance at the solid-liquid interface, we have

$$k_2 \left.\frac{\partial y}{\partial x}\right|_{x=\delta} = \rho_m L \frac{d\delta}{dt} \qquad ...(37)$$

where

k_s = Conductivity of the solidified metal = $\alpha_s \rho_m C_s$,

ρ = Density of the metal (same in the solid and the liquid states),

C_s = Specific heat of the solidified metal, and

L = Latent heat of fusion of the metal.

From equation (34),

$$\frac{\partial \theta}{\partial x} = (\theta_\infty - \theta_s)\frac{d}{dx}\left[\text{erf}\,\frac{x}{2\sqrt{\alpha_s t}}\right]$$

$$(\theta_\infty - \theta_s)\frac{x}{2\sqrt{\alpha_s t}}\cdot\frac{2}{\sqrt{\pi}} \times \exp\left[-\left(\frac{x}{2\sqrt{\alpha_s t}}\right)^2\right] \text{ using equation (18)].} \qquad ...(38)$$

Again, using equations (38) and (37), we obtain

$$\frac{k_s}{\sqrt{\alpha_s}}(\theta_\infty - \theta_s)\frac{1}{\sqrt{\pi t}} \times \exp\left[-\left(\frac{\delta}{2\sqrt{\alpha_s t}}\right)^2\right] = \rho_m L\frac{d\delta}{dt} \qquad ...(39)$$

$$\sqrt{k_s \delta_m C_S}\,\frac{(\theta_f - \theta_s)}{\text{erf }(\xi)}\frac{1}{\sqrt{\pi t}}e^{-\xi^2} = \rho_m L\xi\sqrt{\alpha_s}\frac{1}{\sqrt{t}} \qquad ...(40)$$

or

$$\xi e^{\xi^2} \text{ erf } (\xi) = \frac{(\theta_f - \theta_s)}{\sqrt{\pi}} \cdot \frac{C_s}{L} \qquad \text{...(41)}$$

The solution of equation (41) for ξ can be obtained by trial-and-error. A simpler procedure is to first plot a graph of ξe^{ξ^2} erf (ξ) for various values of ξ. Then, from this graph, the value of ξ corresponding to the value of the right-hand side of equation (49) can be obtained. Thus, once ξ is known, the solidification time t_s, can be obtained, from equation (36) with

$$\delta(t) = h/2, \text{ as}$$

$$2\xi \sqrt{\alpha_s t_s} = \frac{h}{2} \text{ or } t_s = \frac{h^2}{16\xi^2 \alpha_s} \qquad \text{...(42)}$$

This analysis is valid only after the initial solidification stage (0.5 – 1 cm) is over. Similar results for the solidification time of the other shapes can be found from the available literature.

9.3.4 Solidification with Predominant Resistance in Mould and Solidified Metal

Let us now consider the problem discussed in Section 9.2.3 under a little different situation. We assume that the copper mould is quite thick and is not water cooled. Then, the mould-metal interface temperature θ_s, can no longer be assumed to remain at its initial value θ_0. The value of θ_s, still assumed to be constant, is decided by the thermal properties of the mould and the solidified metal. Moreover, after the initial stage of solidification, the interface resistance also becomes negligible. Thus, the only significant thermal resistance is offered by regions 2 and 4 (Figure 9.14) and the resulting temperature distribution at any instant is as shown in Figure 9.20. Assuming the mould to be a semi-infinite medium in the negative x-direction, the temperature distribution in the mould is

$$\frac{(\theta - \theta_s)}{(\theta_0 - \theta_s)} = \text{erf}\left(-\frac{x}{(2\sqrt{\alpha t}}\right) \text{ for } x < 0 \qquad \text{...(43)}$$

And that in the solidified metal is

$$\frac{(\theta - \theta_s)}{(\theta_\infty - \theta_s)} = \text{erf}\left(\frac{x}{2\sqrt{\alpha t}}\right) \text{ for } x > 0, \qquad \text{...(44)}$$

where θ_0 is the initial temperature of the mould, θ_s is the mould-metal interface temperature (θ_0), and θ_∞ is a constant of integration. The other conditions to be satisfied are:

(i) The balance of heat flux at the mould-metal interface, i.e.,

$$k_s \left.\frac{\partial \theta}{\partial X}\right|_{x=0^-} = K \left.\frac{\partial \theta}{\partial X}\right|_{x=0^+} \qquad \text{...(45)}$$

(ii) The rate of energy flow balance at the solid-liquid interface [see equation (37)], i.e.

$$k_s \left.\frac{\partial \theta}{\partial X}\right|_{x=\delta} = \rho_m L \frac{\partial \delta}{\partial t} \qquad \text{...(46)}$$

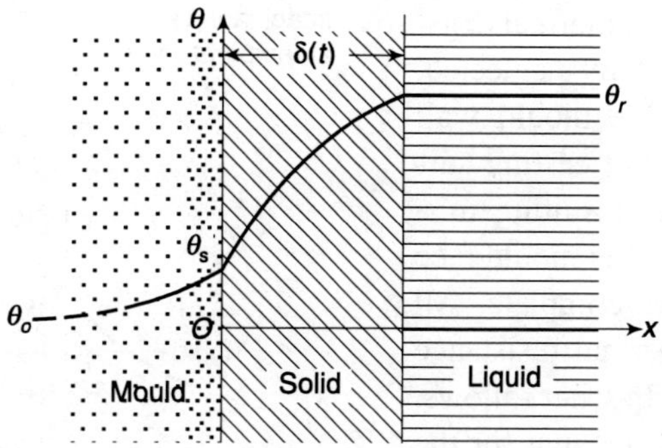

Figure 9.20 Temperature distribution

(iii) Condition is

$$\theta(\delta, t) = \theta_f \qquad \text{...(47)}$$

Combining equations (40) to (44) we get following three equations

$$\frac{(\theta_f - \theta_s) C_s}{L\sqrt{\pi}} = \zeta e^{\xi^2} \operatorname{erf}(\xi) \qquad \text{...(48)}$$

$$\frac{(\theta_\infty - \theta_s) C_s}{L\sqrt{\pi}} = \xi e^{\xi^2} \qquad \text{...(49)}$$

$$\frac{(\theta_s - \theta_0)}{(\theta_\infty - \theta_s)} = \left(\frac{K_s \rho_m C_s}{K \rho C}\right) 1/2 = \phi \text{ (constant)} \qquad \text{...(50)}$$

where ξ is the same as in Section 9.2.3. The three unknowns, namely, θ_s, θ_0, and ξ, can be solved from equations (48)-(50) as follows. Using equations (49) and (50) we can write,

$$(\theta_s - \theta_0) \frac{C_s}{L\sqrt{\pi}} = \xi e^{\xi^2} \phi. \qquad \text{...(51)}$$

Then, adding equations (45) and (48), we get

$$\frac{(\theta_f - \theta_0) C_s}{L\sqrt{\pi}} = \xi e^{\xi^2} [\operatorname{erf}(\xi) + \phi]. \qquad \text{...(52)}$$

Now, the left-hand side and ϕ in equation (52) are known; so, ξ can be determined either graphically or by trial-and-error, as mentioned in Section 9.2.3. In the former approach, a graph of ξe^{ξ^2} ξ versus [erf(ξ) + ϕ] should be drawn for the given value of ϕ, and ξ can then be solved for with the known value of the left-hand side of equation (52). Once ξ is known, the depth of solidification can be computed from equation (36) and the solidification time from equation (42). For such a casting to be feasible, it should be ensured that θ_s works out to be less than the melting point of the mould metal.

In the foregoing analysis, the thermal resistance at the interface was totally neglected. However, in real practice, the mould wall expands due to heating and the casting surface shrinks due to cooling, resulting in a gap. Moreover, the wetting of the mould face by the liquid metal is never complete. All these facts suggest that a thermal resistance at the interface is inevitable. In what follows, we attempt to approximately account for the thermal resistance of layers 2, 3, 4 (Figure 9.14). The temperature profile under such circumstances is as shown in Figure 9.21. Here, two separate surfaces are imagined on either side of the interface, at temperatures θ_{s_1} (on the casting side) and θ_{s_2} (on the mould side). The film heat transfer coefficient, h_f is apportioned on these two sides according to the equations

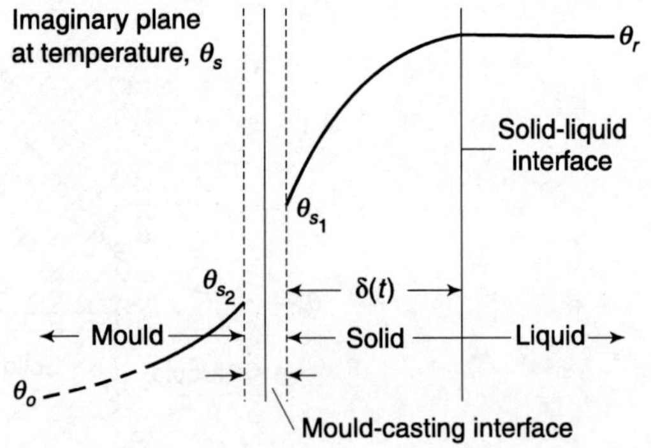

Figure 9.21 Methodology for taking thermal resistance at mould casting interface into consideration

$$h_2 = \left(1 + \sqrt{\frac{k\rho C}{K_s \rho_m C_s}}\right) hf \qquad \text{...(53)}$$

$$h_1 = \left(1 + \sqrt{\frac{K_s \rho_m C_s}{K\rho C}}\right) hf \qquad \text{....(54)}$$

The temperature, θ_s at the imaginary interface is calculated by neglecting the interface resistance (as done in Example 2.6). The solidification depth $\delta(t)$ can then be obtained as

$$\delta(t) = \frac{h_1(\theta_f - \theta_s)}{\rho_m La} t - \frac{h_1}{2k_s} \delta^2 (t) \qquad \text{...(55)}$$

where the factor, a is given by the relation

$$a = \frac{1}{2} + \sqrt{\frac{1}{4} + \frac{c_s(\theta_f - \theta_s)}{3L}} \qquad \text{...(56)}$$

Further, the solidification time t_s can be obtained from equation (55) with $(t_s) = \frac{h}{2}$, h being the thickness of the slab. The mould surface temperature θ_{s_2}, when the casting has completely solidified, is given by

$$\frac{\theta_{s_2} - \theta_0}{\theta_s - \theta_0} = 1 - e^{\gamma} [1 - \text{erf}(\sqrt{\gamma})] \qquad \text{...(57)}$$

where $$\gamma = \frac{h_2^2}{k^2} \alpha t_s \qquad \text{...(58)}$$

and θ_0 is the initial temperature of the mould.

Worked Out Examples

Example 9.1: A large plate of cross-sectional area, A is being cast. Establish a relationship between the time after pouring and the distance of the solidification front from the mould face (Figure 9.22), assuming no superheat.

Solution: During the solidification of this plate-shaped casting, most of the heat is rejected through two side faces, each having the cross-sectional area A. The heat rejected through the other four faces (having very small area as compared with A) is 'negligible'. So, the solidification fronts move from the two sides, as indicated in Figure 9.22.

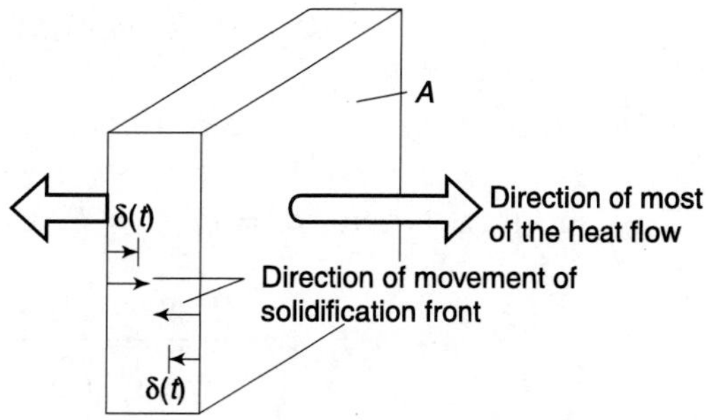

Figure 9.22 Movement of solidification fronts

If the solidification fronts move through a distance $\delta(t)$ at any instant t from the respective mould-metal interfaces, then the heat rejected by each solidified half is

$$Q_R = \rho_m A\delta La \tag{a}$$

Now, the time taken to reject this much of heat through the mould face of the area A is then, from equation (21),

$$Q_R = \frac{2AK(\theta_f - \theta_0)}{\sqrt{\pi\alpha}}\sqrt{t} \tag{b}$$

Equating the right-hand sides of (a) and (b), we get

$$t = \gamma\delta^2$$

or

$$\delta(t) = \frac{1}{\sqrt{\gamma}}\sqrt{t} \tag{c}$$

where γ is given by equation (23).

A smooth progress of the solidification fronts, as envisaged in the foregoing problem, is not observed in real life, especially in the case of alloys. In fact, all our foregoing analyses should be used only to provide an estimate and not the precise answer.

Example 9.2: Determine the solidification time of the following two iron castings when both are poured, with no superheats, into sand moulds at the initial temperature 28°C: (i) A slab-shaped casting 10 cm thick and (ii) a sphere 10 cm in diameter.

The data for iron is θ_f = 1540°C, L = 272 kJ/kg, ρ_m = 7850 kg/m^3, and for sand is c = 1.17 kJ/kg-K, k = 0.8655 W/m-K, ρ = 1600 kg/m^3

Solution:

(i) Let l, b, and h be the length, breadth, and thickness, respectively, of the slab. So, the volume of the casting,

$$V = lbh$$

And the surface area of the casting,

$$A = (lb + bh + lh) \approx 2lb \text{ (as both } l, b \gg h)$$

Hence, $$\frac{V}{A} \approx 5 \times 10^{-2}\,\text{m}$$

As superheat is zero, $\theta_p = \theta_f$, $L' = L$, and $\theta_0 = 28°\text{C}$. Using equation (25), we find that

$$\lambda = \frac{(1540 - 28) \times 1600 \times 1.17 \times 10^3}{7850 \times 272 \times 10^3} = 1.3256$$

So, from equation (27),

$$\beta = \frac{2}{\sqrt{\pi}} \times 1.3256 = 1.4957$$

Now,

$$\alpha = \frac{k}{\rho c} = \frac{\dfrac{0.8655}{1600 \times 1.17 \times 10^3}\,\text{m}^2}{\text{sec}} = 0.46 \times 10^{-6}\,\text{m}^2/\text{sec}$$

$$\alpha t_s = \frac{(V/A)^2}{\beta^2} = \frac{25 \times 10^{-4}}{2.24}\,\text{m}^2$$

So, the solidification time

$$t_s = \frac{25 \times 10^{-4}}{2.24 \times 0.46 \times 10^{-6}} = 2430 \text{ seconds} = 0.675 \text{ hour}$$

(ii) Using equation (24) with $\lambda = 1.3256$, we obtain

$$\beta = 1.4957 + \frac{0.4419}{\beta} \quad \text{or} \quad \beta^2 - 1.4957\beta - 0.4419 = 0$$

If we take only the positive real root, then $\beta = 1.7$. Now for a sphere of radius $R(=5)$,

$$\frac{V}{A} = \frac{(4/3)\pi R^3}{4\pi R^2} = \frac{R}{3} = \frac{5}{3} \times 10^{-2}\,\text{m}$$

$$\alpha t_s = \frac{(V/A)^2}{\beta^2} = \frac{25}{9} \times \frac{10^{-4}}{3.05}\,\text{m}^2$$

or $$t_s = \frac{25}{9} \times \frac{10^{-5}}{3.05} \times \frac{10^6}{0.46} \text{ sec} = 198 \text{ seconds} = 0.055 \text{ hour}$$

Example 9.3: Determine the solidification time of the slab-shaped casting considered in Example 9.2 when

(i) the casting is done in a water cooled copper mould, and

(ii) the casting is made in a very thick copper mould.

In both cases, assume no resistance at the mould-metal interface and use the following data:

For iron	*For copper*
$C_s = 0.67 \dfrac{\text{kJ}}{\text{kg K}}$	$C = 0.376 \dfrac{\text{kJ}}{\text{kg K}}$
$k_s = 83 \dfrac{\text{W}}{(\text{m} - \text{k})}$	$\rho = 8960 \dfrac{\text{kg}}{\text{m}^3}$
$\theta_f = 1540°\text{C}$	
$L = 272$ kJ/kg	$k = 0.398$ W/m – K
$\rho_m = 7850$ kg/m^3	

Solution:

(i) In this case, $\theta_s = 28°\text{C}$ and $h = 0.1$ m. So, from equation (41),

$$\xi e^{\xi^2} \text{erf}(\xi) = \frac{(1540 - 28) \times 0.67}{272\sqrt{\pi}} = 0.21$$

or

$$\xi = 0.98$$

Now,

$$\alpha_s = \frac{K_s}{\rho_m C_s} = \frac{83}{7850 \times 0.67 \times 10^3} = 15.8 \times 10^{-6} \text{ m}^2/\text{sec}$$

The solidification time, t_s is given by equation (42) as

$$t_s = \frac{(0.1)^2}{16 \times (0.98)^2 \times 15.8 \times 10^{-6}} \text{ sec} = 0.0115 \text{ hour}$$

(ii) From the given data, using equation (50), we have

$$\phi = \left(\frac{83 \times 7850 \times .67}{398 \times 8960 \times 0.376}\right) 1/2 = 0.57$$

From equation (52) with $\theta_0 = 28°\text{C}$

$$\xi e^{\xi^2} [\text{erf}(\xi) + 0.57] = \frac{(1540 - 28) \times 0.67}{272\sqrt{\pi}} = 2.1$$

or

$$\xi = 0.815.$$

The solidification time, again from equation (42), is obtained as

$$t_s = \frac{(0.1)^2}{16 \times (0.815)^2 \times 15.8 \times 10^{-6}} \text{ sec} = 0.0165 \text{ hour}$$

The surface temperature of the mould (θ_s) can be found from equation (48).

Thus, $$\frac{(1540 - \theta_s) \times 0.67}{272\sqrt{\pi}} = 0.815e\ 0.664 \text{ erf}(0.815)$$

$$= 0.815 \times 1.943 \times 0.75 = 1.188$$

Or $$\theta_s = (1540 - 854)°\text{C} = 685.4°\text{C}.$$

Since the melting point of copper is 1080°C, the mould will not melt away. Comparing the solidification times obtained in Example 9.2 with those in Example 9.3, we find that the solidification time in the thick copper mould is almost twice that in the water cooled copper mould and both are of an order of magnitude less than that in the sand mould. Thus, the production rate in a metal mould is much higher than that in a sand mould.

Example 9.4: Estimate the solidification time and the mould surface temperature for the casting considered in Example 9.3 (ii) with the total heat transfer coefficient across the casting-mould interface given as 1420 W/m^2°C.

Solution: From Example 9.3 (ii), θ_s = 982°C. Using equation (54), we have

$$h_1 = \left[1 + \sqrt{\frac{83 \times 7850 \times .67}{398 \times 8960 \times .376}}\right] 1420 \frac{W}{m^2} °\text{C} = 2230 \text{ W/m}^2 °\text{C}.$$

From equation (56),

$$a = \frac{1}{2} + \sqrt{\frac{1}{4} + \frac{.67\ (1540 - 982)}{3 \times 272}} = 1.34$$

Now, using equation (55) with $\delta(t_s)$, $= \frac{h}{2} = .05$ m, the solidification time t_s we obtain is

$$0.05 = \frac{2.230 \times (1540 - 982)}{7850 \times 272 \times 1.34} t_s - \frac{2230}{2 \times 83} (0.05)^2 \text{ seconds}$$

or $$t_s = 192.2 \text{ seconds} \equiv 0.0533 \text{ hour}$$

which is significantly more than that estimated in Example 9.3(ii). To compute the mould surface temperature, we first calculate h_2, using equation (53).

Now, from equation (58),

$$\gamma = \frac{3909 \times 3909 \times 0.398 \times 192.2}{398 \times 398 \times 8960 \times 0.37} = 2.19$$

So, using equation (57), the mould surface temperature θ_{s_2}, when the entire casting has solidified, we obtain is (with θ_0 = 28°C and using the error function tables)

$$\theta_{s_2} = 28 + (982 - 28)\ [1 - 8.935\ (1 - .962)]°\text{C} = 658.1°\text{C}$$

which, again, is much less than the value of θ_s estimated in Example 9.3(ii).

Review Questions

1. Explain the cooling curve for a pure metal during casting.
2. Explain the cooling curve for an alloy during casting.
3. Explain microscopic segregation.
4. What are the advantages of fins in casting?
5. Define rate of solidification.
6. Explain solidification with predominant interface resistance.
7. Explain solidification with constant casting surface temperature.
8. Explain solidification with predominant resistance in mould and solidified metal.
9. Explain solidification of a large casting in an insulating mould.

CHAPTER 10

Melting Practices

10.1 INTRODUCTION

Solid metal is converted to liquid state by heating it to a temperature above its melting point and maintaining the temperature at that level. The equipment used for heating is called melting furnace. The metal after melting is tapped or collected from the furnace in vessels called ladles and is poured immediately to the assembled mould. After solidification and cooling to room temperature, the cast product is taken out by breaking the mould. Melting and pouring are considered as the second part of the process of making a casting, while the first part is the production of moulds.

The choice of furnaces for melting metals and alloys for castings is governed by the cost of melting, the flexibility of the unit and the quality demanded on the finished product. Melting is the most difficult problem in foundry and methods of melting should be standardized for all the metals and alloys. There are various types of melting furnaces in use and all of them follow batch type charging except cupola furnace, which is a continuously operating one.

10.2 MELTING OF METALS

When the temperature of a metal reaches its melting point, the strength of the solid metal is reduced. By further adding heat energy to overcome the latent heat of fusion, the metal gradually begins to melt converting to liquid state. This process is done inside an equipment called melting furnace.

Heat Requirements Almost all metals used for engineering products can be cast in foundries. Grey cast iron and cast steel are the metals covering the major share of cast products by weight. Alloys of aluminium and copper come next to them. Zinc, tin, lead, magnesium and nickel are cast comparatively in lesser quantities. The properties and the heat requirement for melting pure metals are given in Table 10.1.

Melting, Tapping and Pouring Temperatures The super heat requirement for melting pure metals is about 30 to 125°C. But the metals melted in foundries for casting products are generally in alloy form in order to get specific engineering properties. This alloying process can be done by adding the alloying materials, while filling the metal charge or in the ladle while the metal is in molten form. To get defect-free castings, the pouring temperature of the metal

Table 10.1 Properties and heat requirements for melting pure metals

Metal	Melting Point °C	Boiling Point °C	Total heat required for melting		
			Latent heat of fusion Cal/g	For 1 kg kcal	For 1 litre kcal
Iron	1537	2900	65.0	279	2204
Nickel	1453	2900	72.1	259	2305
Copper	1083	2500	48.9	161	1449
Aluminium	660	2500	92.7	244	659
Magnesium	650	1100	85.6	268	456
Zinc	420	910	26.3	68	482
Lead	327	1900	5.7	16	187
Tin	232	2600	14.2	27	197

should be sufficiently high to keep the fluidity of metal in the prescribed level. During the tapping of metal from the furnace, handling the metal in the ladle to the mould, and pouring, the temperature of liquid metal drops down about 20 to 40°C per minute for a 50 kg ladle capacity and 5 to 7°C per minute for a 300 kg ladle capacity. That means for large ladles, the temperature drop rate is drastically reduced. The metal tapping and pouring temperatures of commonly cast alloys are given in Table 10.2.

Table 10.2 Melting, tapping, pouring temperatures of commonly cast alloys

Alloy	Density g/cm^3	Temperature °C		
		Melting	Tapping	Pouring
Grey Iron	7.1 – 7.3	1150 – 1260	1360 – 1480	1260 – 1400
Malleable Iron	7.2 – 7.4	1150 – 1350	1410 – 1480	1380 – 1450
Cast Steel	7.8	1420 – 1520	1520 – 1620	1500 – 1600
Bronze	8.6	1000 – 1050	1150 – 1200	1100 – 1150
Brass	8.6	900 – 1050	1050 – 1150	1000 – 1100
Aluminium Alloys	2.55 – 2.83	610 – 660	750 – 830	700 – 780
Magnesium Alloys	1.8 – 1.83	600 – 650	730 – 810	680 – 780

Source of Heat The heat energy source for melting is fuels or electric power. By burning solid, liquid or gaseous fuels in the presence of oxygen, the required temperature can be obtained inside a fuel fired furnace. Coal and coke are solid fuels and can be used as powdered bituminous, anthracites lumps or as briquettes. Fuels in oil or gaseous form are sprayed at pressure through burners and burnt in the presence of oxygen available in a stream of pressurized air supplied by blowers. Various fuels have different burning temperatures. Table 10.3 shows fuels and their use in various furnaces and the metals melted in them.

Electrical energy is found to be a better source than fuel energy for melting process due to its simplicity, neatness, higher temperature and control facilities. But economically, it is much costly both in investment and working. Electrical power is converted to heat by arc, induction or resistance. Table 10.3 gives the details.

Table 10.3 Heat source for furnaces and alloys melted

Heat source		Energy form	Metal melted	Furnace
Fuel energy	Coal	Powdered bituminous	Cast iron	Air furnace
		Anthracite lump or briquettes		Cupola
		Sized lumps		
	Oil	Liquid	Nonferrous metals, iron, steel	Crucible furnaces open hearth
	Gas	Gas		Crucible furnaces open hearth
Electrical energy	Electric power	Direct arc	Steel, Cast iron	I, II, III phase arc
		Indirect arc	Nonferrous, cast iron	Rocking arc
		Induction	All metals	Tilting and lift induction
		Resistance	Nonferrous	Rocking arc

10.3 CLASSIFICATION OF MELTING FURNACES

Metal melting furnaces can be broadly divided into two classes according to the energy used for heating:

1. Fuel fired furnaces
2. Electrical furnaces

These furnaces can also be grouped depending on the type of construction or the method of operation as lift-out type, tilting type, rotary type, shaft type, box type, etc.

10.3.1 Fuel Fired Furnaces

Solid, liquid or gaseous fuel is burnt inside a fuel fired furnace for producing heat to melt metal. Required oxygen is supplied as air blast using blowers. The flue gases formed by burning of the fuel are expelled through chimney fitted over the furnace. The following are the types of fuel fired furnaces used for metal melting:

1. Crucible furnaces
   (a) Coke fired pit type crucible furnace
   (b) Oil or gas fired stationary crucible furnace
   (c) Oil or gas fired tilting furnace
   (d) Oil or gas fired pot furnace
2. Cupola furnaces
   (a) Conventional cupola furnace

   (b) Water cooled cupola furnace
   (c) Coke-less cupola furnace
3. Miscellaneous fuel-fired furnaces
   (a) Coal fired hearth (air or reverberatory) furnace
   (b) Oil or gas fired open hearth furnaces
   (c) Oil or gas fired air or reverberatory furnaces

10.3.2 Electrical Furnaces

Electrical energy is converted to heat by arc, induction or resistance to flow of current using electrical furnaces. These furnaces are better in operation, control and neatness compared to the fuel fired ones. Higher temperatures for melting and super heating are easily obtainable. Close control over composition of melt is also possible. It is to be noted that the initial investment and working cost are high for electrical furnaces. A list of electrical furnaces is given as follows:

1. Arc furnaces
   (a) Direct arc furnace
   (b) Indirect arc (rocking arc) furnace
2. Induction furnaces
   (a) Coreless induction furnace
   (b) Induction furnace—channel type (with core)
3. Resistance type furnaces
4. Special type electrical furnaces
   (a) Electro-slag remelting furnace
   (b) Electron-beam melting furnace
   (c) Plasma melting furnace
   (d) Vacuum induction melting furnace

10.4 CHOICE OF A FURNACE

The choice of a furnace depends on several factors that can significantly influence the quality of castings, the economics of casting operations and the production rate. The important points to be considered are as follows:

1. Initial cost as investment
2. Maintenance and repair cost
3. Composition and melting point of alloy
4. Control of furnace atmosphere
5. Capacity and rate of melting required
6. Power supply and its availability or cost of fuels
7. Ease of superheating the metal

8. Type of charge material that can be used
9. Melting efficiency of the furnace
10. Degree of control on metal purification and refining
11. Environmental considerations such as air pollution and noise

Comparison of different heat sources and the relative performances in a furnace are shown in Table 10.4. It is to be noted that induction furnaces are extremely good in function, but require high investment and maintenance cost.

Table 10.4 Comparison of heat sources for melting furnaces

Item	*Coke*	*Oil*	*Gas*	*Electric (Arc)*	*Electric (Induction)*
Initial cost	Very low	Low	Low	High	Very high
(Relative index)	(1.0)	(2.0)	(2.0)	(20)	(40)
Running cost	Low	Very high	Very high	High	High
Maintenance cost	Low	Low	Low	High	High
High temperature capacity	Good	Low	Medium	Very good	Very good
Heating rate	Low	Medium	High	Very high	Very high
Temperature control	Poor	Good	Good	Very good	Very good
Pollution generation	High	Very high	Low	Low	Very low

10.5 CRUCIBLE FURNACES

Melting in crucible furnaces is one of the oldest and the simplest techniques of melting.

The metal to be melted is placed in a cup-shaped vessel known as crucible and is heated to a temperature above the melting point of the metal. This melts the metal after absorbing sufficient heat from the surroundings to a temperature above its melting point. After melting, the molten metal inside the crucible is poured directly to the mould using crucible holders or taken in a ladle by tilting the furnace and then poured. Crucibles are made of silicon carbide, graphite or other refractory materials. The metal melted in a crucible furnace is usually nonferrous and the quantity ranges from a few kilograms to 150 kg. Crucibles are available in market with a capacity to melt copper of 1 kg to 400 kg and are numbered as 1 to 400.

1. Coke fired pit type crucible furnace

In a pit type crucible furnace, the crucible is kept below the foundry floor level inside a pit. Figure 10.1 gives sectional view of a coke fired pit furnace. Here, a pit of rectangular shape is constructed using brick and concrete. One half of it is converted to cylindrical form using fire bricks as shown in Figure 10.1. This cylindrical portion has an ash pit at the bottom separated by iron bars forming the grate for the fire. Air from the blower enters the coke bed through the ash pit and grate-bars for burning as shown. The crucible placed over a steel base on the grate and the coke is burnt around it. The fireclay-lined lid and the chimney direct the flue gases to atmosphere during burning. The remaining half of the pit is covered by portable steel grill so that the top surface can be used as floor for movement of operators. The use of the free space

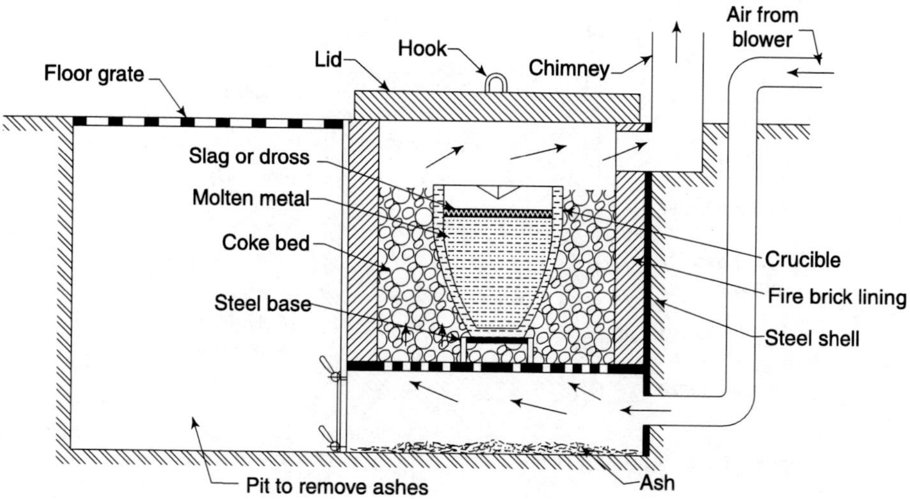

Figure 10.1 Coke fired pit type crucible furnace

under the floor grill is left for the entry of the operator for maintenance of the furnace and the removal of ash from pit.

To melt metal in the furnace, first coke is filled up to the steel base level and is kindled to burn. Then the crucible is lowered into the furnace and coke is added to fill surrounding the crucible. Metal is then charged inside the crucible and the cover is placed in position to facilitate chimney draft. The blower is then run to the full flow level for burning the coke. When the metal melts and reaches the desired temperature, the blower is switched off and the lid is removed for inspection as well as removal of dross or slag from the top of the molten metal. When the metal is ready with the required temperature, the crucible is taken out by means of a special long handled-tong and carried for pouring into the mould.

2. Oil or gas fired stationary type crucible furnace

Figure 10.2 shows sketch of a stationary type furnace in which the crucible is set inside a refractory-lined cylindrical steel shell that has heating burner fitted at the bottom. The crucible is placed at the centre of the furnace on a base of refractory brick. Air from the blower and the fuel enters with pressure through the side of the furnace at the bottom portion. The fuel may be gas or oil, and it should have enough pressure to make a fine spray to the air stream through the burner. The burnt fuel in blue flame circulates around the crucible for heating and the flue gases pass out through the hole of the lid to the chimney.

To start firing, the air blower is switched on to half pressure and the fuel burner is opened to a very low rate. Immediately the burner fire is kindled using a flame fitted on a rod. Then the crucible is kept in position, the metal is charged inside it and the lid is placed. The blower and the burner are then opened to full level for heating the metal till melting. After melting, the metal is poured to moulds as explained for pit furnace.

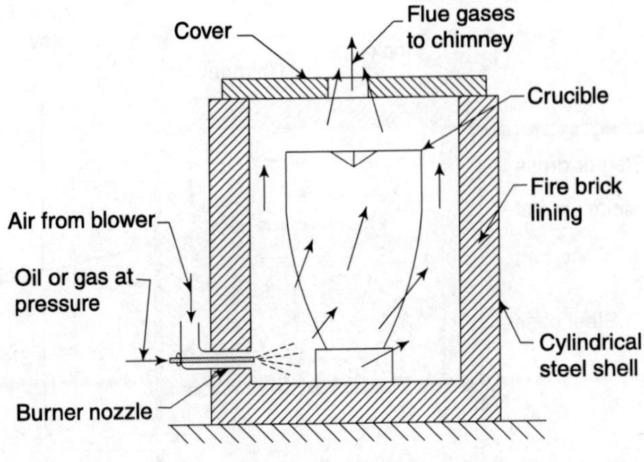

Figure 10.2 Oil or gas fired stationary type crucible furnace

3. Oil or gas fired tilting type crucible furnace

Figure 10.3 gives line sketch of oil or gas fired tilting type crucible furnace. The construction is exactly similar to that of a stationary type explained earlier with a difference that the furnace can be tilted about a trunnion (bearing) to pour the molten metal through the spout. To facilitate this, the air pipeline from the blower is passed through the bearing which acts as the trunnion. An additional support is also given to the crucible at the top in order to keep it in steady position during tilting. The melting process is as explained in the stationary type.

The location of the trunnion (bearing for tilting) can be any of the two positions shown in Figure 10.3 [(a) and (b)]. The centre axis type is a balanced one, hence the tilting is easy, but the lip axis type enables fast emptying of molten metal from the crucible. Tilting type furnaces are commonly used in foundries for batch type melting of nonferrous metals in small and medium rate of production.

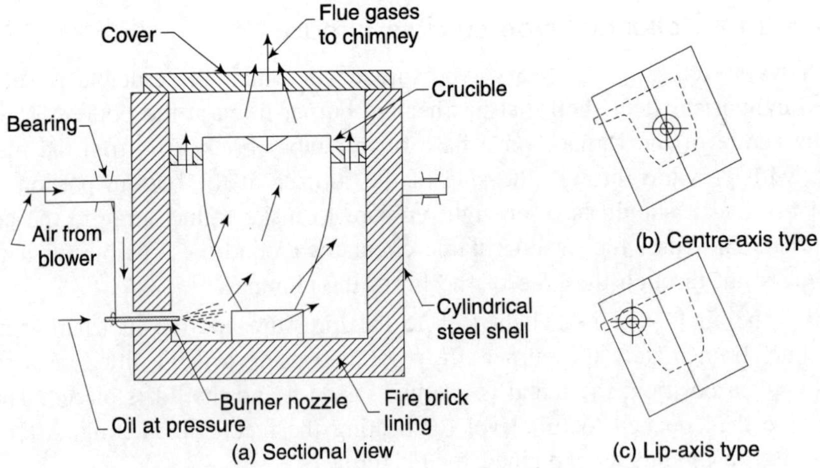

Figure 10.3 Oil or gas fired tilting type crucible furnace

4. Oil or gas fired pot furnace

Figure 10.4 shows line sketch of a gas or oil fired pot furnace. In this furnace, the crucible is replaced by a steel or cast iron pot. The working principle is same as that of stationary type oil or gas fired crucible furnace. The metals melted are magnesium, lead, zinc, tin, lead or aluminium, and their alloys and the quantity is usually small up to 100 kg in a charge. Pot furnaces are generally used for permanent mould casting or die casting of nonferrous metals.

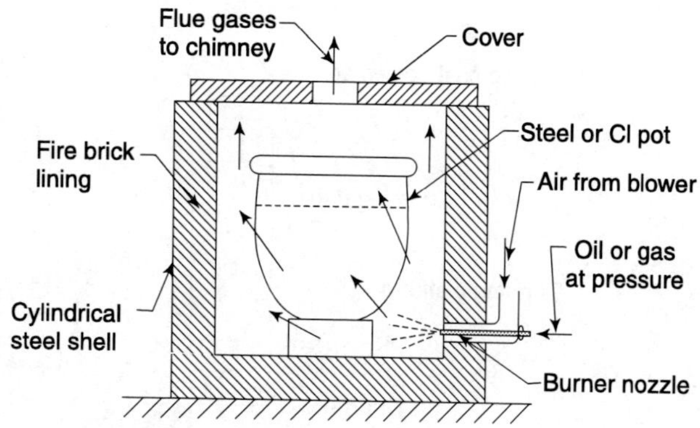

Figure 10.4 Oil or gas fired pot furnace

10.6 CUPOLA FURNACE

Cupola consists of a cylindrical steel shell with its interior lined with heat resisting fire bricks. It is used for melting grey cast iron. The fuel used is coke, and the flux added is limestone. It consists of drop doors at the bottom after closing of which a proper sand bed could be prepared. This sand bed provides the necessary refractory bottom for molten metal and the coke. Immediately above the sand bed is the melting tapping hole which is initially closed with clay till the molten metal is ready for tapping. Above the metal tap hole normally in a position opposite to it is the slag hole through which the slag generated during the melting process is tapped.

Slag generated above is the wind box which is connected to the air blowers supplying the requisite air at a given pressure and quantity. The air enters the cupola through the tuyeres. A little above the charging platform is the charging hole in the shell from where the charge consisting of a combination of pig iron, scrap iron, coke and fluxes, is put into the cupola. The refractory lining above the charge door need not necessarily be as thick as that below, since it is not exposed to much heat, as represented in Figure 10.5. The basic unit consists of:

- The cupola
- The blast delivery system
- The charging system
- The fore-hearth or duplexing furnace
- The slag-handling system
- The emission cleaning system

10.6.1 Types of Cupola

The types of cupola can be classified as

A. Based on lining used
  - Conventional, refractory lined
  - Water-cooled lining-less (water jacketed or external spraying)

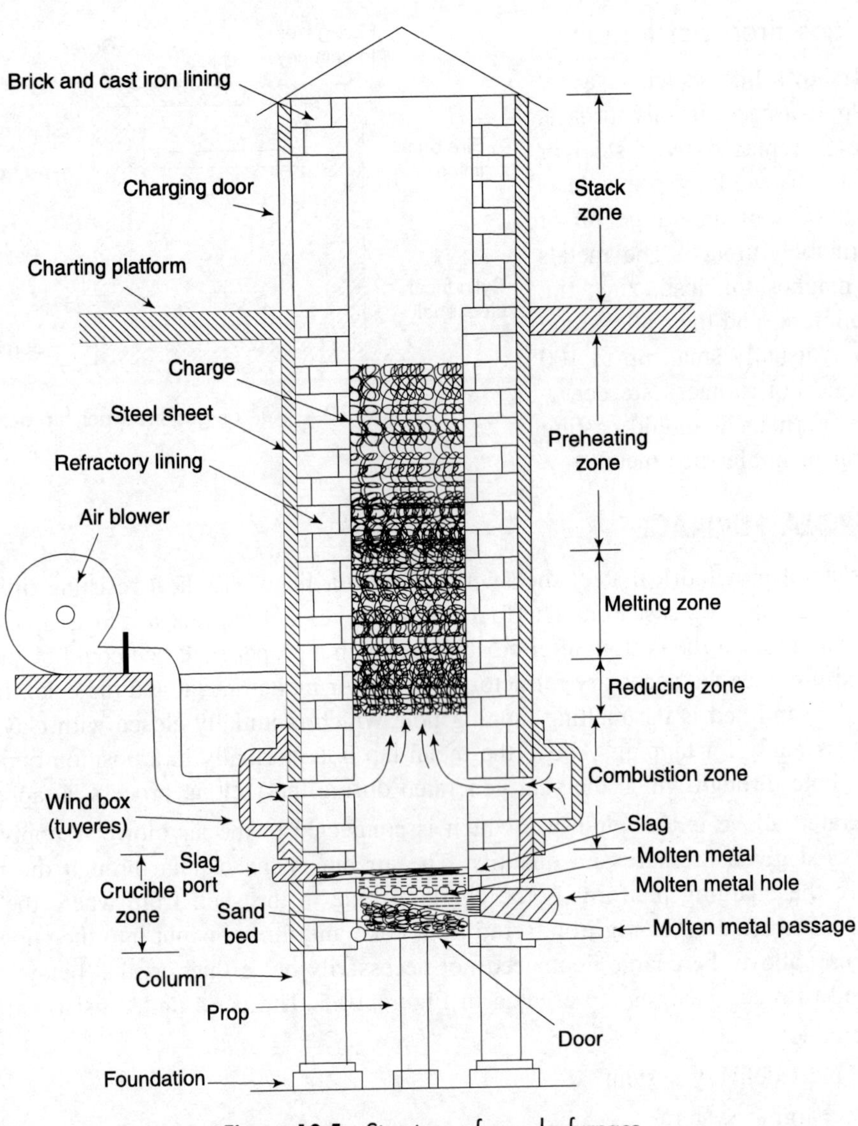

Figure 10.5 Structure of cupola furnace

- Water-cooled partially lined
- Combination water-cooled lined

B. Based upon slag system produced
- Acid-slag cupola
- Basic-slag cupola (to provide low-S and/or high-C iron)

C. Based upon energy conversion
- Hot-blast cupola (air temperature 375-425°C)

- Divided-blast cupola
- Coke-less cupola

I. Zones in a Cupola

The vertical travel distance of the cupola charge from the entry point to the exit as molten metal through the tapping hole, can be divided into five zones:

(a) Hearth or crucible: This is the portion located between the lower edge of the tuyeres and the bottom, of the cupola. No combustion takes place in this zone.

(b) Oxidizing or combustion zone: This zone is located above the hearth and here intensive combustion of the fuel takes place by oxygen of the blast due to which lot of heat is produced in the cupola. Heat is also evolved due to combustion as well as oxidation of silicon and manganese. Due to this high heat, molten drops of cast iron trickles down into the hearth. The reactions taking place in the zone are:

$$C + O_2(g) \rightarrow CO_2(g) + \text{Heat}$$

$$Si + O_2(g) \rightarrow SiO_2(s) + \text{Heat}$$

$$2Mn + O_2 \rightarrow 2MnO\ (\text{Liquid}) + \text{Heat}$$

(c) Reducing zone: The reducing zone is located above the combustion zone to the top of the coke bed. In this zone, the reduction of CO_2 to CO occurs and the temperature drops from 1600°C to 1200°C at the coke bed. Due to the reducing atmosphere, the charge is protected from any oxidizing influence. The following reaction takes place in this zone:

$$CO_2 + C(\text{Coke}) \rightarrow 2CO - \text{Heat}$$

(d) Melting zone: This is the layer of iron just above the coke bed. The charge starts melting here and trickles down through the coke to the bottom of the cupola. A considerable carbon pick-up by the molten metal also occurs in the zone according to the following reaction:

$$3Fe + 2CO = Fe_3C + CO_2$$

(e) Preheating zone: This zone occupies the space between the upper limit of the melting zone to the charging door. In this zone, moisture and volatile matter are evaporated and the materials are heated up. In the preheating zone, the burnt gases coming from the bottom side heat up the charges to a temperature ranging from 400°C to 1200°C. The coke dries first and becomes volatile at higher temperatures. The limestone charged as flux dissociates completely by the reaction:

$$CaCO_3 = CaO + CO_2$$

II. Cupola Lining

Here we discuss the effects that the refractory bricks and materials in the cupola withstand.

(a) Melting zone
- Refractories must withstand high temperature, slag attack, mechanical erosion, fluxing of limestone, coke, ash and sand and abrasion due to charge and air blast.

(b) Crucible zone
- Heat corrosion, abrasion by charge and air blast are to be witheld.
- Less severe than melting zone.

(c) Shaft
- Abrasion mainly happens due to descending charge in the shaft.
- Charging door must withstand the heat of stacked gases.
- Lining is done by fireclay or cast iron to avoid abrasion due to charge.

(d) Lining of acid cupola
- Lining is done by silica bricks, high heat duty fireclay, super duty fireclay and high alumina bricks.
- Super duty fireclay is used mainly in melting zone as it has high refractoriness.
- It has to be fixed hard due to increased porosity by high alumina content.
- The bricks which are hand moulded or machine pressed give strength and low porosity to the brick.

(e) Lining of basic cupola
- Lining is done by magnesite, forsterite, stabilized and semi-stabilized dolomite brick.
- Reduction of sulphur and phosphorus contents is needed for the production of high quality cast iron such as nodular cast iron and high duty cast iron using basic slags.
- Magnesite is the most suitable refractory for basic cupola. However, it suffers from high thermal expansion, high thermal conductivity, poor spalling resistance and high initial costs.
- Sintered dolomite or semi-stabilized tar-bonded dolomite absorbs moisture from atmosphere and is, therefore, advantageous only with continuous cupola.
- Stabilized dolomite is the best among the basic refractories from stability and cost considerations.

(f) Carbon refractories
- Carbon is recommended for lining crucible hearth and tap and slag holes because of the properties like:
    - Very high melting point.
    - Not wetted by molten metal or slag.
    - Chemically neutral to both acids and bases.
    - High resistance to mechanical and thermal shocks.
- But it sometimes gets oxidized above the red heat.

III. Raw Materials Used in Cupola Operation

Cupola melting is a complex processing method occurring at high temperature in which metallic raw materials, the combustion materials, the molten iron products and the gas and slag by-products are all put together.

Materials required for melting cast iron in a cupola are as follows:

(a) Pig iron

- It is a blast furnace product which may have varying amounts of elements like silicon, manganese and phosphorus.
- The type of pig iron depends on the chemical composition of the cast iron.
- Pig iron is the original melting material for iron castings. Until 1950, it was widely believed that, in order to achieve consistent, good quality grey iron castings, it was necessary to retain pig iron as the dominant material in the charge. Typical charge consists of 40-50% pig iron, 25-30% foundry returns and 20-30% ferrous scrap.
- After 1950, open hearth furnaces (large consumers of ferrous scrap) became obsolete. Typical charge consisted of 7% pig iron and 93% scrap.
- In practice, many large foundries use no pig iron at all. Only small plants use about 10-20% pig iron.

Direct reduced iron (DRI)

- Also been referred to as pre-reduced iron, metallised iron and sponge iron.
- DRI is the product of a reduction process using carefully selected, superior quality raw material, especially high quality iron ore.
- Pig iron is a product of total reduction of iron ore. Iron ore completely deoxidized to metallic iron state, melted and superheated about 1595°C, impurities removed as slag, significant percentage of C, Si and Mn are absorbed, which are useful for the foundry men, even though some tramp elements like S and P are introduced.
- DRI is the product of the almost total reduction of iron ore. Iron ore partially deoxidized into the solid state, leaving 5-10% FeO in the product, all impurities remained disseminated through the product, and although it contains up to 0.15% C, no Si or Mn is dissolved in the product. It also does not contain any tramp elements.
- Advantages of DRI is uniformity in composition and low tramp elements.
- Disadvantages are low iron yield due to presence of gangue materials; wasteful oxidation of Si and Mn of charge by reaction with FeO of DRI; high coke consumption; reoxidise and produce heat while in storage by reacting with water and oxygen even at room temperature.

(b) Scrap

- Scrap in the cupola charge may include gates and sprues, rejects from the foundry, turnings and other waste products.
- The use of scrap in the charge reduces the cost of the charge and should be thoroughly cleaned before adding to the charge.

Most important chemical factors determining scarp quality are:

1. Gross chemical analysis of five elements— C, Si, Mn, S and P: To produce grey cast iron, considering all other factors being the same, a tonne of cast iron scrap is worth more than a steel scrap.
2. Residual or tramp alloy analysis: Tramp elements are frequently troublesome, especially for ductile irons. Tramp elements are Cu, Ni, Cr, Mo, Sn, Al and Pb.
3. The melting yield of charge material: Strongly determines the true values of the scrap. Low yield charge often contains non-metallic material which generates gas or slag and is often associated with tramp elements.
4. The by-product disposal effects on the environment.

(c) Alloying additions

- The lack of any particular element is compensated by the addition of suitable amounts of ferromanganese, ferrochromium, ferrophosphorus, ferrotitanium, etc., to the charge or to the ladle to achieve the desired composition of cast iron.

(d) Coke

- It is primarily used as a fuel to supply the necessary heat for melting the iron, and tapping it satisfactorily. It is the source of carbon.
- The oxygen of the air blast comes in contact with the carbon of the coke above the ignition point of the coke, hence the following reaction takes place:

$$C + O_2 \rightarrow CO_2 + \text{heat} \qquad \text{...(i)}$$

- Some of the carbon dioxide produced reacts with excess of carbon (coke) present in fuel bed according to the following reaction:

$$CO_2 + C \rightarrow 2CO - \text{heat} \qquad \text{...(ii)}$$

- Reaction (i) is exothermic while reaction (ii) is endothermic and therefore, temperature in the vicinity of 1700-2000°C is produced at the centre of the cupola.
- For maximum efficiency, the coke should burn to CO_2 with the minimum production of CO. The CO generated, however, is helpful in protecting the metal charge from oxidation during melting. It is, therefore, evident that slow-burning, non-reactive types of coke are preferable for foundry use.

The requirements of suitable coke are:

1. **Moisture:** Moisture in coke results by the water used for quenching and the climatic conditions. It reduces the heating value of a fuel and desirable to have moisture content below a certain limit like 2-4%.
2. **Ash:** It is mixed with coke from the mineral matter associated with the parent coal which would reduce the calorific value of the coke. High ash content means more fluxing material to be added in the cupola operation. In India, the ash content of coke ranges between 18 and 25%.

3. **Volatile matter:** High volatile matter content indicates imperfect carbonization of the coal resulting in coke of poor physical properties, lower ignition temperature, and higher reactivity. The maximum limit of volatile matter in coke has been stipulated as 1.5-2%.
4. **Sulphur and phosphorus:** Both are picked up by the molten iron and are very harmful for the properties of cast iron. Hence these elements in the coke should be low. The maximum allowable limits of sulphur and phosphorus in Indian metallurgical coke are 0.7% and 0.3% respectively.

Besides above, the coke should have the following physical properties:

- **Size:** Large sized coke (generally greater than 2 inches) is desired. Large sized coke offers less resistance to the air blast and exposes smaller surface to combustion, thus retarding the reaction between CO_2 and C. Coke should also be of uniform size. Too much difference in the size of coke spreads the combustion zone, hinders the melting rate and lowers the maximum attainable temperature.
- **Strength and hardness:** Coke for cupola operation should have sufficient hardness and strength to withstand handling and to support the burden without producing an appreciable quantity of breeze.
- **Porosity:** Cokes of low porosity are preferred for foundry use.

Information on the specification of foundry coke in India is given in Table 10.5.

Table 10.5 Specification of coke

Properties of coke	*Grades of coke*		
	Special	*Grade No. 1*	*Grade No. 2*
Ash %	20.0	24.0	28.0
Volatile matter %	2.0	2.0	2.0
Fixed carbon %	78.0	74.0	70.0
Sulphur %	0.7	0.7	0.7
Phosphorus %	0.3	0.3	0.3
Shatter index % on 2″	85.0	85.0	-
Porosity %	40.0	40.0	-

(e) Fluxes

- Slags formed in the melting process contain ash, sand which has adhered to the surface of charge, fused refractory lining and the products of oxidation like SiO_2, MnO and FeO.
- Fluxing materials should be added to increase the fluidity of slag and fluxing agents may be limestone, basic open-hearth slag, fluorspar and soda ash.
- Limestone ($CaCO_3$) on decomposition gives calcium oxide which combines with the products of oxidation to form slag with a relatively low melting point which is not very viscous. It must be dense and homogeneous without contamination of sand or clay.
- Open-hearth slag also contains a considerable amount of CaO which also acts in a similar manner.

- The amount of flux added to the cupola depends on the ash content of the coke. Thus, when the coke contains about 12% ash, the addition of limestone is about 30% weight of the coke.

(f) Air

- In order to achieve uniform combustion of the coke and normal operation, a certain quantity of air per unit of time must enter the cupola.
- In some cases, better efficiency is obtained by using preheated air. About 40,000 cu. ft. of air is required per tonne of iron produced.

Complete guide on selection of cupola melting systems is summarized in Table 10.6.

Table 10.6 Selection of cupola melting systems

1. Process requirement	*2. Equipment requirement*
(a) Type of iron to be produced	(a) Cupola
(b) Chemistry	(b) Blast system
(c) Charge materials	(c) Charging system
(d) Typical charge makeup	(d) Emission cleaning system
(e) Spout metal temperature required	(e) Water system
(f) Metal handling	(f) Controls and instrumentation
(g) Slag handling	
(h) Available utilities	
3. Physical requirement	*4. Production requirement*
(a) Space availability	(a) Melting rate
(b) Access to equipment site	(b) Metal composition
(c) Elevation	(c) Melting schedule
(d) Plan	
(e) Equipment relation	
5. Miscellaneous factors	
(a) Metal transport from cupola without fore-hearth or duplexing furnace	
(b) External desulphurising	
(c) Slag disposal	
(d) Special attachment for collector, gas take-off, top cap or stack burners	
(e) Total weight of equipment to be supported by the cupola stack	
(f) Tool and maintenance equipment	
(g) Personal safety equipment	

IV. Cupola Operation

To melt iron using conventional cupola furnace, there are some systematic operations to be followed before, during and after the process. The furnace is usually run continuously for a period of 12 to 16 hours. On the next day, it is idled for patch work and preparations. For taking castings every day, two cupolas are used side-by-side in a foundry and fired on alternative days. The cycle of events that occur in a heat furnace are given below:

1. Preparation of cupola (the refractory lining on inside vertical surface, bottom, tapping hole and slag hole)
2. Kindling of cupola
3. Charging of cupola
4. Melting by starting the air blast and continuation of charging
5. Tapping and slagging
6. Ending of melting operation and dropping the bottom

The melting practice in cupola includes the following steps:

(a) Preparation of cupola

- The cupola bottom is dropped and running repairs are carried out after each heat treatment.
- The residue is then removed.
- Drop bottom is closed and sand bottom is prepared.
- Moulding sand should contain no more than 50% moisture and a minimum amount of clay.
- If moisture and clay contents are higher, the bottom may bulge and cause difficulties in dropping.
- The sand bottom should have a slope of 7° to 10° towards the tap hole.

(b) Kindling the cupola

- Kindling wood is put on the bottom of the cupola.
- Above that small pieces of fire wood are placed.
- Working door is sealed up with refractory bricks.
- A hole is left for igniting the firewood and air flow.
- When the firewood is burnt, a small quantity of coke is dropped on the wood.
- Some more coke is placed on the top when the coke is burning freely.
- Whole coke of the charge is added in doses.

Coke-bed height

- Primary importance for normal working of the heat and for the production of iron of good quality.
- When the bed height is too low, the oxygen of the air supply will not be consumed completely and will pass through the metal charge causing rapid melting losses due to oxidation of metal.

- The coke height should be such that all the oxygen of the air blast is consumed by the time the gases reach the top of the coke.
- If the coke bed is higher than is necessary for the complete combustion of oxygen, the following reaction will occur:

$$CO_2 + C \rightarrow 2CO - \text{Heat}$$

- Lot of heat is wasted as the above reaction is endothermic.
- As the oxygen is passed through the hot coke bed, it is gradually consumed.
- The coke bed height depends on the air volume and velocity.
- The hole in the working door is sealed up when the coke bed height is reached.

(c) Charging of cupola

- Charge preparation is necessary for smooth running of cupola.
- The size of pig iron pieces should not be greater than 1/3rd of cupola diameter.
- Big pieces of scrap are also broken into pieces weighing not more than 15 kg.
- Charging can be done manually or mechanically.
- The amount of the metallic charge is usually 1/10th of the per hour cupola capacity.
- The coke charge should weigh 10% of the weight of the metallic charge.
- The amount of flux charge is usually 20% by weight of the coke charge.

(d) Air blast melting by starting the air blast and continuation of charging

- The charge is kept for 20 to 50 minutes so that it is heated up with natural drift.
- The process of melting starts about 10 minutes after air blast is started.
- During this period, the tap and the slag holes are stopped with a thin mixture of moulding sand.
- About 40000 ft^3 of air per tonne of iron is required.

(e) Tapping and slagging

- The first tapping can be made 40 to 50 minutes after the air blast is started.
- Sufficient metal is collected in the hearth above the sand bed during this period.
- The slag hole is pierced with the help of a metal bar and the slag is collected.
- The clay plug is then removed from the tap hole.
- Molten metal is carried in ladles and carried to the moulds.
- The tap hole is sealed with clay plug.
- Thus, by repeated slagging and tapping all the metal is tapped and the operation is over.

(f) Ending of melting operation and dropping the bottom

- The blast is shut off as soon as the heat is finished.
- The remaining is allowed to fall on the floor.
- Water is sprayed on the remains.
- When they are cooled, they are removed from underneath the cupola.

V. Difficulties in melting

(1) Hanging

- It is the non-uniform descent of the charge in the cupola.
- It usually occurs in small diameter cupolas due to insufficient repair of the lining.
- Hanging may also occur due to oversized pieces in the charge.
- Hanging may be reduced by proper lining repair and by having a uniform size of charge materials.

(2) Metal run-out

- The molten metal may run out when the hearth has not been properly rammed.
- It can be stopped by cooling with a water jet.
- In case of serious run out, repair can be made with clay and bricks.

(3) Erosion of the cupola shell

- Due to insufficient lining, the cupola shell may be eroded by heat and molten iron.

(4) Blocking of the tap hole

- The tap hole may be blocked by small pieces of coke and due to viscous slag which may freeze on the coke lumps.
- So care should be taken so that the hole is not blocked in the initial stage of melting.
- A thin bar is used to clean the hole.

VI. Specification of cupola

The size of cupola is specified by the inside diameter of the furnace, after deducting the refractory lining thickness from the steel shell diameter. The value ranges from 0.5 m for a small size to 2 m for an extra-large one. They are designated by numbers starting from 0 to 12 respectively. Steel shell outer diameter is also sometimes considered for specifying the cupola size. Table 10.7 gives various dimensions and capacities of cupolas generally in use. The thickness of mild steel cylindrical shell is taken as 6 mm to 10 mm depending on the strength requirements.

Table 10.7 Dimensions and capacities of cupola

Diameter of melting zone, m	*Melting rate tonnes/h, Metal-coke ratio*		*Blast rate, m^3/h*	*Blast pressure, kPa*	*Typical charge, kg*			*Bed height above tuyeres, m*	*Shaft height from tuyeres to charge door still, m*
	10:01	08:01			*Coke*	*Iron*	*Limestone*		
0.50	1.97	1.57	1,340	10.20	20	200	7	1.00	2.50
0.60	2.84	2.46	1,940	10.50	28	284	9	1.00	3.00
0.80	5.11	4.36	3,450	11.20	51	510	17	1.05	3.00
1.00	7.99	6.83	5,380	11.70	80	800	26	1.05	3.50
1.20	11.50	9.79	7,750	12.70	115	1150	38	1.10	4.00
1.40	15.60	13.33	10,600	13.40	157	1570	52	1.10	4.00
1.60	20.44	17.41	13,800	14.40	200	2040	67	1.10	4.50
1.80	25.88	22.05	17,450	15.40	260	2590	85	1.15	5.00
2.00	31.95	27.22	21,550	17.20	320	3200	106	1.15	5.00

VII. Advantages and disadvantages of the cupola

The main advantages of cupola furnace are:

1. Fastest and continuous melting is possible. The furnace can be used for melting for 12 to 16 hours at a stretch.
2. Cost of melting is low.
3. Easy to operate and less floor area is required, since it is of vertical construction.
4. The initial investment required is comparatively low.
5. Operation and maintenance are not much complicated.
6. Cost of operation and maintenance is low.
7. High skill workers are not required to operate cupola successfully.
8. Composition control is easy.

The disadvantages of the cupola furnace are:

1. Certain elements like silicon and manganese are lost from the metal due to burning of coke and some others like sulphur are picked up. This may bring composition change.
2. Control over temperature is difficult to maintain during melting.
3. The waste gases have to be treated for pollution control.
4. Due to oxidation and carbon absorption, there is an excessive loss of alloying elements.
5. Melting to definite chemical composition is difficult.
6. Addition of alloys are made in ladle.
7. Cupola cannot be used to refine metals.
8. As cupola top is open to the atmosphere, a large amount of heat is wasted.
9. Obtaining low carbon percentage ($< 2.8\%$) is difficult.
10. Difficult to melt alloys of cast iron.
11. High temperatures are difficult to obtain in cupola.

VIII. Thermal efficiency of cupola

Thermal or melting efficiency is expressed as:

$$\%\ \text{Thermal efficiency} = \frac{\text{Heat used in melting and superheating the metal}}{\text{Potential heat in coke + Heat from oxidation of Fe, Si, Mn + Heat in air blast}} \times 100$$

IX. Charge calculations

It is very important in the foundry to know the final composition of the metal being obtained, so as to control it properly. The elements in the final analysis are essentially the sum total of what is contained in each of the charge ingredients, with some losses or pick-up in the cupola. Out of the various elements, the ones that are relevant are carbon, silicon, manganese and sulphur.

During cupola practice, the loss or gain of various elements is as follows:

- **Carbon:** It is customary to add from 0.1 to 0.15% to the theoretically calculated percentage

- **Silicon:** A loss of about 10% of silicon may generally be expected due to oxidation.
- **Manganese:** Loss of manganese due to oxidation is about 15%.
- **Sulphur:** It is customary to add from 0.03 to 0.05% to the theoretically calculated percentage.
- **Phosphorus:** There is practically no change in the phosphorus content.
- **Iron:** Loss of iron due to oxidation is only about 4%.

Factors considered in designing the charge makeup

- Size and number of cupolas
- Hours of cupola operation each day
- Iron to coke ratio
- Physical condition and density of scrap
- Maximum tonnage to be melted each hour.

Typical cupola input and output is given in Table 10.8, and Table 10.9 gives the composition changes during cupola melting.

Table 10.8 Typical input and output of cupola furnace

Cupola input	*Cupola output*
1.00 tonne of pig iron, scrap iron and steel	0.98 tonne of molten iron
0.15 tonne coke	0.05 tonne molten slag
0.03 tonne flux	1.35 tonne stack gases
1.20 tonne air	
2.38 tonnes total	**2.38 tonnes total**

Table 10.9 Composition changes during cupola melting

Element	*Changes in analysis*
Carbon	Pick up of about 10-20% of original carbon charged
Silicon	Loss of up to 10% of original silicon charged
Manganese	Loss of up to 15% of original manganese charged
Phosphorus	No change
Sulphur	Gain in total of about 0.03 - 0.05%
Chromium	Loss of up to 10% of original chromium charged
Molybdenum	Loss of up to 5% of original molybdenum charged
Nickel	No change
Copper	No change

Worked Out Examples

Example 10.1: Estimate the final composition of the cast iron produced with the following charge compositions and proportions (Table 10.10).

Table 10.10 Composition of cast iron

	Carbon %	Silicon %	Manganese %	Sulphur %	Phosphorus %	Charge %
Pig Iron 1	3.50	2.50	0.40	0.01	0.40	40
Pig Iron 2	3.20	1.50	1.00	0.02	0.60	35
CI Scrap	3.20	1.20	0.50	0.10	0.40	25

Let us now analyze the total amount of elements present in 1 tonne (1000 kg) of charge assuming carbon pick-up as 0.15%, silicon loss as 10% and manganese loss as 20% (Table 10.11).

Table 10.11 Melting charge analysis

	Charge mass (kg)	Carbon		Silicon		Manganese		Sulphur	
		%	kg	%	kg	%	kg	%	Kg
Pig iron 1	400	3.50	14	2.50	10	0.40	1.60	0.01	0.04
Pig iron 2	350	3.20	11.2	1.50	5.25	1.00	3.50	0.02	0.07
CI scrap	250	3.20	8	1.20	3	0.50	1.25	0.10	0.25
Total	1000		33.2		18.25		6.35		0.36
Charge %			3.32		1.825		0.635		0.036
Change in cupola			+0.15		−0.183		0.127		+0.050
Final Analysis			3.47		1.642		0.508		0.086

The foundry engineer can actually control his final analysis by actually trying with various mixes of charge materials available in his foundry to obtain an economical melt.

Example 10.2: In a foundry, it was required to obtain a cast metal with the following composition: carbon 3.20 to 3.60%; manganese 0.60 to 0.80%; sulphur 0.08% maximum; and phosphorus 0.40 to 0.60%. If the following raw materials are available, estimate the best charge proportions (Table 10.12).

Table 10.12 Composition available

	Carbon %	Silicon %	Manganese %	Sulphur %	Phosphorus %
Pig iron 1	3.50	3.00	1.00	0.02	0.40
Pig iron 2	3.20	1.50	0.50	0.01	0.80
Pig iron 3	3.50	2.50	0.80	0.02	0.50
Scrap 1	3.50	1.80	0.60	0.08	0.50
Scrap 2	3.20	1.20	0.60	0.10	0.40
Ferrosilicon		50.00			

Looking at the raw materials available, pig iron 2 has got very low silicon content, and therefore it should not be used. Similarly, scrap 2 which has low silicon and high sulphur may also be eliminated. So the final choices are pig iron 1 and 3, and scrap. Let the charge be of 40% scrap and pig irons in equal proportions of 30% each.

Let us analyze 1000 kg with the carbon pick-up assumed to be 0.15%, sulphur as 0.05%, the oxidation losses of silicon as 10% and that of manganese as 20% (Table 10.13).

Table 10.13 Analysis of charge

	Charge mass, kg	*Carbon*		*Silicon*		*Manganese*		*Sulphur*	
		%	*kg*	*%*	*kg*	*%*	*kg*	*%*	*kg*
Pig iron 1	300	3.5	10.5	3.00	9.00	1.00	3.00	0.02	0.06
Pig iron 3	300	3.5	10.5	2.50	7.50	0.80	2.40	0.02	0.06
Scrap 1	400	3.5	14.0	1.80	7.20	0.60	2.40	0.08	0.32
Total	1000		35.0		23.70		7.80		0.44
Charge %			3.50		2.370		0.78		0.044
Change in cupola			+0.15		−0.237		−0.156		+0.05
Final			3.65		2.133		0.624		0.094
Required			3.2 to 3.6		2.3 to 2.6		0.6 to 0.8		0.08 max

With the chosen charge composition, the final analysis differs from the required one with carbon, silicon and sulphur percentages. To reduce the carbon and sulphur percentages, we may have to use pig iron 2 which may reduce the silicon percentage but that can be made up by adding suitable amount of ferrosilicon.

The reduction in sulphur that could be achieved by the addition of 1 kg of pig iron 2 for scrap 1 is

$$\frac{0.08 - 0.01}{100} = 0.0007 \text{ kg}$$

The reduction in carbon that could be achieved by the addition of 1 kg of pig iron 2 for scrap 1 is

$$\frac{3.50 - 3.20}{100} = 0.003 \text{ kg}$$

The excess sulphur present in 1000 kg is

$$\frac{1000 \times (0.094 - 0.080)}{100} = 0.14 \text{ kg}$$

The excess carbon present in 1000 kg is

$$\frac{1000 \times (3.65 - 3.60)}{100} = 0.50 \text{ kg}$$

To reduce the sulphur to the desirable limit, the amount of pig iron 2 to be substituted is

$$\frac{0.50}{0.0007} = 200 \text{ kg}$$

To reduce the carbon to the desired limit, the amount of pig iron 2 to be substituted is

$$\frac{0.50}{0.003} = 167 \text{ kg}$$

Hence, substitution of 200 kg of pig iron would be sufficient to give the desired analysis. To confirm this, let us do a fresh analysis (Table 10.14).

Table 10.14 Analysis of melting charge

	Charge mass, kg	Carbon		Silicon		Manganese		Sulphur	
		%	kg	%	kg	%	kg	%	kg
Pig iron 1	300	3.50	10.5	3.00	9.00	1.00	3.00	0.02	0.06
Pig iron 2	200	3.20	6.4	1.50	3.00	0.50	1.00	0.01	0.02
Pig iron 3	300	3.50	10.5	2.50	7.50	0.80	2.40	0.02	0.06
Scrap 1	200	3.50	7.0	1.80	3.60	0.60	1.20	0.08	0.16
Total	1000		34.4		23.10		7.60		0.30
Charge %			3.44		2.41		0.76		0.03
Change in cupola			+0.15		–0.231		–0.152		+0.05
Final			3.59		2.079		0.608		0.08
Desired			3.2 – 3.6		2.3 – 2.6		0.6 – 0.8		0.08 max

It is not possible to increase silicon any further and therefore, the only way is to add Ferro-silicon of requisite amount to make up for the shortfall.

Amount of silicon to be added for 1000 kg is

$$1000 \times \frac{2.300 - 2.079}{100} = 2.21 \text{ kg}$$

This can be made up by adding 5 kg of ferrosilicon in place of scrap 1.

$$\text{Extra silicon added} = 5 \times \frac{50.0 - 1.80}{100} = 2.41 \text{ kg}$$

The final mixture of charge is given in Table 10.15.

Table 10.15 Final mixture of charge

Pig iron 1	300 kg	30%
Pig iron 2	200 kg	20%
Pig iron 3	300 kg	30%
Purchased scrap	195 kg	19.5%
Ferro silicon	5 kg	0.5%
Total charge	1000 kg	100%

In the previous example, we have proceeded with a trial-and-error method and have not considered the cost of various constituents of the charge. A more appropriate analysis would be to take into account the cost of each of the constituents of the charge and the chemical compositions. This can be done by formulating the linear programming problem and solving it by simplex method.

Example 10.3: Compute the probable analysis of the charging mixture so that the cast metal will have the following composition:

Total carbon : 3.3 – 3.5%
Silicon : 1.9 – 2.1%
Manganese : 0.5 – 0.65%
Phosphorus : 0.5 – 0.55%
Sulphur : 0.06% maximum.

The raw materials available are given in Table 10.16.

Table 10.16 Raw material availability

	Ci%	Si%	Mn%	S%	P%
Pig Iron 1	3.5	2.5	0.4	0.02	0.6
Pig Iron 2	3.2	1.5	1	0.01	0.4
Foundry returns	3.2	1.2	0.6	0.009	0.5
Purchased scrap	3.5	1.8	0.5	0.1	0.4
Ferrosilicon		50			

Let us try the following charging mixture (Table 10.17).

Table 10.17 First trial mixture

Pig iron 1	30%
Pig iron 2	20%
Foundry returns	40%
Purchased scrap	10%
Total charge	100%

In this case, let us consider a total mixture weight of 1000 kg.

The analysis of the cast metal on the basis of the above charging mixture is shown in Table 10.18.

Table 10.18 Computing charging mixture of cupola practice

Constituent of the charge	*% of the constituents assumed in the charging mixture*	*Total weight of the constituents in kg per 1000 kg of the charging mixture*	*Analysis of various elements in the constituents of the charging mixture*									
			C		*Si*		*Mn*		*S*		*P*	
			%	*Total wt. in kg*	*%*	*Total wt. in kg*	*%*	*Total wt. in kg*	*%*	*Total wt. in kg*	*%*	*Total wt. in kg*
Pig iron 1	30	300	3.5	10.5	2.5	7.5	0.4	1.2	0.02	0.06	0.6	1.8
Pig iron 2	20	200	3.2	6.4	1.5	3	1	2	0.01	0.02	0.4	0.8
Foundry returns	40	400	3.2	12.8	1.2	4.8	0.6	2.4	0.009	0.036	0.5	2

Purchased scrap	10	100	3.5	3.5	1.8	1.8	0.5	0.5	0.1	0.1	0.4	0.4
Total	100	1000		33.2		17.1		6.1		0.216		5
Analysis of aggregate charge (%)			3.32		1.71		0.61		0.0216		0.5	
Estimated loss or gain in melting (%)			0.15		(10% loss) –0.17		(15% loss) –0.09		0.03		No change	
Estimated analysis of cast metal (%)			3.47		1.54		0.52		0.0516		0.5	
Desired analysis (%)			3.3 – 3.5		1.9 – 2.1		0.5 – 0.65		0.06 max.		0.5 – 0.55	

From the analysis, it is clear that the percentage of the silicon is bit less than the required percentage. If it is necessary that the silicon content must lie within the prescribed limits, it may be achieved by adding ferrosilicon and deducting the same amount of foundry returns according to the following calculations:

Silicon in the final analysis (table): 1.54%

Silicon desired : 2.0%

So, 2 – 1.54 = 0.46% silicon should be added.

i.e., 0.46 × 10 = 4.6 kg silicon should be added to the mixture of 1000 kg.

By adding 1 kg of ferrosilicon, we get 0.5 kg of silicon in the charge.

And by deducting 1 kg of foundry returns we lose 0.012 kg of silicon.

So, total silicon gain = 0.5 – 0.012 = 0.488 kg.

Hence, the total amount of ferrosilicon needed/the amount of foundry returns deducted is:

= 4.6/0.488

= 9.3 kg

Final charge analysis is given in Table 10.19.

Table 10.19 Final charge analysis

Pig iron 1	300 kg
Pig iron 2	200 kg
Foundry returns (400 – 9.3)	390.7 kg
Ferro silicon	9.3 kg
Total charge	1000 kg

Example 10.4: Requirement of Grade 250 cast iron spout composition TC = 3.2%, Si = 1.7%, Mn = 0.7% and P = 0.10%. Composition of charge material is given in Table 10.20.

Table 10.20 Composition of charge material

Material	*C*	*Si*	*Mn*	*P*	*Previous practice*
Low P pig iron	3.0	3.0	1.0	0.10	25%
Grade 250 returns	3.2	1.7	0.7	0.10	35%
Low P iron scraps	3.2	2.2	0.8	0.15	15%
Steel scrap	0.1	0.1	0.3	0.03	25%
FeMn (late addition in spout)				75	As required
FeSi (late addition in spout)			73		As required

Silicon loss is 15% of the charge and manganese loss is 25% of the charge. P changes very little. Total C% in spout is given by Levi equation

$$\text{Levi equation (\%C in spout)} = 2.4 + \left(\frac{\text{total C\% in charge}}{2}\right) - \left(\frac{\text{Si\% + P\% in spout}}{4}\right)$$

Check the suitability of the previous charge make-up to obtain Grade 250 cast iron and determine the amount of FeMn and FeSi to be used (Table 10.21).

Table 10.21 Analysis of charge

	Typical composition,%				*Materials used %*	*Contribution to charge %*				
Material	*C*	*Si*	*Mn*	*P*			*C*	*Si*	*Mn*	*P*
Low P pig iron	3.0	3.0	1.0	0.10	25%	x0.25	0.75	0.75	0.25	0.03
Grade 250 returns	3.2	1.7	0.7	0.10	35%	x0.35	1.12	0.60	0.25	0.04
Low P iron scraps	3.2	2.2	0.8	0.15	15%	x0.15	0.48	0.33	0.12	0.02
Steel scrap	0.1	0.1	0.3	0.03	25%	x0.25	0.03	0.03	0.03	0.01
						Total	**2.38**	**1.71**	**0.70**	**0.10**
Changes during melting (15% Si loss)								−0.26		
Changes during melting (25% Mn loss)									−0.18	
Charge composition							**2.38**	**1.45**	**0.52**	**0.10**
Later on addition in spout (FeSi addition) $\left[\frac{1.71 - 1.45}{0.73} = 0.36\right]$								**0.25**		
Later on addition in spout (FeSi addition) $\left[\frac{0.70 - 0.52}{0.75} = 0.24\right]$									**0.18**	
Final charge composition							**2.38**	**1.70**	**0.70**	**0.10**

$$\text{Levi equation (\%C in spout)} = 2.4 + \left(\frac{2.38}{2}\right) - \left(\frac{1.45 + 0.10}{4}\right) = 3.20\%$$

Final spout charge is given in Table 10.22.

Table 10.22 Final spout charge

TC	3.2%
Si	1.70
Mn	0.70
P	0.10

Calculations such as the above example, should only be used as a guide. The precise carbon pick-up and silicon losses achieved depend on factors such as coke quality, metal temperature, melting rate, etc. Experience will enable more accurate predictions to be made. A number of computer programs are available which carry out the calculations rapidly and enable 'least cost' charges to be selected.

Example 10.5: Requirement of spout composition, C = 3.55%, Si = 2.10%, Mn = 0.75%. Foundry returns and pig iron to be used are 45% and 10% of the charge respectively. Composition of charge material is given in Table 10.23 below.

Table 10.23 Composition of charge material

Material	*C*	*Si*	*Mn*
Cast iron scraps	3.40	1.80	0.60
Steel	0.15	0.20	0.65
Pig iron	4.09	2.08	0.80
Foundry returns	3.55	2.20	0.75
Mn briquettes	0	0	67.0
FeSi briquettes	0	48	0

Silicon loss is 10% of the charge and manganese loss is 15% of the charge. C gain is 20% of the charge. Determine the charge make-up for 1000 kg charge.

Solution: Let the amount of steel and cast iron scraps to be used are A and B kg, respectively.

Charge should contain

$$\text{Si} = \frac{(1000 \times 0.021) \times 100}{90\ (10\%\ \text{loss})} = 23.33\ \text{kg}$$

$$\text{Mn} = \frac{(1000 \times 0.0075) \times 100}{85\ (15\%\ \text{loss})} = 8.82\ \text{kg}$$

$$\text{C} = \frac{(1000 \times 0.0355) \times 80(20\%\ \text{gain})}{100} = 28.40\ \text{kg}$$

Charge balance, total charge = Return + pig iron + steel + cast iron

$$1000 = 450 + 100 + A + B$$

$$A = 450 - B \qquad ...(1)$$

Return used = 45% of 1000 = 450 kg

This 450 kg of return contains

$$\text{Si} = 450 \times 0.022 = 9.90\ \text{kg}$$

$$\text{Mn} = 450 \times 0.0075 = 3.38\ \text{kg}$$

$$\text{C} = 450 \times 0.355 = 15.98\ \text{kg}$$

$$\textbf{Pig iron used} = \textbf{10\% of 1000} = \textbf{100 kg}$$

This 100 kg of return contains

$$\text{Si} = 100 \times 0.028 = 2.08 \text{ kg}$$
$$\text{Mn} = 100 \times 0.008 = 0.80 \text{ kg}$$
$$\text{C} = 100 \times 0.409 = 4.09 \text{ kg}$$

Carbon balance

Total Carbon in charge is given by

$$\text{Total C in charge} = \text{C in returns} + \text{C in pig iron} + \text{C in steel} + \text{C in cast iron}$$
$$28.40 = 15.98 + 4.09 + A(0.015) + B(0.034)$$
$$A = 5553.33 - B(22.67) \qquad \text{...(2)}$$

Using equations (1) and (2) we get,

$$\text{Cast iron} = B = 235.50 \cong 236 \text{ kg}$$
$$\text{Steel} = A = 214.50 \cong 215 \text{ kg}$$

Other elements in cast iron (B)

$$\text{Si} = 236 \times 0.018 = 4.25 \text{ kg}$$
$$\text{Mn} = 236 \times 0.006 = 1.42 \text{ kg}$$

Other elements in steel (A)

$$\text{Si} = 215 \times 0.018 = 3.87 \text{ kg}$$
$$\text{Mn} = 215 \times 0.006 = 1.29 \text{ kg}$$

$$\text{Total silicon in charge} = 9.90 + 2.08 + 4.25 + 3.87 = 20.10 \text{ kg}$$

Total silicon to be in charge = 23.33 kg.

Silicon to be added as FeSi briquets = 23.33 – 20.10 = 3.25 kg

$$\text{FeSi to be added} = \frac{3.23 \times 100}{48} = 6.73 \text{ kg} \cong 7.0 \text{ kg}$$

Total manganese in charge = 3.38 + 0.80 + 1.42 + 1.29 = 6.89 kg

Total manganese to be in charge = 8.82 kg.

Manganese to be added as FeSi briquets = 8.82 – 6.89 = 1.93 kg

$$\text{Manganese to be added} = \frac{1.92 \times 100}{67} = 1.93 \text{ kg} \cong 2.0 \text{ kg}$$

Final charge details are given in Table 10.24.

Table 10.24 Final charge details

Foundry returns	450 kg
Pig iron	100 kg
Steel	215 kg
Cast iron	236 kg

FeSi briquets	7 kg
Mn briquets	2 kg
Total	1010 kg

X. Cupola design considerations

The design of cupola is based on certain mathematical relationships and thump rules. Refer to Figure 10.5. Important relations and values are listed below:

Step 1: Find the inner cross-sectional area of cupola.

$$A_C = \frac{3.14\, d^2}{4} = \frac{Q}{Q_1}$$

d = Inside diameter of cupola in meter

Q = Cupola output required, tonne per hour

Q_1 = Specific output in tonnes/sq. m. area of cupola

Q_1 = 6 to 8 tonnes/hour/sq. m.

Step 2: Find the height of charging door-bottom from tuyeres-centre (H)

$$\frac{H}{d} = 3 \text{ to } 5$$

Step 3: Find the cross-sectional area of tuyeres

Total cross-sectional area of tuyeres is taken as 16% to 20% of A_C, and the size is taken as 50 mm × 150 mm to 100 mm × 300 mm.

Step 4: Find the height of tuyeres

Height of tuyeres from metal tapping hole is determined by the well size required for the heaviest casting to be taken from the furnace. However, it should not exceed 450 mm, but a lower value is preferred.

Step 5: Find the bed coke height

The bed coke height (bed charge) is taken as 900 mm to 1300 mm (refer to Table 10.25) above the tuyeres depending on the size of cupola 1 to 8.

Table 10.25 Sizing of cupola and melting rate details

Cupola size	*Inside diameter*		*Shell diameter in mm*	*Average melt rate in tonne per hour*	*Air through tuyeres in m^3/min*	*Capacity of well in kg*	*Bed coke above tuyeres in mm*
	mm	*inch*					
1	23	584	813	0.9 – 1.8	26	258	900
2	27	686	914	1.8 – 2.5	37	372	950
3	32	813	1170	2.7 – 3.6	51	609	1050
4	42	1065	1420	4.5 – 5.4	88	1030	1100
6	48	1220	1675	8.2 – 10	116	1540	1200
8	60	1525	1980	14 – 15	181	2230	1300

Step 6: Find the weight of metal charge

Weight of metal charge should be from 10% to 15% of the hourly output of cupola.

Step 7: Find the coke to iron ratio

Coke to iron ratio by weight has to be 1/6, 1/8, 1/10 or 1/12 based on the type of coke and the heat content.

Step 8: Find the amount of limestone

Limestone is added as flux about 2% to 3% of the metal charge by weight.

Step 9: Find the required air blast pressure

Limestone is added as flux about 2% to 3% of the metal charge by weight.

The required air blast pressure is calculated as:

$$p = 64\ (q_0 A_C)^{0.5} \text{ mm of water}$$

A_c = Inside cross-sectional area in sq. m.

q_o = Specific volume of air blast/min/sq. m. of cupola cross section (100 to 130 cu. m)

The value of p = 250 mm to 400 mm for small-sized and 400 mm to 750 mm for large-sized cupolas.

Step 10: Find the required thickness of cylindrical shell

The thickness of cylindrical shell is taken as 6 mm to 10 mm and refractory lining thickness as 125 mm to 175 mm.

$$\text{Shell diameter, } D = d + 2b$$

d = Inner diameter of furnace

b = Refractory lining thickness

Step 11: Find the height of stack above the charging door

Height of stack above the charging door-bottom is given as 2 m to 3 m depending on the requirement of chimney height and the size of cupola.

Specific gravity of cast iron is taken as 7.2. The weight of coke as 482 kg/m^3 and weight of air is 1.28 kg/m^3.

Example 10.6: Calculate the following dimensions of a cupola furnace, which is suitable to melt 3 tonnes/hour on an average. Standard thumb rules may be adopted wherever they are required.

1. Inside diameter at the melting zone
2. Shell diameter and thickness of MS plate
3. Tuyere size and numbers
4. The maximum weight of a casting that can be poured by the cupola for a depth of well of 300 mm (bed to slag hole)
5. Height from tuyeres to the charging door-bottom

Solution: Refer to Figure 10.6.

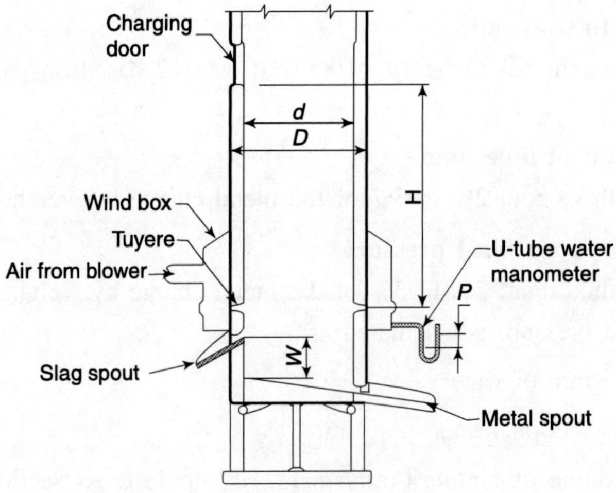

Figure 10.6

Step 1: Find the inner cross-sectional area of cupola

Assuming the specific output of cupola as 7 tonnes/sq. m, (mid value of 6 to 8 tonnes/sq. m), the inside diameter can be calculated by using the equation:

$$\frac{Q}{Q_1} = \frac{3.14d^2}{4}$$

where,

Q = Required output of cupola = 3 tonnes/hour

Q_1 = Specific output of cupola = 7 tonnes/hour/sq. m.

d = Inside diameter in m

$$d = \sqrt{\frac{Q}{Q_1} \times \frac{4}{3.14}}$$

$$d = \sqrt{\frac{3}{7} \times \frac{4}{3.14}} = 0.739 \text{ m}$$

Select the next standard size 813 mm (32 inches) as the inner diameter of cupola 3.

Step 2: Find the shell diameter and thickness of MS plate

2.(a) Shell diameter, $D = d + 2b$

where, b = refractory lining thickness

= 150 mm (125 mm to 175 mm)

= 813 + (2 × 150)

= 1113 mm

(b) Thickness of MS plate may be taken as 8 mm.

Step 3: Find the Tuyeres size and numbers

Cross-sectional area of tuyeres = 18% of A_C

where A_C is the inner cross-sectional area = 3.14 × $(0.813)^2$ = 0.519 cm^2

$$\text{Area of tuyeres} = 18\% \text{ of } A_C$$
$$= 18\% \text{ of } 0.519 \text{ cm}^2$$
$$= 0.18 \times 0.519 \times 104 \text{ cm}^2$$
$$= 934.2 \text{ cm}^2$$

By providing 8 tuyeres, the area of opening of one tuyere = 934.2/8

$$= 116.8 \text{ cm}^2$$

By giving rectangular holes with side height S to length ratio as 1 × 3, the area can be $S \times 3S = 3S^2 = 116.8$. Therefore, S = 6.22 cm, say 7 cm. The allowable size is 5 cm × 15 cm to 10 cm × 30 cm. Hence, the tuyeres size = 7 × 21 cm.

Step 4: The maximum weight of a casting that can be poured by the cupola for a depth of well of 300 mm (bed to slag hole)

For a well depth of 30 cm, the volume of metal that can be collected

$$A \times 30 = 5190 \times 30 = 155700 \text{ cm}^3$$

Taking density of CI as 7.2, the space requirement for slag as 10% and 10% metal loss in gating system, the maximum weight of a single casting that can be made by the furnace = 155700 × 0.9 × 0.9 × 7.2 = 1008 kg

Step 5: Height from tuyeres to the charging door-bottom

Height from tuyeres to charging door, $H = d \times 3$ (3 to 5 times) = 813 × 3 = 2439 mm

So the height, H = 2.5 m.

Example 10.7: A cupola furnace of size 2 has inner diameter of 686 mm (27 inches) and can melt 2 tonnes/hour. Calculate the following values to suit the furnace functioning, if the height of tuyeres from sand bottom is 310 mm.

1. The weight of bed-coke required in kg.
2. The weight of iron, coke and limestone in each charge having coke to metal ratio 1:8.

Solution: Refer to Figure 10.5.

The bed-coke height above the tuyeres is from 900 to 1300 mm. For cupola 2, this may be taken as 950 mm.

Step 1: The weight of bed-coke required in kg

Size 2 (d = 686 mm),

$$A_c = \frac{3.14d^2}{4} = 3.14 \times \frac{(0.686)^2}{4} = 0.369 \text{ m}^2$$

Total height of bed-coke from the sand bottom = 0.95 + 0.31 = 1.26 m

Volume of coke-bed (A_C × ht. of coke)V_b = 0.369 × 1.26 = 0.465 m^3

Weight of bed-coke = V_b × 482 = 0.465 × 482 = 224.13 kg

Step 2: The weight of iron required in kg

For an hourly output of 2000 kg of molten metal, the charge size is taken as 10% of the output.

So, the weight of charge c_m = 2000 × 0.1 = 200 kg

Weight of coke for 1:8 ratio,

$$c_c = \frac{200}{8} = 25 \text{ kg}$$

Weight of limestone (3% of metal charge)

$$c_1 = 200 \times 0.3 = 6 \text{ kg}$$

Hence the charge metal

$$c_m = 200 \text{ kg, coke } c_c = 25 \text{ kg, and limestone } c_1 = 6 \text{ kg}$$

Example 10.8: A cupola of size 3 (0.813 m) can melt 3.2 tonnes of molten metal per hour.

(a) What is the air blast pressure required for the melting?

(b) Calculate the weight of air required in kg/hour, if the specific volume of air blast/minute/sq. m. is 120 cu. m.

Solution:

Step 1: Air blast pressure required for cupola

The air blast pressure required for the cupola is calculated using the relationship,

$$p = 64\ (q_0 A_C)^{0.5} \text{ mm of water}$$

where,

A_c = inside cross-sectional area of cupola

= 3.14 × d^2/4

= 3.14 × 0.813^2/4

= 0.5189 m^2

Now, putting the values in 3.14 = 64 $(q_0 A_c)^{0.5}$ mm of water

3.14 = 64 (120 × 0.5189)$^{0.5}$

= 505 mm of water

Step 2: Weight of air required/hour

Weight of air required/hour = q_0 × A_c × 60 × 1.28

= 120 × 0.5189 × 60 × 1.28 = 4782 kg/hour.

XI. Recent cupola developments

1. Water cooled cupolas: The melting zone is water-jacketed due to which both refractory and overall costs are saved. By proper redesign and control, a conventional cupola layout may be converted to water cooled cupola and can be operated continuously for 2 to 4 weeks, before dropping the bottom for repairs.

2. Hot-blast cupolas: In such cupolas, the blast is preheated.

Advantages:

(a) Coke consumption is reduced.

(b) Cupola outburst is increased.

(c) Poorer quality of coke can be used with preheated blast.

(d) A higher proportion of steel scrap can be used.

(e) Silicon loss is reduced.

Air preheater or sensible heats of waste cupola gases are methods of preheating.

3. Oxygen enriched blast cupolas:

Advantages:

(a) Increases temperature of molten iron.

(b) Output and efficiency of cupola are increased.

(c) Heat losses by waste gases are reduced.

(d) Absorption of hydrogen and nitrogen in the molten iron is minimized.

4. Coke-less cupola: The use of coke as fuel in cupola furnace is replaced by gas or oil in a coke-less cupola. This method of melting was introduced in 1970s and adopted in many countries. Figure 10.7 shows the cross section of the Taft coke-less cupola. This furnace has a vertical refractory lined shaft similar to a conventional cupola. At the middle of its height, there is a water cooled grate using refractory insulated pipes across the vertical shaft. This grate is used to support refractory spheres and the metal charges are loaded above it.

Heat for melting is generated by burning gas or oil through burners (4 to 6 in number) located below the grate in the presence of air blast as shown. The metal and flux are charged through the charging door and the refractory spheres supported by grate restrict the downward movement of solid metal. The heat of the flame melts the charge and liquid metal drops

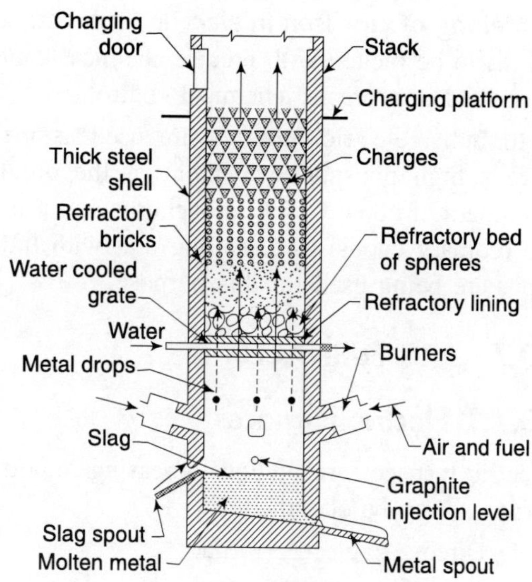

Figure 10.7 Coke-less cupola furnace

through the gaps of the spheres and grate. The dropping metal and slag in liquid form are collected at the bottom well and tapped out as done in the conventional cupola.

Since there is no burning of coke, the molten metal will be having less sulphur and carbon content. In Taft cupola, the iron is carburized to specified carbon content by injecting a fine carburizer (either petroleum coke or graphite) into the metal in the well. A hole is provided for this, just above the level of slag tap-hole.

Pollution from coke-less cupola is considered minimum, since there no coke is burnt. These furnaces are used to melt metal up to a rate of production of 5 tonnes/hr to 10 tonnes/hr. Coke-less cupola can be economical where coke is costly and difficult to get, at the same time gas or oil is available at low cost.

5. Melting of cast iron in reverberatory furnaces: Reverberatory furnace is used to a limited extent for melting cast iron in the manufacture of cast iron rolls which require a substantial quantity of molten metal and a precise control of chemical composition. This furnace is also used in the production of malleable cast iron and alloy cast irons.

The charge consists of iron scrap, steel scrap and pig iron.

(a) The furnace is heated by means of firewood until it reaches a bright yellow heat.

(b) At this stage, charge is placed on the furnace bottom and the furnace is heated by atomized oil.

(c) When meltdown is complete, the slag is removed and a new slag consisting of lime is worked up.

(d) Carbon can be reduced by adding iron ore or mill scale.

(e) The heat is finished when the desired composition of iron is achieved.

6. Melting of cast iron in electric furnaces: This process has an advantage when high grade iron is to be melted with precise chemical control. But the process is costly, with advantage to be melted in a precise chemical control.

In India, electric furnaces are used as melting units for the S.G.-Iron. Indian pig irons have a high phosphorus content of the order, 0.3-0.4%. S.G.-iron specifications call for a maximum of about 0.08% phosphorus and it is not practical to meet the specifications by using the ordinary cupola as a melting unit with Indian raw materials. Thus, electric furnaces of all types are being used for this purpose.

10.7 ARC FURNACES

10.7.1 Electric Furnaces

Electric furnaces are finding increasing importance in the melting practice nowadays. Electric furnaces may be classified as:

1. Direct arc electric furnace
2. Indirect arc electric furnace
3. High-frequency induction furnace
4. Low-frequency induction furnace

Direct arc electric furnace

Principle: An arc is struck directly between the electrodes and the metal to be melted. Heat is produced due to resistance to the flow of electricity offered by metallic charge.

Construction: The furnace consists of a heavy steel cylindrical shell with a spherical or flat bottom (Figure 10.8). The furnace hearth and walls are lined with magnesite bricks. The roof of the furnace can be removed for charging. Three electrodes made of graphite or amorphous carbon pass through the roof and can be raised or lowered. The capacity of such furnace ranges upto about 50 tonnes.

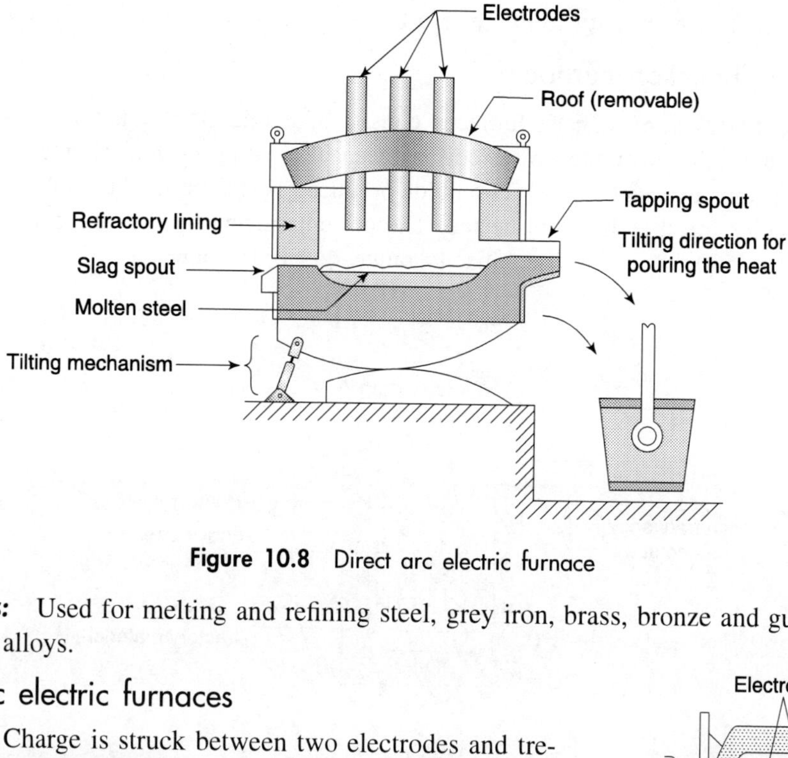

Figure 10.8 Direct arc electric furnace

Applications: Used for melting and refining steel, grey iron, brass, bronze and gun metal and many nickel alloys.

Indirect arc electric furnaces

Principles: Charge is struck between two electrodes and tremendous heat is produced.

Construction: The furnace consists of a cylindrical or barrel shaped refractory-lined shell with two electrodes horizontally mounted (Figure 10.9). The shell is mounted on rollers due to which a rocking motion can be given to the furnace. Such furnaces may be upto 2000 pounds in capacity.

Applications: Used for melting especially brass, bronze, copper and nickel alloys. Small quantities of ferrous alloys may also be produced.

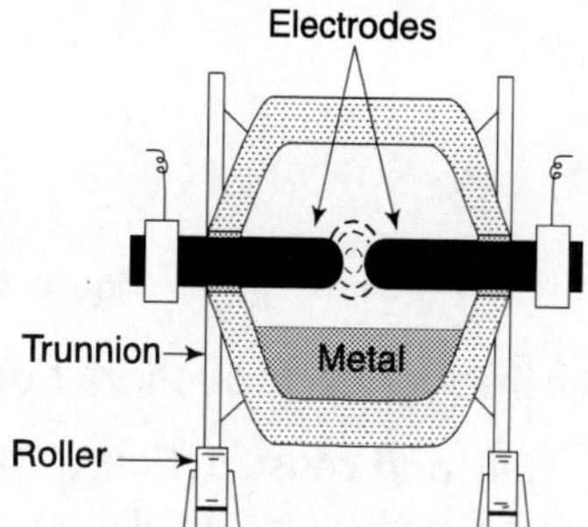

Figure 10.9 Indirect arc electric furnace

Advantages:

1. There is no contamination of metal from external sources and, therefore, they are suitable for melting metals of high purity.
2. Temperature control is easy.
3. Thermal efficiency is high, about 70%.
4. Loss of alloying additions is very less.
5. Furnace atmosphere can be controlled.

Disadvantage:

1. Very high cost of electricity limits their use.

High-frequency Induction Furnace

Principle and construction: These furnaces consist of a refractory crucible containing the metal to be melted and surrounded by a coil with an alternating current (Figure 10.10). The metallic charge is thus exposed to magnetic fields produced by the coil. Alternating currents are set up in the charge and the charge is heated. The current carried by the coil may range from 10,000 c.p.s. up to 500,000 c.p.s. The capacity ranges up to 15 tonnes.

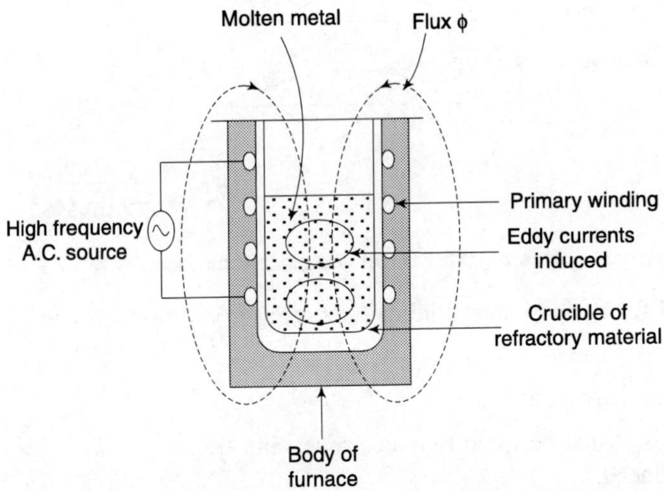

Figure 10.10 High frequency induction furnace

Low-frequency Induction Furnace

Principle and construction: The principle and construction are the same as those of the high frequency induction furnace, except that in this type of furnace, alternating current at 50 or 60 c.p.s. is used.

Applications

1. High frequency furnace: Used for tool steels, nickel-chromium heat-resisting alloys, stainless steels, magnet steels and costly alloys.

2. Low frequency furnace: Used as holding furnaces for die casting and permanent casting of non-ferrous metals and alloys.

Advantages and disadvantages

1. *Induction furnace in general*

Advantages:

1. Possibility of obtaining accurate analysis.
2. No metal contamination by electrodes or fuel.

Disadvantages:

1. As the heat is developed within the metal itself, the slag is cold and has no good reactivity.
2. Needs fine quality raw materials, as refining in the furnace is practically impossible.

2. *High frequency furnace*

Advantages:

1. There is possibility of producing a great variety of products.
2. Cold charge of practically all sizes can be used.

Disadvantage:

1. High initial cost.

3. *Low frequency furnace*

Advantage:

1. Better efficiency.

Disadvantages:

1. The melt can only be started with big lumps or a quantity of liquid metal.
2. Low capacity of hourly production compared with the total capacity of the furnace.

10.7.2 Open-hearth Furnace

Principle: Open-hearth furnace involves the regenerative principle of heating in which the gaseous fuel and air are preheated by the sensible heat of the outgoing products of combustion.

Construction: As shown in Figure 10.11, the furnace consists of a long shallow refractory hearth for holding the metal. Flame and hot gases traverse the bath and pass out at the opposite end of the furnace. The waste gases pass through the checker work which absorbs the sensible heat of the outgoing gases. After some time, the cycle is reversed and the air plus fuel gas mixture is passed through the checker work which has been heated by the waste gases. Now the other checker work gets heated by the outgoing gases. Again after some time, the cycle is reversed. In this way, maximum use of the sensible heat of outgoing gases is made. The capacity of open-hearth furnace varies from 25 to 100 tonnes.

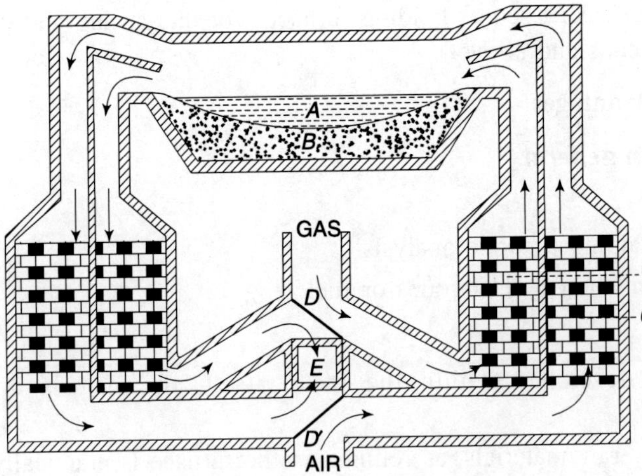

Figure 10.11 Open-hearth furnace

A: Charge of pig iron, scrap iron and flux of limestone

B: lining made of magnesite

C and *C′*: Checkers for gas and air heating

D and *D′* The gas and air unite in a flame over the charge of iron and pass off and then through the firebrick construction which it heats. Changing the direction of the gas and air with the reversing valves, is the special feature of this process.

E: Chimney

Application: Used in the steel foundries for making large quantities of steel.

10.7.3 Air Furnace

Principle: The principle of an air furnace is that of a reverberatory furnace in which heating of the charge is done by the flame deflected from the long low roof of the furnace.

Construction: As shown in Figure 10.12, the construction of an air furnace is almost the same as that of the open-hearth furnace with the only difference that in this case, no regenerative principle is used.

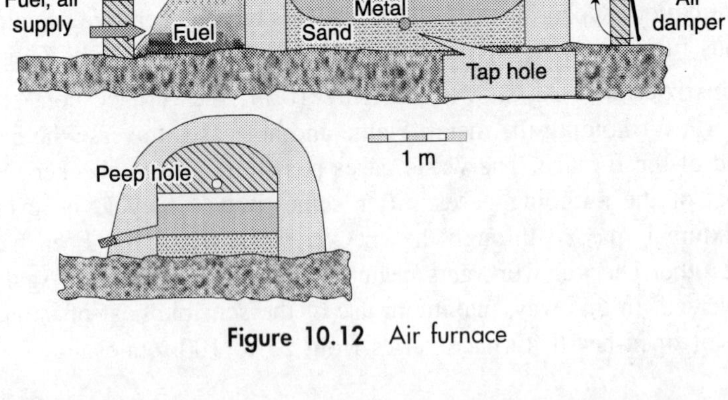

Figure 10.12 Air furnace

The charge is heated by the flame from fuel in a burner or fire box at one end of the hearth. The fuel is usually pulverized coal or oil. The capacity of an air furnace varies from 20 to 50 tonnes.

Application: Used in the production of malleable iron and high quality grey iron castings. This furnace is also used for brass, bronze, etc.

10.7.4 Converter Furnace

Principle: When air blast is passed through molten iron, oxidation of carbon, silicon and manganese takes place which produces lot of heat.

Construction: As shown in Figure 10.13, the converter is a refractory lined vessel mounted on two trunnions so that it can be rotated to any desired position for charging and tapping the metal. Air can be passed through tuyeres located on the sides of the vessel.

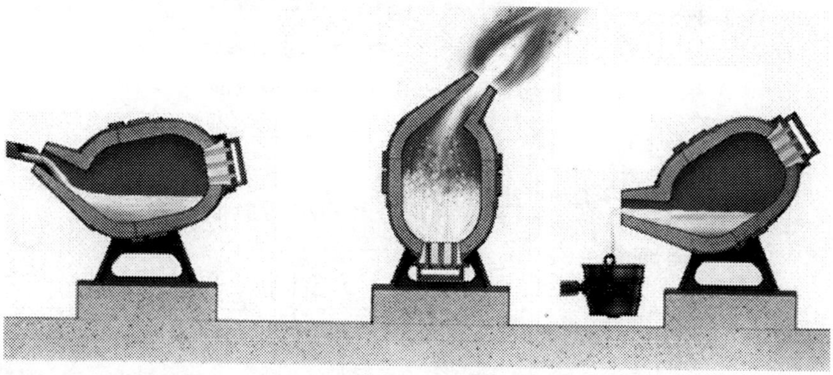

Figure 10.13 Converter furnace

Application: Used in steel production.

Vacuum Melting furnaces

Vacuum melting is a process in which a charge of solid metal is melted and allowed to solidify under vacuum. The object of this process is to avoid contamination by a gas atmosphere and also to eliminate any gaseous or other impurities. (By gaseous impurities are meant those elements or compounds, which are gaseous at atmospheric pressure and room temperature). In the majority of the industrial applications, degassing is the more important process, i.e., every vacuum melting process is also a vacuum degassing process.

The classification of furnaces used for vacuum melting can be done in accordance with the method of heating employed. Basically, all heating for vacuum is done electrically. The following methods of electric heating are used in vacuum melting furnaces:

1. Resistance heating
2. Induction heating
3. Electric arc heating
4. Electron-beam heating

Figure 10.14 shows a vacuum induction melting furnace. The melting pot is enclosed in a vacuum chamber. Speed of melting in a vacuum induction melting furnace is very rapid, and the standard melting rates for air melting can readily be obtained in vacuum. The rapid melting is even more important when combined with vacuum techniques, since it increases the productivity of the high capital investment cost of vacuum installation.

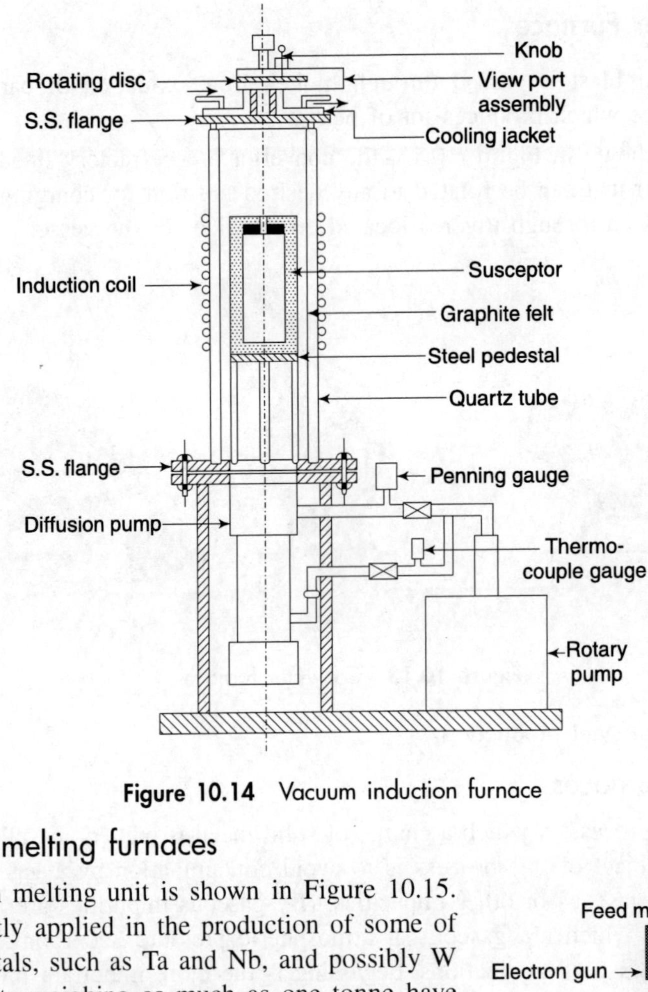

Figure 10.14 Vacuum induction furnace

Electron beam melting furnaces

An electron beam melting unit is shown in Figure 10.15. It has been recently applied in the production of some of the refractory metals, such as Ta and Nb, and possibly W and Mo and ingots weighing as much as one tonne have been produced in the U.S.A. It has been indicated that the process may also have applications in the preparation of very high grade iron and nickel-base alloys. The latest method of melting the refractory metals, electron beam melting represents a completely new departure because unlike arc-melting, it is not just a method of making larger ingots of metals by melting; it is a method of purification

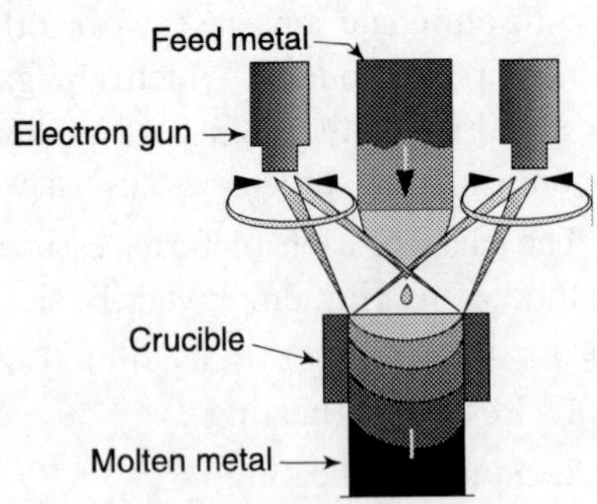

Figure 10.15 Electron beam melting

also. In electron beam melting, fast moving electrons are made to impinge on the feed metal due to which the kinetic energy of the electrons is transformed into heat imparted to the feed metal. This heat developed is sufficient to melt the metal. Some of the kinetic energy of the electrons is also transformed into various types of radiations. The source of electrons is a heated tungsten filament.

Electron Beam Melting (EBM) Process

- EBM furnaces melt the metal by focusing a beam of electrons from an electron gun on the metal. When the electrons strike the material, their energy is absorbed in quantities sufficient to melt the metal. Schematics of EBM is shown in Figure 10.16.
- EBM method was first introduced in 1907 to melt high-melting point metals, such as tantalum and tungsten. It was not until recently that electron beam was recognized for its application in metal-purifying processes.
- This is the principal technology for processing of reactive metal scrap. EB technology is also the only technology that makes it possible to monitor the cleanliness of metals using specially designed button melters.
- EBM is distinguished by its refining capacity. However, electron beam technology is primarily for melting refractory metals and laboratory-scale melts. One of the disadvantages of the electron beam process is its high capital cost due to its requirement of vacuum pumps.

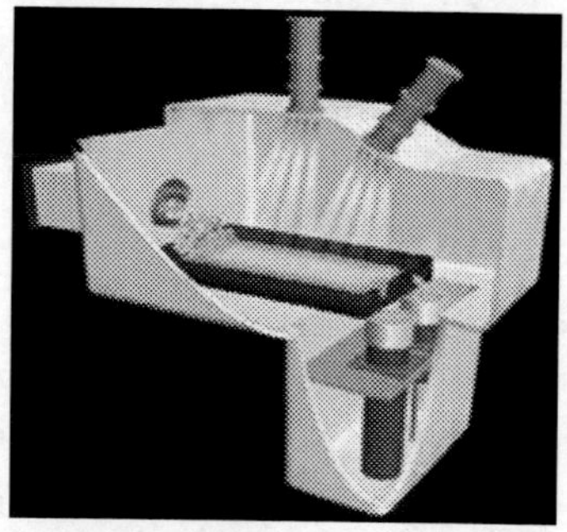

Figure 10.16 Electron beam cold-hearth refining

- In addition, the fitting must be precise and locating the parts with respect to the beam must be perfect. Therefore, the technology is not applicable in typical foundry operations.
- There are two processes used today: drip melting and electron beam cold-hearth refining (EBCHR).
- Drip melting melts the metal in a water-cooled ingot mould enabling removal of entrapped gases and evaporation of all impurities as their vapour pressures exceed that of the metal being processed.
- In EBCHR, the molten metal flows along a water-cooled hearth before entering a water-cooled ingot. While flowing along the hearth, high- and low-density inclusions are removed in addition to evaporative purification. High-density inclusions settle and are removed in the scull, while lighter inclusions either are dissolved in the metal or held back by a dam prior to flowing into the ingot.

Infrared Heating

- Infrared heating is employed by the forging industry; this technology can also be used for die heating and heat treatment in the metal casting industry. It offers a method for reducing time and energy costs of die preheating. Schematics of infrared heating is shown in Figure 10.17.

- A study conducted by ORNL indicated that the surface of the die can be heated to 300°C in less than 20 minutes. An infrared panel heater can bring the furnace to full power in a matter of seconds, saving energy to maintain a hot furnace while idle.
- The rapid heating could also reduce heat-treatment cycle time and increase product throughput.
- Infrared radiation is often supplied by high-intensity quartz heat lamps, producing radiation of around 1 micron wavelength. This radiation melts metals with uniform heat distribution. The high-density infrared technology can achieve a heating rate of 50 to 400°C/sec.

Figure 10.17 Electric infrared flatbed furnace

Microwave Melting

- Originally developed by the DOE's Y-12 National Security Complex for melting of uranium, microwave technology is now capable of melting a wide range of alloys in large quantities. Schematics of microwave melting is shown in Figure 10.18.
- While conventional heating involves transferring of heat through the outer surface of the material to the interior, microwaves can penetrate the mass, enabling rapid transfer of high-intensity heat throughout the bulk. This translates into significant savings in time and energy.
- Three basic elements are required to heat and melt bulk metals using microwaves: a multimode microwave cavity, a microwave-absorbing ceramic crucible, and a thermally insulating casket that is microwave-transparent. Although metal is supposed to reflect microwaves, researchers at Y-12 and Pennsylvania State University found that metal actually starts to absorb the energy and heat at three-fourths of the melting temperature.
- A uniquely designed suscepting crucible used for microwave melting is shown in Figure 10.18. The metal charge is placed in an open (no lid) ceramic crucible, and the insulating casket is positioned to cover the open crucible completely. The heated crucible walls quickly heat the metal charged in the crucible by means of radiation, conduction, and convection. The thermally insulating casket increases the energy efficiency of the microwave system by trapping the heat generated in the crucible.

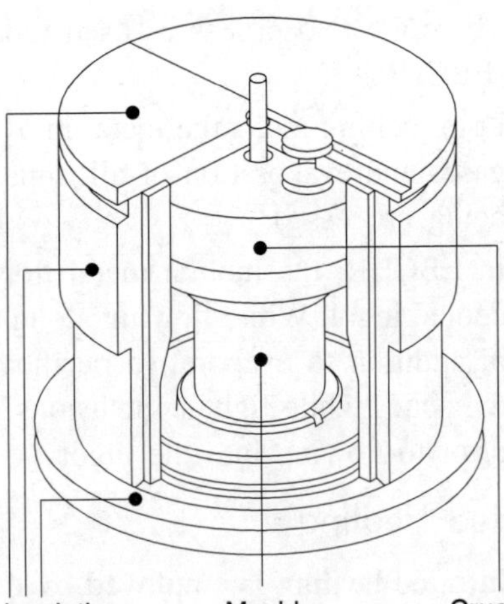

Figure 10.18 Microwave melting

- According to a Y-12 publication, microwave melting could cut melting cost by 30%. Germany and Japan are taking a strong lead in the microwave technology applications in high-temperature processes and microwave-assisted kilns/furnaces.
- Some of the current focus areas in microwave melting technology include use of microwaves for roasting ores, heat-treating glass, detoxifying filter dusts and slag in the metal industry, and decontaminating asbestos containing wastes.
- In addition, research is being conducted by Michigan Technological University using a combination of microwave, electric arc and exothermic reactions to heat and smelt iron.
- Since iron ore and carbon are excellent microwave absorbers, microwave technologies can aid more effective melting in electric arc furnaces used in steel foundries.
- However, bulk metals do not readily couple directly with microwave energy at room temperature because they are electrically conductive, readily reflecting the incident energy.
- In addition, microwave melting requires 2.35 micron radiation which is very expensive. This limits the use of microwave melting to high-end processes and alloys such as investment casting and super alloys.

Plasma Heating

- Plasmas are gaseous collections of electrically charged particles such as electrons and protons. These charged particles carry energy; as the ionized plasma flow hits the metal surface, it releases its energy melting the metal (Figure 10.19).
- The plasma melting process is proved to be more energy efficient than the conventional melting technologies as it can raise the temperature very rapidly.
- Plasma aluminium melting furnaces are reported to impart heat 60% quicker than the conventional high rate melters.
- Due to the fast heat transfer, the melting process is also rapid. The energy consumption rate is as low as 0.198 kWh/lb (of aluminium) compared to 0.345 kWh/lb (of aluminium) in induction melting.
- The plasma melting process also minimizes metal loss due to oxidation and contamination. The plasma keeps a thin oxide film on molten metal intact, protecting the molten metal from absorbing gases.
- In fact, the dross rate of melting aluminium is lowered to less than 1% compared to the conventional 3% to 12%.

Figure 10.19 Plasma heating

- Plasma heating also reduces the cost of fuel. A recent development of plasma heating utilizes ambient air and converts it to nitrogen, eliminating the need to purchase expensive gases such as nitrogen or argon required in conventional plasma heating.

Solar Heaters

- Solar furnaces use large arrays of mirrors to concentrate the sun's energy into a small space and produce high heat flux and very high temperatures (Figure 10.20).
- They can achieve temperatures up to 33,000°C (60,000°F) and produce over 1,000 kWh of energy.
- Solar furnace system consists of a solar concentrator, featuring advanced structural mirrors, and system controls, which provide fully automated operation, fault detection, data acquisition, and communications.
- The sun-tracking mirrors concentrate sunlight onto a receiver that sits on top of a platform extended about 18 feet from the wall mirrors. The receiver collects and transfers the sun's heat to experimental equipment.
- One aspect of the solar system that makes it appealing for electricity production at remote sites is its ability to operate automatically. Without operator intervention or even onsite human presence, it starts up each morning at dawn and operates throughout the day, responding to clouds and wind as needed.
- Finally, it shuts itself down at sunset. The system can be monitored and controlled over the Internet. Although solar furnaces are the ideal metal melting systems in terms of energy efficiency, they are very capital intensive and operate only under a sunny sky.

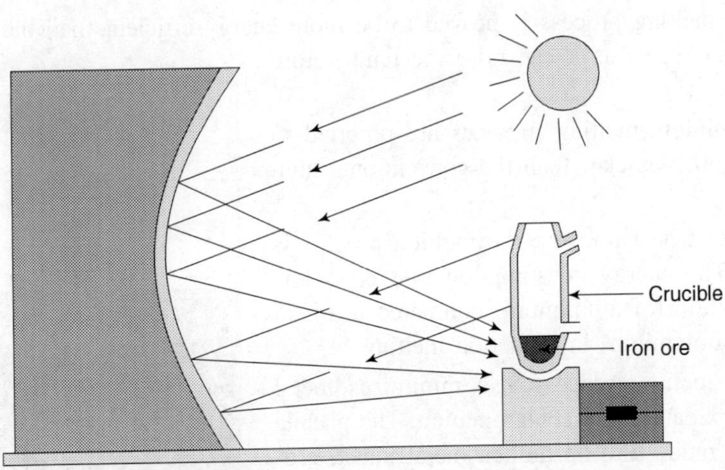

Figure 10.20 Solar furnace

- These furnaces also require acres of land to place the reflective mirrors. Because of these restrictions, solar furnace is used mainly for scientific experiments, lacking the practicality to be applicable immediately in typical foundry operations.

Review Questions

1. Explain the classification of melting furnaces.
2. List all types of electrical furnaces.
3. Explain the working of oil fired crucible furnaces.
4. Give the detailed classification of cupola.
5. With the help of a neat sketch, explain the working of cupola.
6. With the help of a neat sketch, explain high frequency induction furnace.
7. With the help of a neat sketch, explain open-hearth processes.
8. Explain electron beam heating.
9. Explain microwave and infrared heating processes.
10. Explain solar and plasma heating processes.

CHAPTER 11

Casting Defects

11.1 INTRODUCTION

Casting process is also known as process of uncertainty and defects in casting are to be avoided. A casting defect is defined as an observable and unplanned variation of a specification. Casting defects are usually not an accident; they occur because some step in the manufacturing process is not properly controlled. Melting, casting and solidification of metal comprise many intricate operations, and perfect control is impossible. It is not surprising that castings are afflicted with a greater variety of defects than the products of other fabricating methods. Casting defects are often very difficult to characterize due to the intrinsic complexity of the casting process. To study defects rather than sound castings seems a negative approach to learning; however, it is important for a producer, designer, consumer and engineer that such defects as do occur in castings be recognized and their origin and control be properly understood.

Proper classification and identification of a particular defect is the basic need to correct and control the quality of casting.

Casting defects in a controlled process generally fall into one of the two categories:

1. Defects can be chronic, i.e., long-standing, which require a remedy through a process change usually via designed experiments.
2. Sporadic, which are sudden adverse changes in the normal capability of the process. Sporadic defects are probably the most troublesome for casters.

Two distinct journeys must be taken to correct sporadic defects:

1. The diagnostic journey from symptom to cause.
2. The remedial journey from cause to remedy.

There is a temptation to attempt to diagnose a defect by the possible causes; but, an incorrect diagnosis of the cause can lead to an incomplete or incorrect remedy of the problem. It is important to correctly identify the defect symptoms prior to assigning the cause to the problem.

11.2 CLASSIFICATION OF DEFECTS

The International Committee of Foundry Technical Associations (ICFTA) has standardized the nomenclature, starting with the identification of seven basic categories of casting defects:

1. Metallic projections
2. Cavities
3. Discontinuities
4. Defective surface
5. Incomplete casting
6. Incorrect dimensions or shape
7. Inclusions or structural anomalies

(a) Metallic projections: in the form of fins or flash. There are several types of fins or flash defects such as

- Thin fins (or flash) at the parting line or at core prints
- Projections in the form of veins on the casting surface
- Network of projections on the surface of die castings
- Thin metallic projection located at a re-entrant angle and dividing the angle in two parts
- Swells: Excess metal in the vicinity of the gate or beneath the sprue
- Projections with rough surfaces

(b) Cavities:

- Cavities with generally rounded, smooth walls perceptible to the naked eye (blowholes, pinholes)
- Internal, rounded cavities, usually smooth-walled, of varied size, isolated or grouped irregularly in all areas of the casting
- Slag blowholes
- Fine porosity (cavities) at the casting surface, appearing over more or less extended areas (surface pinholes)
- Small, narrow cavities in the form of cracks, appearing on the faces or along edges, generally only after machining (dispersed shrinkage)
- Open, funnel-shaped cavity; wall usually covered with dendrites (open shrinkage)
- Internal, irregularly shaped cavity; wall often dendritic inside (internal shrinkage)
- Porous structures caused by numerous small cavities (micro/macro-shrinkage)

(c) Discontinuities, generally at intersections, caused by mechanical effects (rupture)

- Cold breakage, hot cracking, cold tearing
- Cold shut
- Unfused chaplets
- Separation along grain boundaries of primary crystallization

(d) Defects

- Cope defect, elephant skin, laps

- Wavy fold markings without discontinuities; edges of folds at same level, casting surface is smooth (seams or scars)
- Rough casting surface
- Grooves of various lengths (rat tails)
- Depressions in the casting surface (slag inclusions)
- Plate-like metallic projections (scabs)

(e) Incomplete casting

- Due to loss of metal from mould after pouring (runout)
- Misruns

(f) Incomplete dimensions

- Improper shrinkage allowance
- Shifts
- Castings warpage and distortion

(g) Inclusions or structural anomalies

- Metallic inclusions
- Cold shots
- Slag, dross or flux inclusions
- Slag blowhole defect
- Sand inclusions

Some of the several defects quoted above are discussed below.

1. Blowholes and Open Blows

Balloon-shaped gas cavities caused by release of mould gases during pouring are known as blowholes.

Typical characteristics: Rounded shapes, location always in the upper part of the castings and beneath cores or undercuts. Individual blowholes or large accumulations, then mostly with different sizes.

Preferred defect locations: In parts of the casting in the top of the mould, often above larger cores. In thick parts, but also in thin walls. Beneath undercuts and large cores.

Apart from presence of moisture, they occur due to poor venting and lower permeability of the mould.

Remedial measures

- Adequately dry cores
- Larger proportion of new sand
- Better gas permeability
- Smaller proportion of binder

- Check casting system
- Reduce moisture content of the sand
- Add a little oxide to the melt (e.g., rusty insert)

2. Blowholes due to Wet Sand

Localised patches of round or irregular shaped cavities of significant size. The walls of the casting are smooth and often coloured with oxidation tints.

Causes:

- Excessive moisture in the sand-ramming is too hard.
- Permeability is insufficient.
- Venting is insufficient.

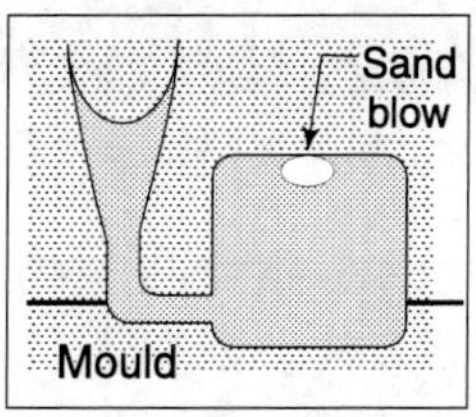

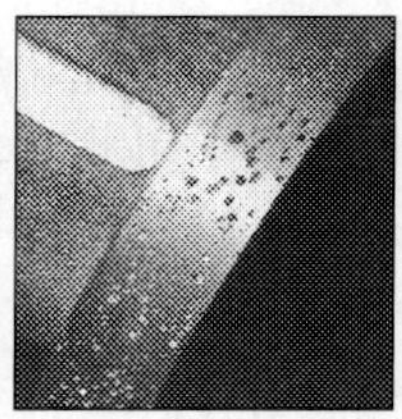

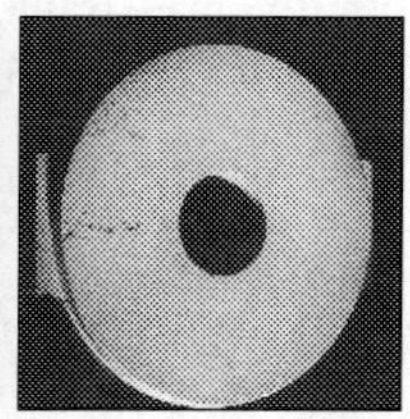

Remedy

- Reduce the moisture content or ensure even mixing.
- Moisture content of moulding sand should be controlled.
- Sand of appropriate grain size should be used.
- Ramming should not be too hard.
- Moulds should be adequately vented.

3. Surface Blowholes

This defect is characterized by localized areas of small cavities or pits on the surface of a casting. The adjacent image shows defects located on the inner diameter (the core side) of a small phosphor-bronze casting containing 10% Sn, 0.3% P: Avoid local overheating of mould or core sand by the utilization of proper running methods. Use of good refractory dressing. Check the pouring temperature.

4. Blowholes due to Sub-surface Slag Reactions

This type of blowhole is unique in appearance as it consists of small holes with rounded or irregular shapes which may at times contain shot/grit at or just below the casting surface. These blowholes mainly occur particularly on top surfaces and below cores. They may be revealed on cleaning or after machining. These holes have discoloured surfaces which are often associated with slag inclusions and sand grains.

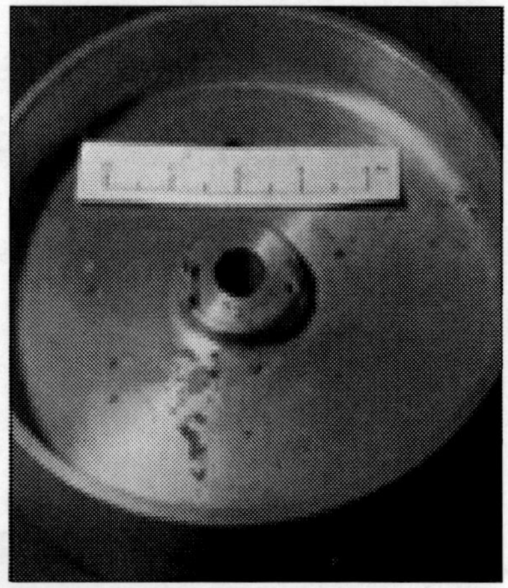

Possible Cause: Fluid containing slag and manganese oxides entering the mould could react with the metal. High manganese content and high sulphur content coupled with low pouring temperatures cause this kind of defect.

Suggested Remedy: Avoid very fluid slags formed by oxidation of the metal in the ladle, or coming from low melting point refractories such as botting clays. Skim the metal thoroughly, use strainer cores, slag traps and good gating practices. Keep the runners full of metal. Reduce the manganese content to a level just adequate to balance the sulphur content (Mn = 1.7 × S% + 0.3), reduce the sulphur content and raise the pouring temperature.

5. Blowholes in Sub-surface

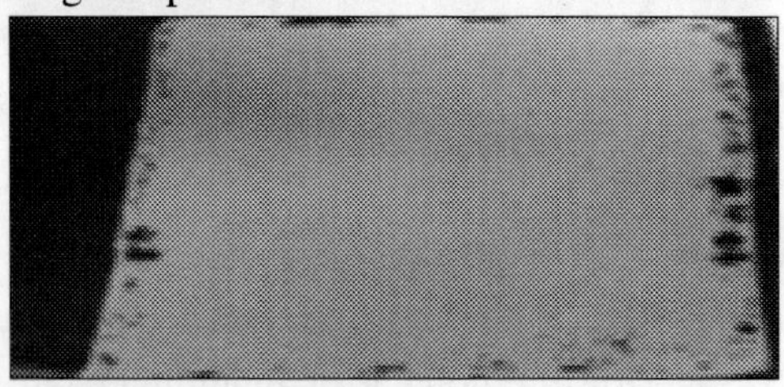

Small cavities adjacent to the surface of the casting (and usually on all surfaces of the casting). In many cases, the surface of the casting appears to be normal (which makes this a difficult defect to identify in the cast condition). In some cases however, surface pitting may be present as illustrated in adjacent figure.

Possible Cause: This type of defect is caused by a reaction of the molten metal with the moisture in the moulding sand. This is mostly associated (but not always) with high pouring temperatures and is encouraged by insufficient de-oxidation, insufficient permeability in the sand and insufficient ventilation.

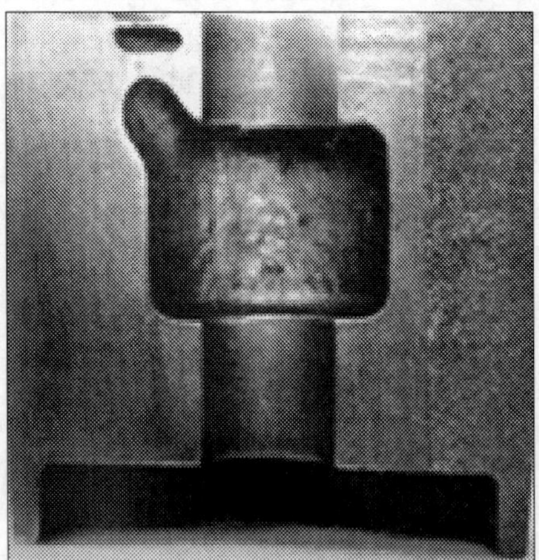

Suggested Remedy: Reduce the volatile content in the mould cores and dressing. Ensure that the material is sufficiently de-oxidized. Increase the level of ventilation and increase the permeability of the sand. Reduce the pouring temperature if possible.

6. Blowholes Due to Mould or Core Gases

Round-shaped holes with bright smooth walls which normally present themselves once the casting is sectioned as shown in adjacent figure. In some cases where blowholes appeared from concentrated pellets of other materials such as clay in the mould, the defect may migrate to the surface of the casting.

Possible Causes: Significant/excessive evolution of gases from the mould or core sand together with low permeability and insufficient venting of the mould or core. It may also be the result of improperly mixed mould and core sand.

Suggested Remedy: Make sure that the sand is properly mixed and milled and that the materials (i.e., bonding) are dispersed uniformly throughout the mix. Ensure that the cores are dry and properly baked. Make use of less organic binders or additions and increase the permeability/ventilation of the sand/mould.

7. Blowholes Due to Entrapped Air in Steel Castings

These blowholes appear to be round with smooth walls found on or just under the surface of the casting.

Possible Cause: Entrapped air

Suggested Remedy: Increase the permeability of the sand as well as venting of the mould. Try and avoid fast and turbulent metal flow into the mould.

8. Entrapped Air

Elongated cavities of appreciable size consisting of smooth walls located on or just below the surface of the casting near the highest part of the cast component.

Possible Cause: Entrapped air

Suggested Remedy: Make use of a vent in the affected area. Wire may be used to increase the permeability of the sand should the surface finish of the product allow for it. Check the degree of ramming or mould hardness at the affected areas.

9. Blowholes Due to Improper Fusion of Chaplets

Smooth-walled cavities adjacent to chaplets which are round or elongated in appearance. This defect is more often than not associated with issues associated to the fusion of the chaplet with the casting and may cause leakage when exposed to pressure.

In the figure, (a) and (b) show blowholes and imperfect fusion of the chaplet in a grey cast iron. (c) shows good fusion of chaplet with metal.

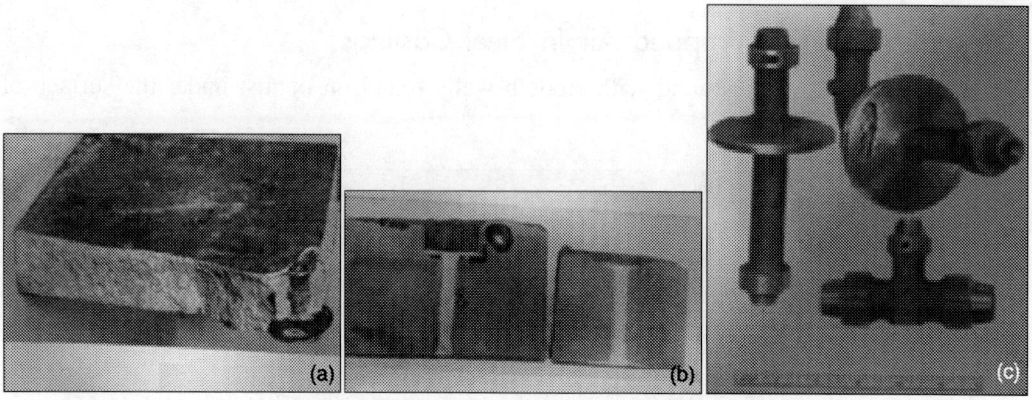

Possible Causes: Rusty, dirty, damp or improperly coated chaplets.

Suggested Remedy: Make sure that the chaplets are clean, dry and free of contamination. Bright iron steel chaplets are deemed to be satisfactory when used immediately. However, they may cause blowing if allowed to stand in the mould (long enough for rust to develop). The use of tin-coated or copper-coated chaplets is recommended to avoid this issue. Zinc or alloys containing zinc are problematic for chaplets as they also cause "blowing".

10. Air Inclusions

1. The atmospheric and other gases absorbed by the molten metal in the furnace or during the flow in the mould, when not allowed to escape are trapped in the casting and weaken it.
2. High pouring temperature increases the amount of gas absorbed.
3. Due to poor gating design such as straight sprues in unpressurized gating, abrupt bends and turbulence causing practices in the gating which increase the air aspiration.
4. Low permeability of the mould itself.

11. Pinhole Porosity

Pinholes are tiny blowholes appearing just below the casting surface.

- Agglomeration of small rounded cavities up to around 5 mm in size. Surface of the cavities is mostly smooth and polished, occasionally with graphite; clearly visible pinholes in the casting skin; surface is often oxidised.

Typical characteristics: All cavities have the same size and the long form is always oriented perpendicular to the surface, arranged at outer edges in the line bisecting the angles.

Preferred defect locations: Outer edges and corners, as well as workpiece surfaces; defects always in the immediate vicinity of the surface. Often at mould parting lines or transitions to cores, these are mostly arranged like strings of pearls/beads.

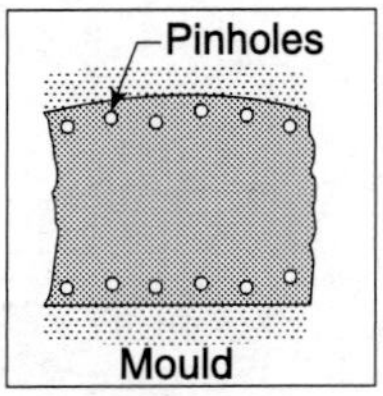

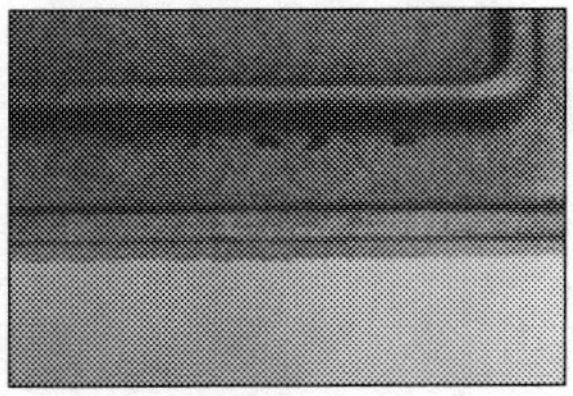

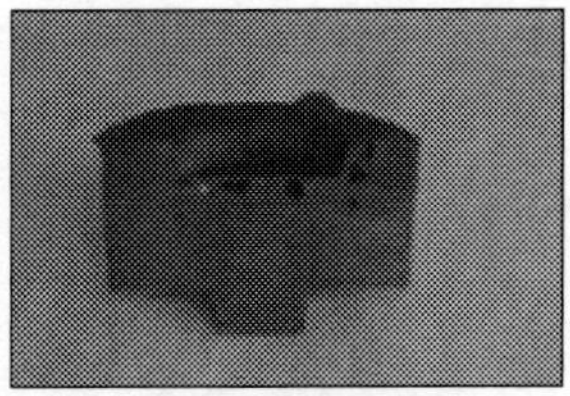

Possible Causes

- Sand with high moisture content.
- Absorption of hydrogen/carbon monoxide gas in the metal.
- Alloy not properly degassed.
- Steel is poured from wet ladles.
- Sand containing gas producing ingredients.

Remedies

- Reducing the moisture content of moulding sand.
- Increasing its permeability.
- Employing good melting and fluxing practices.
- Improving a rapid rate of solidification.
- Optimising mould material.
- Maintain optimum water content.
- Maintain optimum nitrogen content (must be less than 80 ppm).

12. Internal Shrinkage Cavities

- Defect is almost always identifiable during machining. This occurs because specific volume of melt is higher than the specific volume of the solid. Contraction occurs during solidification and cooling. Shrinkage cavities occur if the feed system design is not proper. Shrinkage amounts to 5% by volume and needs design consideration.

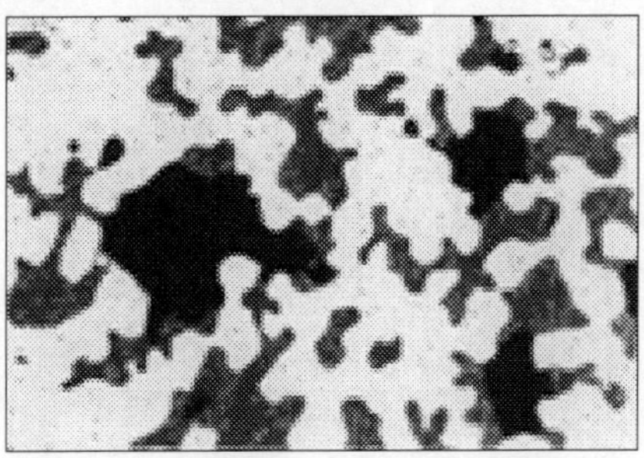

- **Remedial measures**
  - Use optimal gate and feeder system (calculate and simulate).
  - Increase the degree of saturation.
  - Reduce the pouring temperature.
  - Carry out solidification simulation.
  - Use exothermic feeder sleeve.
  - Use proper ramming control.

Table 11.1

Metal or alloy	*Volumetric solidification contraction (%)*	*Metal or alloy*	*Volumetric solidification contraction (%)*
Aluminium	6.6	70% Cu–30% Zn	4.5
Al–4.5% Cu	6.3	90% Cu–10% Al	4
Al–12% Si	3.8	Grey iron	Expansion to 2.5
Carbon steel	2.5 – 3	Magnesium	4.2
1% carbon steel	4	White iron	4 – 5.5
Copper	4.9	Zinc	6.5

Source: After R.A Flinn.

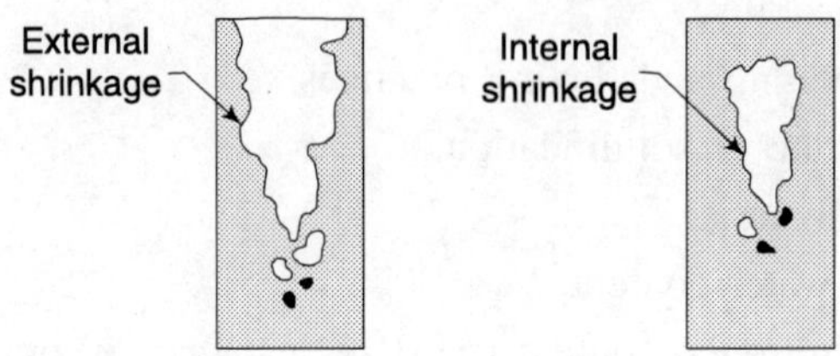

13. Dispersed Shrinkage

- Accumulation of small, crack-like cavities. Only visible after machining. Cross sections up to around 8 mm long, 1-2 mm wide and up to 2 cm deep.
- All cavities are of the same size and always perpendicular to the surface, arranged at the outer edges of the lines bisecting the angles.

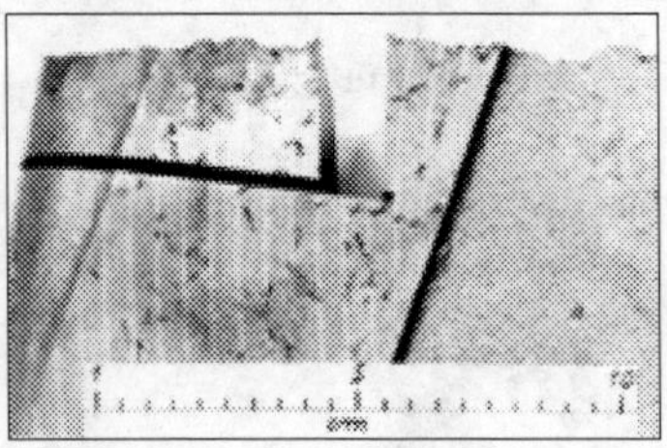

- **Remedial measures**
  - Minimise core sand intake.
  - Reduce water content.
  - Extend mixing time.
  - Increase casting temperature.

14. Moulding Material Defects

Cuts and Washes

Thickening of the casting at one place.

Typical characteristics: Sand washout, combined with sand and/or slag inclusions because of low strength or molten metal flowing at high speed.

Preferred defect locations: Mostly close to the gate or on edges, which tend to heat.

- **Remedial measures**
  - Choose binder with good properties at high temperatures.
  - Greater compaction of the mould sand.
  - Avoid high flow speeds and long pouring times.
  - Improve gate design to reduce turbulence and by increasing size of the gates.

15. Metal Penetration

Penetration occurs when the molten metal flows between the sand particles in the mould. These defects are due to inadequate strength of the mould and high temperature of the molten metal adds on it.

1. More or less thick sand adhesions on the casting, which are held fast on the workpiece by metal penetrations. The metal between the grains cannot always be identified by the naked eye.
2. Widespread, frequently cushion-shaped form.
3. Often cannot be removed by blasting.
4. Penetrating adhesions in blackened moulds and cores can often be cracked off the work-piece using a chisel.

Typical characteristics: Metal between sand grains. Can often only be identified with a microscope.

Preferred defect locations: Thick parts, zones of inadequately compacted mould material and zones, in which cores lie in the hottest parts. Preferably in the lower parts of the mould (higher casting pressure).

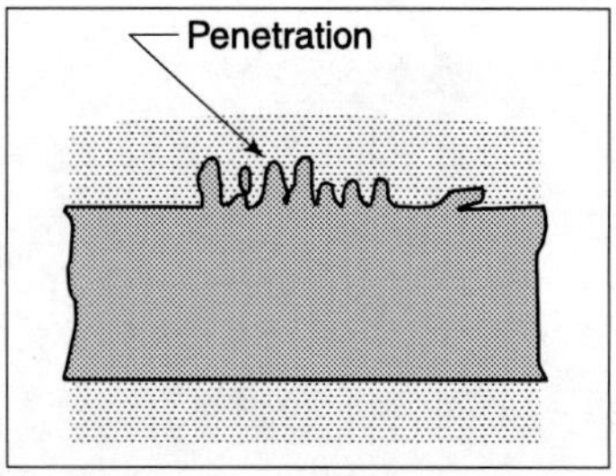

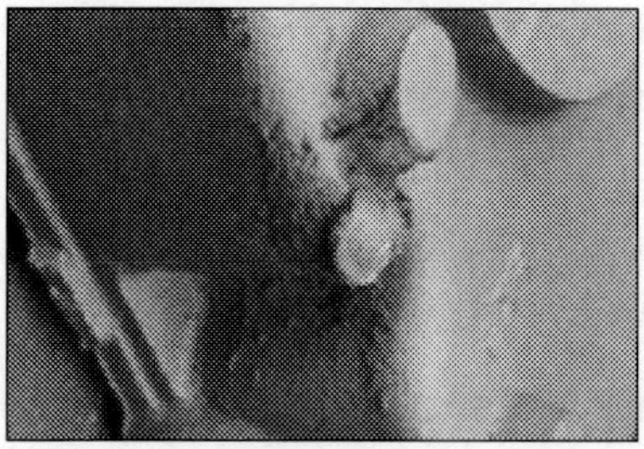

Causes

- Low strength of moulding sand.
- Large size of moulding sand.
- High permeability of sand.
- Soft ramming.

Remedies

- Use of fine grain with low permeability.
- Appropriate ramming.

16. Fusion

- **Manifestation**

  Surface defect.

  Thin sand crusts, which firmly adhere to the casting.

  Typical characteristics: Rough cast surface. Sand sintered on the casting, which can only be removed by grinding.

  Preferred defect locations: Possible on all parts of the casting. Especially in thick-walled castings. Near the gate.

- **Remedial measures**
  - Add more new sand.
  - Reduce dust fractions in sand.
  - Uniform compaction.
  - Reduce pouring temperature.

17. Runout

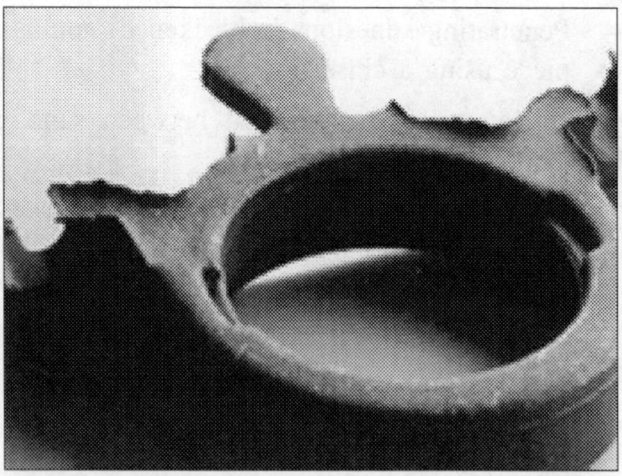

- **Manifestation:** During the casting, metal runs out of the box at the parting.

  Typical characteristics: Casting is incomplete or fully missing.

Preferred defect locations: The upper part of a casting is mostly missing.

- **Remedial measures**
  - Clamp the two halves of the box.
  - Place weights on mould.

18. Rat-tails

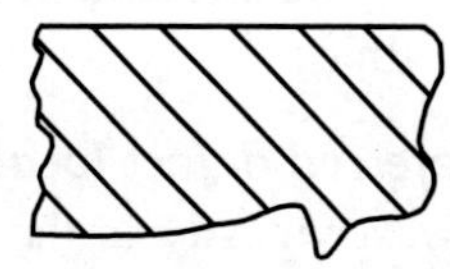

Rat-tail

Rat-tail or a buckle is a long, shallow, angular depression caused by expansion of the sand.

Causes

- Excessive mould hardness.
- Lack of combustible additives in the moulding sand.
- Continuous large surfaces on the casting.

Remedies

- Suitable addition of combustible additives to moulding sand.
- Reduction in mould hardness.
- Modifications in casting design.

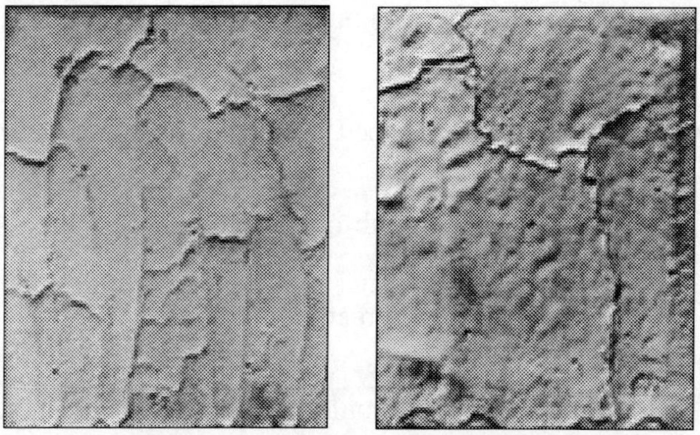

- **Manifestation:** The defect involves a sand expansion defect, which can frequently occur in highly compacted parts of the mould.

Typical characteristics: Scratches appear on the surface of the casting, which can be partly arranged parallelly.

Preferred defect locations: On areas of the mould cavity covers but primarily on the bases. Can be mostly identified on the surface in the cast condition.

Remedial measures

- Optimise mould material preparation with the aim of increasing the wet tensile strength
- Fast mould filling.
- Reduce the casting temperature.

19. Swell

Swells are excessive amounts of metal in the vicinity of gates or beneath the sprue.

Causes:

- Insufficient or soft ramming.

- Low mould strength.
- Mould not adequately supported.

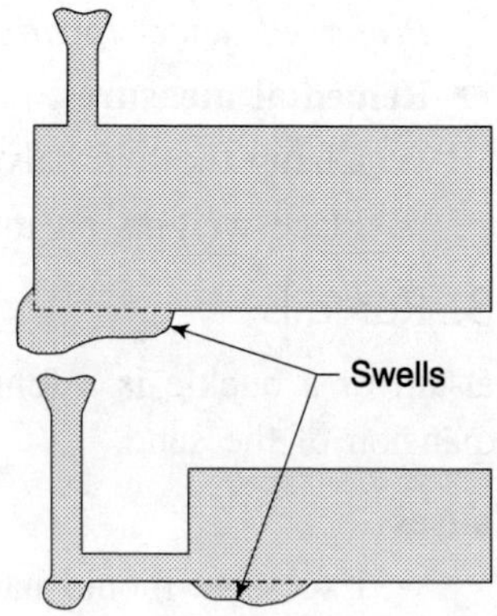

- **Manifestation:** Irregular, widespread thickening of the casting on the outside or inside.

  Typical characteristics: Gradual, unwanted increase in wall thickness.

Preferred defect locations: Possible in all areas of the casting, however mostly in the lower part of the mould.

Remedial measures

- Higher binder content.
- More uniform compaction.
- Reduce casting/sprue height.
- Sand should be rammed evenly and properly.

20. Drop

Drop is an irregularly-shaped projection on the cope surface caused by dropping of sand.

Causes

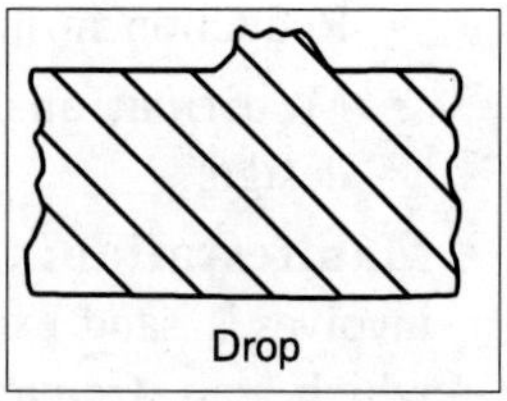

- Low green strength of the moulding sand.
- Low mould hardness.
- Insufficient reinforcement of sand projections in the cope.

Remedies

- Moulding sand should have sufficient green strength.
- Ramming should not be too soft.

21. Washout/Erosion

- **Manifestation:** Thickening of the casting at one place.

  Typical characteristics: Sand washout, combined with sand and/or slag inclusions.

  Preferred defect locations: Mostly close to the gate or on edges, which tend to heat up.

- **Remedial measures**
  - Choose binder with good properties at high temperatures.
  - Greater compaction of the mould sand.
  - Avoid high flow speeds and long pouring times.

22. Misruns and Cold Shuts

1. Metal unable to fill the mould cavity completely is called misrun. A casting that has solidified before completely filling the mould cavity.

2. Cold shut is caused when two streams of molten metal approach each other in the mould cavity from opposite directions but fail to fuse properly, with the result of discontinuity between them.

Causes:

- Low temperature of molten metal.
- Improper gating system.
- Too thin casting sections.
- Slow and intermitted pouring.
- Improper alloy composition.
- Use of damaged pattern.

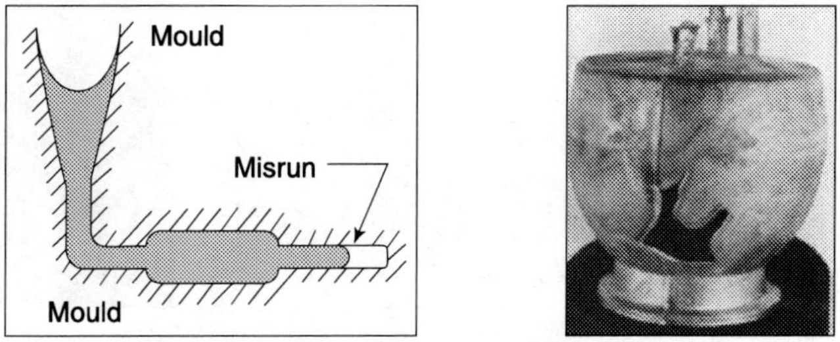

Misrun defect

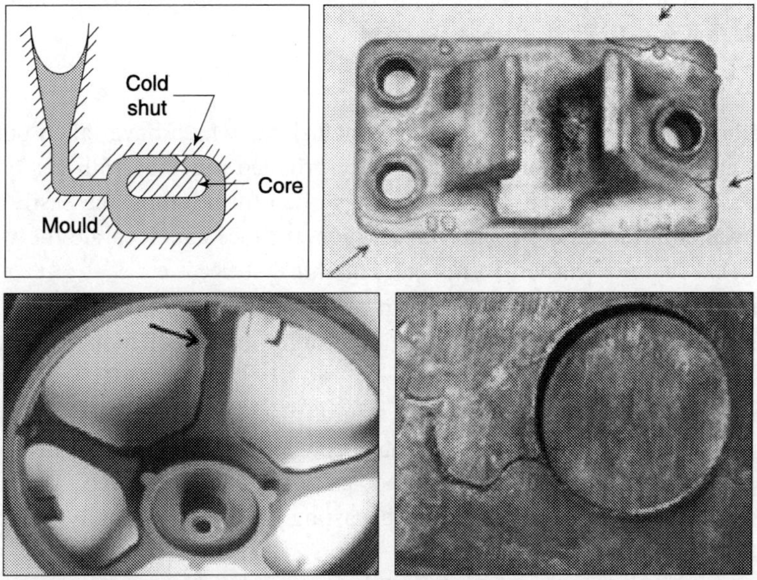

Cold shut defect

Remedial measures

- Increase casting temperature.
- Increase casting speed and make gates larger.
- Smooth pouring with the help of monorail.
- Properly transport mould during pouring.
- Arrange proper clamping arrangement.
- Provide appropriate pouring temperature.

23. Cold Shot Inclusion/Spray Bead

- Spray beads are pearl- or bead-shaped inclusions, which are loosely bonded with the metal. They are partly visible at the surface; they are often not exposed until the machining.

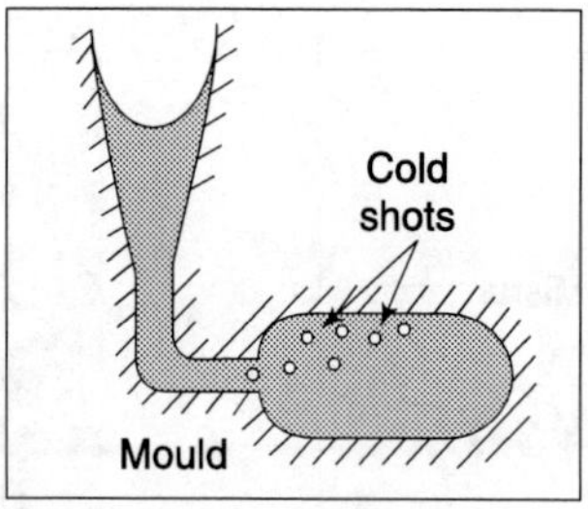

- The beads are created by turbulences during casting or by the effect of spray in the mould; metal particles prematurely solidify in drip or spherical form. These splashes oxidise very easily and the oxide skin prevents the spray beads from being re-incorporated by the subsequent melt flow.

24. Slag Inclusion

- Irregular inclusions or cavities left behind by inclusions, which have fallen out. Mostly on the surfaces at the top of the mould. Frequently grouped together with gas cavities, often with particles of mould material too. Tough slags often form lumps, the easily fusible slags that float upwards in the mould form skins, which at times can separate the whole casting wall. Occasional lustrous carbon forms on slag skins.
- Glassy inclusion, several phases of which can be frequently identified under the microscope. Skins wrinkled or creased, often bonded with a large number of mould material particles (scar like arrangement).
    - ***Preferred defect locations:*** As coarse slag at places which were in the upper part of the mould and under large area cores and core prints. Scar-shaped finely distributed, frequently in the direction of flow on the surface of the workpiece. As slag skin often running in a transverse direction to the casting wall. Occasionally, relatively coarse slags inside castings too.
    - **Remedial measures**
        - Reduce elements which bond with oxygen.
        - Properly dry ladles beforehand.
        - Reduce pouring times.
        - Increase mixing time.

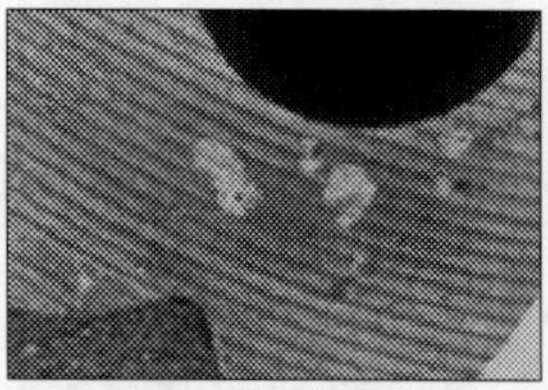

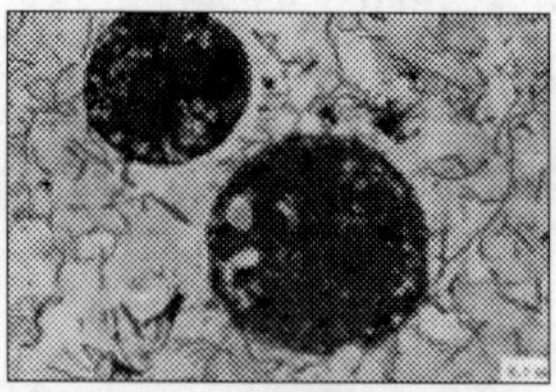

25. Hot Tears

A hot tear is a fracture formed during solidification because of hindered contraction.

- Hot tears are intercrystalline discontinuity and run along the grain boundaries. The risk of cracks at alloys with a high freezing range is higher than with a small freezing range. Hot tears occur due to stresses during solidification.
- Material partition in liquid/solid state, which occurred as the stresses, in the casting in the area of the elastic deformations, were larger than the strength of the material.
- Stresses can occur in case of uneven cooling conditions (differences in wall thicknesses). Special design of the casting can prevent contraction. Shrinkage onto the core, hot cracks primarily occur in steel castings.
- ***Preferred defect locations:*** At sudden, steep transitions in wall thicknesses. At wall thickness transitions with too small radii.

  Remedial measures

  - Design suitable for stresses and prevent residual stresses.
  - Balance out cross sections with large wall thicknesses.
  - Prevent hot sand effects.

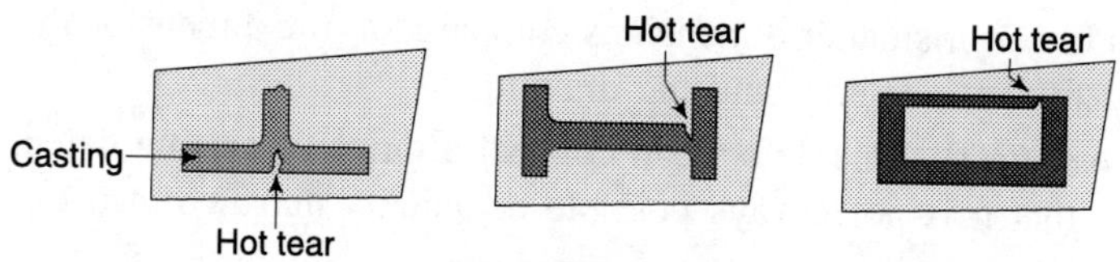

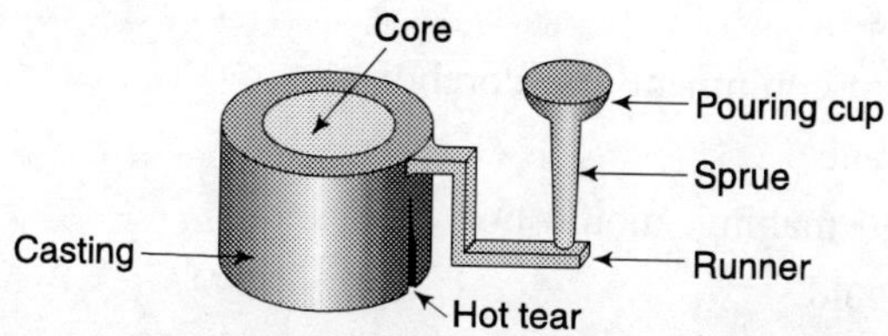

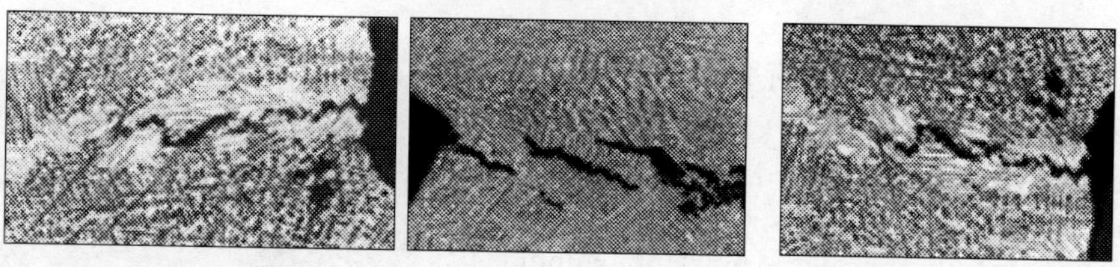

Causes

- Faulty casting design leading to excessive stresses at certain portions in the casting.
- Very hard ramming.
- Too much shrinkage of molten metal.
- Incorrect pouring temperature.
- Improper gates and risers.

- Low flowability of molten metal.
- High sulphur content in molten metal.

Remedies

- Ramming should not be too hard.
- Modification in pattern to take care of residual stresses.

26. Core Damage

- Unintentional protruding (extension) core parts.

  Typical characteristics: Parts of the core are displaced relative to each other.

  Preferred defect locations: At parting lines of the cores.

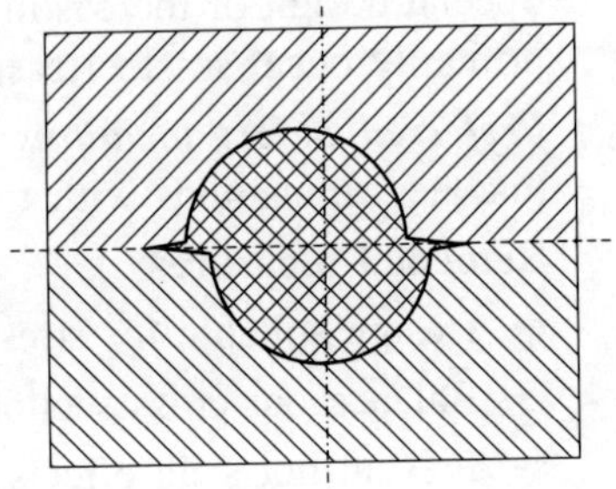

- **Remedial measures**

  Check corebox guides (is there a defect in the core-box?)

27. Dimensional Variation

- The casting has larger or smaller dimensions compared to the dimensionally checked initial pattern released by the customer and the drawing.

  Typical characteristics: This defect can conceal a range of casting defects, starting with pattern wear, so that it is not always possible to quickly narrow down the defect.

- **Remedial measures**
  - Check the pattern equipment and coreboxes.
  - Core print clearance.
  - Check the mould making, mould box guides.
  - Compact the mould.

28. Dross

- Irregularly shaped interruption in the material

  Typical characteristics: Dark scars, foamy dark surfaces, very finely distributed. Dross worsens the mechanical properties. Dross mainly consists of magnesium oxy-silicates and magnesium sulphides, and is a product of the reaction of magnesium with oxygen, sulphur and silicon.

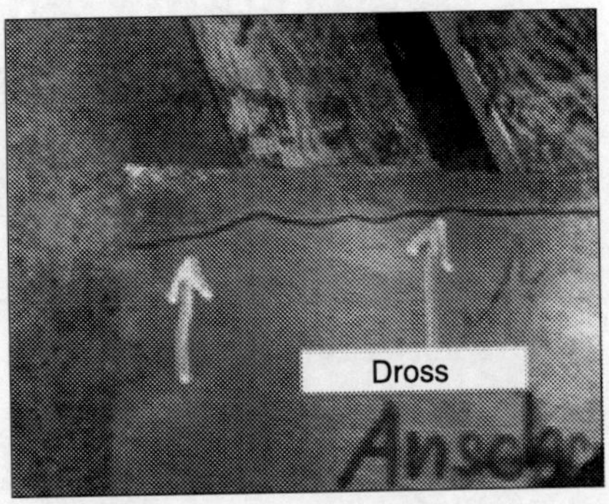

  Preferred defect locations: In the upper casting surface or beneath cores. Mostly in cast iron with nodular graphite, often only visible after the casting skin has been removed.

- **Remedial measures**

Dross cannot be prevented, only reduced.

- **Remedial measures during melting**
    - High temperature
    - Low residual Mg contents
    - Clean melting of raw materials
    - Highly concentrate Mg master alloys
    - Gas purging
    - Refining slags
    - Draw off and restrain treatment slags (large pouring basin, optimum casting system).
    - Filters are very effective.
    - Dry, closed moulds.
    - Hot casting.

29. Elephant Skin

- Irregular scaring and/or creased (bend) surface with net shaped notches.

Typical characteristics: Parchment (skin)-like appearance

Preferred defect locations: Normally in the top part of thick-walled castings on horizontal surfaces. The defect can effectively always be identified in unmachined casting.

- **Remedial measures**
    - Use pure, base iron with low sulphur and oxygen contents.
    - Use a teapot ladle.
    - Adjust the degree of saturation of the melt according to the casting temperature.
    - Possible extension of the dead-melting time after the magnesium treatment.
    - Use refining slags.
    - Use filter.

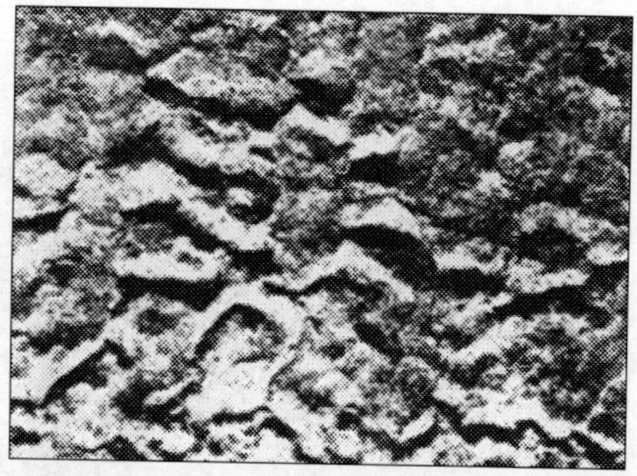

30. Inclusion of Foreign Metal

- **Manifestation:** Material not homogeneously formed
- They can be classified as endogenous and exogenous inclusions. Endogenous inclusions are caused by the reaction products during the melting process (during deoxidation). Exogenous inclusions are caused by other materials in the melt (refractory lining). Thin fluid slag can precipitate at the grain boundaries. Macro inclusions are greater than 20 microns and micro inclusions are less than 20 microns.

Typical characteristics: Visible irregularities in the microstructure.

Preferred defect locations: On the inside of the casting, partly extending out to the surface too. During machining, the inclusions are worn down by the machining tool.

- **Remedial measures**
  - Use clean charge and melt pure raw materials.
  - Optimisation of gating and feeding systems.
  - Draw off treatment slags and hold back (large pouring basin, optimum casting systems).
  - Proper gas purging.
  - Decrease of the dissolved oxygen.
  - Decrease of the overheating temperature.

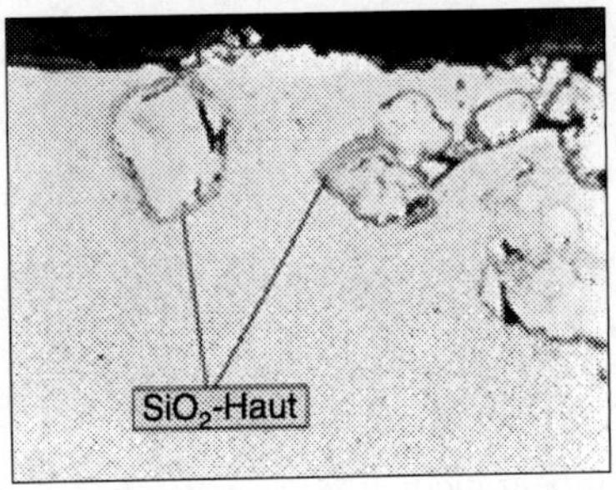

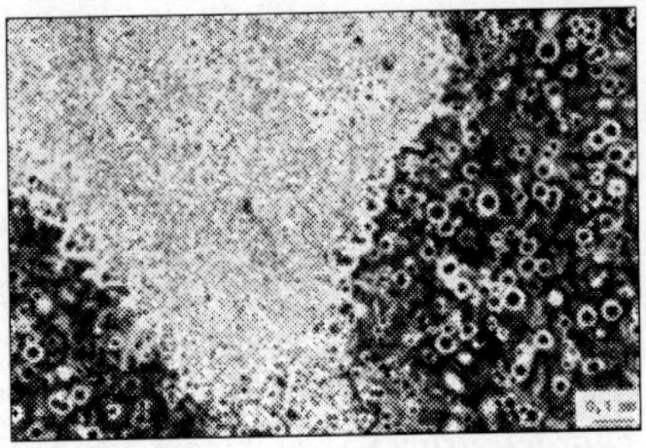

31. Mismatch in Mould

It is a shift /misalignment between two mating surfaces or the top and bottom parts of the casting at the mould joint.

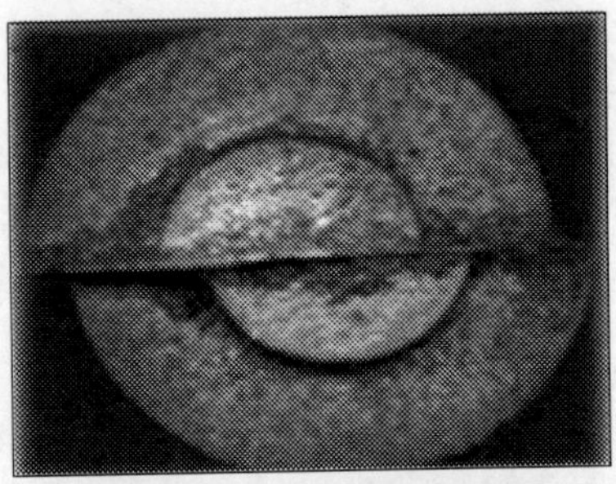

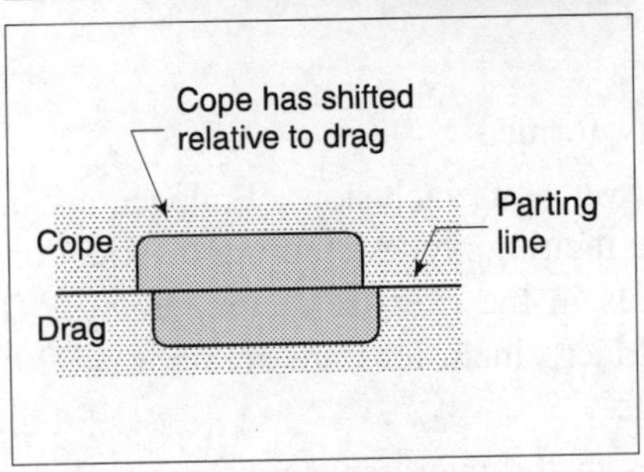

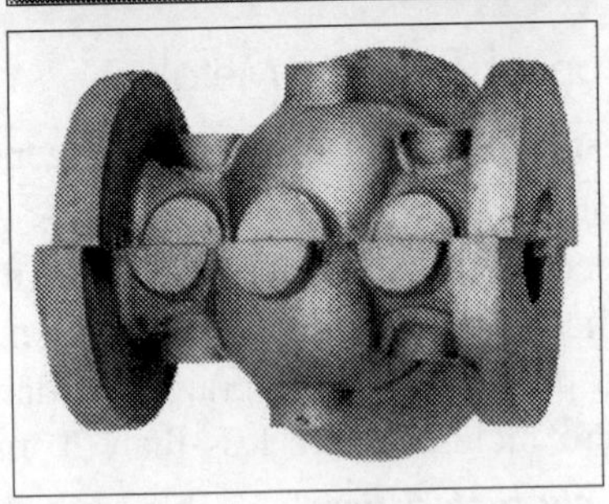

- **Manifestation:** Unintentional protruding (extending) casting parts.

  Typical characteristics: Parts of the casting are displaced relative to each other.

  Preferred defect location: On parting lines.

  - **Causes**
    - Worn dowel in patterns made in halves.
    - Improper alignment of mould boxes due to worn out/ill-fitting mould boxes.
  - **Remedial measures**
    - Check box guides, mould assembly and closure equipment.
    - Check models.
    - Properly move boxes with pins.
    - Prevent box warpage.
    - Use proper clamps.

32. Fillet

- **Manifestation:** Fillets in castings, especially in wall thickness transitions do not often occur according to the drawings, or they contain defects.

  Typical characteristics: Radii at the fillets are considerably smaller in the casting than in the patterns. The surface of the fillets is maculate (unclean).

  Preferred defect locations: Fillets at wall thickness transitions with large agglomerations of mass. At these place the mould material is subjected to extreme thermal stresses.

  Remedial measures
  - Make radii larger than those given in the drawing.
  - Check addition of new sand.

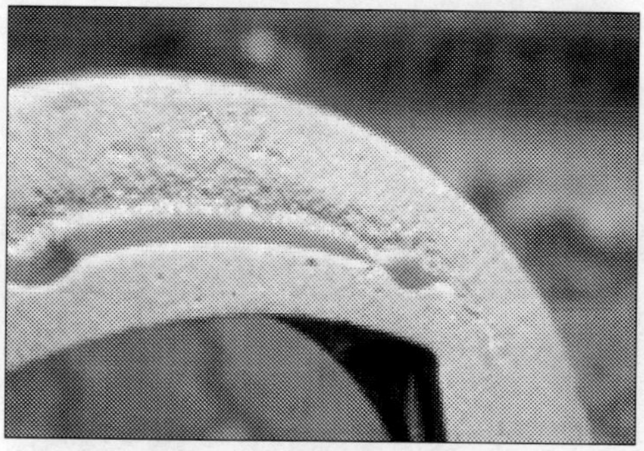

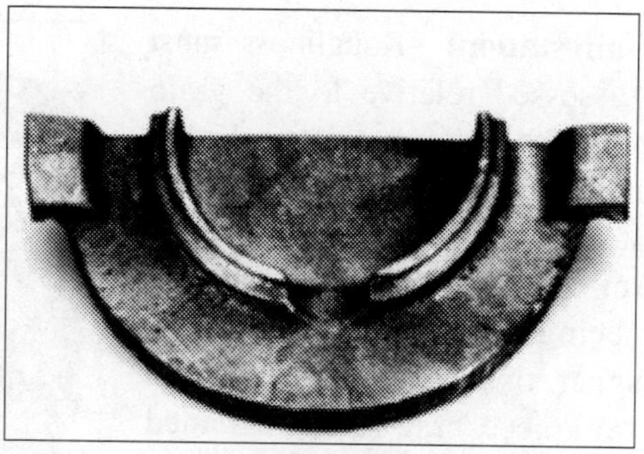

33. Mould Damage

- **Manifestation:** Irregular, widespread thickening of the outside or inside of the casting. After the pattern has been stripped, areas of the subsequent cast contour are broken off in the mould material area of the mould cavity.

Typical characteristics: Enlargement of the wall thicknesses or change in contour, the broken off areas of the mould material are then found in the casting as mould material inclusions, if the defect is caused when the mould is assembled and closed by pressing core prints or in the mould joint.

Preferred defect locations: Joint face or core prints

Remedial measures

- Reopen the mould after assembling and closing and check for pressure points and broken off edges.
- Check pattern and corebox, core print allowance.

34. Rough Casting Surface/Roughness

- **Manifestation:** Roughness must be assessed relative to the grain size of the casting selected. Under certain circumstances, a workpiece cast in coarse sand with fully uniform surface must be assessed as being smooth, although it is rougher than a "rough area" on a workpiece cast in fine grained sand.

  Preferred defect locations: Close to gate, parts deep in the mould, zones with inadequate compaction.

- **Remedial measures**
  - Use finer sand.
  - Reduce water content.

- Increase compaction pressure.
- Reduce casting temperature.

35. Fins

Fins are excessive amounts of metal created by solidification into the parting line of the mould.

Causes

- Inadequately weighted sand as well as incorrectly assembled moulds and cores.
- Over flexible bottom boards.
- Loose plates and improper clamping of flasks.

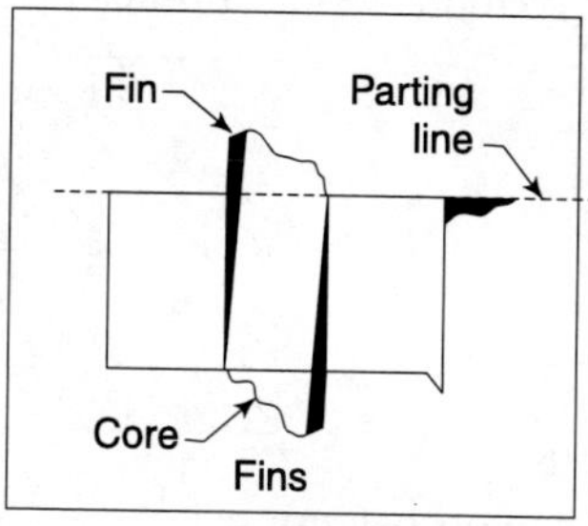

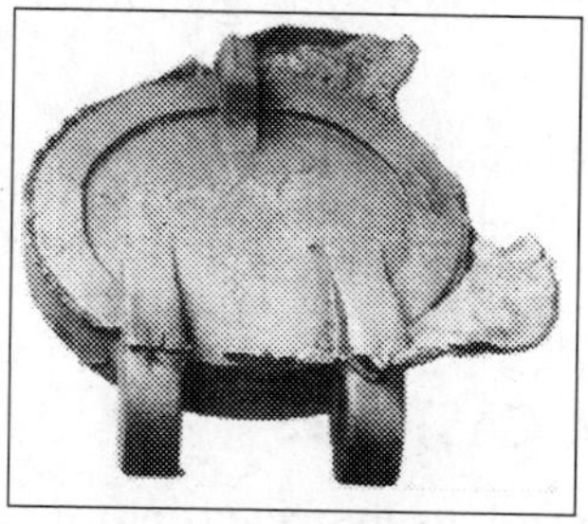

Remedy

- Correct assembly of the mould and cores used for casting.

36. Scabs

Scabs are surface slivers caused by splashing and rapid solidification of the metal when it is first poured and strikes the mould wall.

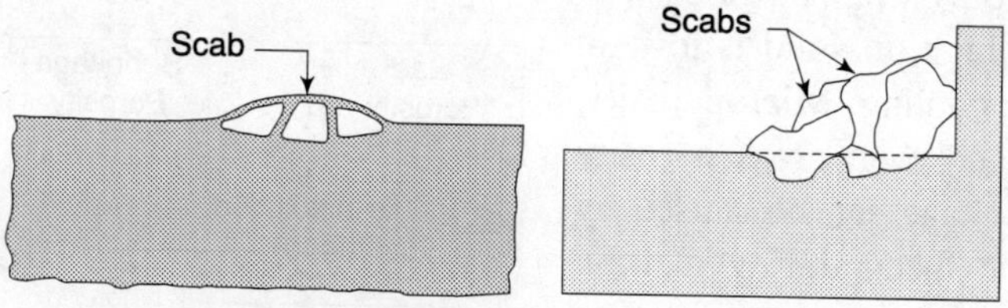

Causes

- Insufficient strength of mould and core.
- Uneven mould ramming.
- Lack of binding material in facing as well as core sand.
- Faulty gating.
- Intense local heating due to slow running of molten metal over sand surface.

Remedies

- Appropriate ramming.
- Improved gating system.
- Addition of sufficient binders in facing and core sands.

37. Hard Spots

These spots are formed due to the local chilling by moulding sand which leads to the formation of white cast iron at those places, rendering them hard.

Causes

- Faulty metal composition.
- Faulty casting design resulting in relatively more rapid cooling of certain spots.

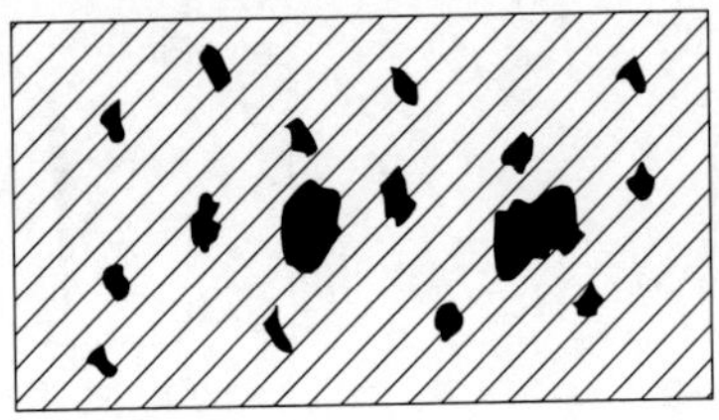

Remedies

- Modification of casting design.
- Modification of casting composition.

11.3 SOLIDIFICATION DEFECTS IN ALUMINIUM CASTINGS

Solidification defects can be subdivided into three main categories:

1. Gas porosity
2. Shrinkage porosity
3. Hot tearing and cracks.

Classification of solidification defects in casting is shown in figure here.

11.3.1 Gas Porosity

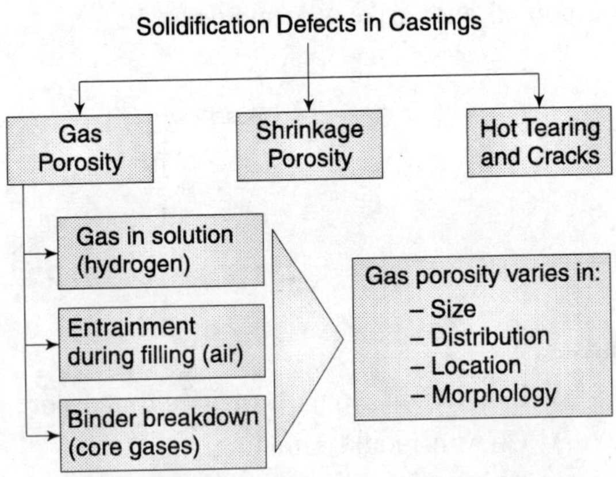

Gas porosity results from entrapped gases. The solubility in liquid is high, in solid is low, so gas is rejected during cooling. Microporosity, centerline shrinkage and even gross porosity resembling gross shrinkage can be caused by gases dissolved in the metal during melting and pouring as well as by solidification shrinkage. Voids in a well fed section must be caused by gas. Rounded voids with smooth walls are generally due to gas, while angular voids, perhaps with dendrite arms protruding into the voids, are usually caused by imperfect feeding. Often, as in the case of the fine microporosity seen occasionally in non-ferrous castings, the distinction is not simple, and both gas content and metal shrinkage must be regarded with suspicion when such microporosity is observed.

Gas held in solution in the molten metal can be precipitated as the metal solidifies due to reduced solubility on freezing. If the mould is filled under very poor conditions, air can be entrained in the metal stream and then trapped as the metal solidifies. The sand binders used to make the moulds and cores often break down when in contact with the molten metal and the gaseous decomposition products can force their way into the solidifying metal, leading to defects which are normally known as 'blows'.

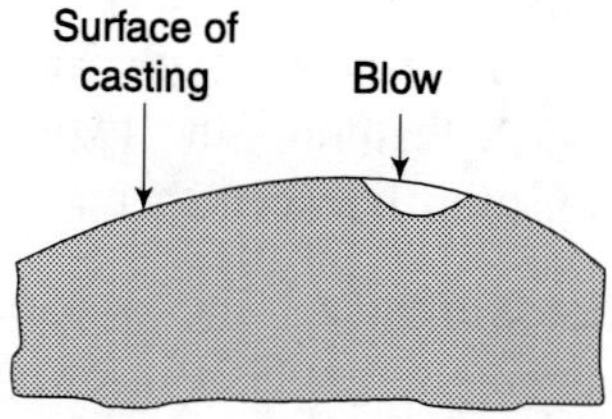

These different types of gas porosity defects vary in their size, distribution, distance below the casting surface and morphology. It follows therefore that the cause of such defects in a real casting can be deduced from a careful examination Characteristics of gas porosity are given in Table 11.1.

Table11.1 Characteristics of gas porosity

Defect	Distribution	Size
Gas precipitation	Uniform, apart from 1-2 mm near surface	0.05-0.5 mm
Air entrainment	Above ingates, especially the first ingate. Concentrated on horizontal ledges. Very close to surface. Only revealed when casting is short-blasted or machined.	1-5 mm
Core blows	At a uniform distance under top of casting	Typically 100 mm in diameter, 10 mm thick

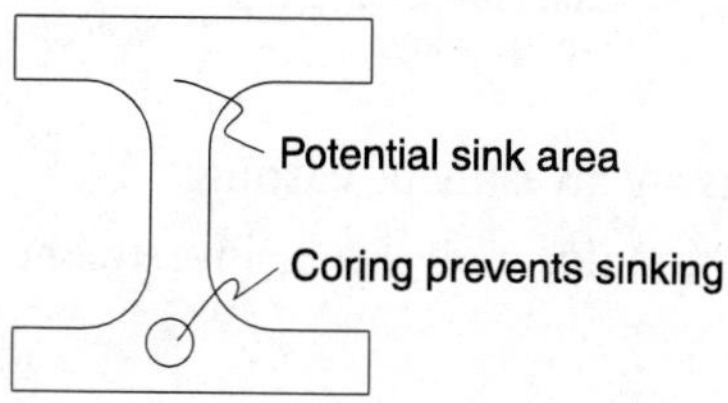

11.3.2 Shrinkage Porosity

Shrinkage porosity is subdivided into macroporosity and microporosity. There is no fundamental difference between these two forms of porosity; one gradually changes into the other as a function of the freezing range of the alloy.

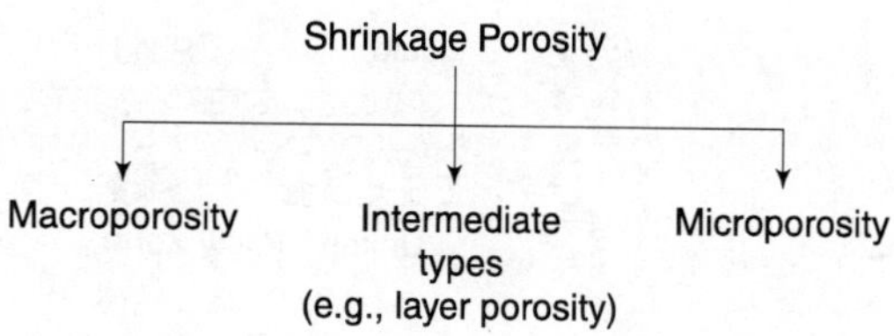

Macroporosity: Solidification starts along the walls and at a slower rate on the top surface. As progressively more metal solidifies, the volumetric contraction is compensated for by the concurrent sinking of the liquid surface, forming a smooth conical funnel or long 'tail' inside the ingot known as shrinkage pipe or piping. It was a commonly-held belief that there are two forms of pipes—primary and secondary—the latter appearing to be discrete islands of porosity below the primary pipe. This is, in fact, an incorrect interpretation of two-dimensional sections of such features—the two are interconnected, constituting the same feature, and there is no distinction between them.

Microporosity This is particularly a problem in long freezing range alloys and/or when the temperature gradient is low. These conditions create an extensive and uniform pasty zone which is favoured by:

1. Metals of high conductivity, such as aluminium alloys;

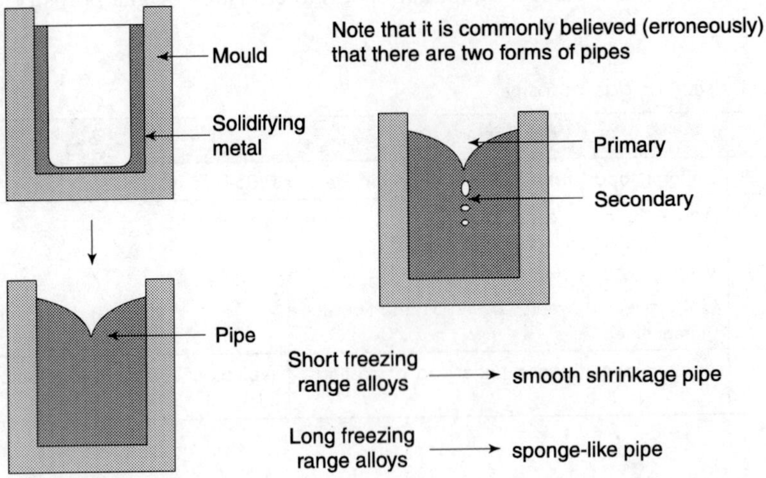

2. High mould temperatures, as in investment casting;
3. Thermal conductivity of the mould, as in sand, investment or plaster low moulds.

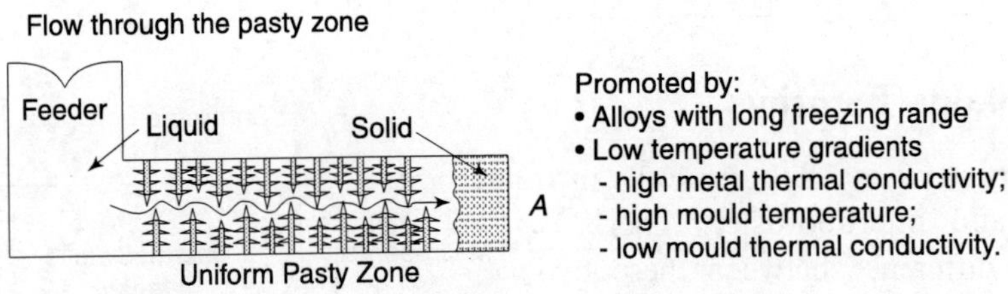

In such cases, towards the end of solidification, there will be a 'pasty' or 'mushy' zone consisting of a forest of dendrites enclosed in the remaining liquid.

Layer Porosity: Figure 11.1 shows a schematic view of the pressure in the liquid metal as a function of time. At time t_0, everything is liquid and so there is no pressure gradient. Once solidification starts, then there will be a parabolic pressure distribution through the pasty zone. As time progresses from t_1 to t_2, the gradual decrease in the capillary diameter r will increase the tensile stress in the remaining liquid.

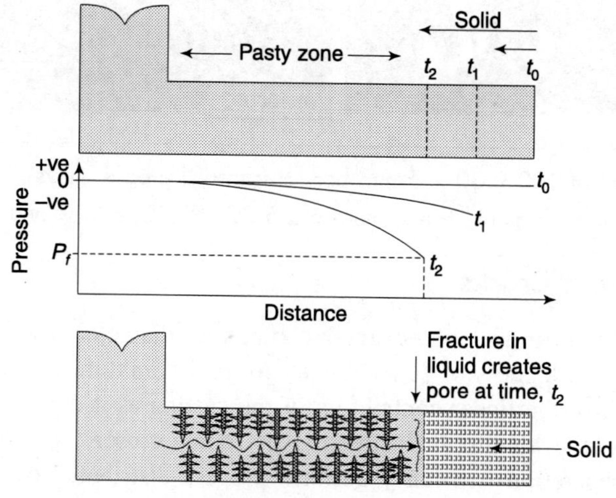

Figure 11.1

The hydrostatic tension in the liquid in the pasty zone continues to increase until, at time t_2, a critical fracture pressure P_f is reached at which a pore will nucleate. This could be an isolated micropore if the shrinkage conditions are not too severe. The generation of small regions of microshrinkage porosity in long freezing range alloys is quite common.

This process repeats until the whole of the casting has solidified (Figure 11.2). Porosity therefore occurs in a sprinkled array of fine pores, or, if the shrinkage conditions are more severe, in an array of layers, as shown schematically here in the final casting.

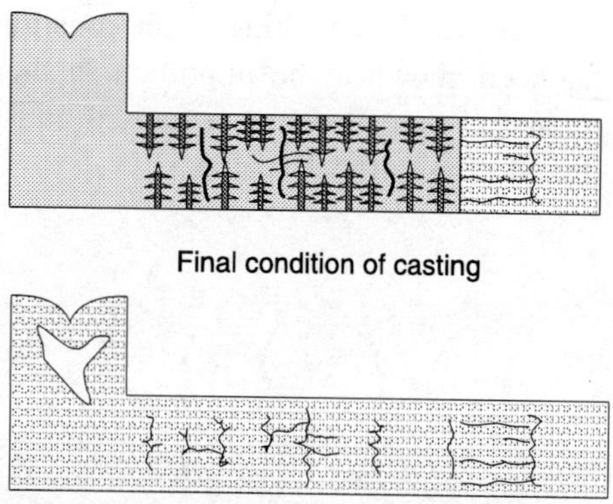

Figure 11.2 Formation of layer porosity

Figure 11.3 show the radiographs of 100 × 30 × 5 mm bars cast in nickel-base alloy at 1620°C in vacuum (15 μm Hg) and with mould temperatures of 250°C, 500°C, 800°C and 1000°C.

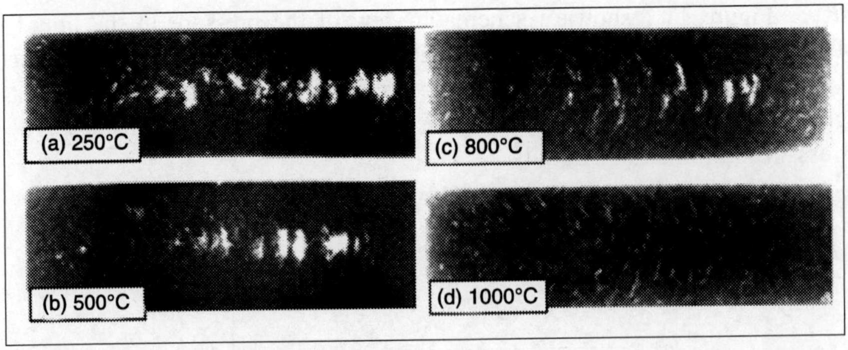

Figure 11.3 Radiographs of 100 × 30 × 5 mm bars cast in nickel-base alloys at 1620°C in vacuum (15 μm Hg) and with mould temperature of 250°C, 500°C, 800°C and 1000°C.

11.3.3 Hot Tears and Cracks

Figure 11.4 shows a simple model of hexagonal grains of diameter *a* separated by a liquid film which initially has a thickness of *b*. At this stage, if the mixture is subjected to a tensile strain, usually as a result of the linear contraction of the casting as it cools, the grains will tend to move further apart in the longitudinal direction but come closer together in the transverse direction. At first this separation is a fairly uniform process distributed over many grains, but later becomes concentrated in one or more branching planes. This causes the residual liquid to rearrange itself somewhat and will effectively create segregation defects in the form of layers of a solute-rich low melting point constituent in the casting. However, such features have not so far been shown to be important in their effect on mechanical or other properties. At this stage, no substantial defect is caused and the grains will still be nicely cemented together by the liquid phase after freezing.

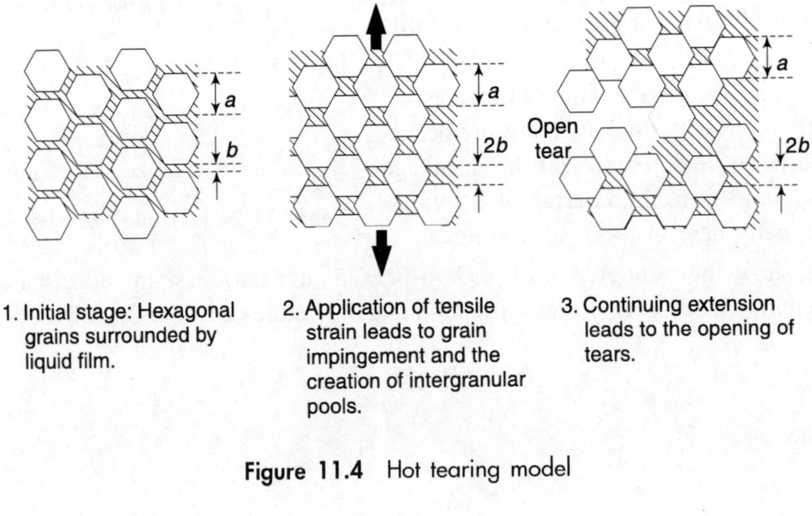

Figure 11.4 Hot tearing model

If strain continues to be applied to the solidifying casting, then the ability of the residual liquid to rearrange itself to fill the volumes left by the rearranging grains is now used up. At this point the further separation of the grains draws air into the space, forming a defect known as a hot tear.

One final aspect of hot tearing is worth emphasising. Hot tears usually have to be nucleated. If no suitable nucleus is present, then it is difficult, or impossible in some alloys, to form a tear. Since it seems that, in aluminium alloys, oxide films are excellent nuclei for hot tears (and excellent for nucleating other volume defects such as the various forms of porosity) then it follows that improving the design of the filling system and increasing the quality of the liquid metal will often cause hot tears to disappear. This non-traditional technique is recommended as the most effective method of all to deal with hot tears. Figure 11.5 gives the summary of parameters involved in hot tearing.

Comments on Hot Tearing

1. Do not confuse hot tears with layer porosity!

• Layer porosity - volume contraction
↓
Separation of liquid
(dendrites not affected)
↓
Pores

• Hot tear - linear contraction of casting
↓
Separation of dendrites
↓
Spread of pore into liquid

2. Increasing residual liquid in solidifying casting reduces hot tears!
E.g., Al-Si- good hot tearing resistance
Al-Cu- poor hot rearing resistance

3. Hot tears require nucleation!
E.g., oxide films in aluminium alloys.

Therefore..... improve running system design
↓
reduce oxide defects
↓
reduce hot tears

Figure 11.5 Summary sheet for hot tearing

Prevention of Hot Tears

1. Alter casting design
2. Chill hot spots
3. Reduce constraint from mould
4. Add brackets and webs
5. Grain refinement
6. Reduce casting temperature
7. Alloying
8. Reduce contracting length

- It is possible to alter the geometry of the casting to reduce stress concentrations and hot spots, for example, by providing generous radii at vulnerable sections.
- Local hot spots can be reduced by local chilling which will strengthen the metal by taking it out of the susceptible temperature range.
- There are various ways of reducing the mould strength so that it provides less constraint to the contracting casting.
- Brackets and webs can be placed across a vulnerable corner or hot spot to provide mechanical support and to enhance local cooling.
- Grain refinement should help to reduce tear initiation since the strain will be spread over a greater number of grain boundaries.
- A reduction in the casting temperature can sometimes help, probably because it reduces the grain size.
- It is sometimes possible to benefit from varying the alloy constituents within the specified composition ranges. In particular, increasing the volume fraction of eutectic liquid may help by increasing the pre-tear extension and by decreasing the cracking susceptibility.
- Finally, it is sometimes possible to site feeders carefully so that the casting is effectively split up into a series of short lengths to reduce the strain concentration.

Cold Cracks

By definition, these form at some temperature below the solidus and often considerably higher than room temperature. Whereas hot tears tend to be rather rugged in nature, cold cracks are straighter and smoother. Cold cracks can be either transgranular or intergranular, depending on the relative strengths of the grains and the grain boundaries. Whereas hot tears are always oxidised, cold cracks can be free from oxidation if they are formed at relatively low temperature and the castings not subsequently heat treated.

Stress is required for a crack to nucleate and grow and there are a number of possible sources. Firstly, the casting continues to suffer strains as it cools to room temperature. This arises simply from the differential cooling rates of different parts of the casting. In addition, of course, the mould or cores may resist the contraction of the casting to such a degree that dangerously high stress is generated within the casting. Stresses can also arise from the volumetric changes associated with phase transformation(s) as the casting cools to room temperature, one example being the transformation from delta ferrite to austenite in steels.

11.4 EFFECT OF DEFECTS ON PROPERTIES

In the past, some castings containing large undetected defects have gone on to serve useful lives in critical applications. Others, however, have failed (e.g., the Tay Bridge disaster in Scotland, UK). The mixed performance arises because of the elementary point that the size of the defect is often of much less importance than its form and position. A large pore in a low-stressed area of the casting may be far less detrimental than a small region of layer porosity in a sharp-corner subject to a high tensile stress. To have blanket specifications requiring the elimination of all types of defect from every area of the casting is therefore, not appropriate. It may result in the scrapping of many serviceable castings. The most logical and effective control over casting performance is achieved by specifying separate designated regions of the casting, and then each separate region being required to contain no defects above the critical size appropriate for that location.

The list of properties which a casting may be required to possess can be long. It might include, for instance, high-temperature properties such as creep resistance, high-temperature fatigue resistance, or oxidation resistance. Alternately, room temperature properties such as resistance to corrosion in specific environments may be required. In this chapter, we shall confine ourselves to the more usually specified properties such as strength, toughness, ductility, and leak-tightness. Different defects have different effects on each of these properties.

11.4.1 Yield Strength or Proof Stress

Because no substantial deformation has taken place, it is logical to assume that the yield stress or 0.2 proof stress will be unaffected by most defects. The only effect will be that due to the reduction in area. But since most defects occupy at most only a few per cent of the area of the casting, this effect is usually hardly detectable. The 1% or so of reduction in 0.2 proof stress because of the 1% or so loss of area of layer porosity cannot be detected in the scatter of the results. Ultimate tensile stress and Young's modulus are, however, noticeably affected. Figure 11.6 shows the effect of layer porosity on the mechanical properties of Mg-Zn alloys.

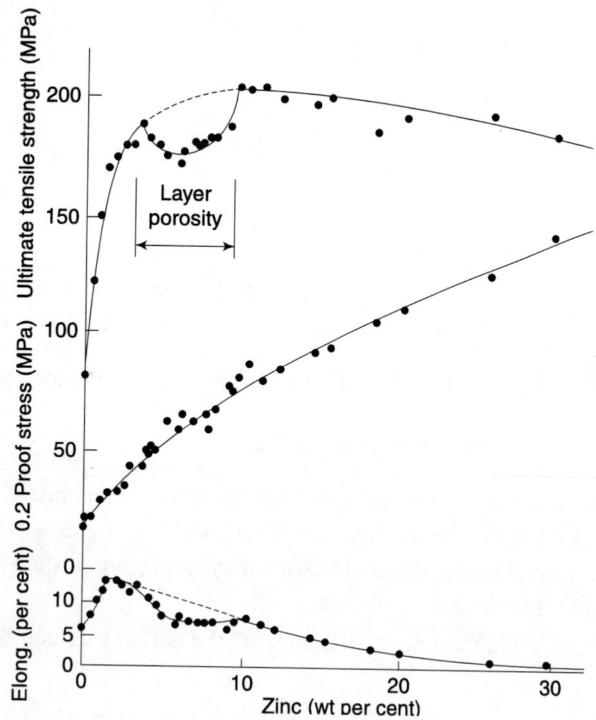

Figure 11.6 The effect of layer porosity on the mechanical properties of Mg-Zn alloys

11.4.2 Steel Embrittlement by Grain Boundary Precipitation of Sulphides

A notable exception to the constancy of the yield stress is found for the embrittlement of steels by grain boundary precipitates of sulphides. The large effect on yield strength is consistent with the large area which the sulphides cover on the grain boundaries, and the weakness of the sulphide phase. Intergranular brittle failure by a fast-running crack through the grain boundaries becomes favoured over extensive plastic deformation of the matrix. Figure 11.7 shows the effect of sulphur on mechanical properties of medium carbon steel.

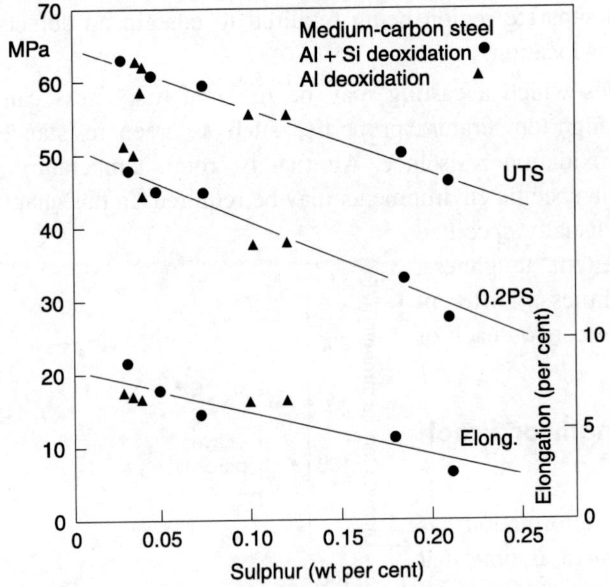

Figure 11.7 The effect of sulphur on the mechanical properties of medium carbon steels

11.4.3 Fracture Toughness

Fracture toughness is a material property, independent of the size of gross defects. It is useful in the prediction of the shapes and sizes of defects which might lead to failure. According to the fracture mechanics, fracture may begin when the stress-intensity factor K exceeds a critical value, the fracture toughness K_{1C}.

For inside crack of diameter d, the critical defect size

$$d = \frac{2K_{1C}^2}{\pi \sigma_y^2}$$

For an Al alloy,

$K_{1C} = 32$ MPa$\sqrt{m}$

$\sigma_y = 240$ MPa

$d = 11$ mm for inside crack

$d = 9$ mm for surface crack

Edge cracks are somewhat more serious than centre cracks. These are comparatively large defects (11 mm/9 mm) that can be detected relatively easily by non-destructive tests prior to the casting going into service. Most current radiography standards which state "no linear defects of any size" or which reject aluminium alloy castings for flaws more than approximately 1 mm in size, may not be logical. When the applied stress equals the yield stress, the greatest resistance to crack extension offered by the materials is controlled by the value of $\frac{K_{1C}}{\sigma_{ys}}$. This parameter is a valuable measure of the defect tolerance of a material. The permissible defect size diminishes with increasing strength. For stronger irons, the permissible defect size is below 1 mm. This is particularly difficult to detect and sets a limit to the stress at which a strong ductile iron casting may be used with safety. Steel has high fracture toughness with correspondingly good tolerance to large defect at low applied stress. However, when they are used in highly stressed applications, the permissible defect size is reduced again. Figure 11.8 shows the permissible edge defect sizes in ductile irons of increasing strength.

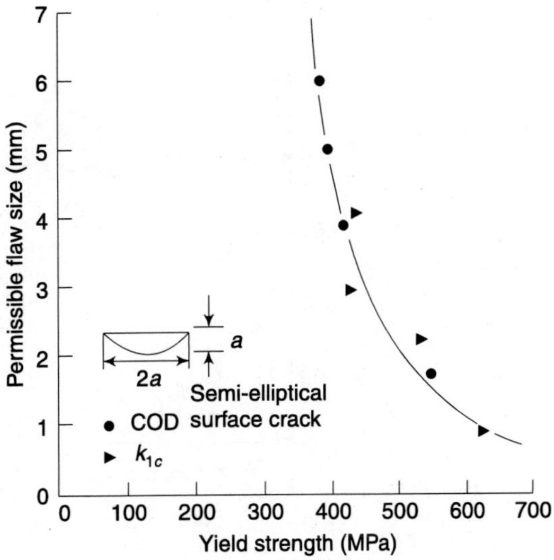

Figure 11.8 Permissible edge defect sizes in ductile irons of increasing strength

For brittle materials, the use of fracture toughness is a more appropriate measure of the reliability of the casting than ductility.

11.4.4 Fatigue

The fatigue performance of cast alloys is generally poor. The initiation of fatigue crack during stage 1 of the fatigue process is short. This is due to the presence of defects (pores, oxide films). The growth and propagation of crack in stage 2 is, however, remarkably slow. This is due to the result of irregular branched nature of the crack, which appear to have followed a zigzag path through the As-cast structure. This contrasts with the wrought alloys where the fine-grained structure allows the crack to spread unchecked along a straight path. Thus, if crack initiation can be slowed, then the fatigue life of cast metals might be extended considerably, perhaps in excess of those of wrought alloys. Figure 11.9 shows fatigue curves for filtered and unfiltered pearlitic ductile irons and for aluminium alloy.

The following factors appeared to have beneficial effect on fatigue performance of cast materials:

- Shot blasting due to the introduction of compressive stress at the surface.
- Filtering the liquid prior to casting and/or improving casting methods (e.g., low pressure, and uphill casting) due to reduced defects.
- Metal mould casting due to reduced size in defects.

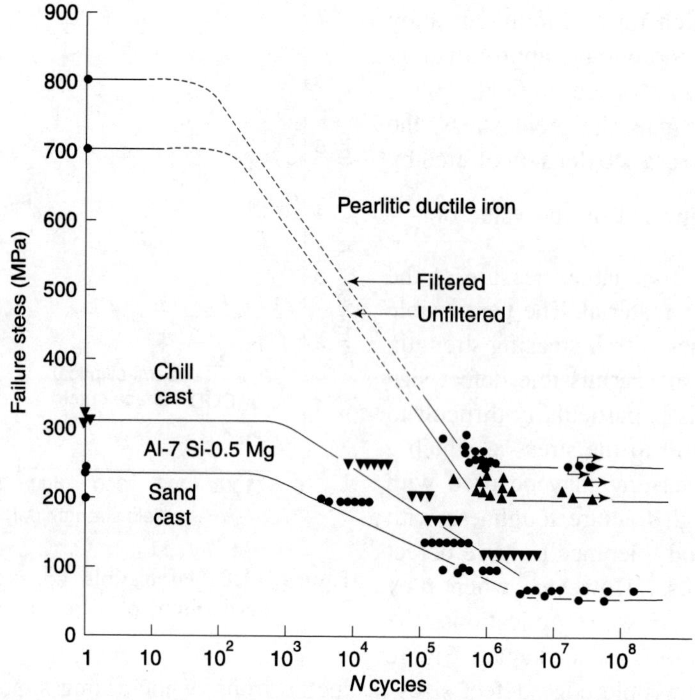

Figure 11.9 Fatigue curves for filtered and unfiltered pearlitic ductile irons and for aluminium alloy

11.4.5 Ductility

All second phase additions (including pore) have deleterious effect on ductility. This is due to the lack of cohesion of the second phase with the matrix when deformation starts. Thus, all particles are soon acting as holes. Figure 11.10 shows ductility of copper containing a dispersion of second phase effect of assembly of holes which affect the ductility. Figure 11.11 shows ductility failure model.

Since failure occurs at 45°, the theoretical elongation is equal to the linear dimension of the cross section, l. For test piece with single pore of size d, the elongation is equal to the half of the remaining sound length, i.e., $(1 - d)/2$. In general, for a spacing s in an array of micropores, we have

$$\text{Elongation} = s - d = \left(\frac{1}{n}\right)^{\frac{1}{2}} - \left(\frac{f}{n}\right)^{\frac{1}{2}} = \frac{1 - f^{\frac{1}{2}}}{(n)^{\frac{1}{2}}}$$

$$n = \text{Number of pores per unit area} = \frac{1}{s^2}$$

$$f = \text{Area fraction of pore on fracture surface} = nd^2$$

Film defects were less important than an equal area fraction of compact inclusions in lowering the ductility of casting. There were between 100 and 1000 times the number of micro inclusions

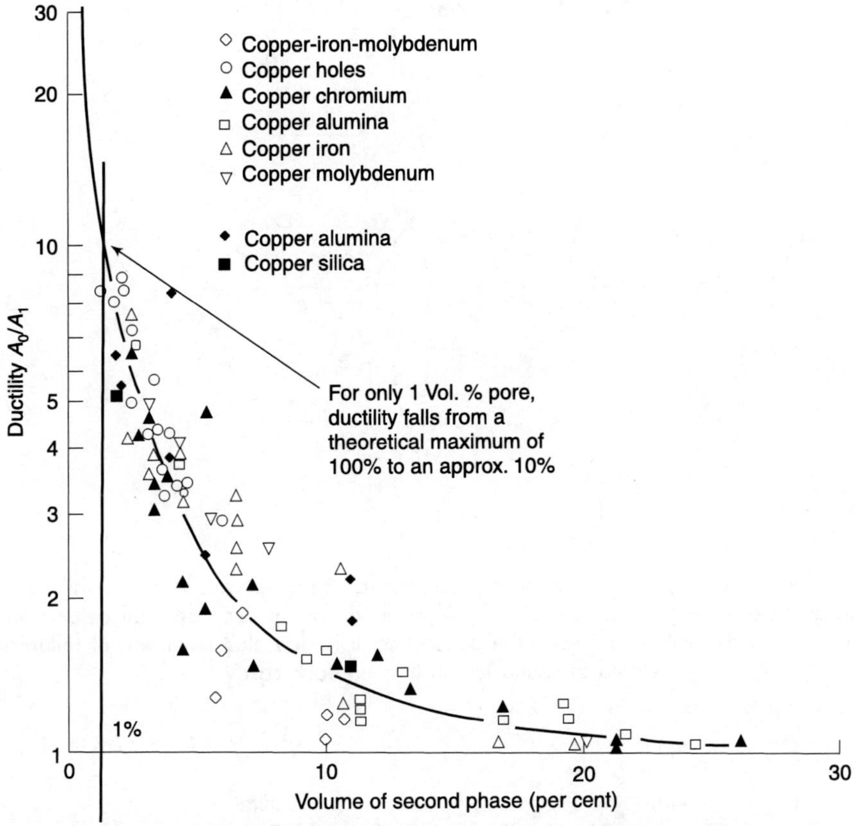

Figure 11.10 Ductility of copper containing a dispersion of second phases

compared to film-type defects in a given area of fracture surface 17. Micro inclusions are more effective than macro inclusions in lowering ductility.

When $f = 1$, ductility falls to zero.

$$\text{Elongation} = s - d = \left(\frac{1}{n}\right)^{\frac{1}{2}} - \left(\frac{f}{n}\right)^{\frac{1}{2}} = \frac{1 - 1^{\frac{1}{2}}}{(n)^{\frac{1}{2}}}$$

This situation can easily happen when turbulence during filling can generate oxide films, which occupy the whole of the cross section of the test piece. Figure 11.12 shows strength and ductility of an Al-4.5 Cu-1.5 Mg alloy as a function of total area of different types of inclusion in the fractured surface.

11.4.6 Ultimate Tensile Strength

Ultimate tensile strength comprises:

- Yield strength plus
- Additional work strengthening during plastic deformation before failure

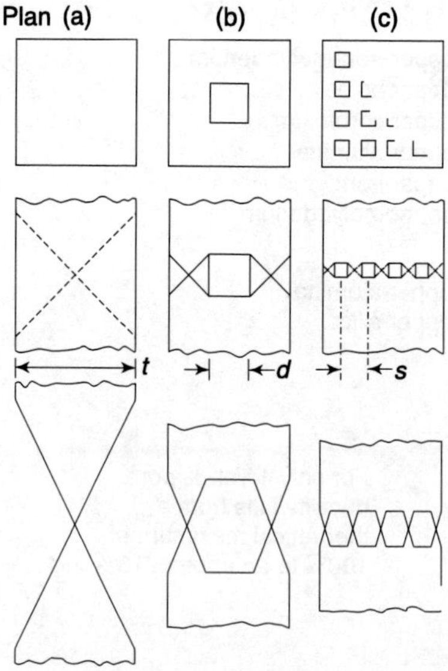

Figure 11.11 A simple ductile failure model, representing a sound specimen in (a) which necks down to 100% reduction area, a single macropore in (b) which leads to cup and cone failure and an array of micropore in (c) which effectively tears along the dotted line. It is clear that extension of failure is directly related to sound length (c-a) in each case

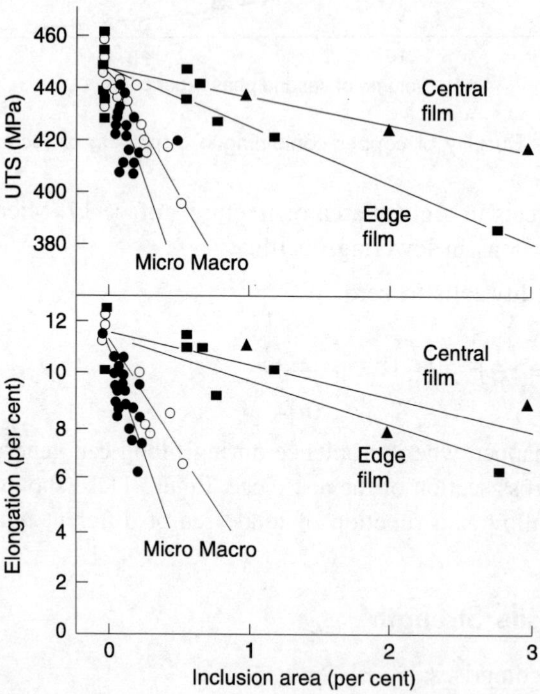

Figure 11.12 Strength and ductility of an Al-4.5 Cu-1.5 Mg alloy as a function of total area of different types of inclusion in the fractured surface

The behaviour becomes more complicated to understand than the behaviour of yield stress or ductility alone.

UTS = proof stress, when

1. Ductility = 0 or
2. Work strengthening effect = 0. This mainly occurs at high temperature when rate of recovery exceeds rate of hardening, ductility can be high for this case.

At room temperature, UTS increases with

1. Increasing ductility (since $\dfrac{\text{yield stress}}{0.2\,\text{proof stress}}$ is relatively insensitive to many variables)
2. Work hardening.

Figure 11.13a shows the mechanical property of Al-4.5 Cu-1.5 Mg and Figure 11.13b, the UTS of a casting as a function of the area of crack.

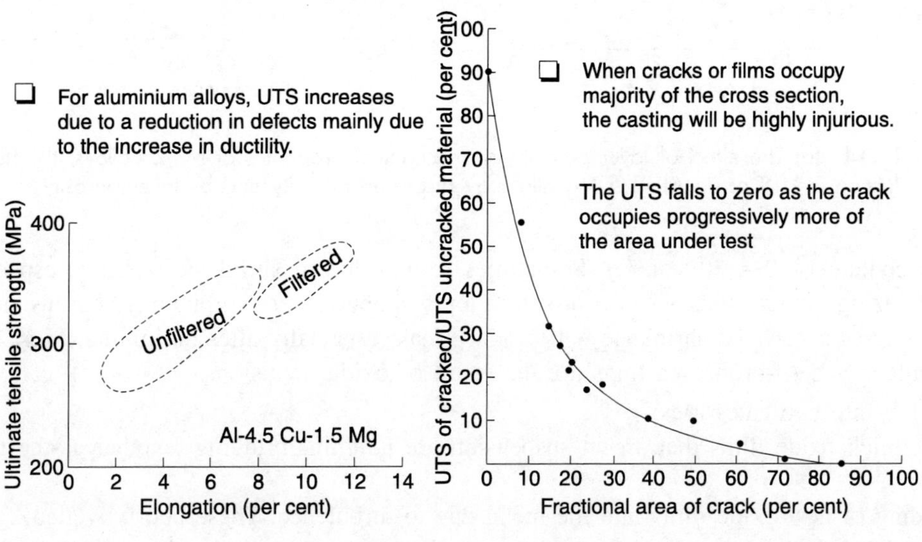

Figure 11.13 (a) Mechanical property of Al-4.5 Cu-1.5 Mg; (b) The UTS of a casting as a function of the area of crack

Layer porosity has similar effect on UTS as it has on ductility. It has larger effect when porosity is layered perpendicular to the applied stress and "stitched" together to form a single crack.

11.4.7 Leak Tightness

Leaks are seldom caused by gas porosity. Because the distribution of gas porosity is not random; they are distributed at specific distances, and are kept apart by the presence of dendrite arms. Leakage due to gas porosity occurs only when the sample contains an impossible amount of

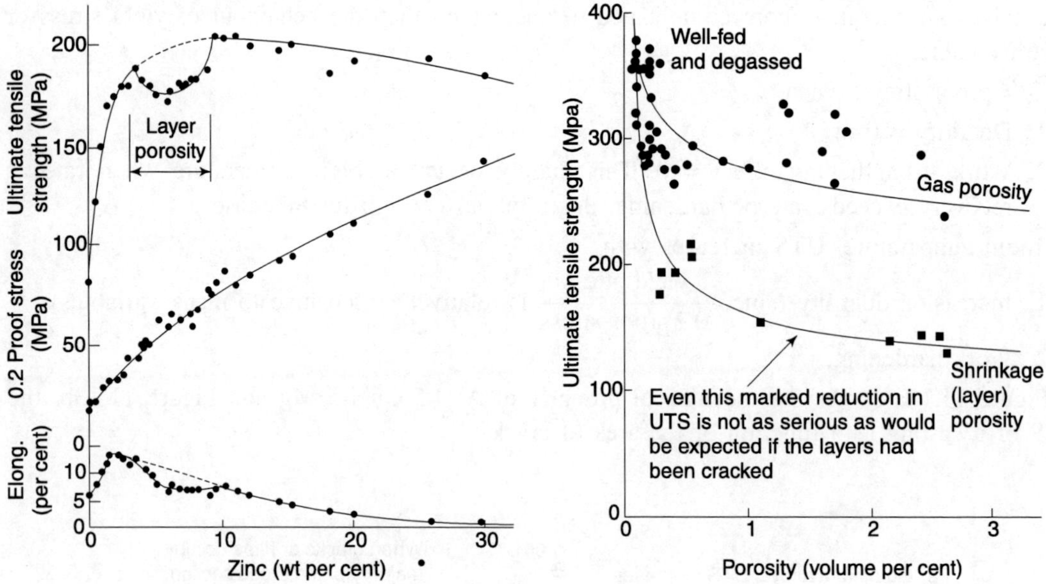

Figure 11.14 (a) The effect of layer porosity on mechanical properties of Mg-Zn alloys. (b) The reduction in UTS of an Al-11.5 Mg alloys by dispersed porosity and by layer porosity.

high gas content (~20 – 30 vol.%.). Sometimes results due to shrinkage porosity, especially in long-freezing-range alloys where porosity adopts sponge/layer morphology; porous metals resulting due to poorly fed shrinkage will produce leak, especially after machining. Most leaks in light alloys and Al-bronze castings are the result of oxide inclusions.

They fall into two categories:

- Old, thick oxide films that are in suspension are jamming/bridging together as the metal rises.
- Folding of new oxide films into the metal due to turbulence. These poorly wetted, folded over dry side to dry side bi-films constitute the major leak paths through the walls of casting.

11.4.8 Residual Stress

Unseen and unsuspected residual stresses can be the most damaging defects of all. The stress can be so large that it can outweigh the effect of all other defects. Residual stress is added up to the applied stress, and put the casting near to its point of failure even at relatively small applied loads:

- Never specified to be low in any standards.
- Practically impossible to measure in any NDT test.

The remedy is, of course, the application of stress relief by thermal treatment.

Appendix A

CATALOGUE OF CASTING DEFECTS AND REMEDIES

DEFINITIONS

1. *Blowholes* or *gas holes*: They are generally rounded cavities spherical, flattened or elongated caused by the entrapped air or gas formed during the casting process. These are generally found inside the castings.

   Porosities are caused by steam or gas passing through the metal. Sometimes pinhole size porosities may be present on the entire surface or just below the surface.

   Blisters are shallow holes on the surface with a thin film of metal over it.

2. *Scars* are generally caused on a flat surface where entrapped gas has prevented the mould cavity from being filled completely. Plates are formed when metal oozes into a scar. Minor defects of the scar or cold shut is known as seam.

   Cold slots or shot iron (in case of grey cast iron) are small globules of metal embedded in due to the action of the entrapped gas or air.

3. A *shrinkage cavity* or *depression*, large or small, results from varying rates of contraction while the metal is changing from liquid to solid. A minor depression is known as sink.
4. *Hot tears* and *cracks* are surface discontinuity or fractures caused by either external or internal stresses or a combination of both acting on the casting.
5. *Mass hardness*: It is caused when the entire casting is too hard. Localized hardness is called hard spot and/or chilled spots.
6. A *misrun* is due to failure of the metal to fill the mould cavity. The cold shut defect is caused due to imperfect fusion where two streams of metal have converged.
7. *Inclusions* are the non-metallic particles of materials embedded in the metal.
8. Sand defects:
   (a) An *erosion scab* occurs where the metal has been agitated, boiled, or has partially eroded away the sand leaving a solid mass of sand and metal at that particular spot. Cuts and washes are caused due to erosion of sand causing rough spots from either excess metals or sand inclusions.
   (b) *Expansion scabs* are rough thin layers of metal partially separated from the body of the casting by a thin layer of sand and held in place by a thin vein of metal. They may be readily chipped off causing indentation known as *buckle*. A *raftail* is a minor buckle occurring as a small irregular line or lines.

   A *pulldown* is a buckle in the cope. A *blacking scab* is caused due to sand expansion. Expansion defects are governed by the extent to which the initial expansion of the quartz can be offered by deformation of the sand mass under compressive stresses. Thus, the relationship between hot deformation and confined expansion test results have been found to give a measure of scabbing tendency.

9. Mould metal reaction defects:
   (a) Sticker or rat is a lump on the surface of a casting caused by a portion of the mould face sticking to the pattern.
   (b) Rough surfaces are caused due to minor mould metal reactions. Metal penetration is a surface defect which looks as though the metal has filled the voids between the sand grains without displacing them. When sand is fused on the surface of the casting, the defect is known as fusion.
10. Defects due to faulty workmanship:
    (a) A mould shift results in a casting which does not match at parting lines. A core shift results from misalignment of cores in assembling. A specialized case of core shift known as "core raise" is caused due to core movement towards the cope causing a variation in wall thickness.
    (b) A *ram-off* or *ramaway* is a defect which results from a section of the mould being forced away from the pattern by ramming sand after it has conformed to the pattern contour. It may be renewable on a localised swell or shifts.
    (c) *Swells, fins, strains* and *sags* result in the castings not being true to the pattern because the metal section is greater or less than required. A swell is an increase in metal section due to the displacement of sand by metal pressure. A *fin* is a thin projection of metal from the casting.

        A *strain* is a type of swell wherein the mould surface has cracked and permitted the metal to form a fin.

        A *sag* is a decrease in metal section due to sagging of the cope or core.
    (d) *Runouts* and *bleeders*: Sometimes called breakouts, these defects result in a casting lacking completeness due to molten metal draining or leaking out of some part of the mould cavity either during (known as runouts) or after pouring (known as bleeders).
    (e) *Crushes, push-ups and damp-offs:* These defects occur as indentations in the casting surface due to displacement of sand in the mould.
    (f) A drop is a defect in a casting due to a portion of the sand dropping from the cope or other overhanging section.
11. Open grain structure is a defect wherein a casting when machined, appears too course grained for the application. The structure may be general or it may be localised.
12. Miscellaneous defects like broken castings are caused during handling the metal and may be due to design defect.

    Inverse chills or *reverse chills* are caused due to inhomogeneity of metal.

    Kish is a defect when in grey cast iron, free graphite has separated out from molten iron.
    Warpage defect is a deformation in a casting other than due to contraction that develops between solidification and room temperature.

Appendix B

DEFECT FACT SHEETS

BLOWHOLES

Appearance

1. Round-shaped holes occurring on the outside of the casting or in the thicker sections.
2. May be found just below the surface on machining. In severe cases, section of casting may be hollow.
3. Cavities may be dull or bright depending on conditions under which they have formed.

Causes

1. Insufficient permeability of moulding or core sand.
2. Insufficient venting in cores.
3. Hard ramming.
4. High moisture content in mould and liquid metal.
5. Rusty or damp chills and chaplets.
6. Incomplete baking.
7. Damp pouring ladles.
8. Too low pouring temperature.

Remedies

1. Increase permeability by use of vent wire or open sand with additions of a coarser silica sand.
2. Avoid excess ramming.
3. Reduce moisture to minimum, consistent with workability.
4. Ensure chaplets are dry and coat chills with oil or proprietary dressing before use.
5. Reduce oil in sand.
6. Ensure vents are clear.
7. Bake until cores are hard.
8. Thoroughly dry all pouring ladles.
9. Increase pouring temperature.

POROSITY

Appearance

1. Casting 'weep' under pressure test.
2. Machined surfaces show cavities in thick sections or a series of fine holes on machined skin.

Causes

1. Wrong composition of metal.
2. Incorrect running and feeding system.
3. Impermeable mould.
4. Oxidized metal.

Remedies

1. Reduce silicon and phosphorus content.
2. Use proper feeding.
3. Improve venting of mould
4. Scavenge the melt before pouring.

SHRINKAGE AND DRAWS

Appearance

1. Rough cavities entering casting on heavy sections at the joint of change of sections.
2. Saucer-shaped depressions of heavy sections, usually with rough edges. Dark areas seen in fracture.

Causes

1. Incorrect gating and feeding.
2. Unsuitable compositions.

Remedies

1. Use risers to feed heavy sections and ensure that they are filled with hot metal. If using open feeders, use feeding flux. If using blind risers, use feeding cores. Embody chills where a heavy section or boss cannot be fed directly with a riser.
2. Adjust silicon and/or carbon equivalent in case of cast iron.

MISRUN

Appearance

1. Holes in thin sections of a casting. Edges are smooth and well-rounded; surface of metal around holes smooth and often shiny.
2. As a line, when two streams of metal have met but not fused together. Fracture may occur along this line. Casting surface may be smooth and shiny.

Causes

1. Low pouring temperature.
2. Low fluidity.
3. Inadequate venting of mould.
4. Faulty pouring practice.
5. Core shift causing uneven thickness.

Remedies

1. Provide hotter metal at furnace spout; reduce heat losses in the ladle by using flux covering.
2. Increase carbon and phosphorus.
3. Avoid excessive ramming and increase vents.
4. Keep runner full of metal during pouring.
5. Take extra care in positioning gates.

SLAG AND INCLUSION

Appearance

1. Pitted surface or inclusions found on machining.
2. Cavities are generally more saucer-shaped and smooth. Slag may be seen before cleaning the casting.
3. Segregations of managanese sulphide.

Causes

1. Dirty metal and dirty ladle lining.
2. Incorrect gating, causing turbulence.
3. Excess of sulphur with high manganese and low pouring temperature.

Remedies

1. Remove all slag from metal before pouring. Thicken slag with sand before skimming. Keep ladle lining free from build-up.
2. Incorporate skim gates or strainer cores in runner systems. Keep runner full while pouring.
3. Restrict sulphur pick-up and avoid pouring 'dull' iron.

DIRT

Appearance

1. Rough cavities and pits in casting surface.
2. If examined before cleaning, the sand may often be seen.

Causes

1. Dirty ladles or shanks.
2. Low sand strength.
3. Loose ramming.
4. Direct wash of metal on sand surface.
5. Poor finish of gating system.
6. Displacement of sand by cores.
7. Disturbed moulds.

8. Insufficient taper on patterns.

Remedies

1. Clean ladle and reline if necessary.
2. Increase green bond with clay additions.
3. Ram evenly.
4. Use well-designed runners and ingates.
5. Finish of running system should be as good as mould. Make bushes and runners with good facing sand.
6. Blow out after placing cores.
7. Place weights carefully; avoid knocking mould.
8. Increase taper to allow clean lift.

ROUGH SURFACE

Appearance

1. Casting surface rough.

Causes

Metal penetration due to:

1. Moulding sand too 'open'.
2. Low coal dust content
3. Uneven ramming.

Remedies

1. Use finer sand and apply mould dressing.
2. Make additions of coal dust.
3. Ram more evenly.

HARD METAL

Appearance

1. Bright areas on machined faces, often at corners or edges of thinnest sections.
2. May occur as scattered hard spots.
3. Shows, when broken, a white fracture.

Causes

1. Wrong composition of metal, with high sulphur and low manganese.
2. High moisture content.
3. Incorrect pouring practice.

Remedies

1. Increase silicon content by:

   (a) altering metal mixture; and
   (b) introducing silicon into ladle.
2. Reduce moisture.
3. Avoid splashing metal down runners and risers.

DISTORTION

Appearance

Casting shows swelling on surface.

Causes

1. Uneven mould hardness due to insufficient ramming to withstand metal pressure.
2. Poor weighting practice.
3. Casting too heavy for greensand mould.
4. Internal strains during cooling.

Remedies

1. Ram evenly and firmly.
2. Increase weight on moulds and ensure it is distributed evenly.
3. Employ dry sand or CO_2 bonded moulds.
4. Improve casting design and moulding.

CRACKS

Appearance

1. Hairline crack showing on casting.
2. When broken, discolouration shows that crack was hot. No discolouration shows cold crack.

Causes

1. High dry strength of sand.
2. Cores too hard.
3. Casting strains.
4. Mechanical reasons.

Remedies

1. Ram softly to allow casting to contract.
2. Reduce oil in cores.
3. Gate evenly; modify pattern design.
4. Take care in breaking off risers. See that risers are provided with correctly designed necks. Allow casting to cool sufficiently prior to knock-out.

SCABS

Appearance

1. Rough 'warty' excretions on surface of casting, mainly on heavy sections.

Causes

1. Hard or uneven ramming.
2. Mould poured too slowly.
3. Sand grain not uniform.
4. Excessive moisture or volatile content.
5. Sand erosion due to: weak bond of facing sand or prolonged metal impingement.

Remedies

1. Avoid hard-rammed areas.
2. Increase ingate(s) areas.
3. Change sand or add a little wood flour.
4. Reduce moisture and coal dust content.
5. Improve sand with clay additions.
6. Use a good refractory mould dressing.

HOT TEAR

Appearance

1. The defect has a characteristic of a tear. Thus, a hot tear is almost certainly a uniaxial tensile failure in a weak material.
2. Form of rugged, branching crack.
3. The main tear and its many faucets follow intergranular path.
4. Failure surface reveals dendrites morphology.
5. Failure surface is heavily oxidized.
6. Its location is often at a hot spot.
7. It does not always appear under apparently identical conditions.
8. Defect is highly specific to certain alloys.

Causes

1. Presence of wetted liquid phase along grain boundary.

   Example: Al-Sn alloy: Liquid Sn wets grain boundary of Al → Hot tear.

   Al-Cd alloy: Liquid Cd does not wet Al, forms compact pool → No hot tear.
2. Strain in the hot spot

$$\varepsilon = \frac{\alpha(\Delta T)L}{l}$$

α = Linear coefficient of thermal expansion

(ΔT) = Temperature difference $(T_m\text{-}T)$

L = Length of casting

l = Length of hot spot

3. Presence of strain concentrator in design
4. Presence of tear initiator → clean melt is difficult to tear.

Remedies

1. Design change
   - remove stress concentrator
   - remove hot spot
   - use chill if necessary
   - reduce mould hardness (increase collapsibility) of mould and core
2. Grain refinement
3. Reduce casting temperature
4. Alloying
5. Reduce contracting length

Review Questions

1. Explain the importance of defect control in casting.
2. Give the classification of casting defects.
3. What are blowholes? Give the causes and remedies.
4. Discuss different types of blowholes.
5. What are hot tears? Give the causes and remedies.
6. Explain various causes of casting distortion.
7. What is gas porosity?
8. Explain the role of residual stresses in casting.
9. Differentaite between layer and surface porosity.
10. What is a rat-tail defect and how to avoid it?

CHAPTER 12

Shakeout and Cleaning of Castings

12.1 INTRODUCTION

Molten metal filled in a sand mould is allowed to solidify and cooled to room temperature to form a casting. The cast solid body is taken out by breaking the mould, and the process is called shakeout. During the casting, refractory used for moulds or cores that still might be sticking to the casting and are to be cleaned. Also, the extra projections like runner and riser are removed to obtain finished casting.

Cleaning methods vary for different castings. See example 12.1 and 12.2. Cleaning is often an expensive operation, and everything possible is done to reduce costs in this department. After cleaning castings are checked for its dimensional accuracy, defects, composition, etc., by various inspection methods.

Example 12.1: The following are the steps involved in cleaning a small aluminium casting:

(a) Gates and risers are removed by band saw.

(b) Rough grinding (snagging) with coarse abrasive wheel.

(c) Finish grinding (polishing) with fine abrasive belt.

(d) Sand blasting.

Example 12.2: The following are the steps involved in cleaning a small steel casting:

(a) Fins are burnt on sand

(b) Gates and risers are removed by flame cutting.

(c) Riser pads and gate stubs are finished with abrasive disk.

(d) Casting shot blasted.

Definition: The cleaning of castings generally refers to all the operations involved in the removal of adhering sand, the gating system, and fins, wires, chaplets, or other metal not a part of the casting. Cleaning operations may also include a certain amount of metal finishing or machining to obtain the required casting dimensions, the savage of casting having minor defects and the inspection of the finished castings.

Reasons for cleaning

Lot of sand and fused particles are clinging to the surface of the casting.

Gates, runners, risers are attached with the main castings which are finally not required.

Hence, the castings need cleaning and trimming operations to be done on them before they can be finally brought to the final shape.

The series of operations performed in the cleaning department may be classified as follows:

1. Removal of gates and risers, rough cleaning.
2. Trimming, the removal of fins, wires and protuberances at gates and riser locations.
3. Finishing final surface cleaning, giving the casting its outward appearances.
4. Inspection

Sometimes heat treatments are involved which necessitate cleaning the castings after the heat treatment. This might be done between steps 2 and 3 above. Often steps 1-4 are carried out simultaneously. Some of these such as, gate removal may occur during shakeout operations.

12.2 COOLING OF CASTINGS

After filling the mould cavity with the molten metal, solidification starts from the surface of the cavity and progresses towards the centre due to cooling. Thin sections solidify first and thicker portions at the end. After solidification, the casting cools to the room temperature. Usually, this takes hours to days for thick medium-sized castings. Hence, the mould is generally broken at a higher temperature at which the casting has enough strength to withstand the shakeout/knockout forces. This early shakeout can shorten the process cycle time, resulting higher productivity with the same floor area. On the other hand, an excessively high temperature of casting at the time of shakeout is undesirable because of the risk of cracking and other casting defects. If a casting is taken out from the mould, the cooling rate in air is high and this may bring uneven cooling for thin and thick sections. This uneven cooling easily brings internal stresses. So to get best results, the high temperature permissible for shakeout is to be determined depending on the metal and shape of casting. Recommended temperature ranges for shakeout of castings in sand moulds are given in Table 12.1.

Table 12.1 Temperature ranges for shakeout of castings

Steel castings	500°C to 700°C
Iron castings	400°C to 500°C
Bronze castings	300°C to 500°C
Aluminium alloy castings	200°C to 300°C
Magnesium castings	100°C to 150°C

The time of holding of casting inside a mould for cooling is determined by the wall thickness of the casting, the properties of cast metal and mould and the shakeout temperature. It can be calculated or determined experimentally. The time for holding in sand moulds may range from a few minutes to hours or even days depending on the metal, shape and size of casting. In some cases, forced cooling like air blowing or water spraying are employed to accelerate the cooling rate. The average rate of cooling of castings in mould may range from 2°C/minute to

150°C/minute. It is chosen by considering the section thickness of casting and the metallurgical properties of metal. Large difference in the cooling rate in various sections of castings can cause high thermal stresses and lead to warpage or cracking of castings. In many instances, the castings shakeout at high temperature and then cool more slowly than in the air. For this purpose, they may be placed into pits or boxes, or left together with shakeout sand.

12.3 SHAKEOUT OF CASTINGS

The process of breaking the sand mould and bringing the castings outside is called shakeout or knocking-out of mould. Since the size of a mould is about five to ten times of the volume of a casting, shakeout process is heavy and requires much material handling. Due to the collapsibility property of mould, the breaking of mould is not much difficult but the removal of dust from the floor is generally a problem. Also, the removal of core sand from the cavities is a difficult task. The shakeout of castings can be grouped into two classes of activities:

1. Shakeout (knocking-out) of moulds.
2. Shakeout (knocking-out) of cores.

12.3.1 Shakeout of Moulds

The breaking of mould is generally done by shakeout (knock-out) grids. Figure 12.1 shows two types of vibrating grids. The cope and drag are placed over a vibrating grid. In each vibration cycle, the mould is thrown up and then drops into the grid or supporting frame. Due to impact, the sand falls through the grid and is transferred by a conveyor system for reclamation. In continuous automatic casting lines, the castings are squeezed out from the flask together with the sand and then freed off the sand on shakeout grids. Thin breakable castings are extracted from the moulds before shakeout.

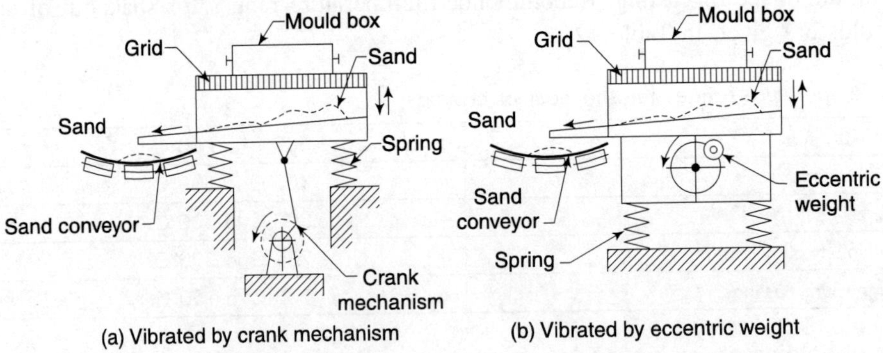

Figure 12.1 Vibrating grid type shaking machines

Since the shakeout process produces much heat and dust, these sections are fitted with efficient draft ventilation.

12.3.2 Shakeout of Core

After extracting the casting from the mould, the cores which are inside the casting, are removed. Cores can be removed by rapping action on the castings. They may also be pushed away from the castings with the help of a bar. Air chisel of vibratory machines, hydraulic boxes or electro-hydraulic installations are used for this. Usually, large steel and iron castings are decored in hydraulic and electro-hydraulic core shakeout installations. In hydraulic core knockout boxes, a casting is placed on a rotatory table and washed out with water jet of 5 to 10 megapascals. This jet removes the core and cleans the internal surface.

Electro-hydraulic spark decoring: In electro-hydraulic decoring installations, the cores are broken by shock waves produced by electric high voltage discharge between two electrodes immersed in water.

Applications: This method is used to remove hard and intricate cores from the castings.

Procedure followed in hydraulic spark decoring is as follows:

- The method makes use of hydraulic shock waves.
- The castings are immersed in a tank of water and subjected to shock waves generated by high voltage electrical discharge across two electrodes submerged in water.
- The shock waves produced in this way are sufficient to loosen cores placed in the vicinity.
- It is possible to break up or loosen internal cores by applying shocks to the solid walls of casting, but best results are obtained with electrodes placed opposite to core opening.

Advantages

1. Removal of very hard and complicated course is easy.
2. Energy requirements are slow.
3. Dust generation is absent.

Disadvantages

1. Contamination of water by core materials makes visual assessment of the process of feeding difficult.
2. Safety devices and isolation of electrical equipment are essential because of high voltages employed.

Although the field of application of the hydraulic spark decoring method is limited but, when decoring by conventional methods is difficult and costly, the method appears to have possibilities.

12.4 CLEANING OF CASTINGS

A casting coming out of a mould has extra fittings like sprue, runner, riser, flow-off, parting line fins, etc., as projections on the casting. The removal of these unwanted projections and the burnt sand sticking on the surface is generally called cleaning of casting. This cleaning process includes the following activities:

1. Removal of sprue, runner and riser
2. Surface cleaning of castings
3. Fettling of castings

12.5 REMOVAL OF SPRUE, RUNNER AND RISER

The extra projections on castings like sprue, runner, riser and fins along parting line have to be carefully removed without destroying the casting. Different methods to remove them and the equipment used are given below.

1. Chipping off the projections using chisel and hammer. Chipping hammers are air driven hammers with a chisel as the working tool. The sharpness of the cutting tool, the material of the blade, the way it is held against the workpiece affects its efficiency. Keeping the cutting tip wet with oil increases the efficiency by 40%. The experience of the operator also plays an important role in the efficiency. Used for low and medium carbon steels and copper base castings. Pneumatic chipping hammers may be used to remove fins, gate and riser pads, wires, etc., and to remove cores.

Applications: This is most suitable for hard and brittle materials like cast iron.

Procedure

- In this method, the extra parts are chipped off either by hand or by pneumatic chisels.
- The use of pneumatic chisels gives rise to the production of dust.
- Removal of dust:
  - The dust can be removed by use of chisel sleeve.
  - Ventilating ducts are incorporated in the sleeve wall in such a manner that there is one pair of ducts above and one below the cutting edge of the chisel. Dust is let away through a 5/8 inch diameter plastic hose.
  - When large quantities of dust are produced during chipping, a conical rubber extension can be fitted over the chisel sleeve.
  - Advantages of sleeve: The introduction of sleeve will also damp the vibration and will provide much more comfortable grip.

Advantages

Useful for cleaning large castings which have large quantities of sand and dust while stripping. For example, stripping of heavy steel castings. In such cases, pneumatic chisel is used.

Disadvantages

- This is a time consuming process.
- Results in rough surface.

2. Flogging (knocking off by hitting with hammer). For more brittle materials, gates and risers are removed by striking with a hammer or sledge. This is called fogging.

Flogging with a hammer or sledge is a positive means of gate removal by impact. A man may be stationed at the shakeout to flog the sprue and risers as the sand falls away from the casting. When the moulds are set out on floors and dumped by hand, men with hammers knock off the gates and toss the castings and gates into separate boxes for transfer to the cleaning room. To sprue the castings is to remove gates in this way. Grey and white iron castings are especially amenable to gate removal by this method. An inherent danger of breaking off the gates is that the break may extend into the casting proper. This condition may be remedied by notching the ingate ahead of the casting. This protuberance remaining on the casting can then be ground flush with the casting wall.

Applications of fogging: Generally used when the casting metal is brittle, the sprues, runners, risers, etc., can be removed simply by giving an impact. This method is suitable for grey and white cast irons.

Advantage

This is a fast and cheap process.

Disadvantage

Sometimes there is a danger that the breaking may extend into the proper casting but this can be avoided by notching the gates of the casting slightly.

3. Shearing of the projections using shearing machines. Shears are heavy, matching steel jaws working past each other by flywheel and cam assembly. They can only be used on materials softer than shear blades such as malleable iron, soft and medium hard steels, brass, bronze, aluminium and magnesium. They are limited to small work, fast and economical. Used to remove stubs left when gates are broken from the castings.

Applications: The shearing machine is most suitable for the removal of extra part of the casting made of ductile materials.

4. Sawing of the projections using hacksaw, circular saw or bandsaw. Bandsaws are used mainly for nonferrous metals. But they are now used for mild and low alloy steels also. It is important to match the type of blade (the number of cutting teeth per inch) to the metal being cut. They are fast and economical when the work is small enough to be held by hand. Automatic handsaws are used for larger works.

Gates may be removed by bandsawing, power sawing, abrasive cut-off wheels or with sprue cutter. A rapid method of removing a number of small castings from a central runner is provided by the sprue cutter.

Applications: Can be used for cutting the extra parts using ordinary hacksaws and band-saws.

Advantages

1. Bandsaws give faster rates of production.
2. This method can be used for ferrous as well as nonferrous castings.

Disadvantages

The ordinary hacksaws are slow and time consuming.

5. Abrasive wheel slitting (grinding). Abrasive cut-off wheels are used when the metal is too hard to be cut by saw, and where cutting and chipping flame is not feasible. Highly effective method, but expensive. New developments have brought down this method into closer economic competition with other methods if careful account of all costs is kept. The cut-off wheel is limited in versatility because cuts must be made on other than flat surfaces.

6. Flame cutting (gas cutting). Flame cutting is an extremely versatile method for removing the metal. Can be used for oxidizable metals and also for non-oxidizable ones. For non-oxidizable ones like stainless steel, iron (in powdered or rod form) is fed into the flame at the tip of the torch to act as an oxidizing medium. This is useful for cleaning steel castings where mould is penetrated by molten metal and layers of sand locked to the surface of castings. Flame gouging and flame scarfing are used to remove excess metal. They find their use in foundry. These methods are used for removing surface defects from ingots, and are used to remove riser pads, cleaning away penetrated sand and preparing castings for welding. Flame cutting is not used for small casings, as the castings would be damaged by heating effect of the torch.

Applications: Mostly for high strength materials like steel castings.

Procedure: In this method, the section to be cut is preheated and a large amount of oxygen at high pressure is blown through the tip of the torch causing the metal to burn vigorously and the metal runs off as a molten slag, exposing more metal to react with oxygen.

Advantages

- Flame cutting is most preferred because of its high speed and economy.
- Other machines are limited to certain metal thickness whereas flame cutting can be used for any size of gates, risers, etc.
- Flame cutting has the ease of moving the cutting tops from one position to other while trimming the casting.

Large risers and gates on steel castings are most conveniently removed by cutting torches. The sprue cutter is limited in metal thickness, whereas the cutting torch and oxygen lance may cut risers of practically any size.

7. Plasma cutting. The gating system can be easily removed by knocking on it with a hammer. On small castings, most of these projections brake during the shakeout operation itself. The gating elements on large-sized castings of carbon and alloy steels are removed by gas cutting process. In the manufacture of high alloy steel castings, machining or grinder cutting is used to remove the excess metal.

Plasma is a mass of ionized gas which is produced by passing a stream of suitable gas to an electric arc discharge. The discharge takes place in a nozzle, with the result that a narrow stream of plasma emerges from the nozzle of the torch at a very high temperature (10,000-15,000°C) and with high velocity. High temperature melts the metal and their high speed of plasma stream blows away the molten metal.

There are two types of plasma torch:

Non-transferred arc plasma torch: (Figure 12.2) In this type of torch, the torch head consists of a tube fitted with water-cooled copper nozzle which is electrically insulated from the tube. A tungsten cathode is contained within the tube and the gas is fed in the tube as shown. A DC arc discharge is made between the tungsten cathode and the copper anode.

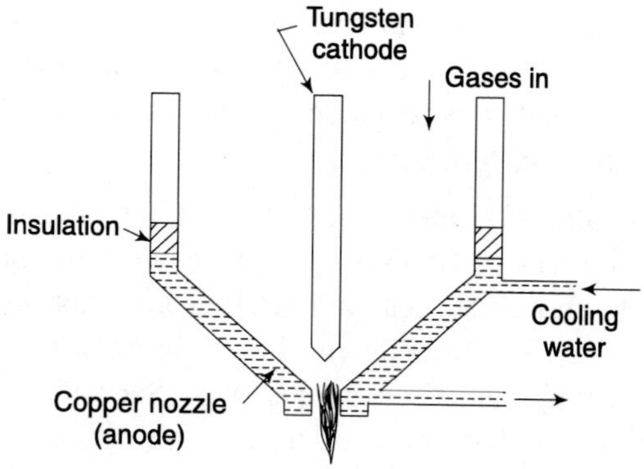

Figure 12.2 Non-transferred arc plasma

Transferred arc plasma torch: (Figure 12.3) As shown the transferred arc plasma torch is similar to non-transferred type, except for the fact that the DC arc discharge takes place between the rear electrode and the workpiece. In this case, the heat transfer is more effective than in former cases.

Plasma techniques have been used in cutting and gouging of nonferrous metals, refractory metals and stainless steels.

8. Air-carbon arc torch cutting

Applications: Air-carbon arc torch is generally used after cleaning of casting has been carried out with pneumatic hammer and chisel or with grinding machine.

Procedure: The principle of operation of the stool is the striking of an arc between carbon electrode and the workpiece, which melts the metal; and the blowing away of molten metal by a jet of compressed air directed along the side of the electrode.

Advantages: It produces a clean and slag-free surface.

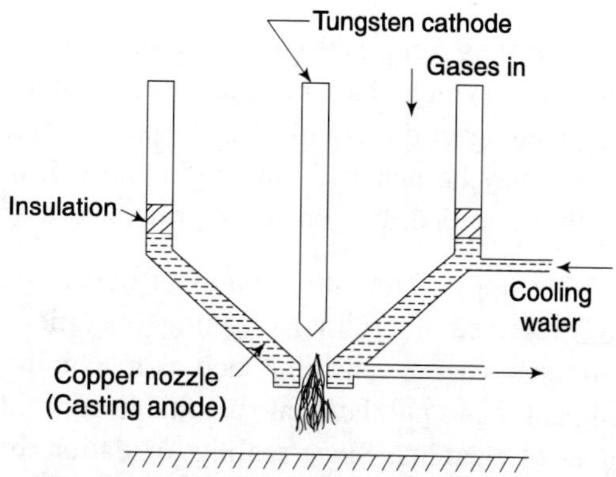

Figure 12.3 Transferred arc plasma

9. Belt saw, circular saw and mechanical presses. These are widely used for cutting and chipping of aluminium, magnesium and copper alloy castings.

10. Grinding. Grinding, or "snagging," of castings is practised to remove excess metal. There are three principal types of grinders, and swing-frame grinders. In addition, specialized machines such as disk grinders, belts, and cut-offs may also be used.

Stand Grinding: Stand grinders are usually of the double-end type and may be of constant or adjustable-speed type, the latter compensating for wheel wear. Low-speed machines operate at speeds up to 6500 surface feet per minute (sfpm), high speed machines at 9500 sfpm. Castings which can be handled manually are ground on machines of this type. The operator presents the areas to be ground to the wheel face under suitable pressure. Traversing of the wheel face is desirable to prevent rounding or grooving of the wheel and the need for frequent dressing of the wheel.

Stand grinding is ordinarily hand work, depending upon the judgement of the operator for the selection of areas to be ground and amount of grinding. When many castings of one type are ground, special guide fixtures or mechanized feeding and positioning fixtures may be used to speed up grinding.

Portable Grinders: Electric or air-operated portable grinders are employed for working over surfaces of castings that cannot be handled by swing or stand grinding. Cone, cup, disk, special shapes, and straight grinding wheels up to about 14 inches. in diameter may be mounted on portable machines. Generally, only light grinding is intended. Disk grinding is sometimes practised in this way. Specially shaped wheels have safe operating speeds which are lower than those of regular wheels, and these must be observed to avoid the hazards of wheel disintegration.

Rotary Tools: Cleaning of softer non-ferrous alloys such as aluminium or magnesium alloys may be done by rotary filling tools or cutters. These may be electrically or air-operated, and are usually of the portable type. The tool operator may then file off excess metal on any casting surface accessible to the tool.

11. Trimming and Sizing: Shearing, punching, coining, and straightening are mechanical operations which may be employed to complete trimming and sizing castings. Coining or straightening is done with dies in presses under hydraulic pressures up to about 1000 tonnes. I-holes may be punched out to eliminate drilling operations. Some castings may have surfaces which are milled, broached, or ground to a specified accuracy as required by the customer.

12. Finishing: The latter stages of cleaning are often referred to as finishing. Many castings have received their final cleaning operation when grinding is completed. Others are given additional surface finishing such as machining, chemical treatment, polishing, hailing, blasting, and painting to put them into a suitable appearance for sale. Heat treatments may come at various stages of cleaning. Since sealing, oxidation, or discoloration occurs during heat treatment, steel and malleable iron casting receive their final surface cleaning after heat treatment. In new ferrous castings, chemical, electrolytic, and mechanical means of making the casting surface attractive may be utilised.

A special salt-bath cleaning operation may be used on ferrous castings to obtain maximum freedom from scale, sand, dirt, and grit. For example, grey-iron castings may be immersed into a molten salt bath of caustic, 95% sodium hydroxide, 5% sodium nitrates and nitrites, at 700°F for cleaning. The castings are then rinsed, dipped in acidified water, 15% hydrochloric acid, hot water-rinsed, and heated with soluble oil. The oil coating provides rust protection. Treatments of this type are utilised only when the additional cost is justified in a particular application or processing sequence.

12.6 SURFACE CLEANING OF CASTINGS

The surface of a casting, after shakeout, contains burnt sand sticking on outside and inside portions. This burnt sand has to be removed and is not good to be mixed again with the moulding sand.

Surface cleaning operations ordinarily follow the removal of gates and risers. However, surface cleaning may facilitate gate removal. For instance, elimination of sand from the casting favours sawing and torch cutting, so that surface cleaning may then be done before gate removal in the case of nonferrous alloys and sometimes steel. When flogging is used on steel or cast iron, it is easier to perform the surface cleaning after gate removal.

The removal of sand particles sticking and the oxides formed on the surfaces is done by using the following methods:

1. Hand methods: This process is manual cleaning on castings of casting surface with the help of wire brushes, files, crow-bar, etc. It is a slow process and hence suitable for production of a few number of castings.
2. Mechanical equipment methods: Here machines are used to remove the sand on the surfaces. The types of mechanical machine cleaning are as follows:
   (a) Tumbling
   (b) Air-blast cleaning
   (c) Air-less shot blast cleaning
   (d) Hydro-blast cleaning
3. Chemical cleaning methods (electrochemical cleaning): The chemical cleaning methods used are:
   (a) Dipping in chemical bath
   (b) Electrolytic method
   (c) Pickling method

Among these different methods, a suitable one is selected according to the size, shape, type of metal, the number of pieces to be cleaned, the finish required, the cost of cleaning, etc.

12.6.1 Tumbling

Tumbling is a surface cleaning and polishing process of metallic components in which the components and star-shaped small-sized hard iron pieces are rubbed together inside a steel barrel by slow rotation of the barrel. The sand particles sticking on the cast surfaces can be successfully cleaned by this simple rubbing process. The tumbling barrel is made of steel and may have cylindrical, hexagonal or octagonal shape with horizontal axis of rotation. The main rubbing material is small-sized hardened star-shaped iron pieces called mill stars. Granite chips and coarse sand grains are added to get smooth finish. Figure 12.4 shows line sketch of batch type tumbling machine. The castings and the hard mill stars are filled inside the horizontal barrel by opening a door and the door is closed and locked.

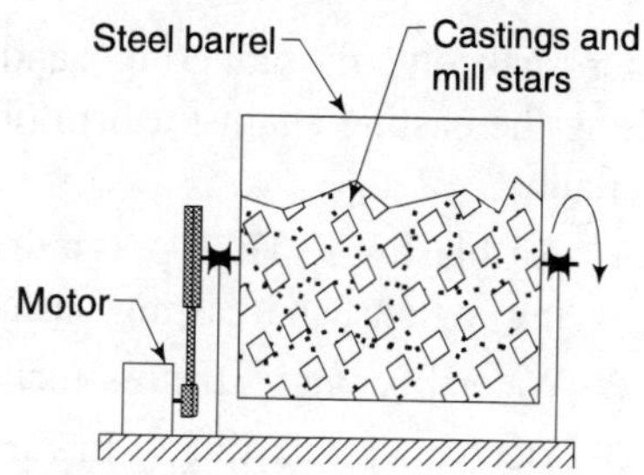

Figure 12.4 Tumbling operation

The barrel is then rotated by a driving mechanism at lower speed of about 30 rpm. To clean a lot, it may take minutes (30 minutes average) to an hour. This process is very effective and economical for small-sized castings of solid shape. Tumbling cannot clean the surface fully, if the shape has undercuts or too many projections. One limitation of the process is too much sound during the rotation of the barrel. For continuous action of tumbling, a double barrel (concentric) type machine may be used. Figure 12.5 gives the line sketch of the mechanism. The inner barrel does the tumbling process and the outer perforated barrel removes the sand particles by sieving.

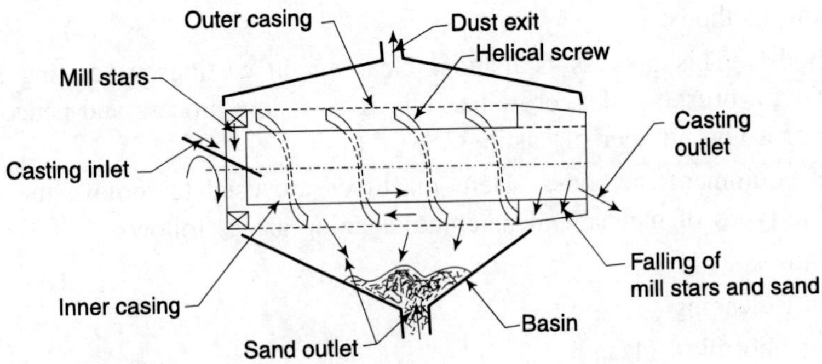

Figure 12.5 Tumbling machine-continuous feeding type

The mill stars are carried back to the feeding end by the helical screw action. The axis of the two concentric horizontal barrels is given a small slope to the exit end so that, when the barrels are rotating, the tumbling materials are moved automatically to that end. Thus, continuous feeding from one end is possible and the finished products are removed from the other end. The sand is moved downwards by gravity and the dust is collected upwards by suction of a blower. The dust problem in tumbling can be eliminated by wet process in which water is added in the barrel.

Tumbling for deburring and brightening of copper-base castings is also practised. The castings may be cleaned and deburred by using water and detergents in combination with sand or pumice stone in the tumbling barrels. Ball burnishing provides a means of imparting high lustre to copper-base castings by tumbling. Steel balls and the castings in a ratio of 2:1 are charged into the barrel along with dilute soap solution. After tumbling, the castings are dried in another mill by tumbling them in sawdust or wood shavings.

12.6.2 Blast Cleaning of Castings

After shakeout, the adhering sand particles are removed with the help of a wire brush by jolting the casting against a firm object, or by *blast cleaning*. Three types of blast cleaning are available:

1. Air blasting: High pressure steam of air propels sand (sand blasting) or shot (shot blasting) against the casting surface.
2. Water blasting: In this technique, water is used instead of air to propel the sand.
3. Mechanical blasting: In this technique, rapidly rotating paddles mechanically hurl shot or sand on the surface to be cleaned.

Different blast-cleaning units are available for different needs.

1. In small cabinet-type sand blasters, the operator goes into the cabinet wearing gloves and manually moves the small parts to be cleaned.
2. In larger sand and shot blast units, a specialized ventilated room is present and the operator (wearing mask and special clothing) directs the blast towards the casting using a flexible hose.
3. In modern blasting units for large castings, complete control of the blast unit is possible from outside the room.
4. Modern automatic units for small castings rotate or tumble the castings under grit blast. The following illustration shows a unit in which the abrasive is blasted onto the castings which are tumbled in an apron conveyor.

Blast cleaning of castings is a process in which abrasive particles, usually steel shot or grit, are propelled at high velocity to impact the casting surface and thereby removing the sand and scales. Centrifugal wheels are mostly used to get high velocity because of their ability to propel large volumes of abrasives. Even though the nozzle blast by air is not efficient as wheel blast, in some applications it may become more efficient, especially in the case of blasting small holes in castings. The material for shot and grit used nowadays is high carbon cast steel that is heat treated to give a desired martensite microstructure and hardness. The hardness of the most commonly used shot and grit is HRC 40 to 50. For faster cleaning or special surface finish requirements, hard abrasive is not often used. While using very hard ones, the wear of the machine parts will be high and the breakdown rate will be more. The diameter of the shot may range from fraction of a millimetre to 2 mm and velocity of the shot may be 2000 m/min to 4500 m/min. The speed of shot blasting wheel is taken as 3000 rpm. There are wide varieties of shot blasting machines in use and the selection depends on the metal cast, size, shape, and number of castings to be cleaned.

Figure 12.6 shows line sketch of barrel type shot blasting machine for small components. Here the components are taken in a barrel or cabinet which is rotating slowly about a horizontal axis, and the shot is accelerated by centrifugal forces formed by rotation of the impeller.

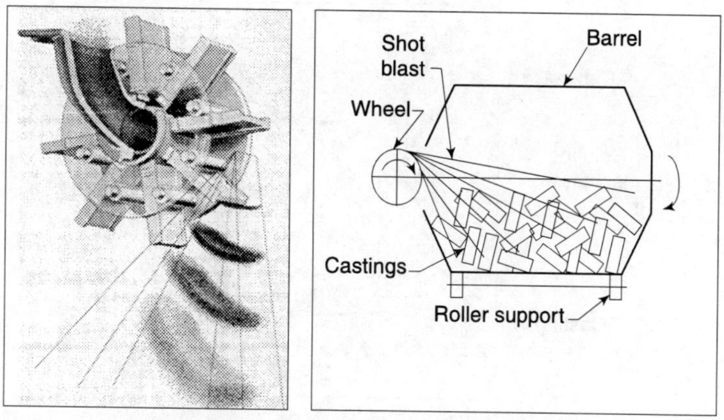

Figure 12.6 Shot blast (rotatory barrel type)

As the shots strike on the surface, the sand particles are removed and the surface is smoothened. The shot blasting barrels are intended for cleaning small or medium-sized castings up to 40 kg. For better cleaning of all surfaces, the castings are taken to move on apron conveyor as shown in Figure 12.7.

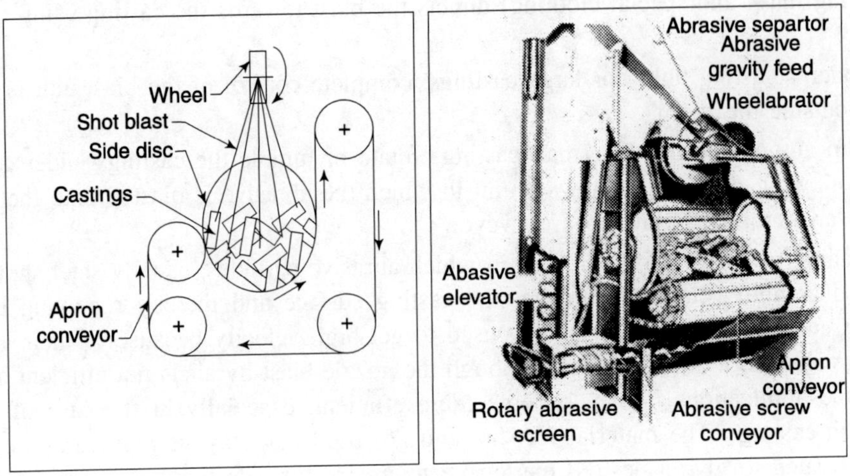

Figure 12.7 Shot blasting (apron conveyor type)

The castings are exposed to the shot blast by the movement of the conveyor over rollers. After completing the cleaning, the rotation of the rollers is reversed for removing the finished components. For large-sized castings (more than 50 kg), better cleaning is obtained by suspending the casting on the rotating hook and moving along a monorail chain conveyor as shown in Figure 12.8. Here the product rotates slowly about a vertical axis and moves horizontally along a monorail through a shot blasting area enclosed by a cabinet. More number of blasting wheels are employed to get fast cleaning of castings.

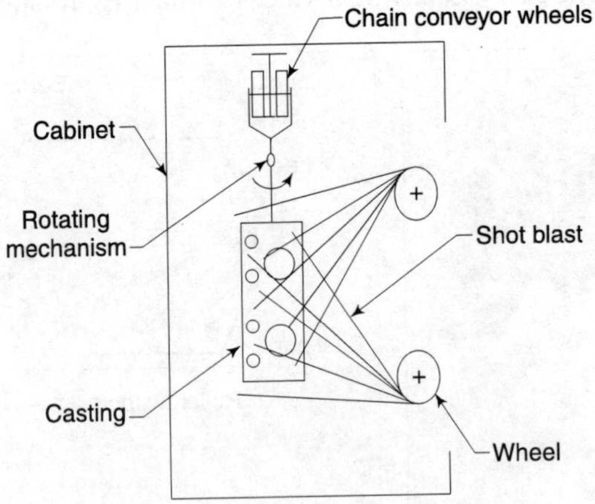

Figure 12.8 Shot blasting (rotating suspension type, inside cabinet)

Instead of shots and grit, sand can also be used as blasting material. Since the sand with compressed air forms dust, it may bring health problems to operator. To avoid the dust formation, air can be replaced by pressurized water and the process is called hydro-blasting. It is an efficient process and is applied successfully for cleaning large castings and removal of their cores. Hydro-blasting is a costly affair compared to others in the case of initial investment.

Blasting: Blasting the surface of castings is the most rapid means of removing sand and scale. Abrasives employed are sand, metal grit, and metal shot. Sand-blasting may be performed using coarse sand, 6- to 30-mesh size, as the abrasive and air as the carrying medium. When air blasting is employed, the blasting must be done in cabinets or rooms provided with a means of handling the dust arising from the disintegrating sand. Water blasting eliminates the dust problem. Table 12.2 gives the recommended size of abrasives.

Table 12.2 Recommended sizes of abrasives

Castings	*SAE Size No.*
Average grey or annealed malleable, with pockets, burnt-in sand, etc.	S390 shot
Light grey iron or annealed malleable	S330 shot
Hard malleable	S330 shot
Brass (all types)	G50 or G120 grit
Die castings	G50 grit
Aluminium	G50 grit
Steel	S390 shot

Grit or shot blasting can be done by the airless-blast method. The metallic particles are thrown by centrifugal force from a rapidly rotating wheel such as that illustrated in Figure 12.9.

12.6.3 Electrochemical Cleaning

Cleaning of internal surfaces of castings is not affected by shot blasting, since there is not much access for high velocity shots to strike the interior. In such cases, electrochemical treatment can be applied. For this purpose, castings are immersed in a bath of molten metal NaOH or KOH at a temperature of 450 to 500°C and connected to a direct current source of voltage 2.5V to 3.5V and current density of 0.03 A/m^2. The castings are connected as cathodes for 7 to 8 minutes, then as anodes, for 3 to 4 minutes, and finally as cathodes again for 3 to 4 minutes. After that, they are washed in hot and cold water. The cleaning effect on the surface is due to the cathode

Figure 12.9 Phantom of airless shot or grit blasting machine head (wheelabrator)

reduction of iron oxides and dissolution of SiO_2 of the sand in the alkali melt with the formation of silicates.

12.6.4 Fettling of Castings

Fettling is removal of small fins and flashes on casting surface or similar projections on the surface. Such projections are removed mostly by grinding using abrasive wheels. Lighter castings, less than 10 kg, are fettled by hand grinding process. Heavier castings are cleaned by suspended, tilting or portable grinders. Semiautomatic and automatic fettling machines and lines are employed for mass production of small or moderate size castings. The castings of tough metals such as steels, malleable cast iron, bronze and aluminium alloys are fettled by corundum wheels, and for brittle alloys carborundum wheels are used. Even though fettling is a kind of surface finishing process by removing unwanted projections, in some foundries this term is used for the removal of all projecting parts on cast body. This includes the cutting of down sprue, runner, riser and flow off of a casting after shakeout from the mould sand.

12.7 FINISH CLEANING

After removal of gates and risers, finishing operations are done to:

1. Smoothen the gate and riser areas of the castings.
2. Remove any excess metal remaining on the casting.
3. Improve surface finish or appearance.

Preliminary metal removal is done by grinding. An abrasive grit of about 25 to 80 mesh is bonded into the shape of a wheel (grinding wheel) or onto a belt of paper or canvas (abrasive belt). Wheels and belts are available in a variety of shapes and sizes, and of different grits, grit sizes and binders. Grinding machines are both portable or stationary. The portable ones are used for large castings. Rotary files are used instead of grinding for softer metals. A rotary file is a hardened steel grinding wheel.

If smooth surface finish is required, castings must be polished, brushed or buffed after grinding.

Polishing: Performed with an abrasive belt machine (similar to abrasive belt grinding) but finer grit is used (80 to 400 meshes).

Brushing: Employs rotary wire and fibre brushes that removes burr and grinding marks.

Buffing: Used to obtain an exceptionally high lustre on cast surfaces. It employs a wheel of buff (disks of muslin sewn together), a very fine abrasive mixed with a grease binder is rubbed on the face of wheel.

Blast cleaning: Blast cleaning of buffing removes excess dirt and grease and smoothens any remaining grinding or polishing marks. In blasting, we should use fine sand or shot grit. Otherwise, it will reduce the surface finish. Sometimes rice hulls, ground corncobs or ground walnut shells are used in blasting operation.

Review Questions

1. What is shake out of casting?
2. Give reasons for cleaning of casting.
3. Write a short note on shaking of cores.
4. Explain electro-hydraulic decoring processes.
5. What are the various ways of cleaning the casting?
6. What are the different methods for removal of sprue, runner and riser?
7. Explain various ways to remove sand particles from casting.
8. Explain the working of tumbling machines.
9. With the help of a neat sketch, explain short blasting processes.
10. Explain electrochemical cleaning.

CHAPTER 13

Inspection and Repair of Castings

13.1 INTRODUCTION

As soon as castings are clean enough, they are inspected by eye, gross defects are repaired if necessary, and the castings are given further inspection. Nearly all inspection methods applicable to metals, in general, can be applied to check and control quality of castings. Radiographic and other critical methods are used to increasing extent, particularly if service performance for the castings is severe. For example, for cast steel valves used in steam systems at 650°C and 100 bar pressure. For such applications, castings must be dependably sound. Very rigid standards have been developed for radiographic inspection. Similar standards have been developed for aluminium and magnesium alloy castings, and it is probable that precise standards will eventually appear for fluorescent and magnetic powder inspection. Originally, radiographic inspection of castings was desired only by the consumer as insurance of casting integrity. Foundrymen have now found that such tests can aid greatly in their efforts to improve their product for quality work; most large foundries utilize every critical test available to them.

The use of statistical quality control methods for inspecting castings is a promising new development for foundries. By statistical methods of sampling, only a calculable number of production castings are checked critically. This saves time and money and often proves to be as adequate as checking 100 per cent of the production.

A defective casting should be identified before adding further operations on it like machining or similar processes. A defect may be internal or external and the person inspecting a casting must know about all the possible castings in a casting.

13.2 DEFECTS IN CASTING

A casting may have large variety of defects formed due to improper processes in the manufacture such as design, pattern making, moulding, mould assembly, metal melting, pouring, cooling, shakeout, cleaning and heat treatments. These defects can be broadly grouped as follows.

13.2.1 External Defects

1. Surface defects—metal penetration, sand fusion, scabs, mould burst, flash, fins, incomplete filling, cold shuts, hot tear, unfused chaplet, pipe formation, etc.
2. Dimensional defects—mismatch, distortion, core shift, dimensional inaccuracy, etc.

13.2.2 Internal Defects

1. Internal cavities—blowholes, gas porosity, pinholes, shrinkage cavity, hot tears, internal cracks, etc.
2. Inclusions—sand inclusion, slag inclusion, unfused internal chills, etc.
3. Metallurgical inclusions—improper composition, defective mechanical properties, internal stresses, etc.

Elaborate study on casting defects and its elimination methods are discussed in the module handling metal solidification and design of castings.

13.3 INSPECTION METHODS

There are many inspection or testing methods available for finding defects in a casting. These can be grouped as follows:

(a) Visual inspection methods
(b) Destructive testing methods
(c) Non-destructive testing methods
(d) Metallurgical testing methods

Visual Inspection Methods

Visual inspection methods are assumed to be simple and fast to find external defects during the casting process, at the time of shakeout or cleaning. If a severe defect is noted on a casting, that component can be removed from the process line immediately so that further wasting of work on it can be eliminated. Most of the external defects mentioned above can be identified by perfect visual inspection after surface cleaning. Dimensional inaccuracy, distortion, etc., can be determined by comparing the geometry of the casting with the drawings. All castings are tested further, only after completing the visual inspection which is the first stage of finding defectives.

Destructive Testing Methods

In destructive casting (DT), the component or a sample specimen is loaded or destroyed to check its capability to withstand the loads or forces. Pressure testing is a semi-destructive testing applied on pipes, vessels, valves, etc. The mechanical destructive testing methods include tensile, bend and impact tests. Specimen is to be made from the casting by cutting it at the required testing area. Hardness at interior portions of the casting is also tested by slicing the casting.

Non-Destructive Testing Methods

Different types of non-destructive testing (NDT) methods are available and they can reveal most of the internal defects precisely.

The following are the popular NDT methods used in foundries:

(a) X-ray radiography
(b) Gamma-ray radiography

(c) Ultrasonic testing methods
(d) Magnetic particle testing
(e) Die penetrant inspection

Chemical and Metallurgical Testing Methods

The composition of a metal is inspected by various chemical testing methods like wet analysis, emission spectroscopy, X-ray fluorescence, automatic absorption spectroscopy, etc. Similarly, metallurgical studies are conducted by microscopic analysis of the grain structure. Specimens taken from the castings are ground, polished and tested using microscope with magnification above 100 times. Latest electronic microscope can show magnifications over 5000×, so that the all properties of the metal can be checked. Such expensive studies are conducted on critical parts of machine components cast from special alloy compositions.

Inspection and Quality Control

Inspection comprises all those operations which check the quality of the casting. The inspection of casting serves the following basic purposes:

1. Inspection ensures that the casting conforms to design and other requirements and thus separates defective castings from the non-defective ones.
2. Inspection provides information required for quality control in the foundry and thus inspection locates defects and flaws at different stages of casting production and use of raw materials. Therefore, the purpose of inspection is also to prevent the occurrence of defects in castings to be produced in future.

Although production and technical staff are concerned in many aspects of quality control, inspection department has a special responsibility for the maintenance of standards laid in contracts and commercial relationship. Thus, systematic collection and interpretation of quality data is generally channelled through the inspection department which is a vital element in an integrated quality control system extending throughout the manufacturing sequence. Close contact and mutual confidence between inspection and production staff are needed for the two-way flow of information required for the maintenance of standards. A concept of organization for inspection and quality control as suggested by Quilter is shown in Chart 1. Modern quality control organization is concerned with a great deal more than the routine testing in relation to specified limits. Much data is obtained by the use of highly sophisticated techniques and more detailed treatment is required in interpreting the results. Adverse trends should be detected before quality deterioration leads to a high proportion of rejects.

13.4 THE COST OF IN-PROCESS INSPECTION

The cost of in-process inspection should be justified by savings in subsequent manufacturing operations or in the quality of the product. Since inspection involves expenses, the purchaser is required to specify the kinds and intensity of inspection. The inspection of castings before shipment is usually the responsibility of the foundry. Unless other inspection methods are specified, usually a visual inspection is given to the standards of that foundry. In case additional inspection procedures are required, detailed information on these should be conveyed to the foundry, since the foundry's inspectors do not know the use of casting and what is critical to its application. Unnecessary tight requirements also increase the cost of castings.

There is also a concern for duplication of inspection, i.e., inspection by the foundry as well as inspection by the purchaser. The inspection at the foundry provides the opportunity to correct process variables in the production schedule. However, inspection by the purchaser is mainly related to the serviceability of castings. Due to different advantages of two inspections, both shipping (foundry) and receiving (purchaser) inspections are carried out.

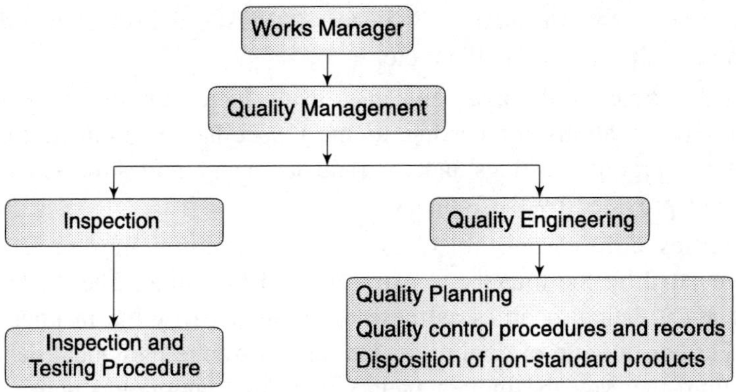

Chart 1 Organization for inspection and quality control

13.5 SAMPLING PLANS FOR INSPECTION

All castings produced in small or moderate quantities are usually inspected by the foundry before dispatching, whereas for castings made in large quantities for a sampling plant inspects only a portion of the castings. Sampling plants should be mathematically employed so that the size of the sample and the frequency of sampling provide the degree of dependability required for each application. The selection of the sample castings should be entirely random as the selection of convenient samples and previewing will make the plan unreliable. Good supervision and auditing is required for successful sampling inspections. The sampling plans are based on the mathematics of probability and in any plan there is a risk of rejecting lots of good castings or of accepting castings of below standard. However, this risk may also occur with the routine inspection of all the casings.

13.6 SPECIFICATIONS

In addition to the quantity of castings and a delivery date, the purchaser is required to supply the necessary information regarding the characteristics of castings such as on surface finish requirements, dimensional tolerances required, properties of metal, etc. The foundry should also fully understand the requirements of purchaser. The purchaser can transmit the requirements to the foundry in a purchase order on an engineering drawing, in a material specification and by the pattern equipment. Sometimes even verbal instructions may be used but all the requirements should be laid on paper to minimize mistakes.

Since castings are used for a wide variety of applications with different combinations of requirements, standard specifications for casting as such are impractical. There are no general standards acceptable for the surface finish, machining allowance, or for dimensional tolerances. Some foundries stabilize guidelines to their own capabilities in dimensional control. However, there are number of standard specifications for metals.

Standard specifications for various types and grades of metals are promulgated by various national and government agencies, for example, American Society of Testing and Materials (ASTM), American Society of Mechanical Engineers (ASEM), British Standard Specification (BSS), Indian Standards Specification (ISS), etc.

The engineering departments of some companies using large number of castings give their own specifications. Specifications for casting to meet specific requirements are particularly written. Some foundries produce castings under a trade name and grades for which the properties and characteristics are provided by the foundry.

The various properties of the castings relating to their serviceability must be established by the purchaser and appreciated by the foundry to avoid misunderstanding. The thickness and shape of the castings influence their properties as these are influenced by the manner in which they solidify and cool. Further the rate of cooling is determined by the individual design of casting, the process of casting, the way the molten metal enters the cavity, and pouring temperature. Thus, it is necessary the specifications be clear to both purchaser and producer and that they agree on referred properties as being in the cast part or in separately cast test specimens. They are standard specifications for both the cases and as and when necessary, special requirements may be added. In general, foundrymen also advise their customers on castings.

13.6.1 Types of Specifications

There are several methods of specifying castings so as to obtain a satisfactory product. Specifications can be written to designate the properties of the metal that are specifically needed, i.e., composition of metal, strength, hardness, finish, accuracy or special characteristics such as resistance to heat or corrosion. There are many casting applications in which the ordinary properties of the designated type of metal/alloy are quite adequate.

When mechanical properties of the castings are important, the most usual procedures are to utilize a standard test bar which is poured separately but along with the specified lot of castings. This test bar should possess the minimum required tensile properties. The relation between properties in the casting and the test bar depends on their relative cooling rates. The properties of iron and steel are particularly sensitive to the section thickness of casting and thus, a series of standard test bars of increasing size should be cast related to various casting thicknesses.

A commonly used method of specifying iron and steels for castings produced in larger quantities is based on the microstructure and hardness of the metal in the castings themselves at a specific location. There is a good relation between hardness and tensile properties but this relation is not used as a part of specifications.

For some applications, the castings can be tested in the manner in which they will be used, e.g., pressure components can be 100 per cent hydraulically proof tested as this test would be non-destructive to satisfactory units. Similarly, load carrying components can be tested by measuring their deflection under load and rotating parts can be spin tested.

Some cast metals can be specified by chemical analysis alone but this is rarely satisfactory to specify iron and steel castings. Limits of the content of some elements are usually included in specifications as these elements can influence properties or characteristics of the metals which are not easily measureable. For example, the thermal shock resistance of grey iron is influenced by its total carbon and corrosion resistance is influenced by alloy content of iron. However, different foundries may obtain the required properties with metals that differ in chemical analysis because their facilities and techniques are different.

Some castings may be required to meet the specific requirements which may include radiographic soundness, stress relief, heat treatment, rust preventing coatings, etc.

13.6.2 Standard Specifications

Standard specifications of most of the castings are laid by various agencies. When metal properties are to be specified, the standard specifications provide the purchaser with maximum number of sources and possible economic advantages because the use of familiar specifications saves time and minimizes the possibility of mistakes. Even with unusual requirements of castings, it is preferable to use an appropriate standard specification as a base and these add the unusual requirements to it as end when possible. However, the additional requirements should be clearly stated and given with the number of standard specifications on the purchase order. Standard specification societies publish a summary of commonly used standard specifications for various types of castings. The summary usually includes specification number, uses, grades or class, yield strength, tensile strength, elongation in tension, chemical analysis, microstructure, hardness, heat treatment, specific requirement, typical applications.

13.6.3 Standards and their Implications

The basic requirement of inspection is the establishment of standards as required by the designer. In a broad sense, the design requires the definition of the casting in terms of physical shape and dimensions, properties, composition, and general quality. Inspection is based on permissible deviation from the ideal standards. Some factors such as dimensions, composition and mechanical properties can be defined quantitatively so that numerical tolerance limits can be prescribed whereas in other cases such as soundness and surface finish, subjective judgements are required. The latter quantities are closely associated with commercial policy and may be set at higher level than actually needed to meet the formal specifications. However, standards must not be set at an unrealistic level in relation to the basic capability of the particular casting technique. Further economic implications of high standards must not be overlooked.

13.7 QUALITY ASPECTS

In general, the quality aspects for casting include surface finish, dimensional accuracy, soundness and dimensional properties.

1. **Surface Finish** The surface condition of castings is influenced by a number of factors such as kind of pattern used, process of moulding, quality of sand and other additives, cleaning process, etc. At one time the kind of moulding sand used was most important to casting finish, but now the abrasive cleaning process may be the most important factor. Smoothness of castings can be evaluated only by visual comparison and there is no other satisfactory way.
2. **Dimensional Accuracy** The basic dimensions of castings are establishment mainly by the pattern equipment but variation from this base may be caused by several other factors in the casting process, i.e., expansion of the mould cavity due to heat and pressure of the metal, contraction of the metal after solidifying and cooling and heat treatment. To compensate all these effects, the pattern is purposely made of oversize. Distortion or warpage of casting may occur due to different cooling rates of different parts. Sometimes, it is necessary to distort or fake the pattern to obtain a true casting. Once a sample casting is approved by the customer, castings are not generally checked for their dimensional accuracy. However, when producing large number of castings, critical dimensions are measured frequently to check for possible drift especially due to pattern wear.
3. **Soundness** Because of the different ways in which various metals/alloys solidify, chances of unsoundness may be more or less depending on the metal to be cast. The metals and alloys showing low shrinkage exhibit less possibility of unsoundness. Often a cavity in the casting will be evident to visual inspection because of an opening or depression at the surface of the casting. Internal soundness can be checked by radiography and ultrasonic scanning. Sample casting may be sawed in half to check for both wall thickness and internal soundness.
4. **Metal properties** The standard specifications for various metals and alloys provide the means for evaluating metals and their properties.

13.8 INSPECTION PROCEDURES

Inspection procedures may be classified into the following categories:

(i) Visual inspection for surface finish and defects.
(ii) Shape and dimensional checking.
(iii) Testing for chemical composition.
(iv) Testing for mechanical properties.
(v) Destructive testing of castings.
(vi) Non-destructive testing for soundness.

13.8.1 Visual Inspection for Surface Finish and Defects

Visual examination of individual casting ensures that none of its features have been omitted by short running, moulding errors, fettling mistakes and errors in other processes of casting. Surface defects observed at this stage include surface cracks, tears, blowholes. Visual examination may be carried out with the naked eye or using a magnifying glass.

Sometimes, numerical standards may be specified which may be evaluated by visual compression of a series of standard surfaces graded arbitrarily to or assessed quantitatively by other means. Alternatively, a style measuring instrument may be used to obtain representations of surface profiles. However, such instruments are usually suitable for distinguishing between relatively smooth surfaces such as those obtained in die casting and other precision casting methods.

Surface roughness methods as a number (in microns), which is an arithmetical mean of the heights of the peaks and depths of the valleys on a casting surface above and below a mean line within a specified sampling length. The surface roughness can be assessed according to IS: 3073, 1967. The approximate roughness values for different types of castings are given in Table 13.1.

Table 13.1 Roughness values for different casting processes

Casting processes	*Roughness value*
Ordinary sand casting	5 – 25
Good quality sand casting	5 – 25
Permanent mould casting	0.8 – 6.3
Die casting	0.8 – 3.2
Shell moulding	1.6 – 6.3
Investment casting	0.8 – 3.2
Ceramic casting	0.2 – 1.6

Presently, a range of optical devices is available to examine the areas inaccessible to the human eye, the bore of a cast cylinder block by using Boroscopes (miniature combined periscope and telescope), light guides or miniature television cameras. A simple and useful method of revealing surface defects is acid etching. The process of etching removes all oxides from the surface of the casting and reveals its crystalline structure and surface defects in sharp relief. The etchant used depends on the type of material, surface condition, and type of defect to be revealed.

Precise measurements of finished surfaces can be carried out by the use of electrical type of direct reading surface measuring instruments or profilometer such as Talsurf. Visual inspection is the simplest and is carried out to ensure casting that looks good. Visual inspection is quite satisfactory for many castings such as manhole covers, drains, counterbalance weights, etc.

13.8.2 Shape and Dimensional Checking

Dimensional inspection is generally required for all types of castings. In some cases, it may not be so critical but in other cases it may be vital. Castings produced by die casting, investment casting shell moulding, etc., and castings requiring adequate machining need close check for the dimensions. All undersized castings requiring machining are rejected as their dimensions will be further reduced after machining. Initially, the castings are made from a new pattern and coreboxes are of correct size. If there is any deviation from the blueprint, the pattern equipment are rectified. In mass production, once the castings are found to be within the tolerances, spot checks of castings, together with regular checks of the pattern and dies being used, may be adequate. However, in case of the jobbing type of foundry, each casting may have to be thoroughly checked for dimensional variations.

Dimensional examination is carried out against the drawing with the help of various measuring instruments such as gauges, jigs, templates, special fixture, etc. Tolerances on cast dimensions are usually included on the drawings. In some cases, marking out for subsequent machining is also included with dimensional checking. If castings are to be machined without individual marking out, specific care is required in respect of dimensions incorporating datum surfaces. In cases where the weight of the casting is critical, a specific weight check is also included. When dimensional errors are detected, their true significance should be determined. Various methods/instruments used for dimensional checking are discussed below.

(a) Measuring instruments such as rule, Vernier's, micrometers, scribing block, combination set, straight edge, squares, spirit level, and dial indicator are commonly used for checking the dimensions. More advanced instruments such as autocollimator, comparator, ultrasonic instruments for wall thickness, projection instruments, etc., are employed for high precision castings or machined castings.

(b) Templates and contour gauges are used for checking profiles, curves, and intricate shapes. Templates having required profile can be easily prepared in mild steel or brass sheet. These save time in measurements and make the entire job easy.

(c) Limit gauges are usually employed for tolerance dimensions on castings produced on a repetitive basis. The various types of limit gauges includes plug, ring, snap, plate, etc., the use of which depends on the shape of the parameter to be checked. The limit gauges should be periodically checked for their correctness.

(d) Special fixtures are employed where dimensions cannot be easily checked by other instruments, e.g., relative dimensions, centre-to-centre distances, angularity of surfaces, etc. Inspection fixtures are specially designed for specific jobs.

13.8.3 Testing for Chemical Composition

Testing for chemical composition provides an actual evaluation of the properties and characteristics of the cast metal. Chemical composition is usually determined by tests carried out on specimens cast separately from the same heat of metal. The samples must be drawn at the same time as the castings are poured, since losses of volatile or easily oxidized elements may cause change in composition and thereby resulting also in difference in mechanical properties. When pouring operation is long or successive ladles are drawn from a single heat, frequent sampling should be carried out at the casting point. Analysis for chemical composition may also be carried on discarded heads and runners. The chemical composition is usually assessed in relation to the specified limits for the major alloying elements and expected impurities, tramp elements and gas content. In the case of iron steel castings, the percentage of C, Si, S, Mn and P are invariably determined. The presence of alloying elements (Cr, Ni, Cu, W, V, etc.) may have to be determined depending upon the composition of alloys. For determination of chemical composition, wet analysis methods are the most accurate and reliable, but take a long time. However, wet methods are unsuitable for determining the composition of the molten metal before pouring to adjust the composition of the final melt. The use of spectrographic, X-ray fluorescence and other physical techniques enables rapid determination of a comprehensive range of elements. The instrumental methods permit more extensive bath sample analysis and thereby provide better control in melting. The composition of last cast should be verified before the further processing of castings minimize loss of production time.

In small foundries not able to afford costly instruments, the elements of critical importance are checked by either cheap instruments designed specifically for a limited number of elements or quick test developed called carbon equivalent (CE) is given by total carbon% + 1/3 silicon%. The instruments work on the principle that CE is directly related to the liquidus arrest temperature of the cast iron. The instrument equipped to hold a small reservoir or molten as it solidifies, is recorded on a chart from which the liquidus arrest temperature be can be easily determined. Conversion table is used to determine the corresponding CE as given in Table 13.2.

Table 13.2 Conversion of temperature to CE

Liquidus	*Temperature*	*C.E.*
1232.2	2250	3.60
1221.1	2230	3.70
1210.1	2210	3.80
1198.9	2190	3.90
1187.8	2170	4.00
1176.7	2150	4.10
1165.6	2130	4.20
1154.5	2111	4.30

Similarly, an apparatus known as Stroheim thermo-electrometer is available for quick determination of silicon in cast irons. It measures the thermal e.m.f. produced on heating a junction of cast iron in which silicon is to be determined and another metal such as copper. Silicon possesses a special position in the thermoelectric series and thus, silicon can be determined by means of empirical calibration curves prepared by using samples of known silicon content.

Spectroscopic analysis is becoming more popular in foundries for quick determination of various alloying and trace elements. Various types of spectroscopic analyzers are available. The selection of a particular type depends on the requirements, i.e., elements to be analyzed, desired accuracy, frequency of tests and the type of cast metal.

13.8.4 Testing of Mechanical Properties

Specimens for mechanical testing are generally prepared by machining of specially designed cast test bars which may be poured separately from the same heat or as directly attached to the heavy casting itself. With certain exceptions, the tests carried out on specimens are representative of the particular heat of metal and not of any individual casting. The measured properties reflect influence of dissolved gas, inclusions, and melt treatments, but may not necessarily be those which are obtained from test pieces cut from actual casting, since there may be difference in section, structure, and degree of feeding. Standardization of test bar conditions permits the use of test results to ensure perfect soundness of the test specimens. Thus, in casting of test pieces, yield is disregarded and most test pieces are over fed.

Test for Tensile properties: Tensile properties are directly determined by testing an especially shaped sample made from separately cast test bars in a tensile testing machine (universal type of testing machine or Hounsfield tensometer). However to establish the actual tensile properties of the metal in a particular section of casting, a tensile test bar can be machined from the actual casting but it may be remembered that this test only provides data on metal that is tested. Tensile properties can be estimated by load testing of the entire casting and also from hardness, microstructure sonic or eddy current data. The tensile results include proof or yield stress, tensile strength, modulus of elasticity, elongation and reduction of area. Casting specifications generally include one or more of these properties.

Microstructure: Microstructure of metals and alloy establishes the type and possible heat treatment in case of iron and steels. Microstructure is identified by viewing a polished and etched surface of the metal under a metallographic microscope. The microstructure can be studied either of a metal piece sectioned out of the casting or on the casting surface itself but latter involves difficult techniques.

Hardness Test: Hardness is the most commonly used test because it is easy to perform and directly provides a numerical value and the test is non-destructive in nature. Hardness tests give the guide to wear resistance, machinability and effectiveness of heat treatment applied to actual castings. In general, correlation of hardness with tensile strength and ductility enables a rough prediction of these properties.

The surface of the casting should be properly prepared for a hardness test. The casting skin should be presented to the indenter to enable the correct measurement of the diameter of the impression with microscope. Brinell test is unsuitable for use at very high hardness as distortion of the ball indenter affects the geometry of the impression. For high quality and precision castings of extreme hardness, Vickers or Rockwell test (gives small indentation) is conducted. The Rockwell test is quick and is used more commonly for control purposes, whereas Vickers pyramid test finds its use in laboratory work.

Indentation tests generally require that the casting be placed on the platform of the hardness testing machine, but certain hardness tests can be applied directly to large castings, where representative samples of heavy castings are difficult to be obtained. Sceleroscope can be used to measure the hardness by rebounding of free falling pointed indenter on any horizontal surface.

Other portable instruments give readings which can be correlated with standard indentation tests. However, these additional tests are used for control purpose rather than for formal inspection. Hardness testers should be routinely checked for accuracy with a calibrated hardness test block.

Sometimes, file hardness test is very useful because it evaluates the hardness of an edge. Particularly iron castings not heat treated can have hard edges which can be destructive to cutting tools. In the file test, a sharp medium cut mill file is held against the edge of the casting and moved just enough to determine whether it is cutting or not. A tendency to slide without cutting under moderate pressure indicates a hard edge.

Impact Test: Though the impact test values do not have direct significance for quantitative design, the important values are included in specifications of some castings. Both Charpy and Izod tests can be used at room temperature (for general castings) or sub-zero temperature (for castings for cryogenic and other low temperature applications). In the case of cast iron castings, notched bar impact testing is hardly of any significance in discriminating between the qualities of different irons, but for an unnotched plain round specimen is held in stepped grips and the fracture energy is determined in the normal Izod test.

Bend Traverse and Compression Tests: These tests are conducted to evaluate the ductility and resistance to shock of the cast metal. All these tests can be conducted on the universal type of testing machine. In the bend test, a bar specimen is supported at its end and load is applied at the middle of the specimen load at which the specimen breaks is noted down. This test is mainly used for grey cast iron.

Fatigue Test: This is carried out where an appraisal of the life of the casting service is to be known. Though frequently the castings are used under fatigue stresses or for high temperature works, fatigue tests and stress rupture or creep tests are seldom used for routine inspection owing to a very lengthy time involved in tests.

Other Tests: Various special tests may be required to establish the response of castings to service conditions. The various tests include: for weld decay as applied to austenitic stainless steel, ballistic tests on plates, weldability tests and other tests based on analogies to the conditions to be encountered by the casting.

13.9 DESTRUCTIVE TESTS ON CASTINGS

In general, variation of properties within a single casting results from inevitable differences in microstructure, and composition produced by local differences in solidification conditions. The normal test blocks (separate cast bars or block attached to castings) are designed to evaluate the metal under ideal conditions of feeding, but tests-sectioned castings can provide a more direct indication of properties within a particular design. Systematic sectioning of casting also aids in direct investigation of structure and internal soundness by metallographic techniques. This also facilitates the accurate interpretation of non-destructive test and radiographic results.

In one form of the destructive testing, direct loading of representative casting is carried out for failure in special fixtures simulating service stresses. The deformation, breaking load and position and nature of fracture guide the quality and behaviour of the design under highly stressed conditions. One simpler form of test consists of fracturing the cast by compression in a press.

Destructive tests can be carried out on rejects from other causes, but in some cases it may be necessary to sample a proportion of castings from regular production. Such techniques can be a vital aid to the interpretation of test results and inspection findings derived by more conventional methods.

13.10 NON-DESTRUCTIVE TESTING ON CASTINGS

Non-destructive tests are conducted to examine the internal soundness of the castings without recourse to sectioning or fracturing. The tests may be carried out for any sub-surface or internal defects, surface defects not detected by visual examination, pressure tightness, and overall soundness which may be required in service. These tests not only assist in maintaining high quality standards, but provide a valuable aid in the development of foundry techniques. These factors have greatly improved the position of the casting in engineering. However in view of increased sensitivity of inspection by non-destructive testing, the techniques should be adopted and interpreted with discretion to avoid unnecessary rejections. Various non-destructive tests can be classified under the following categories:

1. Tests for defects in the surface zone
2. Tests for detecting internal defects
3. Other tests

1. Tests for Defects in the Surface Zone

Cracks and other flaws in the surface of the casting can be detected by a number of methods which include chemical etching, penetrant method, and magnetic flaw detection. All these methods require clean and relatively smooth surface for effective detection.

(a) Chemical Etching Method: Very fine surface cracks, laps and similar defects may be covered by:

(i) Thin layer of metal resulting from superficial plastic deformation in shot blasting, fettling, and machining of casting.

(ii) Oxide scale formed in heat treatment.

In such cases, usual penetrant methods are not effective. Such surface defects can be revealed by a simple chemical etching method in which controlled corrosion attack takes place. The method employs acid or acid salt mixtures developed for particular alloys. The casting is immersed in the etching bath or swabbed down with etching solution. After etching, excess acid is neutralised, the castings are rinsed and dried. This process removes all oxides and superficial metal from the surface of the casting and reveals crystalline structure along with defects. Many etchants are available and the type of etchment to be used depends on the type of the material, surface condition and the type of defect to be revealed. Castings which cannot be prepared readily are pressure tested. Penetrant solution can be applied to one face and the opposite face may be examined for signs of seepage. The penetrant test consists of cleaning of casting, treatment of casting with penetrant, removal of excess penetrant, development and inspection of penetrant indication.

(b) Cleaning: The casting to be tested should be clean and thoroughly free from contaminants such as loose scale, grease oil, rust, etc. Cleaning may be accomplished by using one or more of the following methods:

(i) Vapour degreasing by using trichloroethane.

(ii) Solvent degreasing.

(iii) Pickling in acid or alkaline solutions.

(iv) Steam cleaning.

(v) Cleaning by ultrasonic vibrations given to the casting suspended in a tank of liquid.

Carryover of residues from cleansing solutions, particularly chlorine from degreaser and acid or alkali from pickling solutions should be avoided as these will react and contaminate the penetrant. The physical methods of cleaning such as shot blasting, brushing, etc., should also not be used as these may cause the closing of cracks and holes at the surface of the peening action.

(c) Treatment of casting with penetrant: This consists of allowing penetrant to seep into the surface discontinuities by capillary action for a suitable interval of time (5 minutes to 3 hours). The penetrants employed fall into two categories, i.e.,

(i) brilliant dye penetrants which are solutions of brilliant red dye and reveal red indications on white background and penetrants; and

(ii) fluorescents which are solutions containing fluorescence and emit brilliant green fluorescence when viewed in a darkened room with an ultraviolet light (often known as black light). Brilliant dye penetrants do not require black light and are sensitive to fine defects and very suitable for the inspection of light alloy castings having smooth surfaces. However, these are difficult to be used on rough surfaces and dangerous when used in open tanks.

The choice of penetrant is influenced by various factors such as

- Size, shape and quantity of castings.
- Analysis and surface condition.
- Size, shape and position of expected defects.
- Compatibility of the penetrant with the casting material.
- Sensitivity required.
- Availability of equipment and services.

There are no hard and fast rules for the selection of the best penetrant, but can only be found by experience. The penetrant can be applied to casting by brush or aerosol spray or by complete immersion of casting in a tank of penetrant for a recommended period of time, and the surplus penetrant must be removed.

(d) Removal of Excess Penetrant: The penetrants from either category may be solvent removable, water washable or water emulsifiable (removal of penetrant is possible only after an emulsifier is added). In the case of solvent removable penetrants, operation can be performed by using an aerosol, whereas for water washable penetrant, water hand spray is used with a pressure of 2-2.5 kg/cm^2 at a temperature of 45°C. When using emulsifiable penetrant, paraffin oils an emulsifier is added to the penetrant before water washing. The emulsifier breaks up the continuous oily film into an aggregate of small droplets which can be washed with water spray. Washing of the casting should be carried out from bottom upwards so that water running down does not remove penetrant from the defect. Further, violent wash should be avoided as it will also tend to remove penetrant from the defects.

(e) Development: After the excess penetrant is removed, the casting may be treated either for dry development or wet development. In case of dry development, the casting is dried by either normal evaporation or by forced air circulation (temperature of air should be less than 80°C). After drying, a developer in the form of dry light coloured powder is uniformly applied to the surface by tumbling the casting in the powder or by dusting the powder over the casting or by using an air gun to blow the powder onto its surface.

For wet development, excessive moisture is removed from the casting after washing. Wet developer is applied in the form of dry light coloured powder that is uniformly applied in the form of a powder suspended in water or a volatile solvent (aerosol). For the correct observations, uniformity of developer and thickness of the applied coating are very important (too thick coatings may mask very fine defects). The developer should be vigorously agitated just before its use.

The castings are then allowed to stand for development. The time taken for development may vary from 15 minutes to 36 hours and is known as **bleeding time**.

The penetrant inside the defect will seep out by reverse capillary action (blotting paper action).

(f) Examination of Penetrant Indication: The examination of penetrant flaw indicators must be carried out under adequate illumination. The area to be examined should be equally illuminated without any shadow or glare. The interpretation of the indicators is the most difficult aspect of penetrant flaw detectors. Due to the blotting paper action and porous nature of developer, usually an enlarged outline of defect is obtained. On the other hand if excessive developer is applied to the surface, only a very small amount of penetrant may seep out thus indicating an extremely small defect. Penetrant methods for crack detection are most suitable for nonferrous metals and austenitic steels.

(g) Magnetic Flaw Detection: This method is highly effective and sensitive for revealing linear surface discontinuities (cracks and similar defects) on or just beneath the casting surface in case of magnetic materials (iron, steel and their alloy). The magnetic flaw detection is based on the fact that the magnetic permeability of ferromagnetic materials is greater than that of air. If the magnetic lines of applied field are allowed to pass through a casting of ferromagnetic material containing a defect, they will tend to pass more readily through the material than the surrounding air. Distortion of magnetic lines will be caused in the neighbourhood of the defect. Further, it can be noted that if the defects are positioned across the flux path at 90° to it, maximum distortion of magnetic flux will be observed, whereas minimum distortion of magnetic flux occurs when the defects are in line with the flux. In the latter case, the defects may not be revealed and thus two or more tests will be required to ensure 100% detection of defects (Figure 13.1).

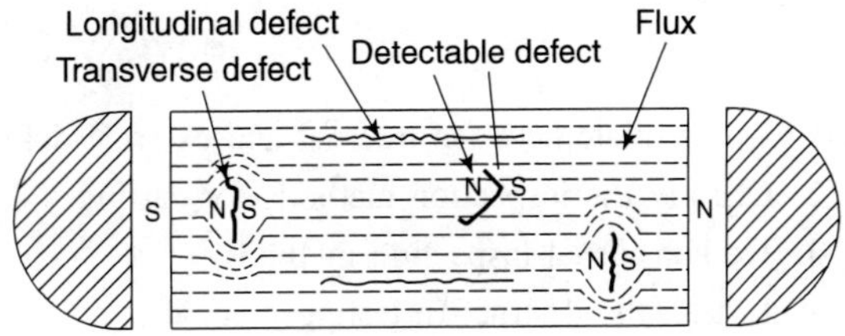

Figure 13.1 Magnetic flux flow around flows in a plate.

13.11 TESTS FOR DETECTING INTERNAL DEFECTS

Radiography

This is a means of obtaining a two-dimensional shadow picture of variations in the density of shapes in metal by using X-rays or gamma rays radiation which are electromagnetic oscillations and share the same properties. A radiograph is a photographic record obtained by passing X-rays or gamma rays through a casting onto a film. The amount of radiation reaching the film depends upon: (a) nature and thickness of material, (b) the number, position, and size of defects, and (0) quality of radiation used.

X-rays are generated from X-ray tube which consists of a highly evacuated glass tube containing an electrically heated filament (cathode) and a target comprising a thin plate of tungsten (anode) embedded in a block of copper. X-rays are generated by applying of high voltage between the cathode and anode causing electrons travelling at high speed from cathode to collide with anode.

Gamma rays arise from the spontaneous and continuous disintegration of atomic nuclei within radioactive substances. The radioactive source itself governs the energy of gamma radiation and cannot be controlled. As the rate of disintegration cannot be changed, the intensity is uncontrollable. For economy, it is desirable to use radioactive isotopes having slow decay rate. An adequate quantity of radiation is required to keep exposure time short and thus, the source must have high activity and be concentrated into a minimum volume.

Natural radioactive sources cannot be obtained with dimensions small enough to give required intensity and thus, artificial radioactive sources are more suitable. The most suitable radioactive isotopes used for industrial radiography include cobalt 60, indium I-92, thallium 170 and caesium 137. It is necessary to select a source (compromising for penetrating power and decrease in contrast) which gives sufficient sensitivity to reveal the defect with the shortest practicable exposure. For the inspection of castings, X-ray radiography is more common as the control of intensity of X-rays is easy.

Review Questions

1. Differentiate between internal and external defects.
2. Classify the inspection methods for casting.
3. Explain visual inspection methods.
4. List all the destructive testing methods.
5. List all non-destructive testing methods
6. Explain metallurgical testing methods.
7. Explain sampling plan for inspection.
8. Write a note on the cost of in-process inspection.
9. Give the roughness values for typical casting processes.
10. Explain radiography in detail.

CHAPTER 14

Heat Treatment of Castings

14.1 INTRODUCTION

Castings after solidification have coarse-grained heterogeneous structure with high hardness and internal stresses. The structure and properties can be improved by heat treatment.

The main objectives are as follows:

1. To receive the internal stresses and remove distortion, dimensional changes, etc.
2. To improve the mechanical properties like tensile strength, toughness, etc., by refining grain size and suitable changes in the structure of the cast metal.
3. To improve the machinability of hand cast metals by annealing.

Heat treatment of metal is conducted at a temperature well below the melting point of the metal, and hence it is a solid state transformation. Three stages covered in all heat treatment processes are:

1. Heating of casting at controlled rate in an oven to the required transformation temperature; this temperature depends on composition, geometry and the type of treatment required for the casting.
2. Holding or soaking of casting of the casting at that temperature for about one hour per each 25 mm thickness of wall, for uniform transformation.
3. Cooling at controlled rate to have the required transformation in microstructure.

14.2 HEAT TREATMENT OF FERROUS METALS

The environmental heat treatment operations applied to ferrous metals are discussed as follows.

1. Annealing: Annealing is the process of heating metal to a temperature that will remove the distortion of grains of the metal and then cooling at a slower rate, so that at room temperature it has no grain distortion or strain. Depending on the temperature, sustaining time, cooling rate, the composition of metal, etc., and the annealing process, it can be grouped as:

1. Full annealing
2. Process annealing
3. Spheroidize annealing

4. Diffusion annealing
5. Isothermal annealing

Figure 14.1 shows a simplified iron-carbon equilibrium diagram showing the temperature ranges for heat treatment of steel at various carbon percentages.

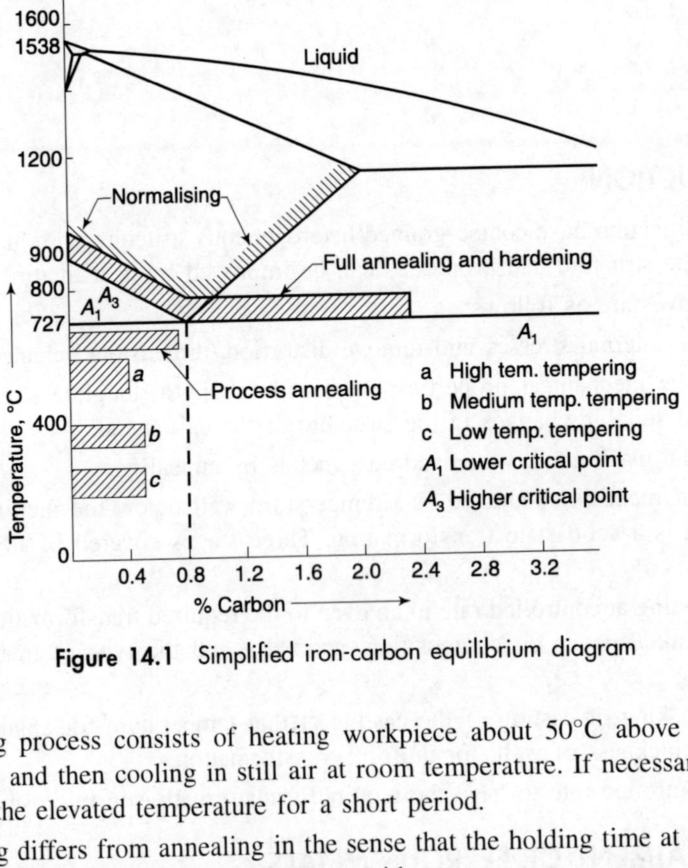

Figure 14.1 Simplified iron-carbon equilibrium diagram

2. Normalizing

- Normalizing process consists of heating workpiece about 50°C above its upper critical temperature and then cooling in still air at room temperature. If necessary, the metal may be held at the elevated temperature for a short period.
- Normalizing differs from annealing in the sense that the holding time at elevated temperature is very short, and the cooling rate is rapid.
- Normalizing process removes the internal stress, refines the grains to smaller size and improves the mechanical properties such as impact strength, yield point, ultimate tensile strength, etc.

3. Hardening

- Hardening is a heat treatment process in which the metal is heated about 50°C above the higher critical point (A_3 line) for steels having more than 0.8% carbon content, this range is taken above the lower critical point (A_1 line).
- Then the metal is held at that temperature for sufficient time to change into austenite structure and cooled rapidly by quenching in a suitable medium like water or oil. This

process gives the maximum hardness for the metal, hence suitable for tools, dies. machine parts, etc.

4. Tempering

- The subsequent heat treatment for reducing the hardness obtained by hardening process is called tempering.
- In this process, the hardened steel is reheated to a temperature below the lower critical point (A_1 line), retained at that temperature for sufficient time and then cooled slowly to the room temperature.

Depending on the holding temperature range, tempering process can be classified as:

1. Low temperature tempering (150°C to 250°C)
2. Medium temperature tempering (350°C to 450°C)
3. High temperature tempering (500°C to 650°C)

In Figure 14.1, the shaded areas *a*, *b* and *c* represent the tempering region in iron-carbon diagram. Tempering relieves residual stresses, improves ductility, and gives toughness by the reduction of hardness.

5. Case Hardening

- Case hardening is a heat treatment process in which the low carbon steel (0.1% C) is heated to red hot and the content of carbon or nitrogen or both on the outer shell of the workpiece is increased by forcing to absorb from the surrounding to the surface for certain depth by diffusion.
- The outer shell is hardened as done in the hardening process. Thus, by getting the hardened surface and a ductile core, a machine part gets high wear resistance as well as heavy dynamic load bearing capacity.

The commonly used case hardening processes are as follows:

1. Carburizing
2. Nitriding
3. Cyaniding
4. Carbo-nitriding

In carburizing process, the percentage of carbon in the outer shell is increased to about 1%. The nitriding process nitrogen is added the outer shell of certain steels containing aluminium and chromium. In cyaniding process, both carbon and nitrogen are added to the surface by using a liquid bath, whereas in carbon-nitriding, the same is added by gas atmospheres.

6. Surface Hardening In surface hardening, only the selected surface is subjected to heating, whereas in case of hardening, the whole object is brought to the hardening property of the surface as obtained to the steel by adding carbon and nitrogen. The heating of surface is done by the following methods:

1. Induction hardening
2. Flame hardening

3. Laser beam hardening
4. Electron beam hardening

Steel Castings

Generally, steel castings are subjected to heavy internal stresses due to the high temperature of pouring. Hence, heat treatment is a must on castings to release stresses and to regulate the properties. The above-mentioned heat treatments are directly applicable to all steel castings and the same can be more understood by referring to the iron-carbon equilibrium diagram.

Grey Iron Castings

Grey iron castings are rarely heat treated. They are treated mainly for the removal of internal stresses. For castings like surface plates, natural ageing is adopted by leaving the castings outside for a few weeks. Critical machine tool castings are, stress relieved by controlled slow heating at a temperature of 200 to 400°C in ovens.

White Iron Castings

Heat treatment is essential for conversion of white iron to malleable iron. This conversion brings structural changes resulting in less hardness, brittleness and achieve desired ductility and tensile strength. The treatment requires heating from 800 to 1000°C, retaining for many hours at that temperature in controlled atmosphere and cooling in a regulated way. The whole process is comparatively an expensive affair.

S.G. Iron Castings

Spheroidal graphite (S.G.) irons or nodular cast irons are heat treated for stress relief as well as for improvement of wear resistance, ductility and toughness. The heat treatment processes are almost similar to steel, and it depends on the original structure and the desired properties.

14.3 HEAT TREATMENT OF NONFERROUS ALLOYS

Nonferrous alloys attain superior mechanical properties by solution-heat treatment and age hardening. The quality of aluminium and magnesium alloy castings can be improved by ageing, annealing or hardening followed with artificial ageing. Copper alloy castings require no annealing except for rare cases, when internal stresses must be relieved by annealing.

The As-cast castings are relatively brittle and possess poor mechanical properties. Heat treatment of the casting is necessary in order to render them serviceable. The main objectives of heat treatment are:

1. To relieve internal stresses which are set up during solidification.
2. To improve the mechanical properties by modifying the structure of the castings.

A. Heat Treatment of Aluminium Alloy Casting

Dimensional changes after casting: Three types of dimensional changes can occur in castings once they have reached room temperature.

Firstly, some alloys undergo dimensional change at their service temperature if they have not been fully stabilised by an appropriate heat treatment. For example, the common structural casting alloy Al-7%Si-0.4%Mg in the solution treated condition will continue to expand up to a total of approximately 0.1% during their first months or years at room temperature. This is because of the precipitation of the compound Mg_2Si. Pre-ageing at a sufficiently high temperature is an effective cure of this dimensional instability.

Secondly, other dimensional changes occur during any heat treatment which involves a change of phase. For example, once again, Al-7%Si-0.4%Mg alloy contracts by 0.1-0.2% during solution treatment, as alloying elements are taken into solution, and then grows by 0.05-0.1% as precipitation occurs.

Thirdly, dimensional changes can occur in service, particularly at elevated temperature. For example, in an analogous way in which the well-known growth of grey irons occurs at temperatures as low as 350°C by the continued precipitation of graphite, growth can also occur in Al-17%Si alloy castings if they are used at a high enough service temperature for silicon precipitation to occur. However, this can be prevented by pre-ageing at 230°C for 8 hours.

Heat treatments are carried out for a number of reasons. The first of these is to improve the mechanical properties and physical properties of an aluminium casting. The strength of some alloys can be improved by a factor of two or three above that which is obtained in the As-cast condition.

The Stages of Heat Treatment

Solution Heat Treatment consists of heating an aluminium alloy casting to a high temperature for a desired amount of time. After the casting is heated to a high temperature it is rapidly quenched, by immersing the entire casting in a water or water glycol solution. This process is usually done to increase the strength of a casting.

Artificial Ageing consists of heating an alloy for an extended amount of time. During the artificial ageing process, the castings are heated for a specified period of time but to a much lower temperature than the solution heat treatment process. By maintaining the heating process for an extended amount of time the grain structure is refined, producing much greater strength properties of the castings. Artificial ageing is used to create a harder casting. Artificial ageing speeds up the process of ageing of a casting that would normally occur naturally over several years.

The objectives for the heat treatment of the aluminium alloys are:

To homogenize the structure which is non-uniform because of selective solidification during freezing.

To relieve stresses induced during solidification, and to produce specific mechanical properties.

The main alloying and magnesium nominal compositions of some of the heat treatable aluminium alloys are as follows:

1. 10% Cu, 1.1% Fe, 0.2% Mg and balance Al.

2. 4.5% Cu, 0.8% Si and balance Al.
3. 4.0% Cu, 1.5% Mg, 2.0% Ni and balance Al.
4. 3.0% Cu, 6.0% Si and balance Al.

The aluminium alloy casting may be subjected to the following heat treatment:

1. Annealing
2. Solution heat treatment
3. Artificial ageing without prior solution heat treatment
4. Artificial ageing after solution heat treatment
5. Stabilizing

1. Annealing

Process: The annealing of aluminium alloy casting is usually done between 325°C for about 12 hours.

Outcomes:

- There may be very little decrease in the tensile strength
- The ductility may be slightly improved
- Relieves macro stresses produced by unequal cooling rates in thick sections.
- The precipitation of the alloying elements which have been retained in solid solution during cooling of the casting may also take place.
- This promotes dimensional stability of parts which are to be used in service at high temperatures.

2. Solution heat treatment

The solubilities of principal alloying elements (copper, silicon, etc.) in aluminium are characterized by sloping solvus line in the respective equilibrium diagrams. Figure 14.2 shows the solubility of silicon in aluminium.

Process

- The alloy is heated to a suitable temperature (above the solvus line but below the solidus) in order to put the alloying elements in solid solution.
- It is then quenched in hot or boiling water.
- The holding time for quenching in hot or boiling water may be about 12 hours for sand casting and about 8 hours for permanent mould castings.

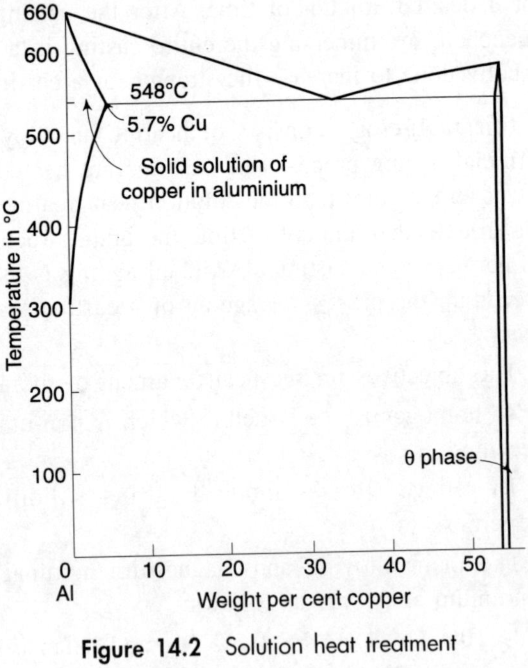

Figure 14.2 Solution heat treatment

Outcomes:

- The tensile strength is improved.
- The yield strength is improved.
- The ductility is improved.

3. Artificial ageing prior to solution heat treatment

Application:

- Used when rapid solidification has retained a considerable amount of the soluble alloying elements in solid solution.

Outcome:

- Results in some increase in strength and hardness.

4. Artificial ageing after solution heat treatment

Process:

- After solution heat treatment, the casting is heated between 250 and 400°F. In some cases the temperature may be as high as 600°F.
- Ageing is done for 4 to 24 hours.

Outcome:

- Maximum strength and hardness are attained by this treatment.

5. Stabilizing

Application:

- This is important for parts to be used at elevated temperatures.

Process:

- Carrying out artificial ageing beyond the point of maximum hardness.

Outcome:

- The precipitate is in its stable form and dimensional stability is achieved.

Example 14.1: Mechanical property improvement

Figure 14.3 shows the effect of solution treatment followed by ageing on the hardness of LM25 alloy (Al-7%, Si-0.3-0.6%, Mg). As would be expected, as the ageing time is increased, the hardness increases as a result of precipitation, reaches a maximum and then decreases due to precipitate coarsening. The time to reach the peak hardness decreases as the ageing temperature increases, but this benefit is offset by a small decrease in the maximum hardness achieved.

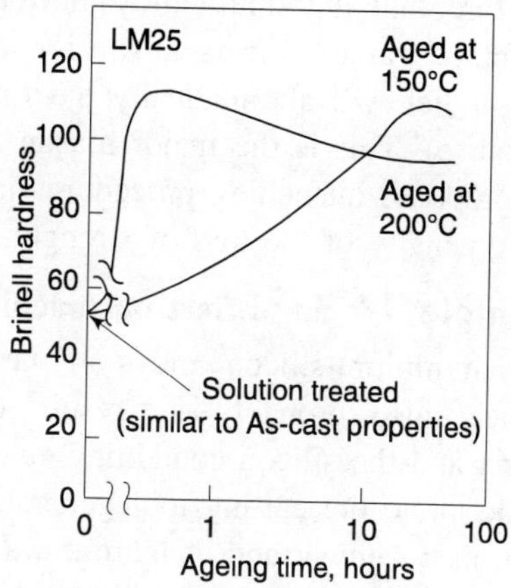

Figure 14.3 Heat treatment of castings improve mechanical properties

Figure 14.4 shows how the 0.2% proof strength and tensile strength vary with ageing time and can be seen to follow the same behaviour as the hardness. It can also be seen that the trend in ductility (measured as % elongation) is the mirror image of the trend in strengths.

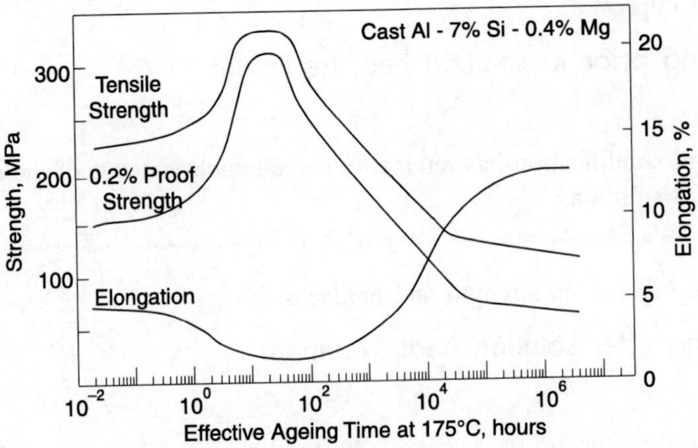

Figure 14.4 Proof strength and tensile strength variation with ageing time

Example 14.2: To relieve stresses

Larger castings are subject to the problem that some parts cool faster than others, so that stresses are developed during cooling in the mould, and particularly during cooling when water quenches after being held at a solution treatment temperature. In fact the quenching stresses usually are much more important than the casting stresses, especially when it is recalled that the casting stresses are normally relieved during the solution treatment, and the quench is the last rapidly cooled that the casting experiences.

It is important to note that for aluminium alloys, the normal ageing treatments are not carried out at a temperature sufficiently high for any useful stress relief to occur. It is unfortunate therefore, that aluminium alloys have to be drastically overaged, thus losing most of the strength advantage gained during any prior ageing treatment. Thus, castings which have to undergo stress relief will automatically have relatively poor strength and hardness (but good ductility, of course). This is the major reason for ensuring that stresses are not put into the casting by inappropriate quenching procedures in the first place. They can be subsequently reduced only at the penalty of the loss of strength.

Example 14.3: Effect on machinability and corrosion resistance

The aluminium-silicon alloys in the As-cast state machine rather poorly, exhibiting a poor surface finish, probably as a result of the large difference in hardness between the aluminium matrix and the silicon-containing eutectic phases. Also, because of the relatively high ductility of the alloy, the cutting tool generates a long curling chip which entangles the machine tool in an inconvenient and dangerous way. Heat treatment leads to a more uniform microstructural hardness and a reduced ductility, which results in a greatly improved machined finish and the swarf from the cutting action generates fine chips in convenient short lengths.

Secondly, it is possible to improve the corrosion resistance of most cast aluminium alloys as a result of solution heat treatment. This is especially noticeable in alloys which contain a small amount of copper. When fully in solution, this is relatively harmless from the point of view of corrosion. However, when present as an interdendritic phase, of composition perhaps $CuAl_2$, then a bimetallic corrosion couple is set up, in which the cathode is the copper-rich phase, and the anode is the surrounding aluminium-rich matrix, in which the rate of local corrosion is high, leading to rapid pitting in an aqueous environment containing oxygen in solution.

Where there is only a limited supply of oxygen in an environment isolated from the air (e.g., in a central heating installation) then the initial oxygen in solution is quickly consumed and not replaced, so that the corrosion is rapidly halted.

The copper content of the relatively pure Al-7% Si-0.5% Mg and Al-11% Si alloys is kept low to reduce these problems. However, a solution heat treatment will ensure that these problems are reduced further.

The aluminium-magnesium alloys also benefit from solution treatment by the removal of the beta phase (Al_5Mg_8) from grain boundaries. If these phases are not removed, the alloy is subject to stress corrosion cracking along the grain boundaries.

B. Heat Treatment of Magnesium Alloy Casting

Magnesium has a density two-thirds that of aluminium and only slightly higher than that of fiber-reinforced plastics and possesses excellent mechanical and physical properties (Table 14.1). When coupled with the inherent advantages of the metal casting process, magnesium alloys yield cost-effective solutions to product needs by allowing for part consolidation and weight savings over other materials and manufacturing methods.

Table 14.1 Properties of magnesium alloy

Property	*Unit*	*AZ91*	*AM60*	*AM50*	*AM20*	*AS41*	*AS21*	*AE42*
Ultimate Tensile Strength	MPa	240 (250)	225 (240)	210 (230)	190 (210)	215 (240)	175 (220)	230 (230)
Tensile Yield Strength (0.2% offset)	MPa	160 (160)	130 (130)	125 (125)	90 (90)	140 (140)	110 (120)	145 (145)
Compressive Yield Strength	MPa	160	130	125	90	140	110	145
Fracture Elongation	%	3(7)	8(13)	10(15)	12(20)	6(15)	9(13)	10(11)
Elastic Modulus, tension	GPa	45	45	45	45	45	45	45
Elastic Modulus, shear	GPa	17	17	17	17	17	17	17
Brinell Hardness		70	65	60	45	60	55	60
Impact Strength Charpy un-notched test bars	J	6 (9)	17 (18)	18 (18)	18 (18)	4 (16)	5 (12)	5 (12)

Note: Values in parentheses show mean property values obtained from separately diecast test bars.

The objectives for the heat treatment of the magnesium alloys are:

- To improve mechanical properties.
- To remove residual stresses produced during solidification.

Most of the magnesium casting alloys respond readily to heat treatment. Aluminium and zinc are the principal alloying elements in magnesium casting alloys. The solubility of these elements in temperature is shown below. It is this fact which makes magnesium alloys susceptible to heat treatment. Phase diagram for magnesium is shown in Figure 14.5.

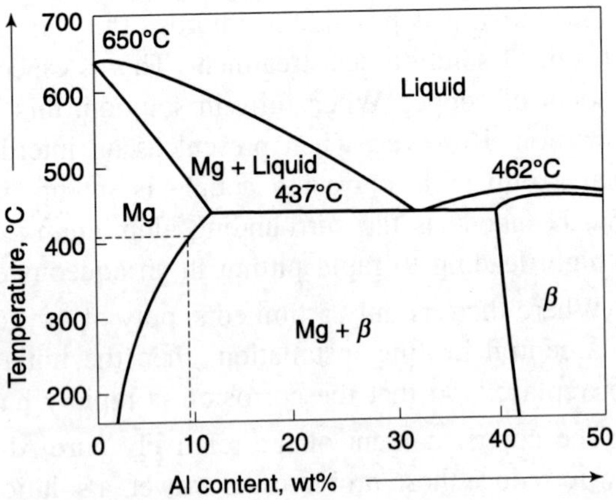

Figure 14.5 Phase diagram for the binary system Mg-Al

Magnesium casting alloys which are heat treatable usually have the following compositions:

1. 8% Al, 0.5% Zn, balance Mg
2. 10% Al, 0.5% Zn, balance Mg
3. 6% Al, 0.2% Mn, 3% Zn, balance Mg
4. 9% Al, 0.1% Mn, 2% Zn, balance Mg
5. 10% Al, 0.1% Mn, balance Mg

There are three types of heat treatment which may be given to magnesium casting alloys:

1. Solution heat treatment.
2. Precipitation hardening or artificial ageing.
3. Annealing or stress-relieving or stabilizing.

1. Solution heat treatment

Magnesium alloy castings are solution heat treated to improve tensile strength, ductility, and shock resistance. This heat treatment condition is indicated by using the symbol –T4 following the alloy designation. Solution heat treatment plus artificial ageing is designated –T6. Artificial ageing is necessary to develop all the properties of the metal.

Solution heat treatment temperatures for magnesium alloy castings range from 380 to 415°C, the exact range depending upon the type of alloy. The soaking time ranges from 10 to 18 hours, the exact time depending upon the type of alloy as well as the thickness of the part. Soaking periods longer than 18 hours may be necessary for castings over 50 mm thickness. Magnesium alloys must never be heated in a salt bath as this may result in an explosion.

A serious potential fire hazard exists in the heat treatment of magnesium alloys. If through oversight or malfunctioning of equipment, the maximum temperatures exceed, the casting may ignite and burn freely. For this reason, the furnace used should be equipped with a safety cut-off that will turn off the power to the heating elements and blowers if the regular control equipment malfunctions or fails. Some magnesium alloys require a protective atmosphere of sulphur dioxide gas during solution heat treatment. This aids in preventing the start of a fire even if the temperature limits are slightly exceeded. Air quenching is used after solution heat treatment of magnesium alloys since there appears to be no advantage in liquid cooling.

After solution treatment, magnesium alloys may be given an ageing treatment to increase hardness and yield strength. Generally, ageing treatments are used merely to relieve stress and stabilize the alloys in order to prevent dimensional changes later, especially during or after machining. Both yield strength and hardness are improved somewhat by this treatment at the expense of a slight amount of ductility. The corrosion resistance is also improved, making it closer to the "As-cast" alloy.

Precipitation heat treatment temperatures are considerably lower than solution heat treatment temperatures and range from 165 to 260°C. Soaking time ranges from 4 to 18 hours.

Process:

- In this treatment, a magnesium alloy casting is heated to temperature between 350°C and 430°C.
- Depending upon the exact alloy composition, it is heated for 16 to 20 hours and then air-cooled. By this temperature heating, the alloying elements will go into solution and give a single phase solid solution structure.

Outcome:

- By this treatment, the tensile strength ductility and toughness are improved whilst proof stress and hardness are little affected.

Precaution:

- During solution heat treatment, the temperature should not exceed the temperature at which the eutectic melts.

Magnesium alloy, after solution heat treatment, are stable; there is no tendency for them to further harden when left at room temperature.

2. Precipitation hardening or artificial ageing

Application:

- This treatment is carried out after solution heat treatment.

Process:

- It consists of heating the casting at temperature of 152 to 20°C for a suitable period (about 15 hours).
- Due to this treatment, the eutectic reappears as fine precipitate.

Outcome:

- This process increases the internal strength and hardness while ductility is somewhat reduced.

3. Annealing stress relieving stabilizing

Application:

- This treatment removes internal stresses which may be produced during casting.

Process:

- In this, the casting is heated at relatively low temperature of 220 to 300°C for 2 to 6 hours.

Outcomes:

- Internal stresses are removed.
- By this treatment, there is no major change in the mechanical properties.

Magnesium alloys are susceptible to oxidization by atmosphere oxygen while heat treatment in the neighbourhood of 400°C and above. The use of sulphur dioxide in atmosphere has been found to be helpful in preventing oxidization. Use of sulphur dioxide gas also minimizes fire hazards.

14.4 HEAT TREATMENT OF STEEL CASTING

The cast steel castings are brittle and possess poor mechanical properties. Heat treatment of steel castings is necessary in order to render them serviceable. The main objectives of heat treatment are:

- Refinement of the coarse structure of the cast metal.
- Relief of internal stresses.
- Development of desired physical and mechanical properties.

The heat treatment of steel casting depends on the following factors:

- Chemical composition
- Grain size
- Shape and size of the section
- Desired mechanical properties

The principal heat treatment operations applied to steel casting are:

1. Annealing
2. Normalising
3. Quenching and tempering
4. Flame hardening
5. Spheroidizing

1. Annealing

Process:

- Annealing consists of heating the casting slowly in a furnace to a temperature slightly above the critical range (about 900°C) for about 1 hour.
- Then it is cooled in the furnace itself.

Certain alloy steels require higher annealing temperature and holding time in order to effect solution and diffusion of alloying elements.

Outcomes:

- It relieves casting stresses.
- Refines coarse structure.
- Makes the product homogeneous in chemical composition.
- Machinability is improved.
- Annealing improved tensile yield strength and increases ductility.

2. Normalizing

Process: This treatment is similar to the annealing process, except for the fact that the castings are removed from the furnace after the holding period and allowed to cool in still air. Normalising and annealing comparison is given in Table 14.2.

Table 14.2 Comparison between normalising and annealing processes

	Normalizing	Annealing
Ductility	Same	Same
Strength	Higher	Lower
Removal of internal stresses	Less	More

3. Double normalizing

- Sometimes, the uniform grain structure is not obtained by single normalizing operation.
- This is achieved by the double normalizing operation which employs a second heating just to the upper critical temperature.
- Double normalizing also improves the ductility of steel castings.
- Sometimes, tempering treatment is given to remove internal stresses or to reduce the hardness of normalized steel castings.
- Tempering treatment involves heating the casting to a temperature below the critical range and allowing it to cool in the furnace.

4. Quenching and tempering

Application:

- It is usually confined to high carbon and alloy steel casting where high strength and resistance to impact and abrasion are required.

Process:

- The quenching and tempering treatment is done after the annealing or the normalizing operation.
- The annealed or normalized casting is reheated and quenched. To eliminate quenching stresses, tempering treatment is necessary.

Outcome:

- It can develop mechanical properties which are not obtained by annealing and normalizing.

5. Flame hardening

Application: Flame hardening is done for those parts which have to withstand extreme wear or abrasion.

Process:

- In the flame hardening treatment, the casting first is annealed or normalized.
- After this, the surface to be hardened is heated to the hardening temperature by torch or induction heating apparatus.
- After heating, the casting is quenched in water.

Outcome:

- Hardness increases on the casting surface.

6. Spheroidizing

Process:

- In this treatment the casting is tempered at a temperature just below the A_{c1} point due which the cementite collects into small spheroids surrounded by ferrite.

Outcome:

- This improves toughness and machinability of steel parts.

14.5 HEAT TREATMENT OF CAST IRON

Cast irons are based on the Fe-C system, and the solid-state transformations on which cast iron heat treatments are based are similar to those applied to steels. Cast iron is a generic term used to designate a family of metals with a wide variety of properties. All cast irons contain more than 2% carbon and an appreciable amount of silicon (usually 1-3%). The high carbon and silicon content means that they are easily melted, have good fluidity in the liquid state and have excellent pouring properties. The basic types of cast iron are best differentiated by their microstructure as opposed to their chemical analysis because the various types overlap (Table 14.3).

Table 14.3 Typical composition of unalloyed cast irons

Element	*Gray iron, %*	*Malleable iron, %*	*White iron, %*	*Nodular iron, %*
Carbon	2.5 – 4.0	2.00 – 2.60	1.8 – 3.6	3.0 – 4.0
Silicon	1.0 – 3.0	1.10 – 1.60	0.5 – 1.9	1.8 – 2.8
Manganese	0.25 – 1.0	0.20 – 1.00	0.25 – 0.80	0.10 – 1.00
Sulphur	0.02 – 0.25	0.04 – 0.18	0.06 – 0.20	0.03 max
Phosphorus	0.05 – 1.0	0.18 max	0.06 – 0.18	0.10 max

The metallurgy of cast iron is more complex than its economics and, in fact, is one of the more complex metallurgical systems (Figure 14.6). Iron-carbon alloys with less than 2% carbon are metastable; the true stable system being iron-graphite (Fe-C). The general term cast iron includes pig iron, grey iron, malleable iron, chilled iron, white iron, and nodular or ductile iron.

If an iron alloy exceeds about 2% carbon, the carbon does not have to nucleate from decomposition of austenite, but instead, it can form directly from the melt by a eutectic reaction. Note that cementite (Fe_3C) can still nucleate at the eutectic more readily than graphite, but on sufficiently slow cooling, graphite itself is able to form and grow.

Consider the solidification of a 3% carbon cast iron (Figure 14.7). At a rapid cooling rate, dendrites of austenite form as the alloy cools below the liquidus and grows until the eutectic temperature is reached.

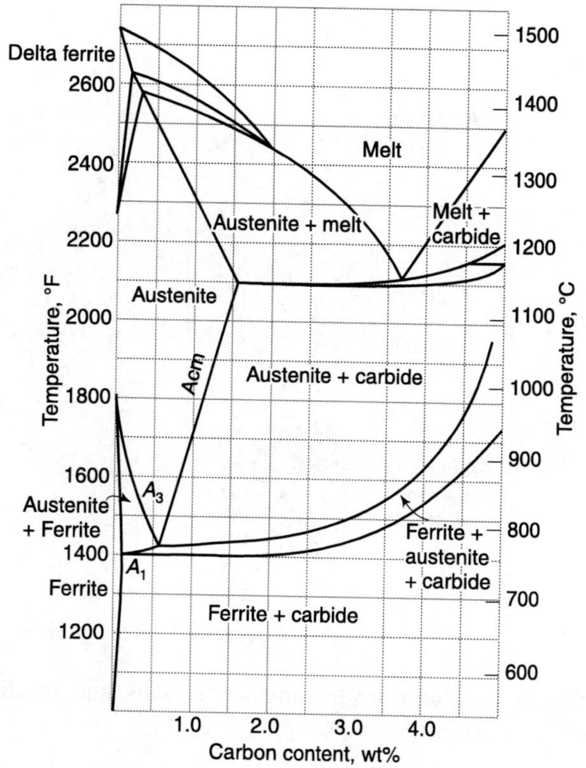

Figure 14.6 Iron-carbon diagram

At the eutectic, graphite formation is suppressed, but austenite and cementite precipitate to form ledeburite, a form of eutectic that consists of spheres of austenite embedded in cementite. Ledeburite forms at the Fe-Fe_3C eutectic (solid line "nm").

On further cooling, the cementite grows as the austenite decreases in carbon content (along the solid line "no") At the eutectic (point "o"), the remaining austenite transforms to pearlite. At room temperature, the iron is hard and brittle and is called white iron because the surface of a fractured piece of iron is white and (somewhat) lustrous.

Upon slow cooling of a 3% carbon cast iron, austenite forms from the melt, but eutectic freezing is now slow enough so the products of the eutectic reaction are austenite and graphite (the reaction takes place at the dotted line "nm"). The eutectic graphite tends to form flakes surrounded by eutectic austenite. As cooling continues, the austenite decreases in carbon content (along the dotted line "no"), while the remaining austentite transforms to pearlite. Because the fractured surface appears dull grey, the material is known as grey iron (or pearlitic grey iron).

Cooling at an extremely slow rate results in phase changes similar to those of a slow cooled component, except the eutectoid cooling is sufficiently slow to permit graphite to precipitate rather than pearlite. No new graphite flakes will form, but the ones present will increase in size. The final microstructure consists of graphite flakes embedded in a ferrite matrix. The resultant material is called ferritic grey iron (cooling of actual castings is seldom slow enough to obtain this structure).

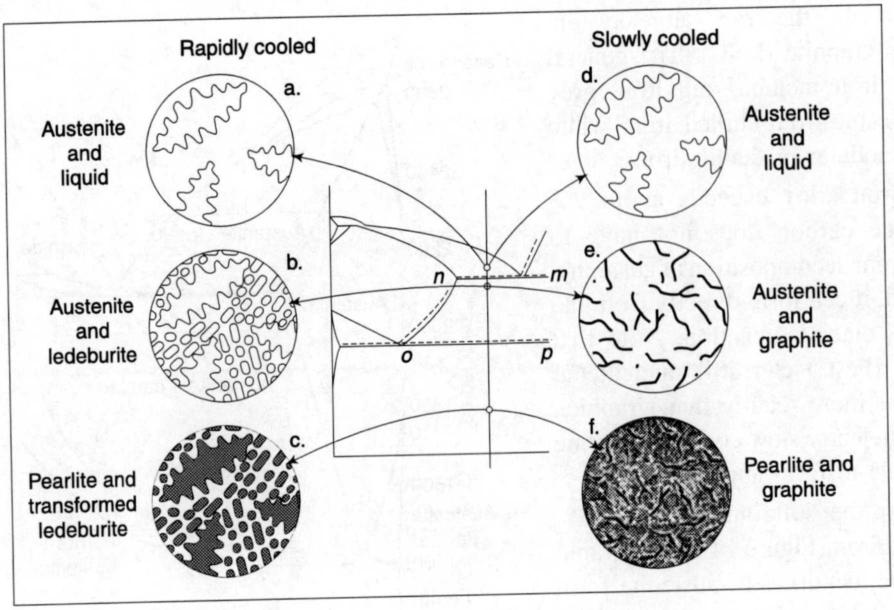

Figure 14.7 Iron graphite phase diagram

Microstructures are obtained by cooling and solidification of cast iron, white iron (left) and gray iron (right).

The cooling rate of a portion of a casting sometimes may vary, resulting in a structure containing patches of both white and grey iron, called mottled iron. Graphite forms in cast iron in several different shapes including flakes, nodules and spheroids (Figure 14.8). Because graphite has very little (cohesive) strength and reduces the effective metallic cross section of the casting, both strength and ductility are affected.

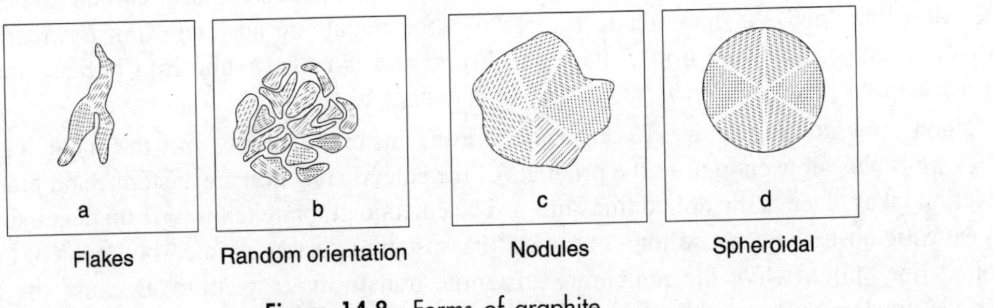

Figure 14.8 Forms of graphite

Pig iron is the term that is generally applied to a metallic product that contains over 90% iron. It contains approximately 3% carbon, 1.5% silicon and lesser amounts of manganese, sulphur and phosphorus. Pig iron along with scarp metal is the base material for both cast iron and cast steel. Grey irons are alloys of iron, carbon and silicon, in which more carbon is present than can be retained in solid solution in austenite at the eutectic temperature. The carbon precipitates as graphite flakes. The grey irons typically contain from 1.7-4.5% carbon and 1-3% silicon as major constituents.

The most common heat treatments applied to grey cast irons are stress relief because of non-uniform cooling of castings and annealing to improve machinability. Subcritical heating is used for both. Stress relief is done at temperatures between 550 and 650°C without significantly lowering strength and hardness. Heating at temperatures between 700 and 760°C lowers the hardness for improved machinability.

Nodular iron, also known as ductile iron or spheroidal graphite iron, is cast iron in which the graphite is present as tiny balls, or spherulites, instead of graphite flakes (as in grey iron), or compacted aggregates (as in malleable iron). The nodular irons typically contain from 3.2-4.1% C, 1.8-2.8% Si and up to 0.80% Mn as major constituents. Several types of matrix structures (including ferritic and pearlitic) can be developed by alloying and heat treatment. The various grades of regular, unalloyed ductile iron are designated by their tensile properties (Table 14.4).

Table 14.4 Common grades of ductile iron

Type (TS-YS-%EL)	*Tensile strength, ksi*	*Yield strength, ksi*	*Elong., %*	*Hardness, BHN*	*Heat treatment*	*Typical microstructure*
60-40-18	60	40	18	137 - 170	Annealed	All ferrite
65-45-12	65	45	12	149 - 229	-	Ferritic
80-55-06	85	55	6	179 - 255	-	Ferrite & pearlite
100-70-03	100	70	3	229 - 302	Normalized	All pearlite
120-90-02	120	90	2	250 - 350	Quenched & tempered	Tempered martensite

Heat treatment of ductile cast iron includes stress relief and annealing, as well as heat treatments used for steels including normalize and temper (for higher strength and wear-resistance), quench and temper (for the highest strength), and austempering. Ferritizing (for the most ductile microstructure) is done by austenitizing at 900°C, followed by holding at 700°C to completely transform austenite to ferrite and graphite.

Malleable irons are relatively soft and can be bent without breaking. They include ferritic (or standard) malleable iron and pearlitic malleable iron. Malleable iron is obtained by the heat treatment of white iron so the hard iron carbide structure of ledeburite is converted to a matrix of ferrite or pearlite and graphite is precipitated within the iron. This form of graphite is sometimes referred to as tempered carbon. A wide range of mechanical properties can be obtained in malleable iron with different matrix structures around the graphite.

Ferritic malleable iron has a ferrite matrix with interspersed nodules of tempered carbon and, depending on how it is produced, some combined carbon. Pearlitic malleable iron is designed to have combined carbon in the matrix, resulting in higher strength and hardness than ferritic malleable iron. The malleable irons typically contain 2.0-2.65% C, 0.90-1.65% Si and 0.25-0.55% Mn as major constituents.

Malleable cast iron can be heat treated to the same microstructures as ductile cast irons. Chilled cast iron is produced by casting the molten metal in such a way as to produce a surface virtually free of graphitic carbon. The surface of chilled iron castings is extremely hard. The depth and hardness of the chilled portion may be controlled by adjusting the composition of the metal.

White iron is a type of chilled cast iron that is virtually free of graphitic carbon, which is achieved by selecting a specific chemical composition to inhibit graphitization for a given section size. The hardness of the casting may be controlled by selection of composition.

In both chilled and white irons, the depth of chill decreases and the hardness increases with increasing carbon content. Carbon varies from 2.5% to above 3.5%. It is also necessary to control silicon content; the chill is reduced by increasing the silicon content. The range of silicon content for white irons is governed by the section size. Silicon content varies from 0.6 to 1.5%. Alloying elements such as nickel, chromium, and molybdenum are used to improve wear resistance.

The following heat treatment procedures are usually adopted for cast iron:

1. Annealing to relieve internal stresses
2. Annealing to improve machinability
3. Malleabilizing
4. Quenching and tempering
5. Surface hardening.

1. Annealing to relieve internal stresses

Process:

- This process involves heating of the castings slowly to a temperature between 500°C and 650°C, holding for a period corresponding to about one hour for each inch of section thickness and then cooling in the furnace.
- During this operation, graphitization takes place which reduces the hardness.

Outcome:

- It renders a high degree of stability and endurance to the casting.

2. Annealing to improve machinability

Process:

- Machinability of cast iron casting may be improved by causing graphitization of hard massive pro-eutectoid cementite.
- This is achieved by heating the casting to above the upper critical temperature (815-983°C) and cooling slowly in the furnace.

Disadvantage:

- Lowers the strength and hardness of the castings to an appreciable extent.

3. Malleabilizing

Malleabilizing is process for producing malleable cast iron. Malleable cast iron nodules of graphite are embedded in the ferrite matrix. There are two main malleabilizing processes as follows:

(a) Whiteheart or European process

(b) Blackheart or American process

(a) Whiteheart or European process

Process:

- In this process white iron castings are packed in boxes with a mixture of fresh and partly used hematite ore, heated gradually to 900-950°C held at this temperature for 5-6 days.
- It is then slowly cooled in the furnace.
- As a result, an oxidizing gas from the iron ore removes carbon from the skin of the castings through oxidation and there is a gradual migration of carbon from the heart of the casting to the skin.
- The process of decarburization continues until the total carbon content of the casting is reduced considerably.

Outcomes:

- This process produces iron having white and crystalline fracture.
- This process produces iron having less than 1 per cent carbon in thick sections.

(b) Blackheart or American process

Process:

- The white cast iron is packed with neutral substances like sand, inert slag, or mill scale to support them during the operation, thus minimizing distortion.
- They are slowly heated to about 900°C for 2-3 days, and slowly cooled.
- Due to inert nature of the packing mixture, little oxidation of the carbon occurs.

Outcomes:

- The blackheart process gives black or grey fracture due to pearlite.
- Carbon is precipitated as tempered aggregate in ferrite.

4. Quenching and tempering

- The quenching and tempering is done to improve hardness and abrasion resistance of grey cast iron.
- Other mechanical properties, however are not improved over the As-cast product.
- Response to hardening is increased by adding certain alloying elements. Addition of 4-5% nickel, for example, develop the martensitic structure in the castings.
- Oil may be employed as a quenching media, but with higher portions of alloying elements, air quenching may be employed.

- An austenitizing temperature in the range of 790-875°C is suitable for most grey iron grades.

5. Surface hardening

Application:

- Surface hardening is done for parts like gears, cams, and lathe beds to produce hard-wearing surfaces.

Process:

- The flame hardening process is quite common in which localized heating of castings is done by gas torches followed by quenching immediately by water jets.

Review Questions

1. What are the objectives of heat treatment?
2. Classify the various heat treatment processes.
3. Describe the effect of heat treatment on properties of castings.
4. What is case hardening?
5. Discuss various case hardening techniques.
6. What is tempering and what are its advantages?
7. What is normalising and what are its advantages?
8. Explain the annealing heat treatment processes.
9. Explain the spheroidizing heat treat processes.
10. Explain the malleabilizing processes.

CHAPTER 15

Casting Design Considerations

15.1 INTRODUCTION

While designing a casting, the responsibility of the engineer goes beyond proper function and service. Economy of production is usually an important factor and consideration of moulding procedure should be constantly incorporated in the development of design. Variations in foundry procedures and in the form and construction of pattern and core boxes should be thoroughly investigated to insure the most economical plan for producing castings of specified quality. Furthermore, operations other than those of the pattern shop and foundry must be considered, and the design of the casting and pattern equipment should be such that cost of all other operations is reduced to minimum. Such operations may include finishing or the elimination of finishing where possible, in whole or in part, assembly of this part to other parts and economical servicing during the useful life of the casting. When a casting is to be produced in large numbers, the final plan of production is determined only after many changes and alterations have been made. During such developments much apparatus is rendered obsolete, with constant monetary loss, but the production of greater quantities of castings assures an overall gain.

15.2 DESIGN RULES

Casting design poses two problems: one for the designer and other for the foundryman. Two designers should foresee the difficulties that might be encountered in the foundry. There are certain basic rules that could easily be followed by the designer:

1. Consult foundryman or pattern maker: explain strength requirement and service functions and where it fits into the total assembly.
2. Investigate the possibility of making part in section which later can be joined together. Rather than attempt too complicated casting. This applies particularly to design which includes large projections or extensions from the main body of the casting.
3. Make casting sections no thicker than essential to the desired unit of strength and functional requirements but of sufficient thickness to insure proper running of the metal in the mould. Avoid sharp angles and use gentle contours.
4. Design sections as nearly uniform in thickness as possible. Avoid abrupt section changes. Make transition gradual, blending the heavier sections into the light ones.

5. Design in a way that all the members of the parts increase progressively in thickness to convenient locations where feeder risers can be placed.
6. Bring the minimum number of sections together. Shrink and porosity troubles most often occur at member junctions. Stagger if possible.
7. In designing adjoining sections, replace corners with radii and avoid heat and stress concentration.
8. Avoid multiplicity of cores. The thickness of inner walls of the casting should be 70-90% of the outer walls, depending on the intricacy of coring.
9. Avoid intersecting ribs. Consider staggering or in the case of large casting, use simple cores at the intersections to equalize cross section of the junctions.
10. Employ ribs only where necessary to increase the strength or reduce the weight or avoid warpage.
11. Reduce bosses, lugs, pads and other projections to minimum. When necessary, design to conform to the thickness of adjacent sections.
12. The form of openings in all castings depends on the size of opening, stiffness or flexibility of the surrounding materials and type and distribution of loads. Contours with large radii of curvature of gradual changes are preferred. An elliptical opening may have advantages if the major axis is oriented correctly.
13. Machining allowances involve many factors and variables. Consult foundryman.
14. In the case of the complex designs, make three-dimensional drawing of model before making the final pattern.

Casting design and reduction of casting strain and distortion:

1. Avoid sudden changes of form producing a corresponding change in the direction of shrinkage.
2. Avoid reentrant angles.
3. Avoid multiplicity of cores. These expand under influence of heat and may offer rigid resistance to free shrinkage with resultant cracking of the casting.
4. Avoid widely differing section sizes, especially those in close juxtaposition, which might give rise to different rates of cooling.
5. Where internal casting strains must be minimum or where dimensional stability is vital, provide stress relief anneal.

15.2.1 Rule 1

Before issuing the final drawing, consult a competent foundryman and pattern-maker

It is all too common to design to suit the engineering department but not the foundry, and result may be failure or disappointment. From the foundryman's viewpoint, it is of first importance to design a casting that can be made rather than one that may be perfect engineeringwise, but which cannot be produced commercially and free from structural weakness.

Foundryman will permit consideration of the foundry problems that are likely to be encountered and will promote the making of sound casting. The time and cost of manufacture can also be considered at this preliminary stage of casting design. Important questions that foundryman answers include type of pattern needed, material shrinkage, and the moulding methods required. Condition necessary to make a dependable casting, machine finish and dimensional limitations.

The engineer must know how to design a casting so that it will actually have the requisite strength and functional properties. The foundryman must be able to make castings, so that they have strength and functional properites the engineer intended to present.

15.2.2 Rule 2

Construct the small model or visualize the casting in the mould

Relatively few engineers or foundrymen can follow all section changes and shapes from blueprint. Create a three-dimensional model using CAD software. This procedure permits the study of how the metal will enter the mould, how solidification proceeds, and shows what parts have to be fed to assure casting soundness.

15.2.3 Rule 3

Design for casting soundness

Most metals and alloys shrink when solidified. Therefore, design so that all members of the parts increase in dimension progressively to one or more suitable locations where the feeder heads can be placed to offset liquid shrinkage. The illustrations shown portray correct and incorrect methods of design. All of the rules set forth here have been proven in service and assure soundness of section.

15.2.4 Rule 4

Always present a cooling surface; avoid sharp angles and corners

Solidification of molten metal always proceeds from the mould face, forming unbalanced crystal grains that penetrate at right angles to the plane of cooling surface. A simple section presents uniform cooling and greatest freedom from mechanical weakness. When two or more sections conjoin, mechanical weakness is induced at the junction and free cooling is interrupted, creating a hot spot. Avoid sharp angles and corners. Engine cylinder parts are often designed to cause local casting weaknesses. Avoid core design which does not present a cooling surface. Streamlining of exterior resulted in heavy section at Y. Thin elongated point of core at A resulted in heat concentration which accentuated heavy section. Resulting hot spot causes leakage due to shrink defect. Redesign eliminates defect and results in solid casting of uniform strength properties.

15.2.5 Rule 5

Bring the minimum number of sections together. Staggered sections minimize the hot spot effects, eliminate structural weakness and reduce distortion. To prevent uneven cooling, bring minimum number of sections together or staggers so that more than two sections conjoin. A cored hole will help to speed up solidification where a number of sections can join. A circular web with adjoining sections is preferred.

15.2.6 Rule 6

Design all sections as nearly in uniform thickness as possible. Failing this, heavy sections should be accessible for feeding.

15.2.7 Rule 7

Proportion dimensions of inner walls correctly. Inner sections of casting, resulting from complex cores, cools much slower than outer sections and causes variations in strength properties. A good rule is to reduce the inner sections 0.9 times the thickness of the outer wall. Avoid rapid section changes and sharp angles. Wherever complex cores must be used, design for uniformity of section to avoid local heavy masses of metals.

15.2.8 Rule 8

The difference in the relative thickness of adjoining sections should be minimum and should not exceed a ratio of 2:1. Where a greater difference is unavoidable consider design with detachable parts, for example, the ways machine tool beds can be bolted, etc. When a change of thickness is less than 2:1 it may take the form of a fillet, where the difference is greater, the form recommended is that of a wedge. Wedge-shaped changes in wall thickness are to be designed with a taper not exceeding 1 inch. Where light and heavy sections are unavoidable, use proper fillets or tapering sections or both. If blending is not permissible, use fillets of fairly large size at junctions.

15.2.9 Rule 9

Fillets are used to reduce stress concentration in the casting in service. They also eliminate cracks, tears and draws at re-entry angle and make corners more mouldable to eliminate hot spots. Number of radii in fillets in one pattern should be minimum, preferably only one. To fulfill engineering stress requirement and reduce stress concentration, relatively large fillets are used with radius equalling or exceeding casting section.

15.2.10 Rule 10

Design ribs and brackets for maximum effectiveness. Ribs increase stiffness and reduce weight. Correct rib depth and spacing is a matter of engineering design. Thickness of ribs should be equal to 80% of casting thickness and should be rounded at edge and correctly filleted. Design preference in average design is for ribs to have a greater depth than thickness. Ribs in compression, in general, offer a greater factor of safety than ribs in tension. However, casting having thin ribs or webs in compression may require design changes to revive necessary stiffening to avoid buckling. Design ribs and brackets for maximum effectiveness; avoid cross ribs or ribbing on both sides of a casting. Avoid complex ribbing. It simplifies moulding procedure, assures more uniform solidification conditions and eliminates hot spots. Casting stresses and stress distribution favour omission of ribbing if the casting wall can be made of ample strength and stiffness itself. Avoid the use of ribs meeting at an acute angle.

15.2.11 Rule 11

Bosses, lugs and pads should not be used unless absolutely necessary

Bosses and pads increase metal thickness, create hot spots and cause open grain or draws. Blend into casting by tapering or flattening the fillets. Bosses should not be included in the casting design when the surface to support bolts, etc. may be obtained by milling or countersinking. A continuous rib instead of a series of bosses permits shifting hole location. Thickness of bosses and pads should preferably be less than the thickness of casting section they adjoin, but thick enough to permit machining without touching the casting wall. Where the casting section is light and does not permit use of this rule, then the following minimum recommended heights can serve as a guide. When there are several lugs and bosses on one surface, they should be joined to facilitate machining if possible, panel of uniform thickness instead of many pads of varying heights simplifies machining.

15.3 FOUNDRY PRACTICE AND REDUCTION OF CASTING STRAIN

General rules followed in foundry to reduce the casting strain are given below:

1. Select the metal having minimum contraction
2. Control pouring temperature
3. Use special moulding procedure
4. Distribute the gates uniformly
5. Avoid rigid cores and restrictions
6. Use moulding devices to obtain uniform contraction conditions

15.4 CASTING DESIGN AND COST REDUCTION

Following rules reduce the overall cost through proper design of castings.

1. Aim to simplify the design as much as possible. It may reduce pattern making cost, eliminate coreboxes and save in moulding expense.
2. Avoid use of loose pieces in pattern; where projections exist, put dovetail slide on pattern or provide corebox.
3. Avoid designs that involve irregular parting lines in the pattern requirement.
4. Avoid the use of deep pockets in the designs. Such design usually involves extra moulding and cleaning shop costs.
5. Do not specify dimensional tolerances closer than are necessary. This costs money.

6. Where lettering and special markings are required, see that they come on surface of casting parallel or nearly so with the mould parting.
7. Consult foundry superintendent if metal inserts or cores requiring the chaplets or stalks are necessary.
8. Dry sand cores are costly; patterns made to leave their own greensand cores in the mould are most economical to manufacture. Any part of the pattern which will not draw from the mould sand requires coring.
9. Where dry sand cores are necessary, make them as few and of simple forms as possible. Provide adequate core prints. In castings that must contain liquid or be pressure tight, design cores such that will not require chaplets.
10. Avoid the use of slender cores surrounded by heavy chunks of metals. Holes can be produced by cheaper and safer methods.
11. Provide vent outlets and clean the openings when using interior cores which are nearly or entirely surrounded by metal.
12. Small cored holes may prove costly, hence consult your foundryman. No core smaller than 25 mm diameter core for casting should not have a diameter less than half of the thickness of the wall.
13. Avoid using chaplets.
14. Patterns with straight parting lines, that is, with parting plane in one plane can be produced easily and at lower cost than those with irregular parting lines. Casting shapes, symmetrical about one centreline or plane readily suggest the parting. Such casting design simplifies moulding and coring and should be used wherever possible. Such design should always be made as spit patters. These require a minimum of hand work in the mould, improve casting finish and reduce costs.

15.5 DESIGN CONSIDERATIONS AND EXAMPLES

Castings are born on a designer's drawing board, but the final shape is often altered during the course of its development. Successful casting practice requires proper control of a large number of variables.

- Characteristics of the metal or alloys to be cast
- Method of casting
- Mould/die material
- Mould/die design
- Various process parameters.

For casting purposes, designing may include the following areas of planning:

- Functional design
- Simplification of foundry practices
- Metallurgical design; selection and optimum use of casting alloy
- Economic considerations

15.6 FUNCTIONAL DESIGN

Sometimes there is only one design which will perform the service intended by the designer. In this case, the foundryman may be required to use all the technical abilities available to produce the casting as it is, whether the design is favourable or an unfavourable one for foundry practices.

1. Mechanical strength: The factual relations between these assumptions and practical castings depend greatly on the casting design, foundry practices, and the nature of the alloy.

2. Accuracy of mechanical property information: Mechanical properties such as tensile, yield, compressive, torsional, shear and impact strengths; endurance limit; notch sensitivity; creep characteristics; hardness; elevated temperature strength; and modulus of elasticity or rigidity all need to have quantitative values for the casting alloys if designers are to employ calculations and empirical design formulas. Unfortunately, many handbooks provide only limited information about the properties of the cast metals.

3. Metal soundness and strength: Lack of metal soundness in a casting is one reason for lower than optimum mechanical properties. The foundryman, by using a sufficient number of foundry techniques such as gating, risering, chills, padding, and thermal gradients, can generally produce soundness even in exceedingly difficult cases of poorly designed castings. If certain design principles are observed, however, the job of producing soundness and uniformly good properties can be made much easier and less costly.

4. Directional solidification: Castings solidify toward areas of large mass or low SA/V ratio. If the last sections to freeze solidify slowly and fully feed those which freeze earlier, then soundness can be obtained by risering. Castings may be intentionally designed to favour directional solidification.

Figure 15.1 shows schematic examples showing correct arrangement to improve casting soundness.

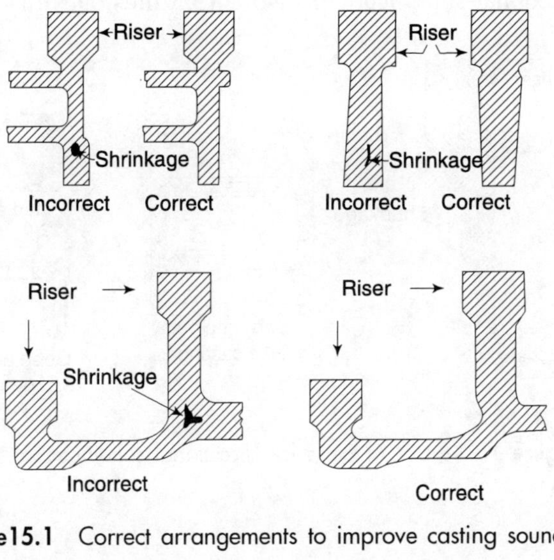

Figure 15.1 Correct arrangements to improve casting soundness

5. Columnar solidification: Because of the columnar growth of dendrites, planes of weakness can be developed during freezing. Figure 15.2 illustrates how columnar freezing develops directional weakness at sharp corners, in rectangular sections, and at perpendicular surface junctions. The use of generous fillets can eliminate this factor as a source of lower than optimum mechanical properties.

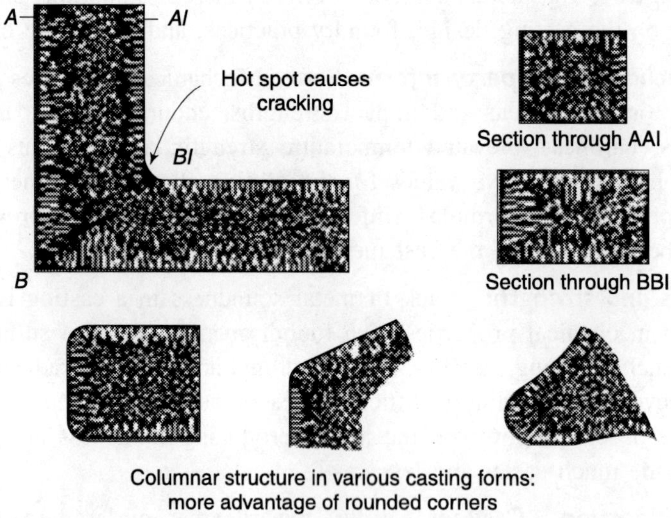

Figure 15.2 Planes of weakness caused by shape effects

6. Padding: Padding is applied to patterns to combat centreline shrinkage and to promote directional solidification. Padding is an excess material added to the casting to develop temperature gradient for directional solidification. Figure 15.3 illustrates the use of padding.

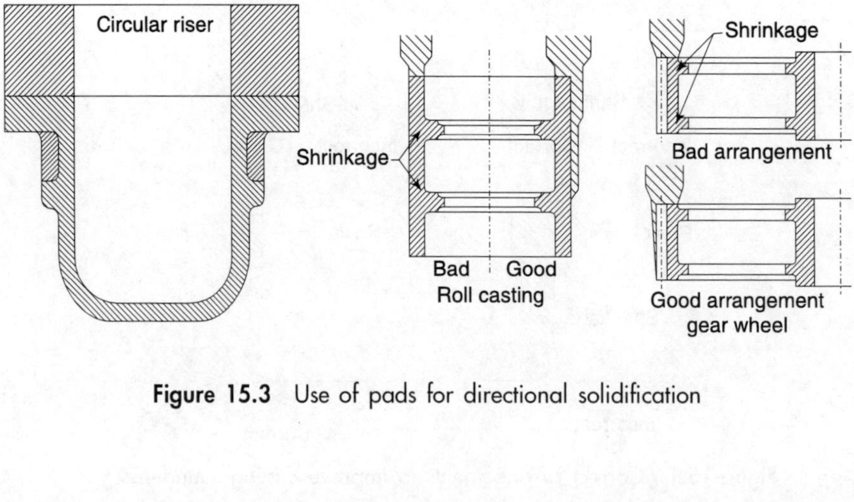

Figure 15.3 Use of pads for directional solidification

7. Mass shape and size effects (hot spots): Shrinkage cavities of any kind of course are harmful for metal properties. Certain shapes, because of their influence on heat extraction during solidification, are likely to cause shrinkage cavities. Whenever solidification is delayed at a particular location, that section will show a shrinkage cavity unless adequate feeding from risers occurs.

The L section shown in Figure 15.4 can be redesigned to improve its heat-dissipating characteristics.

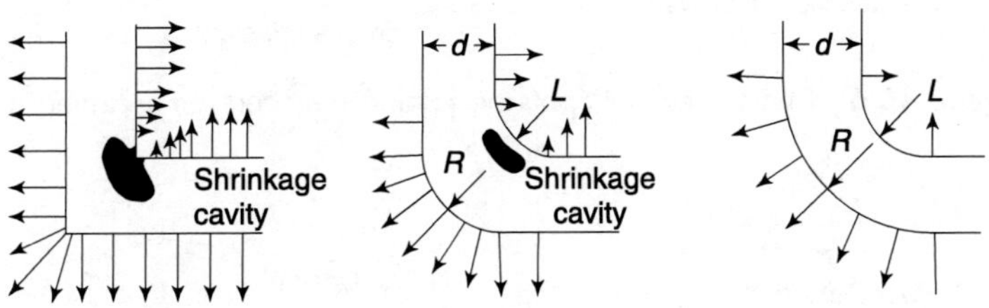

Figure 15.4 Evolution of an L junction to eliminate a hot spot

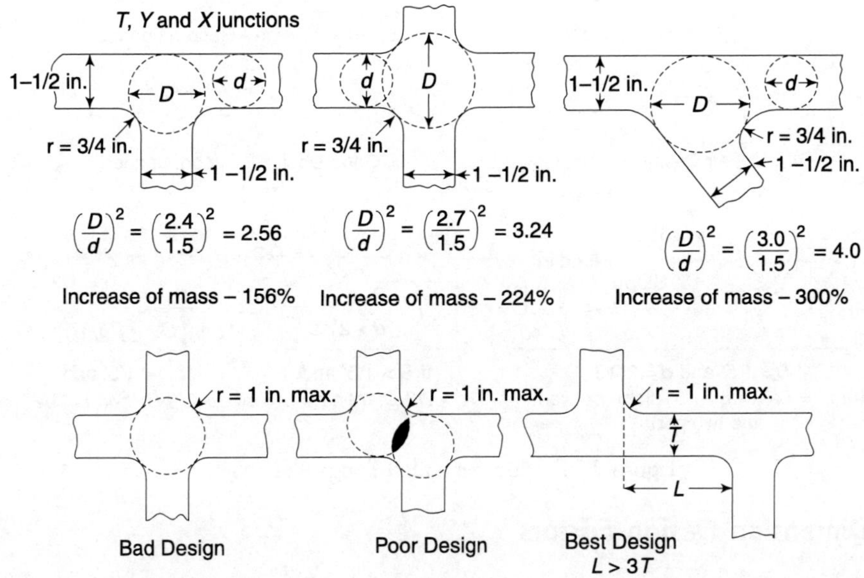

Figure 15.5 Basic design rules of X, Y and T sections

The application of the principle of minimizing hot spots in joining members of a casting is illustrated in Figure 15.6.

8. Section changes: Hot spots problem can be minimized by avoiding abrupt changes in section. Figure 15.7 illustrates section blending and recommends straight uniform section rather than any change. By using tapered sections, however it is possible to achieve soundness more readily than with the perfectly uniform sections.

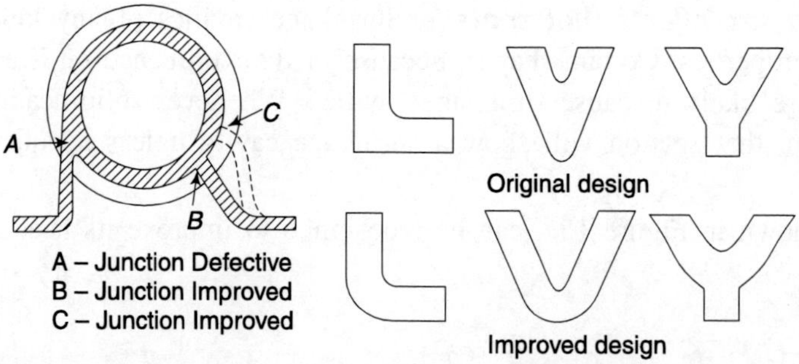

Figure 15.6 Minimizing hot spots on joining members of a casting

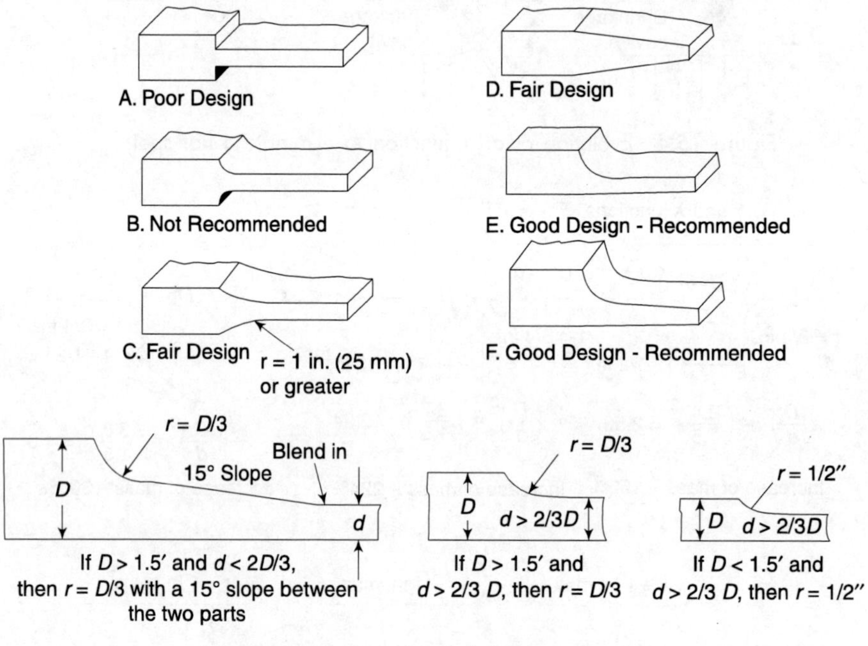

Figure 15.7 Recommended flange designs

15.6.1 Dimension Design Factors

Functional design of castings must take into account the ability of various casting processes to produce the desired dimensions. Certain dimensional characteristics result from the nature of the casting processes.

Minimum section thickness

Minimum section thickness which may be cast in sand should be as per guidelines. For instance, grey irons may chill to white iron if any extended sections of 1/8 inch. are employed. Sections of low-carbon-equivalent irons under 1/2 inch. may cast too hard for machining. Furthermore, the thin section can be run only a certain limiting distance without having misruns appear on the casting.

Cored hole sizes

The minimum size of cored holes which can be cast depends greatly on the accuracy of core location and tolerance required. If a cored hole must be located with extreme accuracy with respect to other surfaces, it is often desirable to drill it rather than core it. However, dimples may be put on the casting surface to locate the drill holes.

Dimensional tolerances

Certain surfaces may necessarily be machined. These require a machining allowance as well as allowances for shrinkage and perhaps drafts as they finally appear on the pattern. In greensand moulding, dimensions ultimately obtained on the casting are the result of the faithfulness with which shrinkage, changes in dimensions of mould cavity, hardness of mould, stability of moulding sand, mechanical alignment of flasks, temperature effects, and so on, are reproduced in the casting process. Of course, many of the variables related to sand moulding are not encountered in die or permanent mould casting.

Surface finish

Surface smoothness of mould cavities made with sand depends on sand fitness number, hardness or denseness of mould, materials in sand, moulding pressure, and other foundry variables.

15.7 SIMPLIFICATION OF FOUNDRY PRACTICE

The steps of moulding, coring, and cleaning may be greatly simplified if the casting is designed with a view to the needs of these operations.

Moulding and coring

Any changes in the casting design which make moulding easier or less costly or result in improved moulds help to make better foundry practice. Some of the important factors are discussed below.

Pattern drawing

Any casting shape that makes patterns more difficult to draw or moulds more complex is to be avoided if possible. Undercuts or protruding bosses, flanges, etc., which are above or below the parting line require the use of cores or loose pieces which might be eliminated with a better design. Of course, the uses of maximum draft on patterns improves drawing. On many

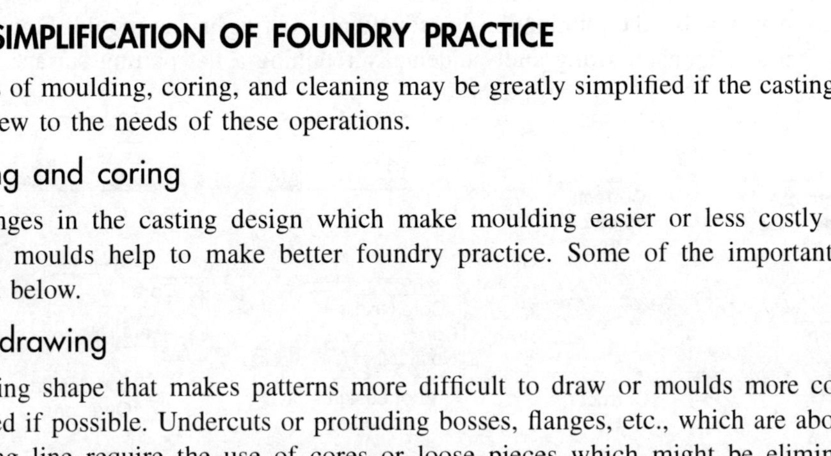

Figure 15.8 Redesign with deep pockets to minimize tearing of mould during pattern drawing

castings, it is completely unnecessary to have vertical inner or other surfaces and generous use of draft is possible. Especially where the casting must have some deep pockets, they can be more easily moulded. Figure 15.8 shows redesign with deep pockets to minimize tearing of mould.

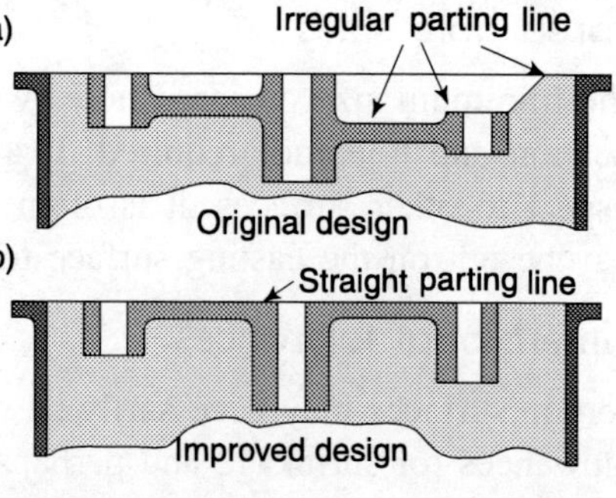

Figure 15.9: Redesign of casting to make straight parting line to avoid defects

Proper selection of parting line

A part should be oriented in a mould so that the large portion of casting is relatively low and height of the casting is minimized. In general, the parting line should be along a flat plane rather than contoured. The parting line should be placed as low as possible relative to casting for less dense material such as aluminium alloys and located at around mid-height for denser material such as steel. Figure 15.9 illustrates the redesign of part to get horizontal parting line.

Elimination of coring

Cost of mould construction with cores is generally greater than with green sand alone. Castings are not often originally designed with the elimination of coring in mind. However, redesign for greater economy may involve the elimination or reduction in coring in the mould. Figure 15.10 illustrates redesign to eliminate coring and incidentally retaining a flat parting surface.

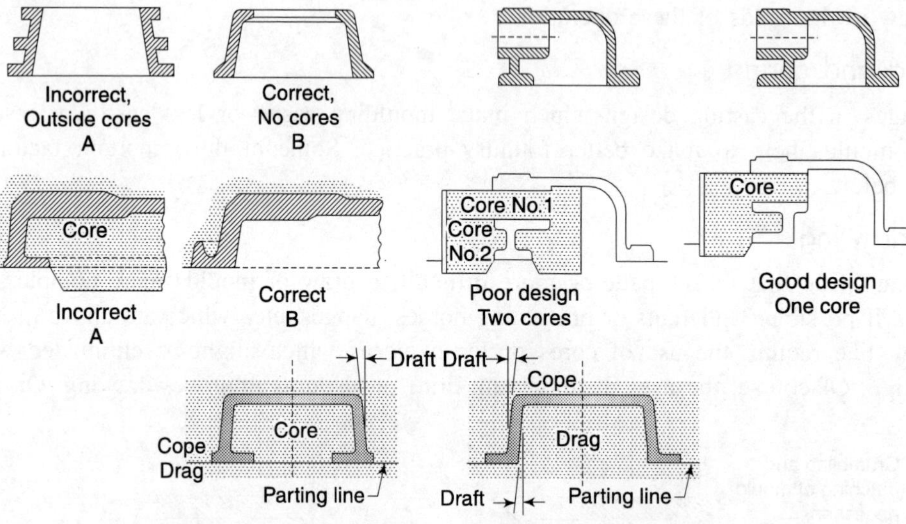

Figure 15.10 Casting design to eliminate/minimize coring

Feeding distance

Ribs are provided to increase rigidity and decrease weight. Effect of feeding distance on defect is shown in Figure 15.11. Directional solidification plays an important role as casting solidifies towards areas of large mass or low SA/V ratio. Ribs will improve the directional solidification and balanced web is still better.

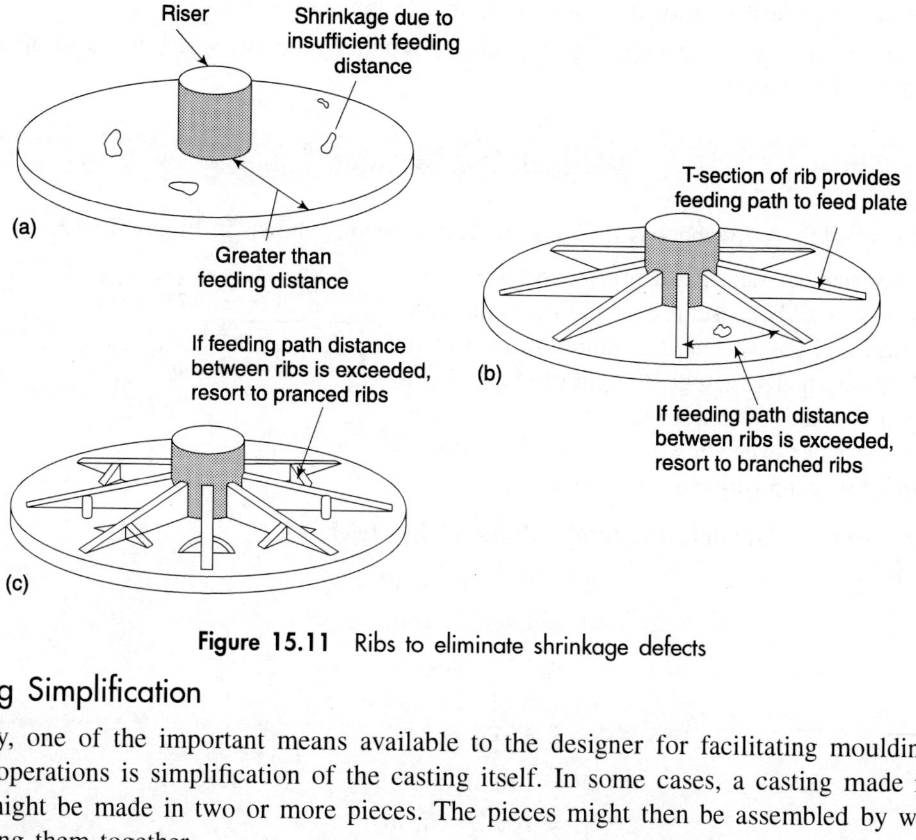

Figure 15.11 Ribs to eliminate shrinkage defects

Casting Simplification

Actually, one of the important means available to the designer for facilitating moulding and coring operations is simplification of the casting itself. In some cases, a casting made in one piece might be made in two or more pieces. The pieces might then be assembled by welding or bolting them together.

Shape and size

Compactness in design favours more economical moulding practices. A casting which is unnecessarily long or has portions which stick out considerably from the main body of the casting requires large flasks, or they may be unduly limited in number of casting per mould. The compact design permits more castings per mould and requires less moulding sand per casting. Compactness of a design does not imply a large concentrated mass, but simply the elimination of unnecessary extensions attached to the main body of the casting.

15.8 METALLURGY DESIGN

Not commonly recognized as a facet of design procedure is the intentional and purposeful application of metallurgical principles for developing optimum properties of alloys in castings.

Areas of metallurgical design may include:

- Selection of the most suitable alloy.
- Effects of the casting process.

  - Importance of thermal effects, heat dissipation, mass, etc.
  - Relationship to metal properties.
- Importance of heat treatment in producing the desired properties.
- Special metallurgical processing, techniques of foundry practice, which assist in obtaining the desired properties.

Worked Out Example

Example 15.1: A flywheel is designed with dimensions shown in Figure 15.12. The designer has decided that a green sand grey iron casting will be used because of cost and machinability consideration and has specified a minimum of 40000 psi tensile strength. The actual strength of each wheel will be proof-tested by requiring that it be spun at 5500 rpm, the latter exceeding any service conditions. What grey-iron composition should be used?

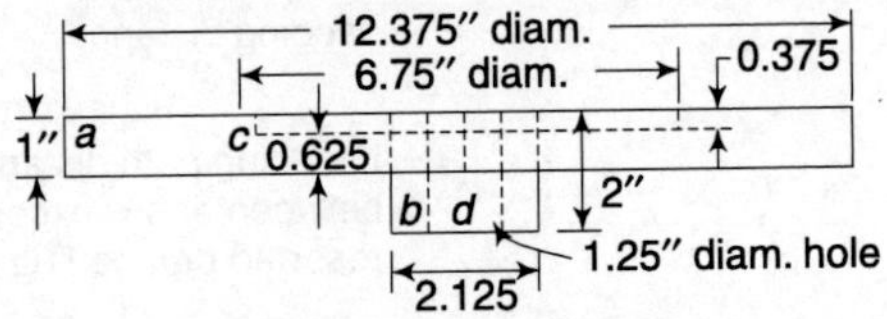

Figure 15.12 Flywheel

Solution: **Step 1 Calculate the total volume of flywheel**

$$V = V^a + V^b - V^c - V^d$$

$$V^a = 120.30 \text{ cu. in.}$$

$$V^b = 3.545 \text{ cu. in.}$$

$$V^c = 12.10 \text{ cu. in.}$$

$$V^c = 2.458 \text{ cu. in.}$$

$$V = 120.30 + 3.55 - 12.10 - 2.46$$

$$= 109.29 \text{ cu. in.}$$

Step 2 Calculate the total surface area of flywheel

$$\text{SA} = \text{wheel} + \text{hub}$$

$$\frac{\pi \times 12.375^2}{4} \times 2 = 240.50 \text{ sq. in.}$$

$$2\pi \times \frac{12.375}{2} \times 1 = 38.90$$

$$2\pi \times \frac{6.75}{2} \times .375 = 7.95$$

$$2\pi \times \frac{2.125}{2} \times .375 = 2.50$$

$$2\pi \frac{2.125}{2} \times 1 = 6.66$$

$$2\pi \times \frac{1.25}{2} \times 2 = 7.85$$

Total SA in sq. in. = 304.36

Ratio SA/V = 304.36/109.29 = 2.78

According to the calculated SA/V ratio of 2.78, a class 40 unalloyed grey iron or class 40 steel can be used.

A final approach to the solution of the problem described above would be to make test castings of the several grey irons selected as possibilities and subject them to mechanical-properties studies. Samples for mechanical property tests could be cut directly from the casting and their properties correlated with separately cast test bars.

Review Questions

1. List all the design rules used in design of castings.
2. What are the general rules followed in foundry to reduce the casting strain?
3. What is functional design?
4. Explain directional solidification and how does it affect quality of casting?
5. What is columnar solidification?
6. Why is straight parting line preferred in casting design?
7. What are the design considerations for pattern to get good mould?
8. Explain the role of feeding distance in solidification of casting.
9. Explain with an example how the core can be eliminated.
10. Explain how ribs can be used to reduce shrinkage.

CHAPTER 16

Foundry Modernisation and Mechanisation

16.1 INTRODUCTION

Foundry layout shows the plan view of the various sections of the foundry. It indicates the activities like material movement, direction of flow, detail of material storage, location of moulding, melting, cleaning, inspection, etc. It also indicates the size of the foundry. It indicates the nature of foundry such as mechanized or unmechanised. A good layout is one which maximizes the production for the given shop floor area that uses least material transportation for a given tonne of casting produced.

- A well laid out foundry is able to carry out the activities with least interference in the material flow.
- It promotes quality in the casting.
- It promotes higher productivity.

16.2 TYPES OF FOUNDRY

The foundry layouts can be generally classified as

- Jobbing foundry
- Captive foundry
- Mechanized foundry

Jobbing foundry

This type of foundry caters to the need of jobs as and when orders are placed. The job items keep on changing every now and then. The quantity required is not very large. Not much stress is laid on the severe quality standards. Simple and general types of castings are produced. Normally, hand moulding is employed. Invariably greensand moulding is used. Production may not be there for all the shifts in a day. The layout will be small and simple. Most of the activities are carried out manually only.

Captive foundry

This type of foundry caters to the need of its own. The foundry is normally a part of a parent organization and will supply the castings against its requirement. A foundry producing automobile castings to meet the requirement of its parent organization manufacturing trucks/buses, etc., is an example of a captive foundry. No external orders are entertained. Orders are repetitive in nature and the production is continuous.

The layout is streamlined. Machines are used to make the moulds and cores, for cleaning the castings, transporting the casting furnaces like arc furnaces or induction furnaces are used in general. Productivity is high.

Mechanised foundry

Mechanization refers to use of machines for carrying out various foundry activities. The human skill or effort or human interference is minimized. Better consistency in the quality of the casting is assured. Mechanization ensures higher productivity. It makes handling of heavy castings and other materials easy. Monitoring of various activities is simplified and made easy.

16.3 ADVANTAGES OF A GOOD LAYOUT

- Equipment, materials, etc., are confined to the correct place, easily accessible to the cranes/handling.
- Material movement will be smooth and crisis crossing is avoided (directional flow of major raw materials realized).
- The distance covered by the raw materials for processing is the least.
- Minimum energy is consumed in material handling.
- Maximum safety is provided for the persons and to the equipment.
- Accidents occur rarely.
- Productivity is high.
- Easy monitoring of the activities.
- Effective use of floor space.
- Effective utilization of manpower.

16.4 SOME CONSIDERATIONS FOR FOUNDRY SITE LOCATION

- Proximity to the power lines or source of power (coke, gas), etc.
- Sufficient space for the layout and for storing large quantities of scrap, sand and other raw materials.
- Away from the populated areas.
- Annual tonnage.
- Type of casting
- Range of casting in terms of size and weight.
- Type of metal handled and product mix.
- Type of moulding, melting, fettling and finishing.
- Provision for future expansion.
- Material storage space.
- Facilities for employees.

- Availability of water source.
- Availability of space for dumping wastes like slag, burnt sand, used refractories, etc.
- Proximity to road/rail transportation.
- Sufficient auxiliary space for future expansion.
- Handling equipment like electric overhead transport (EOT) cranes, conveyors should have adequate capacity.
- Standby arrangement for cranes that handle metals for tapping and pouring is necessary.
- Selection of capacities of production and auxiliary equipment must balance different stages of production process.
- The sand to metal ratio to be handled in a steel foundry is 5:1. Hence, the volume of the sand to be handled is very large. Figures 16.1-16.3 shows layout for different types of foundries.

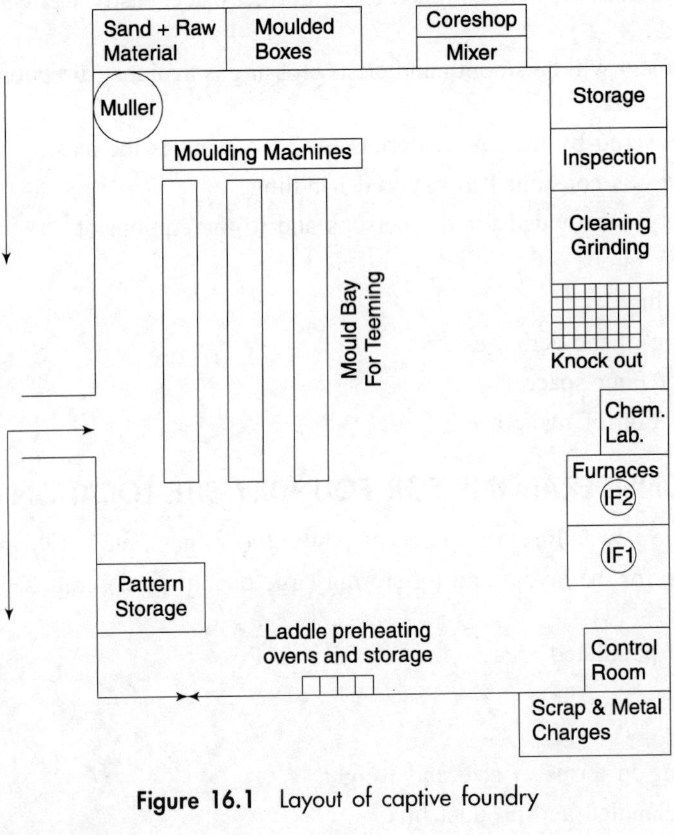

Figure 16.1 Layout of captive foundry

16.5 MODERNIZATION

Modernization refers to updating of technology with respect to machine/equipment, training of personnel/ and processes, etc.

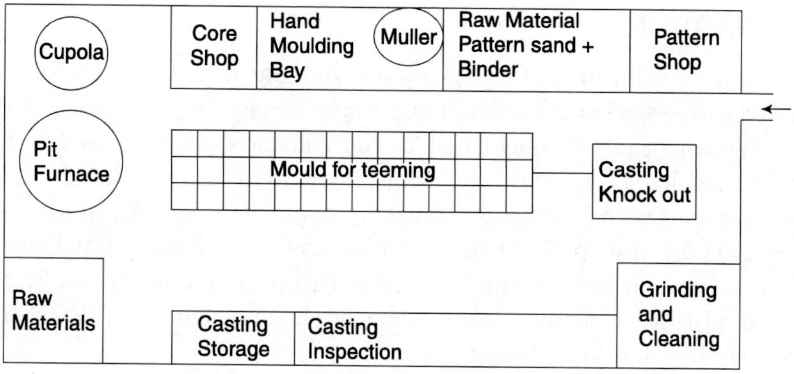

Figure 16.2 Layout of jobbing foundry

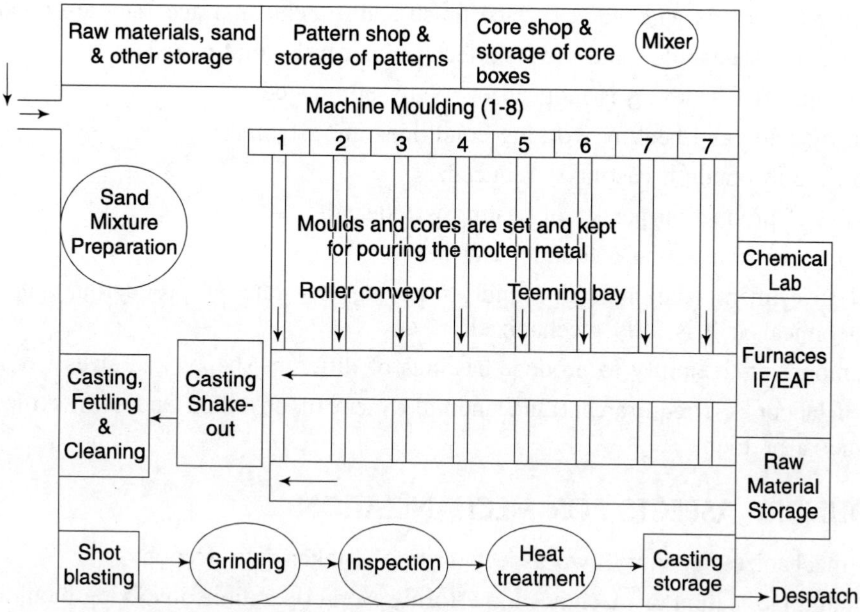

Figure 16.3 Layout of mechanical foundry

The emphasis is on improving the productivity, effective use of machinery and equipment. Economizing the product through energy conservation and streamlining the activities. Hence, need is felt for modernization.

16.6 MECHANISATION

Mechanization means the utilization of machinery for work previously done by hand. Machinery may be used for preparing sand and moulds, transporting the sands and moulds, and transporting and pouring the molten metal. Production of castings in large numbers at low cost can be achieved by mechanization of foundry operations using machineries. Material handling and pollution control are the two primary areas generally considered for this. Modern methods of industrial management are also applicable for optimization of the operations and quality control. By proper mechanization and control with efficient management, a foundry can be converted to a neat, smoke- and dust-free modern industry producing large quantities of defect-free castings at lower cost and reduced unskilled labour content.

Advantages and Limitations of Mechanization

There are many advantages of mechanizing a foundry.

1. The production rate is highly increased for the same floor area of the factory.
2. Defect-free castings with good surface finish and dimensional accuracy are obtained.
3. The number of unskilled and semi-skilled labour can be minimized.
4. Manual material handling is reduced or totally eliminated.
5. Clean, hygienic and healthy working conditions are established.
6. Cycle time to obtain a casting is reduced.
7. The cost of production per kg of casting is reduced.

The limitations of mechanizing a foundry can be:

1. Initial investment is high, and foundries having low rate of production thus become uneconomical, if it is fully mechanized.
2. The amount of flexibility to produce castings of different shapes and sizes is reduced.
3. Skilled labour is a requirement and unemployment of unskilled and semi-skilled labour becomes a problem.

16.7 FOUNDRY ASPECTS FOR MECHANIZATION

In order to mechanize a foundry, two aspects are to be taken into account:

- Design and fabrication of machines to perform several operations involving moulding, core making, mixing, etc.
- These machines have to be matched with material handling equipment so that there is continuity in the process.

Normally, it is estimated that about 50-200 tonnes of material is handled for every tonne of casting produced.

Hence, importance is given to materials handled to accomplish mechanization. Production results from the working of men, material and machinery. One or several raw materials go in making the product with the help of machinery and equipment. To achieve the final product, cost at various stages is incurred. Cost of the product is a function of manufacturing activity, materials handling activity, movement of men and machines, maintenance, etc. It is estimated that handling costs alone could contribute anywhere between 15-75% to the total manufacturing cost. Hence, it is essential that economy of a good arrangement of facilities should be ensured.

16.8 AREAS OF FOUNDRY MECHANIZATION

Mechanization is use of machines, either wholly or in part, to replace human labour. Unlike automation, that may not depend at all on human participation to provide information or instruction. Mechanization began with human-operated machines to replace the hard work of craftspeople. In metal casting industries, following areas are mechanized:

- Preparation of moulding sand
- Moulding and core making
- Melting and pouring
- Shakeout and cleaning
- Material handling

16.8.1 Preparation of Moulding Sand

In a sand casting unit, sand is the major bulk material in volume, which is to be processed and carried on a foundry floor. In a mechanized foundry, all the operations to prepare the mixture are done using machines only. Figure 16.4 shows the flowchart for preparing moulding mixture in a mechanized foundry.

16.8.2 Moulding and Core Making

To meet the requirement of producing large number of moulds of the same design with proper finish and accuracy, the best method followed in a mechanized foundry is machine moulding. This makes it possible to eliminate manual labour in two main moulding operations such as sand ramming and pattern extraction.

Moulding machines

In a mechanized foundry, the following types of moulding machines are usually used. The selection of a suitable machine depends on the size of casting, shape intricacy, production rate, etc.

(a) Classification by sand compaction method:

(i) Squeeze-moulding machines

(ii) Jolt-moulding machines

(iii) Jolt-squeeze moulding machines

(iv) Sand slinging (throwing or shooting) machines

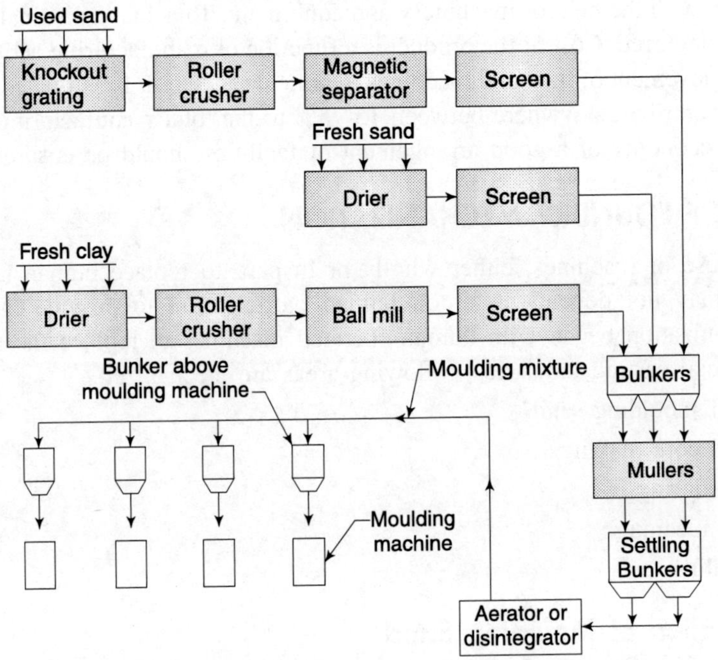

Figure 16.4 Flowchart for preparing moulding mixture

(b) Classification by pattern extraction method:
 (i) Pin-lift moulding machines
 (ii) Pattern-draw moulding machines
 (iii) Roll-over (turn-table) moulding machines
 (iv) Rock-over (turn-over) moulding machines

Core making machines

Core making section is kept as separate block or bay in a foundry shop. The cores are made in coreboxes in which sand is compacted by hand hammers or in machines. Core blowing method is very common for large-scale production of cores.

16.8.3 Melting and Pouring

The moulds, after assembly, are ready for pouring and they are filled with molten metal one by one continuously. This requires continuous supply of molten metal at the pouring temperature. In a mechanized foundry, the following are the functions involved in melting and pouring:

1. Supply of metal, fuel, flux continuously to the furnace using mechanisms.
2. Operation of furnaces and its control.
3. Feeding of raw materials and fuel to the furnace using automatic systems.
4. Collection of molten metal from furnaces and transporting to the pouring area.

5. Operation of holding furnaces to keep the pouring temperature ready.
6. Pouring of moulds using mechanisms.
7. Transportation of moulds from assembly area to the pouring location and then to the shakeout section after cooling.

Raw materials for melting are moved from stores to furnace by belt conveyors, bucket conveyors or overhead cranes. The molten metal obtained from the furnace is collected in high capacity crane ladles (conical, drum type, tilting or bottom-tap) and transferred into intermediate ladles, which is placed on stands in the casting section. The metal is then poured into casting ladles of lower capacity. The capacity of these casting ladles ranges about two to ten times the metal required in one mould. The conical ladles of tea-pot type may have capacity up to 800 kg to make small steel castings.

Automatic pouring machines are used in mechanized foundries for mass production. They can be grouped into electromechanical, pneumatic, electromagnetic and combination of these types.

16.8.4 Shakeout and Cleaning

After filling molten metal, the mould with casting is carried to the shakeout section and on the way it is cooled. The sand moulds are broken down at a temperature below 500°C. The castings taken out from the mould are cleaned and the extra projections like runner and riser are removed. The machines or equipment generally used for various shakeout and cleaning operations are given below:

1. Shakeout of castings:
   (a) vibrating grid type shakeout machines
   (b) hydraulic or electro-hydraulic core shakeout machines
2. Removal of runner and riser:
   (a) machines for abrasive wheel slitting
   (b) shearing machines
   (c) flame cutting machines
   (d) turning, shaping and milling machines
3. Surface cleaning of castings:
   (a) tumbling machines
   (b) air blast cleaning equipment
   (c) shot blast cleaning equipment
   (d) hydro-blast cleaning equipment

Depending on the shape of the casting and requirement of finish of the surface, the above machines are selected and installed in the operation line. Conveyors of various types are employed to carry the castings to the workstations. Heat treatment is done after cleaning the castings. Each casting is then passed through the inspection wing for conforming its quality as per the requirement. Accepted castings are moved to the store, while the rejected ones are taken back to raw material bay for remelting.

16.8.5 Materials Handling

Costs in metal casting operations account for 20 to 40% of the total cost of producing castings. The efficiency of manufacturing system depends upon the ability to move the parts/materials effectively, timely and rapidly. Following are the commonly used material handling equipment:

1. Factory trucks
   (a) manual-wheeled type
   (b) fork-lift type
   (c) pallet life type
2. Conveyor
   (a) belt conveyor
   (b) bucket elevator
   (c) chain conveyor
   (d) roller conveyor
   (e) apron conveyor
   (f) flight conveyor
3. Cranes
   (a) jib crane
   (b) monorail crane
   (c) overhead crane
   (d) gantry crane
4. Hoists
5. Tractors and trailers
6. Robots

16.9 POLLUTION AND ITS CONTROL IN FOUNDRIES

Foundries are among the industrial plants causing environmental pollution, producing substantial quantities of air pollutants. The numerous processes available for moulding, melting and casting are accompanied by evolution of heat, noise, dust and gases. Dust, fines from fly ash, oxides, etc., which form particulate matter are generated in large quantities when preparing core sands and moulds, melting metals, pouring moulds, knocking out poured moulds and loading and unloading raw materials. Gaseous matter like gases, vapours, fumes and smoke are produced during melting and pouring operations. The major pollutants emitted from various work areas in a foundry are given here. The basic means of controlling the emission of pollutants and changing the production process, supplying adequate make-up air, proper aeration and ventilation of the shop, reduction of pollutants at source by taking appropriate control measures, dispersion and dilution of pollutants in the air space and good housekeeping. Areas of pollutions are described here:

- Pattern shop: Where wooden patterns are handled or prepared, saw dust, wood chips are generated which when inhaled possess health problems.
- Sand preparation for moulds and cores: Fine dust from sand, fines from bentonite dextrin, and other additives are generated. Some organic chemicals are used which emit harmful gases/vapours.
- Mould, core-making and mould setting: Sand dust, evolution of gases/vapours and of binders and chemicals.
- Melting shop: Noise of loading/changing of furnace of scrap. Smoke during changing and heating. Noise during furnace operation. Heat radiation, dust, metallic oxides, gases like CO, CO_2 and SO_2, etc. While melting in cupola, the total quantity of emissions that is making pollution is assumed to be 22 kg/tonne of metal cast, and the size of particles may be 1 to 500 microns. The gaseous emissions from the cupola are composed of CO and smaller amount of SO_2 and H_2. The emissions from electric furnaces are less compared to cupola. The amount of dust exhaust from arc furnace is the highest among the electric furnaces, and it varies from 7 kg/tonne to 8 kg/tonne of steel melted. The same in induction furnace is small as 0.12 kg/tonne to 0.7 kg/tonne of metal. A similar important area for consideration is the pouring, shakeout and preparation regions. Gas containing CO_2, CO, N_2, is generated at the time of pouring and is continued till shakeout of the mould. The dust concentration at the shakeout and cleaning section reaches 6 gm^{-3} to 20 gm^{-3} in the nearby vicinity of the points.
- Teeming (pouring and cooling of molten moulds of metal): Gases due to burning of sand binder additives, exothermic compounds, oil vapours, heat radiation.
- Fettling/knock out: Metallic dust, sand dust, gases, noise.
- Heat treatment: Gases like CO, CO_2, SO_2, and oil vapours.
- Finishing: Noise generated due to grinding, welding, chipping operations, etc. Gases are generated during welding. Dust is generated during chipping and grinding. Emission of pollutant gasses/fumes/dust can be taken care of by dust extractors, providing proper ventilation, good housekeeping and modifying the processes. Filters, collectors, cyclones scrubbers, etc., are some of the devices used to control pollution of atmosphere in the foundry.

16.10 APPLICATION OF COMPUTERS IN FOUNDRIES

All industries can benefit from the introduction of computers and the usual area of application is the business and commercial activities. In foundries, added to the usual applications, computer can help in the technical side. The casting design and the production control activities facilitated by computer and its software packages are as follows:

1. Weight estimation of castings and calculation of cost of production.
2. Pattern design giving various allowances and machining of patterns using 3D models.
3. Design of feeding system including riser design.

4. Sand testing and related calculations.
5. Least cost charge calculations for melting and melt analysis.
6. Quality control of castings using statistical tools.
7. Casting stimulation using 3D CAD programmes for estimating the possible defects prior to the production.

The advantages of using computer aided design (CAD), computer aided manufacturing (CAM) and the computer aided engineering (CAE) in foundries can be briefly listed as follows:

1. Increased casting yield per kg of metal poured.
2. Improved casting quality (absence of defect).
3. Enhanced productivity of the casting system.
4. Geometric models provide casting volume, weight, and surface area data, allowing rapid cost-estimation and permitting efficient design.
5. Automated machining of patterns, which, in turn, reduces cost.
6. Fewer prototypes are to be experimentally evaluated. Hence, shorter lead times from design concept to product.
7. Easier implementation and evaluation of engineering changes.
8. Enhanced ability to deal with batch production of castings of different designs.

The software packages developed specially for foundries are very much useful for design, simulation and other casting related processes. The following are some of them:

1. Magmasoft developed by Magma Foundry Technologies Inc., Illinois, is a powerful simulation tool for optimization of castings and foundry processes.
2. CastCAE, developed by CT-Castech Inc., Finland, is a tool for mould filling and solidification simulation coupled with automated risering.
3. CFS Solid 2000, developed by American Foundrymen's Society Inc., a comprehensive software combining thermal and volumetric calculations predicting the shrinkage porosity.
4. ProCast, developed by UES Software Inc., Ohio is based on finite element method and is suitable for sand, shell, investment, die casting, permanent mould and lost foam processes.

AutoCAST, developed by Advanced Reasoning Technologies Private Limited, Mumbai, is a package involving large eddy simulations for combining casting design decisions, casting model creation and process simulation.

16.11 RECLAMATION OF SAND

In a foundry, sand is the major raw material used. It is said that for producing 1T of steel casting at least 3T of sand is required. Hence, sand once used in the mould cannot be thrown out completely. Finding a place to dump huge quantities of used sand is a problem. One way of making use of used sand is by reclamation of sand by restoring its original condition as much as possible.

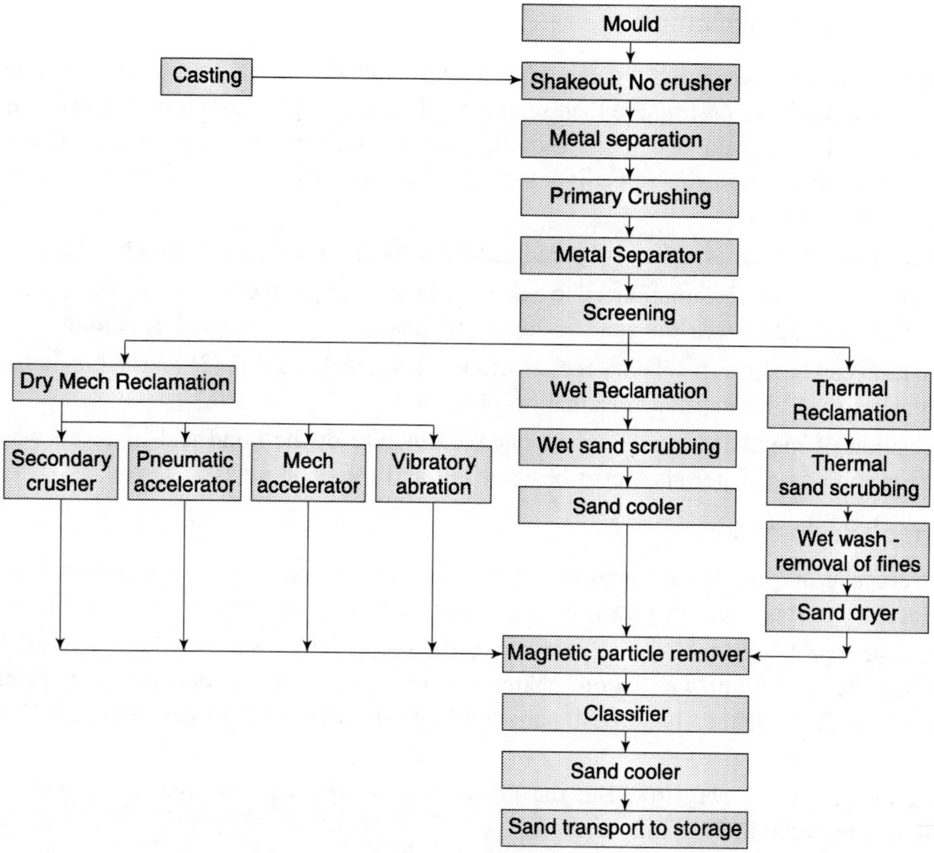

Figure 16.5 Sand reclamation

- If the sand is properly reclaimed, it will give the same performance as that of the new sand.
- Transporting used sand is expensive in addition to not finding dumping space.
- By reclaiming used sand, requirement of new sand can be reduced.

Sand reclamation cost should not exceed the cost of the new sand. In case of a steel foundry, the temperature involved is very high and the sand in the mould which comes in direct contact with the molten metal gets burnt and has to be discarded. This sand is referred to as facing sand. Every time fresh new sand has to be used for facing sand.

The quality of the sand deteriorates faster in steel as compared to other metals.

While preparing sand mix, a number of ingredients like binders, additives, etc., are used which coat the sand grains and form a hard shell around it due to high temperatures. These grains become unsuitable for reuse near the metal mould interface. This is more in case of steel castings as compared to other castings.

16.11.1 Types of Reclamation

The castings are separated from the mould box and sand during shakeout. The sand lumps containing metal pieces of casting are then separated. The lumps are crushed in primary crushers and reduced to original grain size. The magnetic particles are removed by magnetic separation. The sand is passed through the screening section where nonmagnetic particles and oversized sand particles are removed.

The sand thus obtained is called prepared sand and is fed into the declaimer unit.

The purpose of sand reclamation is to remove fines, foreign materials and other organic/ inorganic coatings and restore the sand as nearly as possible to its original condition.

Reclamation can be done by: (1) dry reclamation, (2) wet reclamation, (3) thermal reclamation, or (4) combined wet and thermal reclamation processes.

The selection of method depends on the chemical nature of the bond and its level, whether several organic binding materials are to be removed and the economics.

(a) Thermal Reclamation

Here the screened and metal-free sand is fed into the top section of the claimer (a refractory-lined surface). The sand falls downwards as the hot gases flow upwards (Figure 16.6).

The sand gets preheated in (A) and passes into the second chamber (B) where it is calcined. Later the sand is passed into the bottom chamber where precooling of the sand takes place by passing cool air. The claimer is so designed to allow only the gases to pass through C to B, then to A.

The energy demand is very low. But the initial cost is very high. Normally, organic binder sand systems are reclaimed by this method.

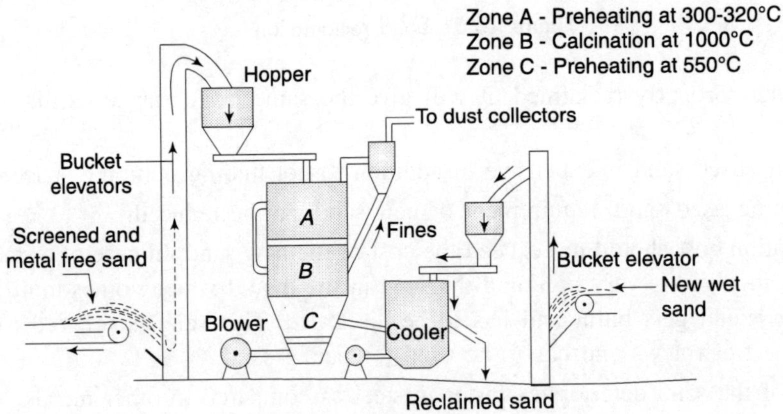

Figure 16.6 Thermal reclamation

(b) Wet Reclamation

This method is used when the binders are soluble in water; normally the inorganic sodium silicate sand and clay systems are reclaimed by this method. Here, the sand lumps after reduction are mixed with water using mechanical stirrers to form a slurry. It is then passed through a primary classifier to remove fines and then sent to wet storage tanks. It enters sand scrubbers (a tank with propeller immersed in it). The propeller rotates at high velocity in the sand slurry. The sand grains rub against each other and the binder is removed.

The slurry then passes through a secondary classifier to remove the fines created in the cleaning operation.

Finally, the water is removed from the sand by a filter and dried in a drier, cooled for reuse. Figure 16.7 shows wet reclamation.

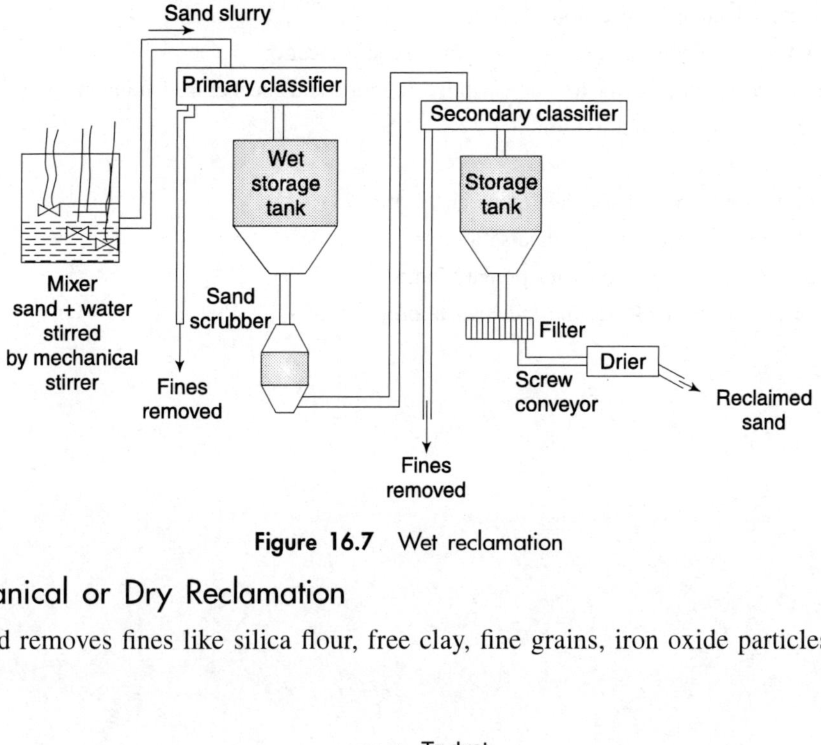

Figure 16.7 Wet reclamation

(c) Mechanical or Dry Reclamation

This method removes fines like silica flour, free clay, fine grains, iron oxide particles and other additives.

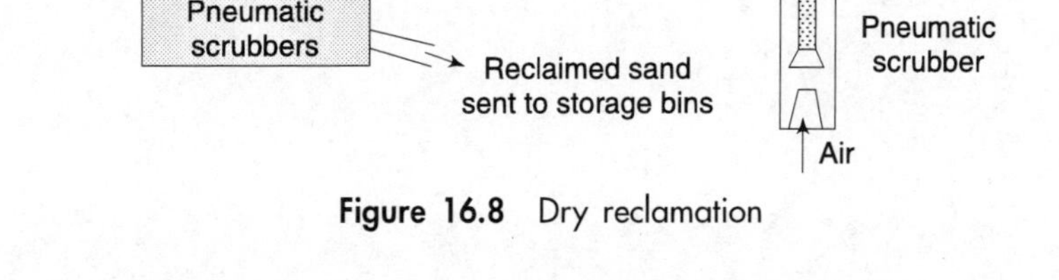

Figure 16.8 Dry reclamation

Iron and steel chips are removed by magnetic separators; lumps of sands are crushed and screened. By screens and air separations, the process is accomplished. A pneumatic scrubber is included to remove tenacious coatings on the sand grain partially. Dry reclamation alone cannot restore the sand to its original quality - as some clay and organic/carbonaceous material may cling to individual sand grains. Figure 16.8 shows a type of sand reclamation.

Review Questions

1. What is foundry layout?
2. What are the advantages of a good layout?
3. Explain what considerations are used for foundry site location.
4. Differentiate between jobbing foundry, captive foundry and mechanised foundry.
5. Explain the need for mechanisation.
6. What is mechanisation?
7. What aspects are to be considered for mechanisation?
8. Identify the areas of foundry mechanisation.
9. Draw a neat flow diagram for sand preparation.
10. Write a brief note on melting line mechanisation.

CHAPTER 17

Centrifugal Casting

17.1 INTRODUCTION

In centrifugal casting, a permanent mould is rotated about its axis at high speeds as the molten metal is poured. The molten metal is centrifugally thrown towards the inside mould wall, where it solidifies after cooling. The casting is usually a fine grain casting with a very fine-grained outer diameter, which is resistant to atmospheric corrosion, a typical situation with pipes. The inside diameter has more impurities and inclusions, which can be machined away. Hence, this method is best suited to cast cylindrical objects with a hole along the axis. Typical materials that can be cast with this process are iron, steel, stainless steels, and alloys of aluminium, copper and nickel. Two materials can be cast by introducing a second material during the process. Typical parts made by this process are pipes, boilers, pressure vessels, flywheels, cylinder liners and other parts that are axi-symmetric.

17.2 PRINCIPLE OF CENTRIFUGAL CASTING PROCESS AND CLASSIFICATION

It works on the basic principle of centrifugal force on a rotating component. In this process, a mould is rotated about its central axis when the molten metal is poured into it. A centrifugal force acts on molten metal due to this rotation, which forces the metal at outer wall of mould. The mould rotates until the whole casting solidifies. The slag oxide and other inclusions being lighter, get separated from metal and segregate towards the centre.

Figure 17.1 shows the principle of centrifugal casting of a cylinder. Here the mould made of metal is fitted horizontally on a faceplate, which rotates along with the shaft. The metal is poured though the central hole of the door or cover while the mould is in rotation. This pouring is done with the help of a spout. Since the mould is rotating, the metal is distributed on the cylindrical surface due to centrifugal force, leaving a horizontal cavity at the centre. The thickness of the cylinder is decided by the volume of metal poured. The principle of centrifugal casting was introduced by a British engineer in 1809. Manufacturing of cast iron pipes using this method was started on large scale by 1920 onwards. Now, a wide range of shapes are cast in almost all alloys using this technique.

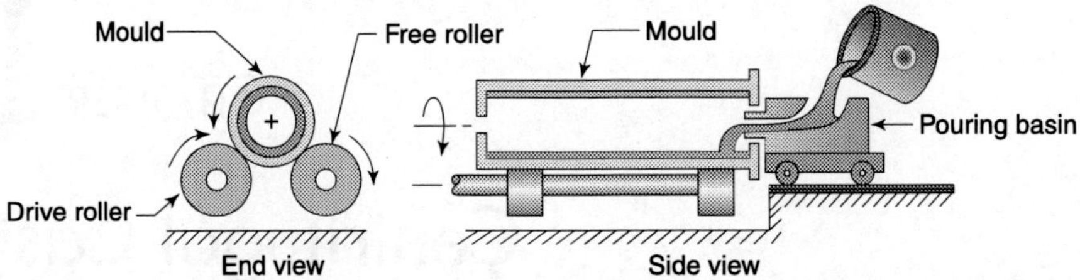

Figure 17.1 Process of true centrifugal casting

Centrifugal casting gives dense metal on the outer ring and separates light floating slag, gas, sand, etc., towards the inner portion. This centrifugal action provides a means for forming hollow cylindrical shape at the centre. The pressure acting radially outwards assists directional solidification towards inside and feeding of metal for shrinkage. The total result is that the process is suitable for cylindrical objects and gives good quality metallurgical products. It is to be noted that, any metal that is cast statically can be cast by centrifugal casting method.

Centrifugal casting processes can be broadly classified as:

1. True centrifugal casting
   (a) Horizontal
   (b) Vertical
2. Semi-centrifugal casting
3. Centrifuge casting (centrifuging or pressure casting)

17.3 TRUE CENTRIFUGAL CASTING

In true centrifugal casting, a cylindrical or tubular product is produced by spinning the mould about its own axis and filling molten metal to get the required casting thickness. The process can be either vertical or horizontal, and the need for a centre core is completely eliminated. Castings produced by this method will always have a true cylindrical bore or inside diameter regardless of shape or configuration. Figure 17.1 is an example for true centrifugal casting of a hollow cylinder with horizontal position of the axis. Similarly, Figure 17.2 shows the method of casting

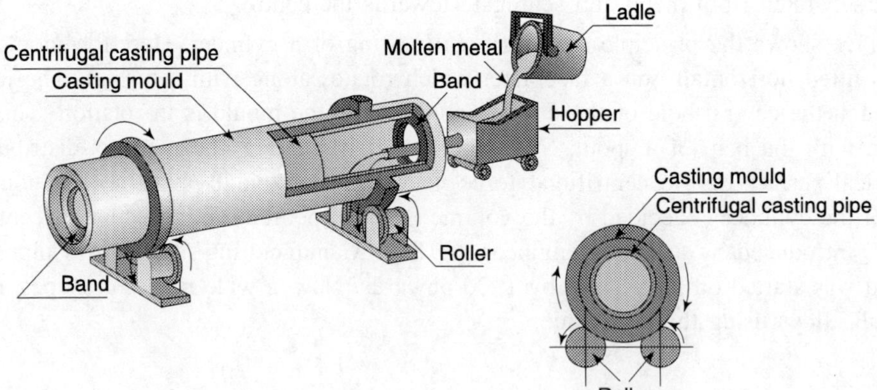

Figure 17.2 True centrifugal casting (horizontal axis)

a gear wheel blank by true centrifugal casting, using a sand mould. It is to be noted that the bore is perfectly cylindrical and no core is required for that. The metal filled in the mould is pushed radially outwards due to the centrifugal force till freezing, so that the cylindrical hole is formed. The quantity of metal decides the diameter of the hole.

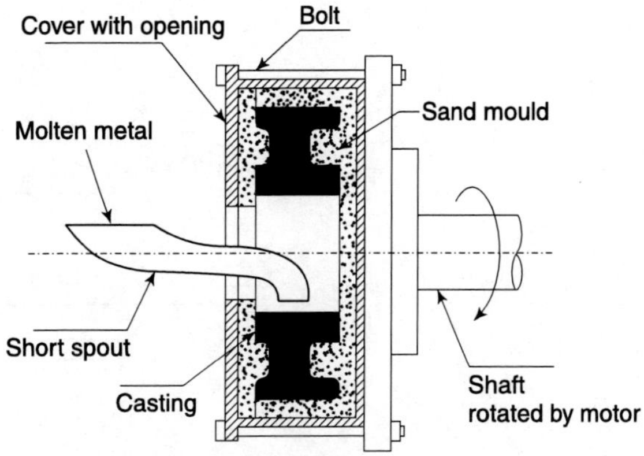

Figure 17.3(a) True centrifugal casting of a gear wheel blank (horizontal axis)

In true centrifugal casting with vertical axis, the cylindrical cavity formed is vertical as shown in Figure 17.3. Rotational speed has great influence on the shape of the product produced. Parabaloid A results from fast spinning and Parabaloid B results from slow spinning.

Castings produced in metal moulds have directional cooling or solidification from the outside of the casting towards the axis of rotation. This directional solidification results in the production of high-quality defect-free castings without shrinkage, which is the largest single cause of defectives in sand castings. Horizontal centrifugal moulds can produce extended-length castings with length three times or more of the outer diameter (L > 3D). The pipes of diameter up to 300 mm and length of 1.8 m are commonly cast by this type of process. Vertical centrifugal moulds are suitable for casting short cylindrical bodies of length less than three times the outer diameter (L < 3D).

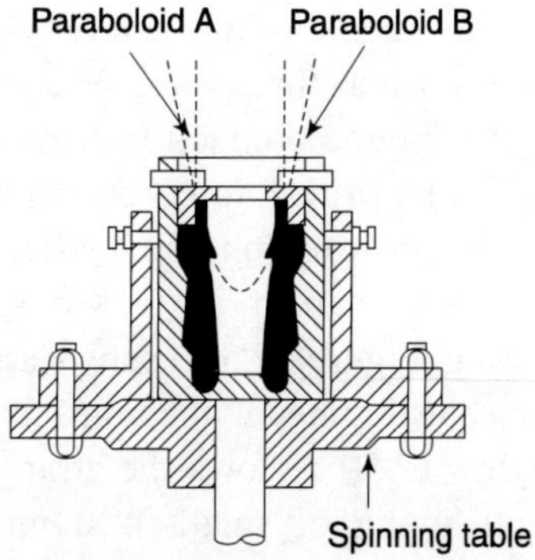

Figure 17.3(b) True centrifugal casting of a cylinder (vertical axis)

17.4 SEMI-CENTRIFUGAL CASTING

Semi-centrifugal casting is used to produce castings with configurations determined entirely by the shape of the mould on all sides (inside and outside), by spinning the casting and mould about its own axis. A vertical spinning axis is normally used for this method. Refer to Figure 17.4. Cores may be necessary, if the casting is to have hollow sections. Directional solidification is obtained by proper gating, as in static casting. Castings that are difficult to produce statically can often be economically produced by this method because centrifugal force feeds the molten metal under pressure to the cavities. This improves casting yield significantly (85% to 95%), completely fills mould cavities, and results in high-quality castings, free of voids and porosity. Thinner casting sections can be produced with this method than with static casting. Typical castings of this type include gear blanks, pulley sheaves, wheels, impellers and rotors of motors.

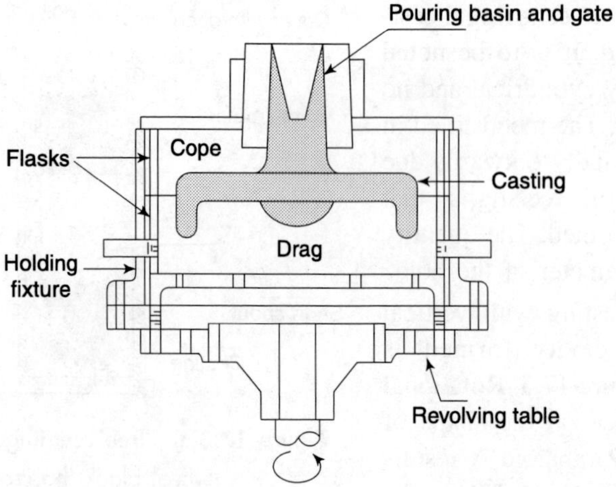

Figure 17.4 Semi-centrifugal casting

17.5 CENTRIFUGE CASTING

In centrifuge casting (centrifuging), the casting cavities are arranged about the centre, (axis of rotation) like the spokes of a wheel, thus permitting the production of multiple castings. Centrifugal force provides the necessary pressure on the molten metal in the same manner as in semi-centrifugal casting. Figure 17.5(a) shows arrangement for centrifuging of medium sized mould. Figure 17.5(b) shows the arrangement for centrifuging of small-sized moulds with individual casting cavities. The moulds are bolted on the periphery of a large-sized face plate in radial position so that the inlet spree of each mould is aligned to the output of the distributor as shown in the figure. The whole unit is rotated at the rated speed and the molten metal is poured to the central pouring basin. Due to the centrifugal force, the metal is pushed through the distributor to the inlet spree of each mould and fills the cavity with pressure. The rotation is continued till freezing. This type of centrifuging is best suited for small components and the yield is as high as above 90% compared to the static method. Centrifuge casting method is typically used to produce valve bodies, plugs, yokes, brackets and a wide variety of various industrial castings.

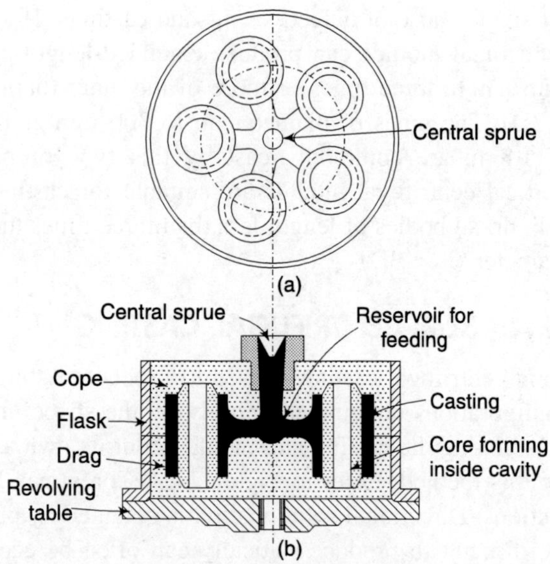

Figure 17.5 Centrifuge casting

17.6 MOULD FOR CENTRIFUGAL CASTING AND ITS SPEED OF ROTATION

Moulds

The selection of mould material for centrifugal casting depends partly on the shape of the casting and partly on the number of castings required. For simple shapes or mass production, metal moulds are usually preferred and that is based on economic grounds as well as for their effect in inducing directional solidification. The metallic moulds may be in steel or cast iron supported in steel shell. The mould is coated for the protection of the cavity surface and easy stripping of the casting from the mould. The coating materials range from refractory slurries to dry powdered dressings containing graphite, ferrosilicon and other substances which serve both to insulate and inoculate.

For complex shapes or for small rate of production, expendable refractory (sand) moulds are normally used. The sand moulds may be in green or dry condition. While using green sand moulds, the molten metal is poured at slow spinning speed. When the pouring is half or full, the spinning speed is increased to the required level. In greensand moulding, the flask is a must to hold the mould, but for dry sand method, snap flask or a special jacket is used to strengthen the mould during the rotation. Whatever may be the type of mould, the casing should firmly be bolted to the machine face plate, which gives rotation to the mould. Graphite moulds are also used for the production of copper alloy castings in small quantities.

Speed of rotation

The centrifugal force acting on a rotating body is proportional to the radius of rotation and to the square of the angular velocity:

$$F_c = mr\omega^2 = \frac{mv^2}{r}$$

where,

F_c = Centrifugal force in N

m = Mass in kg

r = Radius in m

ω = Angular velocity in rad/s

v = Peripheral velocity in m/s

Gravitational force acting on the same mass is given by

$$F_g = mgN$$

where g is the acceleration due to gravity in m/s^2.

The ratio $\frac{F_c}{F_g}$ is called G factor. If N is the revolutions per minute (speed) of the mould, then the G factor can be written as:

$$G \text{ factor} = \left(\frac{r}{g}\right)\left(\frac{3.14}{30}\right)^2 N^2$$

Substituting the known values and simplification,

$$N = 29.9 \sqrt{\left(\frac{G \text{ factor}}{r}\right)}$$

where

N = Rotations per minute of the mould

r = Radius of the mould in m from the axis of rotation

G = Factor to be fixed as per requirement (practical values 10G to 150G)

The highest values of G factor are used for open bore cylindrical components of small diameter and the lower values for semi-centrifugal and centrifuge castings. The preferred values are 60G to 80G for true centrifugal casting. In true horizontal centrifugal castings, if the mould is rotated at slow speed or the pouring of metal is too fast, molten metal may drop like rain during the rotation. This is called raining in centrifugal process.

Safety

Centrifugal casting machines should incorporate all the safety measures possible during handling of molten metal, pouring and solidification. The centrifugal force produced during rotation is heavy and very destructive. The moulds should be symmetrically fitted about the spinning axis to avoid vibration. Static and dynamic balancing may be required sometimes.

The method of attachment of both bottom and top cover plates to the mould is of great importance for withstanding and containing the force of the molten fluid metal while spinning. All clamping arrangements should be designed to tighten with, or to be unaffected by, centrifugal force. Moulds should be firmly clamped to the table because unbalanced moulds may fly off the table during operation. Adequate safety guards with interlocks should be used around all machines to protect workers from molten metal, which can be sprayed from the mould, if too much metal is poured. Moisture in a sand mould or moisture in the mould wash can turn into steam, when the molten metal contacts it. Most of the vertical centrifugal casting machines are installed completely the below floor level for maximum operator safety.

17.7 ADVANTAGES AND LIMITATIONS

The advantages of centrifugal casting compared to conventional casting method can be listed as follows:

1. Very good and defect-free castings with dense metal are obtained.
2. Slag and other nonmetallic inclusions are brought to the internal surface of the casting.
3. Directional solidification from outer to inside is easily maintained.
4. Risers are not usually required in this process.
5. Removal of gating system is simple.
6. High casting yield is obtained and so it is economical.
7. Castings with thin sections can be successfully produced.

8. Centrifugal casting is best suited for producing hollow cylindrical objects of small or medium size.

Centrifugal casting has the following limitations:

1. The mould has to be rotated at high speed, so machines for rotation; strong moulds, perfect clamping and extra power are necessary.
2. All shapes cannot be cast by centrifugal process.
3. Metal spray and mould breaking are hazardous to workers. Hence, extra protection jackets and safety measures are required.
4. Due to additional machines and safety measures, more investment is required.

17.8 CONTINUOUS CASTING

Continuous casting transforms molten metal into solid on a continuous basis and includes a variety of important commercial processes. These processes are the most efficient way to solidify large volumes of metal into simple shapes for subsequent processing. Most basic metals are mass-produced using a continuous casting process, including over 500 million tonnes of steel, 20 million tonnes of aluminium, and 1 million tonnes of copper, nickel, and other metals in the world each year.

Continuous casting is distinguished from other solidification processes by its steady state nature, relative to an outside observer in a laboratory frame of reference. The molten metal solidifies against the mould walls while it is simultaneously withdrawn from the bottom of the mould at a rate which maintains the solid/liquid interface at a constant position with time. The process works best when all of its aspects operate in this steady-state manner.

Relative to other casting processes, continuous casting generally has a higher capital cost, but lower operating cost. It is the most cost- and energy-efficient method to mass-produce semi-finished metal products with consistent quality in a variety of sizes and shapes. Cross sections can be rectangular, for subsequent rolling into plate or sheet, square or circular for long products, and even “dog-bone” shapes, for rolling into I or H beams.

17.9 PRINCIPLE OF CONTINUOUS CASTING AND ITS CLASSIFICATION

The principle of continuous casting was introduced in 1840 by George Seller. He developed a machine for continuously casting lead pipes. The first work on continuous casting of steel was patented by Sir Henry Bessemer in 1890. The process became popular in Second World War.

The principle of continuous casting of molten metal is shown in Figure 17.6. It involves the following operations:

1. Pouring of metal through a distributor (tundish) into the casting mould.
2. Formation of the cast section in a water-cooled copper mould.
3. Continuous withdrawal of the casting from the mould using rolls.
4. Further heat removal to solidify the liquid core of the casting by water spraying.
5. Cutting to length and removing the cast sections.

Molten metal in a ladle is delivered to a reservoir above the continuous casting machine called a tundish. The flow of metal from the tundish into one or more open-ended, water-cooled copper moulds is controlled by a stopper rod nozzle or a slide gate valve arrangement. To initiate a cast, a starter or dummy bar of parent metal, is inserted into the mould and sealed so that, the initial flow of molten metal is contained in the mould and a solid skin is formed by cooling. After the mould has been filled to the desired height, the dummy bar is gradually withdrawn at the same rate that the molten metal is added to the mould. The initial liquid metal freezes onto a suitable attachment of the dummy bar so that the cast strand can be withdrawn down through the machine. Solidification of a shell begins immediately at the surface of the copper mould. The length of the mould and the casting speed are such that the shell thickness is capable of withstanding the pressures of the molten metal core upon exiting from the copper mould. To prevent sticking of the frozen shell to the copper mould, the mould is normally oscillated up and down slightly during the casting operation and a lubricant is added to the mould. The cast metal strand is mechanically supported by rolls below the mould, where secondary cooling is achieved by spraying cooling water onto the strand surface to complete the solidification process. After the strand has fully solidified, it is sectioned into desired lengths by a cutoff torch or shear.

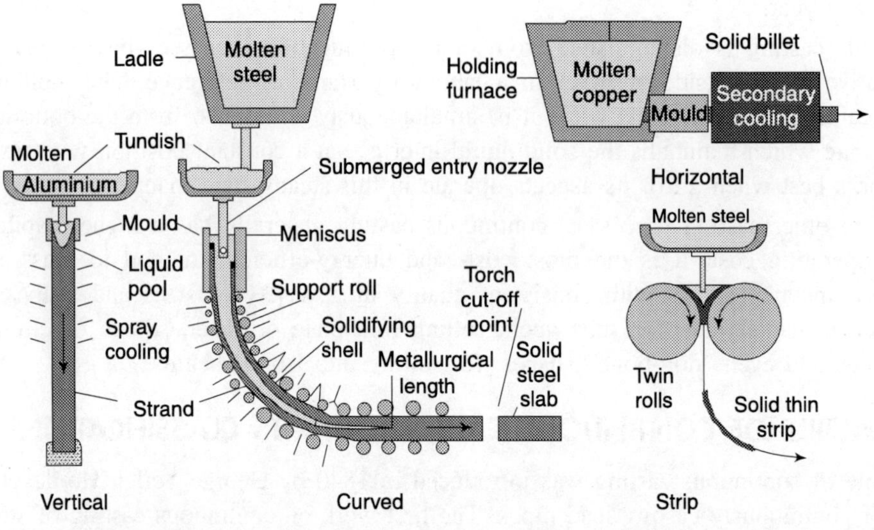

Figure 17.6 Classification of continuous casting methods by position

The above-mentioned casting process can be broadly classified according to the position in which the metal strand is solidified and cooled:

1. Purely vertical strand
2. Straight with bending of strand
3. Curved strand
4. Horizontal strand

The vertical strand type casting method is preferred for thick blooms, billets, cylinders, etc. Straight with bend or curved casting layouts are suitable for slabs and plates having some moderate thickness. Horizontal one is applicable for strips, sheets and wires. It is to be noted that the vertical strand machine requires about 30 m height, whereas for straight with bend layout needs 20 m height from ground. Curved strand requires 5 m to 10 m height, but a horizontal casting machine can work at 3 m.

There are a number of continuous casting designs which are in use under the above categories. They are generally known by either the company name or by the name of the designer. Steel, copper alloys, aluminium alloys and other industrial nonferrous metals are cast in large quantities by the method. Alcova direct chill process, Hazelett process, Williams continuous process, Asarco process are some of the popular methods. If the casting process is not 100% continuous, such a process is called semi-continuous casting. The reason for interruption may be the reaching of the vertical strand on to the ground, need for excess time to cut-off the strand or requirement of more freezing time, etc. Whatever may be the reason, semi-continuous casting is a slow process.

After exiting the spray-chamber, the metal strand passes through straightening rolls (if cast on other than a vertical machine) and withdrawal rolls. There may be a hot rolling strand after withdrawal to take advantage of the metal's hot condition to pre-shape the final strand. Finally, the strand is cut into predetermined lengths by mechanical shears or by moving oxyacetylene torches. In many cases, the strand may continue through additional rollers and other mechanisms which might flatten the roll or extrude the metal into its final shape.

17.10 CONTINUOUS CASTING OF STEEL

The layout of a continuous casting process for steel is shown in Figure 17.7. In this process, molten steel flows from a ladle, through a tundish into the mould. The tundish holds enough metal to provide a continuous flow to the mould, even during exchange of ladles, which

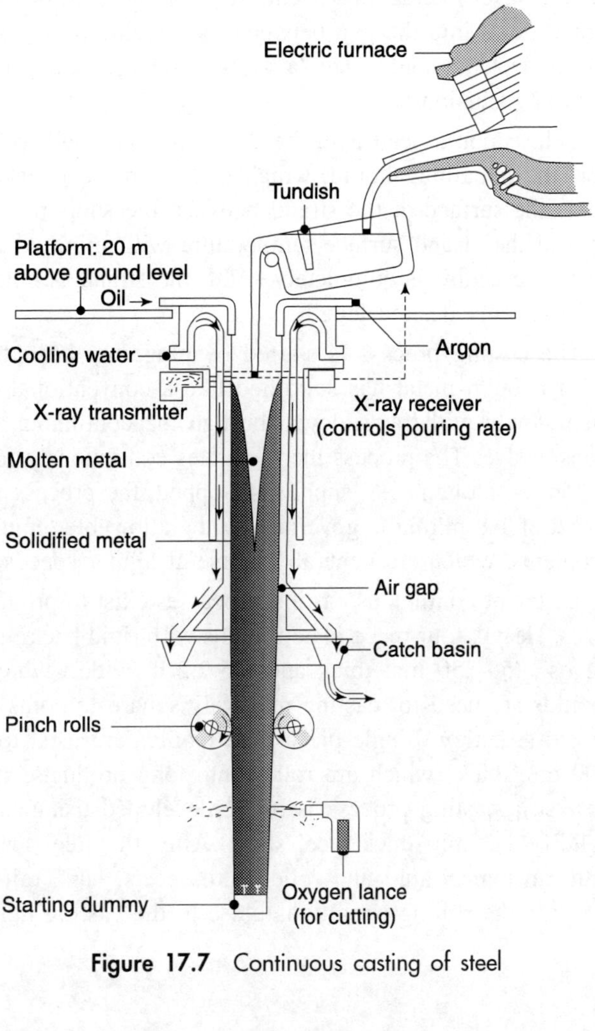

Figure 17.7 Continuous casting of steel

are supplied periodically from the steel-making process. To produce higher quality product, the liquid steel must be protected from exposure to air by a slag cover over the liquid surface in each vessel and by using ceramic nozzles or shrouds between vessels. If not, the oxygen in the air will react to form detrimental oxide inclusions in the steel.

As the molten steel freezes against the water-cooled walls of a bottomless copper mould, a solid shell of the metal is formed. The mould is oscillated vertically in order to discourage sticking of the shell to the mould walls. The pinch rolls lower the shell from the mould at a rate or 'casting speed' that matches the flow of incoming metal, so the process ideally runs in steady state. The liquid flow rate is controlled by restricting the opening in the nozzle according to the signal fed back from the X-ray level sensor near the mould.

The most critical part of the process is the initial solidification at the meniscus, found at the junction, where the top of the shell meets the mould and the liquid surface. This is where the surface of the final product is created, and defects such as surface cracks can form, if problems such as level fluctuations occur. To avoid this, oil or mould slag is added to the steel meniscus, which flows into the gap between the mould and the solidified shell. In addition to lubricating the contact, a mould slag layer protects the steel from air, provides thermal insulation, and absorbs inclusions.

Below mould head exit, the thin solidified shell (6-20 mm thick) acts as a container to support the remaining liquid, which makes up the interior of the strand. Water or air mist spray cools the surface of the strand between the support rolls. The spray flow rates are adjusted to control the strand surface temperature with minimal reheating until the molten core is solid. After the centre is completely solid, the strand is cut with oxyacetylene torches into slabs or billets of any desired length.

The casting process is started by plugging the bottom of the mould with a 'dummy bar'. After enough metal has solidified like a conventional casting onto its head, the dummy bar is then slowly withdrawn down through the continuous casting machine and steady state conditions evolve. The process then operates continuously for a period of one hour to several weeks. When the molten steel supply is stopped, the process must be restarted. The maximum casting speed of 1-8 m/min is governed by the allowable length of the liquid core, and to avoid quality problems, which are generally worse at higher speeds.

Different continuous casting processes exist to produce cross sections of different shapes and sizes. Heavy, four-piece plate moulds with rigid backing plates are used to cast large, rectangular 'slabs', (50-250 mm thick and 0.5-2.2 m wide), which are rolled into plate or sheet. Similar moulds are used for casting relatively square 'blooms', which range up to 400 mm × 600 mm in cross section. Single-piece tube moulds are used to cast small, square "billets" (100 mm × 200 mm thick) which are rolled into long products, such as bars, angles, rails and axles. The new strip casting process is being developed using large rotating rolls as the mould walls to solidify 1-3 mm thick steel sheet. After the steel leaves the casting unit, it is reheated to a uniform temperature and rolled into sheets, bars, rails and other shapes. Modern steel plants position the rolling operations close to the casting unit to save on reheating energy.

17.11 ALCOVA DIRECT-CHILL PROCESS

The vertical direct-chill casting process was patented by Alcova in 1942, and is shown schematically in its present form in Figure 17.8. The process can directly prepare billets for extrusion, blocks for rolling and sheets for fabrication, thus eliminating intermediate mechanical working processes after ingot casting. The process is used for nonferrous metals like copper or aluminium alloys.

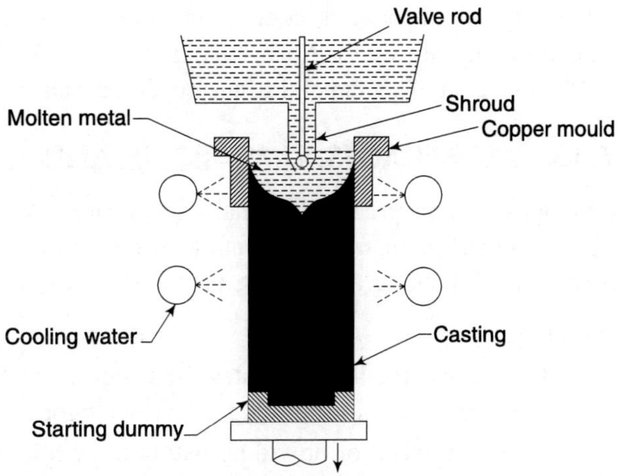

Figure 17.8 Alcova direct-chill process

The layout of the machine is a simple form of the vertical continuous casting of steel. The molten metal supplied by a ladle is first collected in a reservoir called tundish and is poured in the copper mould through a shroud and regulating valve. The cooling is obtained by direct spraying of water (direct-chill) on the mould as well as on the solidified casting surface. To begin the casting, a starting dummy is required to close the bottom of the mould. As the solidification progresses, the platform supporting the dummy is lowered slowly downwards synchronizing the solidification rate.

The key operating requirement in direct-chill casting is that a sufficiently strong shell be developed in the limited time of contact with the mould to retain the interior molten pool. Withdrawal rates of up to 0.2 m/min can be achieved in conventional casting units. Vertical direct-chill casting is used extensively to produce rectangular slabs and cylindrical billets of copper and aluminium alloys and to a lesser extent of pure copper. For copper and copper-base alloys, the liquid metal is poured through a water-cooled, oscillating, graphite-lined mould or collar. The graphite produces a smooth surface on the casting and minimizes oxidation.

17.12 HAZELETT PROCESS

Horizontal continuous casting of copper sheets is schematically shown in Figure 17.9. The molten metal is taken in a reservoir and is supplied through a rectangular spout to the rectangular gap provided by two steel bands (belts). The lower and the upper bands

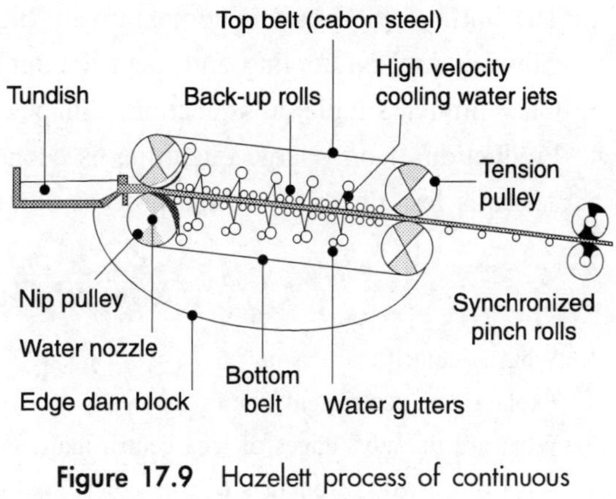

Figure 17.9 Hazelett process of continuous casting

are driven by rollers as shown in the figure and a slight downward slope is given to the exit end. To stop the lateral flow of liquid metal, two flexible side moulds in the form of thick belts, are moved along with the steel band.

The molten copper coming out of the spout flows to the flat steel band and begins to solidify forming the outer skin. The cooling is obtained by spray of water on both sides of the bands. As the belt moves forward, cooling and freezing progress and the solid strip comes out at the exit end along with the belt. The thickness of the strip may be up to 12.5 mm and width can be 1.5 m to 2 m. The same casting procedure can be applied for aluminium alloys also.

17.13 ADVANTAGES, LIMITATIONS AND APPLICATIONS

Continuous casting process has many advantages, and it is very commonly adopted in metal-making industries. In modern manufacturing processes, the location of continuous casting is between metal-making and rolling processes or extrusion.

Advantages

1. This process totally eliminates the rough forming and breaking down rolling operations, to be conducted after casting of heavy ingots.
2. There is no need of providing gating system or risers and the complete metal is obtained as output, so that the yield is 100%.
3. The surface obtained is better than that of the conventional ingots.
4. The metal quality is improved due to reduction of segregation of the alloying elements or impurities.
5. The grain size and its structure can be regulated by proper controlling of cooling rates during the continuous casting.
6. The total cost of production is less and economical than the conventional process.
7. The process is fully automatic and best suited for mass production.

Limitations

1. The initial capital cost is comparatively high to start a continuous casting unit.
2. Close control on cooling and metal feeding is to be maintained.
3. Only limited simple cross-sectional shapes can be cast by this process.
4. Production of only large quantities is economical.
5. There is less flexibility in the process control, shapes, material, etc.

Review Questions

1. What is centrifugal casting?
2. Explain true centrifugal casting process.
3. What are the advantages of true centrifugal casting methods?
4. What are the disadvantages of true centrifugal casting methods?

5. Give classification of continuous casting methods.
6. Explain continuous casting of steel.
7. Explain Alcova direct-chill process.
8. What are the limitations of Alcova direct-chill process?
9. Explain Hazelett process.
10. What are the advantages and disadvantages of Hazelett process?

CHAPTER 18

Die Casting Processes

18.1 INTRODUCTION

The sand moulding technique, though widely used because of its low cost, cannot be reused, which results in lower productivity. The sand moulding often results in rejections and rework due to the formation of blowholes, sand inclusions, dimensional deviations, costly machining and generally lower quality castings. These limitations of sand mould can be overcome with the use of metal moulds.

Die casting is the art of producing accurately dimensioned articles by filling metals dies or moulds with molten metals. There are two types of die casting processes: (i) gravity die casting, and (ii) pressure die casting.

Gravity die casting is a development of the traditional foundry practice, using permanent metal moulds or dies in the place of sand moulds. The molten metal is poured into the die from the ladle and only pressure on the casting during solidification is due to the head of metal in runners and risers.

In pressure die casting, molten metal is forced into the metal mould under a very high pressure applied by means of pistons or by compressed air. In both cases, die is made in two or more parts so that it can be opened to remove the article after casting. In pressure die casting, opening and closing of die and ejection of the cast articles are usually mechanised so that a high rate of production can be achieved.

18.2 ALLOYS FOR DIE CASTING

Most common nonferrous alloys can be gravity die-cast successfully. Many aluminium and magnesium alloys are gravity cast including those which require subsequent heat treatment in order to develop their full mechanical properties. Aluminium, bronzes and various brasses suitable for gravity die casting are also available, while the gravity die-cast iron is now becoming increasingly important.

The choice of alloys for pressure die castings is rather more restricted. Zinc alloys are the most widely used material. Aluminium and magnesium alloys can also be pressure die-cast but subsequent heat treatment is difficult. Both lead and tin alloys are used extensively for pressure die casting.

18.3 SURFACE FINISH AND DIMENSIONAL ACCURACY

The surface finish and dimensional accuracy of pressure die casting is comparable with that of the components manufactured by several machine shop methods. A zinc alloy gives very good surface finish and dimensional accuracy. Accuracy of aluminium and magnesium alloys are not as good as compared to zinc alloys. Lead and tin alloy castings have a surface finish and dimensional accuracy comparable to those of the zinc alloys. In general, dimensional accuracy and surface finish obtained by gravity die castings are much superior to those of sand casting but inferior to those of pressure die castings.

18.4 SELECTION OF PROPER MOULDING PROCESS

The selection of proper moulding process for making a casting from among the sand moulding/ metal mould processes depends upon the following factors:

(a) Metal/alloy being cast
(b) Size and weight of casting
(c) Quantity to cast per order
(d) Quality requirements such as freedom from defects, dimensional tolerances, surface finish, etc.
(e) Overall cost/economy

18.5 PRESSURE DIE CASTING

In pressure die casting, the metal is injected into a die under pressure and solidifies under the same conditions. This enables the production of intricate castings at a higher production rate and at a low cost. The pressure is generally obtained by compressed air or hydraulically and it varies from 20 kPa to 500 MPa. The process is best suited for zinc alloys, copper alloys like bronzes, and aluminium alloy castings. Nowadays, even cast irons can be cast in pressure die casting machine. The pressure die casting methods are classified as high-pressure and low-pressure die casting methods depending on the amount of pressure used.

18.5.1 Principle of Pressure Die Casting Process

Pressure die casting uses a metallic mould called die in two halves and molten metal is admitted to the mould cavity at high pressure and speed. The metal solidifies inside the die at pressure to form the casting and it is ejected, as the die halves are opened. Figure 18.1(a) shows the die in closed position, ready for fitting the metal. The die has one stationary half called cover die and a movable half called ejector die. They are locked in position by means of mechanism of the machine used for the process. These die halves may have internal passages for circulating cooling water. Metal cores, to form internal cavities and holes in the castings, are usually arranged in the movable die half. Solidified castings are removed from the dies by means of pushers called ejector pins, which are rigidly fixed on a plate called ejector plate. Ejector pins and plates are fitted on the movable die half.

In Figure 18.1, the mould cavity is arranged for taking two castings in a shot (filing) and the metal is admitted through the stationary die centre. The metal goes to the castings through the sprue, runner and in gates. A sprue spreader is sometimes provided on the movable die to bifurcate the passage in the sprue. Figure 18.1(b) shows the ejected casting at the time of opening of the die halves. It is to be noted that the ejector die and plate are moving from the locked position to open, while the cover die is stationary.

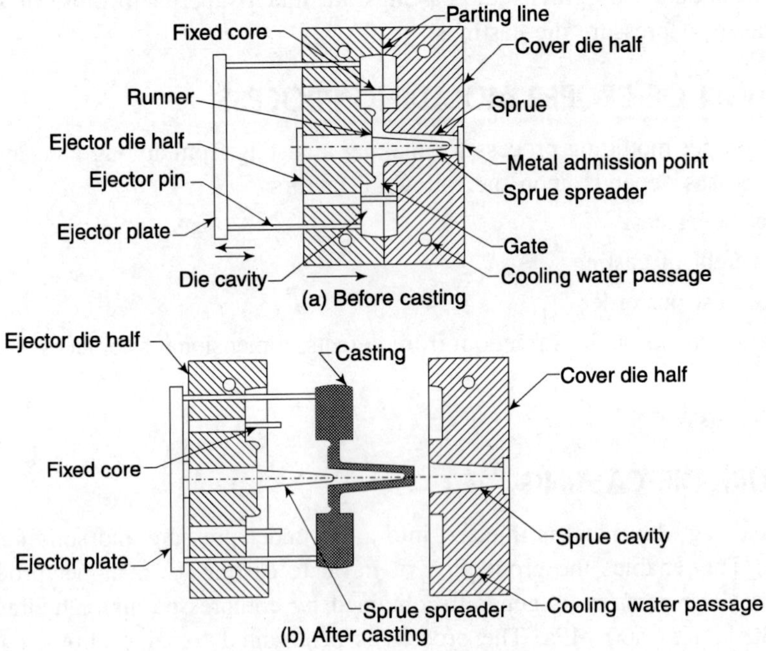

Figure 18.1 Die casting machine

To start the casting cycle, the working surfaces of the open die, i.e., the mould cavity and the metal passages, are washed with a protective liquid composition (lubricant). After that the ejector die half is pressed against the stationary cover die half by means of the locking mechanism of the machine. The molten metal is forced through the sprue of the die at a rate of 0.5 m/s to 120 m/s and pressure of 300 to 3000 bars. The metal is allowed to solidify at that pressure. Just after solidification, the movable half (ejector die) is pulled back together with the casting, runner and sprue. The pusher plate moves back together with the die for certain distance and then stops movement so that, as the die moves back further, the casting is ejected by pusher. The ejected hot casting is taken out at that movement by a manipulator and placed into a press for trimming the gates and fins. The die is then blown with compressed air and washed with the protective liquid for the next casting cycle.

The steps to be followed in pressure die casting process to produce a component can be listed as follows:

1. Application of lubricant on mould cavity.
2. Closing the mould by moving the ejector die towards the cover die.
3. Holding the two die halves securely together.
4. Forcing the molten metal into the die cavity through the inlet passage.
5. Keeping the metal for solidification under pressure.
6. Opening the mould by moving the ejector die backward.
7. Ejection of the casting using ejector pins.
8. Removal of the hot casting using ejector pins.
9. Trimming of the fins and gating system to get the finished casting.

Die casting process was developed in 1838 for the purpose of producing movable letter type for the printing industry. The first die casting related patent was granted in 1849 for a small hand-operated machine for the purpose of mechanized printing type production. Then this process was converted into pressure die casting for better capacities. Recent years have seen the rapid growth of pressure die casting of nonferrous metals, especially zinc and aluminium alloys. The size of the casting has also rapidly increased, following the introduction of heavier machines. Now, there are pressure die casting machines which can take 1000 to 3600 pressing cycles (shots) per hour for smaller objects.

18.5.2 Die Casting Machines

Pressure die casting machines can be broadly classified as:

1. Plunger type cold chamber pressure die casting machines
2. Submerged plunger type hot chamber pressure die casting machines
3. Direct air operated hot chamber pressure die casting machines
4. Vacuum applied pressure die casting machines
5. Low pressure die casting machines

General layout of a die casting machine is shown in Figure 18.2. Here, the injection of metal to the mould cavity is done using a plunger operated in cold state.

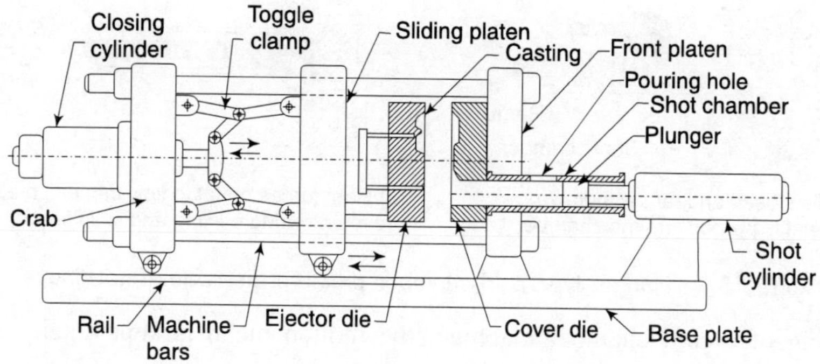

Figure 18.2 Plunger type cold chamber pressure die casting machine

As shown in Figure 18.2, the frame of a pressure die casting machine is designed to support heavy locking loads and to operate the dies in correct alignment. The required locking force is determined by the projected area of the casting in the plane normal to the direction of closing. Very high forces are required for some larger die casting, and this may reach even several hundred tonnes.

The cover die is mounted on a fixed front platen (cover plate) and the moving die on a moving platen (sliding cover plate). The moving die is actuated by a hydraulic ram. The whole structure is held together by four tie bars. The methods used for closing and locking the dies include straight hydraulic or hydraulic with mechanical toggles. With straight hydraulic clamping, the live pressure backs up the die at all times, when it is closed. Machines employing toggles in their clamping mechanism is correctly adjusted, the die can be properly locked and such a system is shown in the figure. The plunger inside the cold chamber is moved by operating the shot cylinder. The whole machine is mounted on a base plate having rails for sliding of the moving platen and the crab.

Figure 18.3 shows the layout and working principle of a cold chamber die casting machine. Here, the shot chamber is horizontal in position and is mounted on the front platen (cover) of the machine. The cover die, which is stationary, is attached to the front platen. Metal for a single shot is loaded in the cylindrical chamber through the top side opening using a hand ladle. The hydraulically operated plunger pushes the metal into the die cavity at very high pressure ranging from 700 to 1400 bars. Because of high pressures, intricate shapes can be made. After solidification of the metal, the pressure is released by backward movement of the plunger. Then the ejector die is moved back, the core block is withdrawn and the casting is ejected from the mould cavity by ejector pins. The metal filling, solidification and ejection operations are completed within a few seconds.

This process gives high production but not usually as fast as hot-chamber machines because of pouring step. Aluminium, brass, magnesium and other high melting alloys can be cast using this method. This process gives less solidification shrinkage losses and less air entrapments.

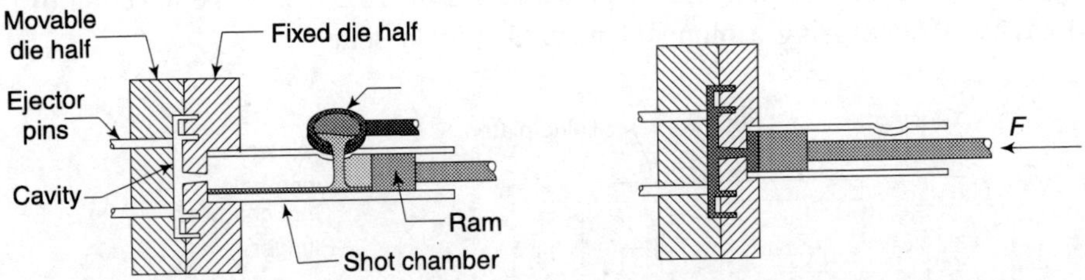

(1) With die closed and ram withdrawn molten metal is poured into the chamber

(2) Ram forces metal to flow into die, maintaining pressure during cooling and solidification

Figure 18.3 Plunger type cold chamber pressure die casting machine

During working of a cold chamber machine, the molten metal is kept readily available by a hot furnace. The filling of molten metal to the shot chamber is accomplished by using hand ladle manually, automatic pumping devices or robotic hand. Cold chamber machine is preferred for casting of relatively higher melting point metals.

Die sequence

The sequence of the die casting process begins when the die is open before the shot. The die is first sprayed with a lubricant, usually an aqueous solution. This spray coats the die with a thin layer of mould release, and the water evaporates, cooling the die surface.

The die is closed, and the metal is either injected into the die (hot chamber) or ladled into the shot sleeve (cold chamber).

The shot proceeds in three stages:

- In the first stage, the piston moves slowly, to fill the shot sleeve so that air is not entrapped in the metal before injection begins.
- In the second stage, the piston moves very quickly, forcing the metal into the mould cavity. The dies are filled in less than 0.15 s. Pressure builds up on the metal during this phase.
- In the third stage, pressure is increased to minimize the formation of porosity. The high pressures require elaborate locking mechanisms to keep the dies closed during the cycle; locking forces of up to 45,000 kN are used in the largest machines.

After the casting is solid, the dies open, and the casting and its runner system are removed from the die. Slides and cores are usually retracted first, and then the dies open. Typical sequence is shown in Table 18.1.

Table 18.1 Machine cycle details for die casting machine

Machine Cycle Details					
Machine Cycle	0	10	20	30	Total
Guard Close					3.50
Die Close					4.50
Slow Stroke					0.70
Fast Shot					0.01
Plunger Dwell					1.50
Coolling Time					4.50
Die Open					3.50
Casting Extraction					5.00
Die Lubrication					4.00
Total Cycle Time					25.00

Submerged plunger type hot chamber pressure die casting machine

Hot chamber die casting machine with submerged plunger is shown in Figure 18.4. Here the plunger and the shot chamber are immersed in a molten metal pot, which is kept well above the melting temperature. Burning of a gas burner below the pot, maintains the temperature at the required level. When the plunger is lifted, molten metal enters from the pot to the vertical shot chamber through the side holes. At the time of injection of metal, the plunger is lowered by hydraulic force. This pushes the metal through the goose neck to the inlet orifice of the die

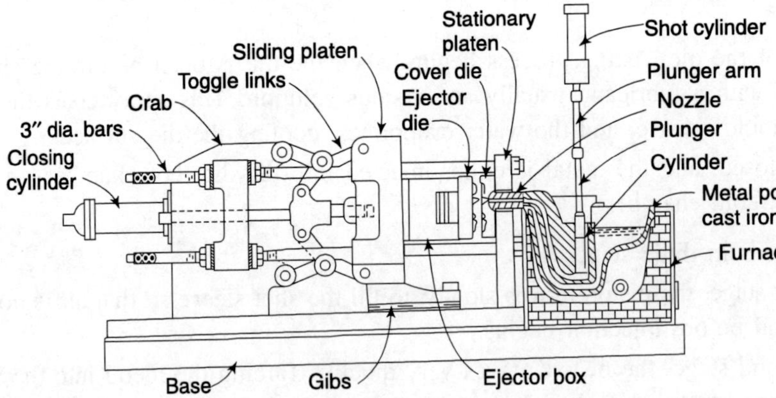

Figure 18.4 Submerged plunger type hot chamber pressure die casting

in closed position. The remaining machine operations are essentially the same as explained for cold chamber process.

Hot chamber machines are extensively used for casting zinc base alloys. Due to the relative high melting point of aluminium as well as similar nonferrous metals, they are not cast by this hot chamber process. Complete sequence is given in Figure 18.5.

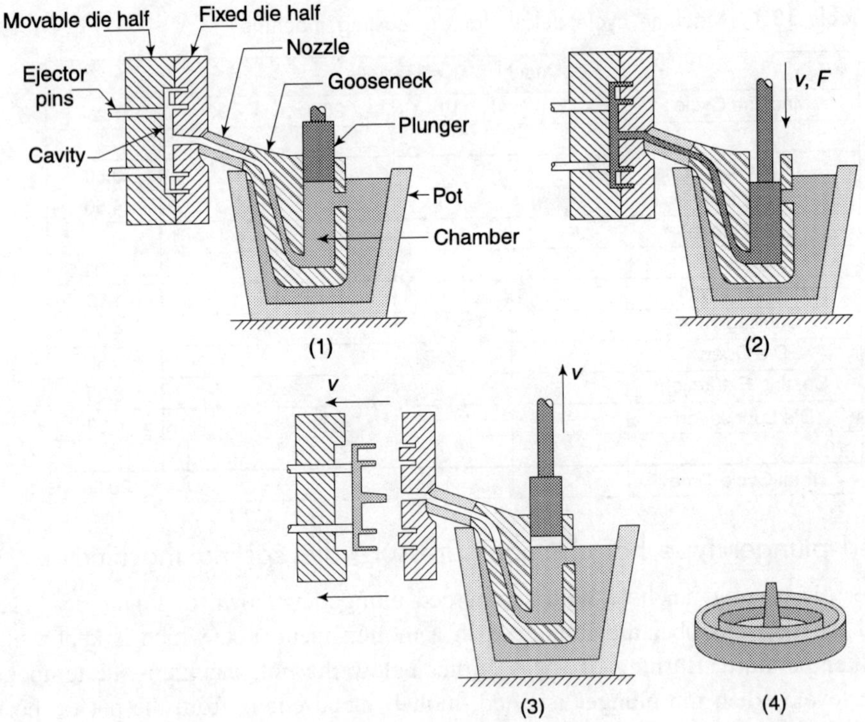

Figure 18.5 Direct plunger operated hot chamber pressure die casting machine

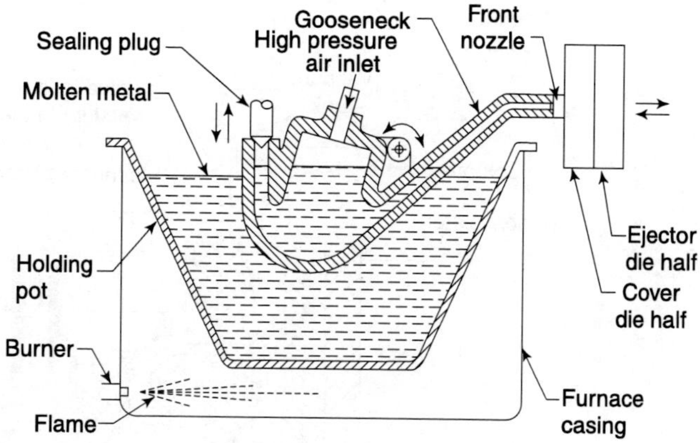

Figure 18.6 Direct air operated hot chamber pressure die casting machine

In this die casting machine, the molten metal is pushed to the die by direct air pressure, instead of a plunger movement in a cylinder. Figure 18.6 shows the working principle of the machine. Compared to the previous hot chamber machine, the vertical shot chamber is replaced by a tilting vessel, which has compressed air connection at the top and a nozzle with sealing pin for the admission of molten metal on one side. The whole vessel unit is tilted anticlockwise about a trunnion so that, the molten metal is replenished to the vessel through the opened nozzle. Then the sealing pin is inserted to close the opening and the vessel is turned back to connect the gooseneck on the metal entry point of the closed die. Then high pressure (about 35 bar) is applied to the vessel through a pipe in order to get the injection of the metal through the gooseneck. The remaining operations are similar to that of a usual die casting process.

Air operated hot chamber machines are simple in construction and there are less moving parts on it. The injection pressure attained by this machine is less than that of plunger type. The production rate is also considered less than that of plunger type. This machine is suitable to cast metals like zinc, tin and lead alloys.

(b) Vacuum applied pressure die casting machines

The presence of air inside the die cavity may bring defects in casting as air bubbles are entrapped in fine compressed air pockets. This problem can be solved by applying vacuum around the die to remove all air from the die cavity. Figure 18.7 gives the method of applying vacuum to the die with the help of a vacuum pump and a tank. As the die closes, it pauses in still position slightly apart and the vacuum valve is opened. In a fraction of a second, the entire air from the compressible seal is evacuated sufficiently and the die completes its closing. The molten metal is admitted at this stage to get a defect-free casting.

Vacuum die casting principle can be applied to both cold and hot chamber processes. In submerged type hot chamber machine, an additional plunger is to be fitted near the main plunger in order to prevent suction of molten metal through the gooseneck during the application of vacuum. Similarly, in cold chamber die casting also, there should be a special mechanism to solve the problem. In some machines, the same vacuum is used for the filling of metal to the shot chamber by suction.

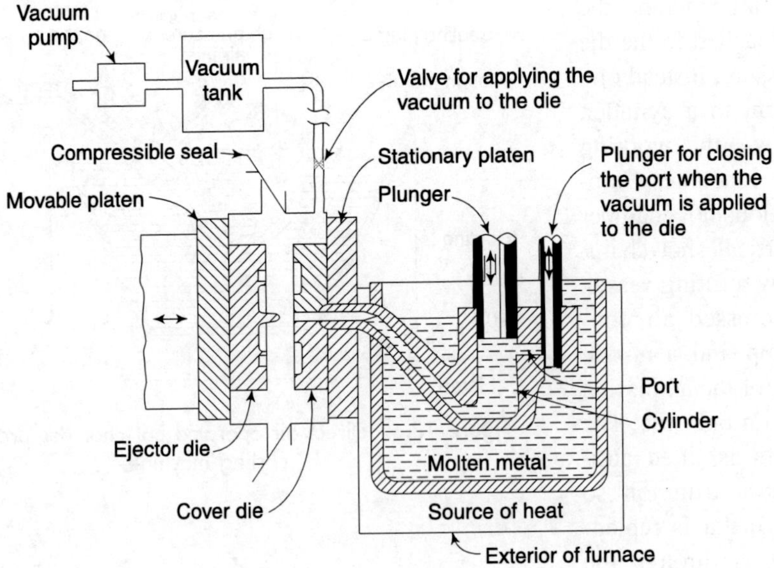

Figure 18.7 Principle of working of vacuum applied pressure die casting

Low-pressure die casting

Low-pressure die casting is a quite different process from high-pressure die casting. By nature of its design, it always involves vertical injection. The pressure involved is normally only in the range of 20-60 KPa compared to up to 200 MPa for high pressure die casting. In this process, sand moulds can be used as well as metal dies.

In low-pressure die casting, the die is placed in a casting device above a sealed airtight chamber that contains a crucible holding molten metal. A fill tube or riser tube extends from the die down into the metal bath, and the casting is made by pressurizing the chamber and forcing metal up the fill tube into the casting. This process usually operates at lower die temperatures and shorter cycle times than conventional gravity die casting. Vacuum casting is similar, except that a vacuum is created in the die cavity and the metal is pulled rather than pushed into the die cavity. These processes are compared schematically in Figure 18.8.

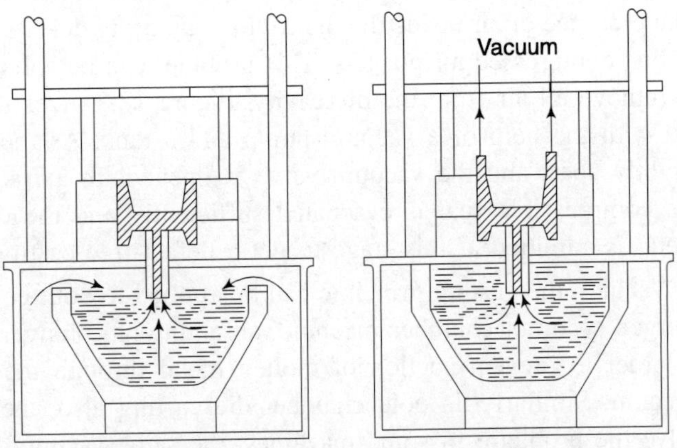

Figure 18.8 Low-pressure die casting

Although the process is most widely used for the casting of aluminium alloys, particularly automotive wheels, and cylinder heads and blocks it has been used for the casting of steel railway wheels; cast iron for the production of baths, and, recently, magnesium alloys for aerospace castings.

The low pressures have in general given low injection velocities, so that, together with the uphill flow of the metal, metal quality is maintained. The process has therefore earned for itself a reputation for delivering castings of constantly high quality.

18.6 DIES FOR PRESSURE DIE CASTING

General features

The mould for die casting is usually constructed from two rectangular blocks of steel forming the die halves. They are joined face-to-face vertically to get the casting cavity. Since heavy pressure is to be applied to the molten metal during casting, the die halves are made of hard die steel. The die half, through which the molten metal is admitted, is called the cover die, and it is kept stationary with the cover platen. The other half is the sliding one and the ejector pins are fitted inside for pushing out the casting after solidification. This half is called ejector die. Fixed core pins or inserts are usually mounted on the ejector die for forming a hole or recess.

For very large dies, instead of two halves, multiple constructions are preferred with die inserts carrying the casting impression. They are supported in heavy blocks or bolsters. For small castings, multiple impression dies are used. This permits mass production in less number of shots. If a die is producing one casting in one shot, such a die is called *single-impression die.* If two or more castings of the same type are taken in a shot, such a die is named *multiple-impression die.* Instead of that, if all the cavities are not alike and different shapes are cast in one shot, such a die is called *combination die.*

Design parameters

The important features to be considered for planning a mould for pressure die casting are briefly discussed as follows:

- Die draft
- Ejector pins and their location
- Air venting
- Cores, slides, loose pieces and inserts
- Die lubricants
- Die cooling
- Gating system
- Metal flow and pressure
- Stroke speed
- Cavity filling time
- Gate velocity
- Final metal pressure

Life and material of a die

Life of a die is determined by its resistance to cracking of the surface due to thermal shock from rapid variation of temperature on the cavity surface. The material for die block is carbon or alloy steel and the die cavity is formed by rough machining followed by spark erosion technique or by precision investment casting process. Dies used for casting low melting point metals like zinc, aluminium or magnesium alloys have long life. The difficulties increase and die life decreases with the rise of melting point of cast metals. Using conventional materials for die and shot chamber, the life can be roughly estimated as:

Zinc base alloys	>100,000 castings
Aluminium and magnesium base alloys	20000 to 100000 castings
Copper base alloys	<10000 castings

The life of a die can be improved much by selecting better composition of die steel. Aluminium alloy castings more than 500,000 can be obtained by using a die steel of composition 5% Cr, 1% Si, 1.3% Mo, 1% V, and 0.37% C, hardened to 425 HB to 460 HB by heat treatment. Die life is increased slightly by light carburizing treatments.

Die temperature

The main die areas to be controlled are those of the cavity face at injection and ejection; face temperature at ejection has an effect on the surface finish of the casting whilst face temperature at casting ejection affects dimensional tolerances of the castings and formation of surface blisters.

The temperature gradient behind the cavity surface determines the rate and pattern of solidification. The cycle time, surface finish and internal soundness are directly related to the die temperature pattern.

Die temperature can be controlled either by regulating the flow of cooling media through the die or using a DTDC (die temperature cycle time) controller, allowing the temperature of the casting instead of the machine timing to control the die cooling cycle.

Dies used for aluminium die casting are water-cooled in service and should be preheated to 175°C. Preheating should be done with water flowing slowly through the die to prevent serious damage that would result from introduction of cold water into a hot die. Preheating should be done with the slides removed since the slides tend to expand sooner than the guide reducing clearance.

Metals for die casting

The metals preferred for die casting are low melting point nonferrous alloys. The following base metals, alloyed with small percentages of other metals, are generally used as die casting material:

1. **Zinc based alloys**: Zinc has several advantages over its rivals. Low melting point and high fluidity of the metal makes zinc based alloys the most easy and suitable for die casting. Thin sections can be cast and dies normally have longer life due to its low melting point. They

exhibit good mechanical properties at room temperature, but not strong at elevated temperatures. Two examples for usually used zinc alloy composition are:

(i) 0.075% Fe, 0.03% Mg, 0.001% Sn, 0.003% Cd, 0.003% Pb, 3.8% Al and balance zinc.

(ii) 0.75% Cu, 0.075% Fe, 0.03% Mg, 0.001% Sn, 0.003% Cd, 0.003% Pb, 3.8% Al and balance zinc.

Automobile parts, door and window fittings, radiator grills, parts of refrigerators, various components of washing machine, television, business machines, etc., are cast by this process.

2. **Aluminium based alloys**: Aluminium is a lighter metal having high electric conductivity and good machining properties. Since the melting point is not high, it can be easily cast by cold chamber pressure die casting process. Two examples for usually used aluminium alloy compositions are:

(i) 2.5% Cu, 11.5% Si, 1.2% Zn, 1% Fe, 0.3% Mg, 0.2% Sn, 1% Ni, 0.3% Pb, and balance aluminium.

(ii) 4% Cu, 9.5% Si, 1% Zn, 0.3% Fe, 0.1% Mg, 0.3% Sn, 0.5% Ni, 0.5% Pb, and balance aluminium.

Automobile parts, components used in aircraft industry, parts of projectors, domestic appliances, components of various instruments, etc., are cast using aluminium based alloys.

3. **Copper based alloys**: Components made of copper based alloys possess good corrosion resistance, better machinability, relatively high tensile strength and wear resistance. Cold chamber process is a suitable casting method and high pressure is to be exerted to produce defect-free castings. Due to high temperature and pressure, the life of the die used is generally less for the metal. Two examples for normally used copper alloy compositions are:

(i) 40% Zn, 1.5% Pb, 1.5% Sn, and balance copper.

(ii) 40% Zn and balance copper.

Aircraft, automotive and marine fittings, impellers for corrosive liquid pumps, electric machinery parts, components of chemical apparatus, etc., are cast by copper based alloys.

4. **Magnesium based alloys**: Extremely light in weight, reasonable strength, good resistance to surface corrosion are some good properties of magnesium based alloys, but they are expensive. Other characteristics are almost same as that of aluminium and cast by cold chamber process. The composition of the most popular magnesium alloy used for die casting is: 1% Zn, 0.5% Mn, 11% Al and balance magnesium.

Aircraft parts, camera components, bodies of portable instruments, laptop frames, etc., are die cast by magnesium based alloys.

5. **Lead based alloys**: Lead based alloys containing 15% antimony, 5% tin and 0.5% copper are good for die casting. Lead base alloys are costlier than zinc and toxic to some extent.

Hence, for specific uses like radiation (X-ray) shielding, battery parts, light duty bearings, etc. only they are selected.

6. **Tin based alloys**: They are excellent for die casting because of the low melting point, but the cost is high and possess poor mechanical properties. The commonly used composition is 19% lead, 16% antimony and the balance, tin. This alloy is used as bearing material called babbit metal and also used in aircraft and automobile industries.

Zinc base alloys	>100,000 castings
Aluminium and magnesium base alloys	20000 to 100000 castings
Copper base alloys	<10000 castings

Table 18.2 Alloys recommended for die casting

Material	Alloy designation					
	Commerical	UNS	Melting poial, °C (°F)	Ultimate tensile strength, kPa (lbf/in^2)*	Castability	Remarks
Aluminium	380	A03800	593 (1100)	317,000 (46,000)	Excellent	Most popular aluminium alloy; best combination of properties and ease of use
	383	A03830	582 (1080)	310,000 (45,000)	Excellent	Higher-silicon version of 380; used when thin walls and pressure tightness are required
	360	A03600	596 (1105)	303,000 (44,000)	Good	Used when better corrosion resistance and ductility are required
Zinc	AG40A (Zamak No. 3)	Z33520	387 (729)	283,000 (41,000)	Excellent	Used for the majority of commercial applications for thin walls and good platability
Magnesium	AZ91B	M11913	596 (1105)	234,000 (34,000)	Excellent	Used for the majority of commercial applications for lightness with strength
Brass	858	C85800	899 (1650)	379,000 (55,000)	Fair	Used when high strength, elongation and corrosion resistance are required

*The typical values indicated are for separately die-cast test bars and do not represent values for specimens cut from die castings.

18.7 ADVANTAGES AND LIMITATIONS

Pressure die casting process has much advantages compared to other special casting methods. The process is widely popular in mass production industries of small- and medium-sized components of nonferrous metals due to its inherent plus points. The list of major advantages is given below:

Advantages

1. Die casting dies retain their accuracy and usefulness for long periods of production. More than 500,000 castings can be taken in zinc alloys using a die set.
2. Pressure die casting is characteristically a high speed production process and a natural choice for high production parts. The rate of production is high as more than 300 shots per hour for submerged plunger machines. In cold chamber machines using hand ladling, this value reaches to 150 shots in an hour.
3. Close dimensional tolerance can be maintained. For zinc alloys, a tolerance of ± 0.075 mm on dimensions up to 25 mm size and ± 0.025 mm for each additional 25 mm can be easily achieved.
4. Intricate castings can be produced in a short time with high repetitive quality and dimensional accuracy.
5. Thin sections (0.5 mm for zinc and 0.8 mm for aluminium) with high tensile strength can be cast without much difficulty.
6. Very good performance is attained in casting of products with complicated details on surfaces. Threaded components and small diameter holes of 0.8 mm can be cast in zinc alloys.
7. Surface finish of 0.80 micron to 3 microns (Ra value) is obtainable in die casting depending on the metal and the controlling parameters.
8. Intricate shapes can also be produced by proper die design.
9. Castings with inserts can be successfully produced by this process.
10. The machine used for casting can be fully automatic; hence the labour content is less relative to other casting methods. The reduction of labour is assumed as 90% to 92% compared to a usual foundry shop producing the same item.
11. Castings produce fine-grained structure, thereby having better mechanical properties.
12. Castings are supplied with the flash trimmed off, and in many cases they are almost ready for assembly. Multitude of operations that would be required with other methods can free productive capacity and space for the other uses.
13. Rapid cooling rate produces high strength and quality.

Limitations

1. Casting size is limited because very large dies are required for larger components. The capacity of machine, the die size and the type of metal decides the maximum casting weight. The reported highest values of cast products are 90 kg in zinc and 30 kg in aluminium.
2. Commercial production is limited to nonferrous metals. Not all metals can be successfully and economically cast by the process.
3. The cost of the die is high; hence the process is economical only if the rate of production is sufficiently large to compensate the expenses. Pressure die castings usually require a few secondary operations.
4. Die cast products generally contain porosity due to entrapped air. This reduces mechanical properties and becomes difficult to heat treat, because the air entrapped expands resulting in blisters on surfaces.

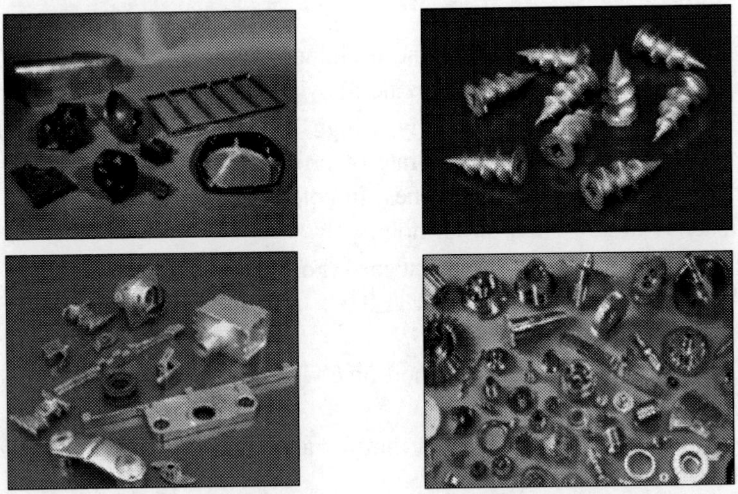

Large to medium size components
Production rates: 120 to 500 shots per hour

Miniature components
Production rates: in excess of 1500 shots/hr

Figure 18.9 Typical components produced by die casting

5. Metal is lost considerably due to badly adjusted burners of hot chamber, poor fluxing practice, and oversized flashes in dies and poor storage conditions.
6. Not much suitable for ferrous castings.
7. Chemical treatment (electroplating, anodising, etc.) after casting is difficult.
8. Intricate casting is not possible because of ejection problem.

Defects in die castings

- Cold shuts: A lapping of solidified metal that occurs due to incoming streams of metal divided into several separate jets that fill different parts of the cavity and meet again after traversing individual parts. During their separate travels, these streams lose some heat and develop thin oxide films in their fronts. To ensure complete fusing, it is necessary that these separate jets retain enough heat to fuse and that they come together under sufficient pressure and velocity to break up the oxide films.
- Excessive heat marks: Excessively blackened portions on the surface of the casting spoil the surface finish.
- Blisters: A surface bubble or eruption is caused by expansion of entrapped gas beneath the plating on the casting.

18.8 CHARACTERISTICS AND APPLICATIONS

High dimensional accuracy and good surface finish are the inherent basic properties of pressure die casting. In these aspects, only investment casting can come near this range. The tolerance attainable for zinc alloys is considered 0.1% to 0.5% of the dimension. For aluminium alloys, the allowance would be increased by 50% of this value. Similarly, thin sectioned castings of about 0.38 mm can be cast for short areas and the general minimum thickness for normal purposes regarded as 1.5 mm for zinc alloys and 2.3 mm for aluminium alloys. This thickness range is comparable to sheet metal products.

Pressure die casting offers reasonable flexibility in design, but not much as in sand casting. Directional solidification is not important here. Most of the complex shapes can be cast, but the restriction is mainly on the cored cavity. The withdrawal of the core has to be planned in a simple way while designing a hollow shape. At the same time, metal or nonmetal components can be used as inserts in the casting for different properties and shapes like threaded components and bearing bushes.

Due to entrapped air, pressure die casting is susceptible to internal porosity. By careful design of the mould and control over the metal injection process, sound and fine-grained castings with good surface layer for a depth 0.02 mm to 0.2 mm (dense and nonporous) can be obtained. Application of the vacuum method and use of wide metal feeders instead of narrow ones in the gating system, can eliminate or reduce the internal porosity. At the end of a shot, a hydraulic shock is experienced for a short time and this moves the ejector die slightly backward, resulting in formation along parting line and slight dimensional change. High locking force and better locking method are the solutions for this kind of error.

Die casting has become the leading process in mass production of castings of zinc, aluminium and magnesium alloys. Automobile parts like carburettor components, aluminium cylinder blocks, brass valves, crank cases, bush bearings, etc., are generally produced by this method. Components used in aircraft industry, parts of projectors, domestic appliances, components of various instruments, marine fittings, electric machinery parts, bodies of portable instruments, laptop frames, etc., all include the items cast by this popular mass production technique. Typical applications for pressure die casting is given in Table 18.3 and comparison of various processes is given in Tables 18.4 and 18.5.

Table 18.3 Typical applications of pressure die casting

Aluminium	*Zinc*
Radiator grills	Lamp bezels
Carburettor bodies	Carburettor and distributor bodies
Distributor bodies	Door handles
Water pump housings	Dynamo end plates
Engine blocks	Starter motor end plates
Cylinder heads	Thermostat housings
Housing for gear boxes, clutches and timing gears	Wiper motor casings
Motor frames	Switchgear housings
Motor rotors	Switchgear covers
Diesel engine oil pans	Gearbox housings and covers
Pulleys and other rotating parts where weight is critical	Pull handles and pulleys
Washing machine paddles	Refrigerator door handles

Table 18.4 Comparison of various processes

Process	*Appearance and finish*	*Cost*
Pressure die casting	Very good, can be finished with variety of mechanical, plated, chemical or organic finishes.	High equiqment cost, thigh tool cost, low piece cost on high activity items.
Gravity die casting	Usually machined or ground but left with base metal surface.	Low equipment cost, medium die cost. Piece cost between die casting and sand casting.
Sand casting	Inferior to die or gravity castings, machined or ground, left with base metal finish.	Low equipment cost, low tool cost, high piece cost.
Powder metal	Good, but porous, usually left with base metal finish.	Medium equipment cost and tool cost, low unit cost on high activity items.
Drop forging and die pressing	Inferior to die casting, usually left in base meta finish.	High equipment cost, high tool cost, reasonably low unit cost on high activity items.
Plastic moulding	Excellent, can be moulded in a variety of colours with excellent finish.	High equipment cost, high tooling cost, low piece cost on high activity items.

Table 18.5 Comparison of die casting with other production processes

Process	*Materials*	*Rate of production*	*Strength of parts*	*Complexity*
Pressure die casting	Lead, tin, zinc, magnesium, aluminium, copper alloys	Very high on certain components up to 2000 per hour	High unit strength	From simple to very complex
Gravity die casting	Iron, magnesium, aluminium, copper alloys	Relatively low	High, particularly when heat treated	In general not so complex as die casting unless sand cores are used
Sand casting	Iron, magnesium, aluminium, copper alloys	Low, unless foundries fully mechanised	Less than that of die castings and gravity die castings	From simple to very complex
Drop forging and die pressing	Steel, magnesium, aluminium, copper alloys	Reasonable	Highest of all	Fairly complex
Plastic moulding	Thermoplastic and thermosetting resins	Very high, equal to die castings	Less than parts mentioned above	As die casting with the advantages that components are self-coloured.

18.9 GRAVITY DIE CASTING

The process is known as gravity die casting in England and as permanent mould casting in USA. The terms are used interchangeably to describe both the process and the product. This process is widely used for aluminium and copper alloys and for a limited output of magnesium alloys and certain grades of cast irons. This process can also cast steels.

The molten metal is poured into the mould under gravity only; no external pressure is applied to force the liquid metal into the mould cavity. So gravity die castings are much thicker in cross section than high pressure die castings, as gravity alone does not provide enough pressure to force molten metal into moulds in sections thinner than 5 mm or less.

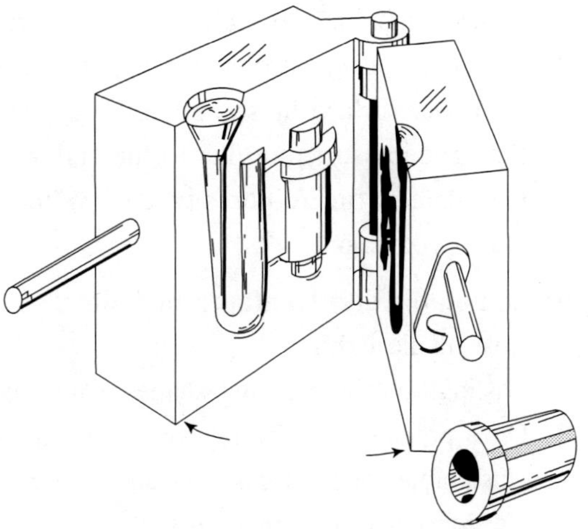

Figure 18.10 Solidified permanent mould casting with gauge and metal core shown in left of mould with completed casting in front

Permanent moulds are made into two halves in order to facilitate the removal of casting from the mould. They may be designed with a vertical parting line, as in Figure 18.10, or with a horizontal line as in ordinary sand moulding. The mould material is usually a good grade of cast iron, although die steel, graphite, copper and aluminium have also been used.

The sequence of operations in gravity die casting includes:

1. Cleaning the mould and maintaining it at proper casting temperature.
2. Painting or spraying the mould surface with a thin layer of refractory wash, or blackening it with carbon soot.
3. Inserting the core, if necessary, and closing the mould by hand or by automatic action,
4. Pouring the metal from a hand ladle.
5. Ejecting the casting from the mould automatically or manually after allowing sufficient time for it to solidify

Permanent moulding is easily automated (Figure. 18.11). Pouring may be automated using tilting pour ladles, or the moulding may be placed in a cradle. As the metal is poured into the moulding, the cradle is lowered at a rate consistent with the desired rate of fill, effectively filling the moulding at a constant level. If the metal source, rather than the moulding, is in the cradle, it is raised, and the moulding remains stationary. This method virtually ensures controlled fill and the elimination of turbulence and dross entrapment in the casting. Another method used to fill large permanent mouldings is to fill reservoirs at one end of the moulding with metal, and then tilt or move.

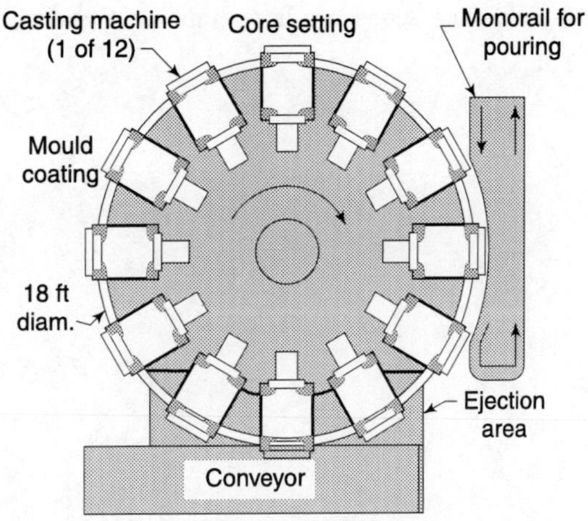

Figure 18.11 Twelve machine turntable for automatic gravity die casting

The most important advantages of gravity die casting over sand casting are.

1. Closer dimensional tolerance and accuracy.
2. Smoother surface finish and better appearance; less machining.
3. Fine-grained structure due to the chilling effect of metal mould.
4. Less defects (micro-porosity, etc.) which are commonly found in sand castings.
5. Mass production.

Some notable disadvantages include:

1. Cost of die is high.
2. Castings of all sizes and shapes cannot be done.
3. Once prepared, the gating system cannot be changed; less flexible operation.
4. Unsuitable for steel castings due to high melting temperature.
5. Uneconomical for small runs.

Review Questions

1. What is die casting?
2. What are the alloys commonly used in die casting?
3. Explain die casting processes with example.
4. What are the steps to be followed in pressure die casting process?
5. Give typical application of pressure die casting.
6. What are the limitations of pressure die casting?
7. What are the advantages of pressure die casting?
8. How do we compare die casting with other production processes?
9. Explain gravity die casting processes.
10. What are sequences in operations in gravity die casting.

CHAPTER 19

Shell Mould Casting

19.1 INTRODUCTION

Shell moulding is a casting process in which the mould is a thin shell (typically 9 mm or 3/8 inch) made of sand held together by a thermosetting resin binder. It was developed in Germany during the early 1940s. As compared to conventional sand casting, this process has better dimensional accuracy, a higher productivity rate, and lower labour requirements. It is used for small to medium parts that require high precision and better surface finish.

The materials required for shell mould preparation are sand, resin-binder and additives. Washed silica sand or zircon sand of A.F.S fineness number 100 to 180 is widely used as the base material for the shell. The synthetic resins used in shell moulding are thermoplastic resins to which setting or polymerizing agents have been added to produce thermosetting characteristics. Phenol formaldehyde resins are the most commonly used because when they are combined with sand, they have high strength, resistance to heat and moisture, and good flowing and curing properties. Additives are sometimes incorporated into resin sand mixtures to improve surface finish and eliminate thermal cracking during pouring. One of the most common additives is vinsol. It is added about 8% to 12% in resin. Other additives are iron oxide, coal dust, treated western bentonite, etc.

19.2 PROCESS PARAMETERS AND STEPS IN SHELL MOULDING

In this process, the moulds and cores are prepared by mixing the dry free flowing sand with thermosetting resins and then heating the aggregate (mixture of fine sand (100-150 mesh) and thermosetting resins) against a heated metal plate. Due to the heat, the resin cures, which causes the sand grains to get bonded with each other and it forms a hard shell around the metallic pattern. The inside portion of the shell is the exact replica of the pattern against which the sand aggregate is placed before heating. The shape and dimension of the inside portion of the shell thus formed is exactly the same as that of the pattern. If the pattern is of two pieces, then the other half of the shell is also prepared the same way. Two halves of the shells prepared are placed together after inserting the core, if any, to make the assembly of the mould. The assembly of the shell is then placed in a moulding flask and backing material is placed all around the shell mould assembly to give its assembly the sufficient strength. Now the shell mould is fully ready for pouring the liquid metal.

Sand

The dry free flowing sand used in the shell mould must be completely free of clay content. The grain size of the sand used in shell moulding is generally in the range of 100-150 mesh, as the shell casting process is recommended for castings that require good surface finish. However depending on the requirement of surface finish of the final casting, the grain size of sand can be ascertained. Also, if the grain size is very fine, it requires large amount of resins, making it expensive.

Resin and Catalyst

The resins most widely used, are the phenol formaldehyde resins, which are thermosetting in nature. Combined with sand, they give very high strength and resistance to heat. The resin initially has excess phenol and acts like a thermoplastic material. In order to develop the thermosetting properties of the resin, the coating of the sand is done with resin and a catalyst (Hexa-methylene-tetramine, known as Hexa). The measure of resin is 4-6% of sand by weight, the catalysts 14–16% of sand by weight. The curing temperature of the resin along with the catalysts is around 150°C and the time required for complete curing is 50–65 seconds. The sand composition to be used in making various castings of different materials can be seen from the relevant standards. The resins available are of water-borne, flake, or the granular types. The specifications of liquid, flakes or powder resins can be obtained from IS 8246–1976, IS 11266–1985, and IS 10979–1981 respectively. The resin-sand mix aggregate can be prepared by the following three ways.

Hot coating process: In hot coating process, the curing of resin takes place due to the combined effect of heat as well as chemical action of the resin with the catalyst. Once curing is done, the cured sand is cooled at 40-50°C to prevent the lumps and agglomerates and to improve the flowability.

Warm coating process: In this process, a different resin formulation (liquid solvent solution) is used and curing takes place at around 80°C. The process is simpler than hot coating but the quantity of resin consumed is more.

Cold coating process: In this process, the sand is first mixed with catalysts, then the resin mixed with alcohol is added to the aggregate. The resin requirement is highest in comparison to the amount required in hot and warm coating processes.

Phenol-Formaldehyde Resins

In manufacturing the shell sand, phenol, formaldehyde resin is used as a binder. The form of resin may be liquid or flake type. Liquid resin is nothing but is a resin dissolved in alcohol. Liquid resin is used for manufacturing shell sand by either warm air process or by ignition process, whereas solid or flake resin is used for hot coating process. Most of Indian manufacturers of shell sand use liquid resin, because of the easiness of resins of the process. The following properties of resins are generally checked as an acceptance criteria.

Liquid Resins

- Clarity

- Viscosity
- Specific gravity
- Solid content
- pH value
- Coated sand properties at certain percentage of the resin.

Solid Resins

- Softening point
- Flow rate
- Particle size

Hexa Catalyst

The phenol formaldehyde resins are thermoplastic in nature and require a formaldehyde donor to cure at a certain temperature. Thus after blending of the resin and the catalyst, it becomes thermoset in nature and the formation of shell moulds and cores is accomplished. The catalyst used is a blend of hexamethylene tetra amine and a lubricant. Lubrication helps in the flowability of shell sands. Hexa catalyst is available in the form of a fine powder.

Use of Additives

Additives may be added to the sand aggregate to further enhance the surface finish of the casting or to improve the strength of the mould or to develop resistance to thermal cracking and distortion. The recommended additives are coal dust, manganese dioxide, calcium carbonate, ammonium boro-fluoride, lignin and iron oxide. To improve the flowability of the sand and to permit easy removal of shell from the pattern plate, some lubricants are added in the resin-sand aggregate. The common lubricant used for such purpose is calcium or zinc stearate.

19.3 STEPS IN PREPARING THE SHELL MOULD

The steps to prepare the shell mould are (Figure 19.1):

1. A match plate metal pattern comprising the cope and drag is heated to the required temperature and fitted over the box containing the mixture of sand and thermosetting resin.
2. Next this box is inverted such that, the sand-resin mixture falls on the hot pattern. This cures a layer of the mixture to a certain extent and forms a hard shell.
3. Once the desired thickness of the shell is achieved, the box is rotated back to its original upside position. The excess sand then falls back into the box, thus forming a shell over the pattern. The obtained thickness depends on the temperature and the time of contact of sand mixture. The required shell thickness for casting depends on the temperature of the pouring metal and complexity of the final casting. It may range from 2-8 mm.
4. To complete the curing process, the sand shell along with the metal plate is heated in an oven for some calculated time.
5. The obtained shell mould from this process is removed from the pattern.

6. Now the two portions (cope and drag) of the shell mould are thereby assembled and some sand particles or metal shots are used in a box for support purpose to add stability to the shell and the pouring is done.
7. After it gets cooled, the finished casting along with sprue is taken out and fettling is done.

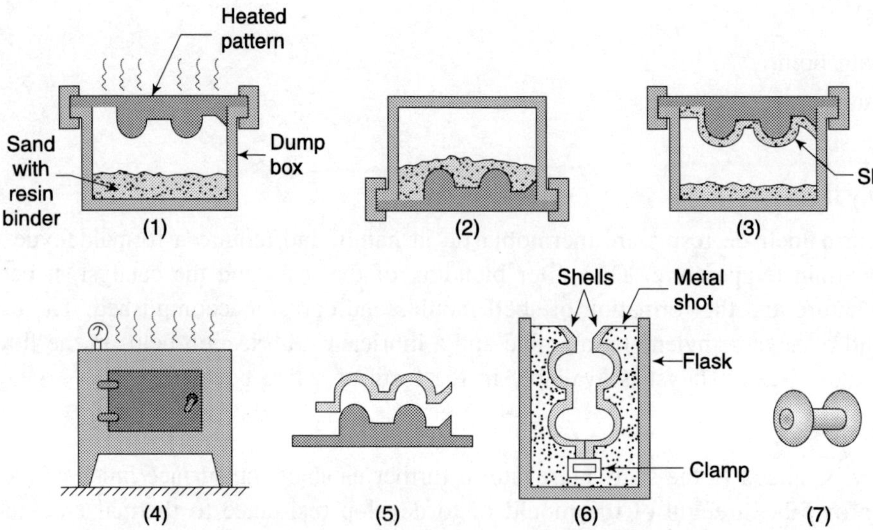

Figure 19.1 Steps in shell moulding processes

19.4 PARAMETERS AFFECTING THE QUALITY OF CASTINGS PRODUCED BY SHELL MOULD PROCESS

To identify the process parameters that affect the quality of the castings produced by shell moulding process, Ishikawa cause-effect diagram has been constructed, as shown in Figure 19.2. It is indicated through the Ishikawa cause-effect diagram that the following process parameters can affect the casting quality (dimensional accuracy, surface roughness, mechanical and metallurgical properties) made by shell moulding process.

- Pattern: Dimensional accuracy, surface finish and draft requirement.
- Additive: Type, function of additive, properties.
- Alloy: Pouring temperature, pouring time and type.
- Sand Type: Silica, zircon; size, size distribution and grain shape.
- Resin and catalyst: Types of resins, catalysts and their costs.
- Curing: Temperature and time.

Advantages and Limitations

The advantages are as follows:

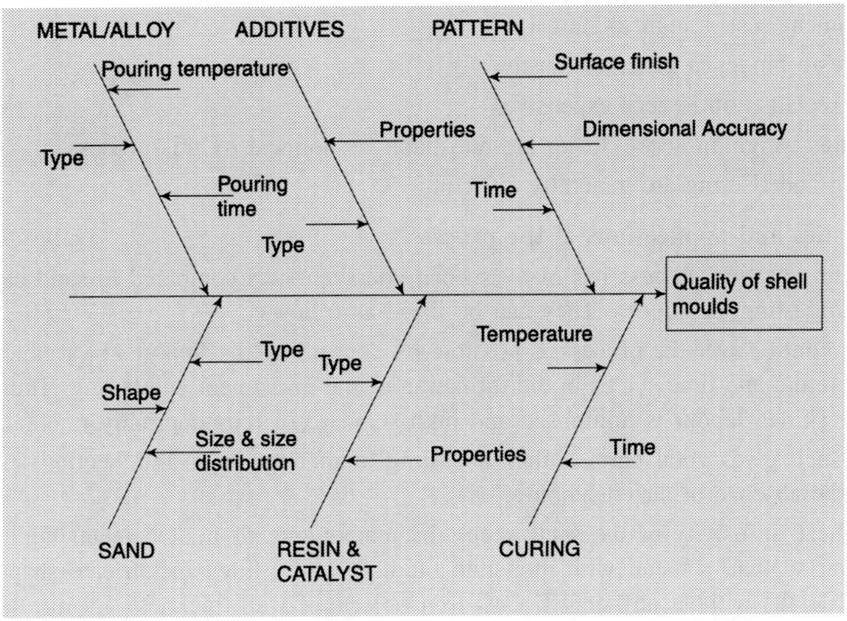

Figure 19.2 Process parameters affecting the shell moulding process

1. The surface of the shell mould cavity is smoother than a conventional green-sand mould, and this smoothness permits easier flow of molten metal during pouring and better surface finish on the final casting. Finishes of 2.5 μm can be obtained.
2. Good dimensional accuracy is also achieved, with tolerances of ±0.25 mm possible on small-to-medium-sized parts. The good finish and accuracy often precludes the need for further machining.
3. Less moulding skill is required because the process is semi-automatic.
4. Sand requirement is about 10% to 12% of conventional sand casting process.
5. Collapsibility of the mould is generally sufficient to avoid tearing and cracking of the casting.
6. This process can be well adopted for mass production of small castings.
7. Less floor space is required.
8. All metals can be cast by this process.
9. Shakeout and cleaning of castings are simple and easy.
10. Shell core are efficient and can be used in other casting processes.

Limitations are as follows:

1. It is a more expensive metal pattern than the corresponding pattern for greensand moulding. This makes shell moulding difficult to justify for small quantities of parts.
2. Pattern cost is high because metal patterns are required.

3. Equipment cost is high to start a unit.
4. The resin binder is generally costly.
5. Sand reclamation is very expensive.
6. Castings weighing above 12 kg are not much economical by this process.
7. Gating and risering are generally difficult.

Characteristics and applications of the process

Shell moulding casting process has many special characteristics compared to sand casting and other special casting techniques. They can be listed as follows:

1. Close control over the process is possible by controlling the important variables such as temperature and time cycles in the formation of the investment and curing. Thus, the part of the skilled labour is minimised and high standard of reproducibility is obtained.
2. Thin shell gives good permeability, so filling of metal even in thin sections is possible with the absence of air cushioning.
3. Low heat diffusivity of the shell keeps the temperature of molten metal high as at the pouring so that, a metal with moderate fluidity can be filled in intricate shapes having sharp details without any defect. Cast iron objects of wall thickness 1.9 mm to 2.3 mm can be successfully cast by this method. Similarly, holes of diameter 3 mm can be cast if required.
4. The draft allowance of 0.5° is the usual value for shell moulding, but this can be reduced to as little as 1 in 1000, if necessary for a particular machine part.
5. Better accuracy and finishing can eliminate machining or reduce to small grinding.
6. Shell mould process is most adaptable to castings which are relatively shallow in a direction normal to the parting line.
7. The process is suited to ferrous and nonferrous alloys and the weight range is usually considered 0.1 kg to 10 kg, but thin walled casting up to 300 kg can be cast.
8. Certain alloys (low carbon and low alloy steels) are susceptible to surface defects called pinholes. The suggested cause is local evaluation of gaseous hydrocarbons at a time, when surface permeability is reduced due to physical changes in the resin.
9. Production costs are only marginally higher than that of sand casting. Initial pattern costs are higher, so a design change is expensive.

Shell moulding process is particularly suited for mass production of small- and medium-sized machine parts in aluminium, copper alloys, and iron, where high quality precision castings are required with a minimum of finishing, and for casting metals that are normally difficult to machine. Air-cooled engine cylinders, valves, rocker arms of engines, camshafts, gears, bearing caps, water meter body, etc., are cast by this process. Shell moulding is now more commonly used for making cores to fit in sand moulds and metal moulds.

Review Questions

1. What is shell moulding?
2. Why is shell moulding used?
3. What are the process parameters in shell moulding?
4. What resins and catalysts are used in shell moulding?
5. What are the uses of phenol-formaldehyde resins?
6. What are the common additives used in shell moulding?
7. Explain the shell moulding process.
8. What are the advantages of shell moulding process?
9. What are the limitations of shell moulding process?
10. What are the characteristics and applications of the shell moulding process?

CHAPTER 20

Investment Casting Process (ICP) and Ceramic Shell ICP

20.1 INTRODUCTION

The investment casting process which is commonly referred to as the "lost wax method", originated in and around the fourth millennium B.C. It is evidenced from the architectural works found in the form of idols, pectorals and jewellery in the remains of ancient Egypt and Mesopotamia. The investment casting process initiates with the production of wax replicas or patterns of the required shape of castings. Each and every casting requires a pattern to be produced. Wax or polystyrene is used as the injecting material. The assembly of large number of patterns are made and attached to a wax sprue centrally. Metallic dies are used to prepare the patterns. The pattern is immersed in refractory slurry which completely surrounds it and sets at room temperature forming the mould. The mould is further heated, so that the pattern melts and flows out, leaving the required cavity behind. After heating, the mould gets further hardened and molten metal is poured while it is still hot. After the casting gets solidified, the mould is broken and it is taken out.

There are two forms of investment casting. One possibility is that the ceramic mixture is poured around the wax in a flask and allowed to harden (the "solid mould" process). In the "ceramic shell" process, the pattern is repeatedly dipped in a ceramic slurry, dusted with refractory, and dried in a repetitive process until a shell of sufficient thickness has been formed around the pattern. Today, the solid mould process is used primarily for jewellery and dental casting; the overwhelming majority of investment castings are made using the ceramic shell process.

The basic steps of the investment casting process are shown in Figure 20.1 and discussed below:

1. Preparing the heat-disposable wax, plastic or polystyrene patterns in a die.
2. Assembly of the prepared patterns onto a gating system.
3. "Investing" (covering) the pattern assembly with a refractory slurry which builds the shell.
4. Melting the pattern assembly (burning out the wax) by firing, for removing traces of the pattern material.
5. The metal in molten state is poured into the mould.
6. Once the metal solidifies, the shell is removed (knocked out).

7. Fettling (cutting off) of the pouring basin and gates followed by finishing operations to get the desired dimensional tolerances and finish.

Steps in Precision Casting Process

Different steps of investment casting (the ceramic shell) process are discussed below:

1. Production of die for making expendable pattern This is known as the master pattern. Generally, die steels or duralumine are used for this purpose which are accurately dimensioned in a tool room. For this purpose, double shrinkage of the expendable pattern and of the die should be taken into consideration. Double shrinkage for a brass casting in a lost-wax process can be calculated as follows:

Shrinkage for wax mixture	1.0%
Shrinkage for brass	1.8%
Total contraction	2.8%

But thermal expansion of mould material somewhat compensates the above shrinkage. This is equal to:

Expansion of mould materials during solidification	0.3%
Thermal expansion of mould during pattern elimination heating	1.0%
Total expansion	1.3%

Thus, a total shrinkage of (2.8–1.3) or 1.5% should be compensated for. This factor again depends on pouring temperature, rate of pouring, restrictive action of mould geometry, and the rate of cooling.

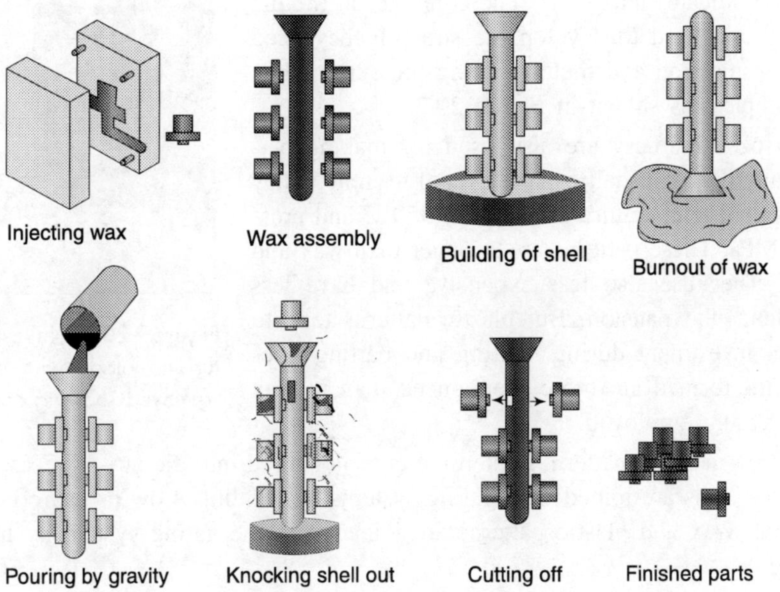

Figure 20.1 Basic steps of the investment casting process

2. Making of expendable patterns and gating system From the master pattern, positive replica of the disposable patterns are made. Most commonly used pattern material is wax from which lost wax process has its name. The pattern is formed by injecting pattern wax into a pattern die. Pattern wax is a blend of natural wax modified with fillers, resins, plastics, plasticizers, antioxidants, and dyes. Waxes used for patterns are most often paraffins, vegetable waxes, and microcrystalline waxes. By blending these waxes with each other and adding modifiers as needed, waxes having very specific properties can be produced. The choice of wax depends on the size and shape of the casting to be produced. Additives are used to increase the strength of the wax and to compensate for its shrinkage on cooling.

The patterns are injected at moderately low temperature and pressure. The temperature and pressure commonly used in wax-injection machines are 70–100°C and 3 MPa respectively. After they harden in the die (which usually occurs in less than a minute for small parts), they are removed from the die and allowed to cool. If they contain preformed cores, they may be X-rayed to be sure that the cores did not fracture during injection. The patterns are inspected visually, die parting lines and the injection feeder are removed, and surface blemishes repaired.

Preformed ceramic cores, if needed, are placed in the pattern die before injection. Alternatively, soluble wax cores may be used instead of ceramic cores; these are dissolved out of the pattern using a weak acid before the slurry is applied. The preformed ceramic cores are primarily made of silica and alumina; they are removed from the casting by dissolving the cores in a caustic solution after casting. The cores are made by injecting the core mixture (usually a slurry of finely divided refractory aggregate in ethyl silicate) into a die. The cores gel in the die and are removed and fired to develop the strength they need to withstand wax injection and metal pouring pressures. They may be very complex, as shown in Figure 20.2.

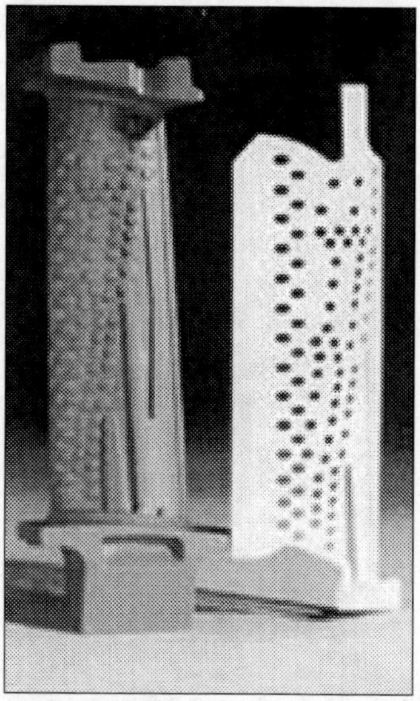

Figure 20.2 Investment cast turbine blade with convex wall removed showing complex core

Plastic, tin, frozen mercury are also used for making patterns. Plastic patterns are usually made of polystyrene. They require higher mould temperature around 150-350°C and pressure about 140 MPa. These patterns are stronger than wax and easy to handle. They are also less expensive and have less thermal coefficient of expansion. But plastic patterns tend to distort the green investment during melting and parting lines and core flash fins formed are more pronounced than in wax due to higher pressure employed.

In case of frozen mercury pattern, molten mercury is poured into die at −40°C as its melting point is −38.83°C. Parts are joined with gating systems by welding. Low pressure is applied at the joints for that. Wax and plastic patterns are joined with the gating system by heating and partially melting.

The gating system is also formed in the same manner as the patterns and attached to the pattern assembly by wax-welding. The assembled clusters are inspected and cleaned by dipping in a solution containing a mild solvent that etches the surface of the pattern so that the slurry adheres to it.

3. Precoating the pattern assembly The pattern assembly is dipped into a slurry of refractory coating material. The slurries are usually mixtures of fine grained refractory aggregate (usually of 325 mesh or less) in an aqueous silica sol (colloidal silica). The refractory sands may be silica, alumina, gypsum, magnesia, zirconium silicate, or a mixture of these. Because of health concerns, ethyl silicate is rarely used as the binder. The common binders used include sodium-potassium silicates, refractory cements, plaster of paris, etc. In addition to the colloidal silica and the aggregate, the slurry may contain a wetting agent, bactericides, antifoaming compounds, and polymers, which are added to give the slurry sufficient green strength, so that it can be redipped without being completely dried between dips. Slurry formulations are usually proprietary. The colloidal silica is a colloidal dispersion of virtually spherical particles of silica in water. The average particle size is 7 to 10 nm. Slurries are controlled by adjusting viscosity, temperature, and solids content.

The first slurry, which serves as coating, consists of a fine refractory aggregate, usually more expensive than that used in later dips. This aggregate contacts the molten metal. The pattern cluster is rotated in the dip tank to coat all pattern surfaces. Automated dipping machines assure repeatability of the dipping process. After dipping, the wet assembly is coated by sprinkling it with a coarser refractory aggregate (40 to 50 mesh). The cluster is dried in ambient air, or heated forced air may be used to accelerate the drying process.

4. Investment of pattern assembly for the production of moulds The precoated pattern assembly is then invested in the mould. Investment mould may be formed by either solid moulding or shell moulding process.

In solid moulding, the pattern assembly is placed in a metal flask and the investment moulding mixture is poured around the pattern. The mould material settles by gravity and completely surrounds the pattern as the work table is vibrated. The moulds are then allowed to air-set. A typical investment moulding mixture consists of 91.2% sand, 33.8% water, 6.5% calcium phosphate and 2.3% MgO (300 mesh).

In shell moulding, the precoated pattern assembly is given a second coat by dipping it in ceramic slurry whose composition is often similar to that of the first slurry. However, the refractory sand that is applied in this coat is usually more coarse and less expensive than that applied in the first coat. The cluster is dried again, and the process is repeated; this time using a coarser, less expensive refractory in the slurry and sanding operations. A number of dips are given in order to build a shell thickness of the order of 6 to 12 mm. The mould is given a final drying and prepared for the dewaxing operation.

5. Removal of pattern from the investment mould The patterns are usually removed by melting or burning them out of the mould. The pattern is melted out of the hardened mould by heating the mould in an inverted position at 100-200°C (for wax pattern) or 60-80°C (for plastic pattern). However, because the pattern material expands about 40 times as much as the mould, care must be taken to avoid cracking of the mould during this operation. Wax drain holes are often added to the cluster to provide a path for molten wax to escape during the dewaxing operation. Wax is a poor conductor of heat, so the outer layers of the pattern melt and run out of the mould before the inner parts of the pattern heat enough to expand to crack the mould.

One method of dewaxing is to use a steam autoclave. In this method, superheated steam is introduced to the patterns at a pressure of 550 to 620 kPa. The steam penetrates the porous mould and gives up its latent heat of vaporization on the wax surface, which heats the surface rapidly and causes the part of the wax in contact with the steam to melt. As that wax runs out of the mould, it exposes the wax underneath to the steam, which melts it, and the process continues. The wax runs into a trough below the moulds and can be recovered and reprocessed. Another method of dewaxing is to flash fire the moulds. In this process, the moulds are placed in a furnace at 870°C to 1095°C. The heat melts and burns out the wax. Wax that is not burnt falls into a trough below the furnace and can be reclaimed. Flash dewaxing units are equipped with afterburners to eliminate smoke emissions.

After dewaxing, the moulds are fired at 870° to 1095°C to remove any residual pattern material and other gas-forming materials, like water and alcohol; completely dry the moulds, and sinter the mould ceramic so that it develops the strength it needs during pouring. This operation may be combined with wax removal in flash dewaxing furnaces and with mould preheating prior to pouring. Investment casting moulds are usually preheated prior to pouring to promote complete metal fill in the thin sections typical of investment castings, and to lessen the thermal shock on the moulds when they are filled with molten metal.

To control solidification, the moulds may be wrapped in refractory blankets prior to preheating. The preheat temperature is determined by the alloy being poured and the casting geometry. When the moulds are completely preheated, they are filled with molten metal. Because the amount of metal poured in investment castings is usually small (<20 kg), the moulds may be poured by hand, or they may be clamped to a furnace just large enough to melt the metal for a single mould. As soon as the metal reaches the pouring temperature, the furnace and mould are rapidly inverted, which fills the mould in a smooth, even manner.

6. Pouring of metal into the mould Melting is usually done using pre-alloyed charges prepared by a custom alloy producer. For some alloys, notably super alloys and titanium alloys, melting, pouring, and solidification is carried out in vacuum (approximately 10-3 torr) to protect the alloys from reactions with oxygen. Induction melting is most often used, although gas-fired furnaces may be used for nonferrous alloys, and vacuum-arc remelting is often used for titanium and refractory metal alloys. In dental and jewellery castings, the moulds are often centrifuged during pouring to assure filling of the intricate designs.

The liquid metal is gravity-poured via sprue into the preheated mould. For thin and intricate castings, centrifuging with compressed pressure of 70 kPa or so may be used. The air pressure is applied through the sprue to force-fill the mould cavity. After the casting is solidified, it is removed from the mould and cleaning operations follow cooling of the casting.

7. Post-casting Operations After casting, the moulds cool, and the ceramic is removed. Clusters are shot blasted, and the casting cut from the sprue. Finishing and inspection operations are similar to those used for sand castings; except, if preformed ceramic cores are used, they must be removed in a caustic bath. If the cores have especially small cross sections, they may be removed in a caustic autoclave, which is pulsed periodically to remove the reaction layer at the core-caustic interface. Caustic (potassium hydroxide) is used to dissolve the silicate binder in the core, allowing the core aggregate material, usually silica and alumina, or aluminosilicate, to fall out of the casting.

Advantages

1. Wax can be recollected and reused.
2. Wax is easy to work and saves time.
3. Surface finish of the products that are cast through investment casting is very good.
4. Due to good surface finish, secondary surface finish is not required and, in turn, material wastage is less. Surface smoothness of 1 to 3 microns, compared to 2 to 4 microns for shell moulding and 5 to 12 microns for sand casting can be obtained.
5. Close tolerance in As-cast condition is obtained.
6. Minimum or no machining is required for completely finished castings.
7. Intricate shapes with good reproducibility can be achieved.
8. Alloys with higher melting points can also be cast.
9. A single process gives good surface finish.

Limitations

1. Patterns are expendable; one pattern is required to produce one casting.
2. Relatively costly process as compared with other casting processes, since it uses the wax, which increases the cost.
3. The cross section of moulds being thin, the process is limited by size and mass of casting.
4. The quality of the pattern determines the quality of the casting.
5. Removal of the wax from the mould again adds up time and cost.
6. Careful handling must be done.
7. Thin and fragile cross sections are obtained and forces are too high to withstand during mould making.
8. There is a size limitation of the part to be cast; majority of castings produced weigh less than 0.5 kg.
9. Use of core makes the process more difficult.

Process capabilities

- Process is capable of producing intricate shapes which are not easily possible through normal process.

- Wax can be easily shaped and formed using a master pattern and finished by intricate tools and skills, making it a robust process.

Applications

- To fabricate difficult-to-machine and difficult-to-work alloys into highly complex shapes.
- For making jewellery and art castings
- In dentistry and surgical implants
- To produce milling cutters and other types of tools
- Defence equipment and accessories
- Automotive equipment
- Aerospace applications
- Marine engineering applications
- Turbine blades
- Cryogenic applications
- Medical applications

Investment casting in dentistry

1. Wax that is made into required shape is further invested into suitable slurry and is made into a pattern.
2. Then the material of the artificial tooth is melted and poured into formed cavity.
3. Again the conventional method for investment casting is repeated.

More issues A typical problem involving investment casting is cleaning and removal of shells. This becomes more important when parts are increasingly complicated and have internal cavities. If some ceramic-sand mixture remains hidden inside the cavities (or) holes, it damages tools like drills and fine grinders during final finishing process and lead to rejection of casting.

Steps taken to correct the above issue:

1. State-of-the-art cleaning systems are required for this purpose.
2. Mainly the following steps in sequence are used:
   (i) Salt baths
   (ii) Rinsing systems
   (iii) Multistage acid brightening systems
   (iv) Efficient material handling devices which are guided by lasers help to pick and place the container containing these investment cast products.
   (v) Automated systems reduce labour costs and increases process efficiency.

Desired properties of the wax (Investment Casting)

1. Wax once injected must get solidified inside the mould quickly as possible.
2. The wax must have low thermal expansion.

3. Melting point must not be higher than the ambient temperature so that the expansion during the injection and energy consumption can be minimized.
4. This improves the cycle time, minimizes the solidification shrinkage which may lead to distortion of pattern.
5. Wax should be sufficiently hard at room temperature so that patterns are self-supported and can be handled without any damages.
6. Wax patterns must have sufficient strength so that breakage is resisted.
7. Velocity of the wax used must be low enough for enabling smooth flow into the thinnest section inside the die cavities.
8. Wax must have smooth and wettable surface so that the same is transferred in the product and also ceramic slurry adheres with its surface.
9. Once the desired mould formation is done, it should be easily released.
10. The wax used should have a lower ash content, such that the shell is eventually clean and ash-free.
11. The wax should be safe from environmental aspects and not form any toxic and hazardous materials after combustion.

20.1.1 Ceramic Shell Investment Casting Process

Ceramic shell investment casting (CSIC) is one of the near net shape casting technologies. The process is based on expendable wax patterns for producing jointless moulds that are required for near net shape castings.

Before we start the discussion on the ceramic shell investment casting process, it is desirable to understand in brief, the process and details of investment casting process, already discussed.

The main difference between investment casting and ceramic shell investment casting is that, in the former process, before dewaxing the wax pattern, it is immersed in a refractory aggregate. Whereas, in the ceramic shell investment casting, a ceramic shell gets built around the tree assembly through repeated dipping of the pattern into the slurry (refractory material such as zircon with binder). After getting the required thickness of cross section, the tree assembly is dewaxed. The shell obtained is further immersed in a refractory coating and the metal is poured into it.

In this process, a wax pattern/assembly is first dipped into a ceramic slurry bath for its primary coating. Thereafter, the pattern is withdrawn from the slurry and is manipulated to drain off the excess slurry to produce a uniform coating layer. The wet layer further stuccoes through sprinkling the relatively coarse ceramic particles on it or by immersing it into such fluidized bed of particles. The ceramic coating is built by successive dipping and stuccoing process. This procedure is further repeated till the shell thickness as desired is obtained. Upon completion, the entire assembly is placed into an autoclave or flash fire furnace at a high temperature. In order to burn out any residual wax, the shell is heated to about 982°C which helps to develop a bonding of high temperature in shell. Such moulds are stored for future use wherein they are preheated for removing the moisture content from it and then, molten metal can be poured into it.

Sequence of Producing Ceramic Shell Investment Casting

The sequence of ceramic shell investment casting is given in Fig. 20.3.

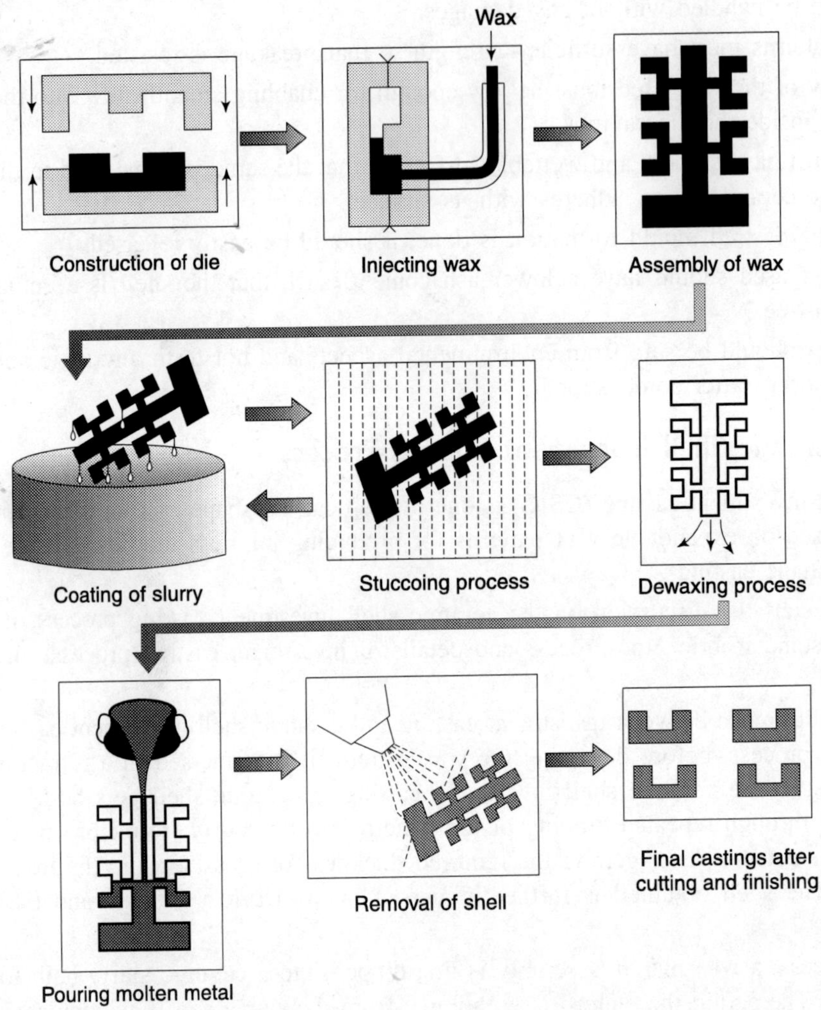

Figure 20.3 Basic steps of the ceramic shell investment casting process

Steps

1. Manufacturing of the master pattern of wax through the master die.
2. Preparation of wax blend and injecting it into the die.
3. Manufacture of wax pattern and assembly of wax pattern.
4. Investment of wax with slurry (coating the slurry).

5. Drying of shell thickness (stuccoing).
6. Dewaxing of raw moulds followed by heating and baking of the shells.
7. Pouring of moulds with molten metal.
8. Once the metal is solidifed, the shells are removed.
9. Cuting off the gates/risers (fettling) followed by finishing operations.

Advantages

- Complex shapes that are difficult to produce by other casting methods are easily produced by this method.
- Thin cross sections and intricacies can be made by this process.
- Finish machining is considerably reduced or eliminated on the castings made by this process, making it economical in cost.
- The process has no metallurgical limitations.
- This process produces castings with excellent surface finish.

Disadvantages

- Expensive process due to the cost of ceramics and pattern (wax cost).
- As the shells are delicate, the process is limited by the size and mass obtained.
- Making intricate and high quality pattern increases the process costs.

Applications

- Aircraft: Turbine blades, carburettor and fuel pump parts, cams, jet nozzles, special alloy valves.
- Chemical industries: Impellors, pipe fittings, evaporators, mixers.
- Tool and die: Milling cutters, lathe bits, forming dies, stamping dies, permanent moulds, etc.
- General and industrial applications: Cloth cutters, sewing machine parts, welding torches, cutter, spray nozzles, metal pumps, etc.

Review Questions

1. What is investment casting?
2. What are the basic steps in investment casting?
3. What are the steps in production of die for making expendable patterns?
4. Explain the process of making of expendable patterns and gating system.
5. Explain the process of precoating the pattern assembly.
6. What are the advantages of investment casting?
7. What are the disadvantages of investment casting?
8. What are the applications of investment casting?
9. What are the process parameters used in investment casting?
10. With the help of a neat a diagram, explain ceramic shell investment casting.

CHAPTER 21

Ferrous and Nonferrous Foundry Practices

21.1 FERROUS FOUNDRY PRACTICES

Ferrous metals or alloys contain iron as the base element with elements such as C, Si, Mn with impurities S and P. Ferrous alloys are very strong, hard and have high density and high melting temperature. Ferrous alloys are used extensively where high strength is required.

They are used in automobile bodies, bridges, IC engines, motor bodies, cranes, etc.

About 75% of the metals/alloys used in most of the engineering applications are made of ferrous alloys.

21.2 CLASSIFICATION OF FERROUS ALLOYS

Classification of ferrous alloys depends on their carbon content:

- Cast iron – C level is 2-4.3%
- Steels – C level is 0.05-2%

(a) Cast Irons

- Wrought Iron: Has a carbon content of less than 0.05% and is the purest form of iron. It is tough and ductile, rusts very slowly and is easy to weld.
- White Iron: Carbon content of up to 2%. It contains iron carbides, is very hard and brittle and is highly wear resistant.
- Malleable Iron: It is heat-treated white iron which contains graphite in the compact form along with pearlite and ferrite in the matrix. They are strong, ductile and malleable.
- Grey Cast Iron: Contains carbon in the range of 2-4.3%. It contains graphite in the form of flakes in a matrix of ferrite and pearlite. Brittle, hard, wear resistant, antifriction and thermal resistant.
- Ductile Iron/SG Iron: Contains the same amount of carbon as grey cast iron (2-4.3%) but it also has sulphur content of less than 0.05%. Graphite will be in the spherical or nodule form in a matrix of pearlite and ferrite. Magnesium is also present to the order of 0.05%. They are hard, have high strength and are ductile.
- Austempered Ductile Iron (ADI): In this material, spheroidal graphite is associated with bainitic matrix. They are tough, strong, wear resistant and can compete with steel.

(b) Steels

Steels are alloys of iron and carbon along with silicon and manganese.

Steels are classified as plain carbon steel which has carbon content of 0.001-2%, with no other special alloying element. They are further classified as:

- **Carbon Steels**
  - Low carbon steel having up to 0.25% carbon, used in general engineering applications.
  - Medium carbon steel having between 0.25% and 0.5% of carbon. Used in hammers and other tools.
  - High carbon steels having more than 0.5% but less than 2% carbon. Used in hand tools, drills, files, etc.
- **Alloy Steels** – Mg, Cr and Ni are used as alloying elements for higher strengths:
  - Low alloy steels contain total alloying elements less than 7%.
  - High alloy steels contain total alloying elements more than 7%.
- **Special Steels**

  They contain specific special alloying elements such as W, Co, Mo, V, Ni, etc.
  - Stainless steel is an alloy of 18 Cr and 8 Ni.
  - High manganese steel is an alloy of 10-11% Mn with 0.9-1.1% C.
  - Heat resistant steel is an alloy of high Cr content used in furnace applications.
  - Die steel contains Cr, Va, Mo and is used for die making.
  - Tool steel is used for making drill bits, milling cutters, punches and dies.
  - High speed steel has a medium carbon content. They are alloyed with W, Cr and Va.

21.3 Fe-C EQUILIBRIUM DIAGRAM

Influence of carbon in iron has the greatest overall effect on its properties as compared to other elements. Only iron with a carbon content of less than 6.7% has any commercial value, since a carbon content above this would give extremely brittle iron.

- The presence of carbon causes a decrease in the coefficient of linear expansion at room temperature from $11.7 \times 10^{-6}/°C$ at 0% carbon content to $8.6 \times 10^{-6}/°C$ at 3.66% C.
- Carbon improves the fluidity of molten iron. As the carbon content increases, so does the fluidity up to 4.3% carbon. Higher levels of carbon tend to reduce fluidity.
- Adding carbon up to 4.3% lowers the melting point of iron from 1534°C to 1147°C for metastable Fe-Fe_3 C system. Further addition raises the melting point.
- The melting point of Fe-C is at a minimum at 4.3% carbon. This is known as the eutectic composition and this temperature 1153°C is the eutectic temperature.
- Iron when solidified can retain in "solid solution" certain amounts of carbon at different temperature levels.
  - At room temperature, only 0.006% C is held in solid solution.

- At 723°C, the solubility of carbon in alpha iron (ferrite) is increased only slightly to 0.02%C at maximum.
- At 1153°C, the solubility of carbon in gamma iron (austenite) is increased. Up to 2% C can be retained in solid solution. The iron is referred to as austenite. Carbon in austenite takes up random interstitial positions in the iron lattice without disturbing their normal face and current locations.

- When an iron containing austenite is cooled very slowly, the austenite transforms into alternate layers or Lamellae of ferrite and cementite. This structure is called pearlite and contains about 0.8% carbon.
- The transformation occurs by a process of nucleation and growth that originates most often at the austenite grain boundaries. If the cooling rate is high, the formation of pearlite is delayed.

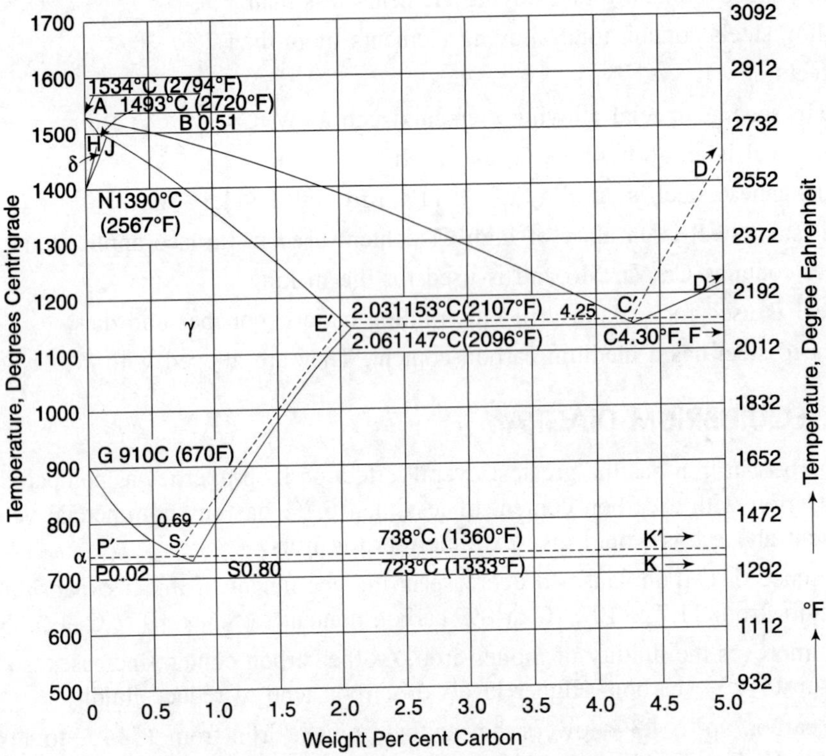

Figure 21.1 Iron-Carbon equilibrium phase diagram

21.4 CONSTITUENTS IN Fe-C SYSTEM

- Ferrite has BCC lattice structure. Soft and ductile. Contains very little carbon content, less than 0.01%C which dissolves in ferrite at room temperature. Referred to as α-iron.
- Austenite – Structure of iron at temperatures higher than 912°C. FCC lattice structure. Not present at room temperature, only at temperatures higher than 910°C.

- Cementite is a compound of iron and carbon (Fe_3C). Hard and brittle. Increases hardness and reduces ductility and toughness of the iron.
- Pearlite has a laminar structure composed of alternate layers of ferrite and cementite. Laminar structure forms a barrier against crack propagation. This gives the iron toughness.
- Martensite has very hard needle-like structure of Fe and C. Formed when steel is rapidly cooled from austenite (above the critical temperature).

21.5 FOUNDRY OF STEEL

Steels contain up to 2% C along with other elements like Si, Mn. S and P are present as impurities. Steels form the majority of alloy material used in the industry. They possess high strength, ductility, wear resistance and so on. Carbon is present in steel in the combined form. Steel has applications in automobile industry, general engineering, cutting tools, screws, bolts, nuts, etc.

Iron can be alloyed with a range of elements to impart some specific properties like corrosion resistance, weldability, impact strength, fatigue strength, and so on. Cr, M, Co, Mo, W, Va are used as alloying elements.

21.5.1 Composition and Classification of Steels and their Applications

Table 21.1 Composition and application of selected steels

Steels	*Composition*	*Applications*
1. Carbon Steel	0.06-0.50% C 0.15-0.80% Si 0.30-0.80% Mn 0.035% max P	General engineering applications, camshafts, axles, etc.
2. High Carbon Steel	0.56-1.10% C 0.15-0.35% Si 0.50-0.90% Mn 0.035% max P 0.035% max S	Cutting blades, bucket tips, wood cutting saws, milling cutters, etc.
3. Low Alloy Steels	0.12-0.23% C 0.15-0.35% Si 0.40-1.2% Mn 0.035% max P 0.035% max S 0.05%-1.3% Cr 0.20-0.85% Mo	These are used in case hardening steels, which are used in drilling machines, spindles, shafts, steering parts, gears, etc.
4. Spring Steels	0.42-0.70% C 0.15-1.80% Si 0.60-20% Mn 0.035% max P 0.035% max S 0.75-1.3% Cr 0.08-0.15% Va	Spiral springs, helical springs, leaf springs, plate springs, etc.

Table contd...

5. High Temperature Application Steels	0.12-0.28% C 0.20-0.45% Si 0.30-0.80% Mn 0.035% max P 0.035% max, S 0.09-5.6% Cr 0.25-1.1% Mo	Boiler components, super heaters, screws, bolts, steam pipe boiler tubes, etc.
6. Carbon Tool Steels	0.70-1.35% C 0.15-0.3% S 0.20-0.35% Mn 0.030% max P 0.030% max S	Heavy forging hammers, sledge hammers, forging dies, chisels, twist drills, milling cutters, threading dies, etc.
7. Dimensionally Stable Steels	0.9-2.9% C 0.3-0.4% Si 0.3-1.0% Mn 0.03% max P 0.03% max S 0.5-12.5% Cr 0.5-1.0% Mo 0.15-0.9% V 0.5-1.5% W	Blanking and punching dies, shear blades, cluster mill cold rolls, cutting knives for wood working, etc.
8. Hot Work Steels	0.3-0.55% C 0.2-1.0% Si 0.3-0.9% Mn 0.03% max P 0.03% max S 1-17% Cr 0.5-1.4% V 0.9-3.5% W	Heavy duty hot work tools, dies, die casting equipment, hot shear blades, extrusion mandrels, etc.
9. Cold Work Steels	0.85-1.45% C 0.2-0.3% Si 0.03% max P 0.03% max S 0.7-1.5% Cr 0.1-1.3% V 0.9-3.5% W	Highly wear-resistant steel, used in spinning tools, band saw, drills, drawing tools, press tools, broaches, gauges, taps, etc.
10. Hot Die Steels	0.42% C 0.03% max P 0.03% max S 1.50% Cr 0.25% Mo 4% Ni	Dies, etc.
11. Ball Bearing Steels	1-1.05% C 0.3% Si 0.3% Mn 0.03% max P 0.03% max S 0.5-1.5% Cr	Balls, rollers, and races, etc.
12. File Steels	1.4% C 0.2-0.3% Si 0.3-0.35% Mn 0.03% max P 0.03% max S 0.7% Cr	Hardest Cr alloy. Used to make files, knurling files, etc.

Table contd...

13. Valve Steels	0.42-0.82% C 2-3% Si 0.4-1% Mn 0.03% max P 0.03% max S 9-19.5% Cr 1.4-14% Ni 1-2.4% W	High temperature steels, resistant to scaling. Used in valves of engines.
14. High Speed Steels	0.75-1.5% C 0.2% Si 0.3% Mn 0.03% max P 0.03% max S 4.3-5% Cr 0.9% Mo 1.1-5% V 9.5-18% W 5-9.5% Co	Taps, milling cutters, broaches, twist drills, reamers, turning and planning tools.
15. Stainless Steels	0.06-0.12% max C 0.03% max P 0.03% max S 18% Cr 8% Ni	For corrosion resistant application, food/dairy brewing industry vessels and equipment.

21.5.2 Effect of Alloying Elements on the Properties of Steel

(i) Carbon – It imparts basic properties like hardness and strength to the steel. 0.05% to 2% carbon is used. It is soluble in iron.

(ii) Silicon – It imparts strength and hardness. It is used in the range of 0.2% to 0.8%. Silicon is also soluble in iron.

(iii) Manganese – Used in the range of 0.5-1.0%. Its effect is similar to that of Si. Mn combines with S to form MnS which will be present as inclusion in steel. Mn is soluble in iron.

(iv) Sulphur – It improves machinability of steel. It is limited to 0.06% as it has a harmful effect on ductility.

(v) Phosphorous – The maximum permissible level is 0.05%. Higher levels cause steel to become embrittled at low temperature leading to "cold shortness".

(vi) Chromium – It makes the steel hard. Causes increase in wear and corrosion resistance. If Cr content is greater than 4% the steels are referred to as stainless steels. It is used up to 18%.

(vii) Tungsten – It ensures the tool edges are maintained at high temperature; used up to 18%.

(viii) Molybdenum – It improves hardness of steel and improves endurance limit; used up to 18%.

21.5.3 Melting Furnace Used

Steel melting – Open hearth furnace (OHF), electric arc furnace (EAF), bessemer converter and induction furnace (IF). (both acidic (SiO_2) and basic refractories ($MgCO_3$) can be used for the lining).

In acid lining – Good grade scrap should be used, since the impurities (S and P) cannot be removed.

In basic lining – S and P can be partially removed.

The selection of melting practice and the type of furnace depends on:

- Tonnage of the metal produced.
- Size of casting and its complexity.
- Type of steel produced, plain-carbon steel (PCS) or alloy steel.
- Raw material available.
- Availability of power and cost.
  - OHF is used for large tonnage production of similar range of composition.
  - EAF is used for smaller heat size with widely changing composition.
  - Special steels and high alloy steels are produced in IF.
  - Converter is used for continuous melting.

21.5.4 Melting of Steel in Electric Arc Furnace and Induction Furnace

(i) Electric Arc Furnace

- First, a thin layer of burnt lime and a few pellets of iron ore are formed at the bottom. On top of this, steel scrap (heavy) is charged. Below the graphite electrodes, small coke lumps are placed. Now the graphite electrodes carrying current will initiate the arcing when they are lowered.
- Slowly, the coke burns and further arcing is smooth.
- Metal charge also starts getting heated and slowly melts.
- The molten metal trickles through the charge, burnt lime and ore.
- Reaction takes place between iron, lime and ore.
- Due to the release of CO_2, vigorous boiling of metal takes place. The elements of iron are oxidized. Oxidation is carried out to bring C level to less than 0.1%.
- When the complete charge melts (melt down), a sample of metal is taken out and analyzed for chemical composition.
- Slag formed on the molten metal bath is removed from the furnace.
- A fresh slag is formed using burnt lime, fluorspar and ore. The slag has lower density and hence floats to the top. In fact, the layer of slag prevents the oxidation of metal and acts as a protective layer.
- Now, based on the chemistry additions of soft coke (for C) ferroalloys are added to make up the composition of steel.

- After adjusting the composition to the required level, the temperature is checked.
- Molten metal is poured into ingots.
- Furnace is again charged and the cycle repeats.
- Any type of charge can be melted in EAF. Scrap with higher carbon content or lower content can be used.
- Oxidation and reduction of the material is easy.
- Temperature control is easy.
- Slag removal is easy.
- Addition of alloying elements is easy.
- Furnace can be used continuously.

(ii) Induction Furnace

- Here selected scrap has to be used as there is agitation in the melt during melting.
- This will not permit separation of the slag and molten metal.
- The furnace is loaded with scrap (mainly return scrap) and solid scrap of known origin is used. Clean scrap is used.
- After melt down, chemical composition is found out.
- Any required additions are made up by using ferroalloys, carbon coke, etc.
- Temperature of the melt is checked. The melt is tapped into the ladle.
- Furnace is charged and recharged for the next melt.

21.5.5 Microstructure of Various Steels

- Ferrite – It is a solid solution of Fe and C. It is also called α-iron. It contains very small percentage of C (less than 0.02%). It is soft and ductile.
- Pearlite – It is a mixture of ferrite and cementite in the solid solution. It consists of ferrite and cementite in the lamellar form, i.e., alternate layers of ferrite and cementite is the feature of pearlite. At the eutectoid, percentage of C is 0.8%. The entire microstructure of steel is pearlite. As carbon is increased slowly from 0 to 0.8% the microstructure varies from ferrite to pearlite. It is hard and strong.
- Cementite – It is a compound of iron and carbon (Fe_3C). It is formed when carbon content exceeds 0.8%. As the percentage of carbon is increased, the amount of cementite also increases.
- Austenite – It is a solid solution of Fe and C. It is also referred to as γ-iron. It is present at higher temperature above the eutectoid temperature 727°C. The maximum solubility of carbon in iron is 2% at 1149°C. It will be present at room temperature under certain conditions.
- Widemanstaetten Ferrite – Ferrite present in the plate-like or a circular form obtained at slightly higher cooling rate. Appears in hypoeutectoid steel.
- Carbon Steel – It is mainly classified based on the eutectoid (i.e., 0.8% C).

- Hypoeutectoid Steel – The carbon content is less than 0.8%.
- Eutectoid Steel – The carbon content is 0.8%.
- Hypereutectoid Steel – The carbon content is above 0.8%.

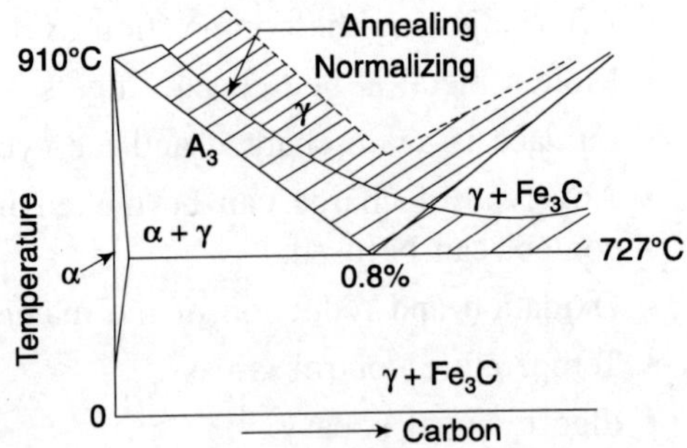

Figure 21.2

From the Fe-C equilibrium diagram, the range of temperature for percentage of carbon can be identified for the above. Above this temperature range, the structure will be austenite or γ-iron. Below this the structure will be ferrite, pearlite or Fe_3C.

If a hypoeutectoid steel is cooled from the austenitic temperature, the microstructure obtained will be different for different cooling rates.

- Very slow cooling rate – Ferrite (major) + Pearlite (small)
- Medium cooling rate – Pearlite (major) + Ferrite (small)
- Fast cooling rate – Bainite + Martensite
- Very fast cooling rate – Martensite

21.6 FOUNDRY OF CAST IRON

Cast iron is an alloy of Fe and C with other elements Si, Mn with impurities S and P. The carbon content varies from 2 to 4.3%. Cast iron is hard and brittle. It has good wear resistant property, good machinability, good thermal shock resistance and good damping property.

Cast iron finds application in all machinery beds, guideways, engine cylinders, slag pots, gear wheels, pulleys, etc.

By varying the carbon content, iron can be solidified as white or grey cast iron. In white iron, the carbon is present as carbides and in grey cast iron, it is present in the free form as graphite.

The presence of carbon in either free form or combined form changes the property of the iron to a large extent.

By treating grey cast iron in the molten metal with Mg and keeping the S level low, spheroidal iron can be produced.

Cast irons are very much sensitive to cooling rates. Cast irons are referred to with a parameter carbon equivalent CE which will dictate the type of cast iron.

Carbon Equivalent (CE): CE is defined as the sum total of percentage of carbon, one-third the percentage of Si and P.

$$CE = \%C + (1/3)\%Si + (1/3)\%P$$

Here Si and P influence the formation of free carbon as graphite. Promotion of graphite formation is referred to as graphitization. The elements promoting graphitization are referred to as graphitizers. Certain alloys like Cr and Ni influence the formation of carbides and are referred to as carbide stabilization. The elements which promote carbide stabilization are referred to as carbide stabilizers. The effect of Si and P is equal to one-third the promotion of graphite, i.e., the amount of graphite formed will be equal to one-third the percentage of Si and P.

Based on the value of CE, a given cast iron is referred to as:

Hypoeutectic cast iron if CE < 4.3%

Eutectic cast iron if CE = 4.3%

Hypereutectic cast iron if CE > 4.3%

CE influences the properties of cast iron.

21.6.1 Composition

Concise details of cast iron family are given below. This will help to identify different irons at a glance.

Table 21.2 Typical compositions of different cast irons

Iron	% C	% Si	% Mn	% S	% P	%Mg	%Fe	Structure
White Iron	2.8-3.3	0.6-1.1	0.25-0.80	0.06-0.25	0.06-0.18	-	Balance	White fracture iron carbide
Malleable Iron	Heat Treated White Cast Iron						Balance	Compact graphite with pearlite
Grey Cast Iron	2.5-4.5	1-3.5	0.4-1.0	0.05-0.25	0.05-0.3	-	Balance	Flake graphite in ferrite and pearlite
Spheroidal Cast Iron	3-4	1.8-2.8	0.5-0.9	0.03 max	0.1 max	0.01-0.1	Balance	Spheroids of graphite with pearlite
Austempered Cast Iron	Heat Treated Spheroidal Cast Iron						Balance	Spheroids of graphite with Bainite
Wrought Iron	<0.05	Minute Traces					Balance	Predominantly ferrite network

The classification of cast iron is mainly based on the percentage of carbon present. The microstructure can be varied by heat treatment, melt treatment or by varying cooling rate in the iron.

21.6.2 Classification of Cast Iron

Grey cast irons can be classified based on the minimum tensile strength as per B.S. 1452-1956 standards. There are seven grades in the standards. Figure 21.3 shows the relationship between CE and tensile strength for various casting diameters and Figure 21.4 shows variation of bar diameter with tensile strength.

As the CE value of iron increases, the tensile strength increases linearly for a given diameter of the casting. Lower diameter yields higher tensile strength and vice versa.

As the diameter of the casting increases, the tensile strength decreases with any grade of iron, non-linearly. The strength value is higher for higher grade irons.

Typical composition of grey cast iron

2.5-4.5% C

1-3.5% Si

0.4-10% Mn

0.05-0.25% S

0.05-0.3% P

Balance Fe

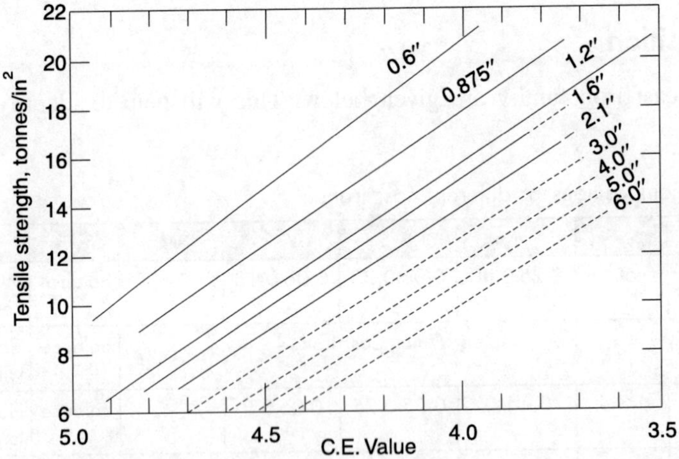

Figure 21.3 Relationship between tensile strength and carbon equivalent value for various bar diameters

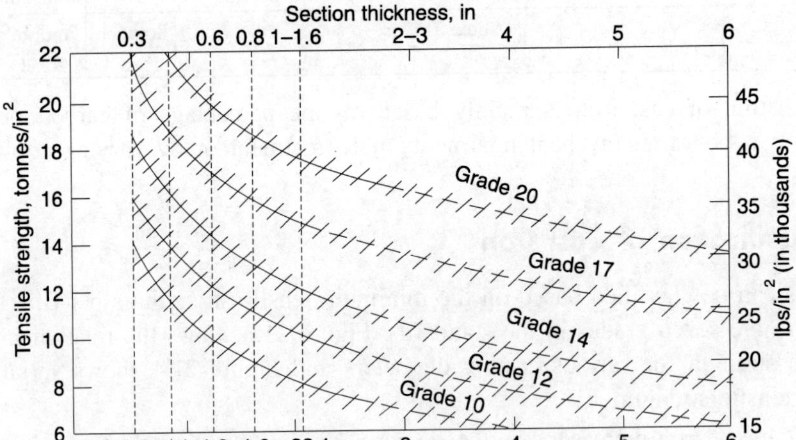

Figure 21.4 Variation of tensile strength with bar diameter or section thickness for several grades of iron

21.6.3 Effect of Elements on Structure and Properties of Cast Iron

Carbon is present from 2 to 4.6% by weight. It is present in the free form as graphite in the combined form as iron carbide (Fe_3C). The total percentage of carbon is the sum total of percentage of free carbon and combined carbon.

Graphite provides good antifriction property, shock resistance and machinability to the iron, but act as discontinuity in the matrix. Graphite will be present as flakes of various sizes. Iron carbide is very hard and imparts wear resistance to the iron, making it brittle.

Carbon at 4.3% results in eutectic formation in the Fe-C diagram.

Silicon is present from 1 to 3.5% by weight. It supports free carbon (graphite) formation, i.e., it acts as a graphitizer and promotes graphitization. Increasing Si content shifts the eutectic to the left in the Fe-C equilibrium diagram.

Hence, eutectic carbon percentage = 4.3 – (1/3)Si. Or carbon equivalent CE = %C in iron + (1/3)Si. If CE = 4.3, then it corresponds (approximately) to eutectic iron. Si in iron increases the hardness and strengthens the ferrite.

Manganese combines with S and forms MnS. It will compensate the bad effect of S. At lower percentage of Mn, it promotes ferrite formation with minimum pearlite. At higher percentage, pearlite is formed. Mn alone acts as carbide stabilization and forms pearlite structure. For correct Mn:S balance in grey cast iron, the percentage of Mn should be at least equal to 1.7 times sulphur content + 0.3%.

Sulphur may be present up to 0.25%. Higher percentages of sulphur favours retention of a completely pearlitic microstructure.

It is a carbide stabilizer. Above 0.25% S results in undesirable hardness and decreased machinability. Sulphur alone will form FeS and segregates into grain boundaries. But when Mn is present, MnS is formed and is randomly distributed.

Phosphorus: Mechanical strength is maximum at about 0.35% P. At about 0.65%, the strength properties fall and continue to do so. It forms a eutectic (Fe + Fe_3P + Fe_3C) as it segregates. It increases the tendency for the iron to be a eutectic type alloy. Its effect is equal to one-third in improving CE. It forms iron phosphide, which is hard and brittle.

21.6.4 Chilling Effect of Cast Iron

Cast irons are sensitive to cooling rate. Depending on the cooling rate, the iron forms grey iron, mottled iron or white iron. Different compositions and types of cast irons show marked difference in the structure when cooled faster from the molten state.

The changes that take place from graphite formation to white iron formation is referred to as chilling tendency, during fast cooling of iron. This chilling tendency is a function of melting conditions, chemical composition (especially C and Si) and section thickness of the casting. Hence, the iron formed may be:

Grey Cast Iron: Normal thickness of the casting. Graphite is present predominantly. The fracture shows grey region. No chilling is observed.

Mottled Iron: For the same normal thickness of the casting. Graphite and iron carbide coexist. The fracture shows partly grey and partly white. Some amount of chilling is observed.

White Iron: When the casting thickness is maintained as before the fracture is completely white. It consists of fully iron carbide or cementite.

Therefore under carefully controlled conditions of parameters, a wedge test is used for determining the degree of chilling tendency.

Wedge Test

A wedge is a triangular cross section specimen.

A sand mould is prepared using a pattern to get a wedge shape cavity.

Figure 21.5 shows the mould setup. Into this cavity, molten cast iron is poured. When the wedge has solidified, it is removed from the mould at a dull red heat at about 600°C and quenched in water. The base side is dipped in water first. The casting is fractured by hammering.

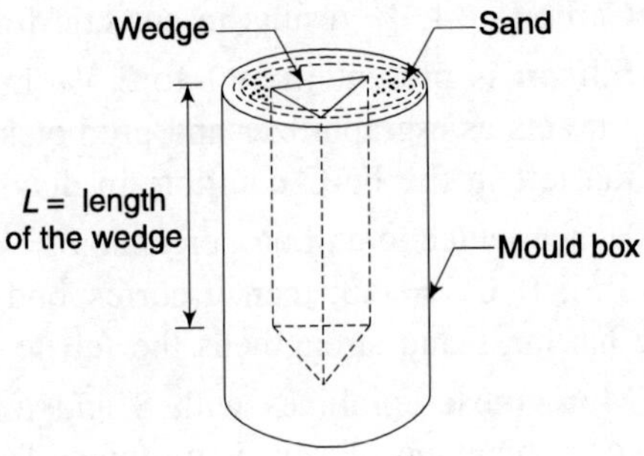

Figure 21.5 Mould with triangular carvity

The cross section of the wedge is observed:

h = height of the wedge

b = base of the wedge

Figure 21.6 shows the fractured specimen with wedge cross section.

- It consists of white region from point portion to a certain height.
- Then a mottled region exists for a certain height.
- There is grey region for the rest of the height.

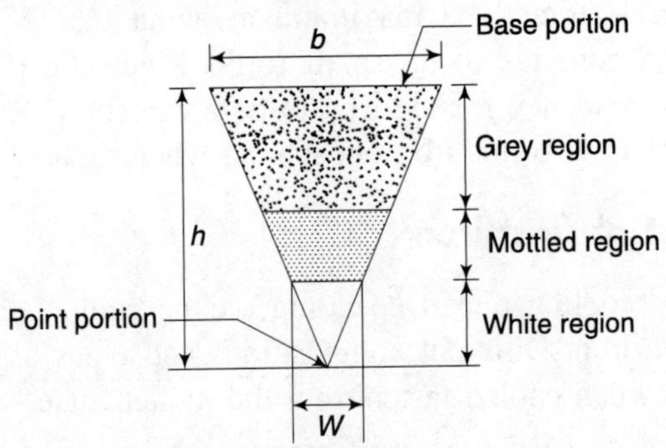

Figure 21.6

Now, the maximum width (W) of the white region is measured in mm. It is taken as the wedge reading. It is referred to as chill depth.

The wedge chosen should be the smallest possible. There are eight standard sizes of wedge (A to H) given in the table below.

Table 21.3 Standard sizes of wedge

Wedge	Base, B	Height, H	Length, L
A	6	11.1	57.1
B	12.7	22.2	101
C	19	38	127
D	25.4	57.1	127
E	38	85.7	152
F	51	120.6	152
G	63.5	165.1	203
H	76.2	215.9	203

The chill depth is a measure of quality iron produced. It is a routine test which will show that a change in the composition has occurred keeping all other variables constant. For a known composition of metal, temperature and type of furnace, chill depth required can be established on a regular basis. Wedge test is carried out for each melt and compared with the established chill depth. If the chill depth is more, there is a tendency for mottling and if the chill depth is less the iron is going to be very soft.

If the chill depth is more in the iron as compared to standard wedge, the composition of iron, is adjusted by adding FeSi. If the chill depth is less, the composition of iron is adjusted by adding steel scrap.

21.6.5 Graphite Shape and Classification

The microstructure of grey cast iron reveals that the graphite shape is flake, but there will be various sizes for the flakes.

Based on this, graphite can be classified as per ASTM A247-47 standards;

Graphite form is indicated by Roman numeral. Ex: Type 1

Graphite distribution is indicated by letters A to E. Ex: Type 1 B

Graphite size is indicated by numbers from 1 to 8. Ex: Type 1 B2

The microstructure is observed at 100x.

Figure 21.7 shows different forms of graphite designed from I to VI.

Form I.... Random graphite flakes

II.... Rosette (Rose petals) type

III.... Small thick flakes concentrated at various regions

IV.... Thin compacted graphite

V.... Thick compacted graphite

VI .... Spherical graphite

The microstructures are very useful in characteristics of the cast iron. It is used as tools for identifying/comparing the microstructure of castings to assess its properties. Usually, A type graphite size 4 is preferred in grey cast iron.

Similarly, graphite can be present in spheroidal form as in SG iron. Accordingly, the graphite size will be used to assess the properties of SG iron.

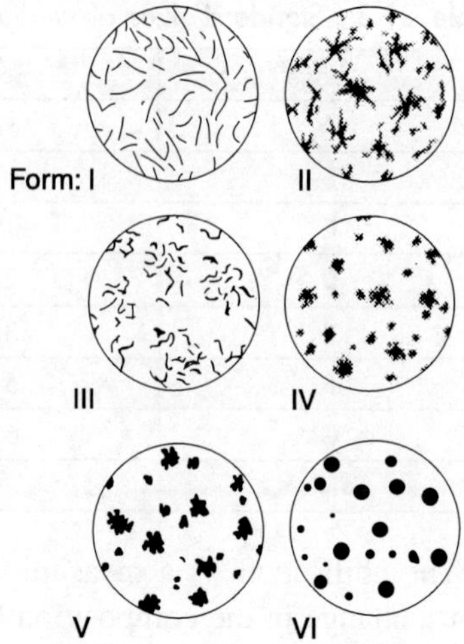

Figure 21.7 Forms of graphite

Figure 21.8 shows the distribution of graphite designed from A to E.

A type graphite: Long random thin flakes associated with small flakes.

B type graphite: Rosette type small thin flakes of graphite with small thick graphite flakes, emerging from a node.

C type graphite: Thick graphite flakes with sharp ends associated with random small thin and thick graphite flakes.

D type graphite: Small specks of graphite flakes (called undercooled graphite) distributed randomly.

E type graphite: Small flakes of graphite uniformly arranged showing dendritic pattern. Again undercooled graphite is present.

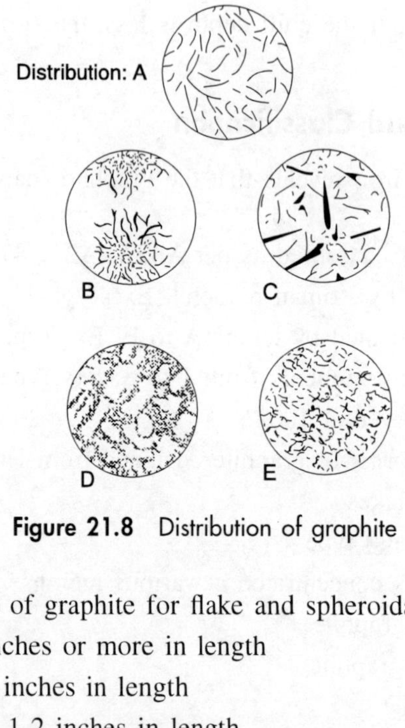

Figure 21.8 Distribution of graphite

Figure 21.9 shows the size of graphite for flake and spheroidal graphite.

Size 1: Longest flakes 4 inches or more in length

Size 2: Longest flakes 2-4 inches in length

Size 3: Longest dimension 1-2 inches in length

Size 4: Longest dimension 1/2 – 2 inches in length

Size 5: Longest dimension 1/4 – 1/2 inches in length

Size 6: Longest dimension 1/8 – 1/4 inch in length

Size 7: Longest dimension 1/16 – 1/8 inch in length

Size 8: Longest dimension 1/16 inch or less

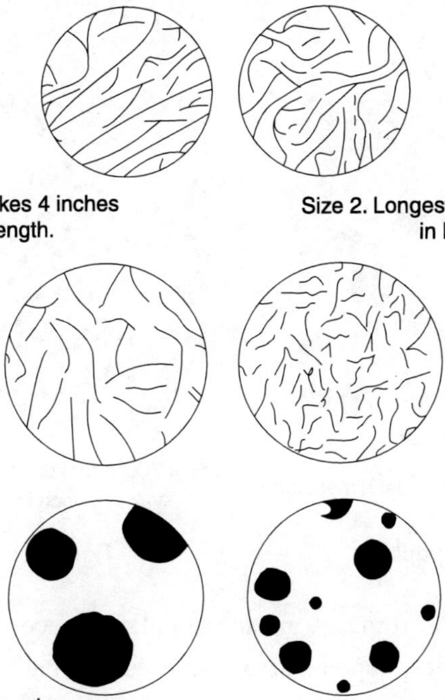

Size 1. Longest flakes 4 inches or more in length.

Size 2. Longest flakes 2–4 inches in length.

Size 3. Longest dimension 1–2 inches in length.

Size 4. Longest dimension 1/2–1 inch in length.

Figure 21.9 Size of graphite

21.7 PRODUCTION OF CAST IRON

For melting cast iron, a number of furnaces can be used:

- In olden days, blast furnace was used to make iron castings.
- Later, air furnaces were used with bituminous coal or briquettes as fuel.
- Open hearth furnace was also used with oil/gas as fuel.
- Reverberatory type furnaces were also used with gas/oil as the fuel.
- Cupola furnace, using coke has become a popular furnace nowadays. Any type of scrap can be used in this furnace. However, cokeless cupola is showing promise.
- Electric arc furnace was also used for making iron casting (both direct and indirect type).
- Induction furnace using electricity is also used nowadays to make iron castings. But selective charge needs to be used.

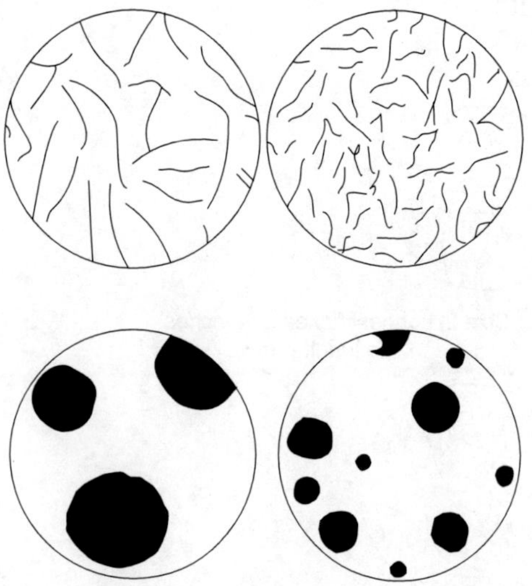

Size 5. Longest dimension Size 6. Longest dimension

Figure 21.10 Size of graphite

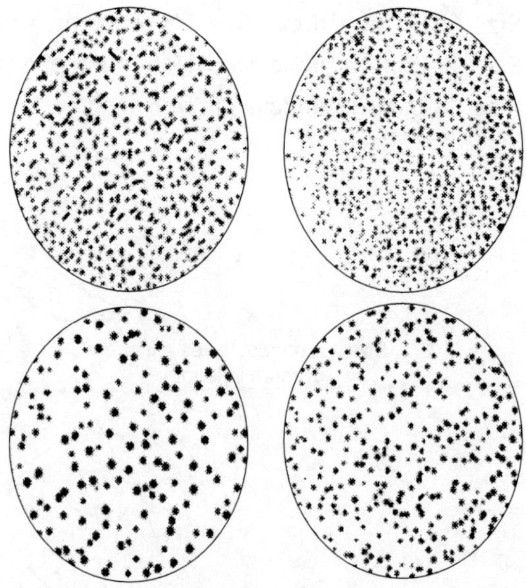

Size 7. Longest dimension $\frac{1}{16}$ – $\frac{1}{8}$ inch in length Size 8. Longest dimension $\frac{1}{16}$ inch in length

Figure 21.11 Size of graphite

- Cupola and induction furnaces are the popular furnaces used for melting cast iron.
- Because of economics, cupola furnace is used in small or big foundries alike.

(i) Melting of Cast Iron (CI)/Production of CI

Melting of CI using cupola furnace is the most popular method. This type of furnace is used by big and small foundries.

Induction furnaces are also used to handle large tonnage of metal to produce bigger castings.

(a) Cupola furnace Melting is based on counter heat exchange principle. Cupola is a tall cylindrical furnace. It makes use of coke as fuel, limestone as flux and pig iron, foundry returns and scrap as the charge material.

Metal Charge: Coke: Metal charge ratio varies from 1:8 to 1:10 based on the quality of coke available. Limestone is used from 15 to 25% of the coke charge, by weight.

Constructional features: The furnace consists of a circular cylindrical steel shell located vertically. The furnace is held on a concrete bed by means of support (legs).

The inside of the shell is lined with refractory bricks. It has two holes in the shell at the bottom lower side. One for tapping the molten metal and the other for removing slag. They are placed 180° apart along the circumference of the cupola. The bottom of the furnace is closed with two semicircular doors which can be closed during melting and opened after melting is over to allow unwanted leftovers to be removed. Above the tap hole, a number of holes are present along the circumference. These holes are surrounded by a wind box. The wind box is connected to the blower. Air through the blower enters the wind box and from there, enters the inside of the cupola radially. This air is used for the combustion of coke. These holes are called tuyeres. There may be 6 to 8 holes along the circumference, depending on the size of the cupola.

Above the wind box, a charging box is located to facilitate operators to stand and keep raw materials for charging. Corresponding to the platform, there is a rectangular hole in the cupola which acts as the charging door. Through the charging door metal charge, limestone and coke are dropped into the furnace periodically.

Cupola size: Cupola furnaces are rated by number from 0 to 12 based on the melting rate varying from approximately 1 to 35 tonnes per hour.

Shell diameter: The diameter of the shell generally varies from 27 to 108 inches. This diameter controls the capacity of the cupola.

Inside shell diameter: It is the inside diameter of the cupola after lining the shell. This diameter actually decides the operating cross section and ultimately the melting capacity of the furnace.

Total tuyere area: Tuyeres are used to introduce air inside the cupola for combustion of coke.

The total area of a tuyere is usually 1/4 to 1/8th of the inside C.S. area of the cupola at the tuyere level. This area is distributed along the circumference of the cupola and the number of tuyeres are decided.

Working Operation

First the refractory lining is prepared, by lining the steel shell with refractory bricks followed by a mixture of granular refractory, clay and water. The bottom doors are closed and the bottom of the furnace is prepared. First moulding sand is placed on the bottom door and rammed well. The bottom is given a taper towards the tap hole.

Tap hole and slag holes may be prepared by using bricks or using a ramming mixture. The vertical distance between the tap hole and slag hole is maintained. The volume of the space inside the cupola between slag hole and tap hole forms the well zone.

Coke bed: When the bottom of the furnace is ready, burning splinters are dropped inside the furnace. Coke lumps are then dropped into the furnace slowly. Once the coke catches fire and starts burning further, coke pieces are added to a preselected height. Now burning coke forms a bed. This is referred to as coke bed height. Coke bed height is the height of the coke inside the furnace above the tuyeres. This decides the actual melting operation, maintains combustion, temperature and melting rate.

Hence, coke bed height needs to be maintained and controlled. This can be measured from the changing door with the help of a rod or chain.

To achieve the correct coke bed height, a burning-in period is adopted. Air at a low blast (half the normal blowing rate) is blown through the bed for a short duration (2-3 min) to hasten the combustion and increase the temperature inside the furnace. Necessary coke may be added after burning of the coke to replenish the burnt coke and to make up the coke bed height. Now charging of cupola can be made.

Charging: Weighed amounts of metal charge, coke and limestone are charged into the furnace through the charging door.

1:8 or 1:10 charge to coke ratio is used for the purpose. Limestone of 20% of coke charge is used as flux. Addition of flux causes formation of slag from the coke ash, refractory erosion and metal oxidation. After charging the cupola completely, it takes a soaking period of 20 to 60 minutes for the entire stock of materials to preheat.

Then the blower is started. After a few minutes, the coke becomes red hot and melting of the charge begins. Within 12 minutes from the start of blowing air, melting progresses sufficiently and trickles of molten iron may be seen through the tuyeres. The molten metal starts collecting at the bottom of the furnace. Tap hole is closed during this period and slag hole is kept open.

The time for the first molten iron at the tap hole is measure of correctness of the coke bed height.

- If the time is < 8 minutes from start of blowing air, it suggests too low a bed height.
- If the time > 10-12 minutes suggests too high a bed height.

It is expected that the molten metal comes out of the furnace consistently at 1510-1593 degrees. To achieve this, air flow rate and coke bed height are controlled. As melting progresses, the stock content settles and new charges are added continuously. As the hot flue gases move upward, the cold charge materials will be descending downwards. This gives rise to efficient heat exchange called *counter current principle.*

Continuous tapping of molten metal and slagging operations are done, with continuous charging of materials. Once the necessary melting is completed and all the molten metal has been tapped, the furnace is ready for shutdown. The blower is switched off after stopping charging of the furnace. Bottom doors are opened and all the contents fall to the floor of the cupola. Water may be sprayed onto the hot materials. The cupola is prepared for the next charging.

(b) Induction furnace melting

Principle

Induction furnace works mainly on the principle of a transformer.

The furnace can be operated at medium and high frequency. Two types of IF are in use, viz., coreless type and core type.

Constructional Features

This furnace has coils wound helically, to get a long cylindrical shell. Mica or asbestos sheets cover these coils from inside. Lining material (pure quartz) is prepared on the sheets of mica/asbestos. The coils are placed on refractory bricks. The lining material is rammed on these bricks to form the bottom of the furnace.

The cavity formed within the refractory lined shell forms the furnace.

The coils are hollow and made of copper. Cooling water is circulated to reduce the temperature of the coil.

A lid is used to cover the furnace at the top. Safety devices are used for protection.

The coils are connected to the mains through which current is drawn at a known value of current and voltage.

When the furnace is having charge, the current passes through the coil and emf is induced in the charge. As the emf is induced in the metal charge a current starts flowing in the charge. But the charge offers resistance to flow of current and thus heat is generated as per Joule's law.

$$Q = I^2Rt$$

where,

I = Current in ampere

R = Resistance of the coil (ohms)

t = Time in minutes

The current flow direction in the charge is opposite to the direction of current in the charge.

Melting takes place quietly. The flow of current in the charge in molten condition results in stirring or churning action. This causes uniform composition and uniform temperature in the melt. As a result of stirring, slag removal becomes difficult and refining is not possible.

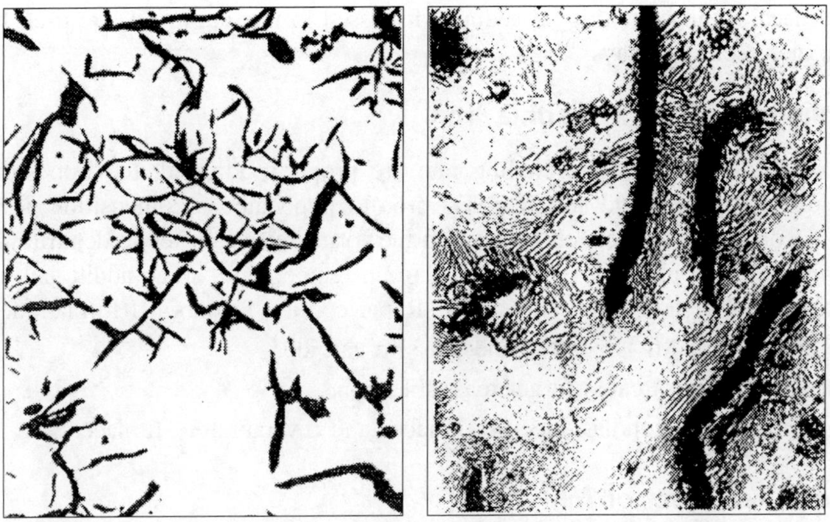

(a) (b)

Representative structure (unetched), of an unalloyed ASTM Class 20 grey iron (left) and ASTM Class 50 high-strength grey iron, etched and magnified 500 times (right). The graphite needles as well as the fully pearlitic matrix can be clearly seen in (b).

Figure 21.12 Micro GCL

Melting of cast iron in induction furnace

Selected scrap consisting of pig iron, return scrap, risers and runners are used in a definite proportion. The chemical compositions of each of these are to be found out. For getting the final composition in the iron produced, charge calculation is done. Accordingly, the charge proportion is obtained for charging into the furnace. The current is passed through the coil which, in turn, induces emf in the charge, then current tends to flow in the charge. This induced current ultimately melts the charge. The composition is checked and any alloys like FeSi, FeMn, C may be added to adjust the composition.

After reaching the correct pouring temperature, the molten metal is poured into the ladle and then into the moulds.

Figure 21.12(a) shows the microstructure of grey cast iron flake graphite and Figure 21.12(b) shows flake graphite in a matrix of pearlite as seen in the etched condition.

21.8 MALLEABLE IRON (MI)

21.8.1 Production of Malleable Iron

MI is produced by heating white iron: It consists of 1.8 – 3.5% C, 0.5 – 1.9% Si, 0.25 – 0.8% Mn, 0.06 – 0.18% P and balance Fe.

The microstructure of white iron is predominantly a massive carbide structure. After heat treatment, carbide is transformed to temper carbon in a ferrite matrix or a pearlitic matrix.

Melting of white iron is generally carried out in air furnace, cupola, induction furnace or electric arc furnace. Similar melting procedures discussed in melting of CI, are used here also. Air furnace is not used nowadays.

21.8.2 Solidification of Malleable Iron

When the composition of iron and cooling rate are properly adjusted, the iron solidifies as massive carbides or carbides and austenite. The iron of hypoeutectic composition provides the necessary C-Si content. Solidification starts with the formation of austenite dendrites followed by the austenite-carbide eutectic. As it cools to room temperature, some additional carbide is formed and the austenite transforms to pearlite. Rapid cooling favours MI formation.

At RT the structure of MI is pearlite and massive carbide.

White iron is to be heat treated to get malleable iron.

White iron is melted in cupolas, electric furnaces and reverabratory furnace.

21.8.3 Heat Treatment of MI

Malleabilization heat treatment of MI results in converting metastable iron carbide to the stable graphite system. The transformation is controlled by nucleation and growth of graphite.

The heat treatment occurs in two stages:

(i) First stage graphitization (FSG)

(ii) Second stage graphitization (SSG).

White iron is heated through the eutectoid range (critical range); the pearlite is transformed to austenite due to the rapid diffusion of the carbon from the cementite into the iron to form a solid solution of carbon in gamma iron. The massive carbides are virtually unaffected by heating.

Above the critical range but below the eutectic, the iron consists of massive carbides and austenite.

When white iron is held at elevated temperatures, above the critical range of graphitization massive carbides will occur. This is referred to as first stage graphitization.

$$Fe_3C \rightarrow 3\ Fe\,(\text{austenite}) + C\,(\text{graphite})$$

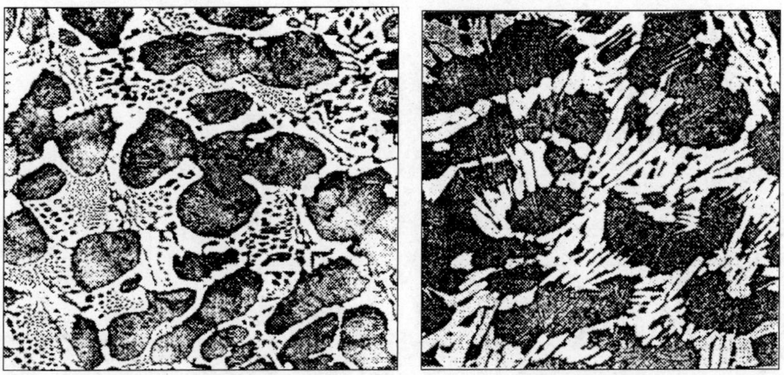

Figure 21.13 Massive carbide (light) with pearlite (dark) ledebutitic eutectic and plate-like eutectic (right) etched in 4% picral, 100x magnification

Higher the temperature, faster the process of graphitization.

Castings are to be scaled tight to prevent excess oxidation and scaling. Nodules of graphite are formed during this process.

In the SSG which involves the precipitation of carbon from austenite as graphite on the nodules of the graphite formed in FSG.

The temperature is lowered after FSG. Due to low solubility of carbon in iron, graphitization proceeds at a rapid rate. Cementite is converted to pearlite.

Cooling after SSG may be carried out at any rate as no further changes occurs.

The malleable iron consists of compact graphite with pearlite structure.

21.8.4 Malleabilization Heat Treatment

Basically, white iron is the starting material for heat treatment. Heat treatment is an annealing process and is often called malleabilization.

White iron consists of massive carbide with pearlite. The cycle involves three stages:

(1) Nucleation: White iron is heated slowly at a rate of 100°C/hr. Pearlite changes completely to austenite. When the maximum temperature of 950-1000°C is reached, the entire structure is converted to austenite.

(2) First stage graphitization (FSG): When the iron is held at maximum temperature for nearly 20 hours, graphitization process which involves a process of solutionizing and precipitation, takes place. The carbides dissolve in austenite and then carbon diffuses to the nuclei and precipitates as graphite. Growth of temper carbon nodules takes place at the expense of carbides.

(3) Second stage graphitization (SSG): Here the temperature is lowered from 790 to 704°C and the castings are held at this temperature for another 30 hours During this period, all the pearlite is converted to ferrite. No change in the compact graphite occurs.

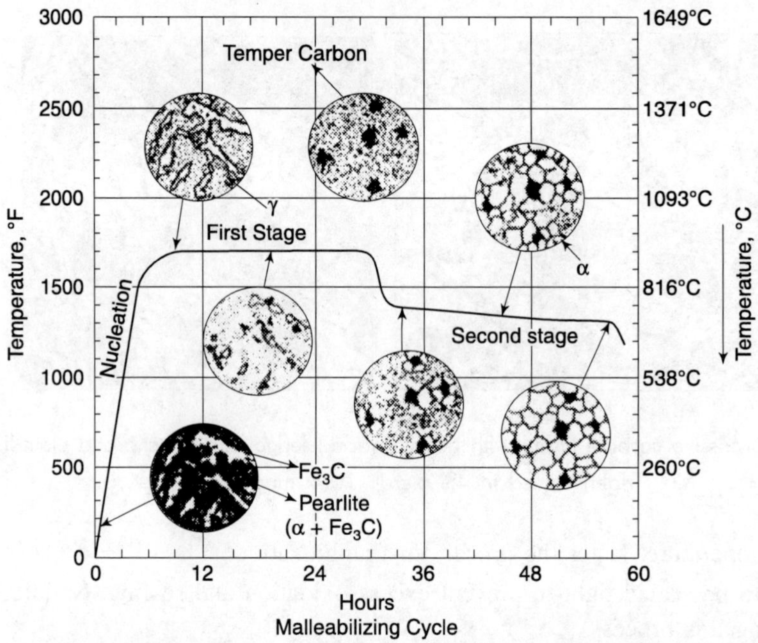

Figure 21.14 Treatment diagram for malleable iron. A typical malleabilizing is shown with the change in microstructure as a function of time and temperature

21.9 SPHEROIDAL GRAPHITE IRON

21.9.1 Production of SG iron

Spheroidal graphite iron or nodular graphite iron has a composition of grey cast iron except that the sulphur level is kept below 0.015% and there will be residual Mg of 0.05%. It consists of graphite in the form of nodules in a matrix of pearlite.

21.9.2 Desulphurisation

Using the calcium carbide powder, the molten metal is treated. As a result of reaction, sulphur is released as slag due to the formation of CaS that goes into the slag.

$$CaC_2 + S \rightarrow CaS + 2C$$

21.9.3 Magnesium Treatment

After desulphurization, the level of S will be less than 0.02%. The melt is treated with nodule forming agents such as magnesium alloys or cerium. This master alloy is kept in the bottom of the ladle and molten iron is poured into the ladle. The master alloy reacts with the molten metal and a vigorous reaction takes place.

21.9.4 Post-Inoculation

Magnesium treatment results in the development of spherical graphite. To produce quality ductile iron, inoculation is required. Since inoculation of the Mg treated iron is done, it is called post-inoculation. This process increases the formation of a number of spheroids. A large number of sites are provided for nucleation of the graphite. Inoculation also prevents chilling tendency in the iron. Inoculation is done with FeSi containing 75% or 85% Si.

21.10 AUSTEMPERED DUCTILE IRON (ADI)

ADI is the new addition to the cast iron family. It is basically SG iron which is heat treated to get biotitic matrix along with spheroids of graphite. Austempering heat treatment is used to bring about the transformation in the matrix.

ADI has a superior mechanical property comparable to that of steel. Hence, it is used extensively in automotive parts.

Production of ADI

The composition of ADI is the same as in ductile iron. The ductile iron, after subjecting it to austempering heat treatment, will have bainitic matrix with graphite nodules.

Austempering heat treatment consists of austenisation followed by austempering in salt bath in the range of 280-370°C followed by slow cooling.

The ductile iron castings are slowly heated to above 910°C, using a heating rate of 110°C/m and held at this temperature for 2-3 hours and then subjected to sudden quenching in salt bath consisting of potassium nitrate and sodium nitrate. The castings are then held in the salt bath at 310°C for 120 minutes. Then the castings are slowly air-cooled. The castings will have graphite nodules in a matrix of bainite.

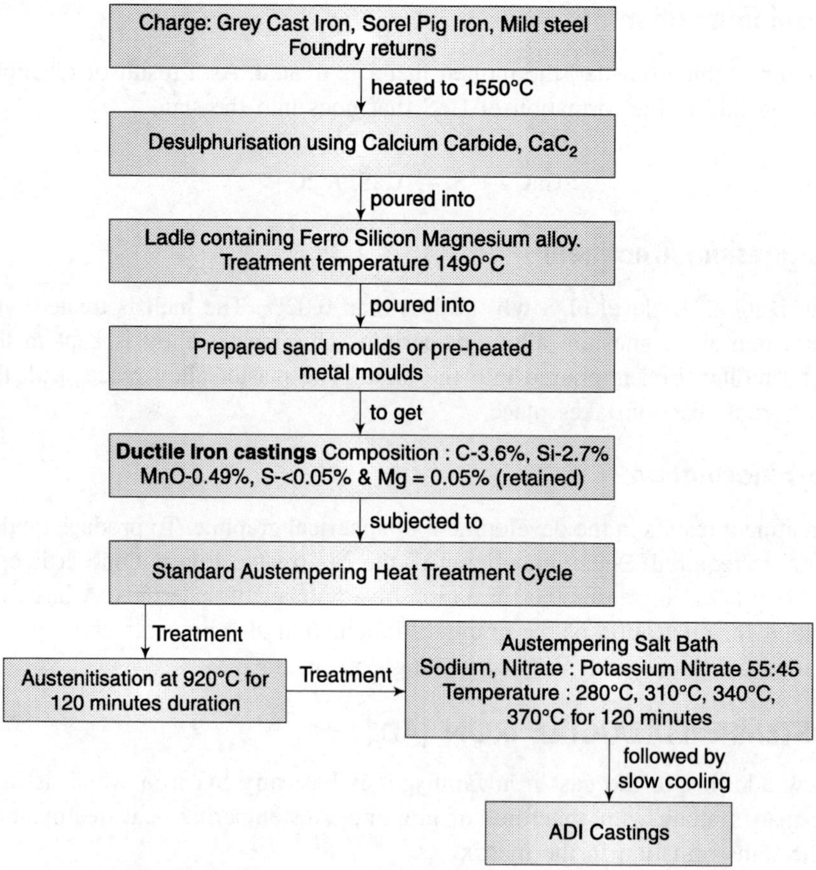

Figure 21.15 Flowchart for the production of ADI castings

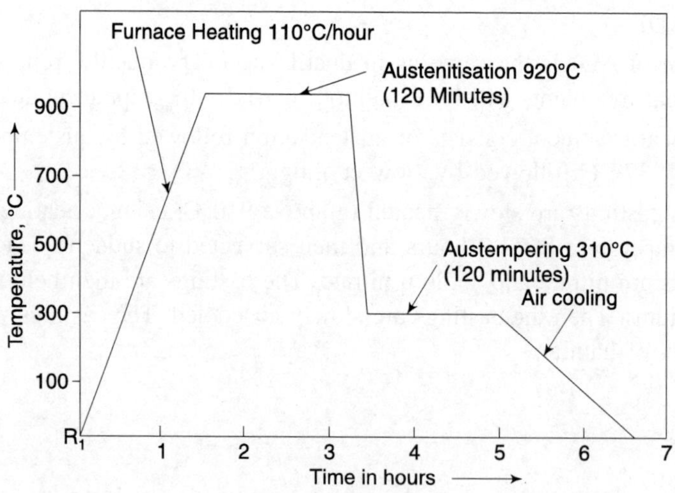

Figure 21.16 Austempering heat treatment cycle employed

21.11 NONFERROUS FOUNDRY

Iron is not a basic or major element in nonferrous metals and alloys even though it may be present as impurity which is not desirable. There is a wide variety of nonferrous metals or alloys like aluminium, copper, magnesium, zinc, titanium, tin, lead, etc. among which Al, Cu, Mg, Ti, Sn and Zn are used to a large extent. Al and Mg alloys are light metal series whereas Cu, Zn alloys form heavy metal series.

Nonferrous alloys are melted in a pit furnace, induction furnace or electric resistance furnace. They are grain-refined to obtain better mechanical properties. Under heat treatment, their mechanical properties are enhanced. Nonferrous alloys have a wide variety of characteristics right from the external colour they possess. Light metals and alloys find extensive applications in aerospace.

21.12 CLASSIFICATION OF NONFERROUS ALLOYS

On the basis of base metal used for making the alloy, these alloys are classified as:

- Aluminium based alloys
- Copper based alloys
- Zinc based alloys
- Titanium based alloys

Various alloying elements are used to make the alloys.

21.13 ALUMINIUM BASED ALLOYS

Aluminium alloys are alloys in which aluminium (Al) is the predominant metal. The typical alloying elements are copper, magnesium, manganese, silicon, tin and zinc. There are two principal kinds, namely, casting alloys and wrought alloys, both of which are further subdivided into heat-treatable and non-heat-treatable alloys. About 85% of aluminium is used for wrought products, for example, rolled plate, foils and extrusions. Cast aluminium alloys yield cost-effective products due to the low melting point, although they generally have lower tensile strengths than wrought alloys. The most important cast aluminium alloy system is Al–Si, where the high levels of silicon (4.0–13%) contribute to give good casting characteristics. Aluminium alloys are widely used in engineering structures and components where light weight or corrosion resistance is required. Alloys composed mostly of aluminium have been very important in aerospace manufacturing since the introduction of metal-skinned aircraft. Aluminium-magnesium alloys are both lighter than other aluminium alloys and much less flammable than alloys containing a high percentage of magnesium.

21.13.1 Applications

Addition of scandium to aluminium creates nanoscale Al_3Sc precipitates which limit the excessive grain growth that occurs in the heat-affected zone of welded aluminium components. This has two beneficial effects: the precipitated Al_3Sc forms smaller crystals than are formed in other aluminium alloys and the width of precipitate-free zones that normally exist at the grain boundaries of age-hardenable aluminium alloys is reduced. Scandium is also a potent grain refiner in cast aluminium alloys, and atom for atom, the most potent strengthener in aluminium, both as a result of grain refinement and precipitation strengthening. However, titanium alloys, which are stronger but heavier, are cheaper and more widely used.

The main application of metallic scandium by weight is in aluminium-scandium alloys for minor aerospace industry components. These alloys contain between 0.1% and 0.5% (by weight) of scandium. They were used in the Russian military aircraft MiG 21 and MiG.

The following aluminium alloys are commonly used in aircraft and other aerospace structures:

- 7068 aluminium
- 7075 aluminium
- 6061 aluminium
- 6063 aluminium
- 2024 aluminium
- 5052 aluminium
- 7050 aluminium

These alloys are used for boat building and shipbuilding, and other marine and salt-water sensitive shore applications:

- 5052 aluminium alloy
- 5059 aluminium alloy
- 5083 aluminium alloy
- 5086 aluminium alloy
- 6061 aluminium alloy
- 6063 aluminium alloy

6111 aluminium and 2008 aluminium alloy are extensively used for external automotive body panels, with 5083 and 5754 used for inner body panels. Hoods have been manufactured from 2036, 6016, and 6111 alloys. Truck and trailer body panels use 5456 aluminium. Automobile frames often use 5182 aluminium or 5754 aluminium formed sheets, 6061 or 6063 extrusions.

21.13.2 Physical Properties

- Low density
- Low melting temperature of 660°C and castings are poured between 700-800°C
- Specific gravity of about 2.7
- Coefficient of linear expansion is 22×10^{6}/°C
- Thermal conductivity is about 0.28 CGS units at 25°C

Heat required to raise temperature and melt aluminium casting alloy is twice that of copper alloy on mass basis and almost same on volume basis.

21.13.3 Chemical Properties

When aluminium is exposed to atmosphere, an oxide film is formed rapidly and this prevents further oxidation. Hydrogen gas readily dissolves aluminium alloys and forms tiny gas bubbles inside, resulting in gas porosity and rejection of casting. To prevent this, molten metal is treated with degassing agents like hexachloroethane or N_2 gas.

21.13.4 Mechanical Properties

Table 21.4 Mechanical properties of aluminium

Properties	Sand Casting	Metal Mould Casting
0.2% proof stress MPa	Varies from 90(As-cast) to 240 MPa (solution treated and precipitation hardened)	Varies from 90(As-cast) to 270 MPa (solution treated and precipitation hardened)
UTS, MPa	Varies from 140(As-cast) to 280 MPa (solution treated and precipitation hardened)	Varies from 220(As-cast) to 310 MPa (solution treated and precipitation hardened)
BHN	60-120	Varies from 60(As-cast) to 150 MPa (solution treated and age hardened)

21.13.5 Alloys of Aluminium

Aluminium is commonly alloyed with Si, Cu, Mn, Mg, Zn and Ti. They provide good of strength and corrosion resistance. Aluminium alloys can be classified as:

- Sand Casting Alloys: LMO, LM4, LM5, LM6, LM9, LM10, LM16, LM18, LM21, LM25, LM27
- Permanent Mould Casting Alloys: LM4, LM6, LM9, LM10, LM12, LM13, LM16, LM18, LM21, LM22, LM27
- Pressure Die Casting Alloys: LM2, LM9, LM20, LM24, LM25
- As-cast: LM0, LM2, LM5, LM6, LM12, LM9, LM20, LM24
- Heat Treatable: LM4(TF) LM9(TF and TE), LM10(TB), LM13(TE), LM16, LM18, LM21, LM22, LM25(TE, TF, TB7), LM26, LM27, LM29 (TE), LM30

where

- TF refers to solutionizing and precipitation
- TE refers to precipitation heat treated
- TB refers to solution treatment
- TB 7 refers to solution treated and stabilized
- LM refers to light metal series (ALAR) (UK)

Table 21.5 shows the chemical composition of the aluminium alloys (ALAR) and Table 21.6 shows the uses of each alloy.

Table 21.5 Chemical composition

Specification B.S.1490	*Cu*	*Mg*	*Si*	*Fe*	*Mn*	*Ni*	*Zn*	*Pb*	*Sn*	*Ti*	*Others B.S.1490*	*Specification*
LM 0	0.03	0.03	0.30	0.40	0.03	0.03	0.07	0.03	0.03	-	Al 99.50 min.	LM 0
LM 2	0.7-2.5	0.30	9.0-11.5	1	0.5	0.5	2	0.3	0.2	0.2	-	LM 2
LM 4	2.0-4.0	0.15	4.0-6.0	0.8	0.2-0.6	0.3	0.5	0.1	0.1	0.2	-	LM 4
LM 5	0.1	3.0-6.0	0.3	0.6	0.3-0.7	0.1	0.1	0.05	0.05	0.2	-	LM 5
LM 6	0.1	0.10	10.0-13.0	0.6	0.5	0.1	0.1	0.1	0.05	0.2	-	LM 6
LM 9	0.1	0.2-0.6	10.0-13.0	0.6	0.3-0.7	0.1	0.1	0.1	0.05	0.2	-	LM 9
LM 10	0.1	9.5-11.0	0.25	0.35	0.10	0.10	0.10	0.05	0.05	0.2[b]	-	LM 10
LM 12	9.0-11.0	0.2-0.4	2.5	1	0.6	0.5	0.8	0.1	0.1	0.2	-	LM 12
LM 13	0.7-1.5	0.8-1.5	10.0-12.0	1	0.5	1.5	0.5	0.1	0.1	0.2	-	LM 13
LM 16	1.0-1.5	0.4-0.6	4.5-5.5	0.6	0.5	0.25	0.1	0.1	0.05	0.2[b]	-	LM 16
LM 18	0.1	0.10	4.5-6.0	0.6	0.5	0.1	0.1	0.1	0.05	0.2	-	LM 18
LM 20	0.4	0.2	10.0-13.0	1	0.5	0.1	0.2	0.1	0.1	0.2	-	LM 20
LM 21	3.0-5.0	0.1-0.3	5.0-7.0	1	0.2-0.6	0.3	2.0	0.2	0.1	0.2	-	LM 21
LM 22	2.8-3.8	0.05	4.0-6.0	0.6	0.2-06	0.15	0.15	0.1	0.05	0.2	-	LM 22
LM 24	3.0-4.0	0.30	7.5-9.5	1.3	0.5	0.5	3.0	0.3	0.2	0.2	-	LM 24
LM 25	0.1	0.20-0.45	6.5-7.5	0.5	0.3	0.1	0.1	0.1	0.1	0.2	Cr 0.6; Co 0.5	LM 28
LM 26	2.0-4.0	0.5-1.5	8.5-10.5	1.2	0.5	0.1	1.0	0.1	0.1	0.2	Cr 0.6; Co 0.5	LM 29
LM 27	1.5-2.5	0.3	6.0-8.0	0.8	0.2-0.6	0.3	1.0	0.1	0.1	0.2	-	LM 30
LM 28	1.3-1.8	0.8-1.5	17-20	0.7	0.6	0.8-1.5	0.2	0.1	0.05	0.2[b]	-	LM 25
LM 29	0.8-1.3	0.8-1.3	22-25	0.7	0.6	0.8-1.3	0.2	0.2	0.1	0.2	-	LM 26
LM 30	4.0-5.0	0.4-0.7	16-18	1.1	0.3	0.1	0.2	0.2	0.1	0.2	-	LM 27

Table 21.6 Aluminium alloys' light metal series: uses and general remarks

Specification B.S.1490	*Uses and General Remarks*
LM 0	Mainly used for sand castings for electrical, chemical and food applications.
LM 2	One of the most widely used alloys for all types of die castings.
LM 4	Most versatile alloy, has very good casting characteristics and used for wide range of applications. Strength and hardness greatly increased by heat treatment.
LM 5	Suitable for sand and chill castings requiring maximum resistance to corrosion, e.g., marine applications.
LM 6	Suitable for large, intricate and thin walled castings in all types of moulds and where corrosion resistance or ductility is required.
LM 9	Used for applications like low pressure die castings that require the characteristics of LM 6 but higher tensile properties following heat treatment.
LM 10	Mainly used for sand and chill castings requiring high strength and shock resistance. It requires special foundry techniques and is heat treated.
LM 12	Mainly used when a very good machined surface finish and hardness are required.
LM 13	Mainly used for pistons.
LM 16	Used when high mechanical properties are desired in fairly intricate sand or chill castings. Requires heat treatment.
LM 18	Combines good foundry characteristics with good ductility. Requires heat treatment.
LM 20	Mainly used for die castings. Similar to LM6 but a little better machinability and hardness.
LM 21	Generally similar to LM4-M in characteristics and applications but better machinability and higher proof strength.
LM 22	Used for chill castings requiring good foundry characteristics with good ductility. Requires heat treatment.
LM 24	One of the two most widely used alloys for all types of die castings.
LM 25	Suitable where good resistance to corrosion combined with high strength is required.
LM 26	Mainly used for pistons as alternative to LM 13.
LM 27	A versatile sand and chill casting alloy introduced as an alternative to LM 4 and LM 21.
LM 28	Piston alloy with lower coefficient of expansion than LM 13. Requires special foundry technique.
LM 29	As LM 28 but lower coefficient of expansion.
LM 30	For unlined die cast cylinder blocks with low expansion and excellent wear resistance.

21.13.6 Melting Procedure for Aluminium Alloys

For melting smaller aluminium alloys pit type furnaces (oil fuel/gas), tilting type crucible furnace, electric resistance furnaces, etc., are used. For large castings, large tilting furnace or electrical induction furnace is used. Bale out furnaces are used for continuous removal of molten metal by small ladle.

21.13.7 Pit Furnaces

Charges below 50 kg are melted using pit furnaces. Metal or alloy or ingots are kept in the crucible which is placed on the pedestal inside the furnace and heated. When the metal reaches molten state, and has attained the desired temperature the crucible is lifted out. A covering flux is used to protect the molten metal against oxidation. When molten metal reaches 750°C, it is treated with solid degassing agent such as hexachloroethane (0.25% by weight) or by bubbling nitrogen gas through the molten metal. Slag is removed. A grain refiner such as Al-Ti (0.25%) in the form of granules is put inside the molten metal and gently stirred to counter-effect the super heating of metal.

(a) Phosphorus unrefined

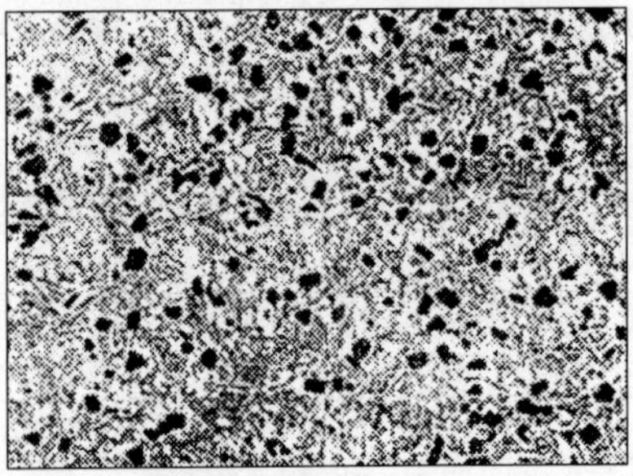

(b) Phosphorus refined

Figure 21.17 Effect of refinement

Slag is removed and molten metal is poured into the moulds. Aluminium alloys containing Si above or near eutectic percentage are modified using phosphorus, strontium or sodium (direct metal or its salts are used). This controls the eutectic Si needles. Modified aluminium alloys show higher strength values. UTS can be enhanced by 13-15%, elongation can be improved by 200%. Chapter on metal treatment may be referred to understand the mechanism of grain refinement and modification. Figure 21.17 (a) shows phosphorus unrefined microstructure and Figure 21.17 (b) shows phosphorus refined microstructure. Figure 21.18 (a) shows non-modified and Figure 21.18 (b) shows modified microstructure of Al-Si alloy.

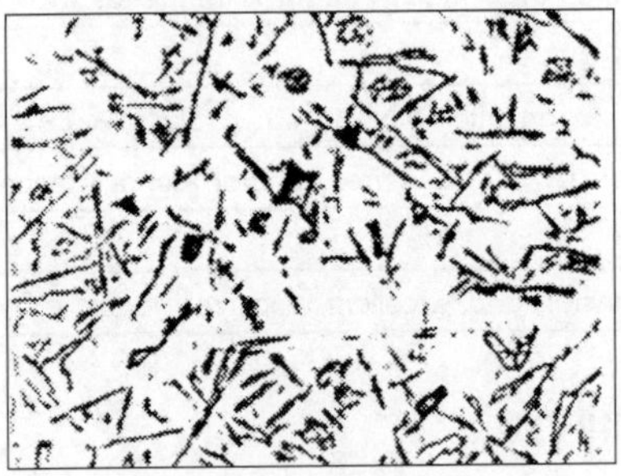

(a) Non-modified

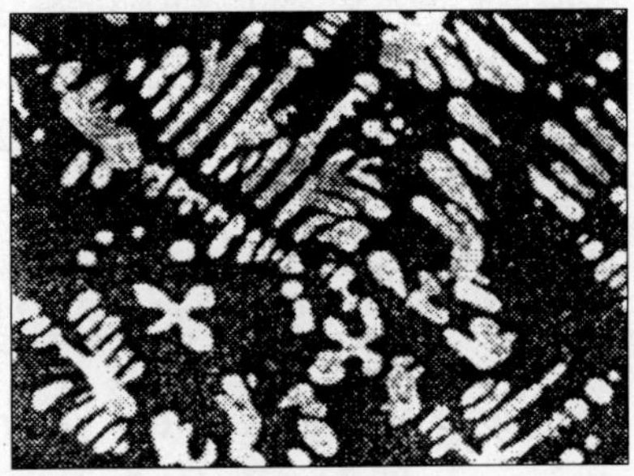

(b) Modified

Figure 21.18 Microstructure of Al-Si alloy

21.14 COPPER BASED ALLOYS

Copper based alloys contain Zn, Sn, Pb, Mn, Ni, Sb as alloying elements with copper as base metal. Their melting point ranges from 800°C to 1080°C, while the density is around

7.83 gm/cc. On exposure to atmosphere, they undergo oxidation which increases with increase in temperature.

Thermal and electrical conductivities are high in these alloys. Their application purview mainly includes heat transfer and heat exchange. These alloys are used in heat exchangers, heat pipes, boiler tubes, etc. They are also used in bearings.

21.14.1 Classification of Copper Based Alloys

They are classified mainly on the basis of the alloying elements used.

- **Brasses:** The predominant alloying element is Zn along with Cu. Zn content varies from 30-40% in the alloy. Small amounts of lead (0.25-3%) along with tin (1-1.5%) may be present. They can be sand cast or die cast.

  Red Brass or 70-30 Brass contains 70% Cu and 30% Zn.

  Yellow Brass or 60-40 Brass contains 60% Cu and 40% Zn.
- **Bronzes:** The predominant alloying element is Sn (4-10%) with small addition of Zn (0.5-2%), lead (0.75-2%). They are highly machinable and corrosion-resistant. Bronze is an alloy of copper and any other element. Various forms of bronzes are Al-bronze, Sn-bronze, P-bronze, Ni-bronze, Si-bronze. They are used in bearings.

  Phosphor bronze: It comprises 9.5% Sn, 0.4% P and remaining is Cu.

  Leaded bronze: It comprises 80% Cu, 10% Sn and 10% Pb.
- **Gunmetal:** They are one of the best materials used in bearing application. The composition of gunmetal is 85% Cu, 5% Zn, 5% Sn and 5% Pb.

21.14.2 Effect of Alloying Elements

Al: It induces oxides and causes porosity. Aluminium should not be present unless specified.

Fe: It helps in increasing hardness and tensile strength. Small quantities of Fe refine the grain structure.

Pb: Improves machinability.

Mn: It is sometimes used as a deoxidiser. It has similar effects as Fe.

Ni: Improves the mechanical properties and increases the resistance to corrosion.

P: It increases hardness and improves fluidity by combining with any trace Fe.

Sn: It improves hardness and strength. It is also used in degassing of copper alloys.

21.14.3 Melting Procedure

Pit furnace, and other conventional furnaces can be used. Copper based alloys containing high percentage of zinc, such as brass alloys, do not dissolve gases such as hydrogen. Vapour pressure of Zn above the melt is higher and prevents dissolution of gases in the metal. Zn also acts as a deoxidant, thereby removing the need for an additional deoxidiser. However, a covering flux is used to prevent excess Zn vapours (glass pieces or proprietary agents are used as flux materials).

21.14.4 Bronze and Gunmetal

These contain high levels of Cu and little or no Zn content. Thereby inducing gas porosity due to hydrogen for which a covering flux (glass pieces or proprietary agents) is used. After melting the top layer, the melt is skimmed off the molten metal and poured into moulds.

21.14.5 High Conductivity Copper

Melted copper, under reducing atmosphere, deoxidises with a deoxidiser containing phosphorus. which is then poured into moulds. Degassing is done using a proprietary degassing agent.

21.14.6 Oxidation-Reduction Technique

It is the main method used in melting copper based alloys. Oxidising atmosphere is created to prevent dissolution of hydrogen gas during melting. Before pouring the molten metal into the moulds, it is deoxidised using phosphorus.

Nitrogen is bubbled through the molten metal to degas it. Alternately, degassing tablets (proprietary) is used to remove dissolved gases.

Table 21.7 Properties of copper based alloys

(i) DTU 160 Aluminium bronze	9-9.8% Al, 0.15% Pb, 0.75%	TS = 451 MPa	Other elements, corrosion resistance, bal copper
(ii) DTU 164 Aluminium bronze	A-9-10% Al, 1-3% Ni, 0.5-2.5% Fe, 0.05% max P 0.2% max Sn, 0.5% max Zn, 0.1% max Si, 0.1% max Mg, 2.5% Mn, Remaining Cu	TS = 565 MPa % elong. = 15%	
Brass 60/40 BS 1400	59-63% Cu, 0.5% max Al, 0.1% max P, 0.75% max others Bal. Zinc	UTS = 248 MPa % elong. = 25%	General purpose, diecast alloy, switchgears, contact parts, brush holders
Naval brass	60% min Cu, 1-1.5% Sn, 0.1% max Pb, 0.75% max others Bal. Zn	UTS = 220 MPa % elong. = 20%	Suitable for hot and cold rolling.

Table Contd...

Leaded 70/30 brass	66-73% Cu 2.5% Pb, 1.5% Sn, 7.75% max Fe, 1.0% max Al, 1.0% max others Bal. Zn	UTS = 151 MPa % elong. = 12%	Suitable for press tight casting, plumbing filling, general hardware.
Guilding Metal 90/10	89-90% Cu 0.3% max Pb 0.6% max impurities	UTS = 220 MPa % elong. = 30%	Tubing, can withstand much work hardening.
Gunmetal 85/5/5/5	4-6% Sn, 4-6% Zn, 4-6% Pb, 2% max Ni 0.5% max others Bal. Cu	UTS = 179 MPa % elong. = 15	For pressure resistance. Has superior machining property. Bearing alloy
Leaded bronze 76/9/0/15	8-10% Sn, 1% max Zn, 13-17% Pb, 2% max Ni, 0.1% max P, 0.5% max others Bal. Cu	UTS = 151 MPa % elong. = 4	Bearing alloy
Leaded phosphor bronze 87/7/2/4	6.5-8.5% Sn, 0.3% max P, 2.5% Pb, 2.0% Zn, 1.0% max Ni, 0.5% max Pb, 0.1% max Si, Bal. Cu	UTS = 165 MPa % elong. = 20%	
Manganese bronze (High Tensile brass)	55% max Cu, 3% max Mn, 2.5% max Al, 0.5-2% Fe, 1.5% max Sn, 1.0% max Ni, 0.5% max Pb, 0.1% max Si, Bal. Zn	UTS = 413 MPa % elong. = 20%	
Phosphor bronze 90/100	10.0% min Sn, 0.05% max Zn, 0.25% max P, 0.3% max others Bal. Cu	UTS = 193 MPa % elong. = 5%	For worm gears, valves, etc.

Table Contd...

Silicon bronze	1.5-5% Si, 5% max Zn, 2.5% max Fe, 1.5% max Mn, 0.5% max others Bal. Cu	UTS = 276 MPa % elong. = 15%	Corrosion resistance, used for bearings.

21.15 MAGNESIUM BASED ALLOYS

Mg alloys are light, strong and have good fatigue resistance. They are used for casting parts of aircraft, and heavy vehicle industry. Melting point is about 650°C and density is around 1.7gm/cc. The main alloying element is Al (up to 10%) with varying amounts of Mn (up to 0.4%), Zn (up to 5%) and Zr (up to 0.7%).

They can be solutionised and precipitation hardened. Ultimate tensile strength of Al, Mn, Zn may vary from 300-430 MPa and elongation 2-5% with BHN varying from 50-90. Table 21.8 gives the details of magnesium based alloys.

Table 21.8 Two important Mg alloys

Element/Design	*L123(AZ 91)*	*L121(A8)*	*AZ 81*
Aluminium	9-10.5%	7.5-9%	8%
Zinc	0.3-1%	0.3-1%	0.5%
Manganese	0.15-0.4%	0.15-0.4%	0.3%
Copper	0.15% max	0.15% max	0.3%
Tin	0.1% max	0.1% max	0.3%
Silicon	0.3% max	0.3%	0.3%
Iron	0.05% max	0.05% max	0.3%
Magnesium	Balance	Balance	Balance

21.15.1 Melting Procedure

Steel crucibles are used as refractory lining of the furnace as molten Mg alloy attack the lining. Ingots are heated in a steel crucible with a proprietary flux placed at the bottom of the crucible. A fluid protective flux is maintained throughout. Then charge is dusted with this flux.

After the molten metal reaches about 750°C, the slag is skimmed off and the flux is dusted on top of the melt layer. The molten metal is gently stirred such that the flux comes in contact with the molten metal. The flux entraps oxides and becomes heavy, hence sinks to the bottom.

The cleansed melt surface is dusted with the flux. Grain refinement is done by using a nucleant (which contains mainly C); molten metal is poured into the mould by dusting sulphur powder. The moulds are made using small percentage of boric acid or ammonium bifluoride to prevent metal-mould reaction. Sulphur may be used with boric acid in moulds but not with ammonium bifluoride. Magnesium alloys have:

- High strength to weight ratio.
- Good fatigue strength near table.
- Good damping capacity.
- Good thermal and electrical conductivity.

21.15.2 Method of Grain Refinement of Mg Alloys

1. Superheating of the metal: The metal is heated to about 900-930°C and held for about 15 minutes in the furnace. The crucible is removed from the furnace and when the pouring temperature is reached, the molten metal is poured into the mould.
2. Carbon inoculation: Disintegration of molten metal is obtained by plunging hexachloro-ethane or hexachlorobenzene tablets. Chlorine gas released removes the dissolved gases. The reaction releases nascent carbon which acts as a nucleus, thereby refining the structure.

21.16 ZINC BASED ALLOYS

Zinc is a metal having a dull silver colour. It has a density of 7.13g/cc and a melting point of 420°C. It is a ductile material. It is used in galvanising as it is electropositive in nature. It is highly corrosion resistant as it forms a thin oxide film, when exposed to atmosphere, which protects it from further oxidation.

It is used in paints, a constituent in brass, to galvanise iron and steel sheets. It is extensively used as a diecast alloy (up to 25 kg). Zinc can mainly be alloyed with Al and Cu but is also alloyed with Mg, Fe, M, Cr, Si, Mn, Pb, Sn in small quantities. Table 21.9 gives the details of popularly used die cast alloys of Zn.

Table 21.9 The two popularly used diecast alloys of Zn

Element	*AG40A*	*Modified AG40A*
Al	3.5-4.3%	3.5-4.3%
Mg	0.02-0.05%	0.05-0.02%
Pb	0.005% max	0.003% max
Cd	0.004% max	0.002% max
Sn	0.003% max	0.001% max
Fe	0.100% max	0.075% max Ni—0.005 - 0.02%
Zn	Balance	Balance

21.16.1 Melting Procedure

Zn alloys are melted and kept in fuel fired open type furnaces or induction furnaces. Grey iron or SG iron melting pots are used. Flux consisting of chloride or fluoride (0.3-0.5%) is used to keep the metal clean.

Molten metal temperature of about 400°C is used for producing die cast alloy castings. Temperature of the molten metal should be controlled within ± 5°C. Excess superheat may result in oxidation of aluminium.

21.17 TITANIUM ALLOYS

Titanium alloys with Al, Sn, Va, Pa. They have high strength and toughness even at high temperature. They are light in weight and extraordinarily corrosion resistant. Being costly, their use is limited to military application, aircraft, spacecraft, medical equipment, consumer electronics and premium sports equipment.

Ti is 7.8% lighter than steel and 60% heavier than aluminium. The density of Ti is 4.32g/cc whereas steel and aluminium are 7.8g/cc and 2.7g/cc respectively. As Ti alloys are highly corrosion resistant, they are used in chemical industry. Ti alloys do not react with human body, hence it is used to fabricate artificial hips and other biological implants. Ti can catch fire when it rubs against other metals at elevated temperatures.

21.17.1 Classification of Ti Alloys

(1) Alpha Alloys: (α-alloy)

Contains Al and Sn as alloying elements although other elements like Mo, Zr, Cu, Ta and Si may also be present.

These alloys do not respond to heat treatment but they are weldable and are used for cryogenic applications, aircraft, and chemical processing industries.

(2) Beta Alloys: (β-alloy)

Have good hardenability, cold formability and can be solution treated. They develop high strength when they are aged. Density of β-alloys vary from 4.84 to 5.06 g/cc. They are weldable and have least creep resistance. Yield strength can be as high as 1345 MPa and is used for heavy duty applications.

(3) Alpha-Beta Alloys

They are heat treatable alloys and are hard and ductile. They are used in aircraft parts, chemical processing equipment, marine hardware, and prosthetic devices. They can be cast, forged, formed. Ti-6 Al-2 Sn-4 Zr-2 Mo and Ti-6 Al-3 Sn-4 Zr-6 Mo are the extensively used alloys.

The strength of the alloy varies from 480-1100 MPa for structural products. Strength greater than 1725 MPa is obtained in wire and spring applications. The properties are given in Table 21.10.

Table 21.10 Properties of Ti alloys

.Material	Characteristics and Applications	TS MPa	% elong
Commercially Pure Ti (99 - 99.5%)	It has superior corrosion resistance. Impurities like O, N, C & Fe reduce. Corrosion resistance used in heat exchanger, piping and valves for chemical industry.	564	15
α-Ti alloys 5% Al, 2.5% Sn, bal Ti	Alloying elements provide solid solution strengthening. These alloys are annealed at the β-range (882°C) and rapidly cooled to a circular structure. Slow cooling after annealing results in plate like β structure which has better creep resistance. These alloys have good weldability. Used in turbine blades, rocket fuel vessels.	880	15
β-Ti alloys 13% V, 11% Cr, 3% Al, bal Ti	High V-content can produce completely α- structure at room temperature, but not used to that extent. Precipitation heat treatment is carried out to get fine α-phase in β matrix. These alloys can be cold worked readily in solution treated and quenched condition and later aged to get very high strength. The alloys have a poor fatigue resistance used in high strength. The alloys have poor fatigue resistance used in high strength fasteners and other fittings for aerospace applications.	1317	5
α-β alloys 6% V, 6% Al, bal Ti	By proper balancing of alloying element, it is possible to get a mixture of α + β phases at room temperature. By solutionising heat treatment followed by quenching, supersaturated β phase is formed which upon ageing provides finely dispersed α-precipitates in a matrix of β. Used in connection rods of cars. Biomedical applications like replacement of hip bones and heart valves, etc.	1057	8

21.18 CASTING OF ALUMINIUM ALLOYS

The aluminium alloy scrap to be melted should be clean. The scrap should be preheated to remove any moisture. The crucible used for melting should be cleaned after each melt. All tools used in melting must be clean and dry to prevent pick up of moisture by the melt. Moisture in the moulding sand causes porosity in the casting. The sand mould must be properly dried and preheated. Combustion gases, sand, tools, etc., containing H_2O may cause hydrogen absorption. Molten aluminium/alloys absorb gases giving a defective casting after solidification. The removal of gases is an important process in the metal casting field. Aluminium is degassed by passing N_2 gas slowly in the melt. A flux is used in melting aluminium. The flux forms a protective cover to minimize oxidation and absorption of the gases. A fine grained aluminium alloy has better strength and this is done by inoculating the melt with elements which form insoluble compound particles which will help to increase the rate of nucleation. Many types of grain refiners and fluxes are available in the market. Preheated aluminium charge is added in the crucible. Once this scrap melts, the remaining charge is added in the melt carefully. The fresh charge should be preheated and added carefully. If any moisture is present on the charge, steam explosion can occur. Grain refinement in aluminium alloys can be refined by adding boron and titanium to the melt. Aluminium alloys can be cast in metallic moulds or sand moulds. The most common casting processes used are sand casting, permanent mould casting and die casting.

21.18.1 Casting Processes

(1) Sand Casting

Sand casting is an important process for casting molten aluminium alloys. It can be used to prepare castings of any shape and size. Molten metal is poured into a mould cavity formed out of sand (natural or synthetic). The sand cavity is formed by using a wooden pattern. The pattern resembles the real casting part. The upper part of sand mould box is called cope and lower part is called drag. The pattern is made slightly oversize to allow for metal contraction as it cools down. The liquid flows into the gap between the two parts, called the mould cavity. Sand castings have a rough surface and are machined later. The metal from the sprue and risers is cut from the rough casting. The riser is a reservoir of liquid metal to supply a contracting, cooling casting with make-up metal. It is used to prevent internal or external voids due to shrinkage.

(2) Casting in Metal Moulds

This uses a metal mould instead of sand. The moulds are made of steel or cast iron. The mould can be of solid type or split type. In the split type, mould casting removal is easy. The moulds are coated with ceramic based coating agent. The mould is dried and preheated before molten metal is poured. The crucible is removed from the furnace and the dross is removed. The melt temperature is taken and the melt is poured into the mould cavity. After solidification, the casting is removed. The shrinkage area is machined away by cutting on a power saw and the rest of the casting can be used. In die casting, molten metal is injected into the mould under high pressure resulting in a better part with good dimensional accuracy.

(3) Centrifugal Casting

The principle originally from a patent, the process used for centrifugal casting is long established, by A. G. Eckhart in 1809. In this process, molten aluminium is poured into a rotating mould, and the metal is held against the wall of the mould by centrifugal force. The mould may be made of cast iron or steel, copper, graphite, ceramic, or dry sand. This process is suitable for the production of hollow parts, such as pipes, bearings, bushings, etc. Most metals like steel, copper/alloys, aluminium/alloys, etc., can be used successfully in centrifugal casting. The impurities which segregate can be machined off. The casting is relatively free from defects. There is less loss of metal in tundish compared to that in gating and risers as in conventional sand casting. The process can be used to manufacture bimetallic pipes. Bimetallic pipe is composed of two different metal layers. Bimetallic pipes are important in applications which demand outer layer to be

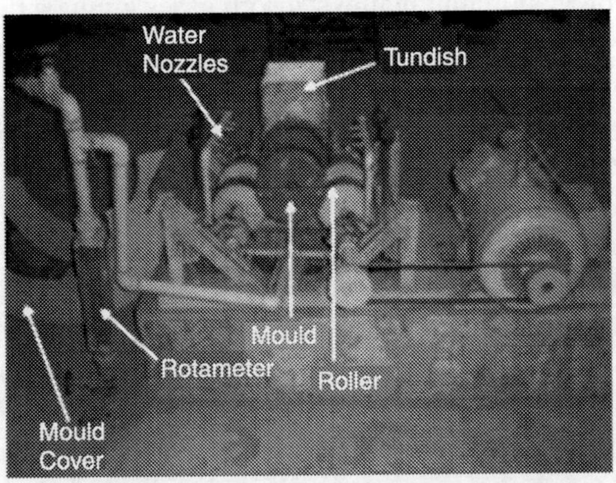

Figure 21.19 Centrifugal casting of aluminium alloys

wear resistant or inner layer to have lubricating properties. This process can be used for prepare a functionally graded metal matrix composite material. The tundish is a vessel for collecting the molten metal from the crucible or ladle. The tundish has a long nose which fits into the centre hole of the mould flange. The tundish transfers the molten metal into the rotating mould. The tundish can be made of graphite or lined with basic lining material.

The mould is driven by a belt driven motor. To avoid any splash of metal, the rotating mould is covered as shown in Figure 21.19.

(4) Shell Moulding

In this process, a heated metal pattern is made to contact sand/resin mixture. This cures the resin in the mixture and it forms a rigid shell. The dump box is then turned back and all the loose sand falls off. The pattern with the sand shell is returned to the oven to harden the shell and is then removed from the pattern. The upper and lower shell moulds are clamped and molten metal is poured inside. The castings made in shell moulds have excellent surface finish, dimensional accuracy and detail. Shell moulding is used for producing simple or complex near net shape castings. The process was developed and patented by Croning in Germany during World War II and is sometimes referred to as the Croning shell process. Details are given in Figure 21.20.

(a) (b) (c) (d)

Figure 21.20 Shell moulding of aluminium alloys: (a) shell part on the heated pattern, (b) preheating of assembled shell parts, (c) pouring molten aluminium into shell, and (d) finished machine casting

(5) Investment Casting

Investment casting or the lost wax process has been known since ancient times. Wax replica of the desired casting is prepared. This is called the pattern. The pattern is attached to a sprue and then dipped in liquid ceramic slurry many times. After the ceramic becomes dry, the wax is melted. Molten metal is poured and after the metal cools, the part, gate, sprue and pouring cup become one full solid casting. The outer ceramic is broken. The gating is machined away from the casting and the final casting is obtained.

(6) Continuous Casting

This process is used to prepare long castings of different cross sections like round, square, etc. The molten metal poured in the preheated tundish passes into the water cooled mould. The mould has the cross section as per the need. The metal is continuously withdrawn and a long casting can be prepared in a smaller mould. The solidified form is cooled and withdrawn at a controlled rate from the water-cooled mould by rollers and the material is cut to length with a travelling saw. Metals such as steel, copper and aluminium are continuously cast.

21.19 CAST BRONZE ALLOYS

The bearing grades of cast bronze can be classified metallurgically into three categories:

- Single-phase solid solution alloys
- Polyphase alloys
- Composite materials

To understand the performance of different alloys, we must first understand what happens to the basic copper structure when small amounts of alloying metals are added. Reactions occur during the solidification and cooling of alloys from their molten state.

In simple terms, the final arrangement of the alloying metals with respect to the normal fcc copper lattice determines the properties of the alloy material.

Alloying metals find their place in the copper lattice in three basic ways:

1. They substitute for copper atoms in the fcc lattice.
2. They combine with the copper and form localized regions (phases) where the crystal structure is of a form which differs from the fcc copper crystal.
3. They are rejected by the solidifying copper lattice but are trapped within the crystals of the alloy as they freeze and grow.

Research has resulted in the graphical representation of how simple binary alloy systems react. This representation is called a phase diagram. The phase diagrams of some binary systems relevant to bronze show the behaviour of alloying elements that typically results in one of the three cases mentioned previously. The copper-tin equilibrium phase diagram is shown in Figure 21.21.

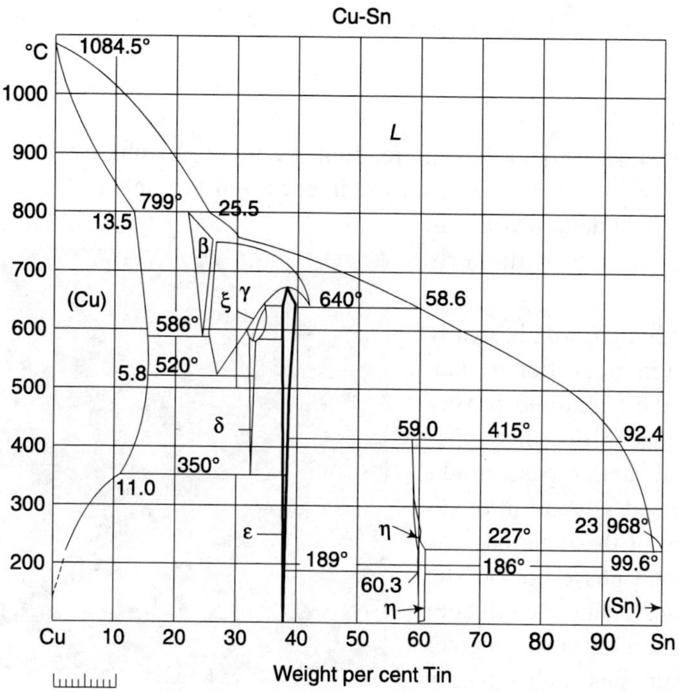

Figure 21.21 Phase diagram of copper-tin

Substitution

Referring to Figure 21.20, at less than 11% tin (at 8% tin for example), an alloy as it cools is seen to solidify over a range of temperature, becoming completely solid as the temperature falls below about 850°C. Under equilibrium conditions of slow cooling, the solid phase that forms are a face-centred cubic crystal (alpha phase). Tin atoms substitute directly into the lattice in place of copper atoms. The tin atoms have the effect of actually strengthening the pure copper because they strain the lattice, that is, they alter the usual distance between the copper atoms. Under commercial conditions of rather slow solidification, nearly all of the metal will solidify as the alpha phase (Figure 21.22). This substitutional phenomenon results in a single-phase solid solution of tin in copper. The crystal structure, though stronger than pure copper due to lattice strain, is still fcc. Consequently, the slip characteristics remain very good. Single-phase solid solution alloys of copper, therefore, retain high ductility despite very significant increases in strength. Such materials find wide use under conditions where the material may be subjected to

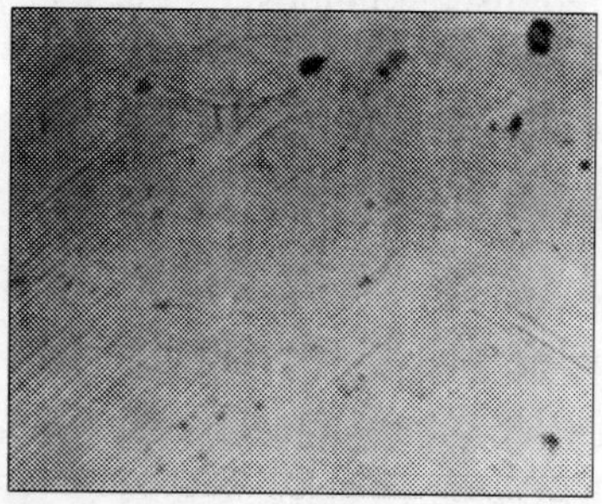

Figure 21.22 Microstructure of a single phase (alpha) copper-tin alloy (88Cu-8Sn-4Zn). Structure shows slip lines. Note also traces of the delta phases (darker islands)

considerable stress but where fracture would be catastrophic (such as fittings in nuclear reactor seawater systems).

Polyphase

If the tin content is increased to 11% or more, some of the alpha phase will transform as the metal cools below 400°C. A new phase appears, interspersed throughout the normal fcc alpha crystals. This phase, called delta phase, can be preserved in the material with fairly rapid cooling.

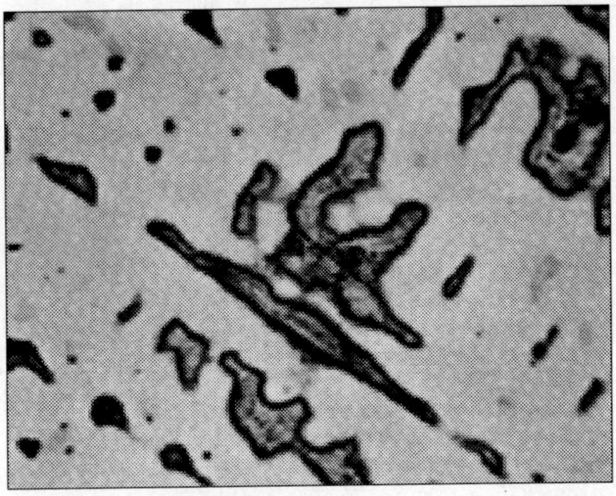

Figure 21.23 Microstructure of continuous cast tin alloy C90500. Delta phase is indicated

The delta phase (though still basically fcc) contains more tin in proportion to copper than is found in the alpha and is very hard and strong but lacks ductility (Figure 21.23). It appears under best conditions as finely dispersed islands throughout the microstructure of the material. The influence of this second phase on the slip mechanism is dramatic, having the effect of pinning the slip planes after small degrees of motion. But the delta phase also greatly increases the wear resistance of the material, as is indicated by the significant increase in hardness. The popular aluminium bronze alloys (C95400 and C95500) and manganese bronzes (C86300 and C86400) gain their high strength and hardness in a similar fashion, although the actors are different. Nonetheless, the properties are the result of the dispersion of another phase (or phases) in the basic fcc lattice with, in nearly all cases, the dispersed phase being much harder and stronger than the bulk of the surrounding matrix material. These "engineered discontinuities" in the material serve to anchor the slip planes and restrict their motion. Details are given in Table 21.12.

These materials are known as polyphase alloys and are characterized by higher strength, hardness and wear resistance than alpha alloys; but they exhibit much less ductility as shown in the tabulation below. As a result, they are best suited for control surface parts where dimensional integrity is most important and for heavy loads and shock loads under slow speeds like in aircraft landing gear bearings.

An additional characteristic of the polyphase alloys is that their properties can vary with temperature to significant degrees compared to the single-phase materials. Such alloys are often heat treatable. Through the manipulation of the microstructure of the alloy as shown on the equilibrium phase diagram by "short circuiting" the equilibrium, certain properties can be obtained which are absent in the As-cast condition.

Table 21.11 Properties of single-phase and polyphase

	Single phase		*Polyphase*	
	C90300	C90700	C92500	C95400
Cu	88	89	87	85
Sn	8	8	11	-
Pb			1	
Zn	4			
Al				11
Ni			1	
Fe				4
TS(ksi)	49	52	50	65
YS	23	29	25	35
% elong.	25	18	20	18
Hardness (BHN)	77	95	83	170

In any case, polyphase materials which do not have significant amounts of lead should only be used as bearings against steel mating surfaces that have themselves been hardened by heat treatment. In the case of aluminium bronze or manganese bronze applications, it is often recommended that the shaft be chrome plated or made from bi-metal material similar to that used for steel mill rolls.

Composite Mixtures

The most widely used bearing materials are really composites. Fundamentally, they can have either single-phase or polyphase structures serving as a matrix around particles or globules of free lead. A look at the copper-lead equilibrium phase diagram (Figure 21.24) indicates that lead is nearly completely rejected by the copper lattice as it freezes. Nonetheless, metallurgical engineers and foundrymen alike have made great efforts to capture the lead between the crystals of copper-based material as these crystals freeze and grow because the resultant alloy makes very fine bearings. A typical microstructure showing the dispersed lead is shown in Figure 21.25.

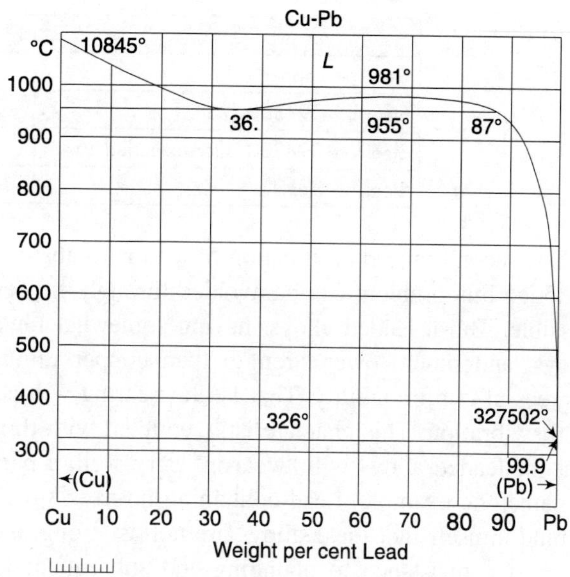

Figure 21.24 Copper-lead equilibrium phase diagram

It is possible to produce copper based material with lead content in excess of 30%, wherein the size of the lead particle is microscopic in scale. On the other hand, if the bearing application indicates larger lead particles are more desirable, it is also possible to produce the alloy in that form.

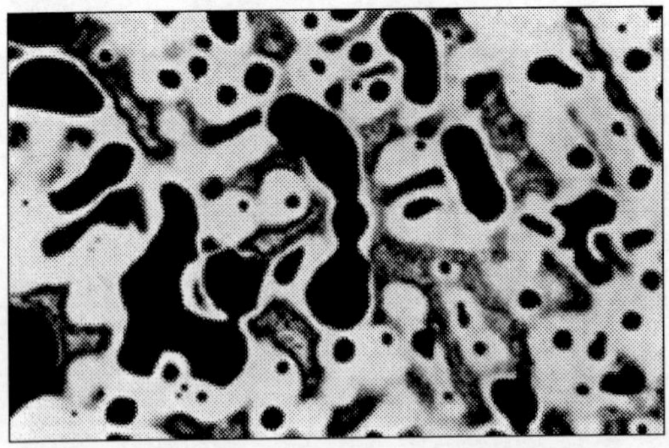

Figure 21.25 Microstructure of continuous cast leaded tin bronze alloy C94100 (20%Pb). Lead particles are indicated

Lead performs three important bearing functions, all of which serve to protect the shaft and improve the performance of the machinery. Of primary importance is the ability of the lead particles to decrease the coefficient of friction between the bearing and the shaft. The mechanism by which this is accomplished is quite interesting. Lead particles are free to be sheared off the bearing surface by microscopic rough edges on the shaft surface. The steel shaft becomes covered with lead which is gradually redistributed to fill in the low spots on the shaft. Once this is accomplished, the coefficient of friction rises only slightly again, as indicated below (Table 21.12). This same phenomenon has a further advantage in that the temperature developed at the points of contact between the bearing and the mated part is limited by the fusion temperature of lead 327°C). Obviously, this property of the leaded alloys is very valuable in the absence of lubrication (planned or accidental) or if the operating environment of the machine is itself subject to wide temperature extremes, for example on aircraft or arctic oil field equipment.

Table 21.12 Coefficient of friction

	Coefficient of friction (sliding)
Steel on copper	0.9
Steel on C94300 (23% Pb)	0.18
Steel on C94300 after extended use	0.30
Steel on Steel	1

The second important function of lead is to absorb dirt which finds its way into the interface, although this problem can be avoided through the design of properly sealed bearings whenever possible. Third, leaded alloys, having somewhat lower strength than the non-leaded copper-tin alloys, and much lower strength than copper-aluminium or copper-zinc alloys, exhibit high degrees of conformability. That is, the bearing will adjust its shape to allow for poor alignment or for vibration. This characteristic, coupled with those previously described, allows one to say that the leaded alloys will "wear-in" very well, a particularly desirable feature for worm gears, to name one example. Lead-containing bronzes are also readily machinable. The engineer must remind himself that these alloys are not as strong as non-leaded materials. They do not exhibit as great a resistance to pounding and subsequent fatigue leading to failure. One comforting consideration, however, is that total failure of the bearing is not likely to destroy the shaft or seize the machine, due to the "softness" of these alloys.

By selecting the proper matrix into which the lead particles will be cast, the engineer can select from a fairly wide range of material strengths compatible with moderate to light loads and high speeds, as shown in Table 21.13. Values are for continuous castings less than 3 inches in diameter.

Table 21.13 Properties of single-phase and polyphase

	C83600	*C93200*	*C93700*	*C93800*	*C94300*
Cu	85	83	80	78	70
Sn	5	7	10	7	5
Pb	5	7	10	15	25
Zn	5	3			
TS(ksi)	45	45	41	34	27
YS	21	24	24	23	13
% elongation	28	16	10	12	15
Hardness (BHN)	72	72	80	62	48

21.20 OTHER CASTING PROCESSES FOR NONFERROUS ALLOYS

The bearing grade alloys of copper available in many forms are produced by various methods of manufacture. Cast production methods are summarized in Table 21.14.

Table 21.14 Sand and chill mould products

Alloy family	*Sand*	*Chill mould*	*Centrifugal*	*Continuous*	*Wrought*
Red brass	X	X	X	X	NA
Leaded Tin bronzes	X	X	X	X	X*
High-Leaded Tin bronzes	S	S	S	X	NA
Tin bronzes	X	X	X	X	X**
Aluminium bronzes	X	X	X	X	X
Manganese bronzes	X	X	X	X	X

* limited to 4% Pb

** Wrought alloys C51000, C52100, C52400

Casting in sand or chill moulds are ideal production methods for small runs or very small parts and are sometimes mandatory for very large parts such as ship propellers.

All of the alloys discussed are available in these forms, although problems with severe lead segregation may occur as the lead content approaches 16%. A wide range of sizes and intricate shapes can be cast. The red brasses, which are very popular as plumbing hardware materials, are produced by these methods, mainly in the form of valve bodies and fittings.

Centrifugally Cast Products

Again, all subject alloys are readily produced by the centrifugal casting method, with the exception of high-leaded tin bronzes in which the lead content approaches 20%. Lead segregation problems are sensitive to the size of the casting. Very large bushings are manufactured by this method. It is likely that most bushings over 14 inches O.D. and up to about 100 inches O.D. are centrifugal castings. Such castings can be made in lengths that exceed 100 inches Nevertheless, small centrifugal castings are also high-volume items. Many of the larger flanged bearings or gear blanks are manufactured by this method. Though sensitive to production quantity, small runs can be highly economical. Stocking distributors maintain inventories of semi-finished centrifugal castings, primarily in standard sizes and especially in alloys C95400 and C93200.

Continuous Cast Products

All alloys are available as continuous cast barstock; lead segregation is generally not a problem. It may be necessary to stress relieve certain castings with very thin walls, particularly if the alloy is C95400, C95500 or C86300, to prevent loss of clearance or tolerance in fabrication and use. A wide size range of solid, tubular and made-to-order cross section barstock is available. Diameters range from less than 0.500 inch up to about 14 inches in O.D., in lengths up to about 13 feet. It is possible to produce very thin-walled bars, at times less than 1/4 inch, depending on the O.D. These products are ideally suited for further fabrication using automatic machine tools.

Larger quantities of continuous cast products are considerably more economical, but again stocking distributors absorb a great deal of this burden, particularly where alloys C95400, C93200 and C90300 are concerned.

21.21 WROUGHT PRODUCTS

The wrought phosphor bronze alloys (C51000, C52100, C52400, C54400) are sometimes used in bearing applications. These alloys are also available as continuous castings in the annealed temper. Wrought phosphor bronze is usually limited to about 3 inches O.D. and under. C54400 has the highest available lead content, about 4%. It is not possible to extrude or roll alloys with higher lead content.

The aluminium and manganese bronze alloys also have wrought equivalents. The wrought alloys have very strong mechanical properties, having been severely worked either by extrusion, drawing, rolling or forging and are widely used in aerospace applications. Some of these alloys are used as weldment materials. These alloys are also available in different extruded shapes, though the variety available is quantity dependent. Heat treatment of cast alloys produces mechanical properties similar to the wrought materials, as does the continuous casting of the aluminium bronzes. Generally speaking, large production quantities are required to make the wrought products economical, although stocking distributors have assumed this burden for the end-user of smaller quantities.

Review Questions

1. List the properties of cast iron.
2. What are the types of cast iron?

3. Explain the production of cast iron castings in foundry.
4. Explain the production of malleable iron in industry.
5. What are ADI castings?
6. Explain classification of alloys of aluminium.
7. Give the classification of copper alloys.
8. What is the melting procedure for copper alloys?
9. What is oxidation-reduction technique?
10. What is grain refinement in Mg alloys and how is it done?

CHAPTER 22

Miscellaneous Special Casting Processess

22.1 INTRODUCTION

There are many special casting processes other than the shell moulding process, die casting process, investment casting, centrifugal casting and continuous casting methods. These special processes are invented for specific purposes and applications.

22.2 THE V-PROCESS SETUP

A photograph of the setup is shown in Figure 22.1. The schematic diagram of the V-process experimental setup is shown in Figure 22.2. The whole setup is divided into the following parts:

- Moulding system: (a) pattern box, (b) moulding box, (c) pattern
- Vacuum system
- Vibrating system: (a) vibration table, (b) vibrator, (c) motor
- Plastic film holding and heating system.

Figure 22.1 Experimental setup in vacuum sealed moulding

1. Moulding System The moulding system comprises the equipment needed to prepare the mould. This includes the pattern box, moulding box and pattern.

2. Pattern Box The pattern box is shown in Figure 22.3. The pattern plate is placed on the straight top open surface of the pattern box, and perfect matching between the mating surfaces is achieved. In order to ascertain the perfect matching, the pattern plate, having equidistant holes, is provided with four support blocks at the four corners in the pattern box. Additional support at the centre has been provided to prevent warping of pattern plate when subjected to vacuum pressure. The pattern box is fixed to the vibrating table.

3. Moulding Box In the design of moulding box (Figure 22.4), the major consideration is to have a uniform distribution of vacuum throughout the body of the sand to form the mould. The design of moulding box for V-process is more complicated than conventional sand moulding process. The box has to be made with annular wall on all the four sides. Inside walls are

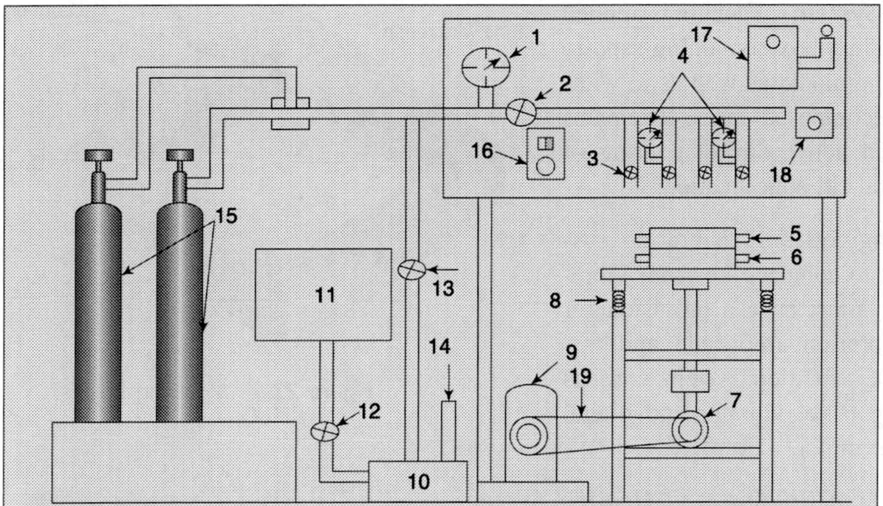

1. Vacuum gauge in main line, 2. Main control valve, 3. Control valves in sub-line for cope and drag, 4. Vacuum gauge in sub-line, 5. Moulding box, 6. Pattern plate, 7. Vibrator, 8. Spring, 9. Motor, 10. Vacuum pump, 11. Water tanks, 12. Water outlet valve, 13. Inlet control valve of vacuum pump, 14. Outlet control valve of vacuum pump, 15. Surge tank, 16. Speed controller box, 17. Main starting switch, 18. Vacuum pump control switch, 19. V-belt.

Figure 22.2 Process setup for V-processes

provided with windows. These windows are covered with very fine mesh and backed by a thin metallic strip having small holes for supporting the fine mesh. This mesh prevents very fine sand particles from being sucked up from the sand voids in the box into the vacuum pump. Two metallic pipes are connected with outside wall of the box. One pipe is used to suck the vacuum from the box and another pipe to release the vacuum to the atmosphere.

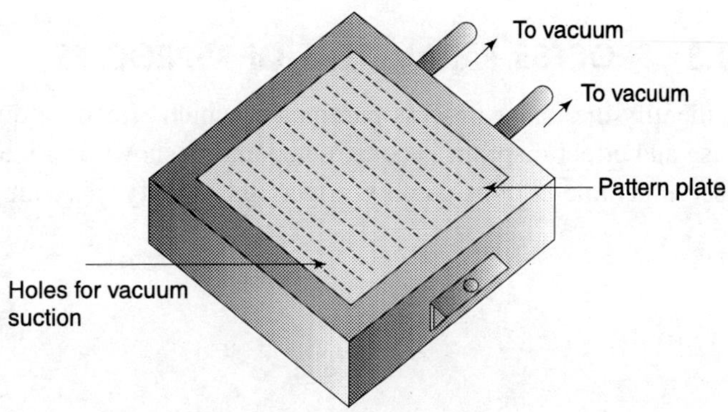

Figure 22.3 Pattern box

4. Pattern The patterns used in V-process are similar to the patterns used in other conventional sand casting processes. Numerous small holes are created on the vertical edges of the pattern, used in V-process, so that plastic film is correctly adhered to the vertical surface of the pattern, thus creating the right cavity in the mould. A simple three-stepped pattern, having small holes at the vertical edges, is shown in Fig. 22.5.

5. *Vacuum System* Vacuum is applied to the moulding box containing sand to ruggedize the mould as well as to withdraw the decomposed gases. It remains constant and active during pouring of molten metal till solidification.

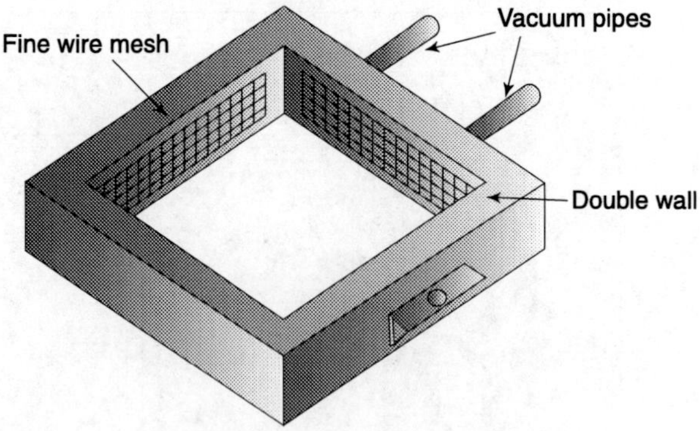

Figure 22.4 Moulding box

6. *Vibrating System* Vibrating system is used to compact the sand. Vibration system must have a provision for varying vibration amplitude and frequency. The vibration system consists of vibrating table, vibrator and digital varied-drive motor.

7. *Plastic Film Holding and Heating System* To prepare a mould, two plastic polyethylene/ethylene vinyl alcohol films are used. The purpose of using two plastic films (one on top and other for bottom of the moulding box) is to seal the box. The film is heated up to plastic *d* formation so that it adheres tightly to the pattern and the top/bottom surface of the mould. Also, heating system is needed to soften the plastic film.

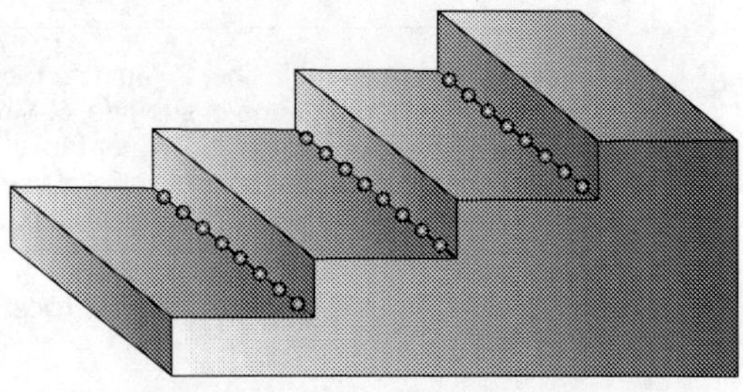

Figure 22.5 Four-stepped pattern

22.3 PROCESS PARAMETERS OF V-PROCESS

To identify the major process parameters which affect the qualities in a V-process, and Ishikawa cause and effect diagram is constructed and is shown in Figure 22.6. It depicts *t* at the following process parameter considerably affects the quality of castings produced by V-process.

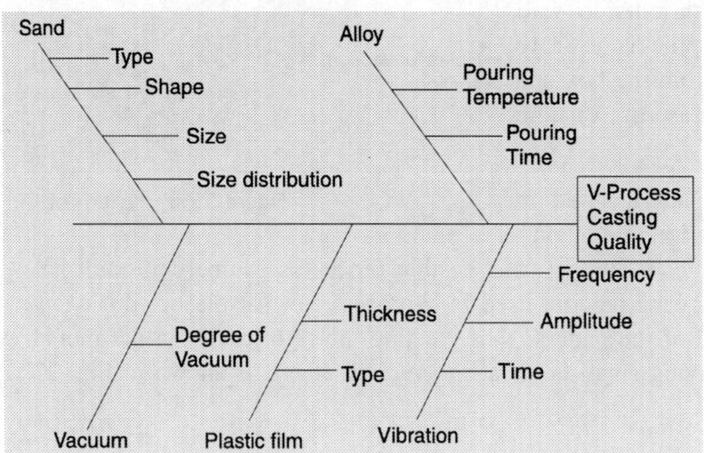

Figure 22.6 Cause and effect diagram for V-processes

- Variables based on moulding sand: shape, type, size and size distribution.
- Variables based on plastic film: type and thickness.
- Variables based on vibration: amplitude, frequency and time of vibration.
- Variables based on vacuum: degree of vacuum imposed.
- Pouring material based variables: pouring time and temperature.

22.4 SPECIFICATION OF V-PROCESSES

Complete V-process outcome specification is given in Table 22.1.

Table 22.1

1.	Tolerances:	One side of the parting line ± .010″ up to 1″.
2.		Over 1″ add ± .002″/in.
3.	Surface Finish:	125 – 150 RMS
4.	Surface Finish: Cored Area	150 – 250 RMS
5.	Draft Requirement:	Nil
6.	Minimum Section Thickness: 2.5 mm	

22.5 INTRODUCTION

The evaporative pattern casting process is also known by several other names such as full mould process, lost foam process, etc.

H.F. Shroyer patented the foam patterns for metal casting in 1958. In this patent, expanded polystyrene (EPS) block was used by him to machine the pattern and during pouring, it was supported by bonded sand. This process is called the full mould process. In this process, the pattern is usually machined from an EPS block and is used to make large one-of-a kind castings primarily. Originally, this process was also known as 'lost foam process'.

M.C. Fleming in 1964, used unbounded sand with the process. This is known as lost foam casting (LFC). With LFC, the foam pattern is moulded from polystyrene beads. LFC is differentiated from full mould with the use of unbounded sand (LFC) as opposed to bonded sand (full mould process).

The evaporative pattern casting (EPC) process is a binderless process and no physical bonding is required to bind the sand aggregates. Foam casting techniques have been known by a variety of generic and proprietary names such as lost foam casting, evaporative pattern casting, and cavityless casting, full mould casting and evaporative foam casting. In this method, expanded polystyrene is used to prepare the complete pattern including the gates and risers. This pattern is further embedded in a sand of no bake type. The molten metal is poured through the sprue, while the pattern is still inside the mould. The heat of the molten metal is sufficient to gasify the pattern and the pattern progressively gets displaced by it.

Using unbounded sand and expandable polystyrene pattern, the EPC process is an economical method for producing complex, close-tolerance castings. The expandable polystyrene can be moulded into numerous complex and rigid shapes. It is basically a thermoplastic material. In the EPC process, expandable polystyrene patterns are attached to an expandable polystyrene gating system and a refractory coating is applied to the entire assembly. Once the coating gets dried, the entire foam pattern assembly is kept on loose dry sand in a vented flask. The flask is vibrated and additional sand is added to it until the pattern assembly gets completely embedded in sand. Molten metal is then poured into the sprue which further vaporizes the foam polystyrene and reproduces perfectly the used pattern to get the required casting.

In this process, a pattern used refers to an expandable polystyrene or foamed polystyrene part which gets vaporized by the molten metal. For every casting process, a new pattern is required.

22.6 ADVANTAGES OF EPC PROCESS

1. In the EPC process, no cores are required making it the most advantageous.
2. No requirement for binders or other additives, as it is a binderless process.
3. Complete sand reclamation is possible using very simple and inexpensive techniques.
4. Sand shakeout is easy as the sand is unbounded.
5. Since the pattern used in EPC process is one piece, hence no parting line and since cores are eliminated, hence no core prints. Also, no mismatch, core shift because of the mentioned reasons.
6. Improved casting quality. Close tolerances are possible.
7. The EPC is an environmentally favourable process.
8. As it is a binderless process, the efforts on cleaning the moulded sand are virtually nil. Therefore, the EPC process is viewed as a value-added process rather than a substitute for sand casting.

22.7 DISADVANTAGES/LIMITATIONS OF EPC PROCESS

1. Since every casting requires a new pattern, it is a costly process.
2. There is a limitation on the minimum section thickness of the pattern.
3. Quality of the casting fully depends upon the quality of the pattern.
4. As the sand is unbounded, during pouring, due to the difference in the evaporation rates of the pattern material and the flow rate of the metal, the sand falls down in the generated cavity generated, thereby leading to a defective casting.

22.8 APPLICATIONS OF EPC PROCESS

1. It is used for making automotive components (cylinder heads, engine blocks, inlet manifolds, heat exchanger, crankshaft).
2. It is used in marine, aerospace and construction industries.

22.9 PROCESSING STEPS, REQUIREMENTS AND VARIABLES IN EPC

Process Description

1. The process of EPC starts with the pre-expansion of polystyrene beads. Once these pre-expanded beads get stabilized, pattern sections are formed by blowing them into a mould. A cycle of steam is used to fully expand and make the beads fuse in the mould itself.
2. Clusters are formed through assembly of the pattern sections using glue. The gating systems are also similarly glued and attached.
3. Ceramic coating is used to cover the foam cluster. The coating acts as a barrier preventing penetration and sand erosion during pouring.
4. Once the coating gets dried, the cluster is placed in a flask along with the backing of the bonded sand. Any suitable sand can be used as long as it resists the temperature of the molten metal being poured. Silica sand, zircon sand, olivine sand and chromite can be used as moulding sand. The EPC process has high degree of sand reclamation hence we can use zircon or chromite sands, which are otherwise considered expensive. The strength of mould can be determined through the frictional resistance between the sand grains. The mould strength with angular grains is higher although higher bulk density is provided by the rounded grains.
5. In order to ensure uniform and proper compaction, a vibrating table is used in the mould compaction. After completing this process, the cluster is packed in a flask and the mould is set ready for getting poured.

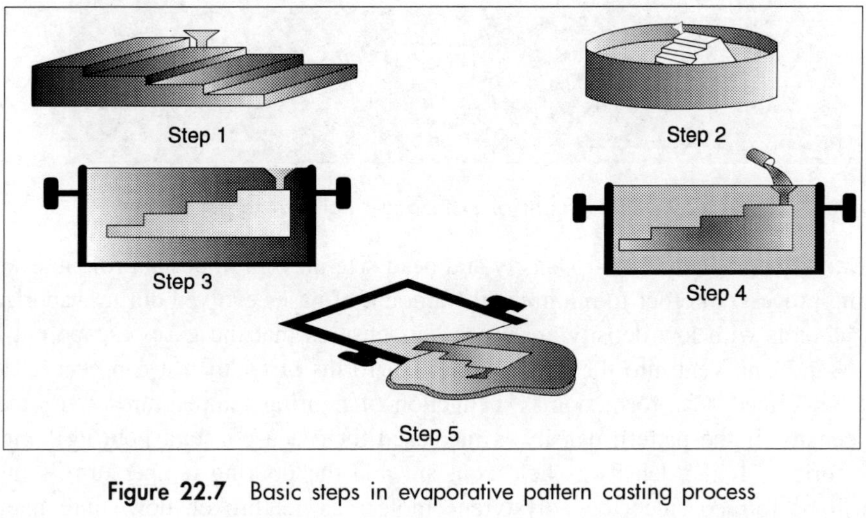

Figure 22.7 Basic steps in evaporative pattern casting process

1. Pattern The patterns used in EPC process are made of expendable polystyrene (EPS). Pattern fabrication starts with the pre-expansion of polystyrene beads. They are blown into die cavity to form pattern sections. When the beads are in the die, a steam cycle causes them to fully expand and fuse together. A photograph of simple pattern made of polystyrene beads is shown in Fig. 22.8.

Figure 22.8 A polystyrene pattern

2. Polystyrene Pattern Die For the fabrication of stepped pattern, die and sprue, a cavity as per the pattern dimensions was prepared. Aluminium was used as the die material. The complete die is fabricated in two parts, called male and female. Three inlet pipes (nos. 1, 2 and 3) and one outlet pipe (no. 4) are attached with male and female parts, as shown in Fig. 22.9. The function of all three inlet pipes is same in male and female parts. The two outermost pipes (nos. 1 and 3) are used to supply water as cooling agent in the die and the middle pipe (no. 2) is used to supply steam into the cavity through pinholes to soften and expand the beads. Pipe 4 is used to draw water and steam from the die. Polystyrene beads are fed into the die cavity through a pipe which is connected at the top of the male part. At the four corners of the die, T slots are fabricated to hold both male and female parts of the die in the press.

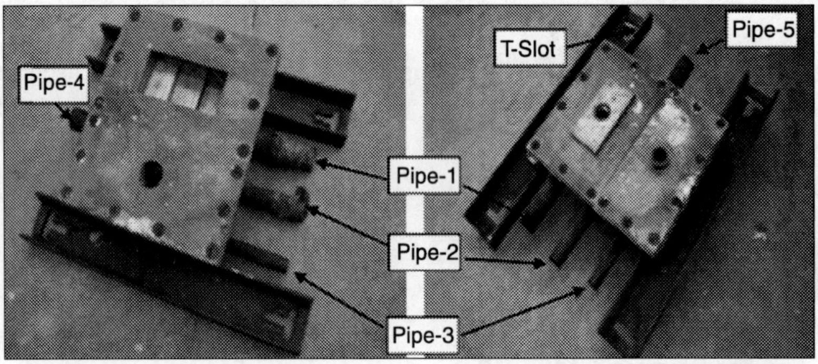

Figure 22.9 Photographs of expandable polystyrene pattern die

3. Pattern Density and Bead Size Density and bead size play an important role in evaporative pattern casting process. In order to minimize the amount of gases evolved during vaporization of the pattern, patterns with low density are used. This ensures that the gases escape out through the coatings, sand, and vent into the atmosphere. If it forms faster than it can vent, a defective casting will be formed. Gas formation is a function of pouring temperature of the metal and the pattern density. If the pattern density is increased then at a constant pouring temperature, more gas is formed. If the density is held constant and the pouring temperature is increased, more gas will be formed since the polystyrene molecules get broken down into many more basic molecules at higher temperatures. The gases formed should pass through the coating on the pattern surfaces. The pattern density generally varies between 1 and 1.5 pcf which further depends on the geometry and the metal that is being poured. Small beads are required for obtaining relatively non-beady, smooth surfaces on the moulded patterns.

4. Requirement of Refractory Coating Material As the coatings are inherent parts of the casting process and also give good quality surfaces to the casting, some specific demands as mentioned below are desired for the refractory coatings in the application of evaporative pattern castings process:

- Highly permeable coating is preferable for rougher sand while medium and low permeable coating is used for finer sand.
- Quick drying is preferred.
- Coating should get easily stuck to the pattern, and there should be possibility of controlling and adjusting the coating layer thickness.
- Appropriate strength, resistance to abrasion, resistance to cracks during storage, resistance to bending and deformation during mould making.
- In cases where rough sands are used for moulding at higher temperatures, thicker refractory coating layers are desired.

5. Coating Material The refractory coating plays an important role in the EPC process. Its function is to provide a refractory protection and to seal any penetration of metals. It should also provide smooth surfaces and act as a membrane between the polystyrene pattern and the sand mass is not bounded. As the sand mass is not bounded, it should be firmly held in place, till the polystyrene foam gets decomposed under the effect of molten metal stream which is entering. This causes a back pressure against the entering molten metal stream resulting in a pressure force that holds the unbounded sand in place. The gases escaping into the sand mass are controlled through an appropriate coating material along with the required permeability characteristics. In order to obtain a variety of properties imperative to the casting process such as thermal insulation, abrasion resistance and liquid absorption, the coating is engineered.

There are several types of refractory coatings available for evaporative patterns. The requirements from the refractory coatings have been enumerated above. Some of the coating materials are zircon sand, kaolin and talc. The coating generally recommended in case of cast iron is iron powder, as it prevents metal penetration. Siliminite, quartz, and aluminium silicate can be used as filler materials to make the coating cost effective.

22.10 EPC PROCESS VARIABLES

To identify the process variables which can have a major effect on the quality of evaporative pattern casting (EPC) process, an Ishikawa cause and effect diagram is shown in Figure 22.10.

The EPC process variables can be grouped in the following categories:

- Variables based on moulding sand: Type of sand, shape of the sand grains, grain size and the size distribution.
- Variables based on pattern: Density and size of pattern beads.
- Variables based on pattern coating: Slurry material-base, filler, additives, and coating thickness.

- Variables based on vibration: Amplitude of vibration, time of vibration and frequency of vibration.
- Variables based on pouring: Pouring time and temperature.

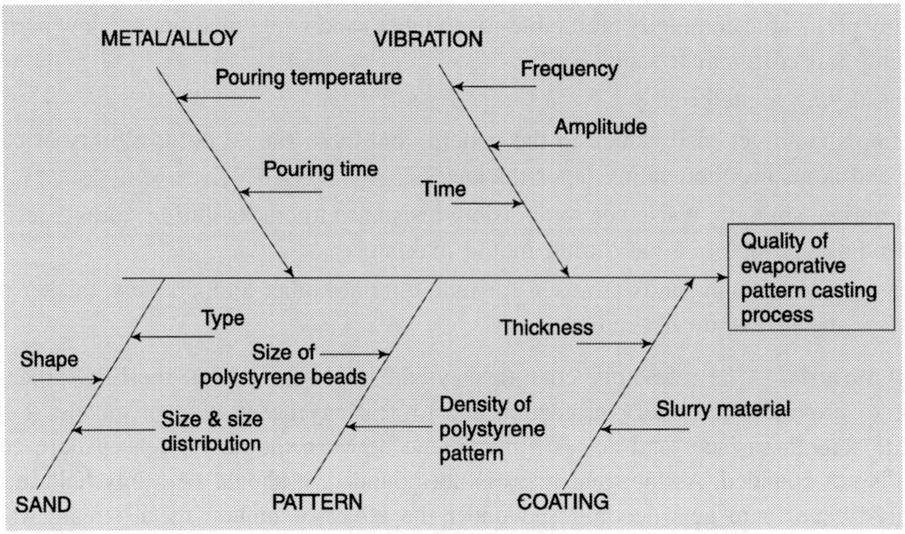

Figure 22.10 Cause and effect diagram of VAEPC process

22.11 EVAPORATIVE PATTERN CASTING

Evaporative pattern casting is also known as "expandable pattern casting" and as "lost foam casting". The pattern is made up of polystyrene (Styrofoam). First it is dipped into gas permeable refractory slurry and dried. During this process, a fine hard shell around the pattern is created. Pack the pattern in the moulding box using sand and support is created by filling sand. Liquid metal is poured into this shell. Foam evaporates as the metal fills the shell and allows to solidify the metal. The polystyrene pattern evaporates and escapes from the mould cavity. Finally, the entire pattern will be burning and escaping the cavity replacing molten metal with pattern material. After solidifying, the shell has a fine surface finish and so solidified casting also has good surface finish. Break the shell and the part comes out (Figure 22.11).

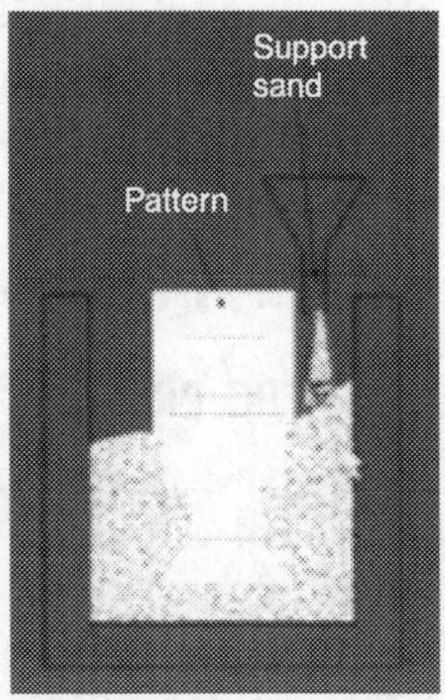

Figure 22.11 Evaporative pattern casting

22.12 HYBRID EPC (VACUUM ASSISTED EVAPORATIVE PATTERN CASTING PROCESS)

One of the limitations of EPC process is that the sand, which is free flowing, falls on the unfilled casting before the final solidification. This happens because the rate of evaporation of polystyrene pattern is higher than the metal flow rate. Pattern vaporizes ahead of the flowing stream of the metal because of the radiant heat generated from the hot metal. As a result, the sand falls in the cavity and, in turn, produces a defective casting. This problem has been overcome by combining the process sequence of EPC and vacuum sealed moulding process.

22.13 VACUUM ASSISTED EVAPORATIVE PATTERN CASTING PROCESS (VAEPC)

The concept of hybrid casting process is introduced with an aim to achieve better performance of the advanced casting processes. The reason for developing hybrid casting processes is to make use of combined or mutually enhanced advantages and to avoid or reduce some adverse effects (if any) of the constituent processes when they are individually applied. The evaporative pattern casting process (EPC) is a binderless process wherein physical bonding is required to bind the sand aggregates. The process however, does not guarantee a sound casting every time as it becomes difficult to control the rate of pouring matching with the rate of polystyrene burning. To overcome this effect, EPC process is combined with the vacuum sealed moulding process (V-process). In the V-process, the physical bonding is achieved by using pressure or vacuum in the moulds. The vacuum system when used with EPC process helps in withdrawing the decomposed gases and it also helps in providing rigidity to the sand mould.

Additional (other than EPC process) requirements for vacuum assisted evaporative pattern casting process (VAEPC) are discussed below.

In VAEPC process, there is a need to have a special pattern and moulding box. The details of these special patterns and moulding boxes are given in the lectures related to vacuum sealed moulding process (V-process). Similarly, as in the V-process, there is a need to have an arrangement for vacuum generating system to apply vacuum in the pattern and moulding boxes. VAEPC process also uses such an arrangement.

1. Pattern Box The pattern box is shown in Figure 22.12. The pattern plate is placed on the straight top open surface of the pattern box and perfectly matching between the mating surfaces is achieved. In order to ascertain perfect matching, the pattern plate is provided with four support blocks at the four corners in the pattern box. Additional support at the centre is provided to prevent warping of the pattern plate when subjected to vacuum pressure. The pattern box is fixed to the vibrating table. A pattern plate, having small holes at equidistant places can be used.

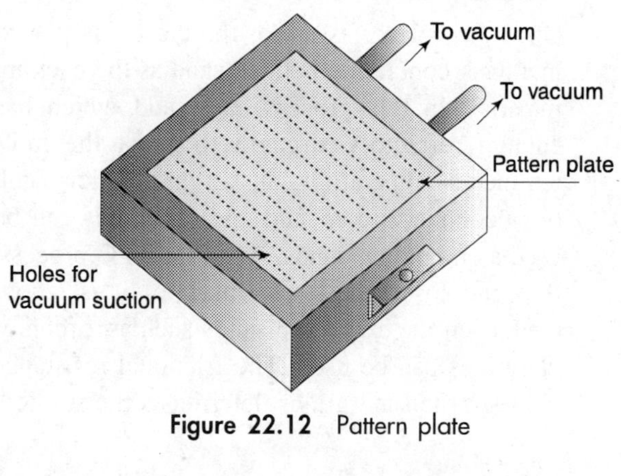

Figure 22.12 Pattern plate

2. Moulding Box In the design of moulding box, the major consideration is to have a uniform distribution of vacuum throughout the body of the sand forming the mould. The design of the moulding box for VAEPC process is more complicated than the conventional sand moulding process. The box has to be made with annular wall on all the four sides. Inside walls are provided with windows. These windows are covered with very fine mesh and backed by a metal strip having small holes for supporting the fine mesh. This mesh prevents fine sand particles from being sucked up from the sand voids in the box and into the vacuum pump. Two pipes are connected with the outside of the wall of the box. One pipe sucks the vacuum from the box and another pipe releases the vacuum to the atmosphere. The moulding box is shown in Figure 22.13.

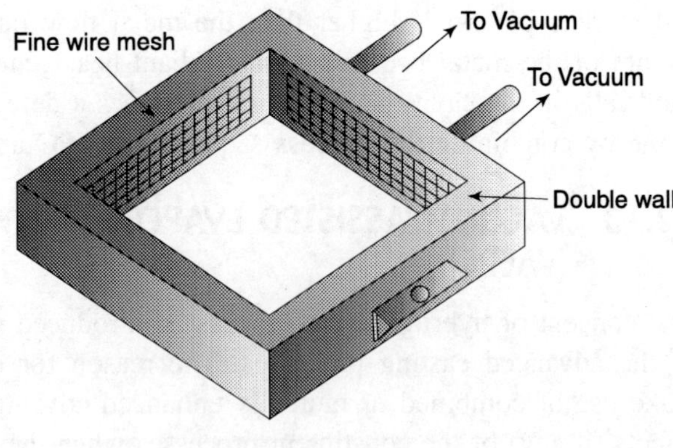

Figure 22.13 Moulding box

3. VAEPC: The procedure of VAEPC is shown in Figure 22.14.

1. The VAEPC procedure initiates with the pre-expansion of beads, usually polystyrene (same as the first step of EPC). After the pre-expanded beads are stabilized, they are blown into a mould to form pattern sections. In the mould, the steam cycle helps in fusing the beads together by full expansion.
2. The different sections of pattern are joined together by gluing and then assembled. The gating system too is glued and attached in a similar fashion.
3. A ceramic coating covers the foam cluster and it forms a barrier such that penetration of molten metal or erosion of sand during pouring does not take place.
4. Once the coating is dried, the cluster is placed in a flask containing bonded sand as the backing material. In this process, any sand which has enough refractory strength to resist the molten metal temperature can be used. Silica sand, olivine sand, zircon sand and chromites can be used as moulding sands. As in the EPC process, the sand loss is less and much of it is recovered, even the expensive sands such as zircon or chromites can be used. The frictional resistance between the sand grains determines the strength

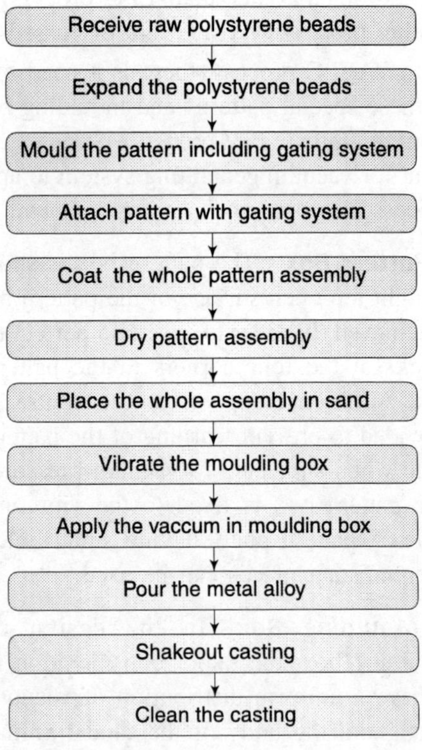

Figure 22.14 Procedure of VAEPC

of the mould. The mould strength with angular grains is higher, although higher bulk density is provided by the rounded grains.

5. The pattern, coated with a suitable refractory wash, is further embedded with dry, unbounded sand. It is then vibrated to produce a rigid mould.
6. Two plastic films are used to encapsulate the mould in-between and vacuum is applied in the sand mould. The vacuum gives rigidity to the mould and greater hardness is thereby achieved. After this process, the cluster is packed and vacuum is maintained thereby making the mould ready for pouring.
7. After pouring, the molten metal occupies precisely the shape and size of the pattern and duplicates all of its features. Hence, the surface finish and dimensional accuracy of the pattern is important as it has a direct relation to the casting quality being produced.
8. Once the casting is removed from the mould by the shakeout method, very little fettling is required, as the process doesn't require mould joint lines and even the cores are entirely eliminated.

4. Parameters of VAEPC To identify the important process variables affecting the quality of VAEPC process, an Ishikawa cause and effect diagram is constructed (Figure 22.15). The process variables of the VAEPC process can be grouped in the following categories:

- Variables based on moulding sand: Sand type, shape of the sand grains, grain size and size distribution.

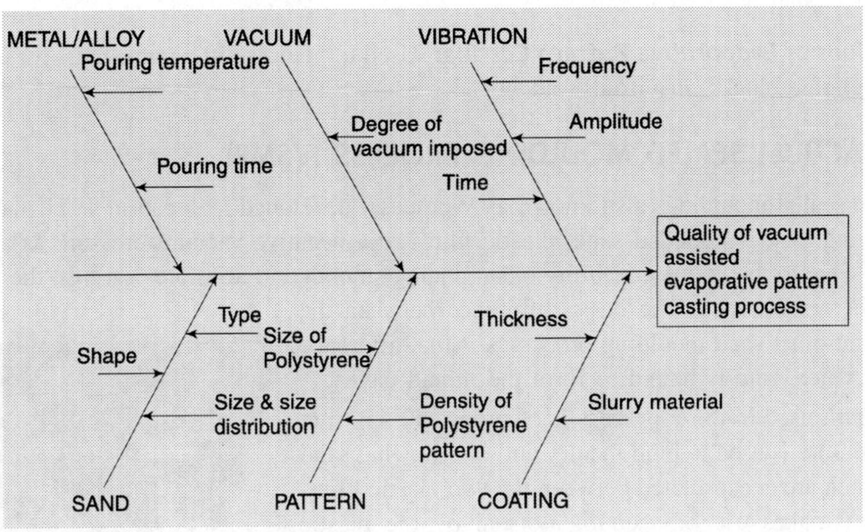

Figure 22.15 Cause and effect diagram of VAPC process

- Variables based on pattern: Density and size of pattern beads
- Variables based on coating of pattern: material used for the slurry and coating thickness.
- Variables based on vibrations: Amplitude of vibration, frequency and time
- Variables based on vacuum: Degree of vacuum imposed
- Variables based on material being poured: Pouring time and temperature

5. Advantages

1. Used for precision casting.
2. Used for both ferrous and nonferrous metals.
3. High dimensional accuracy is obtained.
4. Thin sections can be cast.
5. Complex shapes can be cast.
6. One piece flask is less expensive and easier to cast.
7. Fewer steps are involved compared to sand casting.
8. No need to mix binders or other additives.
9. Multiple castings can be combined in one mould to increase pouring efficiency.
10. The need for skilled labour is reduced.
11. Fettling and machining is minimised.
12. High levels of sand reuse are possible.
13. Lower capital investment.

6. Applications of Evaporative Pattern Casting: Used to manufacture crankshafts for engines, aluminium engine blocks, manifolds, etc.

22.14 VACUUM SEALED MOULDING PROCESS (VSMP)

The vacuum sealed moulding (also known as V-process) developed in Japan in 1971 makes use of free flowing, dry, unbounded sand, plastic film and a negative vacuum pressure as a means for binding. Due to its unique capability of producing smooth and accurate castings, the process gained further importance. The basic difference between the V-process and other sand moulding processes is the difference method by which sand is bound to form the mould cavity.

The vacuum used in the V-process is of order 250-450 mm Hg. This pressure is used to bind the dry and free flowing sand particles which are encapsulated between the two plastic films. This process makes use of vacuum assisted by the plastic film to form a mould cavity over the pattern. Unbounded dry sand is used as a backing material and vibrations are used to compact it. After pouring the molten metal into the mould, the

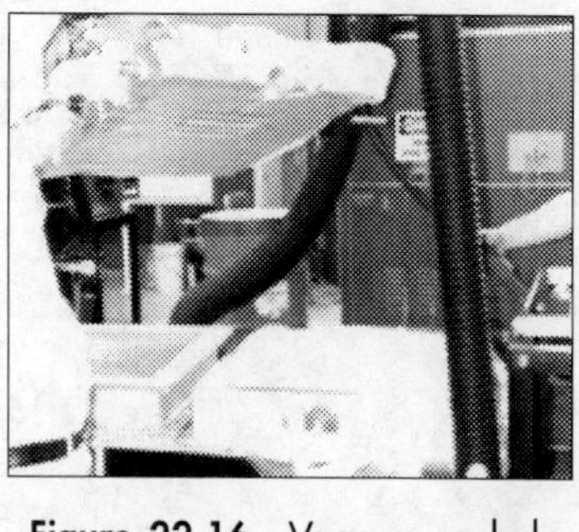

Figure 22.16 Vacuum sealed moulding process

plastic film is sucked inside the sand voids due to imposed vacuum. It further gets condensed and forms a shell-like layer (Figure 22.16).

Vacuum is required to be maintained till the metal is solidified. Then it is released, allowing the sand to drop away, thereby leaving behind the casting with a smooth surface. The process does not require shaking-out methods and equipment to remove the casting and the sand can be reused after cooling without any treatments.

1. Advantages

- Simplified sand control.
- No sand reclamation.
- No mixing of sand and no waste sand removal.
- Good surface finish and dimensional accuracy.
- Patterns have a long life and are inexpensive.
- Reproducibility is good and consistent.
- Draft is not required, thereby reducing material and cost.
- Reduced noise, low cleaning and finishing cost.
- Better general environment.
- No shakeouts and sand lumps.
- Less energy consumption and pattern maintenance.

2. Applications

The size of the product is no limitation in the V-process. However, as found in literature, the application of V-process castings are in the range of up to 8 tonnes for ingots. Some other applications, wherein V-process was used as a preferred casting method are:

- Medical devices
- Computers
- Instrumentation
- Electronic enclosures

3. Sequence of Producing V-process Moulds (Figure 22.17)

1. The pattern is set on the pattern plate of pattern box. The pattern and the pattern plates have numerous small holes, which help the plastic film to adhere closely on the pattern when vacuum is applied.
2. In order to soften the plastic film, a suitable heater is used.
3. The pattern is draped by a softened plastic film. The suction of vacuum takes place through the vents and the plastic film adheres very close to the pattern.
4. The mould box is set on the film coated pattern.
5. The mould box is compacted by filling it with dry sand and providing slow vibrations.
6. The mould is further levelled and a plastic film is covered on top of the box. The suction created due to vacuum helps in stiffening the mould.

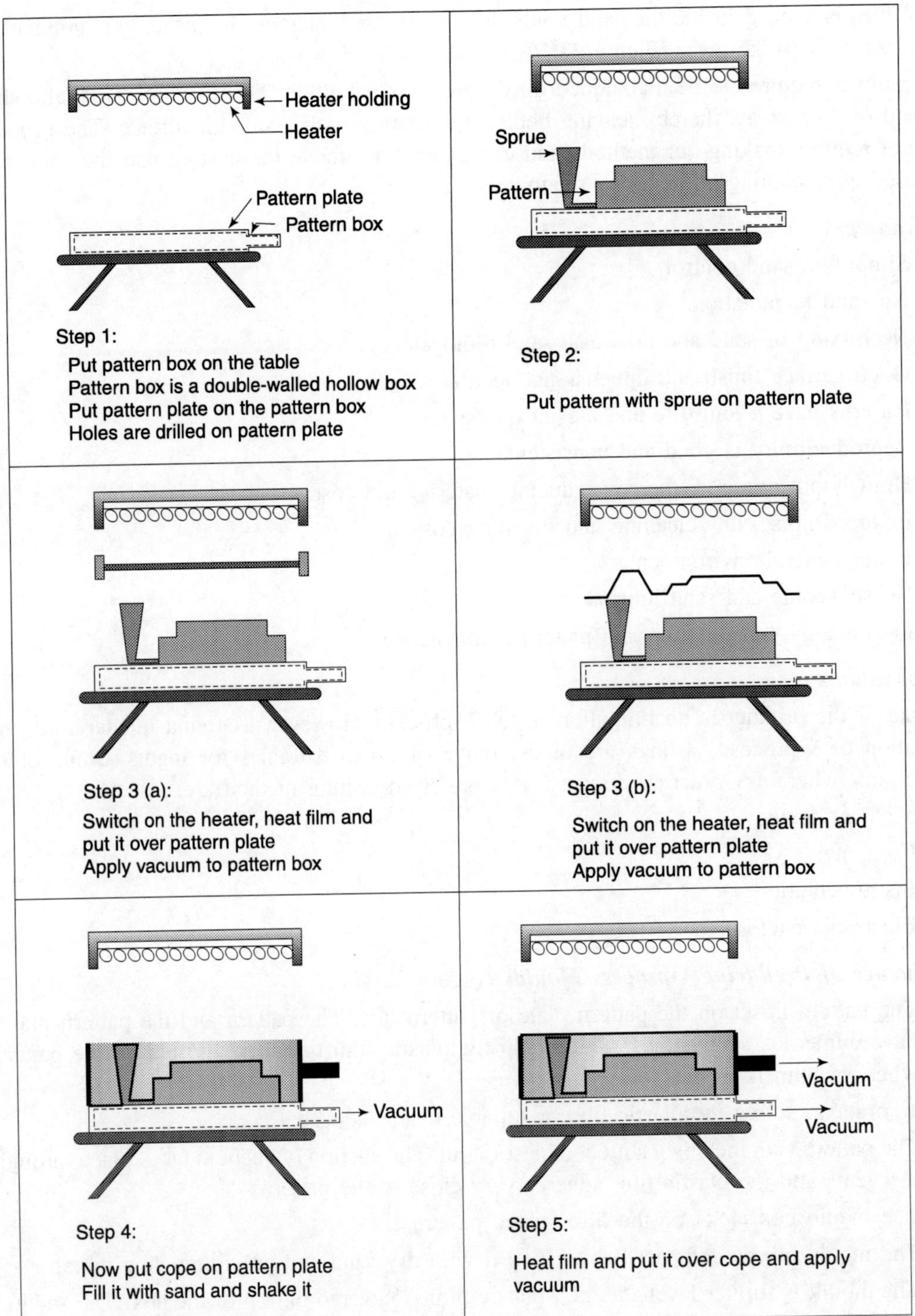
Heater holding
Heater
Pattern plate
Pattern box
Step 1:
Put pattern box on the table
Pattern box is a double-walled hollow box
Put pattern plate on the pattern box
Holes are drilled on pattern plate
Sprue
Pattern
Step 2:
Put pattern with sprue on pattern plate
Step 3 (a):
Switch on the heater, heat film and
put it over pattern plate
Apply vacuum to pattern box
Step 3 (b):
Switch on the heater, heat film and
put it over pattern plate
Apply vacuum to pattern box
Vacuum
Step 4:
Now put cope on pattern plate
Fill it with sand and shake it
Vacuum
Vacuum
Step 5:
Heat film and put it over cope and apply
vacuum

(Contd.)

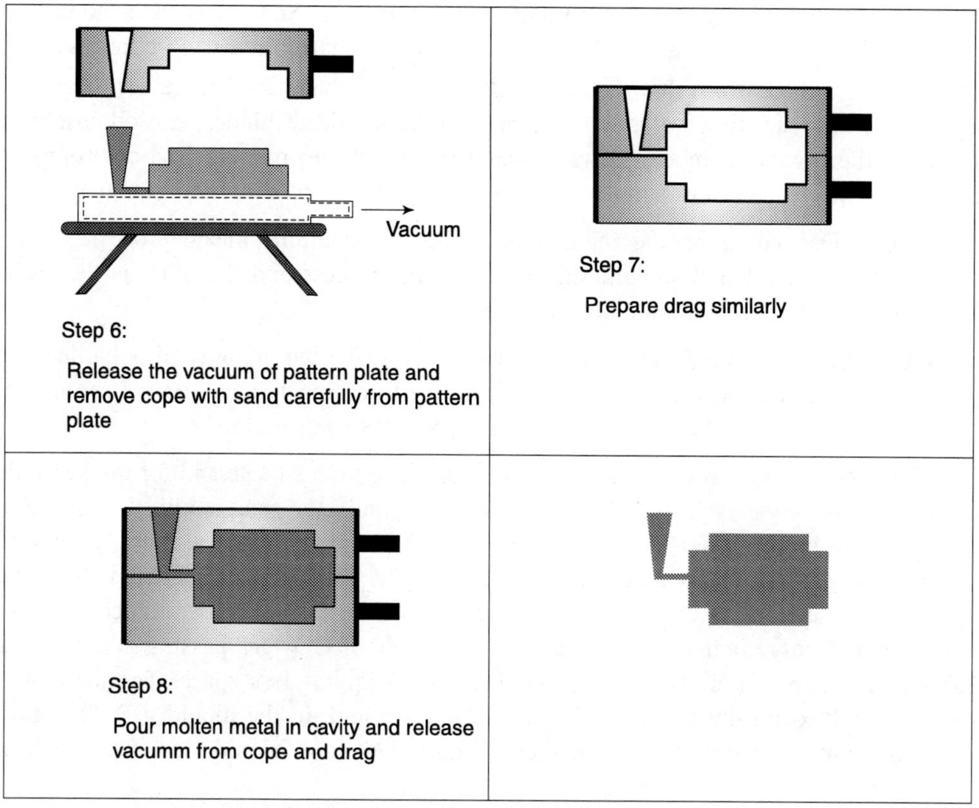

Figure 22.17 Steps in vacuum sealed moulding processes

7. Releasing the vacuum on the pattern box helps in stripping off the mould easily.
8. Cope and drag are both assembled and the metal is poured. Vacuum is maintained during pouring.
9. The vacuum is further released, once the mould boxes get cooled; this allows sand to freely flow back, thereby leaving a clean casting behind.

4. Limitations

1. Skilled workers are required.
2. Polymer film is expensive.

22.15 SODIUM SILICATE PROCESSES

1. Introduction One of the easiest modern moulding and core-making processes for instructional and small foundries to use is the sodium silicate/CO_2 process. In this process, liquid sodium silicate ("waterglass") is mixed with the sand. The sand is rammed into a mould or core box and cured by passing CO_2 gas through the mould/core. Sodium silicate mould/cores are very strong. The bond is so strong that hot tearing and collapsibility can be an issue.

Moulds and cores made from this process produce less gas than other processes. Cleanup is also easy since water can dissolve sodium silicate. Sodium silicate binders are odourless and produce no volatile organic compounds when metal is poured into moulds. The environmental friendliness, ease of cleanup, and simplicity make the process very simple to conduct.

2. *Classification* Depending on the method of hardening, sodium silicate processes are of two types. One is self-hardened sodium silicate moulding process and the other is the carbon dioxide hardened sodium silicate moulding process.

3. *Self-hardened Sodium Silicate Process* In this process, the moulding sand is hardened by mixing up of Portland cement or blast furnace slag or the ferrosilicon, etc., and the setting time will be from 2 hours and it takes up to 24 hours, sometimes even more.

4. *The CO_2 Process* In this process, sand is bonded using sodium silicate and for the hardening purpose, carbon dioxide gas is passed. As in resin-bonded sands, the sand is mixed with sodium silicate (care should be taken not to overmix the sand which will produce a crumbly mixture), placed around the pattern, and vibrated to compact it. The sand is then gassed with carbon dioxide gas (hence the name "CO_2 process"), which hardens the mould almost immediately. The process is rapid and only about a few seconds to 2 minutes of CO_2 gas passing at 1 to 2 bars (140 kPa) pressure is required. The process is cheap and requires less space. The methods of gassing in the corebox are illustrated in Figure 22.18 (a–c) and, in the mould, (f). This gas is dispensed from banks of cylinders or from a bulk liquid container. Moulds are cured 24 hours

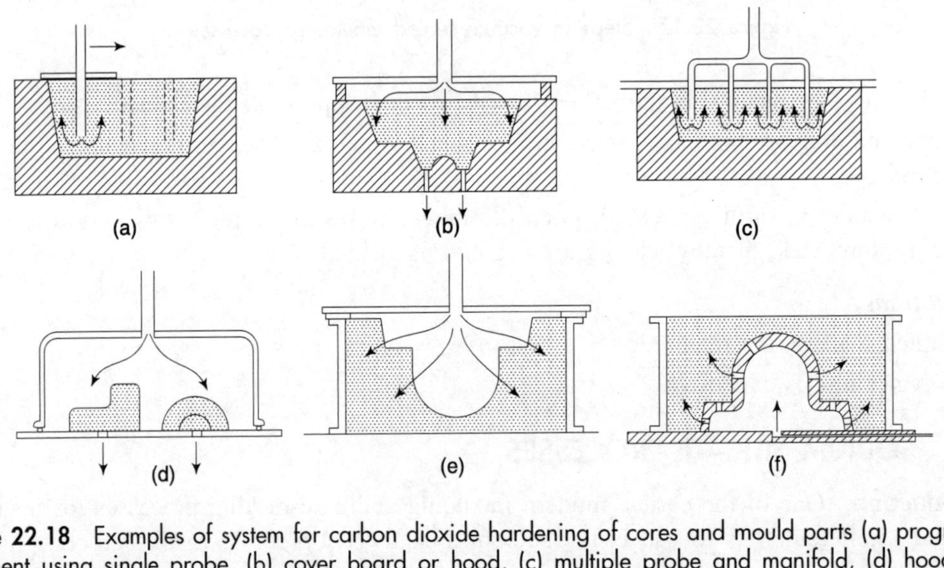

Figure 22.18 Examples of system for carbon dioxide hardening of cores and mould parts (a) progressive treatment using single probe, (b) cover board or hood, (c) multiple probe and manifold, (d) hood over previously stripped cores, (e) treatment of mould after pattern draw, and (f) passage of gas through hollow pattern

at room temperature in dry conditions (high humidity softens the bond). During this time, their strength increases as a result of some dehydration of unreacted silicates and continued gelling of the silicate. They have good shelf life.

Sodium silicate foundry binders are metastable solutions with molecular ratios in the range of 1.5 to 3.1. There are two hardening mechanisms that take place. First, there is dehydration by evaporation. Second, the reaction with CO_2 removes the sodium oxide from the silicate, and the silicate gels. The hardening reaction that takes places is as follows:

$$Na_2O.mSiO_2.(mn + x)H_2O + CO_2 = Na_2CO_3.xH_2O + m(SiO_2.nH_2O)$$
$$\text{sodium silicate} = \text{sodium bicarbonate} + \text{silica gel}$$

SiO_2 thus liberated is amorphous and forms a gel causing increased strength and viscosity. The amount of silicate binder varies from 3 to 6% by weight, depending on the type of sand, its grain fineness, and the amount and type of contaminants in the sand. In foundries SiO_2 : Na_2O ratio of the sodium silicates used is of the order of 2:1 with specific gravities of 1.5 to 1.7.

The influence of gassing time on bond strength is illustrated in Figure 22.19. It is seen that the passing of CO_2 gas is critical and the strength of the mould increases with time and velocity of gas passed but the strength of the bond becomes impaired by overgassing. Overgassing forms sodium bicarbonate and sodium carbonate, which embrittle the gel. Theoretical consideration shows that the ratio of CO_2 gas required to complete bicarbonate stage is 0.22 per unit weight of silicate. However, some excess CO_2 should always be flowed (about 1.5 or even higher ratio than theoretical demand) due to inefficient use of CO_2 gas in actual foundry conditions.

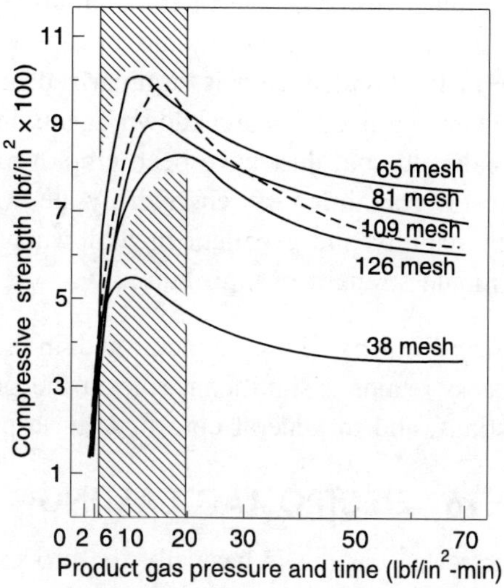

Figure 22.19 Influence of carbon dioxide flow on compression strength, illustrating reduction of strength on overgassing. Sands of varying mesh with constant binder content of 4%

Normal sand handling equipment can be used to prepare the sand. Coreboxes and patterns for sodium silicate moulds may be made of wood, metal, or plastic, and they must be washed thoroughly after use to prevent build-up of sodium silicate. Mulling equipment and ramming tools must also be washed thoroughly after use. Sand mixes that have not been used should be covered with a wet cloth to prevent a hard surface crust from forming because the mixture will slowly cure in air. The sand has reasonably good bench life, but if impurities such as ferrosilicon or lime are present in the sand, the bench life is shortened.

5. *Advantages and Limitations* Moulds and cores prepared by sodium silicate process have adequate strength and hardness to provide good dimensional tolerances and better surface finish

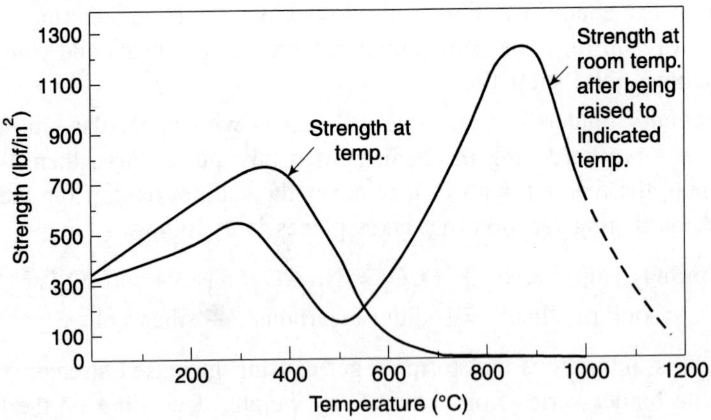

Figure 22.20 High temperature and retained strength of sodium silicate bonded sand

to the casting. The mould and cores can be made well in advance because of their high bench life.

The sodium silicate process is expensive as fresh silica sand is needed every time. Sand reclamation is the most serious limitation to a process otherwise noted for its relatively slight environmental impact. A combination of mechanical reclamation and water reclamation has shown promise, but recovery rates are limited to about 70%.

Another disadvantage of the sodium silicate process is poor collapsibility and additives are often added to improve mould breakdown. The effect of casting temperature on the properties of the bonded sand is summarized in Figure 22.20, where the very high retained strength explains the need for breakdown agents in silicate bonded sands. Molasses, wood flour, starch, cereals and coal dust have been employed for this purpose, but proprietary one-shot high ratio silicates containing starch hydrolysate were later developed to combine high bond strength with stability and excellent breakdown properties. However, these accelerate the deterioration of mould strength in high humidity.

6. Applications Despite the growth in the applications of the organic bonding systems, the CO_2 process retains a significant role in the production of moulds for both ferrous and nonferrous castings, and in general core making, especially for larger castings.

22.16 ELECTROSLAG CASTING

Electroslag casting is basically the process of making casting in water cooled metal mould in which the molten metal is prepared in the mould itself by consumable electrode-electroslag remelting. It is a continuous process, where during the remelting of the consumable electrode, refining and solidification of the steel occur simultaneously. Cast, rolled or forged ingots can be used as a consumable electrode.

The electroslag-remelting process has become an accepted method of producing high-quality ingots in a variety of steels and related alloys. At a very early stage in the process's development, it was recognized that the technique had the potential to cast not only ingots but also simple shapes. From this recognition stems the electroslag casting (ESC) process.

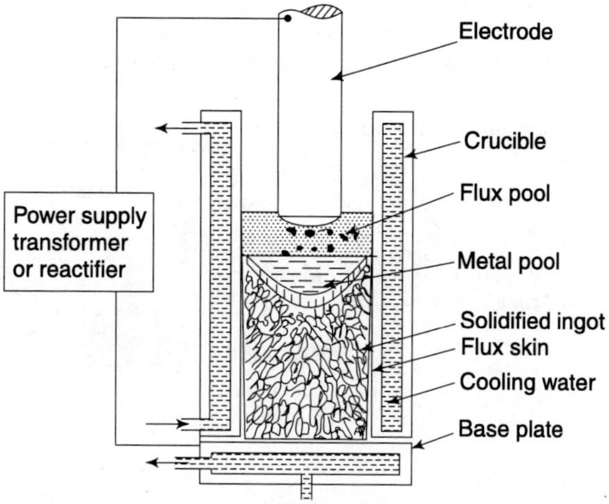

Figure 22.21 Electroslag processes

A schematic of the process is shown in Figure 22.21. The consumable electrode, produced by either rolling or casting, is remelted into a water-cooled crucible resting on a water-cooled base plate. Electrical connections are made to the electrode and to the base plate, using a line-frequency transformer as the power supply. Heat for melting is generated by Joule heating of the liquid slag. The melting process causes a progressive solidification in the casting pool and some chemical refining is also observed due to the contact with a highly basic, superheated slag. The result is a directionally solidified casting, free from shrinkage defects, of very uniform composition, and usually possessing isotropic mechanical properties. A continuous, thin shell of solidified slag on the mould surface also provides for a smooth, defect-free casting finish. Advantages claimed for the product are:

- More reproducible mechanical properties.
- Freedom from conventional casting defects.
- Properties comparable with forged products.

The range of alloys processed by electroslag is very large, covering low-alloy and carbon steels, stainless steels and nickel-base alloys, tool and die steels, and special alloys such as the Inconels, Nimonics, and Hastelloys. The major disadvantage of the ESC process in economic context is that it involves water-cooled copper moulds which represent a substantial cost barrier. One of the remarkable differences between ESC material and conventional castings is the absence of large sulphide inclusions.

In electroslag casting, the parent metal of electrode is remelted and refined in the mould due to slag. During solidification, the high temperature gradient between the slag-metal boundary and the solidification front favours the formation of columnar crystals in the metal structure. The high density and purity of the cast metal ensure high mechanical properties of the casting.

Applications of Electroslag Casting

It is used for making crankshafts, connecting rods of ship engines, critical valves, fittings for nuclear power plants, cast rotors of turbo generators, rolling mill rolls, cement kiln bandings, super high pressure vessels, etc. Some applications are shown in Figure 22.22.

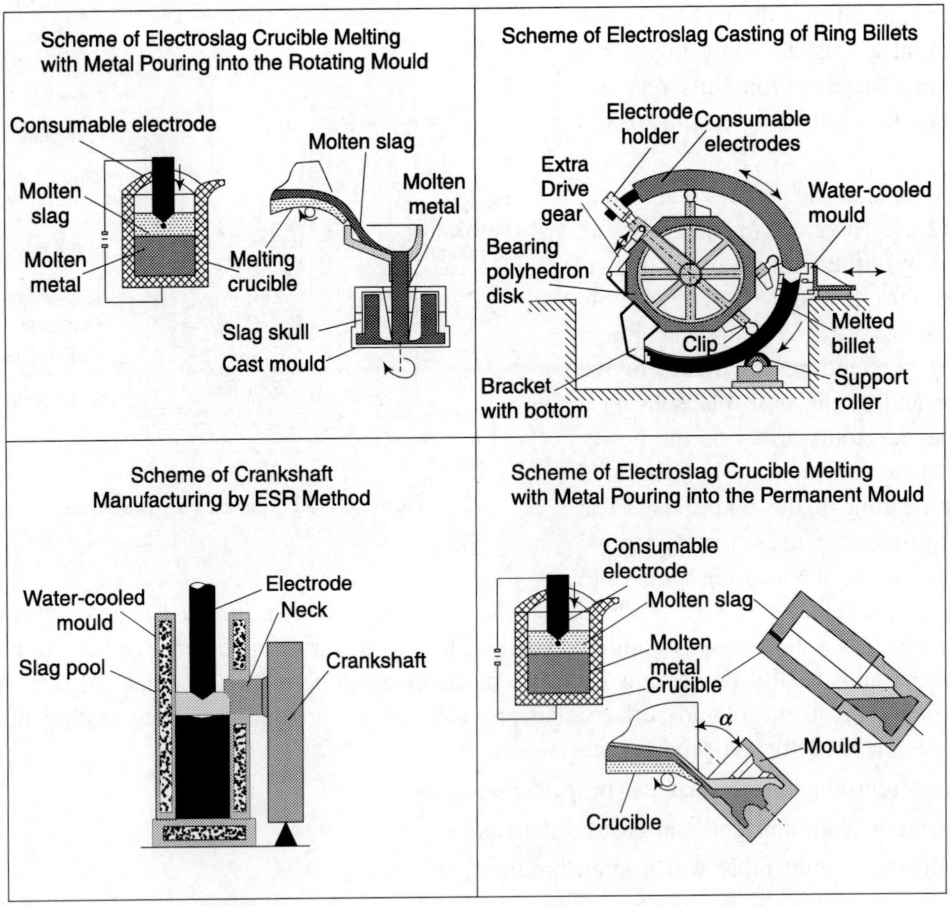

Figure 22.22 Application of electroslag casting

Advantages of electroslag casting

1. There is no requirement of gates, risers, melting and pouring.
2. Castings of good mechanical properties can be produced using these processes. It provides better control of the solidification microstructure and chemical homogeneity;
3. Bulky and heavy casting can be made.
4. The wastage in this process is minimum.
5. Suitable for continuous casting and process is clean.
6. Properties of the cast products are comparable to forgings and stamping.

Disadvantages of electroslag casting

1. Copper metal moulds are required with good water cooling systems.
2. Thin castings are difficult to make.
3. Not economical for batch production.
4. Only simple thick shapes of vertical structures is preferred.

22.17 NEAR NET SHAPE CASTING

Near net shape casting (NSC) is a family of techniques where the initial part production provides a quality surface finish and is close to the final (net) shape. Reducing the need for machining and grinding can eliminate more than two-thirds of the production cost in some industries. NSC techniques include investment casting, spray forming (deposition), polymer blow moulding, vacuum forming, compressed air forming, laser forming, rapid prototyping and nanotechnology manufacture.

With NSC, very close tolerances can be achieved with material waste reduced to a minimum. NSC techniques can be used with ferrous and nonferrous materials such as titanium, aluminium, copper, magnesium, zinc and ceramics. Currently, most steel is continuously cast into slabs, billets or blooms, which later have to be reheated and rolled into final shape. Direct or NSC or thin-strip casting integrates the casting and hot-rolling of steel into one step, reducing the need to reheat the steel before rolling it. With a direct continuous steel casting approach, liquid steel can be directly cast into semi-finished products, which eliminates the need for primary rolling of ingots.

Near net casting can be classified as

(a) Single roll caster
(b) Twin roll caster
(c) Twin belt caster
(d) Block caster
(e) Rotary caster

(a) Single Roll Caster: In a single roll caster, liquid metal is supplied to the surface of a water-cooled rotating roll. The melt contacting with the roll surface cools down and solidifies forming a thin continuous strip.

The advantages of the single roll continuous casting method are:

- high productivity;
- capability to cast very thin strips; and
- low capital cost.

Single roll caster provides very high cooling rate. Liner speed of the roll surface is 15-200 m/min. The method is used for casting thin-gauge (1-3 mm) aluminium strips. Low variation of casting thickness is achieved by uniform cooling of the roll surface.

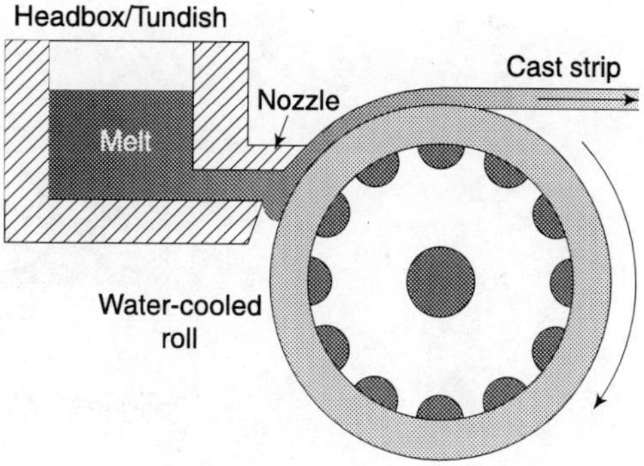

Figure 22.23 Single roll caster

(b) Twin-roll Caster: In twin-roll casting process, a melt is fed through a ceramic nozzle into the gap between two rotating water-cooled rolls. The melt cools down and solidifies between the rolls. Additionally, the solid strip exerts hot rolling with the thickness reduction of about 5-20%. There are three possible configurations of twin roll casters: vertical downward casting, vertical upward casting and horizontal casting (Figure 22.24). The most popular configuration is horizontal and 15° tilt back. The roll dimensions may vary in wide ranges: diameter 300-1500 mm, width 150-2184 mm. The water-cooled roll shells are made of steel, copper or copper-nickel alloys. Sticking between the roll and the solidified strip is prevented by means of a parting agent applied to the roll surface. The agent contains graphite or magnesium hydroxide. Twin roll casting process provides extremely high cooling rate up to 550°C/s. Fast cooling allows to reach high casting rate, which is 0.9-38 m/min for aluminium alloys and 5-460 m/min for steels.

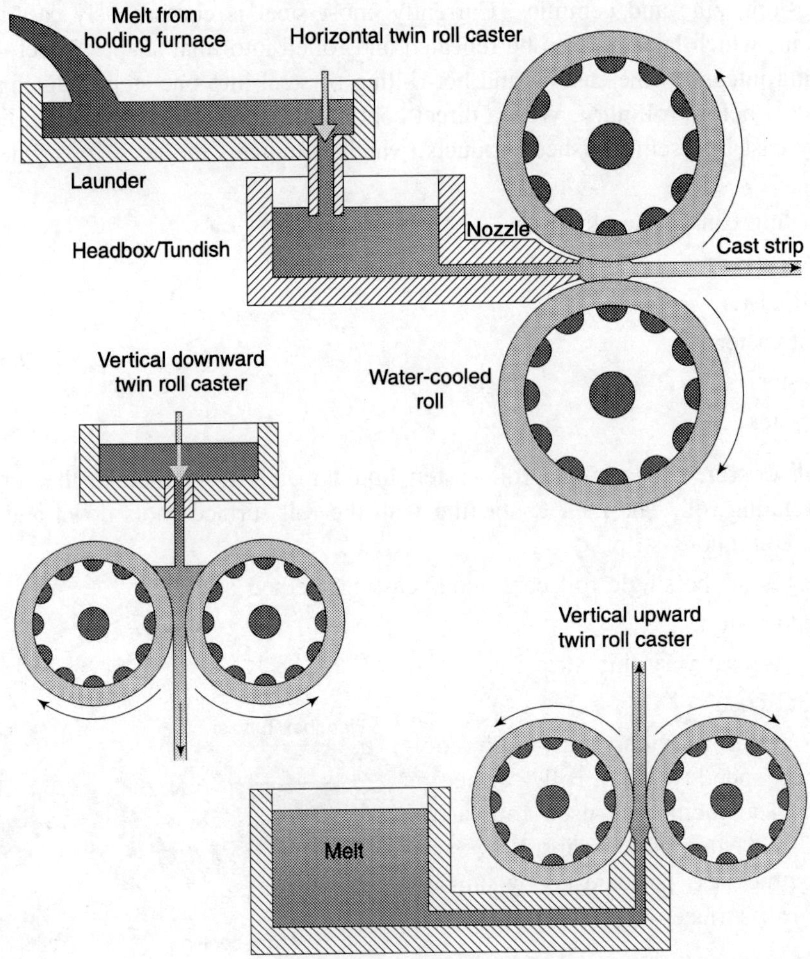

Figure 22.24 Twin roll caster

Thickness of the cast strips may be as low as 0.1 mm. Maximum thickness reaches 5 mm for steels and 10 mm for aluminium. Typical gauge of twin roll cast strips is about 2.5 mm. Twin roll casters are used for casting alloys with relatively low temperature interval of solidification: aluminium and its alloys containing max. 2.5% Mg (wrought aluminium alloys (1xxx), wrought aluminium-manganese alloys (3xxx), wrought aluminium-magnesium alloys (5xxx)), stainless steels, special steels, carbon steels. Twin roll casters are also used for continuous casting of aluminium based bearing alloys.

(c) Twin belt Caster In twin belt casting process, a melt is fed through a ceramic nozzle into the gap between two rotating thin belts held in tension. The belts are cooled by water jets from the side opposite to the surface contacting with the casting. The melt cools down and solidifies between the belts. Width of the cast slab is determined by two chains of steel blocks travelling with the belts and preventing side leakage of the non-solidified melt. The cast slab/strip is pulled out by a withdrawal unit synchronized with the drive of the belts (Figure 22.25).

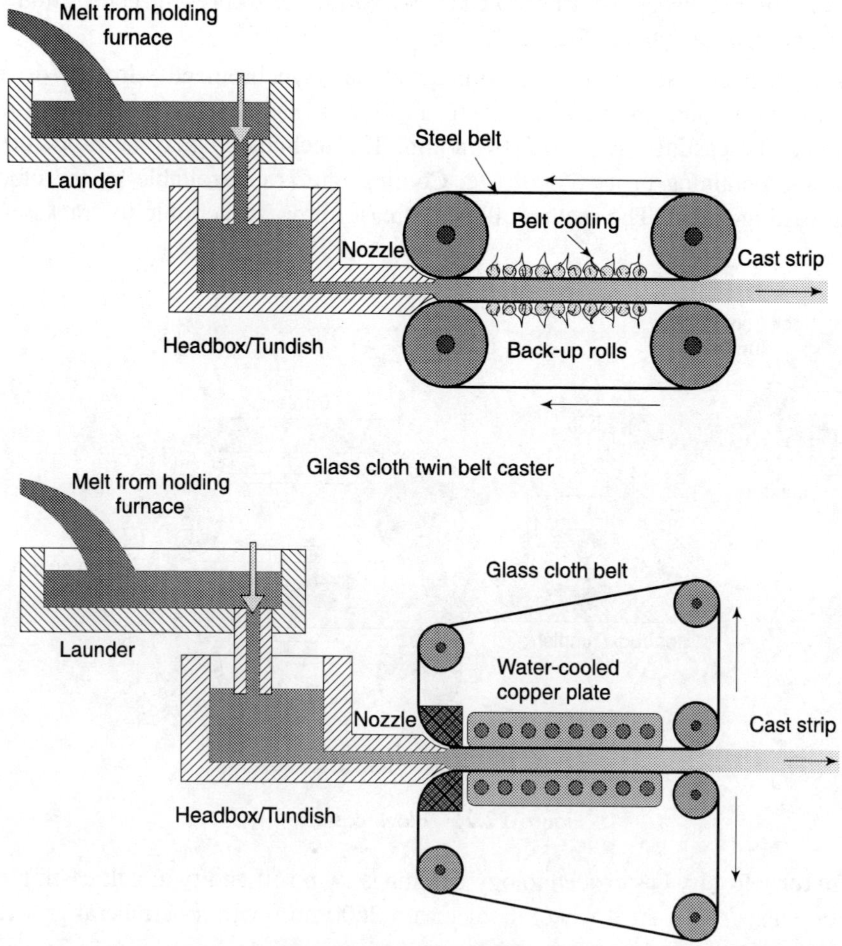

Figure 22.25 Twin belt caster

Most twin belt casters use carbon steel belts of about 1.5 mm thick. The maximum width of the cast slabs is 2000 mm. The strip/slab thickness is within the range 2.5-33 mm. The typical casting thickness is 12.5-25 mm. The maximum casting speed is 9 m/min. The slab leaving the caster is still hot and may be hot rolled in-line to a gauge 2.5-5 mm and then coiled.

These casters are generally used for casting aluminium alloys containing up to 3% of Mg: wrought aluminium alloys (1xxx), wrought aluminium-copper alloys (2xxx), wrought aluminium-manganese alloys (3xxx), wrought aluminium-magnesium alloys (5xxx)), wrought aluminium-magnesium-silicon alloys (6xxx). Twin belt caster is successfully used for continuous casting of aluminium based engine bearing alloys. The belts are supported and chilled by two water-cooled copper plates.

(d) Block Caster In a block caster, molten metal is fed through a ceramic nozzle into the space formed by two chains of steel blocks travelling like caterpillar tracks. The blocks are cooled on their way back by chillers. The melt cools down and solidifies in the space between the steel blocks. The blocks are adjacent to each other in order to prevent penetration of liquid metal into the gap between them (Figure 22.26).

The cast slab/strip is pulled out by a withdrawal unit synchronized with the drive of the blocks. The maximum width of the cast slabs is 1750 mm. The slab gauge is within the range of 10 to 40 mm. The casting rate is 0.5-10 m/min. The technology may be used for most of aluminium alloys containing up to 5% of Mg. Cooling rate is controllable by the blocks temperature's and casting speed. The method allows to cast alloys susceptible to cracks caused by fast cooling.

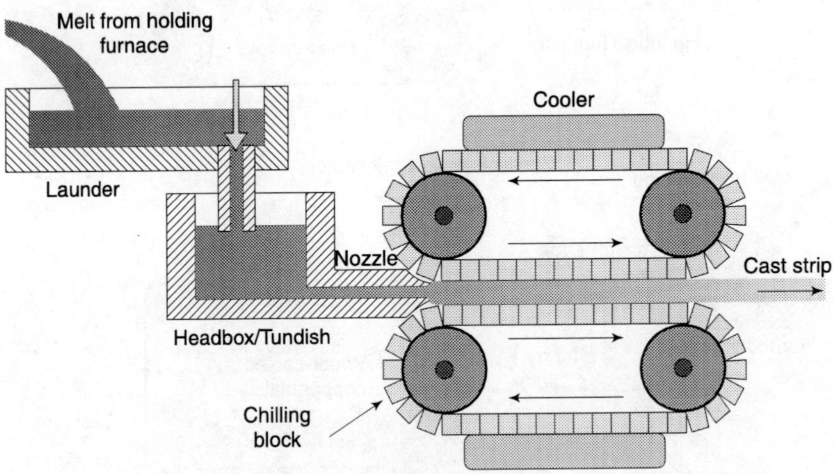

Figure 22.26 Block caster

(e) Rotary Caster Rotary caster technology combines twin roll and twin belt casting methods. A rotary caster consists of a steel wheel diameter up 2600 mm with a peripheral groove and an endless steel belt closing the groove and forming the casting space. The groove has triangular, trapezoidal rectangular cross section. The wheel and the belt rotate with the linear speed of about 14 m/min. The belt is cooled by water jets.

When a melt is poured into the space between the groove and the belt it cools down and solidifies. The slab/strip/bar leaving the caster is still hot and may be hot rolled in-line to and then coiled. The wheel diameter is up to 2600 mm.

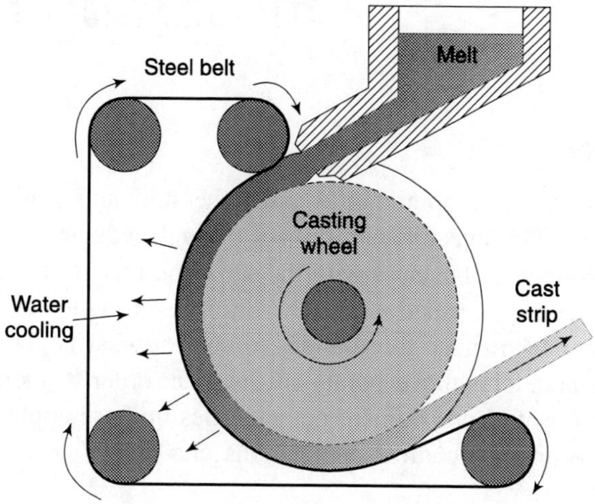

Figure 22.27 Rotary caster

Aluminium alloys (wrought aluminium alloys (1xxx), wrought aluminium-manganese alloys (3xxx), wrought aluminium-magnesium-silicon alloys (6xxx)) and copper alloys for electrical applications (wires, conductors, bus bars) are cast by the rotary method.

Review Questions

1. What is V-process?
2. Explain process setup for V-process.
3. Explain cause and effect diagram for V-process.
4. What are the specifications of V-process?
5. What is the sequence of producing V-process moulds?
6. Explain various evaporative pattern casting processes.
7. What are the advantages of EPC processes?
8. What are the limitations of EPC processes?
9. Explain electroslag casting in details.
10. Write a short note on near net casting process.

CHAPTER 23

CAD and Numerical Simulations in Casting Processes

23.1 INTRODUCTION

The primary purpose of using a computer is to assist the staff and managers to increase the profitability of the foundry. This objective can be met through reducing the cost of performing a task handling a higher volume of business with the same indirect staff and providing the staff and management with the latest information and the information which is not earlier available, which allows them to plan and manage the foundry to control what is about to happen instead of what has already happened. The above points are listed in order to increase profit improvement potential. The highest profit improvement potential lies in those applications which impact on material utilization like process control applications and labour equipment utilization like operation control applications.

The changes can be seen in a number of different aspects in foundry like estimation of cost for manufacturing by estimating the total volume and weight of the casting, calculation of optimum charges in cupola or induction furnaces to achieve the best output regarding mechanical and metallurgical properties. In this a linear programming model could be designed with cost minimization as objective function, material balance, inventory restrictions and the chemical compositions as constraints with the help of a computer. Development of an expert system is another area where computers may be used widely like the selection of best modelling processes, casting defects diagnosis, casting techniques by calculating the gating and risering system, etc. Finally, casting simulation is a process where computers are used as the only tool for any modern foundry. In this technique, a designer can modify the casting parameters along with design and can produce a virtual casting before the actual investment in the shop floor. As a result, the end product becomes less expensive and is quickly delivered to the customer.

23.2 APPLICATIONS OF COMPUTER IN FOUNDRY

1. Estimation of cost
2. Sales and profit data
3. Shipment and invoicing
4. Production planning and control
5. Production reporting and performance analysis
6. Capacity planning and shop loading
7. Quality assurance

Computers have vast capacity to store, process and retrieve data. Nowadays, computers have become a part of Management Information Systems for the effective operation of industry. Old concept of maintaining paper work, big drawing sheets, files in chest box and manual data entry are obsolete and superior computer systems can store all the documents in a better fashion. With development of new graphics software, complicated drawings can be processed by Computer Aided Drafting (CAD) and manufacturing by Computer Aided Manufacturing (CAM) that offers powerful tools for generating drawings with editing facility to create 3D solid models.

23.3 GENERAL APPLICATIONS OF COMPUTERS

Like any other business or industry, computers are used to record data, process and present reports which help managers to know the progress and accomplish objectives. Ordinary PC's can be used effectively for a host of information (PPC), metal charge calculation, orders booking and execution, statistical quality control (SQC), analysis of rejections, etc., that will ultimately reduce the cost of operations. Computers can also be used for effective load control of furnaces. Suitable computer software is used for designing gating and risering system and solidification simulation can detect the poison of hot spot, gas entrapment, etc. which reduces the shop floor trials and certainly cut down the product development period.

Some of the common reports generated through computers are:

1. Pay roll: Employee wage and salary records
2. Financial accounting
3. Material/stores/inventory record, A-B-C analysis
4. Production plan monthly, weekly and daily schedules
5. Progress report, production reports of different departments/sections
6. Keeping track of requirements of sand, scrap, pig iron, mould and pattern
7. Inspection/rejection reports
8. Equipment breakdown data
9. Production costing
10. Sales transaction

Apart from the speed and accuracy, computers help through communication networks across different sections and levels enormously for fast decisions and implementations. Access to information simultaneously to many, helps in better collective action and motivation. Computers help in fast communication not only within the factory but also with the suppliers, customers and others associated with the unit inside and outside the country. Even designs/drawings can be faxed across the world today through computer controlled devices which save time and cost.

23.4 SPECIAL/TECHNICAL APPLICATIONS

(i) Estimation/quotation

The first stage in receiving a new order from customer in quotation in which the foundry estimates its internal cost for manufacture adds suitable profit margin and gives its price. The procedure starts from calculations of volume and weight of the casting from the drawing given by the customer. The estimator divides the casting into many similar shapes like prisms, cylinders, adds allowances for manufacture in different dimensions and calculates the total volume and weight. Estimation of metal required additionally for gating and risers rejection/losses has to be done, in addition to the stagewise costs for the manufacture, depending upon the intricacy and quality requirements. Accuracy and reliability of the estimates make all the difference before getting/losing the order as well as making profit/loss on the order. Computers can be used for applications with much convenience, speed and accuracy. Typical programs in FORTRAN, C, C++ and AUTOSLIP can be developed for casting volume estimation. With a graphic interference like AUTOCAD, the program can also generate a 3D sketch of the part being calculated for convenience.

Cost estimation computes the total cost with the user choosing the variable parameters like casting critically/intricacy, expected rejection, liquid metal, alloy specifications, melting methods, method of moulding heat treatment, overhead costs at different stages through suitable database already built in the program. The additional advantage is fast recalculation with the changing parameters and reliability.

(ii) Generation of Patterns/Process Drawing of Gating and Risering

This is an important stage which influences further production both in quality and cost. Computer Aided Design (CAD) procedures help to clearly visualize complicated drawings in 3D to design patterns and cores for optimum moulding method for high quality. CAD can be defined as the use of computer systems to assist in creation, modification, analysis and optimization of a design. Components could be visualized by sectioning the solid model in different parts of the component. Allowance, parting faces, drafts, core and print details can be added on to the drawing in different layers for easy presentation. Showing in different colours, it is easy to visualize complicated core assembly and proper precautions could be taken before actual production. Calculation of running and gating system through computer helps in correct size of the sprue, runner and ingate considering important parameters like critical wall thickness, fluidity, material, etc. to arrive at optimum values.

Risering/feeding system is vital to manufacture shrinkage-free casting in steel, iron, aluminium and copper alloys. The earlier methods using thumb rules and empirical rules considering only a few obvious cannot be valid today with high demand for quality like non-destructive testing, high completion which necessities optimum but not too liberal safety margins in design at high cost. Computerized system use latest technology in feeding design involving many parameters like cast metal characteristics, mould material characteristic criticality, feeding ranges, etc., to arrive at optimum design. Options of chills, feeding aids like insulating/exothermic sleeves, etc., are made available for practical applications.

With the introduction of powerful characterized Finite Element Analysis techniques, it is now possible to use solidification simulation to visualize the model of casting on computer screen, observe the progress of its solidification to locate the portions that solidify last causing shrinkage. These are made possible by the software which form the thermal data of liquid and heat transfer through the mould, conducts the thermal analysis of the finite element meshed model, developing convergence solution for the unsteady state (involving latent heat generation during solidification) considering effect of conduction in the mould.

Moulds are the best design gating and feeding systems. The computer printout of such a technique using NISA-FEM package to simulate solidification of liquid metal is also available in the market.

(iii) Computer Aided Manufacturing of Pattern

Computer Aided Manufacturing (CAM) techniques are used to manufacture pattern equipment from drawing stage to final product. They are sophisticated solid modelling software to create the component details with allowance by adding machining data. Numerical Control (NC) programs are generated which can be loaded to NC or CNC machine tools to automatically machine the pattern equipment accurately. Typical direct applications of CAM include CNC and adaptive control of manufacturing processes. Indirect applications involve use of computer to support manufacturing operations as in programming, simulation of CNC machines, robots and computer assisted process planning.

(iv) Coordinate Marketing and Measuring Machines (CMMs) for Pattern Inspection

The Coordinate Marketing and Measuring Machines are widely used tools for performing fast and accurate three-dimensional measurements of any component. With the help of sophisticated programming abilities, the need for expensive fixtures and tedious manual operations can be eliminated. The CMM's ability to inspect pattern tooling and repeatability make it ideal for the process of pattern tooling inspection. Inspection can be integrated through computer networks. The exact dimensions of the part as determined in the original CAD/CAM database can be downloaded to the CMM using highly precise linear transducers and a microcomputer based controller to achieve typical measurement accuracies of + 0.005 mm.

Using a reverse engineering system, a model can be produced by a designer and digitized by CMM. The digital information can be brought into CAD package using IGES (Initial Graphics Exchange Specification). CAD system then can generate drawings or NC programs after post-processing software is to read IGES file, validate integrity of the data contained in it and translate the contents of the file into the format required for the CAD system. The typical CAD system has a proprietary internal database that is used to represent geometry, organizational information and quality annotation. With the increasing demand for high standards of quality castings from automotive industries, it is necessary to have patterns of very high standards which can no longer be met by adopting the conventional method of pattern design, inspection and tooling.

(v) Computer Aided Operation Research

Computer helps in optimizing charge calculation to select the best charge composition for making a complicated alloy. Manual methods may ignore important variables like level of impurities in a constituent, melting losses, etc. For example, maximum 40% sponge iron in steel charge, to arrive at the lowest cost charge-mix. Linear programming (L.P.) models are available with easy user interface for this application on computer.

Similar L.P. packages also help to optimize product-mix decisions for different items with varying material and labour requirements to arrive at the mix with maximum contribution in value.

(vi) Statistical Techniques

Computer applications using statistical techniques are very useful in process control, quality control, machine breakdown trend investigation, etc., with easy user interface. Regression analysis techniques help to correlate and predict seemingly unconnected variables from basic data. For example, mechanical properties and graphite morphology of S.G. iron can be predicted from the chemical analysis once a large amount of data is available. These predictions are very useful for decisions, which are difficult to take by normal manual methods.

(vii) Perfect Management

Computer techniques for project evaluation like PERT/CPM prove very useful to plan and monitor time taking activities like new product launching, expansion of production, new technology adoption, etc. Control of such properties involving large number of activities, cost and criticality is very difficult manually, resulting in delays in competitive environment. Many user-friendly computer packages are available with easy uses for this application where there is almost no need of knowledge about High Level Language. Generation of network diagrams, progress reports with time and cost, highlighting areas can be represented easily with the help of related software. Harvard Graphics, Project Manager, TCS PRISM are few leading software packages in this field.

(viii) Casting without Pattern

Rapid casting technology (RCT) was developed in Germany and is also available in USA. In this process, mould cores are made directly from the design stage eliminating the pattern making step. Mould and cores are directly made from CAD files. In this process, 70% time is saved and as result the lead time decreases abruptly. Silica sand and furan binders are used in this process. Castings with complex geometries or lower volume of production suit this process. This system is extremely flexible because any change can be carried out at any stage of the process at the last moment. The issues like gating and feeding design can also be evaluated in this rapid prototyping casting process.

23.5 COMPUTER AIDED MODELLING OF CASTINGS

Today, product development times are reduced from months and years to days and weeks. This has been possible due to advanced computer technology and progress in numerical simulation methods. Gradually, computer technology becoming increasingly user friendly including a number of built-in options and rules as well as supplied databases allow users to perform very complex mathematical computations and engineering analysis. Casting process deals with so many complex issues like gating, risering and solidification, fluid flow, heat transfer, mechanical and thermal stress analysis, computer aided design and tooling manufacture. Due to this, complex processes involved in the casting industry was for long time more an art than science.

However, development of powerful and fast computers as well as optimised computer codes for modelling and simulation of the casting process allowed the transition of metal casting to modern scientific and technological processes, where a little is left for guessing. Nowadays the foundry engineers and technologists successfully deal with complicated issues like optimization of gating and risering, design for manufacturability and assembly, parametric analysis, fatigue analysis, etc. Although a number of things have to be resolved and improved with respect to capability, flexibility, accuracy and capability of the software. Computer aided drafting helps to clearly visualize complicated drawings in 3D to design patterns and cores for optimum methoding. Different allowances can be pattern, core, core print and parting line selection becomes easier after determinig the maximum cross-sectional area. Calculation of gating and risering system through computer helps in selection for correct size of the sprue, runner and ingate considering important parameters like critical wall thickness, fluidity, nature of solidification, pouring temperature of the alloy, rate of pouring, etc.

First of all, we had taken up an industrial component like flywheel (S.G. iron) and then we have designed it using solid works software. It is needless to mention that solid works is software used for designing and assembling. The software is a user-friendly one in which parts can be easily drawn and assembled together.

23.6 GATING AND RISERING

Gating and risering are major parts of casting to provide molten metal to the mould which is a true replica of the actual component. Proper gating system controls the progressive solidification in such a manner that no part of the casting is isolated from active feed channels during the entire freezing cycle. Gating system has four components like pouring basin, sprue, runner and gate. Risers are attached to the casting at the right locations so that they can continuously supply hot liquid metal to the casting during solidification to compensate the shrinkage. If the casting has problems with shrinkage defects, misruns, etc., then the information given below will help to solve the problem. If the pattern is match plate, a gating system is built onto the match plate as well so that it is moulded at the same time as mould cavity.

Main objectives of a gating system are to:

- minimize turbulence to avoid trapping gases and breaking up the sand mould during the entry of the molten metal; therefore, the sprue should be gradually tapered to minimize the vortexing of metal and consequent entrainment of inclusion and gases;
- get enough metal into the mould cavity before the metal starts to solidify. The pouring basin must be full of molten metal;
- prevent the formation of dross, scum, slag and eroded sand entering into the casting, pouring must be done far away from the sprue to allow time for inclusions to float out which can be skimmed with the help of a bar and a turbulence-free molten metal should flow into the sprue;
- prevent casting defects like cold shut, misrun, hot tear, local shrinkage and gas cavities particularly in a steel casting;

- get maximum casting yield with less grinding by proper design of ingates and runners; and
- establish the best possible temperature gradients to achieve directional solidification and minimizing the distortion and shrinkage effect.

In addition to the above-mentioned points, the gating system should be of such simple design as to facilitate moulding at the same time involving minimum fettling cost resulting in maximum casting yield. Many of these requirements and functions are conflicting with each other. Therefore, an effort should be made to optimize these parameters so as to get defect-free castings.

The elements of a gating system are downsprue through which metal enters the runner and from which it, in turn, passes through the ingates into the mould cavity (Figure. 23.1). The part of gating system which mainly restricts or regulates the rate of pouring is the primary choke, more often called the choke. At the top of the downsprue may be a pouring cup or pouring basin to minimize splash and turbulence and ensure the smooth entry of clean metal only into the downsprue. Correct pouring equipment like ladles, crucibles are to be used so that the molten metal should enter into 'the mould cavity gradually through the lip with constant velocity. To prevent entry of dirt or slag into the downsprue, the pouring basin may contain 3 skim cores, a strainer, a delay screen or a sprue plug. To prevent erosion of the gating system when a large amount of metal is to be poured, a splash core may be placed in the bottom of the downsprue or wherever the flowing metal impinges with more than normal 'force'. Gates and risers are often designed to ensure the principle of controlled directional solidification which means that freezing starts from the farthest end of the riser and proceeds towards the riser.

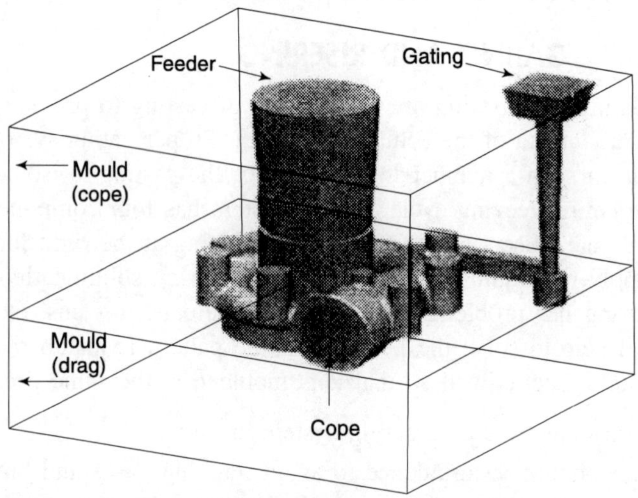

Figure 23.1 Elements of gating system

To accomplish this, castings are riser-gated with metal entering the riser through a downsprue and runner heating both the riser base and riser neck while flowing into the mould cavity. A top riser is placed at the farthest end of the casting to feed the hot metal locally as the side riser may not feed up to that point.

Types of gating: The gating system is named as per the method or location of molten metal entry into casting/mould cavity.

- **Top gate:** Top gates are usually limited to relatively small castings of simple design. The turbulence of metal as it enters the mould cavity causes erosion, which is a major problem in the manufacture of steel castings. Top gates are used in steel foundries only for broad shapes of low height in a casting.

- **Bottom gate:** Bottom gating reduces the turbulence and erosion of the mould to a minimum but creates unfavourable thermal gradients. Whereas local hot spots result at the gate entrance and metal appears in the riser.
- **Parting line gate:** This particular gating is a compromise between top and bottom gating. In this case, metal enters the mould cavity at the same level as the mould joint or parting line. Molten metal enters through the sprue and reaches the parting surface where the sprue is connected to the runner or gates in a direction horizontal to the casting. This type of gate allows easy use of devices like strainer cores which can entrap any slag, dirt or sand effectively during metal flow down the sprue.
- **Side/Step gate:** This type of gating having advantages of bottom gating at the same time allows the hot metal to enter the riser directly. Metal flows through the bottom ingates until the mould is filled to the level of the next higher ingate. At this point metal is expected to start flowing through this ingate and through successively higher ones, as the mould gates get filled. However, in practice step gates do not function in this ideal manner. The inertia of the metal falling through the sprue and the resulting low pressure areas created at the entrance of the top gates shows better distribution of metal but it decreases the yield of the casting. Through experimentation, it has been observed that by slanting the ingates upward at an angle to the casting, step gates can be made to function properly (Figure 23.2).

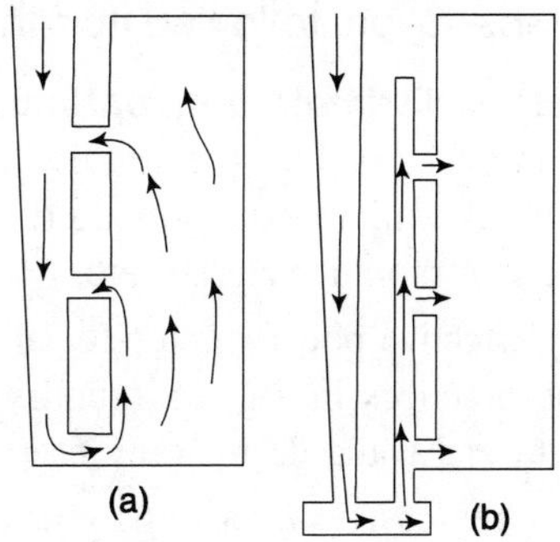

Figure 23.2 Side/Step gate

- **Whirl gate:** This type of gating system is very useful for preventing dirt from entering the mould cavity. Particularly for clean steel castings, it is very important. Following recommendations are very important:
  - (i) Ratio of ingate and out-gate cross-sectional area should be 1.5:1.
  - (ii) Optimum whirl gate basin diameter appears to be between three to five times the ingate width and height about 1.5 times the ingate height.
  - (iii) Performance can be improved by increasing the angular displacement 180 degrees apart.

The gating system may be classified into two types: (i) pressurized (ii) non-pressurized.

Pressurized gating system is kept full of metal and it maintains back pressure throughout the casting; gate and runner areas are equal to or less than the area of sprue base. For example, 1:0.7:0.5 (area of sprue base : area of runner : area of gate). This type of gating system minimizes the danger of the metal pulling away from the mould walls but at the same time it causes aspiration, sand erosion and turbulence problems.

In **non-pressurized** gating system, the area of the runners and the ingates are larger than the sprue area. It provides faster metal flow through the gates farthest from the sprue and slower through gates closer to the sprue due to low pressure in the later case. For example, 1:2:2 or 1:4:4 (area of sprue base: area of runner: area of gate). This type of gating system offers low velocity of metal stream and minimizes mould erosion. Such system favours oxidation of metal and may be partially responsible for the formation of peroxide defects. A slight change in the gating ratio like 1 : 2 : 1.5 will produce steel castings nearly free from sand erosion and oxidation. It also provides uniform flow.

Factors for optimizing gating system

- Control on pouring of metal into basin which should be filled quickly and smoothly
- Selection of correct pouring equipment
- The melt should have sufficient fluidity.

Steps to be followed for the design of gating system

Step 1: Estimation of optimum pouring time of casting: Pouring time is influenced by many factors like the fluidity and dross forming characteristics of metal, casting geometry/ critical wall thickness/modulus, total weight of the casting including runners and risers and type of mould material, etc.

Effective pouring rates for all commercial casting alloys have not been published. Therefore, the practices in a given foundry depend on the expertise available in the particular area. Some data are available for grey cast iron, 8.6 iron, steel, aluminium, etc.

Grey cast iron castings < 1000 kg [according to Nielson]

$$t = 0.32\ (W.G)^{0.4} \text{ seconds}$$

where,

t = pouring time in seconds

W = average wall thickness in mm

G = Gross casting weight in kg

t = M/E [According to Czikel] in seconds

M = modulus of casting = casting volume/casting surface area

E = constant 0.015 to 0.035

The optimum pouring time for grey iron castings is also calculated from the following empirical formula:

$$t = k(G.W)^{3/2} \text{ seconds}$$

where,

k = coefficient dependent on weight of castings

W = average wall thickness of castings in m

G = gross casting weight in kg

The value of k is given in Table 23.1.

Table 23.1 Value of *k*

Type	*Weight*	*k*
Big castings	10-50 tonnes	1.8 to 2.3
Medium castings	1-10 tonnes	1.6 to 1.9
Small castings	Less than 1 tonne	1.4 to 1.6

Pouring time can also be calculated for grey cast iron from the following empirical formulae also:

$$t = Z\ (1.41 + W/14.58)\ G^{1/2},\ G < 400 \text{ kg}$$

$$t = Z\ (1.24\ W/16.65)\ G^{1/2},\ G > 400 \text{ kg.}$$

where, Z = Fluidity factor which depends on the carbon equivalent and the temperature of the hot metal. Value of Z is given in Table 23.2.

Table 23.2 Value of Z

Carbon equivalent	*Value of Z*			
	Temperature °C			
	1425	1400	1370	1315
4.3	1.00	0.93	0.86	0.73
4.2	0.95	0.88	0.83	0.65
4.0	0.86	0.80	0.75	0.58

In the calculation of pouring time, the filling is an important factor. Too low pouring time means very high filling rate which results in a turbulent flow that creates sand erosion which is not desirable. Too high pouring time means very slow pouring rate which results in premature solidification on the walls and some metallic oxides and impurities may solidify and get entrapped on the vertical walls of casting.

The flow rate can be calculated from the following empirical formula

$$V = C/t \text{ cm/sec.}$$

where C = Height of casting from the lowest point to the highest point in cm.
T = Pouring time in seconds.

Table 23.3 gives the permissible flow rate in case of grey iron and steel.

Table 23.3 Flow rate for grey iron and steel

Wall thickness in mm	*Grey iron castings*	*Steel castings*
Up to 4 mm	3-10	-
4-10	2-3	2
10-40	1-2	1
More than 40	0.8-1	0.8

After calculating the pouring time from previous two equations it is necessary to cross-check from the last equation mentioned above; whether the speed of rise of metal in the mould at the time of pouring is lower than the permissible/recommended limit or not. If the value V is too low, we should either change the position of the casting in the mould or make the time of pouring shorter.

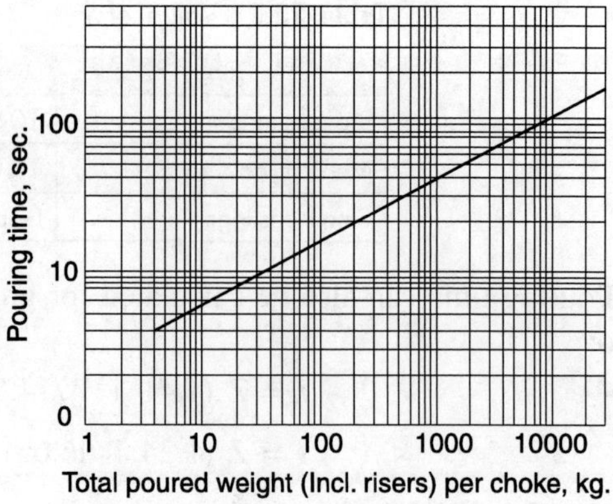

Figure 23.3 Pouring time versus pouring weight curve

S.G. cast iron castings

$$t = 0.97 \times (G)^{1/2} \text{ seconds}$$

$$t = K_1(G)^{1/2}$$

where, K_1 = 2.08 for $W < 9.5$ mm

= 2.67 for $9.5 < W < 25.4$ mm

= 2.96 for $W > 25.4$ mm

$$t = 0.64 \times (G)^{1/3} \times (W + 10)^{1/2}.$$

A simple formula widely used in different foundries.

Another chart is given Figure 23.3.

Steel castings: $t = S\ (W.G.)^{1/3}$ seconds

S = time coefficient varying from 1.2 to 1.6

W = minimum wall thickness in mm.

G = gross casting weight in kg.

Some foundries calculate the following empirical formula in case of steel.

$$t = 1.78 - (2.38 \times 10^{-5}\ G^{1/2}) \text{ seconds}$$

Step 2: Estimation of sprue choke area: This is calculated from the principles of the law of continuity which confirms the constant flow rate in all portions of the fluid system.

$$Q = A_1V_1 = A_2V_2 \text{ (continuity equation)}$$

where, A_1 and A_2 = cross-sectional areas of that channel at two different points 1 & 2.

V_1, V_2 = velocities of metal at two different points 1 & 2

23.7 BERNOULLI'S THEOREM

Total energy comprising potential, kinetic and pressure energy at every point in the gating system is constant.

$$h_1 + V_1^2/2g + P_1/p = h_2 + V_2^2/2g + P_2/p$$

(Heat loss is not considered.)

where, P = pressure in dyne/cm^2

V = velocity of flow = $(2gh)^{1/2}$ in cm/second.

h = effective metallostatic height in centimeters.

g = acceleration due to gravity = 981 cm/sec^2.

p = density of the metal in kg/cubic centimeter.

Note that Bernoulli's theorem is employed here to calculate the velocity of an ideal fluid system, i.e., in the system where the fluid suffers no energy losses. In real gating systems, besides losses due to friction energy losses occur at entrances and exits, bends, enlargements and contractions. The velocity and flow rates obtained from the above equation would therefore be somewhat higher than those found in practice.

The smallest area in the feeding channels (known as choke) controls the rate in the mould cavity and consequently controls the pouring time. Actually, the choke area occurs at the bottom of the sprue to establish the velocity as soon as possible but this is not always the case. If the choke occurs at the base of the sprue, proper area can be calculated by giving a formula based on the application of Bernoulli's theorem. For example, choke area can be determined by using the following formula:

$$S = G/p.t.c.V$$

where, t = pouring time in seconds.

G = total liquid metal poured in kg.

p = density of metal in kg/cubic centimeter.

c = discharge coefficient (0.25-0.4)

S = sprue choke area in square centimeter.

V = velocity of flow = $(2gh)^{1/2}$.

This sprue is normally tapered. As the metal descends through sprue, its velocity increases owing to the acceleration due to gravity. This acceleration has two effects.

(a) It increases the velocity of the metal ($v = (2gh)^{1/2}$).

(b) During the free falling the cross-sectional area decreases as the velocity increases (according to the law of continuity), as a result the metal pulls away from the walls of the sprue with consequence turbulence and aspiration as shown in Figure 23.4. It is advisable to make the sprue tapered so that the metal lies firmly against the wall of the sprue and as a result the turbulence and air aspiration problems are eliminated. This also reduces mould erosion and metal turbulence.

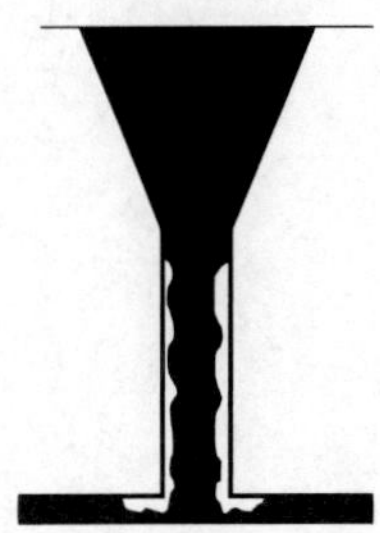

Figure 23.4 Sprue causes turbulence and air aspiration problems

In case of dross forming metals like aluminium and other non-ferrous metals, restricted sprue area is used to reduce the velocity of the metal which may lead to dross formation as it enters into the runners. On the other hand, carbon steels and alloy steels are not so prone to the formation of dross and therefore these precautions are not essential. The pouring temperature of steel is very high and cools rapidly. Therefore, the mould filling must have a fast rate to avoid cooling tendency.

23.8 CASE STUDY

Steps followed:

1. To draw the component, i.e., flywheel with gating and risering systems of the component (flywheel) using solid works software.
2. Finding the maximum area to obtain the parting line.
3. Calculating the gating system.
4. Designing the gating system by solid works.
5. Optimization of gating system.

Design considerations

Some design considerations have been taken for designing the gating system:

1. As the thickness of flywheel is very small (43 mm), so it is placed in the drag only. Instead of using a two-piece pattern, we are using single-piece pattern. This is also useful while considering the economic purposes.
2. The flywheel is placed horizontally to save the space required for the mould box and the eddy is formed due to high speed of the molten metal through the sprue.

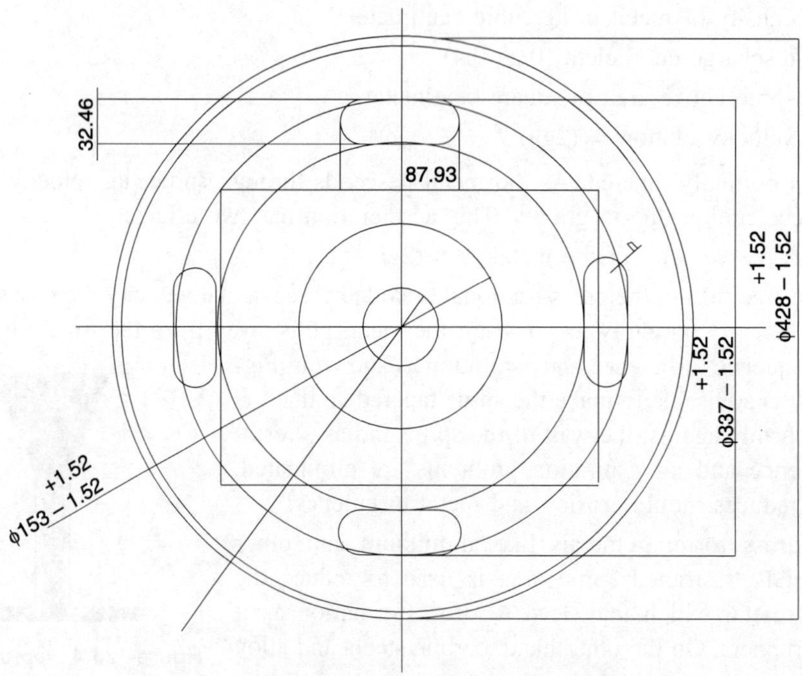

Figure 23.5 Top view of the flywheel

Drawing the gating and risering system

The gating and risering system is drawn using the solid works software. Care should be taken while drawing the dimensions of the flywheel as shown in Figures 23.5 and 23.6.

Outside diameter = 428.5 $\pm$ 1.52 mm.

Thickness of the flywheel = 43.5 $\pm$ 0.76 mm.

Bore diameter = 56 $\pm$ 0.76 mm.

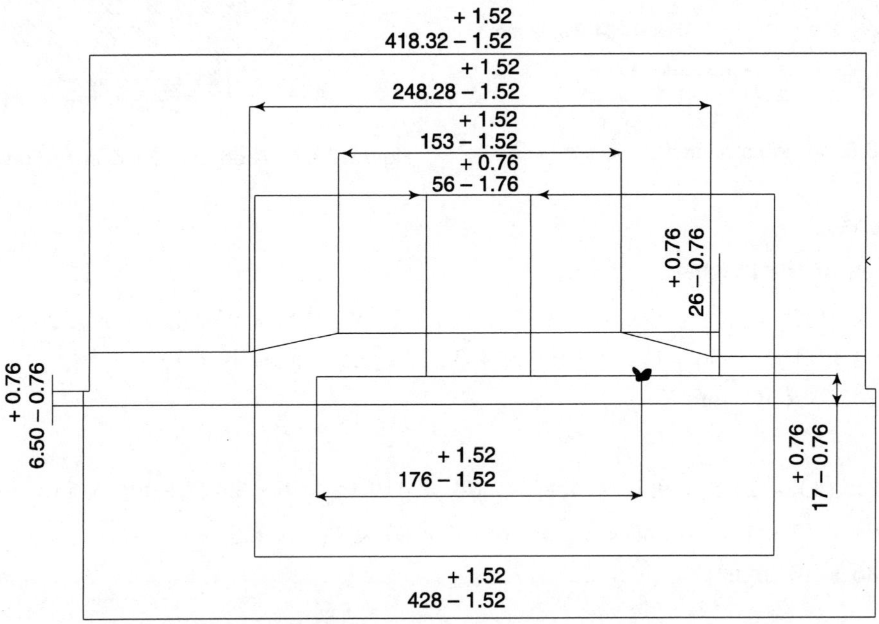

Figure 23.6 Side view of the flywheel

Finding the maximum area to obtain the parting line

Parting line is the dividing line between the two moulding flasks that makes up the sand mould. In split pattern, this is also the dividing line between the two halves of the pattern.

To obtain the parting line, we had different sections of the flywheel. Though we had taken 5-6 sections but due to space constraints, we are giving the snaps of only three sections (Figures 23.7 to 23.9).

So we have seen from the above figures that the maximum area occurs at section 3-3. Hence, the parting line is obtained at section 3-3.

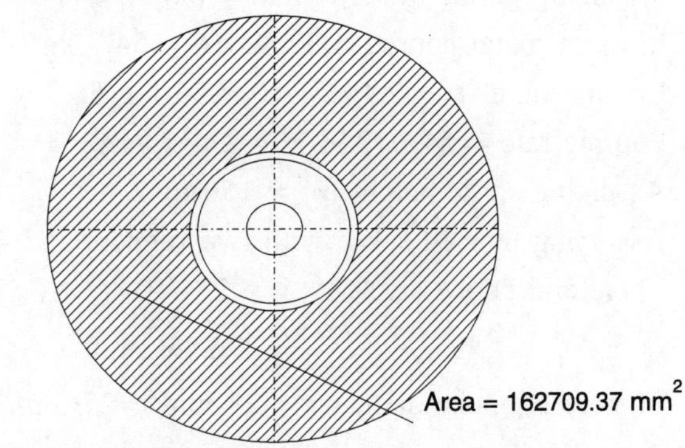

Figure 23.7 Area calculated in section 1-1

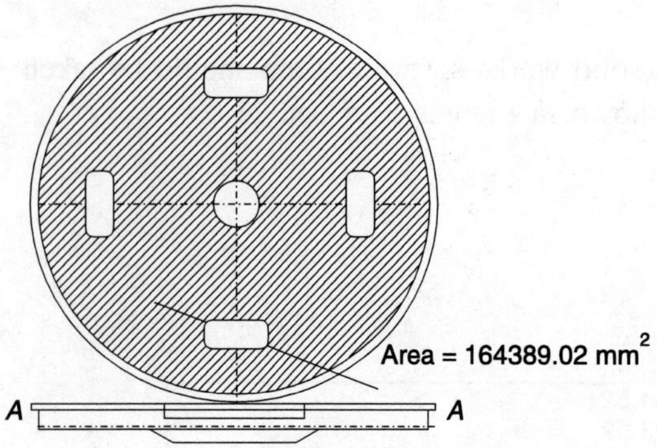

Figure 23.8 Area calculated in section 2-2

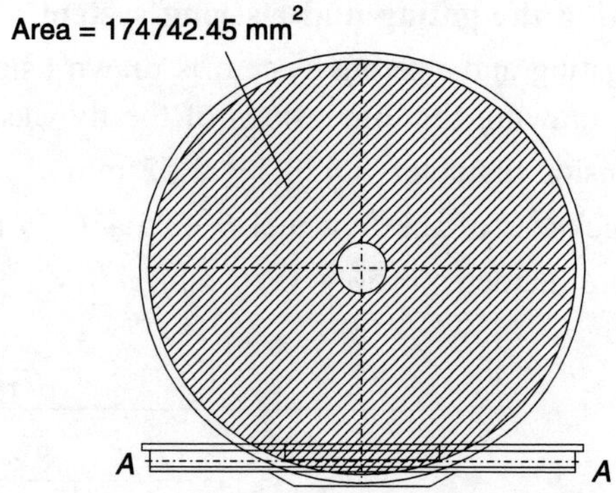

Figure 23.9 Area calculated in section 3-3

Calculations

Surface area of the flywheel

$$= \pi/4 \times (425^2 - 56^2) + \pi \times 176 \times 17 + (43 - 24.5 - 9.5) \times \pi \times 428 + \pi \times 24.5 \times 419 + \pi/4 \times (419^2 - 56^2) + 4 \times 2 \times (89 + 32.2) \times 6.5 + \pi \times 56 \times 76$$

$$= 3.520256 \times 10^5 \text{ mm}^2.$$

Volume

$$= \pi/4 \times (43 - 24.5 - 9.5) \times 428^2 + \pi/4 \times 419 \times 419 \times 24.5 + \pi/4 \times (159^2 \times 9.5) - \pi/4 \times 17 \times 176^2 - \pi/4 \times 56^2 \times 76 - 4 \times 89 \times 32.3 \times 6.5$$

$$= 4.186 \times 10^6 \text{ mm}^2$$

Mass

$$= 4.1864 \times 10^6 \text{ mm}^2 \times 7.84 \times 10^{-6} \text{ kg}$$

$$= 32.82 \text{ kg}.$$

Assuming gating system yield = 0.6

Mass of metal poured = 32.82/0.6 = 54.7 kg.

Pouring time (t) = $2.4335 - 0.3953 \log 54.7 \times 54.7^{0.5}$ = 12.92 sec.

Pouring rate = 54.7/12.92 = 4.234 kg/sec.

Assuming sprue height (h) = 150 mm.

Assuming tapered sprue with two runners and 4 ingates, and efficiency is 0.85.

Therefore, choke area (A_1) = $4.234/\{0.85 \times 7.82 \times 10^{-6} \times (2 \times 9800 \times 150)^{0.5}\}$

$$= 370.55 \text{ mm}^2.$$

Therefore, choke diameter = 23 mm = 25 mm.

Assuming a metal head of 50 mm in the pouring basin,

Therefore, sprue area at the top = $2.2326 \times A_1$ = 828.541 mm^2.

Therefore, sprue diameter at mp = $(2.2336)^{0.5} \times 21.721$ = 32.48 mm (35 mm).

Pouring basin depth = 2.5 × 35 = 87.5 mm

Assuming gating ratio = 1:2:2 (as non-pressurized gating system is adopted)

Runner area = 2 × 1(14 × 252) = 981.75 mm^2.

Now, metal flow of metal through runner = 4.234/(2 × 7.84 × 10^{-6})

= 2.7 × 10^5 mm^3/sec.

Assuming 2 runners having rectangular cross section with 2:1 ratio

Therefore, runner cross-sectional area = 22 × 44 mm^2.

Metal velocity in runner = 2.7 × 105/(22 × 44) = 279 mm/sec.

There are four ingates.

Therefore, metal flow through each gate = 279/4 = 69.73 mm/s

Each gate area = 981.75/4 = 245.44 mm^2.

Now solving the equation (empirical formula for the modulus method)

$$D^3 - 5.46 \times Mc \times D^2 - 0.05093 \times Vc = 0$$

Hence, D = 9L 2 mm ~ 95 mm

Therefore, volume of metal in riser = π/4 × 95^3 = 2.7 × 10^6 mm^3.

Now neglecting the metal flow in runners,

Casting yield = Volume of metal in casting/volume of metal poured (casting + riser)

= (4.1864)/(4.1864 + 2.7) a: 0.607 = 60.7%, which is almost equal to the assumption (60%).

Hence, the design is satisfactory.

Designing the Gating System by Solid Works

The above flywheel has been designed based on the calculations and using the solid works software (Figures 23.10 and 23.11).

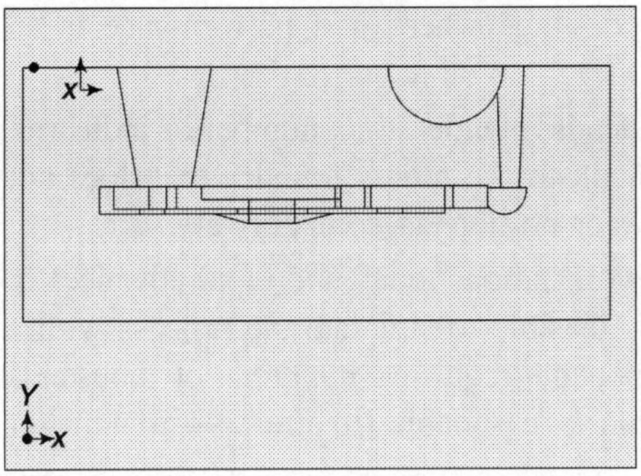

Figure 23.10 Modelling in solid works

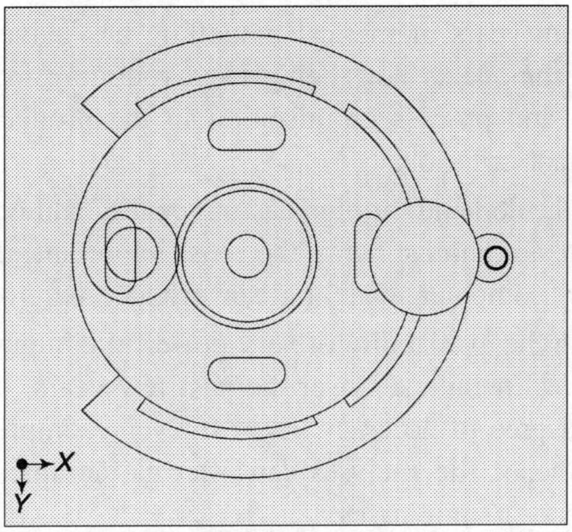

Figure 23.11 Views of gating and risering system

Care should be taken during solid modelling:

1. The design should be safe.
2. The dimensions should be properly given.
3. There should be provision for further alterations and modifications.
4. The design should be easily understood.

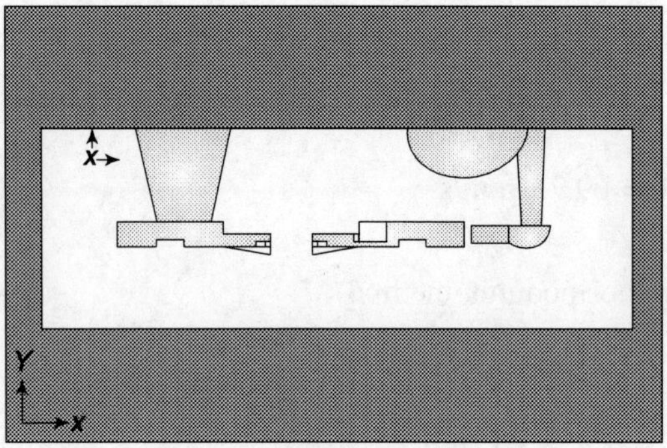

Figure 23.12 Section view of gating and risering system

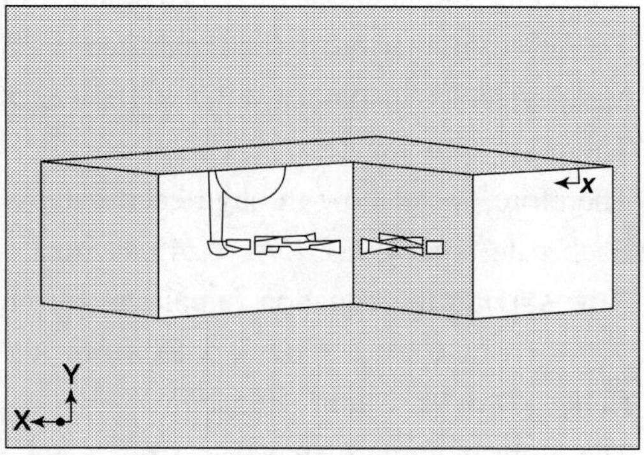

Figure 23.13 Half section view of gating and risering system

It was reported that elimination of defects and increase in the yield of castings is possible by redesigning the feeding system of this particular casting. In this case study, it is claimed that about 16% improvement in yield is possible and shrinkage defects were almost eliminated (which was about 30%) in the successive batches of production with reduced lead time for the production of new castings.

23.9 NUMERICAL MODELLING IN CASTING

In recent years, numerical modelling has gained increasing popularity in the casting industry. The technique is now being used routinely in resolving potential problems related to mould design and process optimization in the production of castings where part performance is an important criterion.

As the demand for high casting quality and performance is relentless in a number of industrial sectors, automotive being the most important, it is difficult to see how a foundry that does not employ computer modeling can produce parts with low or no defects competitively.

With the availability of very powerful computers or workstations today and rapid advances in numerical techniques over the past few years, it is now feasible to simulate the process with a good degree of accuracy. Even for very complex components requiring hundreds of thousands of elements, the analyses can often be completed within a reasonable timeframe—in a matter of hours in some cases.

Although both finite difference and finite element methods are used in casting software packages available commercially, it is generally recognized that the latter method (favoured by developers at IMI) is more suitable for problems that involve parts with a complex geometry, as is usually the case in castings. In addition, the finite element method has a clear advantage in that meshing of the geometry, an extremely time consuming operation in computer modeling, can be performed in most instances in the same CAD system that was isused to design the component.

In terms of modelling the process, the phenomena of interest are filling, solidification, residual stresses and shrinkage. If these can be predicted, it would be possible, at the initial design stage, to establish the size and type of feeding system required, the sizes and locations of the cooling channels and the exact dimensions of the cavity. Furthermore, it would also be possible to use the results to determine certain process or production data such as injection speed, pressure, cooling and cycle times.

In this chapter, the development of a 2D shell finite element model for thin wall pressure dies castings is first described. Simplified governing equations for turbulent flow of a Newtonian fluid are derived. The pressure field is calculated using the Hele-Shaw approximation and a control volume approach is employed to track the flow front advancement. The model can also be used to simulate laminar flow.

Second, for applications where filling cannot be accurately modelled by the Hele-Shaw approximation approach (as in low pressure or gravity casting), a fully three-dimensional solution based on the non-isothermal Navier-Stokes equations is presented. Tracking of the flow front is modelled using a pseudo-concentration technique.

Third, solidification of the metal in the cavity is modelled by solving the energy equation, which takes into account the release of the latent heat of fusion during phase change, employing a macroscopic specific heat model.

23.9.1 Two-dimensional Shell Element Model

Simplified Governing Equation

Components produced by pressure die casting generally have a very small cross-sectional thickness compared to their width and length. Thus, a very complex three-dimensional geometry can be approximated by a series of two-dimensional shell elements with a thickness dimension. Also known as a 2.5D approach, it can be applied to practically all pressure die castings produced industrially. The computational time required for a given filling analysis is considerably less than what would be required for a fully 3D model, often in no more than an hour or so for a component as complex as a transmission housing.

As pressure die casting is a rapid production process, the die being typically filled in less than one-tenth of a second and the part solidified well below a minute, it is reasonable to assume that the heat transfer during the turbulent filling phase is minimal. Assuming isothermal condition in the die, the turbulent flow of a Newtonian fluid which represents the molten metal can be described by the Reynolds-averaged conservation equations:

$$\rho\left(\frac{\partial u}{\partial t}\right) + u\cdot\nabla u = -\nabla \mathrm{p} + \nabla.(S - \overline{\rho u u}) \qquad ...(1)$$

$$\nabla\cdot u = 0 \qquad ...(2)$$

$$S = \eta\ (\nabla u + \nabla u^T) \qquad ...(3)$$

where t is the time, ρ the density, u the mean velocity, P the mean pressure, S the viscous stress tensor, η the viscosity and the overbar denotes a time-averaged quantity.

The quantity $-\overline{\rho uu}$ denoted as τ is known as the Reynolds stress tensor. The simplest turbulence models are based on the Boussinesq eddy-viscosity approximation in which the Reynolds stress tensor is proportional to an eddy viscosity ηe and the mean strain rate tensor

$$\tau = \eta_e (\nabla u + \nabla u^T) \qquad ...(4)$$

Using the lubrication approximation, the governing equations are simplified to the Hele-Shaw equation

$$-\frac{\partial p}{\partial x} + \frac{\partial(\eta + \eta_e)}{\partial x}\frac{\partial U}{\partial z} = 0 \qquad ...(5)$$

$$-\frac{\partial P}{\partial y} + \frac{\partial(\eta + \eta e)}{\partial x}\frac{\partial U}{\partial z} = 0 \qquad ...(6)$$

$$\frac{\partial U}{\partial x} + \frac{\partial V}{\partial y} = 0 \qquad ...(7)$$

Equations (5)–(7) can be combined to yield a single equation for the pressure

$$\frac{\partial}{\partial x} S_1 \frac{P}{\partial x} + \frac{\partial}{\partial y} S_1 \frac{P}{\partial y} = 0 \qquad ...(8)$$

where

$$S_1 = \int_b^a \frac{z^2}{\eta + \eta_e} dz \qquad ...(9)$$

Equation (8) together with the appropriate boundary conditions allow the prediction of the pressure distribution during the filling process.

Formulation of eddy viscosity

Following Prandtl mixing length hypothesis, the eddy viscosity can be expressed as

$$\eta_e = \rho l_m^2 \gamma \qquad ...(10)$$

where l_m is the mixing length and γ is defined as

$$\gamma = \sqrt{\frac{1}{2}[(\nabla u + \nabla u^T):(\nabla u + \nabla u^T)]} \qquad ....(11)$$

In fully developed duct flows, the mixing length distribution is well described by Nikuradse's formula

$$\frac{l_m}{b} = 0.14 - 0.08\left(l - \frac{z}{b}\right)^2 - 0.06\left(l - \frac{z}{b}\right)^4 \qquad ...(12)$$

Since the eddy viscosity depends on the mean velocity, the pressure equation (8) is nonlinear and thus has to be solved numerically.

Numerical implementation

The pressure equation (8) is solved using the finite element method, the control volume approach being employed to track metal advancement in a fixed mesh system. As principal advantage, this approach can handle complex flow fronts of arbitrary shapes and can also incorporate factors such as distortion, flow separation, and front multiplication. A fill factor F, the value of which is between 0 and 1, is associated with each element to indicate the percentage of filling, with a value of 1 representing a completely filled element and 0, an empty element. At each time step, the pressure calculation is performed in order to obtain the velocity distribution in the filled domain. Using the flow rate information based on the velocity field, the fill factor of each element is evaluated by applying the principle of the conservation of mass. Finally, a newly filled domain can be easily defined using an updated fill factor field.

Simulation example

The capabilities of the model to predict flow in a die are demonstrated in two thin wall-die castings in an aluminium alloy. The properties used for the analyses are summarized in Table 23.4.

Table 23.4 Properties of aluminium 356

Part (Aluminium 356)	*Values*
Density	2.685×10^{-3} g/mm^3
Viscosity	2.96×10^{-3} g/mm·s
Specific heat	0.963 J/g·°C
Thermal conductivity	0.151 W/mm·°C
Latent heat	389 J/g
Liquidus temperature	555°C
Solidus temperature	615°C
Poisson's ratio	0.37
Fluidity parameter	10^3 (MPa·s)$^{-1}$
Hardening parameter	0.032
Coefficient of phase change expansion	10^{-4}
Coefficient of thermal expansion	$\alpha_1 = 22.598 \times 10^{6}$°C^{-1}
$\alpha(T) = \alpha_1 + \alpha_2 T$	$\alpha_2 = 0.02387 \times 10^{-6}$°C^{-2}
Young's modulus	$E_1 = 6.93 \times 10^4$ MPa
$E(T) = E_1 - E_2 T$	$E_2 = 106$ MPa/°C
Initial yield stress	$\alpha_1 = 300$ MPa
$\alpha_y^0(T) = \alpha_1 - \alpha_2 T$	$\alpha_2 = 400$ MPa/°C
Gravity	9800 mm/s^2
Mould (H13 Steel)	
Density	7.8×10^{-3} g/mm^3
Specific heat	0.422 J/g·°C
Thermal conductivity	0.0287 W/mm·°C

Comparison between laminar and turbulent flows

The first example provides a comparison between laminar and turbulent flow in an experimental casting, as shown in Figures 23.14–23.17. With a wall thickness over most of the surface of approximately 3.0 mm, the cavity is fed by two tangential gates of just over 1.0 mm in thickness from a runner which tapers from 35 mm wide and 13 mm thick at the entrance to 5 mm × 5 mm at the end. This runner design gives a more uniform pressure distribution at the gate as the cross-sectional area is decreased progressively from the flow origin to the tip. Figures 23.14 and 23.15 are the flow patterns associated with the laminar and turbulent condition respectively, the colour zones representing the fill times for the different parts of the casting. It should also be observed that the patterns produced by these two flow conditions are significantly different in terms of arrival times for the various zones. As viscosity depends on the flow rate, a higher flow resistance is generated in the cavity in turbulent flow. This is reflected in the pressure distributions calculated for the two flow conditions, Figures 23.16 and 23.17. The pressure requirement to fill the part under turbulent flow is significantly higher compared with flow that is laminar.

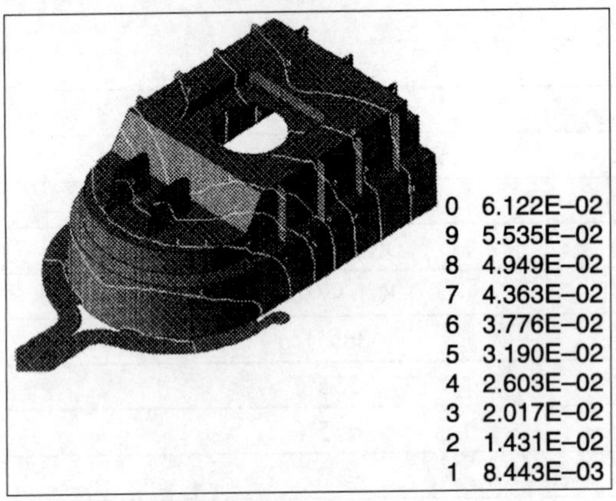

Figure 23.14 Flow pattern for laminar flow

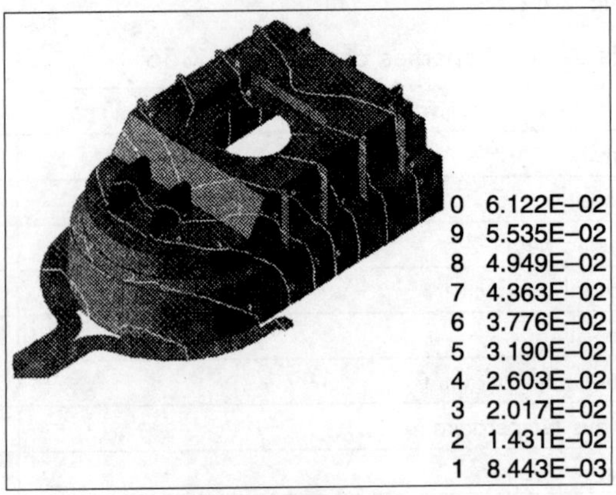

Figure 23.15 Flow pattern for turbulent flow

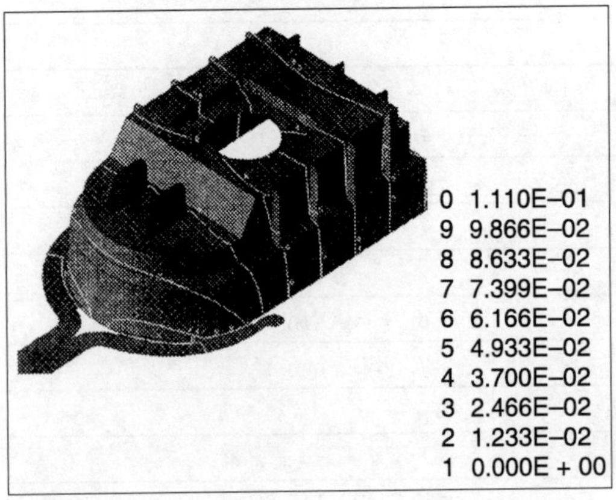

Figure 23.16 Pressure distribution in laminar flow

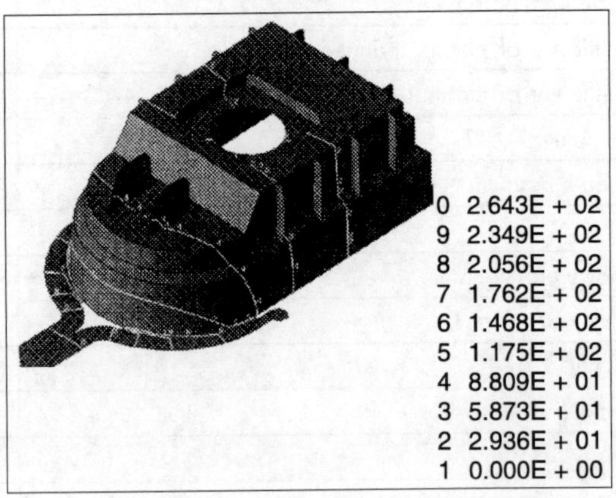

Figure 23.17 Pressure distribution in turbulent flow

Effect of runner design and multi-cavities

Figures 23.18–23.19 show flow in a multi-cavity die for producing a housing component in an aluminium alloy. Two different runner designs are evaluated. In Figure 23.19, the feeding system with a larger cross-sectional area fills the two cavities closest to the origin completely even before the metal reaches the gates in the four end cavities. This is unacceptable in die casting as a high level of casting defects are likely to occur in the end cavities. On the other hand, in Figure 23.19, a more uniform filling of all cavities is observed as the metal reaches each gate more or less simultaneously, an ideal fill condition for a multi-cavity die. These results can thus be applied to optimize the dimension and configuration of the feeding system.

Figure 23.18 Flow fronts at 40% fill (filled region marked is shaded)

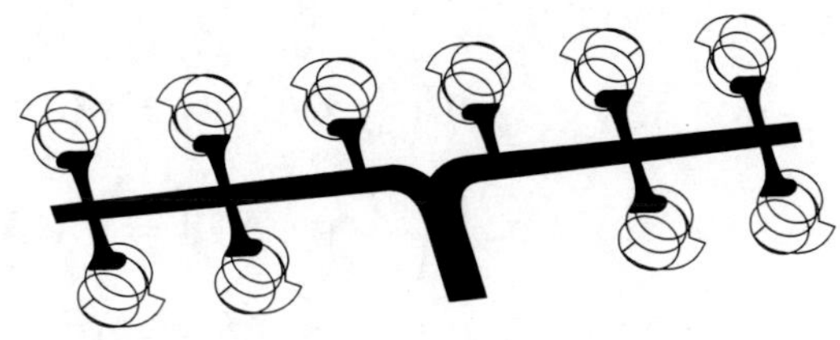

Figure 23.19 Flow fronts at 40% fill (filled region marked is shaded)

23.9.2 Three-dimensional (3D) models

Not all filling simulations in casting can be accurately modelled using the Hele-Shaw approximation approach. Among the difficulties that one could encounter are conditions that relate to thickness of the casting, the fluid behaviour in blunt 3D sections, the behaviour near and at the solid walls, the phenomenon occurring at the merging of two or more fluid streams, waves on the free surface, and the presence of cooling lines and residual stresses.

During mould filling, the melt front impinges against the walls which may or may not have smooth contours. At the point of contact, the flow has to split in order to continue its movement along the wall. Also, two flow fronts may collide with each other and form a "weldline". Because of these and other difficulties, flow can sometimes be best represented by using a fully three-dimensional model.

The main obstacles encountered in the three-dimensional simulation of mould filling from the numerical standpoint may be summarized as follows:

- The computational domain is usually a three-dimensional volume having a complex shape.
- The free surface (liquid-air interface) subjected to large deformations and multiple interfaces may come in contact with each other.

Because of such obstacles, the model should have a multi-step solution algorithm. In the first step, the incompressible Navier-Stokes and energy equations to compute the velocity pressure and temperature fields are solved. Then, based on the pseudo-concentration technique, the position of the flow front is convected using the computed velocity field. Once the cavity is filled, the energy equation and stress equations are solved to predict solidification, residual stresses and deformations.

Flow equations

The Navier-Stokes equations are solved on the whole computational domain, implying that the equations are also solved for the air present in the cavity. It is assumed that the air can exit the cavity without restriction and that its velocity is small compared to the speed of sound. The flow in the empty region of the domain can be assumed incompressible. Therefore, in the numerical analysis, the "air" is referred to as a pseudo-fluid. The equations governing the laminar flow of liquid and pseudo-fluid are expressed as

$$\rho_1\left(\frac{\partial y}{\partial t} + u.\nabla u\right) = -\nabla p + \nabla.(2\eta_l\dot{\gamma}(u) + \rho g$$

$$\rho_1 c_{pl}\left(\frac{\partial T}{\partial t} + u.\nabla T\right) = \nabla.(k_l\nabla T)$$

$$\nabla.u = 0 \qquad \text{on } \Omega_l$$

$$\rho g\left(\frac{\partial u}{\partial t} + u.\nabla u\right) = -\nabla p + \nabla.(2\eta_g\dot{\gamma}(u)) + \rho g$$

$$\rho_g C_{pg}\left(\frac{\partial T}{\partial t} + u.\nabla T\right) = \nabla.(k_g\nabla T)$$

$$\nabla.u = 0 \qquad \text{on } \Omega_g \qquad \text{...(13)}$$

With $$\dot{\gamma}(u) = \frac{1}{2}(\nabla u + \nabla u)^T$$

In the above equation t, u, p, T, $\rho(T, F)$ and $\eta(T, F)$ denote time, velocity, pressure, temperature, density and viscosity respectively. Subscripts l and g refer to liquid and pseudo fluid respectively.

Appropriate boundary conditions complete the statement of the problem

$$u = u_o \qquad \text{on } \Gamma_{\text{inlet}}$$

$$u.n = o \qquad \text{on } \Gamma_{\text{tractions}}$$

$$u.t = \hat{t} \qquad \text{on } \Gamma_{\text{tractions}} \qquad \text{....(14)}$$

The boundary condition on Γ_{wall} is imposed only on the filling material.

Front tracking

The tracking of the flow front in the mould cavity is modelled using a pseudo-concentration technique. This model defines a function $F(x; t)$ in the whole computational domain. The function $F(x; t)$ is basically the distance from the interface. A critical value of F, called F_c, denotes the interface between the alloy and the air. With these definitions, the portion of the cavity where $F > F_c$ represents the filled portion of the cavity. The front tracking equation can be written as

$$\frac{\partial F}{\partial t} + u.\nabla F = 0 \quad \text{on} \quad \Omega \qquad ...(15)$$

with the following initial and boundary conditions:

$$F(x;\, t) = F_c + \text{sgn}(d)\ \Gamma_{\text{inlet}}$$

$$F(x;\, 0) = \text{sgn}(d) \qquad \text{at } t = 0$$

where Γ_{inlet} is the inflow region of $\partial\Omega$ and sgn(d) is the signed distance function. The hyperbolic equation (15) is solved with the element by element technique of Lesaint and Raviart.

Phase change model and solidification simulation

The solidification of the part involves a phase change from the liquid melt to a solid casting. During this solidification phase, the latent heat of fusion is released in the system and the energy equation must take that phenomenon into account. This is modelled using a macroscopic specific heat model:

$$C_p \cong \frac{H(t + \Delta t) - H(t)}{T(t + \Delta t) - T(t)} \qquad ...(16)$$

where H is the enthalpy and is based on different macroscopic solidification models such as Lever and Scheil rules.

Once the solidification simulation is completed, the transient temperature field is then used to provide information on the thermal history and metallurgical condition of the simulated part. Using this information, one can predict the solidification time, microporosity and secondary dendrite arm spacing, as well as the evolution of solid fraction and solidification velocity of the casting.

Stress and shrinkage simulations

During solidification and cooling, metal alloys generally exhibit a time-dependent plastic deformation behaviour. Prediction of stresses in the casting requires the development of a three-dimensional viscoplastic model which would also simulate deformation (shrinkage and warpage). Thus the final shape of the casting, together with residual stresses which arise at the end of the cooling stage, can be subsequently predicted. Effects such as thermal convection, solidification, phase change are calculated independently.

Basic equations

For nonlinear continuum problems, it is assumed that the total strain, ε can be separated into elastic, ε_{el}, thermal, ε_{th} phase change, ε_{tr}, and viscoplastic, ε_{vp}, components, so that the total strain rate can be expressed as

$$\dot{\varepsilon} = \dot{\varepsilon}_{el} + \dot{\varepsilon}_{th} + \dot{\varepsilon}_{tr} + \dot{\varepsilon}_{vp} \qquad ...(17)$$

where a dot represents differentiation with respect to time.

The total stress rate, $\dot{\sigma}$, depends on the elastic strain rate according to

$$\dot{\sigma} = D\dot{\varepsilon}_{el} \quad ...(18)$$

where D is the elasticity matrix.

The viscoplastic behaviour is governed by a scalar yield condition of the form

$$F(\sigma, \varepsilon_{vp}) = f(\sigma, \varepsilon_{vp}) - \sigma_y = 0 \quad ...(19)$$

in which σ_y is the uniaxial yield stress. It is assumed that viscoplastic flow occurs for values of $f > \sigma_y$.

The uniaxial yield stress may be given by

$$\sigma_y = \sigma_y^o + H'\bar{\varepsilon}_{vp} \quad ...(20)$$

where σ_y^o is the initial uniaxial yield stress, H' is the hardening parameter and $\bar{\varepsilon}_{vp}$ is the effective viscoplastic strain.

It is now necessary to choose a specific law defining the viscoplastic strains. The simplest option is one in which the viscoplastic strain rate depends on the current stresses, so that

$$\dot{\varepsilon}_{vp} = \eta\langle\phi(F)\rangle\frac{\partial F}{\partial\sigma} \quad ...(21)$$

in which η is a fluidity parameter controlling the plastic flow rate, $\phi(F)$ is a positive monotonic increasing function for $F > 0$ and the notation $\langle ... \rangle$ implies

$$\langle\phi(F)\rangle = \phi(F) \quad \text{for} \quad F > 0 \quad ...(22)$$

$$\langle\phi(F)\rangle = 0 \quad \text{for} \quad F \leq 0 \quad ...(23)$$

Thermal and phase change rates of strain are given respectively by

$$\dot{\varepsilon}_{th} = \alpha\dot{T} \quad \text{and} \quad \dot{\varepsilon}_{tr} = \beta\dot{f}_s \quad ...(24)$$

in which α is the linear thermal expansion coefficient, T is the temperature field, β is the linear dilatation coefficient due to the total liquid-solid transformation and $\dot{f}_s$ is the rate of the volumetric solid fraction which depends on the solidus and liquidus temperatures.

The equations of equilibrium to be satisfied at any instant of time, t_n, are

$$\int_\Omega [B^n]^T \sigma^n \, d\Omega + f^n = 0 \quad ...(25)$$

where f^n is the vector of equivalent nodal loads due to applied surface tractions and body forces. In (25) the notation B^n is employed to denote the possibility that the strain matrix may not be constant throughout the solution.

As boundary conditions, it is assumed that during the cooling stage, the mould is a rigid body and the part can deform within the cavity.

Results of analyses on a wheel

Filling, solidification and residual stress analyses were conducted on an aluminium wheel produced by the low pressure die casting process. The material properties and operating conditions during production are shown in Table 23.4.

Results of the filling simulation on a section of the wheel are shown in Figures 23.20 and 23.21 indicating the positions of the molten aluminium front in the cavity at different filling times. By observing the direction and shape of flow front, it is possible to define venting and cooling requirements at the mould design stage to avoid the associated casting defects.

Once the mould is completely filled, the solidification and residual stresses simulations follow. These two simulations complement the filling analysis in that potential problems related to microstructure; porosity and part deformation which are not caused by mould filling can also be identified.

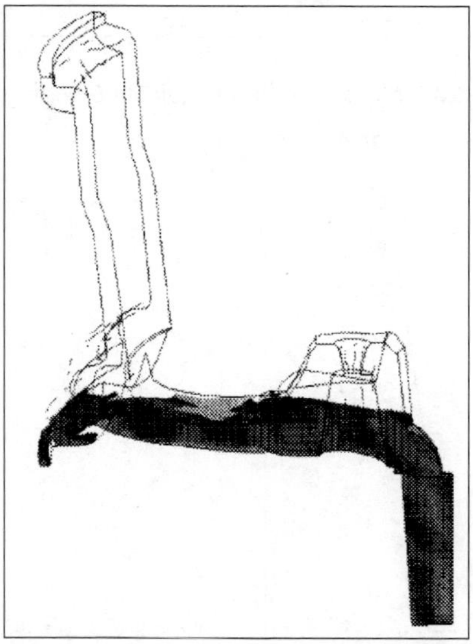

Figure 23.20 Flow front at 12.5 seconds

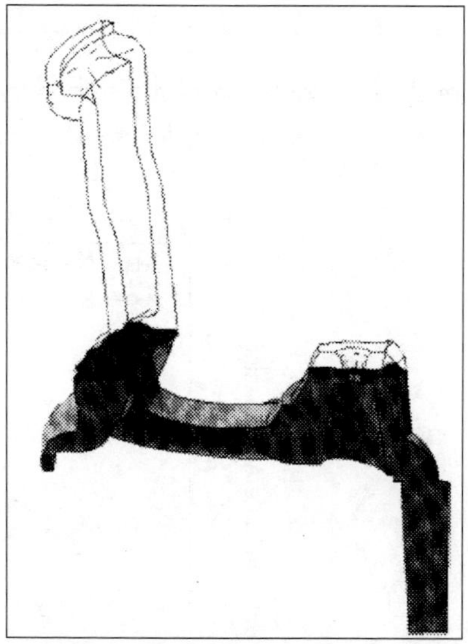

Figure 23.21 Flow front at 20 seconds

Figures 23.22 and 23.23 show the predicted secondary arm spacing and microporosity at different regions through a half-section of the wheel, respectively. For a dendritic alloy, the secondary dendritic arm spacing (SDAS) is the most commonly used parameter to determine the degree of micro-segregation of the casting. In general, the SDAS is directly related to the local solidification time, or average cooling rate during solidification. A more refined SDAS is an indication of better mechanical properties in the part.

The following relationship is used

$$SADS = At_s^n$$

where t_s is the solidification time and the values of parameters A and n depend on the material and s is the solidification time of the material.

When the numerical SDAS values were compared with those obtained experimentally, a reasonable fit between these two sets of data was obtained, as shown in Figure 23.24.

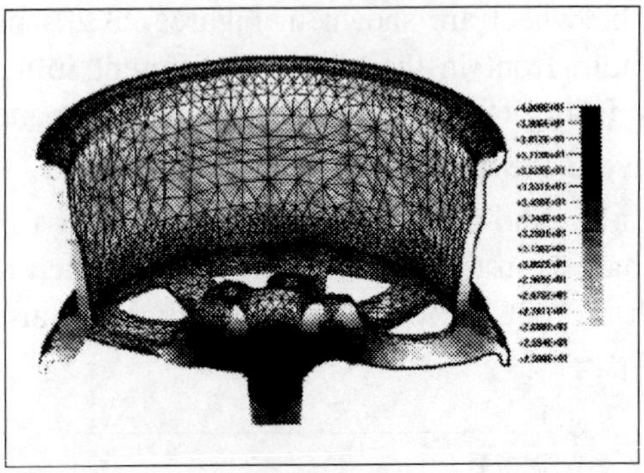

Figure 23.22 Prediction of the secondary dendrite arm spacing (microns)

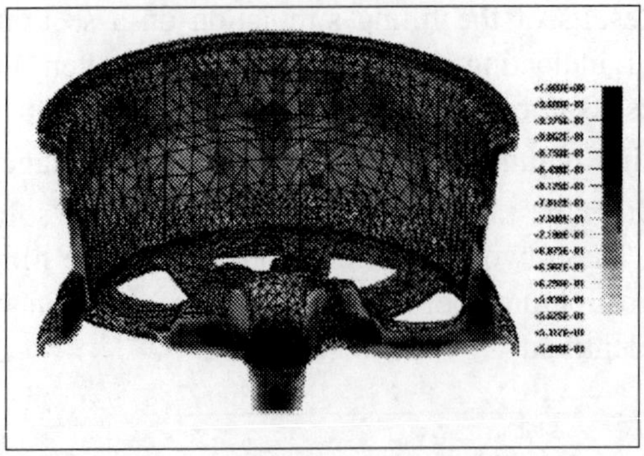

Figure 23.23 Niyama criteria indicating regions of high porosity

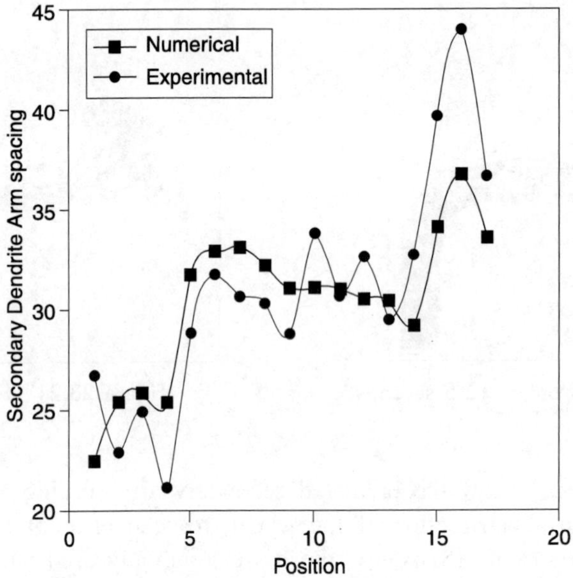

Figure 23.24 Dendrite arm spacing at different positions on the wheel, comparison of experimental and numerical

A convenient way to predict microporosity numerically in castings is the use of a porosity criterion function which relates microporosity to the thermal condition and other parameters in the mould, such as part geometry. When the well-known Niyama criterion $N_{Iyama} = G/\sqrt{R}$ (with G: thermal gradient and R: cooling rate) was applied to predict micro-shrinkage towards the end of solidification (near the solidus temperature), the zones susceptible to microporosity (Figure 23.23) were found to correspond well to what was observed in the casting.

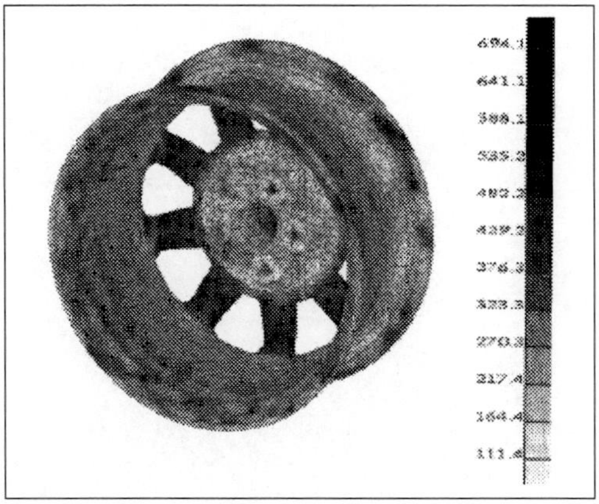

Figure 23.25 Residual Von-Mises stresses of the wheel (MPa)

Figure 23.26 Shrinkage of the wheel amplitude by a factor of 20.

Figure 23.25 shows residual Von-Mises stresses throughout the wheel at the end of the cooling stage. The maximum residual stresses are located at the spokes and at isolated parts of the wheel geometry where contraction of the metal during solidification and cooling is constrained by the wall of the mould. Subsequently, the degree of shrinkage at different sections of the casting depends, to a great extent, on the degree of constrain experienced by the material during solidification, as highlighted in Figure 23.26. The air gap which forms as a result of this shrinkage has a bearing on the thermal conductivity between the solidifying casting and the mould and could affect the final microstructure of the part. A maximum gap of approximately 1.5 mm was calculated.

Review Questions

1. List the application of computers in a foundry.
2. Explain the use of CMM in pattern making.
3. What are the statistical techniques used in a foundry?
4. Write a brief note on computer aided modeling of castings.
5. What are the design considerations in CAD of castings?
6. How is the riser and gating design done using computers?
7. Explain the computer aided design procedure used in a foundry.
8. How is solidification simulation done?
9. Discuss two-dimensional shell element model.
10. Write a short note on numerical modelling in casting.

Bibliography*

1997. Ed. Beech, J. and Jones, H., Dept of Eng. Materials, University of Sheffield, 45 (1997)

Anon. Br. Foundrym., **80**, 20 (1987).

Apelian, D. and Entrebin, C. H. Jr. (1986). *Int. Met. Rev.* **31**, 2, 77.

Ashton M. C., Sims B. J. and Sharman, (1985). Steel Founders' Society of America, Illinois.

Ashton, M. C. (1991). *Metals and Materials* **7**(1), 12.

Ashton, M. C., Sharman, S. G. and Sims, B. J. (1983). *Br. Foundrym.* **76**, 162.

Balewski, A. T. and Dimov, T. (1965). *Br. Foundrym.*, **58**, 280.

Ballmann, R. B. (1988). *Trans. Am. Fndrym. Soc.* **96**, 465.

Batyshev *et al.* (1972). *Russ. Cast. Prod.* 220.

Beeley, P. (2001).. Foundry Technology, 2nd Edition, Standard Media.

Beeley, P. R. and Smart, R. F. (1995). (Eds) *Investment Casting*. Institute of Materials, London.

Beeley, P.R. (1986). *Br. Foundrym.* **79**, 441.

Begg, J. and Clegg, A. J. (1996). *Trans. Am. Fndrym. Soc.* **104**, 1159.

Blackwood, R. and Perkins, J. (1944). *Trans. Am. Fndrym. Ass.*, **52**, 273.

Bobrov, V. I., Batyshev, A. I. and Bidulya, P. N. (1967). *Russ. Cast. Prod.*, 153.

Bralla, J. Handbook of Manufacturing Processes, Industrial Press Inc., U.S.A.

Bridges, P. J. and Hauzeur, F. (1991). *Cast Metals* **4**, 3, 152.

Brown, J. R. (1992). *Metals and Materials* **8**, 10, 550

Bunshah, R. F. (1958). *Vacuum Metallurgy*, Reinhold, New York.

Campbell, J. (1981). *Engineering* **221**, 3, 185.

Campbell, J. (1988). *Cast Metals*, **1**, 3, 156.

Campbell, J. (1991). *Castings*, Butterworth-Heinemann, Oxford

Campbell, J. (2010). Casting, Reed Elsevier India Pvt. Ltd.

Campbell, J. (2011). Metal casting handbook. Butterworth-Heinemann; 1st Edition.

Campbell, J. and Wilkins, P. S. A. (1982). *Br. Foundrym.* **75**, 233.

** This book is an in-depth compilation of the workings of more than 1500 different authors and researchers in Metal Casting and Manufacturing Processes in Metal Casting Industries. Author is pleased to acknowledge with gratitude the valuable information obtained from the following individual's work.*

Carless, P. and Tingle, A.B. (1987). *Fndry Trade J*, July 2nd 561.

Carreker, R. P., Jr. (1963). *J. Metals*, **15**, 774.

Carrington, F. G. (1944). *Trans. Am. Fndrym. Ass.*, **52**, 333.

Chadwick, A. and Yue, T. M. (1989). *Metals and Materials* **5**, 6.

Chakrabarti, A. K. (2005). Casting Technology and cast alloys, Prentice Hall India Learning Private.

Chandley, G. D. (1983). *Modern Casting* **73**, 10, 29.

Chandley, G. D. (1989). *Cast Metals*, **2**, 1, 2.

Chandley, G. D. and Cargill, D. L. (1990). *Trans. Am. Fndrym. Soc.* **98**, 413.

Chandley, G. D., Mikkola, P. and Shah, R. C. (1996). *Trans. Am. Fndrym. Soc.* **104**, 903.

Choi, S. H., Kang, C. S. and Loper, C. R. (1989) *Trans. Am. Fndrym. Soc.* **97**, 971.

Clegg A. (1985). J. *Fndry Trade J*, **158**, 472 and 486.

Clegg, A. J. (1986). *Fndry Trade J. International* **9**, 30, 51–61, 69.

Crepeau, P.N. (2007). Shape casting, John Wiley Publishers.

Crow, T. B. (1947). *Metallurgia*, **35**, 141.

Cumberland, J. (1963). *Br. Foundrym.*, **56**, 26.

Dalton, R. F. (1957). *Trans. Am. Fndrym. Soc.*, **65**, 483.

Das, A. A., Yacoub, M. M., Zantout, B. and Clegg, A.J. (1988). *Cast Metals*, **1**, 69.

Dietrich, W. and Stephan, H. (1988). *Metals Handbook*, Vol. 15 Casting, 410, ASM International, Metals Park.

Dowsing, R. J. (1975). *Metals and Materials*, July/August, 26.

Dowsing, R. J. (1981) Metals and Materials, July/August, 27.

Dunlop, A. (1986). *Fndry Trade J.* **123**, 124.

Ebisu, Y., *Trans. Am. Fndrym. Soc.* **85**, 643 (1977).

Else, G. E. (1983). *Br. Foundrym.* **70**, 9, 145.

Emley, E. F. (1958). *Br. Foundrym.*, **51**, 501.

First Report IBF Working Group T20 (1994). *Foundrym.* **87**, 6, 223.

Fleetwood, M. J., *Metals and Materials* **3**, 1, 14 (1987) 672 Foundry Technology.

Flemings, M. C. (1979). Solidification and Casting of Metals. The Metals Soc. London, 510.

Flemings, M. C. Solidification and Casting of Metals. Proceedings of 1977

Flemings, M. C., Adams, C. M., Hucke, E. E. and Taylor, H. F. (1956). *Trans. Am.Fndrym. Soc.*, **64**, 636.

Fndry International, **16**, 3, 224 (1993).

Fndry Trade J. Int (1991). **14**, 4, 145.

Fndry Trade J. International (1989). **12**, 2, 20.

Fukusako, T., Ohnaka, I. and Yamauchi, I. (1979). *Br. Foundrym.* **72**, 26.

Gangadhar, S. A., Mahajan, B. K. (2010). Casting Defects: Its Analysis & Remedies, CBS.

Ghosh, A. (2010). Manufacturing Science, Pearson India.

Glazoff, M. (2018). Casting Aluminium alloys. Butterworth-Heinemann Inc.

Goddard, D. M. (1986). *Trans. Am. Fndrym. Soc.* **94**, 667

Goulette, M. J., Spilling, P. D. and Arthey, R. P. (1984). Proceedings Int.conf.on Superalloys, Am. Soc. Metals.

Gouwens, P. R., Watmough, J. and Berry, J. T. (1959). *Trans. Am. Fndrym. Soc.*, **67**, 577.

Griffin, J. C., Patton, K. D. and Bates, C. E. (1991). *Trans. Am. Fndrym. Soc.* **99**, 203.

Groover, M. P. (2014). Principles of Modern Manufacturing. John Wiley.

Groover, M. P. (2016). Automation, Production Systems and CIM. Pearson Publications.

Grote, R. E. (1986). *Trans. Am. Fndrym. Soc.* **94**, 181.

Hall, J. H. (1948). *Foundry*, **76**, 76

Heine, R. W., Loper, C., Rosenthal, P.C. (2017). Metal Casting. Tata McGraw Hill.

Herlach, D. M., Cochrane, R. F., Egry, I., Fecht, H. J. and Greer, A. L. (1993). *Internat. Mat. Rev.* **38**, 6, 273.

Herrera, A. and Campbell, J. (1997). *Trans. Am. Fndrym. Soc.* **105**, 5.

High-Temperature Steels and Alloys for Gas Turbines (1952). Specia Report 43, Iron Steel Inst., London.

Jain, P. L. (2015). Foundry Technology, Tata McGraw Hill Publishing Co Ltd.

John, K.C. (2015). Metal Casting, Prentice Hall of India.

Jones, H. (1991) *Metals and Materials* **7**, 8, 486.

Kalpakjian, S. (2008). Manufacturing, Engineering and Technology. Pearson India.

Kapranos, P. (1994). *Materials World*, **2**, 9, 465.

Kasai, H. (1980). *Trans. Am. Fndrym. Soc.* **88**, 535.

Khanna, O. P. (2016). Foundry Technology, Dhanpat Rai Publications (P) Ltd.

Kirkwood, D. H. (1994). *Int. Mat. Rev.* **39**(5), 173.

Kirkwood, D. H. and Kapranos, P. (1989). *Metals and Materials*, **5**, 1, 16.

Kono, R. and Miura, T. (1974). *Br. Foundrym.* **67**, 319.

Kono, R. and Miura, T. (1975). *Br. Foundrym.* **68**, 70.

Kumar, D. and Jain, S. K. (1999). Foundry Technology, CBS Publishers.

Lavernia, E. J., Ayers, J. D. and Srivatsan, T. S. (1992). *Internat. Mat. Rev.* **37**, 1, 1.

Lavington, M. H. (1986). *Metals and Materials* **2**, 1, 713.

Leatham, A. (1996). *Materials World* **4**, 6, 319.

Leatham, A., Ogilvy, A., Chesney, P. and Wood, J. V. (1989). *Metals and Materials* **5**, 3, 141.

Lee, O. R. J. and Northcott, L. (1947). *J. Inst. Metals*, **73**, 491.

Lipson, S. and Ripkin, F. (1965). *Trans. Am. Fndrym. Soc.*, **73**, 194.

Littleton, H. E., *et al.* (1996). *Trans. Am. Fndrym. Soc.* **104**, 335.

May, R. P. (1979). *Br. Foundrym.* **72**, xxxviii.

McIntyre, J. B. (1961). *Fndry Trade J.*, **110**, 459.

Mehrabian, R., Riek, R. G. and Flemings, M. C. (1974). *Metall. Trans.*, **5**, 1899 674 Foundry Technology.

Method of Precision Coring in the Casting of Metallic Articles (1957). British Patent 782, 553.

Mitchell, (1983). Solidification in the Foundry and Casthouse.273, 147, The Metals Society, London.

Molding Methods and Materials, Am. Fndrym. Soc., Des Plaines, Ill., 536 (1962)

Morgan, E. and Milnes, H. H. M. (1959). *Br. Foundrym.*, **52**, 240.

Morley, J. I., *Ref. 22*, 195

Morton, J. R. and Barlow, J. (1994). *Foundrym.* **87**, 23.

Mykura, N. (1991). *Metals and Materials* **7**, 1, 7.

Nolte, M., Neussl, E. and Sahm, P. R. (1994). *Trans. Am. Fndrym. Soc.* **102**, 949.

Northcott, L. and Dickin, V. (1944) *J. Inst. Metals*, **70**, 301.

Northcott, L. and Lee, O. R. J. (1945). *J. Inst. Metals*, **71**, 93.

Northcott, L. and Mclean, D. (1945). *J. Iron Steel Inst.*, **151**, 303.

Pannen, H. and Sick, G. (1988). *Metals Handbook*, Vol. 15 Casting, 419, ASM International,Metals Park.

Pichler, F. (1992). Heinz Schwartzel Cast methods in Modelling, Springer.

Plyatskii, V. M. (1965). *Extrusion Casting*, Primary Sources, New York.

Poirier, D. R. (2010). Heat Transfer Fundamentals for Metal Casting. John Wiley & Sons.

Portcullis P. (1982). Fundamentals of Foundry Technology.

Production Techniques 3 Further casting techniques 673.

Radhakrishna, K. (2015). Foundry Technology, Sapna Book house.

Rao, P. N. (2015). Manufacturing Technology, Volume 1, Tata McGraw Hill Publishing Co Ltd.

Rao, T. V. R. (1996). Metal Casting, New Age Publishers.

Ravi, B. (2013). Metal Casting Computer Aided Design & Analysis. Prentice Hall of India.

Reed, D. S. and Jones, M. L. (1991). *Trans. Am. Fndrym. Soc.* **99**, 697.

Reynolds, D. J., Shamsuzzoha, M. and El-Khaddah, N., Solidification Processing

Roberts, R. J. (1996) *Trans. Am. Fndrym. Soc.* **104**, 523.

Rohatgi, P. K., Asthana, R. and Das, S. (1986). *Int. Met. Rev.* **31**(3), 115.

Rose R. L. (1973). *Trans. Am. Fndrym. Soc.* **81**, 233.

Roy A. (1989). Lindberg, Processes And Materials of Manufacturing, Pearson.

Rubtsov, N. N. and Gini, E. CH. (1962). *Russ. Cast. Prod.*, 281.

Ryzhikov, A. A. and Timoveev, G. I. (1962). *Russ. Cast. Prod.*, 4.

Sahoo, M. and Sahu, S. (2017). Principles of Metal Casting, McGraw Hill Education. 3rd Edition.

Samuels, M. L. and Schuh, A. E. (1951). *Foundry*, **79**, July, 78; Aug., 84.

Second Report of Working Group T20 Casting Processes (1994). *Foundryman* **87**, 386.

Shroyer, H. F. (1958). *Br. Pat.* 850, 331.

Singer, A. R. E. and Evans, R. W. (1983). *Metals Technology* **10**, 61.

Sinha, K. P. and Goel, D. B. (2011). Standard Publishers Distributors.

Smith, R. A. and Wilkins, P. S. A. (1986). *Trans. Am. Fndrym. Soc.* **94**, 785.

Spencer, D. B., Mehrabian, R. and Flemings, M. C. (1972). *Metall. Trans.* **3**, 1925.

Suryanarayana, C., Froes, F. H. and Rowe, R. G. (1991). *Internat. Mat. Rev.* **36**, 3, 85.

Taylor, I. C. (1988). *Castings Buyer* **1**, 3, 12.

Taylor, P. B., Gettings, M., Eyre, N. J. and Bushell, G. C. (1986). *Metals and Materials*, **2**, 8, 494.

Terry, B. with Jones, G. (1990) *Metal Matrix Composites*, Elsevier Adv. Tech. Oxford.

Thornton, A. E. (1958) *Br. Foundrym.*, **51**, 559.

Thornton, A. E. (1959). *Br. Foundrym.*, **52**, 80.

Thornton, A. E. (1963). *Br. Foundrym.*, **56**, 63.

Tsvetnenko, K. U. (1961). *Russ. Cast. Prod.*, 273.

Typical Microstructures of Cast Metals. (1966). 2nd Edition. Inst. Br. Foundrym. London.

Walling, R. P. and Dantzig, J. A. (1994). *Trans. Am. Fndrym. Soc.* **102**, 849.

Watmough, J., Berry, J. T. and Gouwens, P. R. (1961). *Trans. Am. Fndrym. Soc.*, **69**, 701.

Wightman, P. and Hengsberger, E. (1991). *Metals and Materials* **7**, 11, 676.

Wilkins, P. S. A. (1983) *Br. Foundrym.* **76**, 72.

Williams, G. and Fisher, K. M. (1983). Solidification Technology in the Foundry 137 London.

Wittmoser, A. (1972). *Br. Foundrym.* **65**, 73.

Wood, J. V., McCartney, D. G., Dinsdale, K., Kellie, J. L. F. and Davies, P. (1995). *Cast Metals* **8**, 1, 57.

Index

K

L

M

N